Other Prentice-Hall Books by Dr. KAMILO FEHER

TELECOMMUNICATIONS MEASUREMENTS, ANALYSIS,
AND INSTRUMENTATION
1987

DIGITAL COMMUNICATIONS:
Satellite/Earth Station Engineering
1983

DIGITAL COMMUNICATIONS:
Microwave Applications
1981

# ADVANCED
# DIGITAL
# COMMUNICATIONS

# ADVANCED DIGITAL COMMUNICATIONS
## Systems and Signal Processing Techniques

### Dr. KAMILO FEHER
**Editor**

*Professor, Electrical and Computer Engineering*
*University of California, Davis*
*Davis, California 95616*

*Director, Consulting Group, DIGCOM, Inc.*

# PRENTICE-HALL, INC.
*Englewood Cliffs, New Jersey 07632*

*Library of Congress Cataloging-in-Publication Data*

FEHER, KAMILO.
    Advanced digital communications.

    Includes bibliographies and index.
    1. Digital communications.   2. Signal processing—
Digital techniques.   I. Title.
TK5103.7.F428   1987        621.38′0413        86-17074
ISBN   0-13-011198-8

Editorial/production supervision and
    interior design: Reynold Rieger
Cover design: Ben Santora
Manufacturing buyer: Gordon Osbourne

Printed in the United States of America

10  9  8  7  6  5  4  3  2  1

ISBN 0-13-011198-8   025

Prentice-Hall International (UK) Limited, *London*
Prentice-Hall of Australia Pty. Limited, *Sydney*
Prentice-Hall Canada Inc., *Toronto*
Prentice-Hall Hispanoamericana, S.A., *Mexico*
Prentice-Hall of India Private Limited, *New Delhi*
Prentice-Hall of Japan, Inc., *Tokyo*
Prentice-Hall of Southeast Asia Pte. Ltd., *Singapore*
Editora Prentice-Hall do Brasil, Ltda., *Rio de Janeiro*

# CONTENTS

## 4   ECHO CANCELLATION IN SPEECH AND DATA TRANSMISSION   182

*Dr. David G. Messerschmitt*

## 5   DIGITAL SPEECH INTERPOLATION SYSTEMS   237

*Dr. S. J. Campanella*

Contents

# 8 CORRELATIVE CODING: BASEBAND AND MODULATION APPLICATIONS 429

Dr. Subbarayan Pasupathy

Contents

**11   ADVANCED CONCEPTS AND TECHNOLOGIES FOR COMMUNICATIONS SATELLITES     573**

Dr. Douglas Reudink

## 12  ADAPTIVE EQUALIZATION       640

*Dr. Shahid U. H. Qureshi*

## INDEX       715

# FOREWORD

Applications of digital signal-processing techniques appear around nearly every corner in the labyrinth of recently developed methods for solving the demanding technological challenges arising in the design of modern telecommunication systems. The potentials of digital communications for providing high transmission rates and integrated audio, video, and data services are realized, in many cases, only by using innovative digital signal-processing techniques to achieve economical, reliable, and robust systems. The topics and techniques addressed in this book promise to remain as important and active areas of research and development in coming years, stimulated by demands for new and better telecommunication systems.

One major contribution of this book is to provide in-depth coverage of the major problem areas within digital communications where digital signal-processing techniques have (or will have in future systems) decisive importance. These specialized chapters are written by recognized experts from around the world, and the result is a book whose perspective and scope is unmatched by any other single book on digital communications.

But the contribution of the book goes well beyond the sum of the individual contributions, thanks to the efforts of Dr. Kamilo Feher to coordinate the presentations and to provide a perspective of the topics covered from an overall systems standpoint. The introductory and tutorial sections appearing in later chapters is complemented in essential ways by the overview of his first chapter. In terms of the analogy mentioned earlier, the reader of this book can come away with an appreciation of the maze from both the Minotaur's view as well as that of Daedalus.

Student, professor, researcher, and practicing engineer will find this to be a valuable book, both for technical details as well as for insightful discussions. Dr. Feher and his coauthors have made an important contribution to the literature of the digital communications field, and I applaud the results of their efforts.

BRADLEY W. DICKINSON
*Professor of Electrical Engineering*
*Princeton University*

# PREFACE

This book is for engineers, students, and professors interested in the fascinating fields of Digital Communications (DIGCOM) and Digital Signal Processing (DSP). An introductory-overview chapter and the tutorial sections, presented in the first part of each chapter, should facilitate the comprehension of the advanced material.

I do not know of another book in the digital communications and signal processing field of comparable scope and depth. Most modern DIGCOM and telecommunications related DSP topics are thoroughly covered. In our study of DSP techniques we highlight new techniques and applications of DSP in telecommunication systems and networks. However, to keep the size of this book reasonable we decided to omit the description of a number of important topics, including switching systems, optical fiber communications, error-correction coding, and spread spectrum systems. These topics are adequately covered in a number of other books.

If you have a general, fairly broad knowledge of the subjects covered in this volume, then I hope that it will motivate you to study the advanced topics in depth and that you will acquire a solid foundation of DIGCOM and DSP system principles, techniques, and applications. Even if you do not have the time and energy to study all of the derivations and concepts, you could benefit from the many graphs, tables, hardware photographs, and original laboratory measurement results that highlight the text. Thus you could use this book for your in-depth studies, as a comprehensive text, or as an encyclopedia/reference sourcebook.

If you are a specialist and have up-to-date, advanced knowledge of most of the topics covered in this book, then the newest discoveries, research and development results, new principles, and ever-increasing DIGCOM and DSP applications in the United States, Canada, Japan, Europe, and throughout the world should be of interest to you.

Perhaps the single predominant reason that I did not even dare to attempt to write this complete volume by myself is that I recognize that I am not an expert and a leading authority in all the closely interrelated topics covered in this book; there is probably no single individual who could be considered so.

All coauthors are internationally recognized authorities in the fields addressed in their respective chapters. Their extensive research and development, management, and teaching experience, and their significant contributions to the professional literature, combined with their strong theoretical background, practical design, and operational experience are reflected in this book. I believe that all readers of this book, including expert research engineers and professors, will find a wealth of new information in the following chapters. Innovative and original techniques discovered by the authors are also highlighted.

The close relation between various chapters and topics is highlighted by many flowcharts in the introductory chapter, Chapter 1.

I would be pleased to hear from you. If you have comments, questions, or suggestions related to the many challenging research and development topics covered in this book, then please feel free to write to me.

DR. KAMILO FEHER
*Professor of Electrical Engineering*
*University of California, Davis*
*Davis, California 95616*

# ADVANCED
# DIGITAL
# COMMUNICATIONS

# 1

# DIGCOM AND DSP: DIGITAL COMMUNICATIONS AND DIGITAL SIGNAL-PROCESSING OVERVIEW

**DR. KAMILO FEHER**

*Professor, Electrical Engineering*
*University of California, Davis*
*Davis, California 95616*
*and*
*Director, Consulting Group, DIGCOM, Inc.*

## 1.1 OBJECTIVES OF THIS INTRODUCTORY CHAPTER

To motivate you to read and study the many advanced and fascinating **digital communications** (DIGCOM) and signal-processing subjects covered in this book, I believe that an introductory and overview chapter is well justified.

In Fig. 1.1, the outline of the material covered in this volume is presented in a block diagram form. **Digital signal processing** (DSP) plus transmission techniques and network and technology principles, techniques, and applications are presented in depth in Chapters 2 through 12. In this introductory chapter, we present an overview of the later advanced chapters and highlight the close *relationship* among the chosen topics. Numerous block diagrams (flowcharts) highlight the synergisms between the topics presented in various chapters.

We attempt to point out to you the benefits that could accrue from an in-depth study of this book. A sampling of such benefits is indicated by the questions below, which are answered in this introductory chapter:

1. Why do we need ISDN?
2. Why digital speech?
3. What is wrong with PCM?
4. What are the echo-cancellation requirements?

1

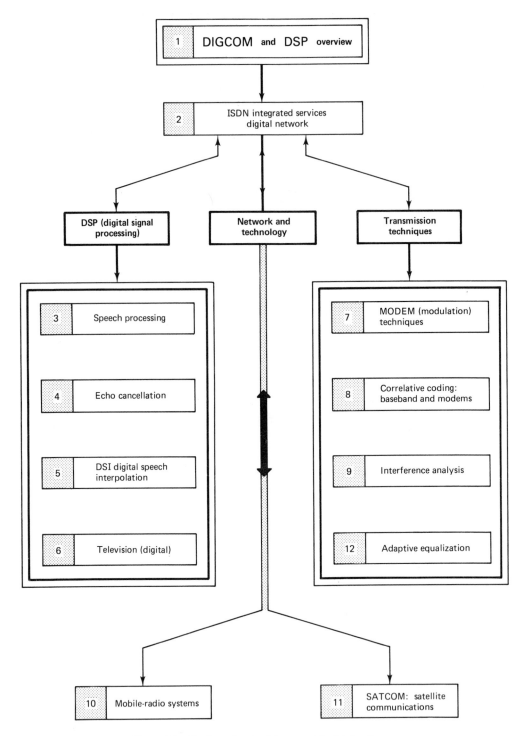

**Figure 1.1** Outline of material covered in this book.

5. How can circuit capacity be multiplied with DSI?

6. Why should we study or develop new digital TV-processing techniques?

Brief tutorials are included in several sections of this chapter to complement the material presented in the respective chapters. To expedite comprehension by readers unfamiliar with a topic, nonmathematical, simple physical explanations of later-described principles and techniques are given. Illustrative results and applications are also highlighted.

The large number of digital communications and signal-processing topics covered in this book, the depth of the material, and the size of this volume warrant an introductory overview chapter. However, I want to admit that writing this relatively short chapter has been a more difficult challenge than writing more advanced and longer chapters on modern communication techniques. This challenge came from one of this book's reviewers, who did not see this chapter. He made the following comment:

> *I am interested in seeing how Dr. Feher will convince the reader in Chapter 1 that the marriage of these chapters, for which he is the matchmaker, is a marriage made in heaven.*

After reading the following sections and chapters, you will be in a good position to reach your own conclusions.

## 1.2 ISDN: INTEGRATED SERVICES DIGITAL NETWORK

### 1.2.1 International IDN and ISDN Developments

**Integrated digital networks** (IDN) have been established worldwide during the last two decades. Architectures and standards for digital transmission networks and switching systems dedicated to voice and/or data have been developed. These networks are called integrated digital networks, due to the commonality of digital techniques used in transmission and switching systems. These IDNs are evolving toward the ISDN.

The **telephony IDN** is based on 64-kb/s **pulse code modulation** (PCM) coding of speech signals. As a result, this standard has been applied to both transmission and circuit switching systems. Circuit switched systems are controlled by software programs stored in reliable processing units. In advanced telephony IDN structures, call control (signaling) information is exchanged among offices through a **common channel signaling** (CCS) network. The CCS network is a *computer communication* network employing packet switching techniques to transfer signaling messages among office control computers.

For *data applications,* two IDN modes have emerged as best suited to handle bulk and bursty data: circuit switching and packet switching, respectively. **Circuit switched IDNs** dedicated to data applications are designed according to the same principles as those used for telephony applications, especially for the case of synchronous communications. The basic user data rates for synchronous data networks are 2.4, 4.8, 9.6, 19.2, 48, 56, and 64 kb/s. **Time division multiplex** (TDM) techniques are used to interleave lower rate channels, thus forming a basic 64 kb/s channel. Circuit switching can also be performed for lower-rate digital data channels. **Packet switched IDNs** represent an evolution of message-switching techniques. Data information is assembled in short messages (packets) that are transmitted by users at the same channel rates as specified for synchronous circuit switched networks. Packet switched data networks and offices use **statistical multiplexing** (SM) of data *packets over transmission links* and employ the store-and-forward principle.

### 1.2.2 Emerging Communications Services

The fast-growing demand for responsive communication services along with recent advances in communications, electronics, and computer technology create strong incentives for the development of integrated *services* digital networks. Several parameters that characterize communication services are

—Market (residential, business)
—Information type (voice, sound, video, data)
—Statistical traffic behavior (traffic intensity, terminal activity, . . . )
—Channel capacity requirements (expressed in bits/s)
—Performance requirements (error-free seconds, bit error rate, delay, . . . )

Some examples of channel capacity required to support specific types of services are presented next.

Telemetry applications such as meter readings, security or alarms, telecontrol, and opinion polling require 10- to 100-b/s data rates. Services classified in the 1- to 10-kb/s range include text, graphics, and data communications. These services (teletex, home computer, videotex, facsimile, and low-speed switched data) are frequently interactive and thus are suitable for the packet switched mode of operation. Various digitized voice services between 10 and 100 kb/s find applications. The traditional 64-kb/s rate, required for PCM telephony voice transmission, is being reduced (without loss of voice quality) to 32-kb/s and even lower bit rates. Chapter 3 describes modern reduced bit rate voice-encoding techniques, which enable intelligible voice transmission at a reduced rate of only 1 kb/s. Due to the bursty nature, i.e., the 35 to 40% activity factor, of one-way telephone conversations, **digital speech interpolation** (DSI) is used to enhance further the efficiency of the ISDN. High-speed switched data, digital facsimile, and slow-scan video are bulk services which also require a 10-kb/s to 100-kb/s data rate. High-speed facsimile, high-quality music and videoconferencing require a 100-kb/s to 2-Mb/s capacity. Broadcast-quality TV programs require rates between 20 and 100 Mb/s, whereas high-definition TV requires transmission rates of over 100 Mb/s.

From the previous list of service requirements, it is clear that ISDN will have to cater to a large variety of service demands and applications.

### 1.2.3 The ISDN Concept

The ISDN concept is characterized by its three main features:

1. End-to-end digital connectivity
2. Multiservice capability (voice, data, video)
3. Standard terminal interfaces

The ISDN provides a network transport capability for a variety of services (ranging from telemetry up to broadband video application) using a variety of digital communication modes (from leased and semipermanent connections to circuit and packet switched connections). Essential to the functions of an ISDN is not only its integrated operating capability, but also its administrative and maintenance functionality, such as network management, and billing. The key element of service integration in the ISDN is the provision for a limited set of standard user-network interfaces.

### 1.2.4 Performance Objectives of ISDN

The **bit error rate** (BER), **Error-Free Second** (EFS), and **availability objectives** of digital connections which may form part of an ISDN are highlighted in this section. ISDN parameter specifications have a direct impact on the transmission and signal-processing hardware and software subsystem and overall system design philosophy.

In a complex, long-distance national or international data or speech signal connection, such as an ISDN, many regenerative transmit-receive sections, signal-processing, multiplex, and switching subsystems may be required. For this reason we highlight the overall ISDN performance objectives (EFS and BER) for a hypothetical 27,500-km reference connection of a **64-kb/s data stream** and also the objectives of a 2500-km radio section, which may form part of the connection. Detailed analysis and description of performance measurement techniques and of international ISDN standard requirements (CCITT) are presented in [CCITT, 1.3].

*Error performance* on an international digital connection, forming part of an ISDN, has been defined by the CCITT. The rationale for the development of error performance objectives may be summarized as follows:

1. Services in the future may expect to be based on the concept of an ISDN.
2. Errors are a major source of degradation in that they affect voice services in terms of distortion of voice and data-type services in terms of lost or inaccurate information or reduced throughput.
3. Although voice services are likely to predominate, the ISDN is required to transport a wide range of service types and it is therefore desirable to have a unified specification.

Error performance objectives in international ISDN connections are summarized in Table 1.1. Since the objectives defined in Table 1.1 address the overall

**TABLE 1.1** CCITT error performance objectives for an international 64-kb/s ISDN connection*

| Part | Performance measurement | Objective |
|------|-------------------------|-----------|
| (a) | $BER < 10^{-6}$ for $T_0 = 1$ min | > 90% of 1-min intervals to have 38 or fewer errors |
| (b) | $BER < 10^{-3}$ for $T_0 = 1$ s | > 99.8% of 1 s intervals to have less than 64 errors |
| (c) | $BER = 0$ for $T_0 = 1$ s (EFS) | > 92% EFSs |

Total measurement time: $T_m = 1$ month

*27,500 km hypothetical reference connection. Based on Table 1/G.821—CCITT draft revision. With permission of the International Telecommunications Union. See also [CCIR 594, 1.3].

Notes

1. The time intervals mentioned should be derived from a fixed time pattern, i.e., not specifically related to the occurrence of errors.
2. Total time, $T_L$, has not been specified since the period may depend upon the application. A period of the order of any one month is suggested as a reference.
3. The alternative use of the *error-free decisecond* (EFdS) is currently under study. Should this take place, the objective for classification (c) would need to be based on the EFdS equivalent value of the EFS figure, together with an assessment of the customer's perception of circuit quality.
4. In a practical measurement a small number of 10-min integration periods may include periods when the connection is judged to be unavailable and/or contain seconds where the number of errors exceeds 63.
5. Further study needs to be carried out to determine whether the 10-min average period is acceptable (possible network implications). An alternative value of 5 min is also under consideration.

ISDN 27,500-km connection, it is necessary to subdivide the table into constituent parts and even into individual equipment and systems. Apportionment strategies for the 1-minute measurement interval, EFS requirements and for services utilizing higher bit rates have been studied recently by the CCITT.

Here we highlight only one apportionment, namely, the allowable BER at the output of a 2500-km hypothetical reference digital path for *radio systems* [CCIR 594, 1.3]; see Table 1.2.

### 1.2.5 Low Error-Ratio Value

The CCITT has proposed an error-ratio design objective of 1 *in* $10^{10}$ *per kilometers* for the transmission system in a 25,000-km hypothetical reference digital connection. For a 2500-km digital radio-relay path, this gives an error ratio objective of $2.5 \times 10^{-7}$ (this figure excludes the contributions due to the multiplexing equipment). In digital radio-relay systems, this error ratio will, according to present design criteria, be achieved for at least 99% of the time. Therefore, a low value of error ratio of about $10^{-7}$ would seem appropriate for a 2500-km hypothetical reference digital path.

**TABLE 1.2** BER and errored seconds in the redrafted CCIR-G.821 recommendation (May, 1984) for a 2500-km hypothetical reference digital path (HRDP)

| BER not to exceed | For more than X% of any month | Integration time |
|---|---|---|
| $1 \times 10^{-6}$ | 0.4% | 1 min |
| $1 \times 10^{-3}$ | 0.54% | 1 s |

Errored seconds: Total should not exceed 0.32% of any month*

*We interpret this as:

| Number of errored seconds not to exceed | Per "average" period of |
|---|---|
| 8294 | 1 month (30 days) |
| 276 | 1 day |
| 11 | 1 hour |

### 1.2.6 High Error-Ratio Value

The proportion of time for which the high error ratio may be exceeded significantly influences the design of a system. For example, the requirement that the higher error ratio for a 2500-km circuit not be exceeded for more than 0.01% of any month would severely penalize the economics of a practical system, whereas a corresponding error ratio of 0.1% for any month, which compares favorably with existing frequency division multiplexed–frequency modulation, or FDM-FM, radio-relay systems, would not penalize costs severely.

### 1.2.7 Error Free Second (EFS) Objectives

The CCITT has proposed that for data transmission, a 95% EFS objective should be attained on the local exchange—or a local exchange segment of a 25,000-km hypothetical reference connection. A corresponding criterion of 99.5% EFS for the 2500-km hypothetical reference digital path seems appropriate.

In general there is no simple relationship between BER performance and the long-term EFS performance. For this reason and in order to assure that the BER and EFS objectives are simultaneously satisfied, both of these system requirements have to be specified.

### 1.2.8 Relation of the ISDN Chapter to Other Chapters

The material of Chapter 2, "Integrated Services Digital Networks," is related to all other chapters of this book. This relationship is illustrated in Fig. 1.1 as well as in the following sections. Evolving international standard transmission rates, network configurations, computer communications, and network protocols all have impacts on the communications subsystems and communications systems analyses presented in later chapters. Almost every block (chapter) in Fig. 1.1 could be considered as a subsystem of evolving ISDN, although topics like theoretical interference analysis or satellite radiation patterns are not part of ISDN networks. However, we must carefully analyze our overall system interference and, establish our satellite radiation patterns and our overall system-design guidelines to satisfy the requirements of emerging ISDN.

## 1.3 SPEECH-CODING ALGORITHMS AND VECTOR QUANTIZATION

### 1.3.1 Why Digital Speech (Voice)?

Analog speech (voice) signals must be converted into a digital format in order to enable their digital processing, including storage, compression, forwarding, time division multiplexing, switching, and transmission over emerging digital networks such as the IDN and ISDN.

PCM is one of the oldest and conceptually most simple analog-to-digital (A/D) conversion processes used in speech and video signals. A detailed description of the principles of conventional PCM systems is given in most books on telecommunications engineering including [Feher, 1.1; Feher, 1.3; Bellamy, 1.9; Martin, 1.11]. Here we present a brief review of the major steps used in PCM encoders. An essential step in PCM conversion of analog signals is **sampling.** The sampling theorem requires a signal to be sampled at a frequency which is at least two times greater than the higher edge of the analog signal bandwidth. For "toll-grade quality" telephony signals, the speech signal is bandlimited to a nominal 3.4-kHz bandwidth. Including guard bands, the signal bandwidth is 4 kHz. Thus a sampling rate of $f_s = 8$ kHz is used. At the output of the sampling device, a sequence of pulses is obtained. The amplitude of each pulse is proportional to the amplitude of the analog input signal at the sampling instant. This step is known as **pulse amplitude modulation** PAM; see Fig. 1.2.

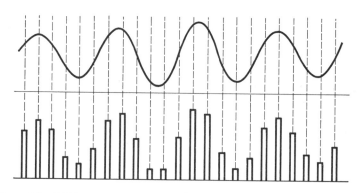

**Figure 1.2** PAM stream of an analog input speech signal pattern. [Martin, 1.11]

The heights of the pulses in Fig. 1.2, which contain all the information in the original analog waveform, still carry the information in analog format. Sampling at an $f_s$ = 8-kHz rate assures that the PAM signal is a *discrete time* signal, that is, it is fully defined in integer multiples of $\frac{1}{8}$ kHz = 125 μs; however, it is not yet a digital signal.

As a second step, we convert this discrete time analog signal (having a continuous range of possible levels) into a *quantized* discrete time signal that reduces the number of possible levels from infinity to some finite number. In this process, a **quantization error,** also know as **quantization noise,** is introduced (see Fig. 1.3). In this figure, in order to have a simpler conceptual and visual presentation, we illustrate eight possible quantization levels. Thus in this case the maximum error

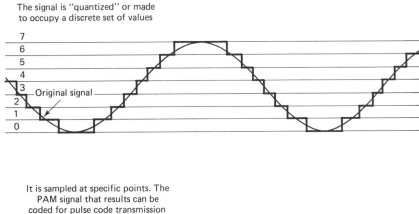

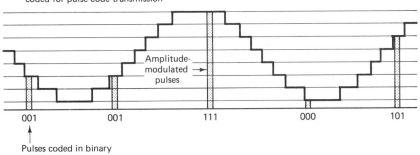

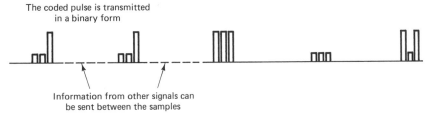

**Figure 1.3**  Basic steps in PCM conversion and TDM of a speech signal. (With permission of Prentice-Hall, Inc. [Martin, 1.11].)

voltage or quantization-noise component is approximately 6%. It has been found experimentally that for a practical toll-grade quality telephony signal, a ratio of **signal to quantization noise** $(S/N)_q$ in the 30-dB to 40-dB range is required. Thus the number of quantization levels must be increased to 256.

In the third step, the *quantized PAM signal is converted into* a *binary data stream,* which in turn may be *time-division-multiplexed* (TDM) with other PCM data streams. For the unique identification of 256 possible quantization levels, 8 bits per quantization level are required, as $2^8 = 256$. Thus the resultant bit rate requirement for a PCM-encoded 3.4-kHz analog speech signal is 8000 samples/ second × 8 bits/sample = 64 kb/s.

In the described process—namely, sampling—quantization and quantized PAM to binary conversion, $(S/N)_q$ depends on the rms value of the analog input signal and the number of quantization levels. In this process, the larger input signals have a larger (better) $(S/N)_q$ and weaker signals have a lower $(S/N)_q$. This would be an undesirable effect in telephony systems, as some users speak far more softly than others. For the listener, it would be unpleasant and annoying to listen to a very low amplitude signal corrupted by a relatively high amplitude quantization error (low $(S/N)_q$). To remedy this problem and achieve approximately the same value of $(S/N)_q$ for a small-amplitude signal as for a large-amplitude signal, a quantizer with a nonuniform step size is required. Nonuniform step quantization is frequently implemented with uniform step-size quantizers preceded by **nonlinear (logarithmic) compressor** or **compression** devices. Compressors followed by uniform step-size quantizers amplify the low-volume signals more than high-volume signals. In the PCM decoder (receiver) the inverse nonlinear *expansion* of the PCM encoded signal is performed. The overall system is called **companded** PCM.

### 1.3.2 Problems with PCM: Alternative Solutions

For a 3.4-kHz analog speech signal, a PCM encoder transmits a data rate of $f_b = 64$ kb/s. For the same source (analog) speech signal in a simple binary baseband system, the minimum theoretical bandwidth requirement for the transmission of this digital signal is 32 kHz. Thus bandwidth almost ten times as great is required if binary digital transmission is used instead of analog transmission. The main drawback of PCM-encoded signals is that they require much more transmission bandwidth than conventional analog speech signals. Fortunately, spectrally efficient modulation techniques, such as 16 QAM, 64 QAM, and 256 QAM may be used to resolve this drawback and use the radio-frequency spectrum more efficiently. For example, currently operational high-capacity 64-QAM digital microwave systems have a practical spectral efficiency of 4.5 bits/s/Hz. With such systems a PCM 64-kb/s speech signal is transmitted in a reduced bandwidth of 64 kb/s ÷ 4.5 bits/s/Hz = 14.22 kHz. (High-capacity microwave systems transmit TDM PCM voice channels. For example, a 90 Mb/s, 64-QAM radio system transmits 1344 multiplexed voice channels in a radio bandwidth of 20 MHz.) However, 14.22 kHz is still a large bandwidth requirement. In conventional 64-kb/s PCM systems the correlation between samples is not exploited; in other words the redundancy inherent in the speech signal is not removed. Each sample is quantized independently, regardless of whether it contains any new "information" (rapidly changing amplitude) or no new "information" (a continuous and constant dc level).

Modern speech coding algorithms transmit, store, and synthesize speech at a given quality using *many fewer bits than conventional PCM*. Adaptive signal prediction, linear least-mean-square (LMS) estimation techniques, scalar quantization, vector quantization, and a number of other recently developed techniques enable a bit compression of more than a factor of two for toll-grade quality speech transmission. In other words, a toll-grade quality telephony system requires only 10 kb/s to 30 kb/s (instead of 64 kb/s) if modern digital speech processing techniques are employed.

In Table 1.3, the approximate bit rate and bandwidth requirement of toll-grade and of acceptable-communication quality telephony speech signals are listed. Here we assume binary (not multilevel) digital baseband transmission.

For many telephony systems applications, toll-grade quality is not essential. For "intelligible and acceptable communication quality" telephone-signal transmission, a dramatically reduced bit rate in the 0.5-kb/s to 5-kb/s range has been achieved with modern speech coding and vector quantization algorithms, described in Chapter 3.

In Fig. 1.4, the relation of the speech-coding (processing) material, covered in Chapter 3, with that of other chapters in this book is highlighted.

**TABLE 1.3**  Bit rate and bandwidth requirements of toll-grade and of "intelligible and acceptable communication quality" telephone signals. Binary baseband transmission is assumed. Further spectral compression of the digitally encoded speech signals is achieved if signal transmission by multilevel modulation is employed.

| Technique used | Approximate requirement for toll-grade | | Approximate requirement for acceptable-communication quality | |
|---|---|---|---|---|
| | Bit rate | Minimum bandwidth (binary) | Bit rate | Minimum bandwidth (binary signaling) |
| PCM[1] | 56–64 kb/s | 28 kHz | 45 kb/s | 22.5 kHz |
| ADPCM[2] | 30–40 kb/s | 15 kHz | 6–30 kb/s | 3 kHz |
| ADMOD[3] | | | | |
| LPC[4] | | | | |
| . | | | | |
| . | | | | |
| . | ≤20 kb/s | ≤10 kHz | 0.5–5 kb/s | 0.250 Hz to 2.5 kHz |
| Vector quantization | | | | |
| Analog voice (baseband) | — | 4 kHz | — | 2.7 kHz |

[1]PCM: Pulse coded modulation

[2]ADPCM: Adaptive differential PCM

[3]ADMOD: Adaptive delta modulation

[4]LPC: Linear Predictive Coding

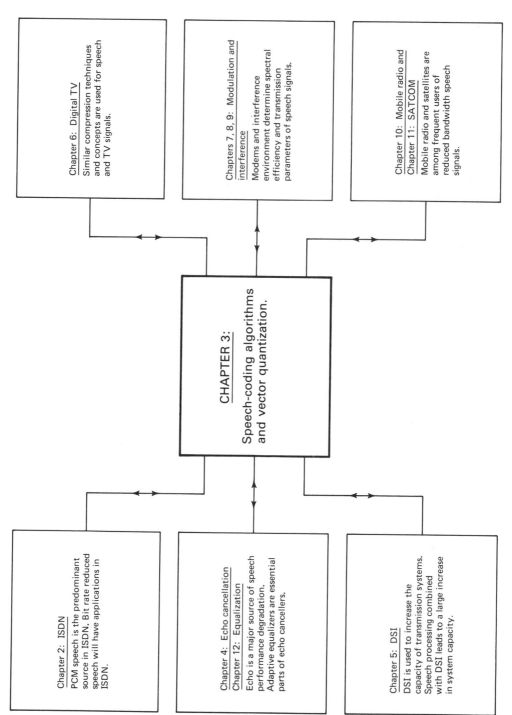

**Figure 1.4** Relation of Chapter 3 to other chapters in this book.

Chapter 6: Digital TV
Similar compression techniques and concepts are used for speech and TV signals.

Chapters 7, 8, 9: Modulation and interference
Modems and interference environment determine spectral efficiency and transmission parameters of speech signals.

Chapter 10: Mobile radio and
Chapter 11: SATCOM
Mobile radio and satellites are among frequent users of reduced bandwidth speech signals.

CHAPTER 3:
Speech-coding algorithms and vector quantization.

Chapter 2: ISDN
PCM speech is the predominant source in ISDN. Bit rate reduced speech will have applications in ISDN.

Chapter 4: Echo cancellation
Chapter 12: Equalization
Echo is a major source of speech performance degradation. Adaptive equalizers are essential parts of echo cancellers.

Chapter 5: DSI
DSI is used to increase the capacity of transmission systems. Speech processing combined with DSI leads to a large increase in system capacity.

## 1.4 ECHO CANCELLATION IN SPEECH AND DATA SIGNALS

### 1.4.1 Echo-Cancellation Requirements

The telephone network generates undesired echos at points within and near the ends of telephone connections. Echo-cancellation and/or suppression subsystems are required to combat this echo for both speech and data transmission. Typical parts of a telephone system include two-wire subscriber loops, hybrids, and a four-wire trunk. On the two-wire subscriber loop, both directions of transmission are carried on a single wire pair. In the four-wire trunk (transmission, switching) the two directions of transmission are segregated on physically different facilities. This is required where it is desired to insert carrier terminals, amplifiers, or digital switches. In cases where carrier transmission is used, these are not necessarily four physical wires, but the terminology is still useful and descriptive.

Echo is introduced in the feedback loop around the four-wire portion of the connection. Without sufficient loss in the round-trip path, there is a degradation of signal quality or, in extreme cases, oscillation (called singing). The hybrid provides a large amount of attenuation around this round-trip loop without inserting significant loss in the two-talker speech path. The longer the speech echo is delayed, the more disturbing it is and the more it must be attenuated before it becomes tolerable. Synchronous satellites (most commercial satellites are *synchronous* (i.e., geostationary) hover roughly 40,000 km above the earth's surface. Due to this large distance, the round trip of a telephone conversation relayed via satellite, including the terrestrial segment, takes about 500 to 600 ms. For a double-satellite loop, the round-trip delay exceeds 1 s. The large round-trip delay combined with echo is one of the most serious drawbacks of satellite communication systems.

A typical long-distance telephone circuit is illustrated by Fig. 1.5. At any location in this circuit, if the transmitted signal encounters an impedance mismatch, a fraction of this signal gets reflected as an echo. Telephone sets in a geographic area are connected by a two-wire line to a hybrid transformer (frequently called a hybrid) which is located in a central office. For both transmission and reception to and from the central office, only two wires are used. This provides considerable savings of wire, local switching equipment, and maintenance.

The circuit diagram of a hybrid located at $B$ is shown in Fig. 1.6. If the two transformers are identical and the balancing impedance $Z_n$ equals the impedance of the two-wire circuit, the signal originating on the "in" side gets transferred to the two-wire circuit of $B$ but produces no response at the "out" terminal. On the other hand, if the signal originates in the two-wire circuit (talker $B$ is active), this signal is transferred to both paths of the four-wire circuit. This signal has no effect on the in signal path, since the amplifiers shown in Fig. 1.5 amplify signals only in the opposite direction. Echoes are generated whenever the hybrid in side is coupled (has a leak-through) to the out side. Unfortunately, this occurs in almost all networks, as the $Z_n$ network is not identical to the distributed time-variable impedance of the two-wire circuit. Also, we note that a four-wire circuit may be con-

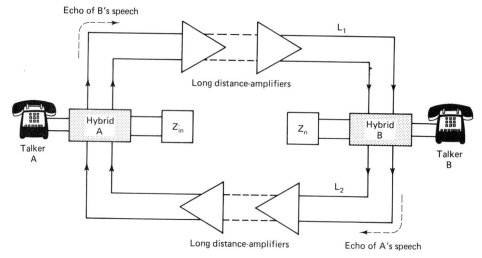

Echo of B's speech

$L_1$

Long distance-amplifiers

Hybrid
A

$Z'_{in}$

$Z_n$

Hybrid
B

Talker
A

Talker
B

$L_2$

Long distance-amplifiers

Echo of A's speech

**Figure 1.5** Typical long-distance telephone circuit. The hybrids are located in central offices $A$ and $B$, respectively.

nected to a large number of two-wire circuits. Thus the need for echo control (suppression or cancellation) in long-distance systems is evident.

### 1.4.2 Echo Suppressors

A simplified block diagram of an echo suppressor (echo canceler) is shown in Fig. 1.7. Most of the time $A$ and $B$ speak alternately; thus the echo can be prevented from returning to $A$ by disconnecting the path $L_2$ during the intervals when only $A$ is talking. This is achieved by detecting the presence of the signal originating

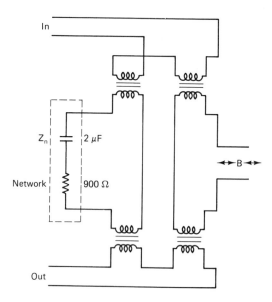

In

$Z_n$

$2\ \mu F$

$\blacktriangleleft\!\!\blacktriangleright B\blacktriangleleft\!\!\blacktriangleright$

Network

$900\ \Omega$

Out

**Figure 1.6** Circuit diagram of hybrid located at $B$ (see Fig. 1.5).

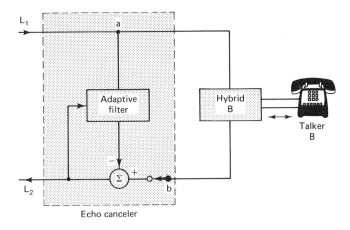

Figure 1.7 Echo canceler: conceptual diagram.

by talker *A* and activating (opening) switch *S*. However, interruptions are an important part of conversations, particularly of somewhat heated discussions. A statistical analysis of on-off patterns in numerous conversations demonstrates that there is a 20% probability for interruptions, that is, *simultaneous speech*. During simultaneous speech the switch is prevented from opening. Thus during interruptions (simultaneous speech), echoes are not eliminated.

Chopping the initial portions of the interrupter's speech is unavoidable. When telephone conversations are conducted over satellite systems that have a round-trip delay on the order of 600 ms, the interruption rate increases and the subjective effect of echoes during interruptions worsens. For these reasons, another generation of echo-control devices, known as echo cancelers, have been developed.

Note that the speech detector and comparator circuit design are fairly complex if good accuracy of control and comparison activities is required.

### 1.4.3 Echo Cancelers

The conceptual diagram of an echo canceler is shown in Fig. 1.7. The basic idea of echo cancellation is to generate a synthetic replica of the echo and subtract it from the leaked-through echo signal returned through hybrid *B* (Fig. 1.7). If an adaptive filter that perfectly matches the transfer function of the echo path were designed, complete echo cancellation could be achieved. Adaptive filtering is required to match the time-variable distance and device-dependent complex impedance characteristics of the two-wire circuit. Transversal adaptive filters and lattice filter and continuous time filter techniques described in Chapters 4 and 12 are frequently used for the implementation of synthetic echo signals.

The widespread application of echo cancellation has recently been stimulated by advances in large-scale integrated microelectronics, bringing the computational requirements of echo cancelers within the reach of inexpensive chip implementations. A detailed study of the principles and applications of echo-cancellation techniques for speech and data communications applications is presented in Chapter

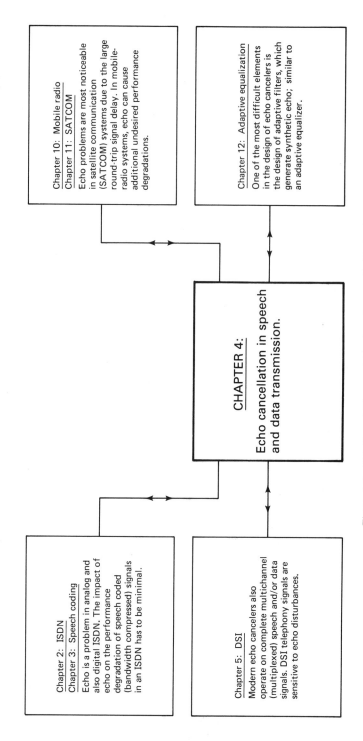

Chapter 10: Mobile radio
Chapter 11: SATCOM
Echo problems are most noticeable in satellite communication (SATCOM) systems due to the large round-trip signal delay. In mobile-radio systems, echo can cause additional undesired performance degradations.

Chapter 12: Adaptive equalization
One of the most difficult elements in the design of echo cancelers is the design of adaptive filters, which generate synthetic echo; similar to an adaptive equalizer.

CHAPTER 4:
Echo cancellation in speech and data transmission.

Chapter 2: ISDN
Chapter 3: Speech coding
Echo is a problem in analog and also digital ISDN. The impact of echo on the performance degradation of speech coded (bandwidth compressed) signals in an ISDN has to be minimal.

Chapter 5: DSI
Modern echo cancelers also operate on complete multichannel (multiplexed) speech and/or data signals. DSI telephony signals are sensitive to echo disturbances.

**Figure 1.8** Relation of Chapter 4 to other chapters in this book.

4 by Dr. Messerschmitt. Echo cancellation is directly related to a number of other topics presented in this book. In Fig. 1.8 this relation (interdependence) is highlighted.

## 1.5 DSI SYSTEMS

### 1.5.1 DSI—An Effective Circuit Multiplication Method

The DSI technique exploits the fact that in telephone conversations one person is usually speaking while the other listens. There is usually a channel inactivity between calls; pauses, hesitations, and intervals of silence are parts of normal conversations. On the average, speech of a particular customer is present 40% of the time. Modern DSI equipment exploits these relatively low voice-activity properties by compressing the number of outgoing channels used to transmit a given number of incoming telephony channels. Voice detectors are used to detect speech activity on each incoming channel input to the DSI system. The idle time between calls and conversation pauses is used to accommodate additional incoming calls. In a number of references, incoming channels are designated as *terrestrial channels* and outgoing channels as *satellite channels*.

In Section 1.3 we mentioned the main disadvantage of the conventional (logarithmic) PCM technique for voice—namely, its large bandwidth requirement of 32 kHz. However, even with conventional PCM-encoded telephony signals, significant overall system capacity advantages can be obtained on a number of satellite systems. Application of DSI systems to PCM encoded voice signals in **time division multiple access** (TDMA) satellite systems leads to a major capacity advantage of digital TDMA systems over **frequency division multiple access** (FDMA) analog satellite systems. An examination of Table 1.4 indicates that the *capacity,* expressed in terms of number of toll-grade quality 3.4-kHz channels (analog or PCM) of a digital TDMA satellite system, is two to three times larger than the capacity of a corresponding analog FDMA satellite system. This almost three-to-one capacity advantage justifies the additional hardware and software complexity required for digital TDMA-DSI satellite systems.

**TABLE 1.4** Approximate capacities for the INTELSAT-V hemispheric beam 72-MHz transponders for INTELSAT standard A (30-m antenna diameter earth stations)

| Transmission mode | Maximum capacity expressed in terms of number of 3.1-kHz telephony channels |
|---|---|
| FDMA (analog system) | 1100 |
| TDMA (digital system) | |
|     Without DSI | 1600 |
|     With DSI | 3200 |

Similar system-capacity advantages can be achieved in terrestrial line-of-sight (LOS) microwave system (for short, microwave system) applications particularly if DSI subsystems are utilized in conjunction with highly spectral-efficient modulation techniques, such as 64 QAM or 256 QAM.

### 1.5.2 DSI Principles and Performance Limitations

Theoretically, the DSI gain can be adjusted over a fairly wide range. If the number of satellite channels (outgoing channels) for a given number of terrestrial channels (incoming channels) is reduced, the DSI gain increases. However, the gain increase is limited by increased clippings in terrestrial channels, which cannot be transmitted for given periods of time due to nonavailability of satellite channels. Thus the terrestrial channels that do not have a satellite channel assignment are *frozen out* until one of the terrestrial channels that was previously assigned a satellite channel becomes inactive. Thus front-end clipping of speech is evident during *freeze out*. The DSI gain that can be obtained with a subjectively acceptable performance for **time-assigned speech interpolation** (TASI) and for the **speech predictive encoded communications** (SPEC) systems is about 2:1. The development of freeze-out specifications was preceded by numerous subjective tests. These specifications require that the percentage of clipping be less than 2% of the speech spurts that experience clips greater than 50 ms. (Each period of time occupied by a caller's speech is called a *speech spurt*.)

In the SPEC digital speech-interpolation system, the competitive clip problem experienced by the TASI-DSI system is completely avoided. The speech predictor algorithm used in the SPEC system eliminates unnecessary samples in the instantaneous speech waveform or in the short intersyllabic pauses. The SPEC predictor removes more than 25% of the PCM samples during voice spurts 25% of the time. On the other hand, the major performance degradation in the SPEC digital speech-interpolation system is the production of prediction distortion.

The principle of compression of a larger number of terrestrial channels (or incoming channels) into a smaller number of satellite channels (or outgoing channels) is illustrated in Fig. 1.9. The PCM encoder accepts $N$ analog telephony inputs and converts them into $M$ binary bit streams. The output of the DSI module provides $M$ binary channels for satellite transmission, where $M < N$. One of these channels, the assignment channel, is used to transmit the satellite channel assignments to the receiving earth station for proper reconnection by means of assignment messages, which consist of the terrestrial channel number and associated satellite channel number.

### 1.5.3 Phenomenal Circuit Multiplication with Modern DSI and Efficient Speech-Coding Algorithms

Digital TASI without and with PCM overload channels, variable-rate delta modulation used with speech interpolation and SPEC techniques are described in Chapter 5. These techniques exploit the speech activity properties to reduce the information rate needed to handle a multiplicity of speech-carrying telephone channels.

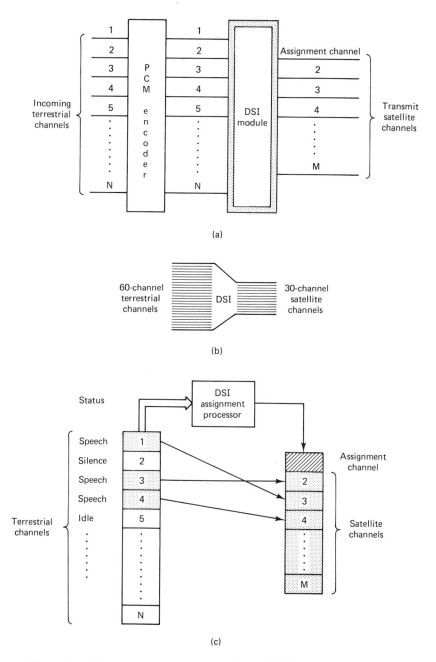

**Figure 1.9** DSI channel-compression capability. (a) DSI transmit equipment: $N$ incoming terrestrial channels are compressed into $M$ outgoing satellite channels through DSI operation, where the ratio $N/M$ is the DSI gain. (b) 60 incoming channels are compressed into 30 satellite channels using DSI with a gain of 2.0. (c) Mapping of terrestrial channels into satellite channels.

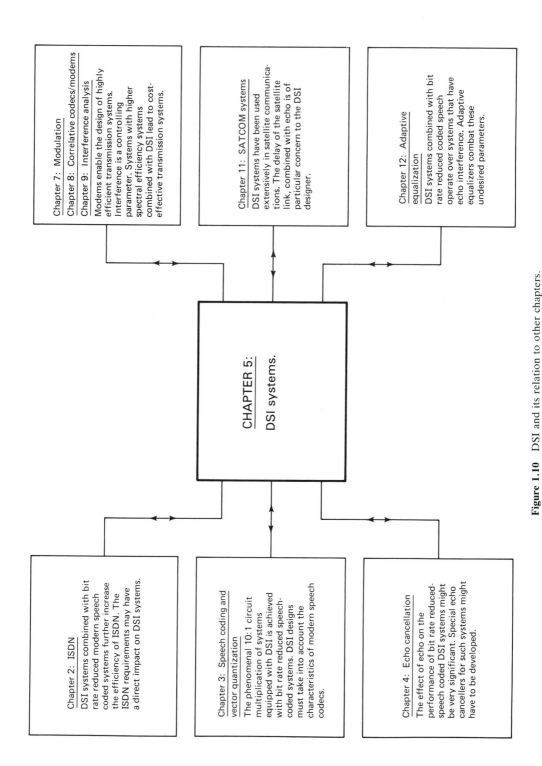

**Figure 1.10** DSI and its relation to other chapters.

In Chapter 5, Dr. Campanella presents techniques (modern speech encoders and digital speech interpolation) that achieve a net circuit multiplication of 4:1 to 5:1 compared with conventional noninterpolated PCM telephony. In this case, off-the-shelf, large-scale integrated circuit, 32-kb/s rate, toll-grade quality digital speech encoders are used. During the 1990s efficient algorithms for speech encoders are expected to provide toll-grade quality telephony signals at a 16-jkb/s rate (or even lower rates). When combined with digital speech interpolation these yield a *net circuit* (channel) multiplication ratio of between 8:1 and 10:1. This circuit multiplication or effective increase in system capacity is truly phenomenal!

Digital speech interpolation is related to a number of other topics described in this book (see Fig. 1.10).

## 1.6 DIGITAL TV-PROCESSING TECHNIQUES

### 1.6.1 Analog TV for the 21st Century?

Color TV sets with high-quality video and sound reception are part of our daily lives. We are accustomed to live programs, which require transmission from nearby studios as well as from remote locations in other countries and continents. These remarkable technical services have been achieved by means of analog signal processing, transmission, distribution, and reception.

Most source signals are analog. A free-for-all in a hockey rink, an explosion on the evening news, the facial expressions of a cat rejecting that other brand of cat food—all these images and motions from your favorite television programs are examples of analog source signals, which come to you via analog transmission and reception. Most importantly, the human eye and ear are analog receivers and are expected to stay analog (at least throughout the active lifetime of this book). In summary, analog processing and broadcasting of TV signals has led to high-quality (toll-grade or broadcast quality) cost-efficient TV distribution to hundreds of millions of homes. It is very unlikely that the majority of home analog TV receivers will be replaced with digital receivers during this century. Thus we might ask ourselves the following question: *Why should we study and develop new digital TV signal-processing techniques?* This question is answered in this section.

### 1.6.2 Motivation for Digital Television (Analog RF Bandwidth 18 to 30 MHz)

Land and satellite link rentals for transmission of analog broadcast-quality TV signals are fairly expensive due to the relatively large information bandwidth (about 7-MHz baseband bandwidth for the combined video and accompanying analog signals) and high signal-to-noise ratio ($S/N$) requirement (requires high frequency FM deviation). Many TV signals are transmitted through analog terrestrial FM microwave systems and through FM satellite systems. An FM-modulated TV signal may require a **radio-frequency** (RF) bandwidth of 18 to 30 MHz. Digitized and bandwidth compressed broadcast quality video could be transmitted in an RF bandwidth of only 3.7 MHz (see Table 1.5).

### 1.6.3 ISDN Evolution—Digital Transmission, Switching, Multiplexing

As stated earlier, there is a worldwide trend toward the ISDN. During the transition period, between now and a truly operational ISDN, more and more new digital multiplexing, switching, and transmission facilities will be installed. In addition to digitized voice and other digital services, these facilities will be suitable for digitized TV signal transmission. Modern high-capacity digital fiber optic systems and TDMA satellite systems can more efficiently transmit digital TV signals than analog TV signals.

Digital switching and multiplexing is considerably more cost-effective than analog switching. In the digital approach, conceptually, a simple logic gate provides the switching and TDM functions. In analog FDM, frequency translators with frequently expensive filters are required (very strict amplitude and phase specifications raise filter costs). Furthermore, analog switches require high isolation, low insertion loss, high return loss (impedance match), and many other parameters that are expensive to implement and maintain and are not required in the digital approach.

### 1.6.4 Video Teleconferencing

During the 1980s, a large increase of video-teleconferencing service requirements, particularly in the United States, occurred. High-quality visual communication as well as audio communication at a reasonable cost are essential elements of video teleconferencing. Easily accessible video-teleconference rooms, preferably located at the customer's or subcontractor's premises, are required for two-way teleconferencing systems. In a number of applications, the video and audio signals are

**TABLE 1.5**  Practical RF spectral requirement for the transmission of DSP television signals

| | Approximate bit rate requirement | Practical radio-frequency spectral requirement | | | |
|---|---|---|---|---|---|
| | | QPSK | 64 QAM | 256 QAM | Analog FM |
| Conventional PCM broadcast quality | 90 Mb/s | 60 MHz | 20 MHz | 15 MHz | — |
| Interframe coding broadcast quality | 20 Mb/s | 15 MHz | 5 MHz | 3.7 MHz | — |
| Interframe coded teleconference quality | 600 kb/s | 450 kHz | 150 kHz | 110 kHz | — |
| Analog broadcast quality | 5.5 MHz (baseband) | — | — | — | 18–30 MHz |

broadcast in only one direction, and audio (telephone grade 3.4-kHz voice channel) is broadcast only in the return direction. For example, during a seminar I presented at Stanford University, California, we broadcast full video (visual) and audio to the attendees, who asked questions through the returning conventional telephone network.

Despite the advantages of video teleconferencing, the costs for video transmission sometimes discourage its use. The more bandwidth required, the higher the cost. Since the bandwidth requirements of analog or digital full-motion broadcast-quality video signal transmission are more than a thousand times wider than a voice telephone channel, the video costs far exceed the audio costs. Accordingly, the TV signal long-distance transmission and local distribution cost (assuming no digital signal processing is used to compress the required bandwidth) could be prohibitive for many applications.

### 1.6.5 Digital Processing For Bandwidth Compression of TV Signals

One of the oldest digital TV signal-processing techniques is conventional PCM. Since the analog TV, full-motion, broadcast-quality video signal is bandlimited to approximately 5.5 MHz, the sampling theorem requires this signal be sampled at an $f_s = 11$-MHz sampling rate. To obtain an acceptably low quantization noise (or, in other words, a sufficiently high signal-to-quantization-noise ratio), experiments determined that 8 to 10 bits/sample are required. Thus the transmission rate requirement for a PCM-encoded full-motion TV baseband signal is in the 88-Mb/s to 110-Mb/s range. Assuming use of the digital modulation most frequently used in current satellite links, QPSK, an RF bandwidth of about 60 MHz, would

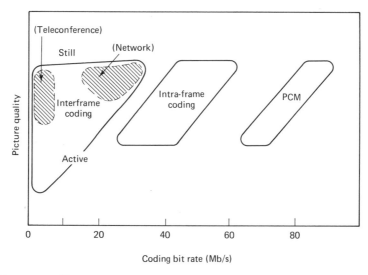

**Figure 1.11** Picture quality as a function of coding and bit rate for typical TV digital signal processing (coding) methods.

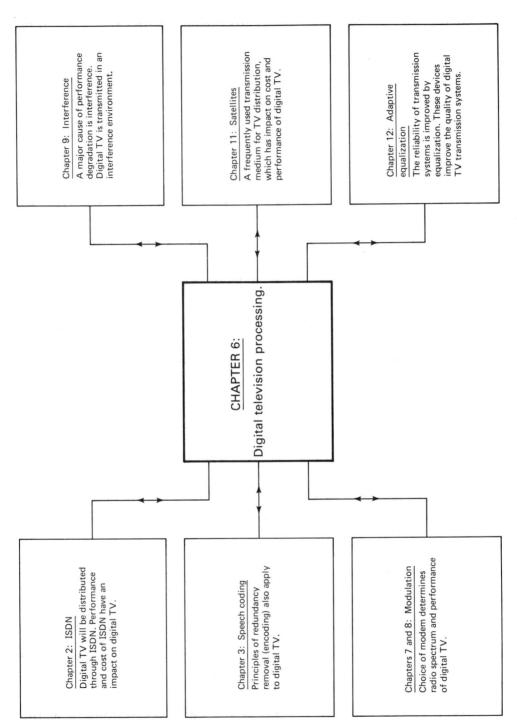

**Figure 1.12** Relation of Chapter 6 to other chapters in this book.

Chapter 9: Interference
A major cause of performance degradation is interference. Digital TV is transmitted in an interference environment.

Chapter 11: Satellites
A frequently used transmission medium for TV distribution, which has impact on cost and performance of digital TV.

Chapter 12: Adaptive equalization
The reliability of transmission systems is improved by equalization. These devices improve the quality of digital TV transmission systems.

CHAPTER 6:
Digital television processing.

Chapter 2: ISDN
Digital TV will be distributed through ISDN. Performance and cost of ISDN have an impact on digital TV.

Chapter 3: Speech coding
Principles of redundancy removal (encoding) also apply to digital TV.

Chapters 7 and 8: Modulation
Choice of modem determines radio spectrum and performance of digital TV.

be required. Thus we note that the digitized PCM video signal (without compression) requires a bandwidth two to three times as larger than its FM-modulated counterpart. Therefore, to be bandwidth-competitive with its FM-modulated counterpart, the requirement for bandwidth compression is essential for most digital TV applications.

· Intraframe and interframe coding techniques combined with differential PCM and other digital signal-processing techniques exploit the inherent redundancies present in video signals, particularly in signals that represent relatively slow movement images, as is the case in video teleconferencing. In Fig. 1.11 the coding bit rate requirement as a function of picture quality for typical digital processing methods is illustrated. With sophisticated video digital signal-processing (encoding) techniques, a compression ratio of 40 or even more has been achieved. For video teleconferencing, video encoders requiring a 2-Mb/s rate (or even the lower rate of about 500 kb/s) are commercially available. The required radio-frequency spectrum of the transmission of various TV signals is illustrated in Table 1.5. In Fig. 1.12, the relationship of Chapter 6 to the other chapters in this book is highlighted.

## 1.7 DIGITAL MODEM (MODULATION-DEMODULATION) TECHNIQUES

### 1.7.1 Modem Requirements

To convert a binary or a multilevel digital baseband signal to an **intermediate frequency** (IF) or an RF signal, digital modulators are employed. The modulated IF signal may be frequency translated to an RF signal or transmitted as an IF signal. In the receiver the inverse signal processing steps are performed, that is, the received modulated RF signal is downconverted to a suitable IF frequency and is demodulated. The demodulated baseband signal is regenerated and the source bit stream is restored.

For the ideal **additive White Gaussian noise** (AWGN) channel, **coherent** demodulators are required for theoretically optimal performance. For coherent demodulation, the unmodulated carrier frequency and the symbol timing clock of the received signal must be recovered. Nonlinear signal processing is required to obtain these synchronization signals.

For performance below the theoretically optimal limit, **differentially coherent** and **noncoherent** demodulators can be used. In principle, these have a simpler structure as they do not require a coherent carrier recovery circuit. Since their performance is not as good as that of coherent modems, for a specified BER or **probability of error** ($P_e$), noncoherent modems require a higher **carrier-to-noise ratio** (CNR) than coherent modems.

For a number of system applications, such as digital satellite and mobile radio communications systems, **power efficiency** is perhaps the most critical system parameter. For other system applications, such as digital LOS microwave systems, coaxial cable, and telephony systems, a **high spectral efficiency** is a mandatory requirement.

### 1.7.2 Power-Efficient Modems

We consider a modem to be power efficient if it has a good $P_e$ performance in a relatively low CNR environment. More specifically, a power-efficient modem satisfies the following normalized performance objectives:

|  | Theoretical $E_b/N_O$ | Practical $E_b/N_O$ |
|---|---|---|
| $P_e = 10^{-4}$ | 8.4 dB | 10.8 dB |
| $P_e = 10^{-8}$ | 12 dB | 15 dB |

Where $E_b/N_O$ represents the normalized CNR of the system. Note that $E_b$ is the average energy of a received bit and it equals $CT_b$, that is, average power multiplied by the unit bit duration. The received noise density $N_O$ is average noise power in a normalized 1-Hz bandwidth.

For a power-efficient transmission system, the **high-power amplifier** (HPA) of the transmitter should operate at its full power, known as **saturated power**. In this case the modulated signal is nonlinearly amplified and may suffer a significant distortion unless specific signal characteristics are chosen.

Most power-efficient modems, with the exception of combined coded/modulated systems, have an RF spectral efficiency of less than 2 bits/s/Hz.

### 1.7.3 Spectrally Efficient Modems

We consider a modem to be spectral efficient if it has an RF spectral efficiency of more than 2 bits/s/Hz. Modern microwave, coaxial cable and telephony systems require spectral efficiencies in the 2- to 7-bit/s/Hz range, and it is anticipated that spectral efficient modems in the 7- to 10-bit/s/Hz range will be required toward the end of this century. Spectral efficient modems require a considerably higher CNR than power-efficient modems. Combined coded/modulated systems reduce the overall CNR requirement without loss of spectral efficiency. However, they do this at the expense of increased hardware and/or software complexity.

### 1.7.4 Classification of Modems

Many different modem techniques have been discovered and used in various systems applications. Some of the earliest noncoherent binary FM modems were used for the transmission of 50- to 200-bit/s information rates in a 3.4-kHz telephony channel. During the 1980s, data rates of up to 19.2 kb/s have been transmitted through the same telephony systems.

In earlier power-limited satellite systems, binary **phase-shiftkeyed** (PSK) modems equipped with half-rate **forward-error-correction** (FEC) devices have been used. Recent satellite systems give promise of spectral efficiencies of more than 2 bits/s/Hz of RF. More importantly, this increased spectral efficiency could be achieved with fully saturated high-power amplifiers.

Digital 64-QAM LOS microwave systems operate with a practical spectral efficiency of 4.5 bits/s/Hz. Recently developed 256-QAM modems, which have applications in microwave systems, have a practical spectral efficiency of about 6.6 bits/s/Hz.

Modems are built for bit rates from 10 bits/s to 1 Gb/s, that is, $10^9$ bits/s. The technology, implementation, and design constraints depend on the specified bit rate (*s*).

Modem techniques and applications are highlighted in Fig. 1.13. A detailed study of these modems is presented in Chapter 7. The material in Chapter 7 is particularly related quite closely to material in several other chapters (see Fig. 1.14).

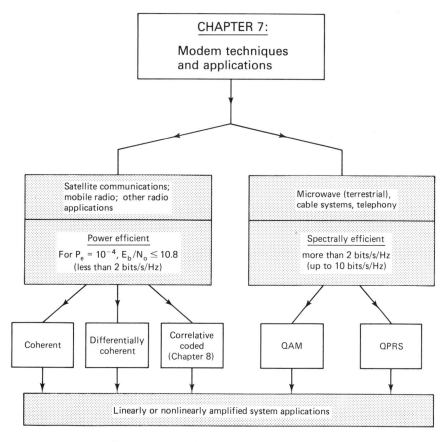

**Figure 1.13**  Modem techniques and applications.

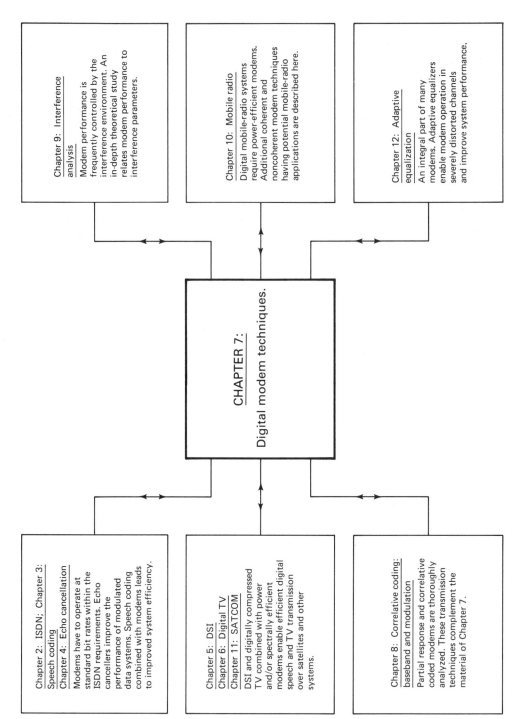

**Figure 1.14** Relation of Chapter 7 to other chapters.

Chapter 9: Interference analysis

Modem performance is frequently controlled by the interference environment. An in-depth theoretical study relates modem performance to interference parameters.

Chapter 10: Mobile radio

Digital mobile-radio systems require power-efficient modems. Additional coherent and noncoherent modem techniques having potential mobile-radio applications are described here.

Chapter 12: Adaptive equalization

An integral part of many modems. Adaptive equalizers enable modem operation in severely distorted channels and improve system performance.

CHAPTER 7: Digital modem techniques.

Chapter 2: ISDN; Chapter 3: Speech coding
Chapter 4: Echo cancellation

Modems have to operate at standard bit rates within the ISDN requirements. Echo cancellers improve the performance of modulated data systems. Speech coding combined with modems leads to improved system efficiency.

Chapter 5: DSI
Chapter 6: Digital TV
Chapter 11: SATCOM

DSI and digitally compressed TV combined with power and/or spectrally efficient modems enable efficient digital speech and TV transmission over satellites and other systems.

Chapter 8: Correlative coding: baseband and modulation

Partial response and correlative coded modems are thoroughly analyzed. These transmission techniques complement the material of Chapter 7.

## 1.8 CORRELATIVE CODING: BASEBAND AND MODULATION

### 1.8.1 Why Use Correlative Coding?

**Correlative coding**, also known as **partial-response signaling** (PRS), was introduced in the 1960s by Dr. Adam Lender to attain certain beneficial signal-transmission effects such as convenient spectral shaping. Lender's PRS also possesses, in general, the property of being relatively insensitive to channel imperfections and to variations in transmission rate.

Correlative coded baseband and modulated systems found numerous applications. One of Lender's earlier inventions, the conventional duobinary system, enables transmission up to 43% above the binary Nyquist rate. Wire, cable, and radio systems engineers have been using duobinary and modified duobinary (partial-response) systems because these systems are not sensitive to group-delay variations near the edge of the available bandwidth and to the high-pass effect of transformer coupled or, in general, ac coupled networks.

Chapter 8 presents a brief tutorial introduction to PRS. Following this material, an in-depth study of structures and the performance and applications of various PRS decoders is presented. Bit-by-bit PRS decoders as well as Viterbi receivers for these signals are studied. Correlative coded and continuous phase-modulation systems are described. The emphasis throughout Chapter 8 is on concepts not treated in detail in previous tutorial surveys or textbooks.

Partial-response and quadrature partial-response modulated systems are also covered in Chapter 7 (Section 7.12). The treatment in Chapters 7 and 8 is quite different: One is more intuitive and practical (experimental), whereas the other more analytical. If you are interested in further extending the research frontiers of partial-response systems, you will want to study both these chapters in depth.

Chapter 8 highlights partial-response baseband and modulated systems. Its relation to other chapters in this book is practically the same as illustrated in Fig. 1.14.

## 1.9 INTERFERENCE ANALYSIS AND PERFORMANCE OF DIGCOM SYSTEMS

### 1.9.1 Is Interference Analysis Essential?

Interference is present in all practical systems. With the advent of low-noise receivers, power and spectrally efficient modem/transmission techniques, and congestion in the radio frequency bands, the impact of **cochannel interference** (CCI) and **adjacent-channel interference** (ACI) on the performance degradation of digital transmission systems can be very significant. Interference may very well limit the performance of our present and future communication systems.

In most modern transmission-system designs, in addition to the classical thermal noise assumption, interference must also be taken into account. Thermal noise has an approximately Gaussian **probability density function** (pdf), whereas the interfering signal may have another type of pdf or be deterministic in nature. Therefore, the method of analysis and system-performance prediction that is used

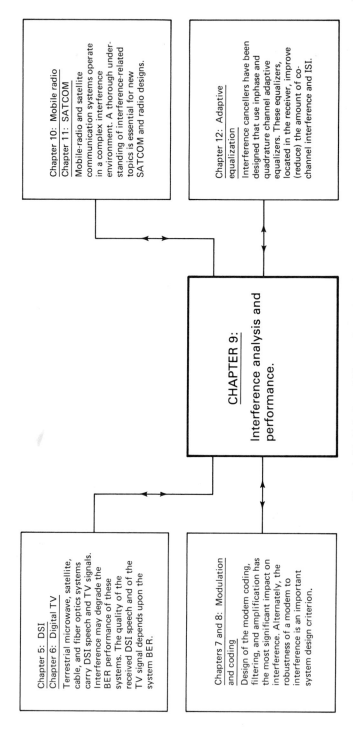

**Figure 1.15** Relation of Chapter 9 to other chapters.

to determine the effect of Gaussian noise alone on the performance of communications systems should not be used.

A new methodology is required to determine the performance of communication systems subject to non-Gaussian interference and to optimize their performances. Since thermal noise, however small, is almost always present, it is necessary that we consider the combined effect of Gaussian-distributed thermal noise and non-Gaussian-distributed interference. This makes the analysis, performance prediction, and practical specification of the transmission parameters more difficult. For example, after the choice of basic system parameters, such as required bit rate, RF bandwidth, power amplifier (linear or nonlinear), type of modulation technique (16 QAM, 49 QPRS or 64 QAM), and the frequency plan, the total interference should be calculated. Based on the initial interference calculations and its predicted impact on system performance degradation, one or more iterations in the original design parameters may be required.

Consider the design of a 49-QPRS modem for a microwave system application having a transmission rate of 90 Mb/s in a 20-MHz bandwidth. After preliminary study, we find that the 20-MHz RF bandwidth limitation requires very steep filters, which may introduce excessive **intersymbol interference** (ISI). Thus in the second iteration of our system analysis and design, we assume a 64-QAM modem for the same microwave system application and find that more practical filters (having a Nyquist-shaped raised-cosine roll-off factor of 30 to 50%) can be used. Now we further analyze the interference and "optimize" our system design. Here system optimization assumes that practical cost and implementation trade-offs are taken into account.

Power-efficient modulated systems, such as QPSK, have a better interference immunity than spectrally efficient modems, such as 64 QAM and 256 QAM. A 15- to 25-db interference power (below the modulated carrier power) degrades the $P_e$ performance of a QPSK system by about 1 dB. In a 256-QAM system the interference must be about 40 dB below the carrier for a 1-dB performance degradation.

Chapter 9 presents a comprehensive analysis of new methods to determine the joint effect of thermal noise and non-Gaussian interference and presents new insights into their performances. Some of the material covered in this chapter is very new, having been reported in the literature only during the last few years, and is being adopted and applied by system designers.

The material of Chapter 9 is closely related to the material presented in several other chapters (See Fig. 1.15).

## 1.10 MOBILE RADIO COMMUNICATIONS

### 1.10.1 An Ultimate Objective of Communication Systems

An ultimate objective of communications systems is to enable anyone to communicate instantly with anyone else from anywhere. In this ultimate scenario you (or your computer) should be able to establish a high-quality, reasonable-cost two-way communications link from your house, office, automobile, airplane, train, or bus. This can be achieved only by mobile radio communications.

A well-known historical event that clearly showed the importance of mobile radio communications was the distress of the Titanic in 1912. Morse-coded digital modulation was used in the early days of telegraph-mobile communication systems. Afterwards, analog and digitized voice transmission requirements increased.

Since the advent of mobile radio telegraph, various technological advances brought the appearance of many other mobile radio communication systems (radio telephone, radio paging, emergency dispatch, navigation control, and status reporting, for example). Demand for mobile radio communication services has steadily increased. However, mobile radio communication systems were limited to services for specialized groups because of the limited frequency spectrum allocated for mobile use.

In order to meet increased user demands, the new higher-frequency 900-MHz mobile radio band has emerged. Since this higher-frequency band has sufficient bandwidth to accommodate the increased demand, advanced high-capacity **cellular** mobile radio-telephone systems using this band have been developed in various countries.

Cellular mobile radio systems enable high-density geographical cochannel reuse and effectively achieve efficient spectrum utilization. Although digital techniques have been effectively applied for nonmobile radio systems to achieve high-speed, highly reliable data-transmission and high-grade, highly flexible system control. In general, digital voice transmission has not yet been adopted for mobile radio systems.

ISDNs are being designed on a global scale. It is expected that future mobile radio communication systems will be integrated into the digital telecommunications network and a variety of digitized voice and data services will be provided. For efficient implementations of digital mobile radio systems, a number of modern digital signal-processing and modulation/transmission techniques can be used.

*Efficient spectrum utilization* including *frequency, time,* and *space* is a basic requirement. Narrow-band transmission, multichannel access, and small cell layout are the three major solutions for conserving spectrum. Narrow-band transmission is achieved by low bit rate coding, bandwidth-efficient modulation, and stabilized carrier source techniques. Power-efficient modulators require fully saturated output RF power amplifiers. Multichannel access, which enables time-shared use of channels, requires a central processor using a stored-program-control scheme and microprocessor-controlled frequency-variable mobile radio transceivers. Small cell layout, which makes possible cochannel reuse with high geographical density, is achieved by a high-grade system control. Diversity and error control, which enable improvement of the cochannel interference performance, are effective for achieving small cell layout.

Chapter 10 presents a comprehensive study of modern digital mobile radio communication techniques. Mobile radio system requirements, propagation characteristics, digital modulation techniques suitable for mobile radio applications, diversity techniques, and radio link design/optimization techniques are highlighted. The material contained in the mobile radio chapter is closely related to the material in a number of other chapters (see Fig. 1.16).

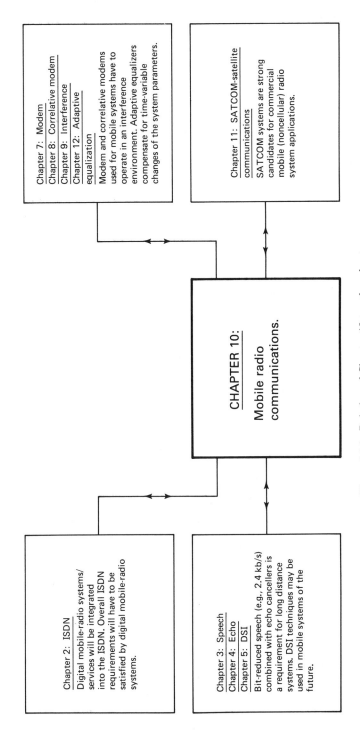

**Figure 1.16** Relation of Chapter 10 to other chapters.

## 1.11 SATCOM: SATELLITE COMMUNICATIONS: CONCEPTS AND TECHNOLOGIES

### 1.11.1 Unique Ability of SATCOM Systems: Phenomenal Growth

The unique ability of telecommunication satellites to service the globe has opened a new era for regional and global communications. Systems using a **single satellite** offer the flexibility to interconnect thousands of pairs of users separated by great distances up to 17,000 km (approximately one-third of the circumference of the earth). Systems using three satellites can provide a *global coverage* with multiple-access flexibility.

> *The growth of satellite communications capacity and capability has been revolutionary, a result of the flexibility provided by multiple-access, global-coverage digital satellite systems.*

Since 1965 more than 100 communications satellites have been launched. Predictions are that the trend will continue to grow, with some studies showing a need for 250 or more 1980-capacity satellites serving the United States by the year 2000. This truly unique multiple-access flexibility includes communication links between satellites and fixed points on earth, ships at sea, airplanes, trains, automobiles, other moving space vehicles, and "person-pack" terminals carried by a person and installed in 5 min or less. There are no other communication vehicles that can approach this flexibility.

The variety of data formats and services that can be provided by satellite links include: telephony signals; TV (visual and audio) signals; computer-generated signals (computer communications); broadcast data for computer communications; teleprinter; large-screen teleconferencing; interactive education; medical data; emergency services; electronic mail; newspaper broadcast; control data for power systems and utilities; traffic information; weather and land surveillance; navigational data for ships and airplanes; and military strategic data.

This list is far from complete and, as each year passes, it will become even less complete as new requirements are created and accommodated. The flexible multiple-access long-and-short-distance satellite systems offer more and more reliable, cost-effective solutions.

### 1.11.2 Topics Covered

Advanced SATCOM concepts and technologies are described in Chapter 11. A brief tutorial entitled "Basics of Satellite Communications" is followed by in-depth technical considerations of some of the most interesting topics of satellite communications. Multiple-access and demand assignment, the scanning spot beam concept, multiple scanning systems and spacecraft antenna technology are described. For the advanced systems engineer, the "transmission capacity" and

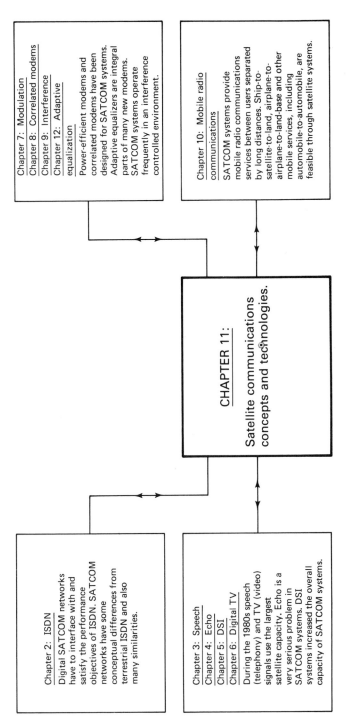

**Figure 1.17** Relation of Chapter 11 to other chapters.

Chapter 7: Modulation
Chapter 8: Correlated modems
Chapter 9: Interference
Chapter 12: Adaptive
equalization

Power-efficient modems and correlated modems have been designed for SATCOM systems. Adaptive equalizers are integral parts of many new modems. SATCOM systems operate frequently in an interference controlled environment.

Chapter 10: Mobile radio
communications

SATCOM systems provide mobile radio communications services between users separated by long distances. Ship-to-satellite-to-land, airplane-to-airplane-to-land-base and other mobile services, including automobile-to-automobile, are feasible through satellite systems.

CHAPTER 11:

Satellite communications concepts and technologies.

Chapter 2: ISDN

Digital SATCOM networks have to interface with and satisfy the performance objectives of ISDN. SATCOM networks have some conceptual differences from terrestrial ISDN and also many similarities.

Chapter 3: Speech
Chapter 4: Echo
Chapter 5: DSI
Chapter 6: Digital TV

During the 1980s speech (telephony) and TV (video) signals use the largest satellite capacity. Echo is a very serious problem in SATCOM systems. DSI systems increased the overall capacity of SATCOM systems.

"resource sharing of digital satellites" sections are of particular interest. Satellite and earth station frequency-reuse techniques and interference considerations complete the coverage of this chapter.

### 1.11.3 Relation to the Previously Published SATCOM Book and to Other Chapters

The advanced SATCOM topics described in Chapter 11 complement the material presented in my previously published book on digital satellite communications systems [Feher, 1.1]. The material in Chapter 11 is related to material presented in most of the other chapters of this book (see Fig. 1.17).

## 1.12 ADAPTIVE EQUALIZATION

### 1.12.1 Traditional and New Applications

Adaptive equalizers have frequently been used for voice-bandwidth "telephony" channels required to transmit digital information. The analog passband of these channels is nominally in the 300-Hz to 3000-Hz range. Data rates in the 50-b/s to 19.2-kb/s range have been transmitted through time-variable telphony channels. As the data rate transmission requirements over the analog telephony plant increased, the complexity of adaptive equalization techniques increased as well. Large-scale integrated circuits enable the implementation of fully digital adaptive equalizer circuits.

The problem of undesired echo has been noticed in systems that have a particularly long transmission delay between the transmitter and receiver. In very long long-distance terrestrial and satellite systems, echo effects frequently annoy the user. To cancel the echo efficiently, *echo cancelers* described in Chapter 4 have been developed. One of the most challenging circuits of modern echo cancellers is the adaptive equalizer, which adapts its transfer function (time domain impulse response) to the time-variable echo characteristics.

Modern terrestrial LOS microwave systems have increased spectral efficiency. For example, current operational 90-Mb/s and 135-Mb/s rate radio systems have a practical spectral efficiency of 4.5 bits/s/Hz. The spectral efficiency of a recently developed 1.544-Mb/s system having an encoded 256-QAM modem is 6.7 bits/s/ Hz. These highly spectral-efficient radio systems operate in a time-variable "selectively faded" system environment, which is caused by multipath fading. To compensate for time-variable propagation distortions as well as for hardware (filter) imperfections, adaptive equalizers are used. For the 90-Mb/s to 135-Mb/s radio systems, equalizers with about 5 to 10 taps are used. For the more efficient 1.544-Mb/s case, adaptive equalizers with up to 50 taps (per rail) are used. Modems used for these systems and overall system considerations are described in Chapter 7 and in [Feher, 1.2].

Satellite and terrestrial microwave systems frequently use a cochannel frequency plan to increase the overall system efficiency. Crosspolarization discrim-

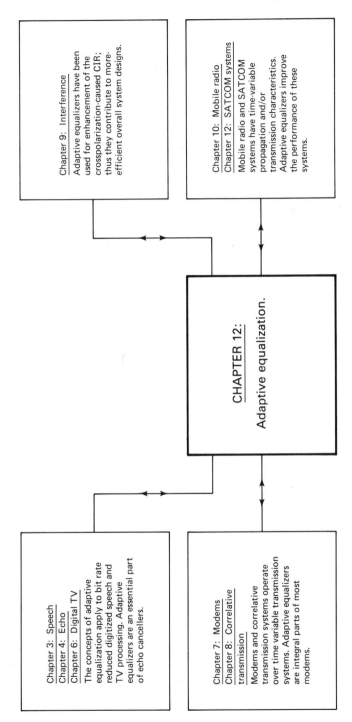

**Figure 1.18** Relation of Chapter 12 to other chapters.

ination between the vertically and horizontally polarized radio channels is a function of time variable multipath propagation conditions. Adaptive equalizers have been successfully employed for enhancement of the crosspolarization-caused **carrier-to-interference ratio (CIR)**. SATCOM systems, particularly high-speed TDMA satellite communications systems operated at 10-Mb/s to 180-Mb/s transmission rates also require adaptive equalizers for a number of applications.

### 1.12.2 ISI Cancellation

In spectrally efficient communication systems, the effect of each transmitted symbol extends far beyond the time interval used to represent that symbol. The **distortion** caused by the resulting overlap of sequence of received symbols is known as ISI. Transmitted symbols are recovered (regenerated) in the receiver in consecutive sampling instants. To achieve optimal performance, it is required to minimize the ISI in consecutive sampling instants. Time-variable distortion, introduced by less-than-perfect transmission systems, may be reduced by adaptive equalizers.

### 1.12.3 Adaptive Equalizer Structures

Modern equalizers are very powerful and efficient. A large class of equalizer structures has been developed to enable an efficient application for various system requirements.

Chapter 12 presents an in-depth description of some classical and some new, advanced equalizer structures. Linear and nonlinear equalizer structures, optimum and suboptimum receive filters, LMS adaptation and fast-converging equalizers are described in this chapter.

The material presented in Chapter 12 is useful for a number of systems applications described in other chapters (see Fig. 1.18).


## REFERENCES

This is a *partial* list of references. Comprehensive reference listings are provided at the end of Chapters 2–12.

[1.1] Feher, K. *Digital Communications: Satellite/Earth Station Engineering*, Prentice-Hall, Englewood Cliffs, N.J. 1983.

[1.2] Feher, K. *Digital Communications: Microwave Applications*, Prentice-Hall, Englewood Cliffs, N.J. 1981.

[1.3] Feher, K., and Engineers of Hewlett-Packard, *Telecommunications Measurements, Analysis and Instrumentation*, Prentice-Hall, Englewood Cliffs, N.J., 1987.

[1.4] Feher, K. "Digital Modulation Techniques in an Interference Environment," in Vol. 9, Encyclopedia on EMC, Don White Consultants, Inc., Gainesville, Va., 1977.

[1.5] Gowar, J. *Optical Communication Systems*, Prentice-Hall, Englewood Cliffs, N.J., 1984.

[1.6] Viterbi, A. J., and J. K. Omura. *Principles of Digital Communications and Coding*, McGraw-Hill, New York, 1979.

[1.7] Dixon, R. C. *Spread Spectrum Systems,* John Wiley, 1976.

[1.8] Oppenheim, A. V., and R. V. Schafer. *Digital Signal Processing*, Prentice-Hall, Englewood Cliffs, N.J., 1975.

[1.9] Bellamy, J. C. *Digital Telephony,* John Wiley, New York, 1982.

[1.10] Ziemer, R. E., and W. H. Tranter. *Principles of Communications-Systems, Modulation and Noise,* 2nd ed., Houghton Mifflin, Boston, 1985.

[1.11] Martin, J. *Communication Satellite Systems*, Prentice-Hall, Englewood Cliffs, N.J., 1978.

[1.12] Proakis, J. G. *Digital Communications*, McGraw-Hill, New York, 1983.

# 2

# ISDN: INTEGRATED SERVICES DIGITAL NETWORK: ARCHITECTURES AND PROTOCOLS

*DR. MAURIZIO DECINA\* and DR. ALDO ROVERI*

*Professors of Electrical Communications*
*INFOCOM Department*
*University of Rome*
*Rome, Italy*

## 2.1 INTRODUCTION

An introduction to the emerging integrated services digital network (ISDN) is given in this chapter, which is organized in three parts.

The first part gives a framework for the ISDN concept. International worldwide developments promoting the introduction of such networks are also described. This first part is a self-contained contribution for readers interested only in a general overview.

The other two parts expand the concepts introduced in the first part. They deal with network architectures and protocols for the ISDN with an emphasis on terrestrial network implementations.

In particular, the second part considers the general trade-offs among various communications switching techniques necessary to enable voice, data, and video information transfer. Also described are the main ISDN approaches chosen in North America and Europe, with reference to their planned implementations.

The protocol aspects, described in the third part, cover both dedicated network standards and current trends for ISDN applications. Both user-access protocols and interexchange protocols are briefly considered.

*On leave of absence at the ITALTEL Central Research Laboratories, Milan, Italy.

## 2.2 THE EMERGING ISDN: AN OVERVIEW

In this first part of the chapter an overview of current implementations of dedicated digital networks for voice and data is first presented (Section 2.1.1). Then, a wider range of communication services and technological advances providing the synergistic element for the development of the ISDN is considered (Section 2.1.2). The evolving ISDN scenarios are then outlined in terms of both network architectures and protocol structures (Section 2.1.3).

### 2.2.1 Dedicated Digital Networks

During the last two decades, architectures and standards for digital transmission networks dedicated to voice and/or data have been established worldwide. Such networks are called integrated digital networks (IDNs), where the term *integrated* refers to the commonality of digital techniques used in transmission and switching systems. The two applications (voice and data) are considered next.

**Voice Applications**

The telephony IDN is based on 64-kb/s pulse code modulation (PCM) coding of speech signals. As a result, this standard has been applied to both transmission and circuit-switching systems. We distinguish between **PCM multiplexing** and **digital multiplexing** equipment shown in Fig. 2.1; both operate according to the time division multiplexing (TDM) principle, that is, a multiplexing in which two or more channels are interleaved in time for transmission over a common channel.

PCM-multiplexing equipment (channel banks) use 8-bit time slot interleaving of 64 kb/s channels to form primary PCM multiplex signals at 2.048 Mb/s or 1.544 Mb/s. This is done in accordance with the European or the North American multiplexing hierarchy.

Digital multiplexing equipment uses bit-interleaving of digital tributary signals at a given hierarchical bit rate to form a single signal at a higher bit rate.

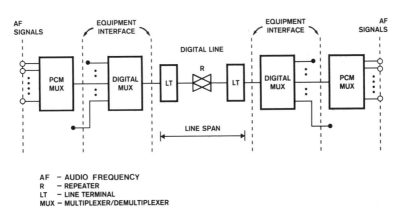

AF  − AUDIO FREQUENCY
R   − REPEATER
LT  − LINE TERMINAL
MUX − MULTIPLEXER/DEMULTIPLEXER

**Figure 2.1**  Typical digital transmission system.

The digital multiplexing hierarchy establishes the sequence of hierarchical bit rates as follows:

| | |
|---|---|
| Europe | $2.048 \rightarrow 8.448 \rightarrow 34.368 \rightarrow 139.264$ Mb/s |
| North America | $1.544 \rightarrow 6.312 \rightarrow 44.736$ Mb/s |

Such bit rates apply to equipment interfaces, as shown in Fig. 2.1.

Figure 2.2 presents an illustrative example of a digital transmission chain, which provides digital connectivity between two PCM offices. This chain includes multiplexing equipment and line transmission equipment interconnected at various hierarchical bit rates according to the European hierarchy.

**Circuit switched offices** are controlled by software programs stored in reliable processing units [Joel 2.33; Joel 2.34]. Switching systems (connection networks) provide inlet-outlet interconnections at 64 kb/s and operate on time-slot multiplexed signals at bit rates greater than or equal to the primary ones.

In advanced telephony IDN structures, call control (signaling) information is exchanged among offices through a common channel signaling (CCS) network. The CCS network is a computer-communication network employing packet switching techniques to transfer signaling messages among office control computers. CCS switches are called **signaling-transfer points** (STPs) and are geographically duplicated to enhance interoffice signaling reliability. The packet switching mode employed in CCS networks is datagram-oriented (see Section 2.3.1). Each **datagram** (DG) is a message that contains sender and receiver office addresses in its header; it can be routed as a lone message through the network. As a peculiar feature of the CCS networks, DGs belonging to a single signaling transaction between offices are routed via a single path through the CCS network.

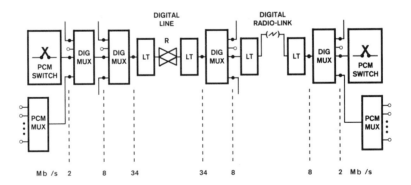

R   – REPEATER
LT  – LINE TERMINAL
MUX – MULTIPLEXER-DEMULTIPLEXER

**Figure 2.2** Typical digital transmission chain for the European hierarchy.

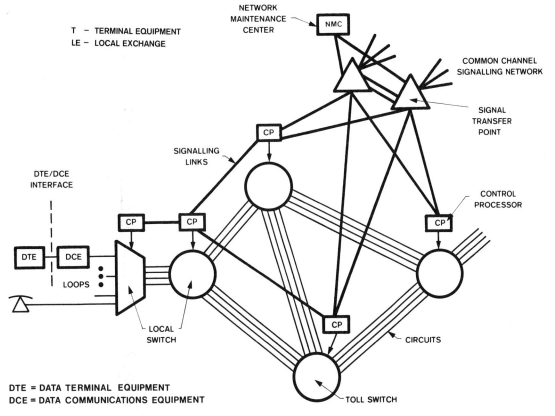

**Figure 2.3** Synchronous circuit—switched network (for voice and data applications) using CCI.

The model of a circuit switched network suitable for both voice and data applications is also shown in Fig. 2.3. Here, we can distinguish between the **trunk network** and the **CCS network**. **Local switches** are composed of **subscriber stages** (with interface and concentration functions) and **group stages** (with distribution functions). Subscriber stages can be colocated with group stages or they can be remotely placed, while **toll switches** are composed of group stages only. Trunk circuits are provided by 64-kb/s paths implemented via transmission systems.

Both PCM multiplex and switching equipment operate *synchronously* according to a common network reference frequency. Hence, a suitable timing distribution must be implemented according to a selected network **synchronization strategy**.

Figure 2.3 presents an ultimate scenario for the telephony IDN that has been evolving since its origin in the second half of the 1960s. The evolution of telephony IDN followed a path dictated by technological advances subject to some cost-effectiveness criteria. The principal steps were

1. Point-to-point PCM transmission in the exchange plant
2. Short-haul PCM transmission

**3.** Toll PCM switches with CCS capabilities

**4.** Local PCM switches

Although 64-kb/s PCM techniques have achieved a high degree of penetration into the telephone networks, economics still favor analog techniques in at least two areas. These are long-haul terrestrial or satellite links and subscriber loops. Moreover, cost-effectiveness is not apparent when application of CCS to local switching is compared exclusively to the basic requirements of **plain old telephone service** (POTS). These considerations have different impacts on different national network geographical layouts. For example, in most European countries, 64-kb/s telephony is cost-effective over long-haul trunks due to their limited length. However, in North America, long-haul digital transmission is more attractive when using **low bit rate voice** (LBRV) techniques* to encode speech at rates less than 64 kb/s [Aaron, 2.1]. On the other hand, digital subscriber loops and local CCS are both expected to be stimulated by the multiservice capability in the evolution towards the ISDN.

### Data Applications

Data applications can be classified according to the *terminal activity* during a call. **Terminal activity** refers to the fraction of time during which the data terminal is active during the data transfer phase; it will be denoted by $\alpha$ with $0 \leq \alpha \leq 1$. Two types of data can be broadly identified, **bulk** and **bursty,** depending on the amount of terminal activity. Bulk data, such as facsimile and batch data transfer, are characterized by high terminal activity ( $\alpha > 0.5$), whereas bursty data, such as interactive start-stop terminal-to-computer communications, present low terminal activity during the call (as shown in Table 2.1).

Two digital communication modes have emerged as best suited to handle bulk and bursty data. These are **circuit switching** and **packet switching**, respectively. Dedicated data networks are implemented on the basis of these two switching principles.

**Circuit switched IDNs** dedicated to data applications are designed according to the same principles as those used for telephony applications, especially for the case of synchronous communications. The basic user data rates for synchronous data networks are 2.4, 4.8, 9.6, 19.2, 48, 56, and 64 kb/s. These rates apply to **data terminal equipment** (DTE) and **data circuit-terminating equipment** (DCE) interfaces. TDM techniques are used to interleave lower rate channels, thus forming a basic 64-kb/s channel. Circuit switching can be performed for lower-rate digital data channels (**data subrate switching**) either on a subrate backbone switch or by using a common 64-kb/s connection network. In this latter case, appropriate bit stuffing-destuffing techniques should be provided to handle the lower bit rate channels.

Figure 2.3 illustrates a model that is also appropriate for the **synchronous** circuit switched data networks. In this case, a full digital facility should be available from the very early stages of network development.

---

*A detailed description of LBRV coding techniques is presented in Chapter 3.

**TABLE 2.1** Classification of data in accordance with terminal activity

| | |
|---|---|
| Low Activity (<0.5) | **Bursty Data** Remote interactive Remote file management Remote entry: on-line Videotex Telemetry |
| High Activity (> 0.5) | **Bulk Data** Remote entry: off-line CPU-CPU: file transfer Remote batch Teletex Facsimile |

Circuit switched data networks have also been implemented for **asynchronous** (start-stop) data communications. Such dedicated networks were originally designed to handle telex and start-stop data up to 300 bits/s. At the present time, a number of countries have established circuit switched data IDNs suitable for both asynchronous and synchronous communications [Staudinger, 2.50].

**Packet switched IDNs** represent an evolution of message-switching techniques [Rosner, 2.46]. Data information is assembled in short messages (packets) that are transmitted by users at the same channel rates as specified for synchronous circuit switched networks. Packet switched data networks employ statistical multiplexing (SM) of data packets over digital transmission links. These links generally operate at a rate of 64 kb/s.

Packet switching offices operate on the basis of the **store-and-forward** principle. They are realized by a suitable interconnection of stored program-processing units. Users can communicate by lone data packets (DGs) or by establishing a **bidirectional call** involving multiple data packets (**virtual call**).

Virtual calls are established (or cleared) by means of call setup packets (or clear request packets) that create (or disconnect) a virtual circuit through the network to be followed by the subsequent data packets. Virtual call control packets have the same format as data packets and dynamically share with these the same communication network resources, such as transmission links, link buffers, switching paths, and so on.

**Virtual circuit routing** permits less packet header overhead than **DG routing**. In the first case data packets are routed through the network by using short logical channel numbers instead of rather long DTE addresses. A second advantage is that virtual circuit routing minimizes the probability of out-of-sequence delivery of packets belonging to the same call. On the other hand, DG routing is potentially more suitable for the adoption of adaptive mechanisms on a per-packet basis rather than on a per-call basis.

The model of a packet switched data network is illustrated in Fig. 2.4. In this system both packet-mode DTEs and non-packet-mode DTEs (e.g., start-stop terminals) can access packet switched services. **Packet assembling/disassembling**

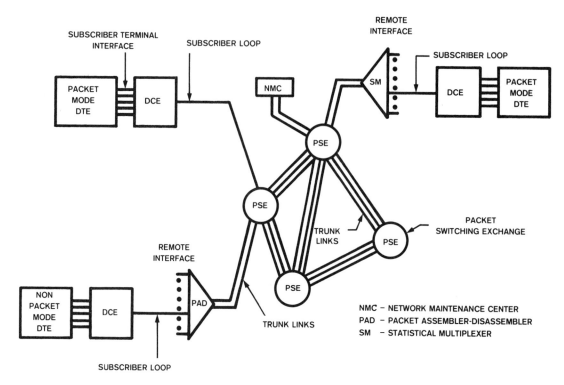

**Figure 2.4** Packet switched data network.

(PAD) devices are used to interface non-packet-mode DTEs, usually via dial-up connections over the analog telephone network.

### 2.2.2 Emerging Communications Services

The scenarios depicted for dedicated digital telephony and data network are today evolving towards the ISDN scenario. The fast-growing demand for new communication services along with recent advances in technology create strong incentives for the development of the integrated services networks. A review of emerging communication services is presented next.

Communication services may be characterized by several parameters and thus may be classified in accordance with any of them. For example, they can be classified according to **market** (residential, business); **information type** (voice, sound, video, data); **statistical traffic behavior** (traffic intensity, terminal activity, message length distribution); **performance requirements** (delivery delay, errors); or required **channel capacity** (in bits per second). Here we choose a simple unidimensional classification based on the latter parameter. Table 2.2 assigns channel capacities for present and potential digital services, covering a broad range of bit rates ranging from less than 1 kb/s up to about 1 Gb/s.

The group of services requiring rates in the 10- to 100-bit/s range encompasses **telemetry applications**. These are typical of home and business applications and involve data sources with low activity and burst mode of operation.

**TABLE 2.2** Required capacity (in b/s) for emerging digital communication services

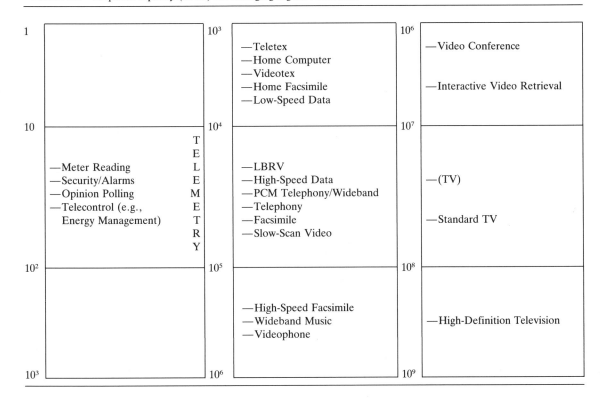

The group of services classified in the 1- to 10-kb/s range finds application in home and business and include *text, graphics,* and *data communications.* These services (*teletex, home computer, videotex, home facsimile, low-speed switched data*) are largely interactive and suitable for the packet switched mode of operation. In particular, Teletex and Videotex are CCITT-defined **telematic services.** Teletex is devoted to remote word processing. Videotex addresses interactive alphanumeric or alphageometric text retrieval.

Between 10 and 100 kb/s, we find **voice services.** In addition to 64-kb/s PCM telephony, **wideband voice** (up to 7 kHz) at 64 kb/s and LBRV at rates of 32 kb/s or less are included. At 32 kb/s an **adaptive differential PCM** (ADPCM) coding scheme has been adopted as an international standard [CCITT, 2.10]. Such a coding algorithm uses both adaptive quantization and adaptive prediction techniques [Petr, 2.43]. At 16 kb/s and below, more-sophisticated coding schemes should be adopted in order to obtain telephone quality voice over an end-to-end digital connection. It is expected that further advances in technology will lead to new LBRV standards in the near future.

Voice communication is bursty in nature. The average voice activity is estimated to be in the order of about 0.35 to 0.40 and is largely dependent on particular speech detector characteristics such as the hangover time [Gruber, 2.25]. Such a low-voice activity enables suitable speech interpolation on long-haul trunks in circuit switched networks and presents attractive possibilities for **speech packetization** within the evolving ISDN.

**High-speed switched data, digital facsimile**, and **slow-scan video** are bulk services that also belong to the group requiring between 10 and 100 kb/s. They can be conveniently handled by circuit-switching techniques using a single 64-kb/s time slot.

A multislot ($n \times 64$ kb/s) circuit switched transport capabilities system is required to handle bulk services between 100 and 1000 kb/s, such as **high-speed facsimile** and **high-quality music**.

The last column of Table 2.2 comprises **bulk video services** using capacity between 1 Mb/s and 1 Gb/s. Dedicated broadband digital transmission and switching facilities are needed to provide such services. There is a big difference in the digital capacity required to encode *business* as compared to *entertainment* video signals.

Among business applications, the **videoconferencing** service is in wide use. It requires an approximate rate of a few megabits per second in order to encode a standard color TV signal: Specifically, rates of 2 Mb/s for Europe and 3 ($2 \times 1.5$) Mb/s for North America are possible choices. The quality reduction resulting from the use of sophisticated redundancy reduction algorithms is tolerable when considering the loose service requirements in terms of movement reproduction.

On the other hand, a **broadcast-quality TV program** may require about 100 Mb/s to cope with the stringent transmission requirements set up by international bodies. The three levels in the digital hierarchies are points of reference for current development of TV codecs. Data rates of about 34, 68 ($2 \times 34$) and 102 ($3 \times 34$) Mb/s are potential bit rates for Europe, whereas 45 and 90 ($2 \times 45$) Mb/s are most likely for North America.

Entertainment video services can be offered in a variety of modes, such as *basic TV, pay TV, video on demand, video shopping*, and *two-way video*. All these modes can be implemented in a flexible manner when associated with sophisticated high-capacity digital subscriber signaling capabilities. However, we note that there is a growing worldwide interest in the introduction of **high-definition TV** standards needing up to 35 MHz for the analog baseband signal. This implies the need for several hundred megabits per second for conversion to a digital format [Ninomiya, 2.41].

### 2.2.3 The ISDN Concept

Service-dedicated networks have been conceived and implemented to provide, in a cost-effective manner, a limited range of services centered around the main service application to which they are devoted. The proliferation of different types of digital networks (e.g., a circuit switched IDN for telephony, a circuit switched IDN for bulk data, a packet switched IDN for bursty data) imposes a burden on both network users and providers. This is particularly true when taking into account the multiplicity of access interfaces and access facilities. Access arrangements represent a very significant portion of the overall system cost.

These circumstances, together with the rapid growth of communication system capabilities and technologies, have stimulated the emergence of the ISDN concept [Decina, 2.15; 2.16]. An ISDN can be characterized by its three main features:

1. End-to-end digital connectivity
2. Multiservice capability (voice, data, video)
3. Standard terminal interfaces

With reference to features 1 and 2, the ISDN provides a network transport capability for a variety of services (ranging from telemetry up to broadband video applications) using a variety of digital communication modes (from leased and semipermanent connections to circuit and packet switched connections). Essential to the functions of an ISDN are its integrated operating capability as well as its administrative and maintenance functionality, such as network management and billing.

For feature 3, the key element of service integration in the ISDN is the provision for a limited set of standard user-network interfaces. Four types of ISDN users are shown in Fig. 2.5. These are: (1) a single ISDN terminal, such as a **multiservice station**; (2) a multiple ISDN terminal installation, such as a **multidrop arrangement** directly controlled by the local office; (3) a user network, such as a **local area network** (LAN), a **private branch exchange** (PBX) or a network of PBXs; (4) a service vendor, such as a **data base** for information retrieval. Through access interfaces, users can select the requested communication facilities for single or multiple simultaneous services.

An ISDN is recognized by *the service characteristics* (protocol, performance, etc.) *offered at its access interface* rather than by its internal architecture, config-

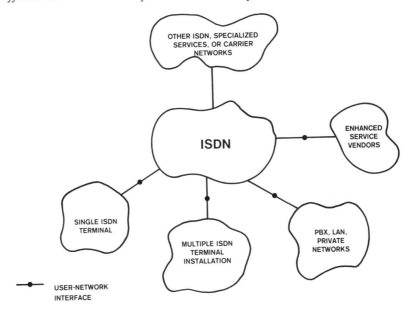

**Figure 2.5** ISDN access interfaces.

uration, or technology. This concept plays a key role in permitting user applications and network technologies to evolve separately. Essential to the ISDN concept is indeed the integration over access interfaces and loops. Hence, the ISDN may be implemented in a variety of configurations according to the particular telecommunication environment from which it emerges, in terms of initial network status and progressive evolution.

In general, the ISDN should be based on and evolve from the 64-kb/s telephony IDN, including digital subscriber loop facilities. The telephony IDN will provide interconnection with current service-dedicated facilities (such as data packet switching and wideband switching). It will progressively incorporate, according to technology evolution and economic considerations, additional network functions and features, including those of any other dedicated service. As the ISDN evolves, the trend is to proceed with integration from the access arrangement toward the core network equipment. To improve cost effectiveness, current service-dedicated facilities will be incorporated into common network equipment.

The transition from the existing network to a comprehensive ISDN will require a long period of time. In the initial stages of evolution, new network and system architectures will arise to improve the efficiency of integrated voice, data, and wideband communications.

The following two subsections are devoted to the evolving ISDN architectures, including service capabilities, and to the access or internal network protocols, respectively.

### Network Architectures

Three main stages of the evolving ISDN architecture are here identified: (1) an early ISDN architecture for **voice and data** (V&D) capability; (2) an advanced stage for an enhanced integrated transport capability also for V&D; and (3) an ISDN architecture with broadband capability.

***Early ISDN Architecture for V&D.*** In the early stages of development, **hybrid switching systems** (i.e., circuit and packet switching) characterize the ISDN. These systems are implemented by integrating current carrier network facilities for voice and data. Bulk data and voice are treated by 64-kb/s **circuit switching** (CS) and CCS, while bursty data are handled by **packet switching** (PS). These integrated facilities can share common network equipment with an increasing degree of commonality.

There is a need to exploit existing resources in the subscriber loop plant essentially to provide voice-data integration over copper loops. This medium is adequate for the provision of digital access capability to small terminal clusters requiring capacities up to $n \times 64$ kb/s, $n$ being a small integer. Existing copper plant can also conveniently provide bit rates up to 1.544 or 2.048 Mb/s to connect large terminal clusters supported by PBXs or LANs.

The primary PCM multiplex bit rates can also be applied to broadband services, such as videoconferencing. However, the use of existing copper plant at such bit rates involves very short loops, thus restricting the service penetration. Therefore, broadband video services for both business and entertainment purposes may

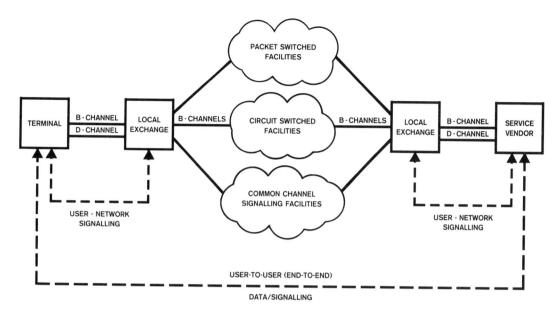

**Figure 2.6** Early ISDN architecture for V&D capability.

more effectively be deployed by using new fiber distribution plant. The actual switching of broadband signals via terrestrial facilities implies an advanced technology that is not yet mature. From the outset, the switching of 1.544- or 2.048-Mb/s videoconferencing channels through the ISDN will be provided on demand by satellite connections. However, it is expected that voice and data service will be dominant in the first stages of the ISDN evolution, whereas broadband video services will be deployed at later stages.

To set the ISDN concept in motion, various access types are foreseen, the most significant being the **basic access**. This access is intended to serve single or multiple ISDN terminal installations (e.g., up to 8 terminals). To facilitate this, two types of digital communication channels are defined, the **B-channel** and the **D-channel**. The B-channel operates at 64 kb/s, whereas the D-channel operates at 16 kb/s.

The basic access structure consists of a TDM assembly of two B-channels and one D-channel. The D-channel is message-oriented and carries the **signaling information** (*s*-information); this controls circuit switching of the B-channels through the ISDN in cooperation with CCS facilities. **Telemetry** (*t*-information) and **low-speed bursty data** (*p*-information) are multiplexed into the D-channel together with signaling messages. The B-channel is devoted to circuit switching through the ISDN for a variety of transmission applications, typically for PCM voice.

An early ISDN architecture is shown in Fig. 2.6. Here, the terminal gains access to an ISDN local exchange through the B- and D-channels. The toll facility is comprised of circuit switched 64-kb/s channels, common channel signaling and data packet switching channels. The D-channel and CCS facility transport *s*-information, whereas *p*-information is routed through the PS facilities by virtual circuits. For the case of *t*-information, this could be routed through either PS or CCS facilities. Here, it may be handled either by a **permanent virtual circuit** (PVC)* or by DGs. Typical uses of B-channels include the transmission of

1. PCM speech at 64 kb/s;
2. Data information corresponding to CS or PS data user classes of services at bit rates adapted to 64 kb/s;
3. LBRV combined with data information, both directed towards the same destination;
4. Wideband digital speech at 64 kb/s.

Alternate selection among the stated uses may be provided by one B-channel on a call-by-call basis or by a change during an established call.

For application 2, the B-channel may also be used as a support for PS data. In this case this channel provides a digital dial-up connection to a PS equipment port that handles the standard in-band PS protocol. The PS equipment may either be integrated into the local exchange or it may be placed in the PS facilities common to the whole ISDN.

For application 3, in general B-channels may be built up as a TDM assembly of subrate channels (i.e., at 8, 16, or 32 kb/s) and each subchannel could be routed to a different destination. Handling of individual subchannels through the ISDN may be implemented in various ways. For example, circuit switching of subrate channels through 64-kb/s connection networks can be performed by employing rate adaptation at digital office terminations.

The ISDN services currently considered for B- and D-channel access capabilities are summarized in Table 2.3.

The network architecture just described may be used by any type of access based on a TDM assembly of multiple B- and D-channels. The D-channel bit rate may also be extended up to 64 kb/s and a multislot transport capability for up to six 64-kb/s time slots (384 kb/s) may also be included.

More advanced network architectures should eventually be developed to offer an enhanced integrated transport capability of voice, data, and broadband information.

***Advanced ISDN Architecture for V&D.*** V&D integration can be enhanced by using a **common PS** technique [Decina, 2.17]. Packetized voice requires the definition of *low functionality protocols*, simpler than those used today for data communication (see Section 2.4.1). Also required is the provision for rather complex *voice-interface* functions. Packetized voice and data protocols should avoid error recovery and window flow control functions on a link-by-link basis through the network. These functions should be allocated only for data com-

---

*A PVC is a virtual circuit permanently set up between end users for a preassigned period of time.

**TABLE 2.3** ISDN services for B- and D-channels

---

**D-Channel Services (16 kb/s)**
  Enhanced telephony
  Low-speed data (PS)
  Videotex
  Teletex
  Telemetry
    Emergency services
    Energy management

**B-Channel Services (64 kb/s)**
  Voice
  High-speed data (CS & PS)
  High-quality voice
  Voice and Data End-to-End
  Assembly of subrate channels
  Facsimile
  Slow-scan video

---

CS = Circuit switching
PS = Packet switching

---

munication and should be initiated at the interfaces of the packetized voice and data facilities. The voice interface to packet switched facilities should include

1. LBRV coding (e.g., at 32 kb/s)
2. Speech detection for talkspurt identification
3. Echo control (e.g., by echo cancelers)
4. Packet assembly and disassembly
5. Transport packet delay equalization
6. In-band signal classification to handle telephone tones or voice-band data alerting tones

For packetized voice, the **delay performance** is a critical parameter that depends largely on the carrier-channel bit rate. The use of high-speed links at 1.544 or 2.048 Mb/s throughout the network may result in acceptable delays (a few hundred milliseconds) for voice. Also, a 1.5- or 2-Mb/s packetized digital pipe could carry (assuming statistical multiplexing) not only voice and bursty data but also a certain amount of bulk data traffic having a limited peak rate (e.g., 64 kb/s).

To make effective a packetized voice and data network, it is essential to develop *high-capacity PS systems,* characterized by very low transit delay. This capacity should be in the order of 1,000,000 packets/s and exhibit transit delays of no more than about 1 ms. Hence, in order for packetized voice and data facilities to become economical, substantial progress is necessary in signal processing and switching technology implemented with **very large scale integration** (VLSI) circuits.

An advanced ISDN scenario is shown in Fig. 2.7, in which the local section represents the same hybrid overlaid structure as in Fig 2.6. The communication

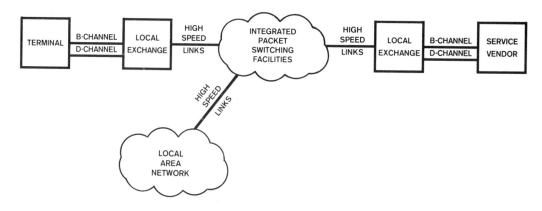

**Figure 2.7** Advanced ISDN architecture for V&D capability.

facilities in the toll section are integrated in a single packetized facility using high-speed links and high-capacity switching systems. In this toll scenario, signaling, voice, and data dynamically share the same transmission and switching facilities. Interface conversion functions for voice, data, and signaling occur at gateways to and from local exchanges. We note that the access to the integrated packet-switching facilities by LANs can be directly provided, thus bypassing intermediate local exchanges.

*ISDN Architecture With Broadband Capability.* The provision of broadband services requiring large analog (5 to 35 MHz) or digital (up to several hundreds of megabits per second) capacities appears feasible. This is so in the light of the deployment of **fiber distribution plant** and **wideband switching systems**. The fiber medium is well suited to carry information flows, each in either digital or analog form, via **wavelength division multiplexing** (WDM), including

1. Digital signals for narrowband applications (e.g., signaling, voice, data, and slow-scan video) up to about 2 Mb/s;
2. Broadband video signals in digital format;
3. Broadband video signals in analog format.

Hence, the provision of one or two fibers to each subscriber allows a tremendous multiservice capability with transmission in both directions. Interface procedures for communication control and service selection may be supported by the D-channel protocol, adequately extended to cover the requirements of video services.

Present trends in structuring wideband distribution networks allow for several architectural alternatives including *bus, ring,* and *star* layouts. An attractive alternative considered is the double-star network centered around the central office and a number of **remote switching units** (RSUs). Such RSUs would be placed near (within 1 to 2 km) the subscriber premises and serve a few hundred subscribers, each with a fiber drop of moderate quality (low cost). RSUs are connected to the local exchange via high-quality (low-loss) fiber feeders. These feeders supply the RSUs with a number of TV channels, thus providing selective channel distribution

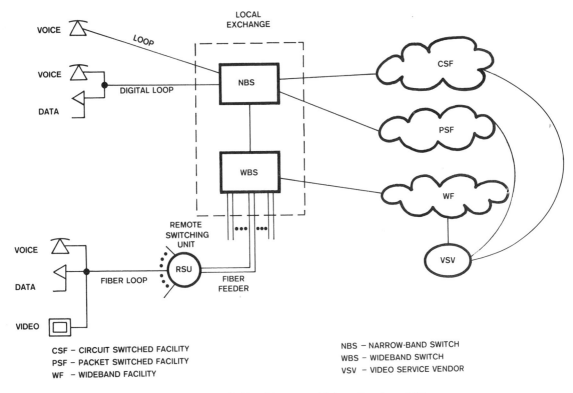

CSF – CIRCUIT SWITCHED FACILITY
PSF – PACKET SWITCHED FACILITY
WF  – WIDEBAND FACILITY

NBS – NARROW-BAND SWITCH
WBS – WIDEBAND SWITCH
VSV – VIDEO SERVICE VENDOR

**Figure 2.8** ISDN architecture with broadband capability.

to requesting subscribers. The double-star solution seems to meet the important objectives of ease of expansion and enhanced maintainability [Kneisel, 2.36].

An ISDN architecture with wideband service capabilities is shown in Fig. 2.8. The local ISDN exchange presents an adjunct **wideband switching system** (WBS) [Cardwell, 2.6], which constitutes the center of the double-star fiber distribution network. It provides the gateway to the wideband toll transmission and switching facilities. The WBS also provides the gateway to the narrow-band ISDN facilities via the local **narrow-band switching system** (NBS).

As far as the *role of satellites* in ISDN is concerned, it is recognized that satellites are certainly suitable as multiservice toll communication media. They may be an alternative to be used in conjunction with terrestrial media [Guenin, 2.26]. It is also agreed that satellites can provide cost-effective connectivity for digital wideband services. The capability for provision of worldwide videoconference services is significant. However, satellite transmission is characterized by a high transit delay (about 280 ms) that in the case of multihop connections may cause unacceptable performance for voice and data applications. This refers specifically to subjective voice quality and data throughout resulting from a particular protocol.

Satellites can be used not only as toll communication facilities but also as means to implement specialized communication networks. These networks can provide end-to-end digital connectivity and multiservice capability to users on a private basis. Specialized satellite networks have emerged to serve the large business community in the absence of widespread terrestrial ISDN capabilities.

### Network Protocols

The definition of a suitable protocol architecture is the key issue in partitioning processing and communication resources among users and networks in order to establish information services [Pouzin, 2.44; Tanenbaum, 2.51; Zimmerman, 2.53; des Jardins, 2.18].

At first, principles of protocol layering for data networks are here considered by introducing a suitable **reference model**. A distinction is also made between access and internal network protocols. Then, problems related to ISDN protocol architecture are focused and the corresponding standard activities are summarized with reference to the present trends of CCITT.

*Protocols For Data Networks.* The **open system interconnection** (OSI) reference model is a model for developing compatible protocols for communication among heterogeneous "systems" (terminals, computers, networks, processes, etc.). The OSI model proposes a structured approach to protocol specification based on a seven-layer architecture. Layering is employed in order to

1. Subdivide the complex problems of protocol implementation into a workable number of less-complex pieces;
2. Ensure a high degree of independence of each piece, or layer.

This allows changes of functions within a layer to be made, thus taking advantage of new advances in hardware or software technology without affecting other layers.

The key aspects of protocol layering are: (1) the definition of homogeneous functions and procedures within each layer, which allow entities operating in the same layer to communicate correctly; and (2) the creation of boundaries between layers, so that each layer has direct interaction with only the two adjacent layers. With a sequential number scheme from 1 through 7, each layer uses a set of "services" offered by the next-lower-numbered layer and provides a larger set of services to the next-higher-numbered layer. Exceptions are layers 1 and 7, which offer services to and receive services only from their adjacent layers, respectively. Moreover, layer 1 receives services from the *communication medium* that supports the exchange information, whereas layer 7 offers services to the *application process* that is the ultimate source or sink of information.

An **application** may be composed of a number of cooperating application processes, which perform the processing of information. Applications or application processes may be of various kinds—for example, manual (a person operating an automated banking terminal) or computerized (a FORTRAN program executing in a computer center and accessing a remote data base). Communication among application processes is accomplished in accordance with the application protocols.

Layer 7 of the OSI model is called the **application layer**. This layer comprises both the communicating portions of application processes (application entities) and

**TABLE 2.4** OSI layers

---

**Layer 7**: Application layer
**Layer 6**: Presentation layer
**Layer 5**: Session layer
**Layer 4**: Transport layer
**Layer 3**: Network layer
**Layer 2**: Data link layer
**Layer 1**: Physical layer

---

the protocols by which they communicate. The application processes are self-contained and are outside the layered model. The other OSI layers are presented in Table 2.4.

The lower three layers are those typically used in carrier communication networks. These layers deal with the physical and logical characteristics of the local connections between the communicating system (user) and the terminating network (local exchange).

The **physical layer** (layer 1) is concerned with mechanical, electrical, functional, and procedural characteristics required to establish, maintain, and release the physical connection (*data circuit*). It offers to layer 2 a set of services such as transparent bit transmission, physical fault detection, and performance monitoring (e.g., error rate).

The **data link layer** (layer 2) provides the functional and procedural means to activate, maintain and deactivate the logical connection (*data link*). It performs functions such as data unit (frame) delimiting and synchronization (data transparency), data unit sequencing, error detection, error recovery, and window flow control and offers the corresponding services to layer 3. Data link protocols present features similar to those of popular data communication protocols, such as **high-level data link control** (HDLC) [Carlson, 2.7].

The **network layer** (layer 3) provides the means to establish, maintain, and terminate network connections (i.e., a connection path across the network) between end systems, which include communicating application entities. Signaling and routing protocols belong to this layer. Layer 3 also performs a set of functions on data units (packets), such as segmenting and blocking, sequencing, flow control, expedited transfer, and reset and offers the corresponding services to layer 4.

While protocols at layers 1, 2, and 3 are operated between user and network termination, protocols at layer 4 and above are operated instead end-to-end across the network between end users.

The OSI reference model has been useful mainly in establishing **access protocols**, that is, protocols belonging to all layers and appearing at user-network interfaces. On the other hand, Fig. 2.9 indicates that among network nodes, other types of protocols are operated in general on a link-by-link basis. These are the **internal network protocols**. At each network node, protocols at layers 1, 2, and 3 are regenerated. Local exchanges provide conversion between access and internal network protocols, whereas transit exchanges provide regeneration of internal network protocols. Fig. 2.9 also outlines the **peer-to-peer procedural flow** between entities belonging to the same layer (i. e., the protocol interaction at that

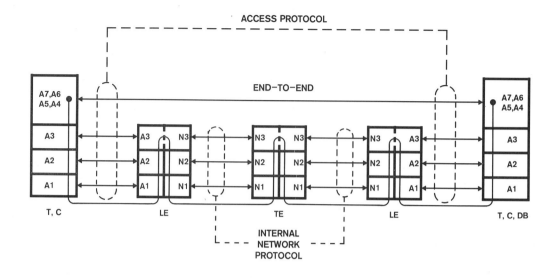

Ai (i=1,2,3,...,)  –  Access Protocol, Layer i

Ni (i=1,2,3,  )  –  Network Protocol, Layer i

&rarr;   Peer-to-Peer flow

&bull;—&bull;   Level-to-Level flow

LE  –  Local Exchange
TE  –  Transit Exchange
T  –  Terminal Equipment
C  –  Controller
DB  –  Data Base

**Figure 2.9**   Distinction between access and internal network protocols.

layer) and the **layer-to-layer** (or level-to-level) **procedural flow** through interlayer interfaces. Standards apply to peer-to-peer protocols and specify the exchange of services between layers. They are not intended to describe interlayer physical interface arrangements.

Concerning the end-to-end protocols belonging to layers 4 and above, the **transport layer** (layer 4) exists to provide to the upper adjacent layer a comprehensive set of transport-service capabilities. These service capabilities are in association with the underlying service provided by the lower layers. The transport layer also guarantees transparent data transfer from the source interface to the destination interface. It relieves the transport users from any concern with the detailed way in which reliable and cost-effective transfer of data is achieved in the various types of communication networks.

Once the data-transport capability is guaranteed, the **session layer** (layer 5) provides the control (organization, synchronization, and management) for the dialog between communicating systems.

The **presentation layer** (layer 6) provides the data formats, codes, and representation used in that dialog in a way that preserves meaning while resolving syntax differences.

***ISDN Protocols.*** The OSI reference model was developed to harmonize protocol standards for distributed applications in computer-communication networks

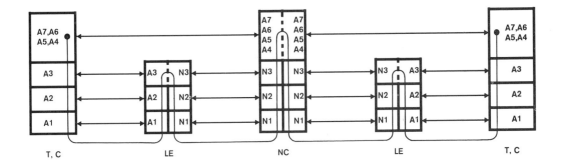

Ai (i=1,2,3,...) − Access Protocol, Layer i

Ni (i=1,2,3, ) − Network Protocol, Layer i

◄──────► Peer-to-Peer flow

●────────● Level-to-Level flow

LE − Local Exchange
T  − Terminal Equipment
C  − Controller
NC − Network Center

**Figure 2.10**  Network-provided features to handle CCITT-defined services at higher layers.

(i.e., data networks). In such networks a single physical channel is used for both control signaling (a layer-3 protocol) and data transfer. This circumstance does not always apply in ISDN applications. Data applications over the D-channel ($p$-information) certainly can be fully characterized by using the OSI model. However, circuit switched applications over the B-channel are controlled by the combined use of signaling messages over the D-channel and of data transfer protocols (including further possible signaling messages) over the B-channel. An extension of the OSI model is required to describe such ISDN applications (see Section 2.4.2).

By referring to Fig. 2.9, a description of ISDN data communications over the D-channel only can be seen. It illustrates the case in which the ISDN is offering a basic transport capability (i.e., the network equipment handles protocols at layers 1, 2 and 3 only).

Depending on national regulatory situations, storage and information-processing capabilities for certain services might also be incorporated within the ISDN (see Fig. 2.10). This might be the case for CCITT standard services such as teletex and videotex at layers 4 and above. This may be true only for particular applications such as telephone directory or mailbox facilities. In this case, access protocols for layers 4, 5, 6 and 7 may be exercised between terminals and network-allocated resources.

In structuring the ISDN protocol architecture, both access and internal network protocols play important roles. Internal network protocols should not necessarily be structured in the same manner as access protocols. Thus functions can be partitioned differently among layers provided that services offered to the end-to-end layers remain unchanged. To achieve cost-effectiveness, the internal network layering may be arranged as shown in Fig. 2.11. Layer 3 is structured into

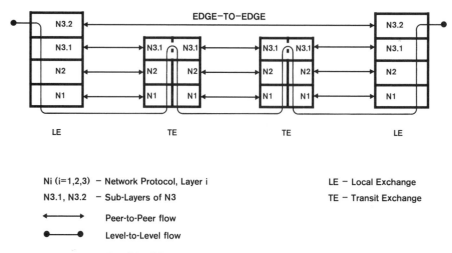

**Figure 2.11** Possible differences between access and internal network protocols.

two sublayers: 3.1 and 3.2. The protocols for sublayer 3.1 are exercised on a link-by-link basis within the network, whereas those for sublayer 3.2 are exercised from network edge to network edge. Some complex procedures (e.g., layer 3 sequencing and window flow control) may be allocated to sublayer 3.2. This simplifies functions at lower layers operated link-by-link and has an impact on the complexity of each switch port internal to the network [Fraser, 2.21].

Figure 2.12 illustrates another application of internal network protocols. Here they allow network processor-to-processor communication for network manage-

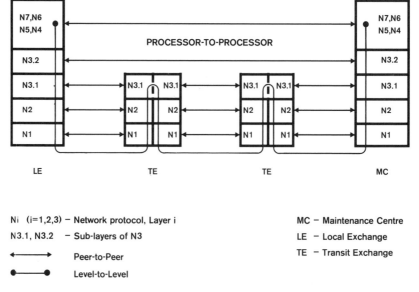

**Figure 2.12** Network processor to network processor communications.

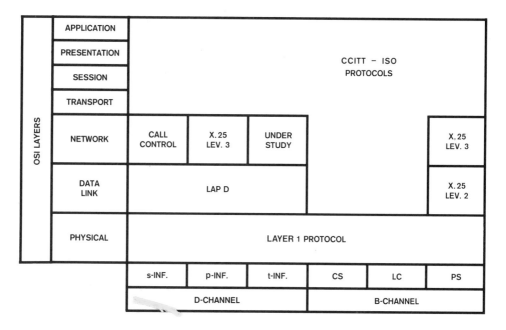

**Figure 2.13**  Protocol layers at ISDN user-network interfaces.

ment functions (i.e., operation, administration, and maintenance). Layers 4, 5, 6, and 7 support internal network applications.

*ISDN Standard Activities.*   Activities related to ISDN protocol standards are focused on *access protocols* via D- and B-channels [CCITT, 2.11]. A status summary of these activities is given in Fig. 2.13, which shows the correspondence between OSI layers and ISDN user-interface layers.

Layer 1 of the ISDN access protocol is common to both B- and D-channels. It specifies the TDM frame format for these channels, together with the necessary mechanical, electrical, and procedural characteristics of the physical connection at user-network interfaces.

The ISDN layer 2 protocol applies differently to the D- and B-channels. The **link access protocol** (LAP) on the D-channel (LAP D) is largely based on balanced LAP (LAP B) of Recommendation X.25* (see Section 2.4.1). It comprises the standard HDLC frame format and the error-recovery procedures for single-frame and multiple-frame operation.   In addition, two address bytes of LAP D are used to support **multiple logical links** (i.e., to provide independent multiple LAPs between exchange terminations and logical end points at user premises). Logical links may be used either to access different layer 3 protocols for *s*-type, *p*-type and

---

*In the text the term *Recommendation* (Rec.) always refers to CCITT standards.

*t*-type information or to address different terminal end points (*multipoint capability*).

The layer 3 format on the D-channel is different for various information types. For the *s*-information, the call control required to support communication on the B-channel is DG-oriented. The *p*-information uses the same protocol specified by Level 3** of Rec. X.25 (i.e., the standard packet level, including in-band virtual call setup and clearing procedures). For the case of *t*-information, the layer 3 protocol is still under study; it may finally follow either *p*-type or *s*-type information procedures [Leth, 2.37].

Above layer 3 on the D-channel, the *p*- and *t*-information can exploit any suitable set of end-to-end protocols according to particular applications. Also, for *s*-type information, end-to-end protocols apply. In Fig. 2.6, end-to-end information between terminals and service vendors may well be considered as signaling and be conveyed over CCS facilities. This is particularly true where they are executed in conjunction with a circuit switched service over B-channels.

Also shown in Fig. 2.13 is the protocol layering on the B-channel. Here, the B-channel may be used for circuit switched and leased transparent connections as well as for a dial-up connection carrying packet switched data. In this latter case, the dial-up signaling to the PS equipment port is conveyed over the D-channel while protocols on the B-channel comply with level 2 and level 3 of Rec. X.25.

Concerning *internal network protocols* in the ISDN, we have a brief comment with respect to signaling and, in particular, the correspondence between *s*-information and CCS interexchange procedures. These latter procedures are based on the CCITT **Signaling System No. 7** (SS No. 7).

Figure 2.14 illustrates the correspondence between access signaling and interexchange signaling, with reference to the OSI layers. SS No. 7 is a standard for communication among network processors and has been structured with four levels. Here, the **message transfer part** (MTP) occupies levels 1, 2, and 3 and the **user parts** (UPs) occupy Level 4. The right half of Fig. 2.14 shows the actual correspondence between the SS No. 7 levels and the OSI layers. In SS No. 7, level 3 adopts a DG-oriented procedure. OSI layer 4, the transport layer, is not implemented. An adequate mapping between the *s*-information at the D-channel protocol layer 3 and the SS No. 7 UP has been developed in order to guarantee satisfactory operation of the ISDN.

Further development activities on internal ISDN network protocols are expected. Areas such as integrated packet switched voice and data facilities for the toll network will be explored. Packetized voice and data protocols, once established for toll applications, are expected to be made available also as access protocols.

---

**Note that early CCITT specification for protocols are organized in *levels*. The levels of a given CCITT recommendation specify a particular set of functions. These functions are selected from the larger set contained in one or more layers of the OSI model. A careful distinction should, therefore, be made between the OSI layers and the levels in which a particular CCITT recommendation is organized.

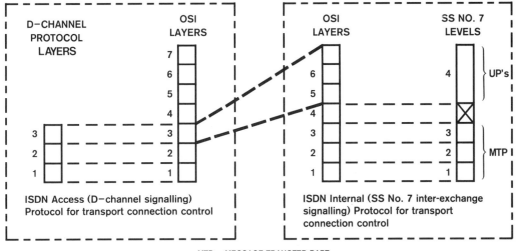

MTP = MESSAGE TRANSFER PART
UP = USER PART

**Figure 2.14** Correspondence between D-channel signaling procotol layers, SS No. 7 levels, and OSI layers.

## 2.3 NETWORK ARCHITECTURES

Various switching alternatives for providing communication services are described in this section. This description is done in the context of networks that cover large geographical areas. Emphasis is given to voice and data integration alternatives, whereas broadband video applications are only briefly reviewed. In Section 2.3.1 CS, PS, and **hybrid switching** (HS) concepts are further described in order to introduce evolving ISDN architectures. The overlayed hybrid switched network approach, as was proposed for the initial implementation of the ISDN, is then thoroughly described in Section 2.3.2.

### 2.3.1 The Evolving Network Architecture

This section first faces basic switching modes for voice and data; a comparison is also made between circuit switching and packet switching techniques.

*Basic Switching Modes.* Table 2.5 shows switching modes suitable for voice and data communications. To introduce the basic switching modes, we refer to Figs. 2.15 and 2.16. In particular, Fig. 2.15 is a space-time diagram, which illustrates circuit and message switched communications between terminal equipment 1 and 2 via a number of intermediate switching nodes, A, B, C, and D.

CS is characterized by an initial call **setup phase**, followed by a bidirectional **information-transfer phase**. A **teardown phase**, which may be initiated by either party, terminates the call. The call setup delay is accounted for by the signaling processing required at each node in order to select the route, the signaling message-transmission delay, and the transfer delay through STPs in case of nonassociated

**TABLE 2.5** Switching integration alternatives for V&D

---

**Circuit Switching (CS)**
  Subrate/multislot
  Fast
  Enhanced (DSI & ADM)

**Packet Switching (PS)**
  Virtual circuit
  Data/voice gram
  Cut through
  Fast

**Hybrid Switching (HS)**
  Basic CS for voice and bulk data
  Basic PS for bursty data
  Enhanced (DSI and ADM)
  Advanced CS for voice and data
  Advanced PS for voice and data

---

DSI = Digital speech interpolation
ADM = Adaptive data multiplexing

CCS (see Section 2.4.1). During the information-transfer phase, the two parties enjoy a dedicated physical connection (implemented by time division or space division arrangements) characterized by a constant transport capacity (fixed analog bandwidth or digital bit rate) and a constant delay.

In principle, circuit switched connections are *transparent to user information*. However, in practice these connections may include systems or devices that limit transparency. Some such devices are line coders, echo controllers, and digital pads. Transparency is twofold: it offers a dedicated channel to meet any user demand and it forces two users to verify their compatibility (in terms of speed, transport protocols, etc.) before establishing their connection.

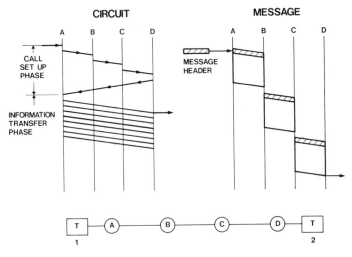

**Figure 2.15** Network delay for circuit and message switched communications.

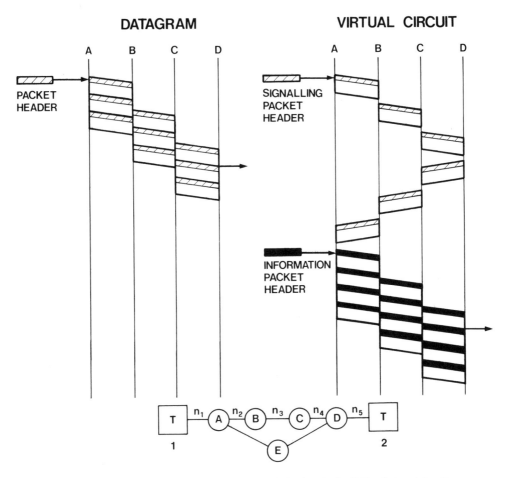

**Figure 2.16** Network delay for packet switched communications in the DG and virtual circuit modes of operation.

Message switching provides unidirectional message transfer according to the store-and-forward principle. Messages are routed through the network using message headers. These headers specify source and destination terminal equipment addresses. Message switching is characterized by a variable transport capacity and variable delay in accordance with the characteristics of the statistical multiplexing and of the traffic load. The delay in message delivery depends on the bit rate of interconnecting links, the message length, the number of intermediate hops, the header-processing time, and the message-queueing time at each intermediate node. Message switching permits two users to communicate via a standard transport protocol, even if they operate at different speeds.

Message-oriented communications have evolved toward *PS* techniques, as illustrated in the space-time diagram of Fig. 2.16. Two frequently used modes of PS are the datagram (DG) and the virtual circuit modes.

The **DG mode** represents a direct evolution of message switching, in which messages are split up into smaller chunks (packets). Packets are provided with headers containing source and destination addresses. This allows them to become self-contained entities in the transmission network. In this manner the shorter the packet length is, the shorter the packet delivery delay becomes. This technique contains two main drawbacks. First, the information overhead is increased, since headers (with length on the order of a few tens of bytes) are repeated for each packet of useful information (with length on the order of a few hundreds of bytes). Second, packets can be delivered out of sequence to their destinations. Thus the order in which they originate from the source may not be preserved over the network. In Fig. 16 some DGs of a message may be routed from node $A$ to node $D$ via intermediate nodes $B$ and $C$, whereas others might be routed via an intermediate node $E$, thus likely encountering a shorter delay.

The **virtual circuit mode** of operation employs the best features of circuit and message switching. Thus it provides a bidirectional information transfer through the network. A virtual call is characterized by *call setup, information transfer,* and *teardown* phases. Signaling packets are used to seize a path (virtual circuit) and are followed by subsequent information packets. Signaling packet headers are structured as DG headers. They contain source and destination addresses, as well as logical numbers—for instance, according to a numbering scheme that uses 2 bytes.

With reference to the network at the bottom of Fig. 2.16, in order to identify the virtual circuit for a given call, a logical circuit number $n_i$ ($i = 1, 2, \ldots, 5$) is assigned to each link crossed by the call setup packet in both directions. Intermediate nodes store the mapping between incoming and outgoing logical circuit numbers ($n_i$ and $n_{i+1}$) assigned to the same call. Once the call setup phase is successfully terminated (i.e., a call-accepted packet is received by the calling terminal equipment), the information packet headers need to contain only the logical circuit numbers. This permits routing to their destination, thus avoiding overhead for terminal equipment addresses. Hence, virtual circuit routing overcomes the drawback of the DG mode: It implies less overhead for the information packets and it also guarantees packet sequencing, since all the packets of a virtual call follow the same route through the network.

*CS Versus PS Technique.* CS and PS techniques are cost-effective for different service applications. Their cost-effectiveness depends on source-traffic characteristics and on relative switching or transmission costs with reference to given network topologies [Frank, 2.22; Kummerle, 2.37; Parodi, 2.42; Rosner, 2.45]. Figure 2.17 shows the qualitative trade-off between the transferring efficiency of CS and PS techniques. We propose to serve a source with terminal activity $\alpha$ ($0 \leq \alpha \leq 1$) by a channel of a given peak bit rate using either CS or virtual circuit PS. During the information-transfer phase of the call, the CS channel is used $\alpha \times 100\%$ of the time. The same information can be transferred by messages, provided that appropriate message headers are added. An average overhead $\varepsilon$ is employed in the information transfer during the virtual call, thus resulting in channel utilization $\alpha (1 + \varepsilon) \times 100\%$ of the time.

Moreover, we note that a PS channel cannot be utilized 100% of the time. This is because it operates as the server of a queueing system and it has to maintain a certain channel vacancy rate (unoccupancy) in order to reduce the queueing delay

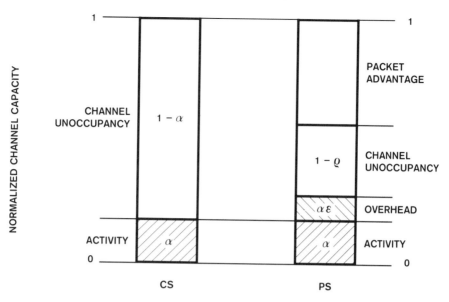

**Figure 2.17**  CS versus PS: trade-off on channel utilization.

to acceptable values. Figure 2.17 indicates a maximum PS *channel occupancy* $\rho$ $(0 < \rho < 1)$ so that, for the particular example shown, PS still offers a certain advantage in channel utilization over CS. In other words, while $(1 - \alpha) \times 100\%$ is the percent disadvantage of CS, two drawbacks apply to PS: These are the overhead and the channel unoccupancy $1 - \rho$ (statistical disadvantage).

The overhead factor $\varepsilon$ depends on the packet size and is roughly of the order of 0.1 when using virtual circuit routing. The $1 - \rho$ statistical disadvantage depends on traffic engineering criteria and the required service delay performance. In present PS data networks, it can range up to 0.4 due to the limited knowledge of data source traffic characteristics and to the relatively slow speed (64 kb/s) of the transmission links. In conclusion, Fig. 2.17 shows that CS is favored in the case of high-activity factors, whereas PS is more attractive for the case of low-activity factors.

However, to compare CS and PS techniques fairly, switching costs and packet-processing costs should also be taken into account. Several simplified data network models have been developed to describe cost trade-offs between CS and PS as applied to data communications. Key parameters in such models that are used to characterize source traffic refer not only to terminal activity but also to message fragmentation during the calls (i.e., message length and intermessage interval length) and intensity of call traffic. Other parameters refer to the cost of various network resources such as transmission, processing, and storage. Physical distances between terminals assume an important role in defining breakpoints between areas, which determine the application of CS or PS.

Although it is difficult to summarize the specific results obtained with such models, some rough conclusions can be drawn. Disregarding distance, CS always outperforms PS for the case of very high activity (continuous messages). For very low activity (intermittent short messages with long intermessage intervals), PS is preferable. For the intermediate range of activities, CS is favored for short transmission distances, whereas PS is superior for long distances.

In the following we describe advanced CS and PS techniques and we introduce the combined HS mode of communication, according to the classification given in Table 2.5. Each main entry of this table is intended to apply to the whole variety of voice and data (bulk and bursty) applications.

### Advanced Circuit Switching

CS systems today use digital techniques to provide a 64-kb/s rate service. Both **multislot** and **subrate** switched connections can be implemented (to match voice and bulk data requirements) using bit rates that are multiples and submultiples of 64 kb/s, respectively. However, to handle bursty data, in a cost-effective manner, *fast CS* should be used.

The **fast CS** technique is similar to the conventional CS, but it ensures that the circuit setup and teardown process is implemented very quickly throughout the network. Thus each data burst (one or more packets) can be serviced by the established circuit at the time of its occurrence.

Fast CS is effective only if the connection setup and teardown delays are small (less than a few hundred milliseconds). This objective is difficult to achieve in a large carrier network using nonassociated CCS. This results from the additional signaling delay introduced by STPs (on the order of a few ten milliseconds per STP crossing) [Harrington, 2.31].

In general, the main drawback of fast CS occurs either in the imposing of stringent requirements on the call-processing resources (one of the most critical resources in a switching system design) at each network node, or in providing priority in setting up and tearing down circuits for each data burst. Fast CS is not an attractive solution at the carrier network level. However, it presents favorable features when applied at the switching system level.

Improved *transmission link efficiency* can be realized in CS networks by taking advantage of voice and data burstiness. DSI [Campanella, 2.6] and Adaptive Data Multiplexing (ADM) may be used on long-haul trunks to convey only the active parts of voice and data calls, respectively.

The DSI system is dynamic digital TDM equipment, which uses talkspurt and pause patterns in speech conversation to transmit a number of input digital speech channels (possibly together with some voiceband data channels) on a lower number of TDM channels (DSI channels).

This objective is reached by a system operation providing (1) a DSI channel allocation that can be dynamically changed depending on the active-nonactive status of input channels; (2) a speech-encoding word whose length can be adaptively controlled (i.e., by reducing it from a standard to a lower value) depending on both the number of input channels issuing talkspurts and the possible amount of voiceband data; (3) transparency of voiceband data ensured by data encoding with fixed code word length and by assigning a DSI channel for data transmission with highest priority.

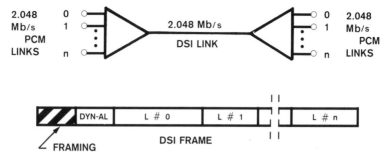

**Figure 2.18** DSI link and frame layout.

About feature 1, a DSI channel can be allocated to an input channel only when it is actually active, i.e., when the corresponding information source is issuing talkspurts (as recognized by speech detectors) or voiceband data. When all DSI channels operate at standard bit rates (*normal conditions*), the speech signals transferred by the DSI system are not affected by any degradation due to system operation, but only a limited number of input speech channels can be served at a given time. This constraint would give rise to the potential for partial or entire talkspurt loss when additional input speech channels request service (*overload conditions*).

Talkspurt front-end clipping or loss causes an annoying subjective effect called the **freeze-out** effect. Subjective speech quality can be significantly improved if, when approaching overload conditions, variable bit rate voice-coding schemes [Cox, 2.14; Goodman, 2.23] (e.g., an embedded coding*) are adopted (feature 2).

As a consequence of this operation, the number $m$ of available DSI channels is time variable from a minimum to a maximum value. The minimum value, $m_1$, occurs in normal conditions and corresponds to the adoption of standard bit rates in speech encoding (e.g., 64 kb/s or 32 kb/s for PCM or ADPCM encoding, respectively). The maximum value, $m_2$, refers to overload conditions in which talkspurt samples are encoded with a minimum length* of code word (e.g., less than 8 or 4 bits for PCM or ADPCM encoding, respectively).

The ratio between the total number of input channels and the minimum value $m_1$ of DSI channels is a measure of the compression capability of a DSI system and is called **DSI gain**; this parameter depends on the speech activity (see Sec. 2.2.1) and the adopted encoding scheme.

A DSI system operating on a 2.048-Mb/s point-to-point DSI link serving $n + 1$ tributaries is shown schematically in Fig. 2.18, in which each tributary is supposed to operate with the European primary PCM format ($32 \times 64$ kb/s time slots). In the same figure the **DSI frame** is shown.

---

*An embedded code is a digital code that admits simple bit dropping and insertion (as distinct from elaborate code conversion); standard 8-bit PCM is an example of embedded code.

*The selection of the tolerable minimum value of word length is suggested by voice-quality constraints.

In general the DSI frame includes two main fields. The first one (O-field) is dedicated to overhead information and, in particular, includes a frame-alignment word (framing) and a DSI allocation message (DYN-AL, or **dynamic allocation field**), specifying the different assemblies of DSI channels carried by each frame. The second field (I-field) is reserved for speech and voiceband data information issued by each input channel; it is the support of DSI channels. In Fig. 2.18 the I-field includes $n + 1$ L-subfields, each associated to an input tributary, meaning that each L-subfield shows a variable length, which can span from 0 to the whole I-field, depending upon the active-nonactive status of the input channels.

Figure 2.19(a) shows how the DSI multiplexing technique can be modeled by a multiple-server loss system in the particular case in which a buffering is not required for talkspurt processing; in the case of use of buffering, the model would be a multiple-server waiting system. The service is requested from active input channels, whereas the $m$ servers represent the available DSI channels. The characterization of the model in Fig. 2.19(a) requires the description of talkspurt interarrival and service processes. In particular, as far as the distribution of talkspurt duration is concerned, the corresponding average value is about 350 ms for speech detectors having a 16-ms hangover time.

An embedded code scheme for variable bit rate encoding is considered in Fig. 2.19(b). This scheme is similar to the one representing standard PCM coding. Here each speech sample is coded by $k$ bits, where $k \le 8$. By deleting some of the least significant bits, the remaining bit configuration still constitutes a valid codeword. This codeword represents the speech sample with a coarser quantization.

Taking into account that voice activity is in the range of 35 to 40%, DSI gains from 2 to 3 are obtainable using embedded codes (together with an appropriate overload-control strategy). Hence, in the DSI arrangement shown in Fig. 2.18, $n$ can span a range up to 3.

As with DSI, ADM techniques may also be used to take advantage of data burstiness over long-distance links. ADM implies the a priori knowledge of the particular data protocol operated over the links. This enables transmission of only the active parts (dataspurts) of the data calls. When approaching overloads, the dataspurts can be either lost or buffered while waiting for the next free server.

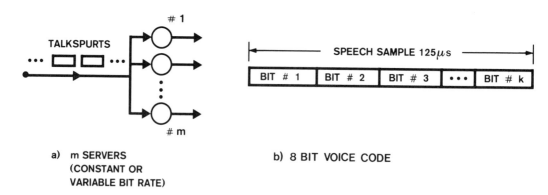

a)  m SERVERS
(CONSTANT OR
VARIABLE BIT RATE)

b) 8 BIT VOICE CODE

**Figure 2.19**   DSI multiplexing (a) and embedded voice codes (b).

## Advanced Packet Switching (PS)

Conventional PS is available today for loops or trunks operating at rates of up to 64 kb/s. It is operated in public data networks to provide DG and/or virtual circuit capabilities. Virtual circuit network implementations are more popular than DG network implementations.

As a refinement of the store-and-forward principle, PS may also employ the **cut-through principle** [Kermani, 2.35]. In the former technique, data packets are entirely stored, processed, and then forwarded to their destination. In the latter technique, packet headers (placed at the front end of the packet) are always stored and processed, whereas packet information need not be entirely stored if the outgoing channel is idle. Once the route selection via header processing has been accomplished, packet forwarding can start even if the entire packet has not been yet received. If packets are blocked due to a busy output channel, they are entirely stored. **Cut-through switching** aims to shorten packet delivery delays, since it avoids the delay due to unnecessary buffering while there is an idle channel. This technique turns out to be more effective than store-and-forward only if link loads are small, implying that a packet encounters an empty queue with high probability. Cut-through switching is useful in local networks where link occupancy can be maintained at a low level without severe cost penalties.

In data networks, **access protocols** and **internetworking protocols**, such as CCITT Recs. X.25 and X.75, respectively (see Section 2.4.1), have been designed to meet bursty data requirements. This is particularly necessary in terms of error recovery and flow control of data packets. These rather complex protocols are also used as internal network protocols in several network implementations (see Fig. 2.9). However, actual trends favor the simplification of interexchange protocols by allocating the most-complex functions to the network edges, away from the network core. Figure 2.20 shows how protocol conversions between access and internal network protocols can be performed at local exchanges. As an example of the radical difference between access and internal protocol structures, complex functions, such as sequencing, error recovery, and window flow control (see Section 2.4.1), could be avoided at layers 2 and 3.1 of the internal network protocols (see also Fig. 2.11). These functions could indeed be transferred to layer 3.2, which is used by the originating and destination local exchanges. In this manner switching ports internal to the network are not burdened by processing resources being activated on a packet-by-packet basis.

Simple protocols are key elements that allow PS of voice traffic [Cohen, 2.12; Decina, 2.17]. Voice cannot tolerate error retransmission or window flow control, due to its stringent delay requirements [Gruber, 2.24], and voice accepts packet losses much more easily than data. Moreover, packetized voice transmitters and receivers operate at the same speed, which may not be the case for data.

On the other hand, packetized voice sources generate a very large amount of traffic as compared to that produced by data sources. For example, with 35% voice activity and packets each containing 16-ms voice segments, an average 180-s call corresponds to about 4000 packets for each transmission direction. For a data call of similar duration, a few packets per data source can be expected.

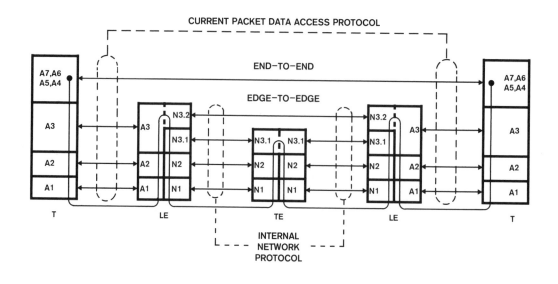

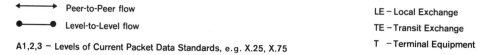

**Figure 2.20** Protocol conversion between access and internal network protocols.

A large digital local CS office can today handle up to 100,000 subscribers. With 180-s calls and an average traffic intensity of 0.1 Erl/subscriber, such offices have to handle about 55 calls per second. Packetized voice local offices of comparable capacity should have a throughput of about 0.5 Mpacket/s. This throughput is two orders of magnitude greater than that of operational PS systems. Hence, a simple protocol should ensure a minimum processing time per packet in order to make feasible high-throughput PS systems. It should also meet stringent transit-delay objectives such as 1-ms packet cross-office time (this figure has to be compared with the few tens of milliseconds for current PS systems).

Another key element, necessary in order to permit packetized voice transmission in large carrier networks, is the use of high-speed links (e.g., at bit rates 1.544 or 2.048 Mb/s) to limit link transmission delays [Coviello, 2.13]. For example, with a 560-bit-long voice packet (resulting from 16-ms voice chunks employing 32-kb/s ADPCM coding with 4 bits/sample and about 10% packet overhead), the transmission time at 2.048 Mb/s is less than 0.3 ms. This value is about 1000 times less than an end-to-end voice-delay objective of about 300 ms. This fact provides further incentives for the development of viable packetized voice networks.

As far as the implementation of high-throughput–low-delay PS systems is concerned, various solutions are obtained when simple protocol processing (no error recovery or window flow control) and high-speed links are used. A simple conceptual model for such a system is shown in Fig. 2.21. This system is composed of a centralized control for virtual call processing and administration purposes, trunk controllers for interfacing high-speed links, and an interconnection network to switch packets among trunk controllers. Fast CS offers a possible solution for implementing the interconnection mechanism. To switch each packet between trunk controllers, a circuit is set up and torn down within the circuit switched–oriented interconnection network. In contrast to applications in large carrier networks, fast CS presents attractive features at the switching-system level, due to the relatively small delay involved in establishing the link [Hardy, 2.27]. The use of self-routing interconnecting networks (see Section 2.3.2) is favored, as particularly small delays can be achieved.

**Fast PS** is an emerging technique based on the use of simple protocols, high-speed links, and high-capacity–low-delay PS systems [Decina, 2.17]. However, in order for fast PS to be implementable, substantial progress in signal-processing and switching technologies is required.

Congestion control in packetized voice networks can be accomplished by employing flow-control techniques based on embedded codes, as described for DSI systems [Bially, 2.4]. The single-server queue model used to represent packet multiplexing on an outgoing link is shown in Fig. 2.22(a).

Voice packets can be appropriately structured by using embedded voice codes, as shown in Fig. 2.22(b) or 2.22(c). In Fig. 2.22(b) the packet information field (see Section 2.4.1) is structured into $k$ subfields ($k$ being the embedded codeword length), each containing a group of $m$ bits having the same rank in the embedded coding of $m$ speech samples. With this arrangement, link congestion problems are eased by deleting one or more of the least significant subfields of the incoming packet before entering the queue.

Figure 2.22(c) illustrates another solution, in which $k$ voice packets of ranked priority are formed. Each packet contains a group of $h$ bits having the same rank

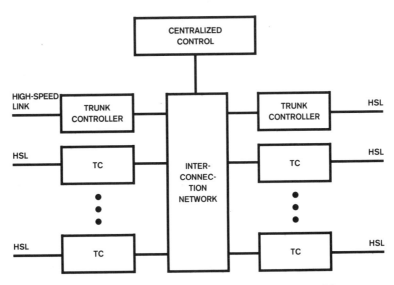

**Figure 2.21** Conceptual layout of a high-throughput/low-delay PS system.

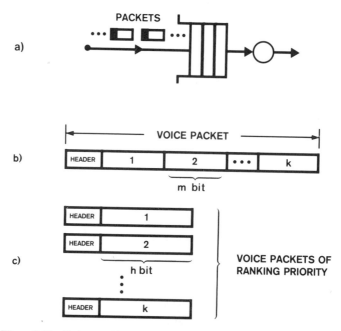

**Figure 2.22** Packet multiplexing (a) and embedded voice packets (b), (c).

as in the case of embedded coding. Under congestion conditions, lower-priority packets are discarded. This solution is less attractive than the previous one, since it implies a larger overhead if $h < mk$, as it is required to limit the number of speech samples processed together. This constraint is needed to prevent excessive degradation in quality due to possible packet losses during network transport.

### Hybrid Switching (HS)

HS techniques were implemented in early ISDN systems. Conventional CS (including some multislot and subrate capability) is used for voice and bulk data, whereas conventional PS is used for bursty data (see Fig. 2.6). CS and PS facilities are overlaid functionally. In the access and in the local sections, these facilities are integrated into common network systems; in the toll section they are implemented by dedicated equipment. Section 2.3.2 describes the hybrid overlayed approach of the ISDN and presents planned system implementations.

In Table 2.5, the CS portion of the basic hybrid network can be enhanced by using DSI and ADM over long-haul digital trunks. HS can also be based on more-advanced implementations. In an advanced scenario, CS of voice and data might be available in local areas, whereas fast PS for integrated voice and data communications might be used for toll connections (see Fig. 2.7). The basic scenario, shown in Fig. 2.6, calls for overlayed CS/PS facilities through the entire network. However, in the advanced scenario of Fig. 2.7, overlayed CS/PS facilities are initiated locally and integrated unconventional PS facilities are deployed in the toll network. In the latter case, access to fast PS facilities might also be gained directly from user premises, thus bypassing local facilities.

ISDN: Integrated Services Digital Network     Chap. 2

Concerning hybrid-overlaid facilities, different degrees of integration can be envisioned for both switching and transmission systems.

The different arrangements to merge CS and PS communications in the same digital carrier link by using the European standard primary PCM multiplex frame format are shown in Fig. 2.23. Figure 2.23(a) shows a conventional 32-time-slot format. Each time slot can support various types of communication such as a CS application, a PS application, or a CCS bearer channel, all operating at a rate of 64 kb/s. We note also that time slot 0 provides frame alignment and alarms and will incorporate in the near future an error-checking sequence. Also, multislot ($n \times 64$ kb/s) and subrate ($n \times 8$ kb/s) channels can be provided. Finally, the information about the frame format subdivision (CS, PS, multislot, subrate) is transferred by a CCS channel (e.g., at 64 kb/s), either incorporated in the same frame at a preassigned time slot or carried by a parallel 2.048-Mb/s link. Such an arrangement implies essentially fixed boundaries among overlaid facilities by using as building blocks the 64 kb/s time slots.

The case of a 2.048-Mb/s link fully employed for packetized voice and data applications is shown in Fig. 2.23(b). Framing and error check are included in packet headers, whereas signaling, voice, and data share the same medium by statistical multiplexing. Here, there is no TDM frame format.

Figure 2.23(c) illustrates the case in which one part of the frame is devoted to CS and CCS, whereas the other part is allocated to PS [Miyahara, 2.40]. The boundary between these parts may be *fixed* or *movable*. The fixed-boundary case

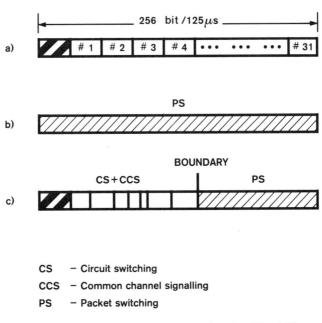

CS — Circuit switching

CCS — Common channel signalling

PS — Packet switching

**Figure 2.23** 2.048-Mb/s link configurations for CS and PS.
(a) Multiple fields arrangement.
(b) Full PS arrangement.
(c) Two-field arrangement.

is similar to that shown in Fig. 2.23(a); however, it employs bits as building blocks and defines just two contiguous portions in the frame. Any fixed-boundary strategy presents drawbacks, since it might not match the dynamically changing traffic requirements. In this case, blocking may occur on one facility, while the other is idle. The movable-boundary approach aims to cope with these problems. However, two major factors make this approach cumbersome; these are (1) the task of finding an effective algorithm that controls the boundary movements under any changing traffic pattern, which is not trivial; and (2) the implementation complexity of that approach, which is significant and has a relevant impact on switching system architectures and cost. Thus the potential gain in transmission efficiency may well be negated by high switching costs.

### 2.3.2 Basic Hybrid Switching (HS)

As explained in Section 2.2.3, the ISDN may be implemented in a variety of configurations. These depend on the particular network environment from which the ISDN emerges and what stage in the evolutionary process the network has achieved. Essential to the early ISDN implementations is the integration of CS and PS communication facilities over the access loop. Two main alternatives for ISDN access are considered. These are the **full digital** and the **combined (analog-to-digital, or A/D)** techniques. The digital one was preliminarily described in Section 2.2.3.

To introduce these two approaches it is convenient to define at first the anticipated configurations for user premises equipment and the location of standard equipment interfaces. Then the full digital approach is illustrated, as is the combined A/D approaches.

#### Reference Configurations for ISDN User Access

The first step in defining and understanding the reference configurations and interfaces is to group together functions which may exist on the user's premises. The generic groupings of functions and the corresponding reference points to be used in describing standardized interfaces are shown in Fig. 2.24. The functions described here need not all exist and the list of functions is not exhaustive. Functions in different groupings can be merged into the same physical equipment.

In Fig. 2.24(a), the **network-termination 1** (NT1) functions include those necessary to terminate appropriately the transmission line into the network. The NT1 functions broadly correspond to those included at layer 1 (physical interface functions) of the OSI reference model. The **network-termination 2** (NT2) functions include a wide variety of communication functions corresponding to those at layers 2 and above of the OSI model. Particular implementations of NT2 functions are PBXs, terminal controllers, and LANs. The **terminal 1** (T1) functions correspond to individual ISDN terminals for a variety of services.

Reference points *S* and *T* define appropriate locations for standard ISDN user-network interfaces. Depending on local regulations, some ISDNs will terminate at reference point *T,* whereas others will have reference point *S* at the end

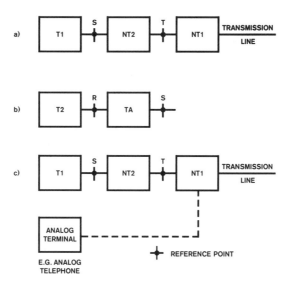

**Figure 2.24** Reference configurations for ISDN user-access interfaces.
(a) Case of ISDN terminals.
(b) Case of non-ISDN terminals.
(c) Combined access arrangement.

of their network. An important configuration is the case in which the NT2 functions are absent, and hence reference points *S* and *T* coincide.

The case in which a terminal does not meet ISDN interface standards is depicted in Fig. 2.24(b). **Terminal 2** (T2) corresponds to such a terminal and requires a **terminal-adapter** (TA) function to convert the non-ISDN interface at reference point *R* into an ISDN standard interface at *S*. Thus reference point *R* may correspond to any current physical interface, such as EIA RS-232 or CCITT Rec. X.21 (see Section 2.4.1).

The reference configuration for the combined (A/D) access arrangement is shown in Fig. 2.24(c). In this case digital information originates at T1, and analog information, such as analog telephony, originates in the analog terminal and merges with the digital information at NT1.

**The Full Digital Approach**

The full digital approach is illustrated conceptually in Fig. 2.25. This approach requires that a digital loop capability be provided for the two types of ISDN communication channels (the B-channel and the D-channel; see Section 2.2.3).

As shown in Fig. 2.25, the various information signals offered by the full digital access are split up at **exchange terminations** (ET). B-channels and the associated *s*-information are routed over CS and CCS facilities. The *p*-information is routed to the PS facilities by means of a **statistical multiplexer** (SM), which concentrates virtual circuits toward the PS equipment. The interface between the SM and PS devices in Fig. 2.25 is likely to be operating at a rate of 64 kb/s with an X.25 type of protocol (see Section 2.4.1). The *t*-information could be handled

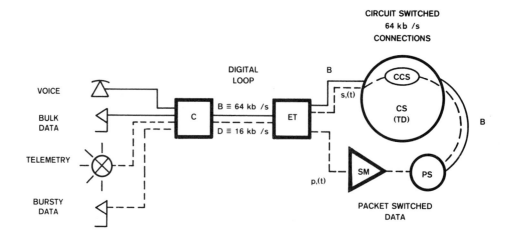

CIRCUIT SWITCHED
64 kb /s
CONNECTIONS

C – TERMINAL CONTROLLER
ET – EXCHANGE TERMINATION
SM – STATISTICAL MULTIPLEXER

PS – PACKET SWITCHING FACILITY
CS/TD – CIRCUIT SWITCHING/TIME DIVISION FACILITY
CCS – COMMON CHANNEL SIGNALLING

**Figure 2.25**  Full digital approach: Merging of digital voice and data networks.

by either CCS or PS facilities.   In the former case it is treated as a DG, whereas in the latter case it is transported via virtual circuits or permanent virtual circuits.

Also illustrated in Fig. 2.25 is the case in which the B-channel is used as a digital dial-up support for PS data.   The right part of the diagram shows the interconnection between CS-CCS and PS facilities.

*Full-Duplex Transmission Techniques.*   To obtain full-duplex transmission on two-wire subscriber loops, two techniques are available [Ahamed, 2.3].   The first one may be referred to as **time compression multiplexing** (TCM), burst mode, or Ping-Pong mode.   It is achieved by partitioning the user bit streams in each direction of transmission into segments (bursts) of $n$ bits.   The burst duration is given by $\Delta = n/F$, where $F$ is the user data rate.   Each burst is transmitted on the loop at a bit rate $F_0$ that is at least twice the user data rate $F$.   Bursts from each direction of transmission are forwarded in alternate time intervals as shown in Fig. 2.26a.   The extent to which $F_0$ exceeds $2F$ depends on the burst duration $\Delta$, the transit time $\delta$ for a pulse to travel down a line of maximum length, and the guard time $\tau$.   This is the time during which both transmitter and receiver are idle; it is required for the line transients to subside.   As an example, suppose that the user data rate is $F = 64$ kb/s and that the maximum loop length is 9 km.   If the burst duration is 3 ms, then $F_0$ is given by $2.25F$.   In general $F_0$ decreases for increasing values of $\Delta$, but $\Delta$ should be kept small to limit signal delay over TCM loops.   A TCM terminal block diagram is shown in Fig. 2.26(b).   As shown here, the TCM technique has a simple configuration.   It may be very convenient during the early stages of loop digitalization.

The other technique is called the **hybrid balancing** (HB) method with echo cancellation and is shown in Fig. 2.27. This method provides transmission in both directions simultaneously at the user rate $F$. Hybrids provide directional isolation with the aid of adaptive echo cancellation to improve channel separation. The advantages of the HB method over that of TCM are the smaller transmitted signal bandwidth ($F$ instead of $F_0$) required and the nonaccumulating delay in long repeated loops. The main problem in implementing HB systems is the complexity

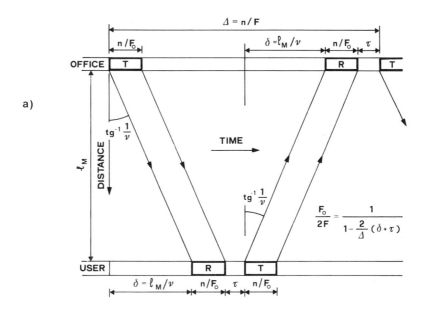

a)

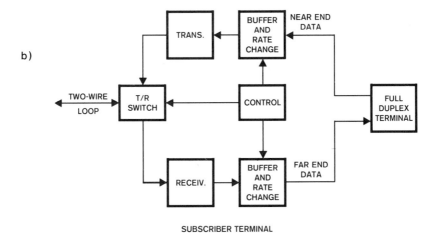

b)

SUBSCRIBER TERMINAL

**Figure 2.26** TCM.
(a) Space-time diagram of signal transfer.
(b) Subscriber terminal.

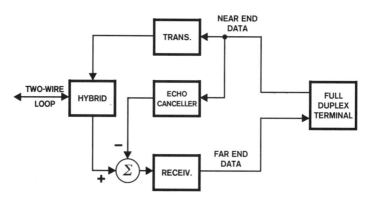

**Figure 2.27**  Hybrid balancing with echo cancellation.

involved in design and implementation of the circuits necessary for near-end signal suppression. Cost-competitive HB systems will likely become feasible by means of VLSI technology and will be deployed in the evolving ISDN.

*HS Office.* The integration of overlaid PS and CS facilities in the local exchange can assume various forms depending on the progressive evolution of the switching technology. The conceptual layout of an **ISDN HS office** centered around a distributed control architecture is shown in Fig. 2.28. We can distinguish the **interface modules** devoted to analog and digital loops and trunks and the **centralized modules** devoted to **operation, administration, and maintenance** (OAM) functions, including operator interfaces. Centralized modules can perform other centralized functions (depending on the degree of decentralization and multiservice capability), such as *call processing, recorded announcements,* and *packet handling.* Connection between these two types of modules is provided by an interconnecting network suitable for circuit and message connections. Message connections allow communication among the module processors. The layout of Fig. 2.28 shows the ISDN capability as spread between interface and centralized modules.

The *ISDN loops module* interfaces with digital loops and incorporates an ET, SM, and a part of the CS/CCS functions considered in Fig. 2.25. The *packet-*

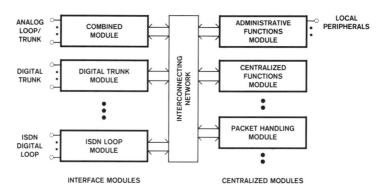

**Figure 2.28**  Conceptual layout of an HS ISDN office.

*handling module* does not always need to be locally present; it performs the PS functions shown in Fig. 2.25. Therefore, the ISDN loop module provides a gateway for concentrated CS and PS traffic toward centralized CS and PS facilities that may be available either in the same local exchange or in other sites.

There are several alternatives to the conceptual layout shown in Fig. 2.28. ISDN loop interface functions may be incorporated in combined modules, or they can be integrated in a single module together with packet handling functions.

As far as interconnection networks are concerned, they can be built up using a variety of structures. Topology and operational principles of such structures can be based on

**1.** Different building elements, such as buses, rings, or switching matrices;

**2.** Different path-tracing criteria, such as space division or time division;

**3.** Different information transfer modes, such as dedicated paths (circuit-oriented) or statistically shared paths (message-oriented).

Two significant examples of topological structures based on switching matrices are given in Fig. 2.29. These matrices are suitable for either time or space division implementation and are open to either circuit- or message-oriented solutions.

Figure 2.29(a) shows the popular three-stage **Clos network** that has been for many years the point of reference for telephony switching systems [Marcus, 2.38]. This network presents $m$ paths for any inlet-outlet pair, $m$ being the number of

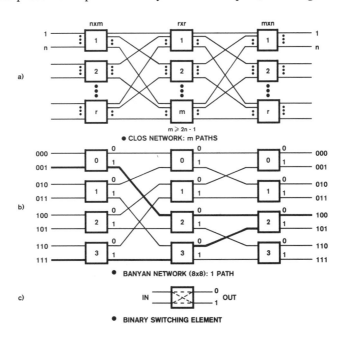

**Figure 2.29** Interconnection networks.
(a) Clos network: $m$-Paths.
(b) Banyan network ($8 \times 8$): 1-Path.
(c) Binary switching element.

matrices in the intermediate stage. When $m \geq 2n - 1$, the Clos network is *strictly nonblocking*. Thus, under any load condition, any idle inlet-outlet pair can be connected.

Figure 2.29(b) gives the layout of a three-stage *Banyan network* (the $\Omega$ network), composed of binary switching elements shown in Fig. 2.29(c) [Wu, 2.52]. There is only one path for any inlet-outlet pair, and the path selection can be simply performed on the basis of the binary address of the required outlet. For example, outlet 100 can be reached by reading sequentially at each stage the corresponding binary digit composing its binary address. This is true when starting from any inlet, from 001 to 111 in the figure. Because they have this property, Banyan networks are often referred to as **self-routing** networks. The drawbacks of such networks lie in the measures required to reduce blocking.

*Various Access Types.* Referring back to the full digital approach, note that the B- and D-channels can give rise to different assemblies at user-network interfaces. Some assemblies have been internationally standardized. The **basic access** structure B + B + D consists of a TDM assembly of two B-channels and one D-channel. This structure applies to physical interfaces corresponding to $S$ or $T$ reference points. An essential feature of the basic access is that some ISDN services do not require the use of one or even both B-channels. Therefore, the subscriber loop can be equipped to support a subset of B + B + D, such as B + D or D only. Thus loop transmission can operate at a net rate of 144, 80, or 16 kb/s. However, all loops utilize the basic terminal interface running at the net rate of 144 kb/s (see Table 2.6).

Another standard access structure is called **primary rate multiplexed access**, which operates at 1.544 or 2.048 Mb/s. It consists of a TDM assembly of $n$ B-channels ($n = 23$ to 30) plus one D-channel operating at 64 kb/s. Access with multiple D-channels for a single 1.544- or 2.048-Mb/s interface or one D-channel for several 1.544- to 2.048-Mb/s interfaces is also possible.

There is also a **primary rate broadband access**, in which user-network interfaces operate at 1.544 or 2.048 Mb/s. In this case the TDM frame format always includes capacity for housekeeping functions (framing, alarms, etc.). Additionally, one 64-kb/s D-channel and one or more time slots for voice and other narrow-band

**TABLE 2.6**  ISDN user-access types

**Basic Access**
    * B + B + D, interface net rate: 144 kb/s
      B + D        ⎫ Possible loop net rates:
      D            ⎬ 80 kb/s; 16 kb/s
                   ⎭

    B = 64 kb/s; D = 16 kb/s

**Combined Access**
    * A + C
    A = analog; C = 8 or 16 kb/s

**Primary Rate Multiplexed Access**
    * nB + D, interface rates: 1544 or 2048 kb/s
    B = 64 kb/s; D = 64 kb/s

services (e.g., facsimile) may be included. The remaining capacity is assigned to the **wideband channel**, H. Typical applications on the H-channel include video-conferencing supported by circuit switched operation, either on demand or in real time. The present standards for the H-channel include two alternatives: the $H_0$ channel at 384 kb/s (6 time slots) and the $H_1$ channel at 1536 or 1920 kb/s (24 to 30 time slots).

The previously described ISDN architecture can allow such broadband access by using the multislot switching capability. However, leased satellite capacity, which could be on a demand-assigned basis, will likely provide the initial broadband capability in the ISDN. The development of ubiquitous real-time broadband digital switching facilities will occur when justified by technology and market demand. In the later ISDN stages, this development will be in conjunction with the deployment of fiber optic distribution plant.

### The Combined Analog/Digital Approach

Two modes of implementing the combined (analog/digital) approach are shown in Fig. 2.30: the **simultaneous** and the **alternate** solutions.

The *simultaneous* provision of the analog and digital capability at user interfaces is depicted in Fig. 2.30(a). Here the A-channel corresponds to the conventional analog voiceband channel, whereas the C-channel is similar to the D-channel discussed in the full digital approach.

The C-channel operates at 8 or 16 kb/s and carries message-interleaved *s*-, *p*-, and *t*-information. The *t*- and *p*-information are routed via the C-channel to the PS facilities, whereas the A-channel is switched through space or time division exchanges. CCS facilities and *s*-information, when present in this approach, may be used to provide enhanced voice services.

Combined loops use a digital data above analog voice transmission technique applied on two-wire copper pairs. The digital data capability (the C-channel) is full duplex. **Frequency division multiplexing** (FDM) is used to separate the voiceband and the duplex data channels.

The compatibility of services provided via the D-channel and the C-channel in the two ISDN approaches, illustrated in Figs. 2.25 and 2.30(a), is a key issue in the establishment of protocol standards. We favor development of compatible protocols over these channels, so that the C-channel protocol is a compatible subset of that of the D-channel.

In the arrangement shown in Fig. 2.30(a), a local space division exchange may be equipped with ET and SM functions, whereas PS functions will likely be placed within preexisting PS facilities.

Another implementation of the combined approach is shown in Fig. 2.30(b). Here voice (A-channel) and data information are presented *alternately* to the user interfaces. The digital data channel currently operates at 56 kb/s for bulk data transfer. Signaling to support this bulk data channel is in-band and is provided via the A-channel. Analog voice and digital data are addressed to the same destination and are switched alternately during the call. The circuit switched 56-kb/s connections may be implemented through space division local offices and time

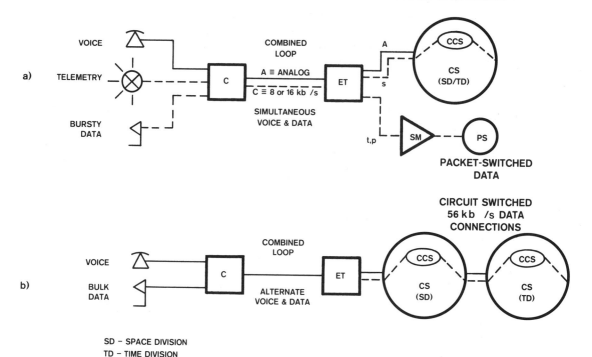

**Figure 2.30** Combined analog and digital approach.
(a) Simultaneously at the user interface.
(b) Alternately at the user interface.

division toll offices. CCS facilities are important in the toll section, but they may not be present in the local section. The configuration shown in Fig. 2.30(b) represents a pragmatic approach to the early provision of end-to-end digital connectivity by exploiting the widespread use of analog facilities.

The ISDN standard user-network interfaces will employ the simultaneous combined approach only (see Table 2.6), whereas, the alternate combined approach can be regarded as an interim solution.

In terms of service offerings, the combination of the two approaches, illustrated in Fig. 2.30, corresponds to the approach given in Fig. 2.25. The choice of one approach versus the other depends on a variety of factors. These include the marketplace addressed (residential or business), the actual penetration of space division or time division local offices, and the degree of digitalization that the toll network has achieved. Taking these factors into account, the Europeans have chosen the full digital approach for their ISDN [Ackzell, 2.2; Hardy, 2.30; Irmer, 2.32]. The North Americans have found more attractive both combined approaches [Felts, 2.19; Handler, 2.28; Handler, 2.29].

## 2.4 NETWORK PROTOCOLS

This section presents a review of protocols developed for dedicated digital networks. Also, trends on protocol developments to be used in service integrated carrier networks are described. In particular, Section 2.4.1 deals with customer and interexchange signaling in circuit switched synchronous networks for voice and data. Access and internal network protocols in packet switched data networks are also presented. Section 2.4.2 describes the perspectives for overall protocol handling in the early ISDN. Access arrangements and their protocol features at layers 1, 2, and 3 of the OSI model are briefly introduced.

### 2.4.1 Protocols in Dedicated Networks

Traditionally, protocol structuring followed different approaches in CS and PS networks. Hence we describe separately the criteria for protocol layering in these two types of networks. We outline differences and controversial issues with respect to the OSI reference model. Unresolved issues are expected to be clarified in the development of protocol layering for HS networks, such as the ISDNs.

This section is subdivided in five parts. The first two are devoted to protocol approaches in CS and PS networks. Then, two other parts are dedicated to the description of access protocols and interexchange protocols in CS networks. Finally the last part faces PS protocols.

#### CS Protocol Approach

In CS networks we may distinguish between **call control information** (signaling) and **user transport information** (user data), such as voice, video, or data information. These two types of information can be transported either over a unique channel or separated (by space, frequency, or time division) channels.

Single-channel transmission techniques are illustrated in Fig. 2.31(a). Signaling and user data are forwarded in disjoint time intervals over a call-dedicated channel. This method is typically used on POTS subscriber loops and trunk connections, by employing dial pulses or tones for signaling. This method is also used on digital data loops, where signaling uses an 8-bit character format as specified by CCITT Rec. X.21. Call control may be employed not only for call setup and teardown but also to allow network recalls for changing call conditions (e.g., to extend the call for conferencing).

The case of separated channels for access and interchange links is shown in Fig. 2.31(b). The signaling-dedicated channels can be used to support either a single-user data-transfer channel, **channel-associated signaling** (CAS),* or multiple-user data channels, CCS. CAS is typically employed on interexchange PCM trunks. In TDM systems a digital signaling channel is provided in association with each preassigned voice channel. CCS, according to CCITT SS No. 7, is typically op-

---

*The signaling transfer according to Fig. 2.31(a) is also called CAS: In this case signaling and user data share the same channel.

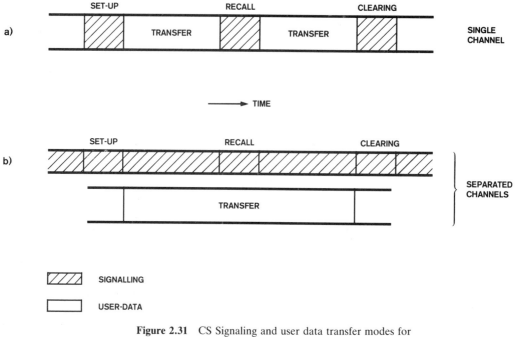

**Figure 2.31** CS Signaling and user data transfer modes for
(a) Single channel.
(b) Separated channel.

erated for voice and data applications on interexchange trunks. Another form of CCS is that envisioned for application on ISDN access loops (the D-channel protocol).

Different protocol-handling methods apply in general to signaling and user data transfer in CS networks, as shown in Fig. 2.32. During the data-transfer phase, **user-data protocols** are exercised end-to-end through the network and only the protocols at layer 1 are regenerated by intermediate network equipment. Figure 2.32(a) shows the interconnection, during the data-transfer phase, between two terminals via two intermediate local exchanges. Physical-layer protocols find a breakpoint at local exchanges, since access and interexchange links present different physical (mechanical, electrical, and functional) characteristics.

The purpose of **signaling protocols** in CS is to provide the physical connection between end systems. Once the connection is established as a concatenation of link-by-link circuits, end systems can exercise higher-level protocols. A possible interpretation would state that signaling is exercised between application entities; thus signaling protocols comprise in principle all the seven OSI layers. Also, signaling protocols allow an interaction among signaling-application entities, which is concurrent with the primary interaction among user data application entities. This interpretation is depicted in Fig. 2.32. In particular note that layers 1 to 7 of the signaling protocols are considered as embedded in layer 1 of the user-data protocols. Signaling protocols apply to both access and interchange links. Their protocol layers refer to the secondary communication function needed to interconnect signaling entities that are allocated in terminal and network equipment (local exchanges in the figure).

Note that some functions are absent in both user-data and signaling protocol layers. Considering user data, the network-routing functions of layer 3 are absent because the layer 3 protocol is operated end to end. Other layer 3 functions, such as virtual circuit multiplexing, may be present, since they can also be found useful in end-to-end applications. As far as the signaling protocol layers are concerned, functions at layers 3 and 4 can be avoided on access links. Moreover, with reference to SS No. 7, layer 4 functions are absent over interexchange links.

### PS Protocol Approach

Signaling and user-data information can also be distinguished in PS networks. In **virtual circuit** (VC) switching, signaling and user data are structured, in principle, in separated messages. They share dynamically a common medium (also with other virtual call information) by means of statistical multiplexing as shown in Fig. 2.33(a). In DG switching, signaling (DG header) and user data are embedded in a common message, carried together with other messages over the transmission medium as illustrated in Fig. 2.33(b).

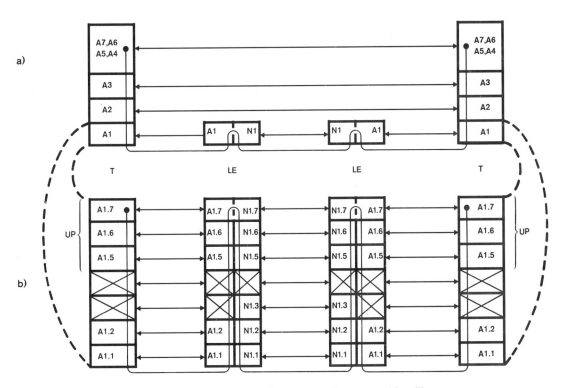

**Figure 2.32** CS Signaling and user-data protocol handling.
(a) User data.
(b) Signaling.

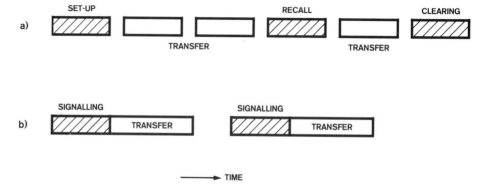

**Figure 2.33** PS Signaling and user-data transfer modes for
(a) Switching.
(b) Datagram Switching.

In PS networks common protocols apply to both signaling and user data at layers 1, 2 and 3. This is typical of protocols such as CCITT Recs. X.25 and X.75. Protocols at these three layers are regenerated by intermediate network equipment, as shown in Fig. 2.34(a). According to the previous interpretation of CS protocols, signaling is also performed between signaling-application entities in order to support the concurrent primary communication process between user-data application entities. This concept is illustrated in Fig. 2.34(b). Here, layers 3 to 7 of the signaling protocols are considered as embedded within layer 3 of the user-data protocols. Access and interexchange protocols for PS communications can be identified with reference to both signaling and user-data transfer. In this case, as before, some layer functions are absent in the signaling protocol layers, such as functions at layer 4 on both access and interexchange links.

Our interpretations of Figs. 2.32 and 2.34 are not necessarily shared by other authors. While user-data transfer is well understood as being modeled by Fig. 2.32(a) and 2.34(a) for CS and PS, respectively, the modeling of signaling transfer as given by Figs. 2.32(b) and 2.34(b) is *controversial*. In particular, considering Fig. 2.32(b), there is agreement on the interexchange protocol layering (e.g., SS No. 7); however, some authors include signaling protocol functions just up to layer 3 on access loops. Considering the interpretation of Fig. 2.34(a), most authors agree that signaling resides within layer 3, without outlining the further structuring given in Fig. 2.34(b). We believe that our interpretations offer attractive features for the user interconnection model in hybrid switching networks. A generalization of our interpretation is given in Section 2.4.2.

### Circuit Switched Access Protocols

We distinguish at first three cases, that of **POTS loop signaling, digital data loop signaling** (Rec. X.21) and **digital telephony loop signaling**. The first two cases are described here, whereas the third case falls in the ISDN domain and is presented in Section 2.4.

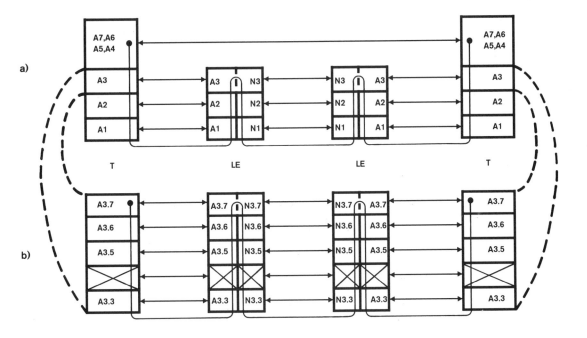

a)

T — TERMINAL EQUIPMENT
LE — LOCAL EXCHANGE

**Figure 2.34** PS Signaling and user-data protocol handling.
(a) User data.
(b) Signaling.

*POTS Loop Signaling.* On POTS loops, call control is achieved by means of a particular set of signals (pulses and tones), which are exchanged between the customer and the local exchange as illustrated in the space-time diagram of Fig. 2.35. The signal flow for a typical successful call is illustrated in this figure.

The calling customer alerts the local exchange (originating exchange) with an **on-hook** to **off-hook** transition. Upon reception of the audible **dial tone** indication (proceed to select), the calling party sends the selection signals that code the called party address. Once the address signals have reached the terminating exchange, **ringing** is initiated at the called customer and **ringback** indication is forwarded to the calling party, both actions being taken by the terminating exchange. When the called customer answers, his or her off-hook condition causes the termination of both ringing and ringback signals from the terminating office. This condition is forwarded to the originating office, where **call charging** is initiated. **Call clearing** can be initiated by either party, but usually only when the calling customer hangs up does the total path through the network disconnect. Unsuccessful calls are characterized by reception of various tones or announcements by the calling party during or after the selection phase.

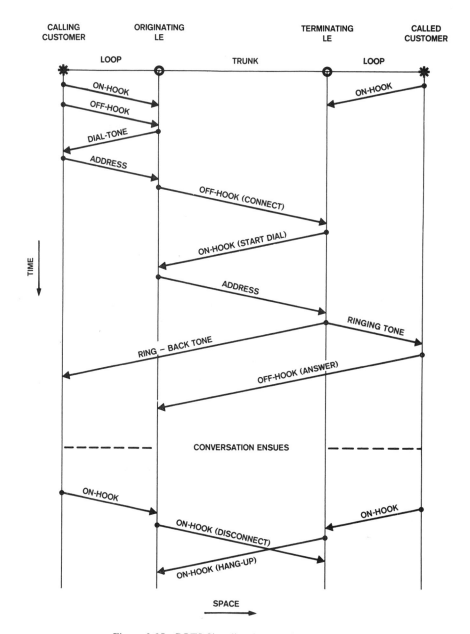

**Figure 2.35** POTS Signaling for a typical connection.

The interpretation of this procedure in the OSI model framework appears controversial and of little practical interest.

***Digital Data-Loop Signaling.*** Concerning CS digital data, we make reference to the *CCITT standard X.21* interface procedure [Folts, 2.20]. In this case customer signaling is obtained by 8-bit characters according to the **international alphabet 5** (IA5) together with the use of additional control information carried out-of-band.

To describe the X.21 standard, we refer first to the physical arrangement at DTE-DCE interfaces illustrated in Fig. 2.36. This figure shows six leads at this interface plus a common return. The T- and R-leads convey data and character signaling, whereas the C- and I-leads provide out-of-band control functions by means of on-off indications. The S-lead provides signal element (bit) timing from the network, whereas the B-lead is optional and provides an 8-bit byte alignment with the network. On subscriber loops the T and R data are transferred together with the on-off status sampling of the C- and I-leads.

The sequence of events necessary for a successful call is shown in Fig. 2.37. Here the evolution of the calling DTE-DCE interface states are described with reference to the sequence of signals appearing on the T-, R-, C-, and I-leads. The calling DTE alerts the DCE by changing the (T-, C-) leads conditions from (1, off) to (0, on). Then the DCE sends on the R-lead a continuous sequence of "+" IA5 characters as a *proceed to select* indication. On reception of this indication, the calling DTE forwards IA5 selection signals. Character alignment in Rec. X.21 is provided by using two or more contiguous "SYN" characters preceding the character sequence. Call-progress signals, similar to POTS ringback or busy tones, can appear on the R-lead to describe call status. Change of I-lead from off to on, together with steady 1s on the R-lead, indicates the *ready for data* condition (i.e., the availability of the end-to-end data connection). DTE clear request is signaled by turning off the C-lead and is acknowledged at the far end by turning off the I-lead. The same sequence of events at calling and called DTEs is shown in Fig. 2.38 in terms of signal flow as given in Fig. 2.35.

The layer 1 **data format** employed on X.21 loops is called an **envelope** and is comprised of 8 or 10 bits. For example, the 8-bit envelope consists of an envelope-framing bit, a status bit that conveys C- or I-lead status sampling and 6 bits that correspond to T- or R-lead data chunks.

As far as the application of the OSI model to X.21 signaling is concerned, the physical and functional interface lead arrangement belongs to layer 1 and the SYN character alignment falls within layer 2. The remaining signaling procedures

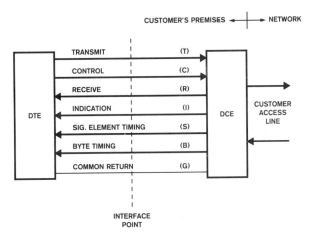

**Figure 2.36** DTE/DCE Interface according to Rec. X.21.

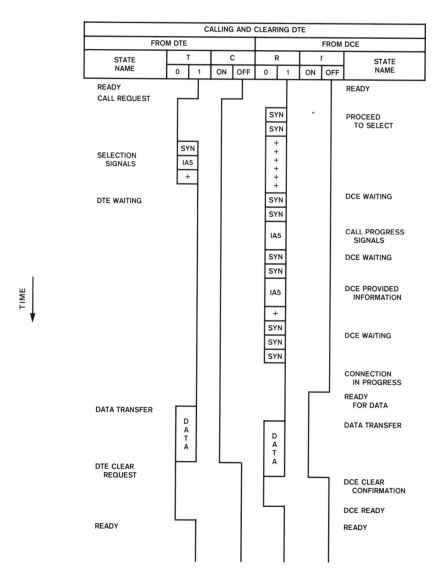

**Figure 2.37** Recommendation X. 21: example of sequence of events for the successful call setup and cleardown at calling and clearing DTE/DCE interface.

can be regarded as structured among layers 5, 6, and 7, whereas functions at layers 3 and 4 are absent. Thus the signaling-session control, the IA5 character codes, and the signaling-procedure application may correspond to Layers 5, 6 and 7 functions, respectively.

### Circuit-Switched Interexchange Protocols

In this section we describe the SS No. 7 standard for interexchange CCS [CCITT, 2.8]. In a CCS network we distinguish terminal points and switching

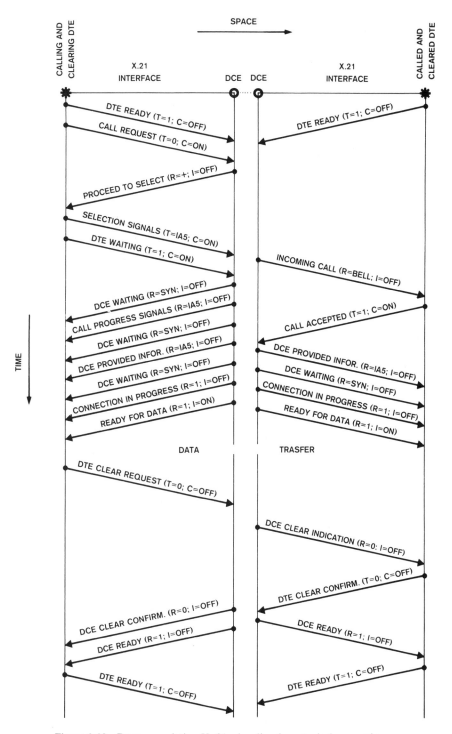

**Figure 2.38** Recommendation X. 21: signaling for a typical connection.

nodes. Terminal points, also called **signal points** (SPs), represent the source and sink of signaling information and correspond to control units of SPC exchanges. Switching nodes are called STPs. The branches of a CCS network are called **signaling links**. The CCS network is overlaid on the trunk network that provides physical end-to-end circuits between the calling and called parties.

Circuit setup in the trunk network from the originating to the terminating office occurs according to a link-by-link path search-and-seizure process. Thus in a tandem connection involving a number of intermediate offices, the signaling message related to the connection should be received and processed at each transit node.

The main **modes of operation** in a CCS network are illustrated in Fig. 2.39. In the **associated mode** depicted in Fig. 2.39(a), a signaling link is dedicated to serve circuits belonging to a given trunk route only. This mode is characterized

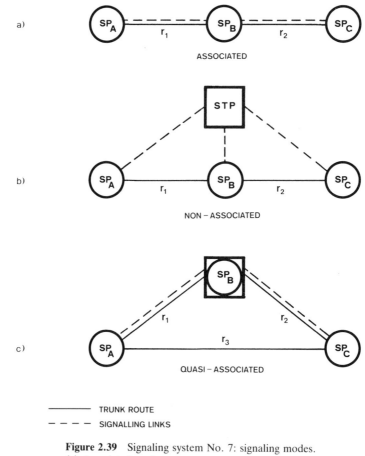

**Figure 2.39** Signaling system No. 7: signaling modes.
(a) Associated mode.
(b) Nonassociated mode.
(c) Quasi-associated mode.

by a large number of signaling links. Some of these links have a low utilization factor because of the limited amount of traffic carried by the corresponding trunk routes.

This drawback can be overcome by using the nonassociated mode of operation, based on the introduction of STPs shown in Fig. 2.39(b). This figure shows the fully **nonassociated mode**, in which each SP is connected in principle by a unique signaling link to an STP. This link serves all trunk routes departing from the office that corresponds to the SP. In this mode the number of signaling links is greatly reduced and a high link utilization is realized. To set up a physical circuit between $SP_A$ and $SP_C$, the signaling messages should travel link by link. Thus the message travels at first between $SP_A$ and $SP_B$ (via STP) and then between $SP_B$ and $SP_C$ (via STP).

In the **quasi-associated mode** shown in Fig. 2.39(c), some SPs incorporate STP functions so that the CCS network operates under a mixture of associated and nonassociated modes. In this example, the trunk routes $r_1$ and $r_2$ are served in the associated mode, whereas trunk route $r_3$ is served in the nonassociated mode. Hence the signaling links $SP_A/SP_B$ and $SP_B/SP_C$ carry associated signaling messages between the corresponding end points. They also carry nonassociated signaling messages between $SP_A$ and $SP_C$ via the STP colocated with $SP_B$.

These operational modes have application in the evolving development of CCS networks. In the early stages, the associated or quasi-associated modes are easier to implement, whereas in the more mature stages the nonassociated mode is more attractive for deployment [Spring, 2.49].

A CCS network is a computer-communication network among office-control processors. The OSI model is well adapted to describe protocols of CCS networks. The SS No. 7 standard is organized in 4 levels. The lower levels 1, 2, and 3 correspond to OSI layers 1, 2, and 3, respectively, and are called all together the MTP. OSI layer 4 functions are absent, and level 4 comprises functions of OSI layers 5, 6, and 7. Each group of these functions, referred to a specific application, is called a UP. The general structure of the SS No. 7 functions and their nomenclature are shown in Fig. 2.40.

Table 2.7 summarizes the SS No. 7 level functions. At level 1 we have the *signaling-data-circuit* functions, comprising transmission up to 64 kb/s and the line protection switching functions. Level 2 includes the *signaling-data-link* functions, such as message-frame formatting, sequencing, and error recovery. At level 3 we distinguish between *message handling* and *network management*. Message handling comprises routing according to DG switching. Network management refers to the additional procedures to be adopted for CCS network operation under abnormal conditions such as faults and overloads. The UPs at level 4 correspond either to various CS services (telephony, TUP, and data, DUP) or to applications of the CCS network other than signaling for CS services, i.e., OAM applications (OAMUP). This latter application allows for an efficient trunk network management service via the **network-management centers** (NMCs) and **traffic service position** (TSP) systems.

Figure 2.41 illustrates communications protocols between two SPs via an

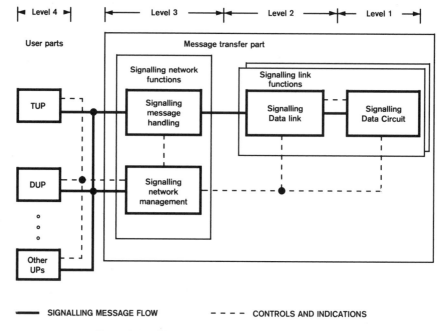

**Figure 2.40** SS No. 7: general structure of SS functions.

intermediate STP. Level 4 protocols are exercised end to end through the CCS network, since STPs operate functions up to level 3 only. In the following some features of levels 2 and 3 are introduced.

*SS No. 7 Level 2.* Figure 2.42 illustrates the format of signaling messages, which are also called **signaling units** (SUs). Opening and closing **flags** (F) delimit SUs, while a two-byte **frame-check sequence** (FCS) serves for error detection. Both

**TABLE 2.7** Signaling system No. 7 level functions

---

**Level 1—Physical**
    Medium, code, error performance, protection
        switching, etc.
**Level 2—Data Link**
    Delimitation, error recovery, sequencing,
        error-rate monitoring, link initialization
**Level 3—Network**
    Discrimination, distribution, routing
        traffic, route and link management
**Level 4—User Parts**
    TUP, DUP, OAMUP

---

TUP = telephony user part
DUP = data user part
OAMUP = operation, administration, and main-
    tenance user part

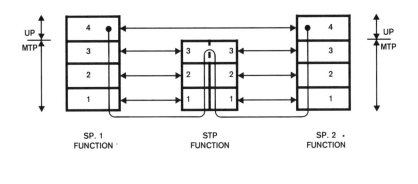

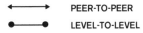

**Figure 2.41**  SS No. 7: protocol levels.

the F and FCS fields comply with the basic HDLC frame format specification given in level 2 of the X.25 protocol (see Section 2.4.1).

The HDLC protocol provides *data transparency* by means of a *zero insertion-deletion* mechanism that prevents simulation of the F pattern (01111110) by the remaining bits in the frame. The mechanism consists in inserting a dummy 0 bit after five consecutive 1 bits in the transmitted bit stream (excluding F). Dummy bits are then removed at the receiving end.

Sequence numbering and error recovery are controlled by a 2-byte field composed of a **forward sequence number** (FSN), a **forward indicator bit** (FIB), a **backward sequence number** (BSN), and a **backward indicator bit** (BIB).

Two additional fields are present in the SU format. The first one is the 8-bit **length indicator** (LI) that codes the SU length by the number of 8-bit bytes it contains. The LI provides redundant information in view of the closing flag and discriminates between **message SUs** (MSUs) and **link status SUs** (LSSUs). The latter are used in the link-initialization procedure, whereas the former are exchanged once the data link has been initialized. The second field is the **service information octet** (SIO), which is included in MSUs only. It provides, at SP locations, message discrimination and distribution to the appropriate UP.

Included in the MSUs is a **signaling information field** (SIF) with length $n \times 8$ bits ($n > 2$), whereas LSSUs contain an information field of 1 or 2 bytes. Error recovery is applied to MSUs only.

The error-recovery mechanism is based on the combined use of sequence numbers FSN and BSN, together with the two status bits FIB and BIB. The principle of this mechanism is illustrated in Fig. 2.43, whereas Fig. 2.44 illustrates an operational example of such a mechanism.

*Error recovery* is based on positive **acknowledgements** (ACKs) to indicate correct transfer of SUs and **negative acknowledgements** (NACKs) to request retransmission of SUs received in a corrupt form, as detected by the FCS field. Transmitted but not positively acknowledged SUs remain available for retransmission. To maintain the correct SU sequence when a retransmission is made, all

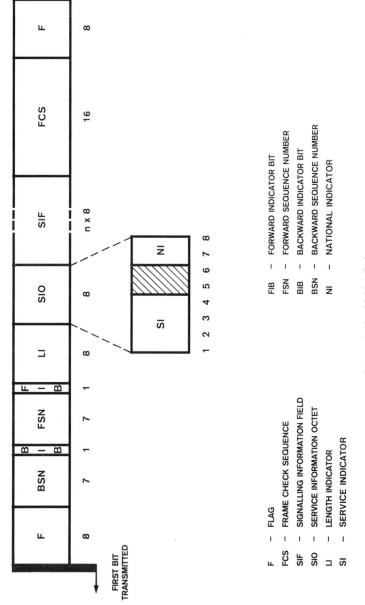

**Figure 2.42** SS No. 7: format.

F   —  FLAG                             FIB   —  FORWARD INDICATOR BIT

FCS   —  FRAME CHECK SEQUENCE              FSN   —  FORWARD SEQUENCE NUMBER

SIF   —  SIGNALLING INFORMATION FIELD        BIB   —  BACKWARD INDICATOR BIT

SIO   —  SERVICE INFORMATION OCTET          BSN   —  BACKWARD SEQUENCE NUMBER

LI   —  LENGTH INDICATOR                     NI   —  NATIONAL INDICATOR

SI   —  SERVICE INDICATOR

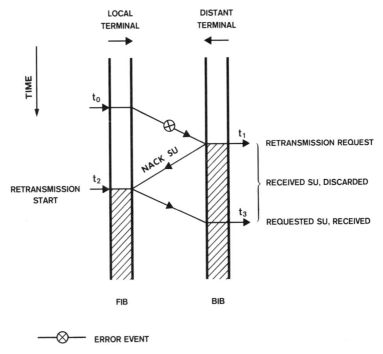

**Figure 2.43** SS No. 7: principle of the SU error recovery.

SUs are retransmitted, starting from the one requested for retransmission. Sequence control is performed by means of the FSN, and the acknowledgement function is performed by means of the BSN. Numbering of FSN and BSN is cyclical modulo 128.

For ACK, the receiving signaling terminal accepts a received SU by assigning its FSN value to the BSN of the next SU sent in the opposite direction. For NACK, the receiving terminal (distant terminal) inverts the BIB value of the next SU sent in the opposite direction (NACK SU), according to the scheme shown in Fig. 2.43.

A NACK SU is generated at the time $t_1$, when the distant terminal detects a discontinuity in the received FSN sequence. When the local terminal receives the NACK SU, it initiates retransmission at time $t_2$ and inverts the FIBs of all subsequently transmitted SUs. Continuous retransmission starts from the SU numbered by the BSN contained in the NACK SU. The distant terminal discards all correctly received SUs in the time interval $(t_1, t_3)$. This is because these SUs, transmitted by the local terminal in the time interval $(t_0, t_1)$, have a FIB value different from the BIB value of the SUs that are being transmitted by the distant terminal. In the example of Fig. 2.44 we have assumed that, during idle periods (i.e., when there are no new SUs to be transmitted), transmitting terminals repeat continuously the last transmitted SU. Therefore, the first error burst does not cause NACK of SU number 9 emitted by the local terminal, since this SU is repeated. NACK is generated for SU number 11 transmitted by the local terminal when the distant terminal receives the SU number 12 from the local terminal.

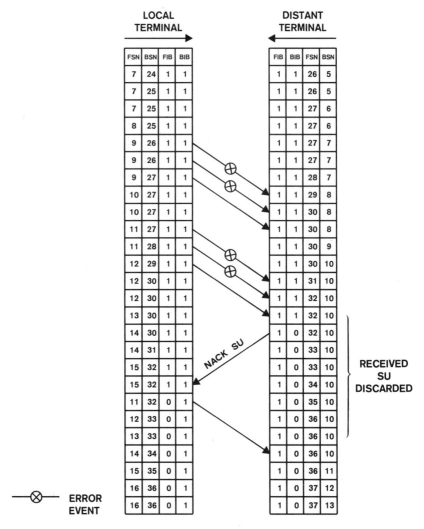

**Figure 2.44** SS No. 7: operational example of the SU error recovery.

The error-recovery mechanism adopted in SS No. 7, based on sequence numbers and status bits, is an efficient technique that avoids the use of time-outs as envisioned, for example, by the corresponding mechanism of Rec. X.25.

*SS No. 7 Level 3.* The protocol at this level is based on DG switching. This implies that each SU is marked with the addresses of the originating and destination SPs. This is true for all SUs even if they belong to the same call to be controlled (about 10 SUs per call are used in SS No. 7).

The *SU label* placed at the beginning of the SIF field is shown in Fig. 2.45. The label comprises three main fields. These are **destination point code** (DPC) of 14 bits, the **originating point code** (OPC) of 14 bits and the **circuit-identification code** (CIC), with length depending on the application, typically 12 bits. The CIC field serves to identify the one-trunk circuit engaged in the call among all the circuits in the trunk route between origination and destination SPs. The first 4 bits of the CIC field are called the **signaling link selection** (SLS) code. Together with DPC and OPC they compose the **routing label**. In fact, the SLS code is used to route the SUs in addition to DPC and OPC, since the CCS network takes advantage of enhanced link and node redundancy in its topology.

Due to its essential role in establishing communications, the CCS network has to be very reliable. STP offices duplicate the control and interconnection equipment, as is usually done for any telecommunication switching node. Additionally, a geographical area is served by two STP offices, which are geographically separated and called **twin STPs**. Figure 2.46 considers two regions A and B, each served by twin STPs, and displays the redundant signaling links among SPs and STPs. Twin STPs are linked together and both are connected to each SP included in their region. All STPs of the CCS network are fully mesh interconnected. The basic configuration shown in Fig. 2.46 is popular in many CCS network implementations. However, other interconnection topologies are also used in some networks. We describe *CCS routing,* distinguishing between the absence or presence of link and node failures.

In *absence of failures* the SLS code is used to provide a uniform distribution of traffic throughout the network and the identification of a unique path through the CCS network. This path is from the originating to the destination SP and is followed by all SUs of calls engaging a given trunk circuit between the two SPs. The uniform traffic distribution allows the minimization of SU transmission delay, and the unique path permits the elimination of out-of-sequence occurrences of SUs serving the same call. This latter feature is typical of CCS networks. Although these networks employ DG switching, this feature allows a minimal out-of-sequence

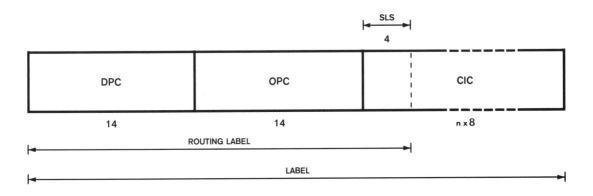

DPC – DESTINATION POINT CODE
OPC – ORIGINATING POINT CODE
SLS – SIGNALLING LINK SELECTION
CIC – CIRCUIT IDENTIFICATION CODE

**Figure 2.45** SS No. 7: Label structure.

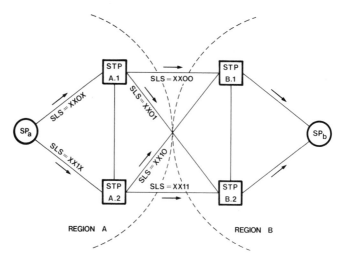

**Figure 2.46** CCS Network: an example of routing in the absence of failures.

probability, even in the case of failures, as in the case of virtual call switching. Referring again to Fig. 2.46, the normal message routes from $SP_a$ to $SP_b$ are the following, according to the particular SLS code:

$$SP_a - STP_{A1} - STP_{B1} - SP_b \quad \text{for} \quad SLS = XX00$$

$$SP_a - STP_{A2} - STP_{B1} - SP_b \quad \text{for} \quad SLS = XX10$$

$$SP_a - STP_{A1} - STP_{B2} - SP_b \quad \text{for} \quad SLS = XX01$$

$$SP_b - STP_{A2} - STP_{B2} - SP_b \quad \text{for} \quad SLS = XX11$$

For the topology shown in Fig. 2.46, only 2 bits of the SLS code are necessary. The other SLS bits (indicated by the character $X$) can be used either to provide additional load sharing on multiple-data channels implementing a single signaling link or to permit more complex CCS network topologies based on the hierarchical partitioning of STPs. An example of the two-level STP hierarchical network, in which lower rank STPs serve metropolitan areas, is shown in Fig. 2.47.

The *network-management procedures* at level 3 ensure the most efficient network utilization in the case of multiple link and node failures. Here we describe just one of these procedures, since it allows a low out-of-sequence probability in the presence of failures. Let us suppose that in the network of Fig. 2.46, the link between $STP_{A1}$ and $STP_{B2}$ fails. According to a protection switching mechanism, the traffic on this link is diverted to the link between $STP_{A1}$ and $STP_{B1}$. An out-of-sequence event can occur if $STP_{B2}$ is congested, and it delays an SU received before the link failure as long as the next SU diverted via $STP_{B1}$ reaches the destination $SP_b$. To minimize the probability of occurrence of this event, a *time-controlled traffic diversion* is activated, such that the diverted traffic is buffered in $STP_{A1}$ for an adequately long time period (a few hundred milliseconds).

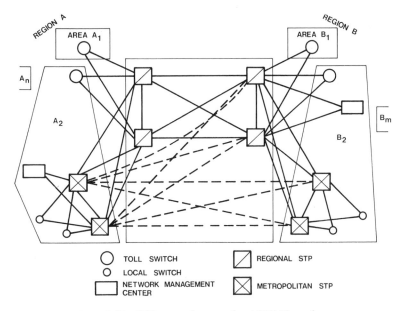

**Figure 2.47** CCS network: a two-level STP hierarchy.

**Packet Switched Protocols [CCITT, 2.9]**

With reference to Fig. 2.4, we identify the following main equipment interfaces:

1. Packet-mode DTE-DCE interface, which complies with the Rec. X.25 standards.
2. Non-packet-mode DTE-DCE interface; in the case of asynchronous terminals Rec. X.28 applies, whereas for synchronous terminals a variety of procedures (such as **binary synchronous communications** (BSC), **synchronous data link control** (SDLC), etc.) are commonly adopted because of the lack of an international standard.
3. PSE-PSE (**packet-switching exchange**) interface (i.e., interexchange interface).

In the case in which PSEs belong to different networks, the internetworking standard protocol complies with Rec. X.75, whereas for PSEs belonging to the same network a variety of procedures can be used, although X.25-X.75 type protocols are commonly adopted. Concerning this latter situation, internal network protocols should not necessarily be structured in the same manner as access or internetworking protocols (also see Section 2.4 and Fig. 2.11).

Here, we limit our consideration to access protocols according to Rec. X.25, and we outline the strict commonality of these protocols with those specified by Rec. X.75. In particular we refer to levels 1, 2, and 3 of Rec. X.25.

The physical DTE-DCE interface (level 1) adopts the lead arrangement specified for Rec. X.21 as given in Fig. 2.36. The C- and I-lead conditions are used in a simple procedure (the same as for an X.21 leased-circuit service) to activate, maintain, and deactivate the physical connection.

Levels 2 and 3 are referred to as the **frame** and the **packet** levels, respectively. Figure 2.48 shows frame and packet formats. The **frame format** given in Fig. 2.48(a) complies with the original HDLC specification and includes *Flags* (F), *Frame Check Sequence* (FCS), *Address Field* (A) and *Control Field* (C); the *Information Field* (I), if present, contains one packet only. The packet format, shown in Fig. 2.48(b), comprises a common packet header (3 bytes) and additional packet header information and/or user data. The common packet header includes a **general format identifier** (GFI), a **logical channel indicator** (LCI), and a **packet identifier** (PI), whose last bit (*S/D*) discriminates between control (including signaling) and data packets. The use of the previously mentioned frame and packet fields is explained next.

*Level 2 of Rec. X.25* [Carlson, 2.7]. For level 2 of Rec. X.25 the F and FCS fields have the same function and coding as in level 2 of SS No. 7.

The A-field is used in the original HDLC applications to provide multipoint capabilities. This use is kept in ISDN applications (see Section 2.4.2) but is not specified in the X.25 recommendation. Here one bit only of the A-field is employed to discriminate between *command* and *response* frames, as specified in data link procedures. The remaining A bits are not used.

The C-field coding is shown in Table 2.8. We distinguish three types of frames: the **information transfer** frames (I-frames), the **supervisory** frames (S-frames) and the **unnumbered** frames (U-frames). Once the data link has been initialized by means of the U-frames, the I-frames are transferred and the S-frames are employed to control the I-frame transfer, as explained later.

The I-frames are labeled with two **sequence numbers**, N(S) and N(R), that code, modulo 8, the transmitter send and receive sequence numbers, respectively. N(S) and N(R) play the same role as FSN and BSN in SS No. 7. The main difference between them is that N(R) codes the next frame expected to be received, while BSN indicates the last correctly received SU. The **poll/final** (P/F) bit is used to provide a *check-point mechanism*. This allows a response frame to be logically associated with the appropriate initiating command frame. This bit is considered to be a **poll bit** (P-bit) if the frame is a command and a **final bit** (F-bit) if the frame is a response, as specified by the procedure.

Four types of S-frames exist according to the S-bit code. Typically the **receive ready** (RR) and the **reject** (REJ) frames are used to provide positive and negative acknowledgements. S-frames contain only a receive sequence number, N(R). The RR frames are used for positive acknowledgements when I-frames are not available for such a task. REJ frames provide an explicit negative acknowledgement message, different from SS No. 7, in which such information is provided implicitly by FIB and BIB bit status.

Up to 32 different U-frames can be specified according to the **more data bit** (M-bit) code. Typically, in the **balanced link access protocol** (LAP B), two U-

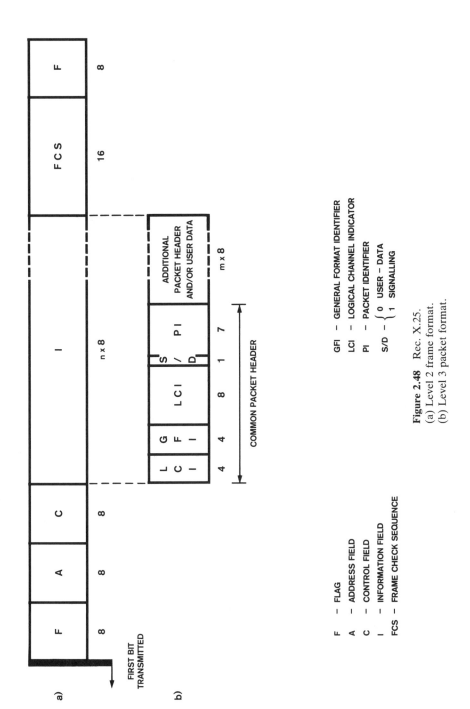

**Figure 2.48** Rec. X.25.
(a) Level 2 frame format.
(b) Level 3 packet format.

105

**TABLE 2.8** Recommendation X.25: C-field format

| Control field for | Control field bits | | | | | | | |
|---|---|---|---|---|---|---|---|---|
| | 1 | 2 | 3 | 4 | 5 | 6 | 7 | 8 |
| Information transfer (I-Frame) | O | N(S) | | | P/F | N(R) | | |
| Supervisory (S-Frame) | 1 | O | S | S | P/F | N(R) | | |
| Unnumbered (U-Frame) | 1 | 1 | M | M | P/F | M | M | M |

N(S) = Transmitter send sequence number
N(R) = Transmitter receive sequence number
S = Supervisory function bit
M = Modifier function bit
P/F = Poll bit when issued as a command, final bit when issued as a response

frames, the **set asynchronous balanced mode** (SABM) and **disconnect** (DISC) command frames, are used to initialize and disconnect the data link. U-frames do not contain sequence numbers and are not subject to error recovery.

Some examples of the error-recovery procedure used in LAP B are shown in Fig. 2.49. These are with reference to the time sequence for frame transmission and reception at a DTE-DCE interface. We assume that the data link has been established and that there is zero processing and transmission delay. Each frame is labeled with the frame type (I, RR, REJ) by the sequence numbers N(S), N(R) for the I-frames or N(R) only for the S-frames. Figure 2.49(a) illustrates the case of *absence of transmission errors* and shows the use of RR frames for positive acknowledgements in the absence of I-frames.

Figure 2.49(b) shows the *effect of a transmission error* on the (I, 4, 1) frame from DTE to DCE. Upon reception of the (I, 5, 3) frame the DCE detects an "out-of-sequence," discards the (I, 5, 3) and (I, 6, 3) frames, requests retransmission by the (REJ, 4) frame and returns in sequence upon reception of the requested I-frame, (I, 4, 4).

Figure 2.49(c) refers to the case in which errors occur on a S-type frame, the (RR, 3) frame. This example shows that, according to the LAP, the sending terminal, after transmitting each I-frame, initiates counting of a timeout $T_1$, allowing for reception of positive acknowledgement. When such a time expires, the I-frame is retransmitted. In the example the (I, 2, 0) frame is retransmitted, thus causing a duplication at the DCE that is recovered by means of the (REJ, 3) frame.

*Level 3 of Rec. X.25* [Rybczynski, 2.47]. With reference to Fig. 2.48(b), the LCI enables the statistical multiplexing of virtual channels (the can range up to 4095 since the all-zero LCI code is reserved for procedural needs). Each virtual channel can be used to support VC or DG communications. In addition, a particular VC facility called **fast-select facility** (FSF) can also be implemented over

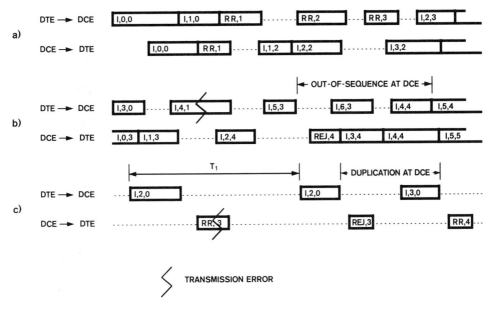

**Figure 2.49**   Rec. X.25: examples of level 2 error recovery

virtual channels.   In the following, we briefly describe how such types of communications are implemented in Rec. X.25.

Some typical packet formats are shown in Fig. 2.50 and 2.51.   These are the **call-request** packet (as a particular signaling packet) and the *data* packet, respectively.   The formats are represented by an octet bit-sequence numbered in order of transmission.   The first three octets are always present in the packet header and they contain, besides the LCI field, the GFI and the PI fields.

The first bit (*S/D* bit) of the PI field (octet 3) discriminates between control and data packets.   The remaining 7 bits in data packets contain the **send** and **receive sequence numbers**, P(S) and P(R), together with a **more data bit** (M-bit)* (see Fig. 2.51).   P(S) and P(R) serve for data packet sequence control to provide flow control, procedural errors, and abnormal conditions at level 3.   They are not intended to control transmission error conditions.   The M-bit is used to link together a sequence of packets.   Since the user-data field in each packet is normally limited to 128 or 256 octets, DTEs wishing to form data messages of greater length can link the desired number of packets by the M-bit.

As far as the *control* packets are concerned, we further distinguish between **signaling (call-control)** packets and **data-transfer control** packets.

An example of call control packet is given by the **call-request** (CR) packet. The format of the CR packet marked with a particular PI code is shown in Fig. 2.50.   This packet includes calling and called DTE addresses, a number of user facilities requested by the calling DTE (e.g., reversed charging, throughput class)

---

*The M-bit (more data) at level 3 should be distinguished from the M-bits (modifiers) at level 2.

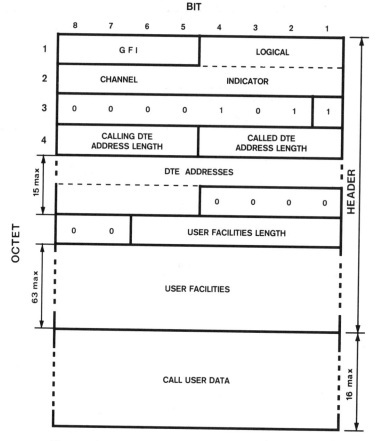

**BIT**

**Figure 2.50** Rec. X.25: format of the call-request packet.

and some user data (16 octets maximum) normally reserved for the higher-layer protocols (e.g., for the exchange of system password).

The data-transfer control packet can be of various types such as *interrupt, reset, restart,* and *supervisory*. Two supervisory packets are labeled as RR and REJ with the same meaning as the corresponding S-frames at level 2. Supervisory packets contain in their PI field only P(R), while the other control packets are marked with a fixed PI code.

The GFI is composed of 4 digits and is intended to convey information about a number of features such as the numbering modulo (8 or 128) of P(S) and P(R) and whether the packet-delivery confirmation is given by local network equipment or by the far-end DTE.

Figure 2.52 schematically shows the evolution of a successful virtual call (**switched virtual circuit**, SVC). For call establishment, two signaling packets are exchanged. These are the CR packet that appears as an **incoming call** (IC) packet to the called DTE, and the **call-accepted** (CA) packet that appears as a **call-connected** (CC) packet at the calling DTE. The call-clearing phase is characterized by the **clear-request** (CRR) packet appearing as a **clear-indication** (CRI) packet to

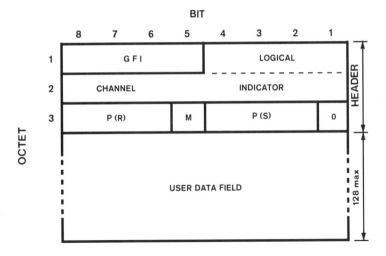

**Figure 2.51** Rec. X.25: format of the DATA packet.

M — MORE DATA BIT
P(S) — PACKET SEND SEQUENCE NUMBER
P(R) — PACKET RECEIVE SEQUENCE NUMBER

the distant DTE. The CRR and CRI packets are acknowledged by **clear-confirmation** (CRC) packets from the local network equipment. Call establishment and clearing at the calling DTE-DCE interface are depicted in Fig. 2.53 by means of a state transition diagram.

A PVC consists of a permanent association between two DTEs, analogous to a point-to-point leased physical circuit, that requires no call-setup or call-clearing actions by the DTEs. Thus in Fig. 2.52 a PVC evolution is characterized by the data transfer portion only.

As already stated, level 3 of Rec. X.25 includes two further types of communication, DG and FSF. Illustrated in Fig. 2.54 is the coexistence of SVC, PVC, FSF, and DG modes of communication over a single DTE-DCE interface. Figure 2.55 shows the procedural characteristics of DG (Fig. 2.55(a)) and FSF with *call-accepted response* (Fig. 2.55b), or *immediate-clear response* (Fig. 2.55c).

A DG is a self-contained packet, including calling and called DTE addresses, user facilities, and user data (up to 128 octets). If a DG cannot be delivered to the destination or is detectably lost, the network attempts to advise the source DTE via a **DG service signal** (DSS) packet (see Fig. 2.55(a)).

An FSF can be activated as a particular facility of a virtual call and is signaled explicitly in the user facility field of the CR packet. In the FSF CR packet, the user data field can range up to 128 octets. The destination DTE can respond with a CA packet containing also up to 128 octets of user data. Subsequently the call evolves as a normal virtual call; Fig. 2.55(b) shows, for example, the skipping of the data-transfer phase. Alternatively, the DTE can respond with a clear request

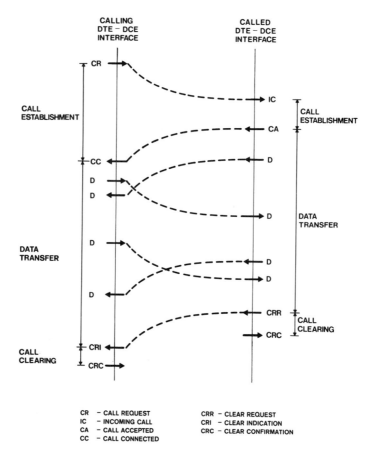

Figure 2.52 Rec. X.25: evolution of a virtual call.

containing up to 128 octets of user data, thus completing a two-way transaction and terminating the call, as illustrated in Fig. 2.55(c).

Inquiry-response applications, characterized by a small number of transactions, can be efficiently served by any of the techniques shown in Fig. 2.55. The selection of the appropriate technique depends highly on the specific application.

To conclude this brief summary of Rec. X.25 level 3 features, we explain the window flow control procedures, based on the use of P(S) and P(R) sequence numbers. We refer to Fig. 2.56, where the DCE window flow control mechanism is illustrated by assuming a P(S) and P(R) modulo 8 numbering. The DCE **transmit window** controls the packet flow from the DCE to DTE, while the DCE **receive window** controls the packet flow from the DTE to DCE. The transmit window is characterized by a lower limit $L$ and a width $W$. In the DCE-to-DTE packet transfer, this window imposes the following constraint on the P(S) of the transmitted packets: $L \leq P(S) \leq L + W - 1$. The lower limit $L$ is assumed to be equal to the last P(R) number that crossed the interface from DTE to DCE. The maximum value of $W$ is equal to $N - 1$, if $N$ is the numbering modulo of P(S) and P(R). The receive window operates in a similar manner but is completely independent

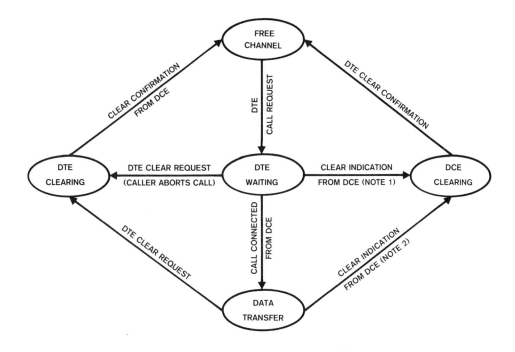

NOTE 1  :  EITHER CALLED DTE REFUSED CALL OR CALL ATTEMPT HAS FAILED

NOTE 2  :  EITHER CALLED DTE CLEARED DOWN CALL OR CALL
           CLEARED DUE TO NETWORK FAILURE

**Figure 2.53**  Rec. X.25: call establishment and clearing at the calling DTE/DCE interface.

of the transmit window.  The lower limit is equal to the last $P(R)$ number sent from DCE to DTE.

Both transmit and receive windows slide clockwise according to the $P(R)$ numbers crossing the interface on both directions.  The DCE does not transmit to the DTE packets with $P(S)$ values not contained in the DCE transmit window.  Also, it does not accept packets from DTE with $P(S)$ values not contained within the DCE receive window.  The number $W$ can be established in the user facility field of the CRR packet; otherwise, $W = 2$ is normally adopted.

### 2.4.2 ISDN Protocols Approach

This section presents the approach followed in the international activities for the standardization of the ISDN protocols.  In the mid-1980s, these activities were still to be completed.  It is envisioned that a comprehensive standard will be published by CCITT in the next few years.

Therefore, the considerations given below are limited to the basic guidelines followed in the standardization process.  Hence, user-access arrangements, pro-

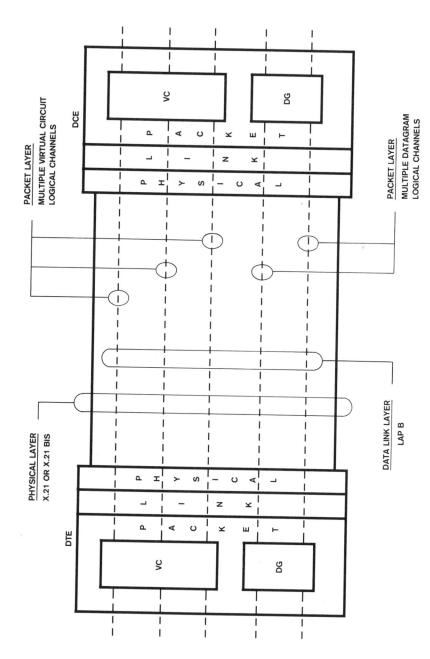

**Figure 2.54** Rec. X.25: Model of DTE/DCE interface supporting VC and DG operation.

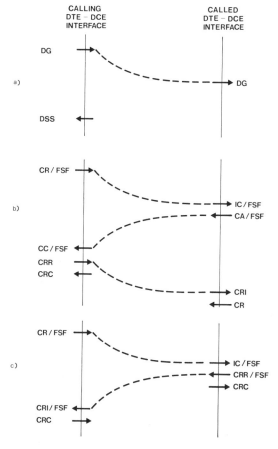

**Figure 2.55** Rec. X.25: comparison between

(a) Datagram.

(b) Fast select with call-accepted response.

(c) Fast select with immediate clear response.

tocol features at layer 1, layer 2, and layer 3, and the ISDN protocol reference model are only briefly introduced.

### User-Access Arrangements

We now consider possible user-access arrangements. Figure 2.24 shows functional groupings and reference points, and Fig. 2.57 illustrates physical implementations and their relationships with functional groupings. There are several significant issues about these arrangements.

It is noteworthy that a *direct connection* (i.e., by wires) is a possibility between points $S$ and $T$ (see Fig. 2.57(a)). This implies that a device intended to connect to an interface at reference point $S$ could also work in an interface at reference point $T$.

Another important configuration is the case in which several interfaces at point $S$ connect to a single connection at point $T$, using type NT2 functions (see Fig. 2.57(c)). This is the classical *star* distribution case illustrated in Fig. 2.58(a).

Multidrop connections, illustrated in Fig. 2.57(b), are also being considered in which several devices are directly connected to the same termination, either at

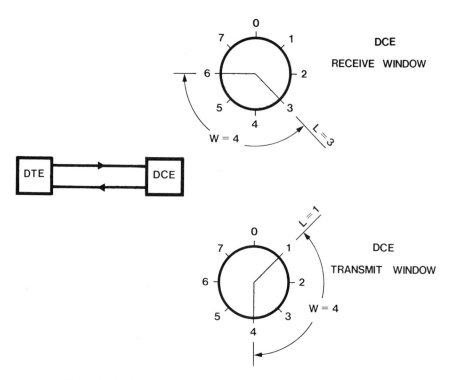

**Figure 2.56** Flow control: example of DCE receive and transmit window.

reference point $S$ or at reference point $T$.  This is not intended to require that individual terminals can talk to each other, as in an LAN, but rather that each of the terminals can communicate with a single "host" (the exchange termination). This configuration is also known as **passive bus** distribution (see Fig. 2.58(b)).

**Ring**, or **active bus**, configurations are not considered to operate across the interface at reference points $S$ or $T$.  However, the NT2 functional grouping could itself include a LAN active bus or ring distribution, as shown in Fig. 2.58(c) and 2.58(d), respectively.

Other configurations in Fig. 2.57 illustrate that either $S$ or $T$—but not both— need correspond to a functional or a physical interface in a particular configuration. Thus the functions of categories T1 and NT2 or NT2 and NT1 could be merged. If neither reference point $S$ nor $T$ is identifiable, then the configuration is not considered to be an ISDN configuration.

### Level 1 Protocol

Here we describe the physical layer with reference to the basic $2B + D$ access (see Section 2.2.3).  Two basic configurations have to be considered.  These are the point-to-point and the point-to multipoint, or passive bus, illustrated in Fig. 2.59.  There is a strong tendency towards an 8-wire transformer-coupled user interface.  This interface will have electrical characteristics suitable to cover the

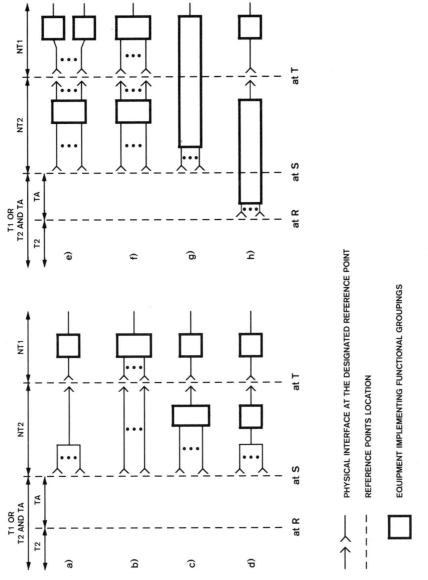

**Figure 2.57** ISDN user access: examples of physical configurations employing multiple connections.

115

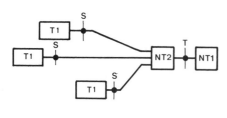

a)    STAR  DISTRIBUTION

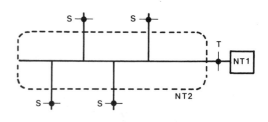

b)    PASSIVE  BUS  DISTRIBUTION

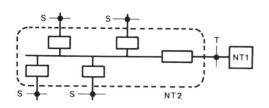

c )    ACTIVE  BUS  DISTRIBUTION

d )    RING  DISTRIBUTION

**Figure 2.58**   In-house distribution schemes.

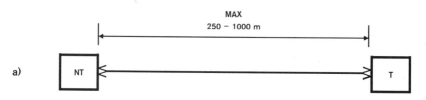

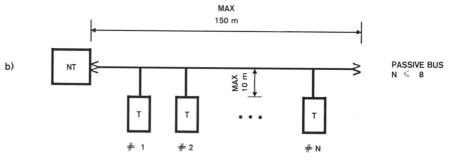

**Figure 2.59**   Basic configuration of ISDN user access.
(a) Point-to-point configuration.
(b) Point-to-multipoint configuration.

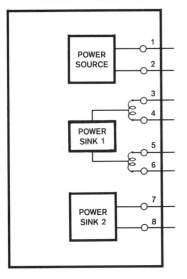
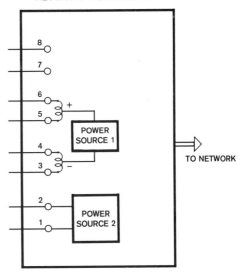

**Figure 2.60**   ISDN user access: Reference configuration for power feeding.

maximum interface length for the two configurations.   In a passive bus arrangement the maximum number $N$ of terminals is equal to 8.

The interface pin allocation for signals and power transfer is shown in Fig. 2.60.   Pins 3 and 4 are devoted to the signal transmitting pair, whereas pins 5 and 6 are allocated to the receive pair.   Power can be delivered either by using signal pairs or by using a dedicated pair.   Power from NT to terminal can be made available either by remote feeding from the local exchange or by a local NT source. This latter source will be needed, since the fairly complex user equipment in an ISDN environment requires a large amount of power, and there are strong limitations on the power-distribution capabilities through copper or fiber loops.

The most relevant aspect of the ISDN layer 1 protocol is represented by the *contention resolution* mechanism adopted for the passive bus distribution.   It is understood that such a mechanism should not impose an undue burden on terminals, taking into account that simple point-to-point interface configurations not requiring contention resolution will occur.   The layer 1 contention mechanism is indeed based on the TDM frame (Layer 1-frame) format used to assemble the B- and D-channels in the basic access.   The layer 1-frame for the basic access interface is shown in Fig. 2.61.   The line code is the **alternate mark inversion** (AMI) with 100% duty cycle.   Here a binary 0 is represented by alternative positive and negative pulses, whereas a binary 1 is represented by no pulse signal.   The frame structure is composed of 48 bits, yielding a total duration of 256 μs, assuming an interface bit rate of 192 kb/s.

With reference to Fig. 2.61, it is noteworthy that the pseudoternary line code has been chosen to permit contention resolution in the presence of a transmission delay corresponding to the maximum bus length (300 m).   Also, it allows a secure framing signal composed of a sequence of two alternate marks (F and L bits)

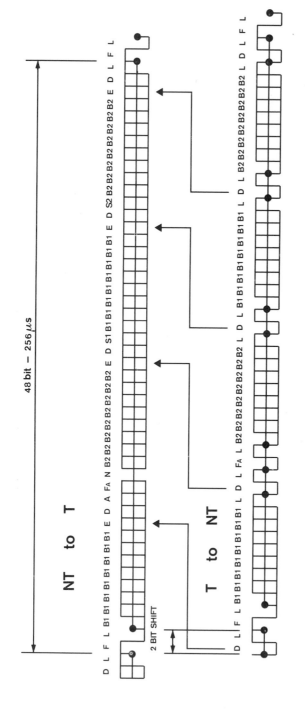

**Figure 2.61** ISDN user access: layer 1 frame structure at reference points S and T.

DOTS MARK THOSE PARTS OF THE FRAME THAT ARE
INDEPENDENTLY DC-BALANCED.

F   –  FRAMING BIT

L   –  DC BALANCING BIT

D   –  D-CHANNEL BIT

E   –  D-ECHO-CHANNEL BIT

$F_A$  –  AUXILIARY FRAMING BIT

B1   – BIT WITHIN B-CHANNEL 1

B2   – BIT WITHIN B-CHANNEL 2

S1, S2 – SPARE BITS

N   – BIT SET TO "$\overline{F_A}$"

A   – ACTIVATION BIT

followed by a bipolar violation. In the case of the AMI code, a DC balancing bit (L bit) is required which works in association with groups of bits that can be handled together by any terminal attached to the bus.

The frame comprises, in the direction from network termination (NT) to terminals ($T$), two B-channels (B1 and B2), the D-channel and the E-channel. The **E-channel** (echo channel) is used for contention resolution in the access control of the D-channel. This channel conveys the echo signals of the D-channel in the $T$ to NT direction. In fact the NT, on receipt of the D-channel bit from a $T$, reflects this bit condition toward the terminal in the next available E-channel bit time. The D-channel access procedure enables a number of terminals, connected in a passive bus configuration, to gain access in an orderly fashion. This procedure ensures that in cases in which two or more terminals attempt to access the D-channel simultaneously, one terminal will always be successful in completing the transmission of its information.

D-channel sensing is based on the all binary 1s pattern applied in the idle conditions. This is also based on the frame format adopted at layer 2 (layer 2 frame), delimited by flags (consisting of the binary pattern 01111110) and using the zero insertion-deletion mechanism (see Section 2.4.3).

Hence a terminal that needs to transmit information on the D-channel monitors the condition on the E-channel. After it counts a sequence of consecutive binary 1 equal to or longer than a present value $X$, a terminal is allowed to start transmission of a layer 2 frame. Detection of a binary 0 resets the counting process. While transmitting information in the D-channel, the terminal monitors the received E-channel and compares the last transmitted bit with the next available bit. If the transmitted bit is the same as the received echo, the terminal continues its transmission; otherwise the terminal ceases transmission immediately and returns to the D-channel sensing state.

The value of the counter $X$ is preset during the manufacturing or installation process in order to provide an access priority mechanism. Lower values of $X$ correspond to higher-priority classes. Two standard values of $X$ ($X = 8$ and $X = 10$) are envisioned for two classes of priority.

Moreover, to provide a fair allocation of the D-channel capacity among terminals belonging to the same priority group, the preset value $X$ of a terminal is increased to $X + 1$ (i.e., the priority is lowered) once a terminal has successfully completed the transmission of a layer 2 frame. This $X + 1$ value should be kept until the terminal has detected in the sensing state $X + 1$ consecutive binary 1s on the E-channel, thus returning to the original $X$ value.

### Layer 2 Protocol

The LAP at layer 2 of the D-channel (LAP D) uses a frame structure including frame delimiters (flags), zero insertion-deletion, and cyclic redundancy check. This provides secure alignment, transparency, and error detection. The error-recovery and flow-control procedures should be based on those of LAP B of Rec. X.25. This solution is preferred over use of the Level 2 procedures of SS No. 7. Full

LAP B and an appropriate subset of the LAP B procedures are envisioned to cope with PBXs and simple terminal configurations, respectively.

In addition, two *address* bytes of LAP D are used to support **multiple logical links**, providing independent multiple LAPs between exchange termination and logical end points at user premises, as shown in Fig. 2.62. Logical links may be used either to access different Layer 3 protocols for *s*-type, *p*-type, and *t*-type information (classes of service) or to address different terminal end points. The support of different Layer 3 procedures for *s*-type and *p*-type information is justified by taking into account that signaling does not require certain features needed by data (such as multiple VCs, packet sequencing, and flow control). Support of a multipoint capability is justified by referring to the passive bus configuration. However, the multipoint capability can also be applied to a star distribution characterized by the allocation of simple layer 2 functions at NT2. By using the address field to discriminate terminal end points, the NT2 functions can be limited to queueing for D-channel access at *T* and for distribution towards the end points, without handling error-recovery or flow-control procedures. Thus the same level 2 protocol appears at both *S* and *T* interfaces ("transparent" NT2).

In the case of PBXs or complex NT2 functions, end-point addresses may not be required, thus leaving the use of logical links just to discriminate information types.

LAP D should also be designed to handle properly the framed message traffic composed by various types of information. The signaling and data-delay requirements over the D-channel should dictate adoption of adequate delay control mechanisms. Some of these are limitations of the *p*-information frame size or adoption of a priority scheme for *s*-information (e.g., a preemptive priority).

LAP D also includes a **single-frame** procedure to be used for signaling and data transfer by simple ISDN terminals. This procedure is provided in addition to that currently adopted by LAP B and is referred to as the **multiple-frame** procedure. The single-frame procedure is based on the exchange of unnumbered frames (**sequential information** (SI) frames) in a send-and-wait mode. In fact the single-frame operation is characterized by data transfer with unity-width window.

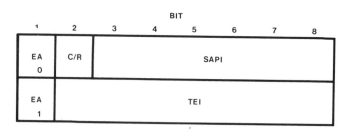

EA — EXTENDED ADDRESS BIT
C/R — COMMAND/RESPONSE BIT
SAPI — SERVICE ACCESS POINT IDENTIFIER
TEI — TERMINAL ENDPOINT IDENTIFIER

**Figure 2.62** LAP D protocol: address field format.

### Layer 3 Protocol

For *s*-type and *p*-type information, multiple layer 3 procedures are required. For logical links supporting signaling, a DG-oriented layer 3 procedure (such as that at layer 3 of SS No. 7) seems appropriate. However, procedures for *p*-information may take several forms. To provide compatibility with existing X.25 terminals, the procedures at level 3 of Rec. X.25 should be supported by the appropriate logical links over the D-channel. This includes the in-band call setup and clearing procedures defined in Rec. X.25.

In addition, there may also be a need for other classes of *p*-information. This aspect is particularly critical, since it involves the definition of new procedures for packet switched data. Data would be transferred over virtual channels in a format compatible with level 3 of Rec. X.25. However, the procedures for virtual channel setup and clearing would be different, since the *s*-type information would be used to set up and clear these virtual channels. This may make it possible to create calls consisting of both circuit switched and packet switched information, using a single, integrated signaling mechanism (**multimedia calls**).

The *s*-information procedures require capabilities that enable common telephony features such as conferencing, call forwarding, call hold and call transfer. In addition, the *s*-procedures should be defined to allow an easy mapping to both the circuit switched data (e.g., according to Rec. X.21) and telephony signaling. The procedure should also allow simple interworking with the future ISDN user part of SS No. 7.

A functional message signaling architecture seems appropriate to cope with these requirements and is also consistent with the long-term trends favoring increasing intelligence in equipment placed at user premises.

Layer 3 procedures for call control in case of point-to-point and point-to-multipoint connections are shown in Fig. 2.63.

Telemetry information at layer 3 may follow *p*-type procedures (e.g., by using a permanent virtual circuit) or *s*-type information procedures.

Finally, note that procedures at layers higher than 3 refer primarily to *p*- and *t*-type end-to-end applications across the transmission network. However, end-to-end information (e.g., between terminal and service vendors) may well be considered as signaling (*s*-type) and conveyed over CCS facilities, in particular when they are executed in conjunction with a circuit switched service over B-channels.

### Toward an ISDN Protocol Reference Model

An ISDN protocol reference model is here tentatively introduced as a means of describing the interchange of information between ISDN user and network elements and also between pairs of network elements. In particular, from an application point of view, we are interested in the communication and control aspects exercised between an ISDN user terminal equipment and

1. Another ISDN user terminal equipment;
2. Some network-control facilities inside the network that perform such functions as, for example, closed user-group registration;

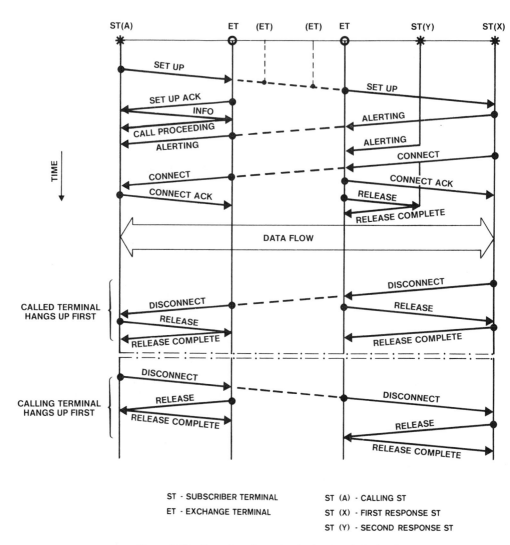

SPACE →

TIME ↓

ST(A)    ET    (ET)    (ET)    ET    ST(Y)    ST(X)

SET UP
SET UP
SET UP ACK
INFO
ALERTING
CALL PROCEEDING
ALERTING
ALERTING
CONNECT
CONNECT
CONNECT ACK
CONNECT ACK
RELEASE
RELEASE COMPLETE

DATA FLOW

DISCONNECT
CALLED TERMINAL
HANGS UP FIRST
DISCONNECT
RELEASE
RELEASE
RELEASE COMPLETE
RELEASE COMPLETE

DISCONNECT
CALLING TERMINAL
HANGS UP FIRST
RELEASE
DISCONNECT
RELEASE COMPLETE
RELEASE
RELEASE COMPLETE

ST - SUBSCRIBER TERMINAL        ST (A) - CALLING ST
ET - EXCHANGE TERMINAL          ST (X) - FIRST RESPONSE ST
                                ST (Y) - SECOND RESPONSE ST

**Figure 2.63**  Procedure for a simple circuit switched call.

**3.** Some information processing-messaging facilities that may reside within or outside the network and include database facilities.

The ISDN protocol reference model has a layered structure, based on the OSI reference model (see Section 2.2.3), and is a logical representation of all types of the communications and controls described.

From a modeling point of view, the various elements in the ISDN user premises and the network—that is terminal equipment, network termination, exchange termination, signal point, signal transfer point—may be modeled by a generic

**fundamental building block** (FBB). Such an FBB is comprised of three types of **information functional groupings** (IFGs). These are the **user** (U-IFG), the **signaling** (S-IFG) and the **management** (M-IFG). The interaction between these information groups can be described by the three-dimensional representation shown in Fig. 2.64.

The U-information (voice, data, etc.) functional group is the key focus of a communication service. It may be transferred transparently through the network or it may be processed within the network. The various activities involved in this group are described by a seven-layer structure.

The S-information functional group refers to all signaling and control aspects related to user-to-exchange connection control (setup, supervisory, clear down), nonconnection related control, control, and transfer of user information, access to network control facilities, user-to-user signaling, and so on. As just outlined, the distinction between U- and S-information is not explicitly made by the OSI model but is here important from a conceptual point of view. According to the considerations in Section 2.4.1, the S-IFG is represented by the seven OSI layers.

The M-IFG refers to all the local management aspects with the transfer of U-S information and also includes network management and traffic control to optimize utilization of network resources in normal and fault conditions. Note that the M-IFG is not partitioned into OSI layers.

Various interactions between the three IFGs within an FBB are possible, including interaction between U- and S-IFG taking place via the M-IFG, direct

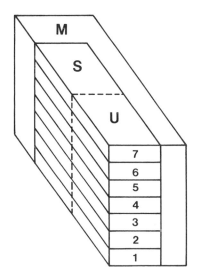

U – USER INFORMATION FUNCTIONAL GROUPING
S – SIGNALLING INFORMATION FUNCTIONAL GROUPING
M – MANAGEMENT INFORMATION FUNCTIONAL GROUPING

**Figure 2.64** Protocol fundamental building blocks.

interactions between U- and S-layers from 4 to 7, and sharing or separation of some functions at layer 1 regarding U- and S-functional groups.

Considering the external interactions of FBBs, we can distinguish between physical information transfers between two FBBs and interactions at the upper face of an FBB. The physical transfer can take place over separated physical media attached to both U- and S-IFGs or over a common physical medium shared by both U- and S-IFGs. The interactions at the upper face of an FBB are intended for application processes and include system-management, user, and signaling applications.

The general **ISDN protocol reference model** is shown in Fig. 2.65, in which

1. The network elements that perform end-system functions (such as those providing information processing facilities) are treated as a user system;
2. The network elements that perform the relaying functions (such as switching exchanges and STPs) are treated by a mirror-image version of an FBB;
3. The access between a user system and the network is represented by the *S-T* reference plane.

Some applications of the ISDN protocol reference model are given in Figs. 2.66 to 2.69. Figure 2.66 considers the case of a **circuit switched connection** and can be interpreted in the same manner as Fig. 2.32. Figure 2.67 refers to the case of a **packet switched connection via D-channels** and represents the direct access to packet network facilities (see also Fig. 2.34). Figure 2.68 is the model of a **packet**

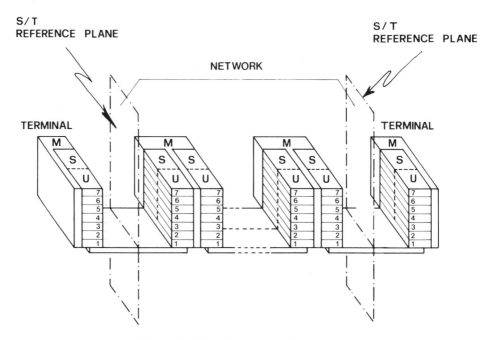

**Figure 2.65** ISDN protocol reference model.

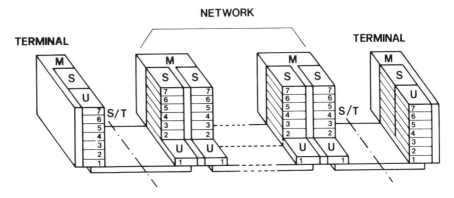

**Figure 2.66** ISDN protocol reference model for circuit switched connections.

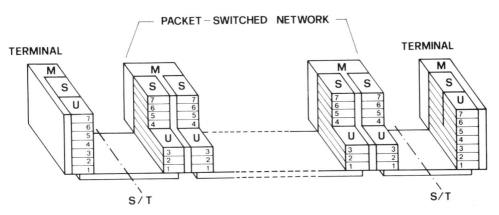

**Figure 2.67** ISDN protocol reference model for packet switched connections via the D-channel.

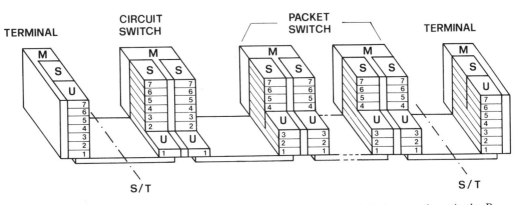

**Figure 2.68** ISDN protocol reference model for packet switched connections via the B-channel.

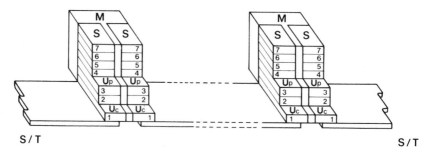

**Figure 2.69** ISDN protocol reference model in multi-media applications: Circuit switched connection type $U_c$ and packet switched connection Type $U_p$.

**switched connection via B-channels** and can be justified by taking into account the access to packet network facilities via a circuit switched network element.

Finally, Figure 2.69 shows the reference configuration for multimedia applications. Here user information is switched by different media in the two directions of transmission. Both circuit switched and packet switched connections are considered. As a consequence, we have two versions of U-IFG. These are the $U_c$ and $U_p$ versions, that refer to circuit and packet switched connections, respectively. An example of multimedia service applications is given by the bulk data delivery from a service center to a subscriber via a circuit switched channel, whereas user-to-center control information is implemented via a packet switched channel.

In conclusion, Figure 2.70 [Scace, 2.48] summarizes the protocol layering and interactions that might take place at an ISDN terminal equipment operating in accordance with a basic access standard.

## 2.5 ACRONYMS AND ABBREVIATIONS

| | |
|---|---|
| A-channel | Analog channel |
| A-field | Address field |
| ACK | Positive acknowledgement |
| ADM | Adaptive data multiplexing |
| ADPCM | Adaptive differential pulse code modulation |
| AMI | Alternate mark inversion |
| B-channel | 64 kb/s channel |
| BIB | Backward indicator bit |
| BSC | Binary synchronous communications |
| BSN | Backward sequence number |
| C-field | Control field |
| CA packet | Call-accepted packet |
| CAS | Channel-associated Signaling |
| CC packet | Call-connected packet |
| CCITT | International Telegraph and Telephone Consultative Committee |

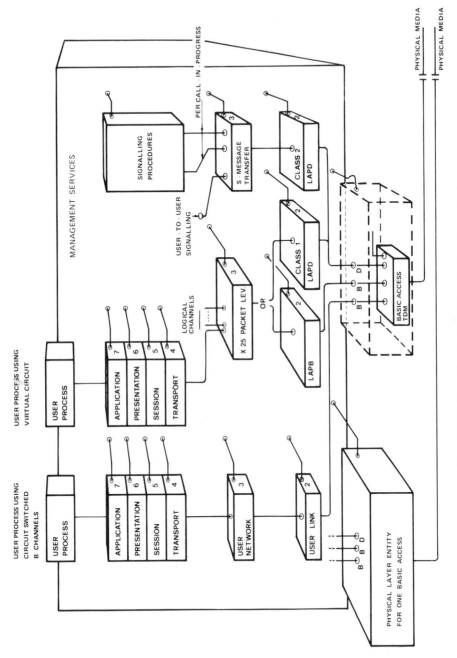

**Figure 2.70** Example of an ISDN terminal equipment.

| | |
|---|---|
| CCS | Common-channel Signaling |
| CIC | Circuit-identification code |
| CRC packet | Clear-confirmation packet |
| CRI packet | Clear-indication packet |
| CR packet | Call-request packet |
| CRR packet | Clear-request packet |
| CS | Circuit switching |
| | |
| D-channel | Signaling and data channel |
| DCE | Data communication equipment |
| DG | Datagram |
| DPC | Destination point code |
| DSI | Digital speech interpolation |
| DSS | Datagram service signal |
| DTE | Data terminal equipment |
| DUP | Data user part |
| DYN-AL | Dynamic allocation field |
| DISC frame | Disconnect frame |
| | |
| ET | Exchange termination |
| | |
| F | Flag |
| FBB | Fundamental building block |
| F-bit | Final bit |
| FCS | Frame check sequence |
| FDM | Frequency division multiplexing |
| FIB | Forward indicator bit |
| FSF | Fast-select facility |
| FSN | Forward sequence number |
| | |
| GFI | General format identifier |
| | |
| H-channel | Wideband channel |
| HB | Hybrid balancing |
| HDLC | High-level data link control |
| HS | Hybrid switching |
| | |
| I-field | Information field |
| I-frame | Information transfer frame |
| IA5 | International alphabet 5 |
| IC packet | Incoming call packet |
| IDN | Integrated digital network |
| IFG | Information functional grouping |
| ISDN | Integrated service digital network |
| | |
| LAN | Local area network |
| LAP | Link access protocol |
| LAP B | Balanced LAP |
| LBRV | Low bit rate voice |

| | |
|---|---|
| LCI | Logical channel indicator |
| LI | Length indicator |
| LSSU | Link status signal unit |
| | |
| M-bit | More data bit |
| MSU | Message signaling unit |
| MTP | Message-transfer part |
| | |
| NACK | Negative acknowledgement |
| NBS | Narrow-band switching system |
| NMC | Network-management center |
| NT | Network termination |
| | |
| OAM | Operation, administration, and maintenance |
| OAMUP | Operation, administration, and maintenance user part |
| OPC | Originating point code |
| OSI | Open system interconnection |
| | |
| P-bit | Poll bit |
| *p*-information | Low-speed bursty data |
| PAD | Packed assembling/disassembling |
| PBX | Private branch exchange |
| PCM | Pulse code modulation |
| PI | Packet identifier |
| POTS | Plain old telephone service |
| PS | Packet switching |
| PSE | Packet-switching exchange |
| PVC | Permanent virtual circuit |
| | |
| *R* | Reference point in ISDN local access arrangement |
| Rec. | CCITT recommendation |
| RR frame | Receive ready frame |
| REJ frame | Reject frame |
| RSU | Remote switching unit |
| | |
| *S* | Reference point in ISDN local access arrangement |
| *s*-information | Signaling information |
| S-frame | Supervisory frame |
| SABM | Set asynchronous balanced mode |
| SDLC | Synchronous data link control |
| SI frame | Sequential information frame |
| SIF | Signaling information field |
| SIO | Service information octet |
| SLS | Signaling link selection |
| SM | Statistical multiplexer (or multiplexing) |
| SS No.7 | CCITT signaling system No. 7 |
| STP | Signal-transfer point |
| SU | Signaling unit |
| SVC | Switched virtual circuit |

| | |
|---|---|
| T | Reference point in ISDN local access arrangement |
| t-information | Telemetry information |
| TA | Terminal adapter |
| TCM | Time compression multiplexing |
| TDM | Time division multiplexing |
| TSP | Traffic service position |
| TUP | Telephony user part |
| TV | Television |
| U-frame | Unnumbered frame |
| UP | User part |
| VC | Virtual circuit |
| VLSI | Very large scale integration |
| V&D | Voice and data |
| WBS | Wideband switching system |
| WDM | Wavelength division multiplexing |

## REFERENCES

[2.1] Aaron, M. R., and N. S. Jayant. "Special issue on bit rate reduction and speech interpolation," *IEEE Trans. Commun.*, Vol. COM-30, No. 4, April, 1982.

[2.2] Ackzell, L. "Evolution of ISDN in Europe," *ISSLS* 82, Toronto, September 1982, pp. 104–8.

[2.3] Ahamed, S. V., P. P. Bohn, and N. L. Gottfried. "A tutorial on two-wire digital transmission in the loop plant," *IEEE Trans. Commun.*, Vol. COM-29, No. 11, November, 1981, pp. 1554–64.

[2.4] Bially, T., B. Gold, and S. Seneff. "A technique for adaptive voice flow control in integrated packet networks," *IEEE Trans. Commun.*, Vol. COM-28, No. 3, pp. 325–33.

[2.5] Campanella, S. J. "Digital speech interpolation," *COMSAT Tech. Rev.*, Vol. 6, Spring, 1976, pp. 127–58.

[2.6] Cardwell, R. E., and H. R. Lehman. "Experimental wideband switching system capability," *ISS* 81, Montreal, September, 1981, Vol. 2, pp. 21B5/1–7.

[2.7] Carlson, D. E. "Bit-oriented data link control procedures," *IEEE Trans. Commun.*, Vol. COM-28, No. 4, April, 1980, pp. 455–67.

[2.8] CCITT. "Yellow Book—Q Series of Recommendations," ITU, Geneva, 1981.

[2.9] CCITT. "Yellow Book—X Series of Recommendations," ITU, Geneva, 1981.

[2.10] CCITT. "Red Book—Recommendation G721," ITU, Geneva, 1985.

[2.11] CCITT. "Red Book—I Series of Recommendations," ITU, Geneva, 1985.

[2.12] Cohen, D. "A protocol for packet-switched voice communications," *Comput. Networks*, Vol. 2, September/October, 1978, pp. 320–31.

[2.13] Coviello, G. J. "Comparative discussion of circuit vs. packet-switched voice," *IEEE Trans. Commun.*, Vol. COM-27, No. 8, August, 1979, pp. 1153–60.

[2.14] Cox, R. V., and R. E. Crochiere. "Multiple user variable rate coding for TASI and packet transmission systems,"*IEEE Trans. Commun.*, Vol. COM-28, No. 3, March, 1980, pp. 334–37.

[2.15] Decina, M. "Progress towards user access arrangements in integrated services digital networks," *IEEE Trans. Commun.*, Vol. COM-30, No. 9, September, 1982, pp. 2117–30.

[2.16] Decina, M. "Managing ISDN through international standards activities," *IEEE Commun. Mag.*, Vol. 20, No. 5, September, 1982, pp. 19–25.

[2.17] Decina, M., and D. Vlack. "Special issue on packet switched voice and data communication," *IEEE J. on Selected Areas in Commun.*, Vol. SAC-1, No. 6, December, 1983.

[2.18] desJardins, R., and H. C. Folts. "Special issue on open system interconnection (OSI)—Standard Architecture and Protocols," *PIEEE*, Vol. 71, No. 12, December, 1983.

[2.19] Felts, W. J. "Bell's concept of the ISDN," *Telephony*, Vol. 203, No. 18, October, 1982, pp. 43–61.

[2.20] Folts, H. C. "Procedures for circuit-switched service in synchronous public data networks," *IEEE Trans. Commun.*, Vol. COM-28, No. 4, April, 1980, pp. 489–96.

[2.21] Fraser, A. G. "Delay and error control in a packet switched network," *ICC*, 1977, pp. 22.4/121–5.

[2.22] Frank, H., and I. Gitman. "Economic analysis of integrated voice and data networks: a case study," *PIEEE*, Vol. 66, No. 11, November, 1978, pp. 1549–70.

[2.23] Goodman, D. J. "Embedded DPCM for variable bit rate transmission," *IEEE Trans. Commun.*, Vol. COM-28, No. 7, July, 1980, pp. 1040–46.

[2.24] Gruber, J. G. "Delay related issues in integrated voice and data networks," *IEEE Trans. Commun.*, Vol. COM-29, No. 6, June, 1981, pp. 787–800.

[2.25] Gruber, J. G. "A comparison of measured and calculated speech temporal parameters relevant to speech activity detection," *IEEE Trans. Commun.*, Vol. COM-30, No. 4, April, 1982, pp. 728–738.

[2.26] Guenin, J. P., F. Lucas, and P. Montaudoin. "The role of satellites in achieving ISDN," *ICC* 81, Vol. 1, Denver, June, 1981, pp. 19.5/1–5.

[2.27] Hardy, D., J. L. Dauphin, O. Louvet, and G. Pays. "PALME: a switching system integrating voice and data," *ISS* 81, Vol. 2, Montreal, September, 1981, pp. 21B2/1–6.

[2.28] Handler, G. J. "Circuit switched digital capability," *ICC* 82, Philadelphia, June, 1982, Vol. 3, p. 6A.5/1–5.

[2.29] Handler, G. J., and R. L. Snowden. "Planning packet transport capabilities in the Bell System," *ICCC* 82, London, October, 1982, pp. 121–25.

[2.30] Hardy, J. H. M., and C. E. Hoppitt. "Access to the British Telecom ISDN," *ICCC* 82, London, October, 1982, pp. 43–48.

[2.31] Harrington, E. A. "Voice/data integration using circuit switching networks," *IEEE Trans. Commun.*, Vol. COM-28, No. 6, June, 1980, pp. 781–93.

[2.32] Irmer, T. "The international approach to the ISDN," *Telecommun. J.*, Vol. 49, No. 7, July, 1982, pp. 411–15.

[2.33] Joel, A. E., Jr. "What is telecommunication circuit switching?" *PIEEE*, Vol. 65, No. 9, September, 1977, pp. 1237–53.

[2.34] Joel, A. E., Jr. "Digital switching—How it has developed," *IEEE Trans. Commun.*, Vol. COM-27, No. 7, July, 1979, pp. 948–59.

[2.35] Kermani, P., and L. Kleinrock. "Virtual cut-through: a new computer communication switching technique," *Computer Networks*, Vol. 3, September, 1979, pp. 267–86.

[2.36] Kneisel, K. E. "Goals and strategies for the introduction of broadband optical communication systems into the Deutsche Bundespost Telecommunication networks," *ICC* 82, Vol. 2, Philadelphia, June, 1982, pp. 4D.1/1–5.

[2.37] Kummerle, K., and H. Rudin. "Packet and circuit switching cost/performance boundaries," *Computer Networks*, Vol. 2, February, 1978, pp. 3–17.

[2.38] Leth, J. W., and P. E. White. "A level 3 signaling architecture for ISDN subscriber access," *GLOBECOM* 82, Vol. 2, Miami, December, 1982, pp. 762–5.

[2.39] Marcus, J. "The theory of connecting networks and their complexity: a review," *PIEEE*, Vol. 65, No. 9, September, 1977, pp. 1263–71.

[2.40] Miyahara, H., and T. Hasegawa. "Performance evaluation of modified multiplexing technique with two types of packet for circuit and packet switched traffic," *ICC* 79, Boston, June, 1979, pp. 20.5/1–5.

[2.41] Ninomiya, Y. "An accurate 8-bit A/D converter sampling at 100 Mhz," *IEEE Trans. Commun.*, Vol. COM-29, No. 9, September, 1981, pp. 1353–56.

[2.42] Parodi, R., N. Corsi, and L. Musumeci. "Circuit and packet switching in an integrated digital network," *ISS* 79, Paris, May, 1979, pp. 765–72.

[2.43] Petr, D. W. "32 kbps ADPCM-DLQ coding for network applications," *GLOBECOM* 82, Vol. 1, Miami, December, 1982, pp. 239–43.

[2.44] Pouzin, L., and H. Zimmermann. "A tutorial on protocols," *PIEEE*, Vol. 66, No. 11, November, 1978, pp. 1346–70.

[2.45] Rosner, R. D., and B. Springer. "Circuit and packet switching—a cost and performance trade-off study," *Computer Networks J.*, Vol. 1, No. 1, January, 1976, pp. 7–26.

[2.46] Rosner, R. D. *Packet Switching*, Lifetime Learning Publications, Belmont, Calif., 1982.

[2.47] Rybczynski, A. M. "X.25 interface and end-to-end virtual circuit service characteristics," *IEEE Trans. Commun.*, Vol. COM-28, No. 4, April, 1980, pp. 500–10.

[2.48] Scace, E. Private Communication.

[2.49] Spring, P. G. "The evolving call handling CCIS interprocessor communications networks," *GLOBECOM* 82, Vol. 2, Miami, December, 1982, pp. 721–5.

[2.50] Staudinger, W. "Digital data networks: activities of the Deutche Bundespost," *ISS* 79, Vol. 2, Paris, May, 1979, pp. 530–34.

[2.51] Tanenbaum, A. S. *Computer Networks*, Prentice-Hall, Englewood Cliffs, N.J., 1981.

[2.52] Wu, C. L., and T. Y. Feng. "On a class of multistage interconnection networks," *IEEE Trans. Comp.*, Vol. C-29, No. 8, August, 1980, pp. 694–702.

[2.53] Zimmermann, H. "OSI Reference model—The ISO model of architecture for open systems interconnection," *IEEE Trans. Commun.*, Vol. COM-28, No. 4, April, 1980, pp. 425–32.

# 3

# SPEECH-CODING ALGORITHMS AND VECTOR QUANTIZATION

**DR. JEAN-PIERRE ADOUL**

*Professeur Titulaire, Génie Electrique*
*Université de Sherbrooke. Québec, Canada J1K 2R1*

## 3.1 INTRODUCTION

The implementation of sophisticated speech-coding algorithms is increasingly cost-effective in numerous applications as a result of recent advances in large-scale integration. The purpose of these algorithms is to transmit, store, or synthesize speech at a given quality using fewer bits. This reduction is achieved by removing the redundancy of the speech signal.

It is possible to find common traits in the many coding algorithms that have been developed over the past 20 years. It is the purpose of this chapter first to pinpoint these common traits (fundamental coding functions and goals) and then to illustrate their operation in basic coding schemes. In subsequent sections we deal with the concept and applications of the recently developed technique of vector quantization. Table 3.1 gives a list of the fundamental coding functions, which are discussed together with the algorithms considered. Several good tutorial [Crochiere, 3.1] review papers [Jayant, 3.2; Flanagan 3.3; Gibson, 3.4] and books [Rabiner, 3.5; Jayant, 3.72] published in recent years provide complementary insights and overview of this vast field. Preceding the discussion of the various coding functions we outline the overall characteristics of coding algorithms and briefly state the basic properties of the speech signal.

### 3.1.1 The Rate Quality Complexity Referential

There are three major parameters in the design of a digital speech coding technique: (1) the (average) binary rate (typically expressed in bits per second), which is directly related to the bandwidth requirement; (2) the overall transmission quality;

**TABLE 3.1** Classification of coding functions and speech-coding techniques discussed in this chapter. The chapter further concentrates on the vector quantization technique and provides interpretations when this technique is used to quantize speech spectra.

| Coding function | Function and/or purpose | Associated concepts | Algorithm considered |
|---|---|---|---|
| Quantization (2.1) | Living with approximation | Scalar Quantization<br><br>(Vector quantization developed in Section 3.3) | Log PCM |
| Prediction (2.2) | Removing "blatant" redundancies | Linear prediction and least mean square estimation | DPCM |
| Adaptation (2.3) | Taking full advantage of speech short-term stationarity | Energy estimate<br>Syllabic/instantaneous<br>Switched/continuous adaptation<br>Forward/backward<br>Sequential/block processing | ADPCM (Adaptative quanti- zation<br>Delta Modulation (CVSD/ Jayant)<br>NIC, BAR<br>ADPCM (Gradient adaptive prediction) |
| Readiness and foresight (2.4) | Living with the unexpected | Spreading the "information bursts"<br>Delayed coding | Pitch compensation<br>Time-amplitude coding<br>Tree coding |
| Interpitch processing (2.5) | Tapping the redundancy of speech pseudoperiodicity | Pitch extraction<br><br>Pitch prediction | AMDF (Pitch extraction algorithm)<br>APC<br>TDHS |
| Noise control (2.6) | Paying attention to speech perception | Auditory masking<br>Mel Scale | Subband coding<br>ATC<br>Noise shaping |
| Speech-production modeling (2.7) | Trimming the "excitation signal" to the essentials | "Vocal track" filter<br>Residual signal | Pitch-excited LPC<br>Baseband RELP |
| Vector quantization (3.) | Tapping the redundancy of speech production constraints | K-means algorithm for quantizer design (3.3)<br>Parametric representations of the spectrum<br>Spectral distances | Conditional (3.1) quantization<br>VQ (3.2) of the LPC coefficients (3.3)<br>Waveforms VQ (3.4) |

and (3) the hardware complexity, an important—albeit evasive—parameter. Other considerations such as robustness to channel error and tandem behavior are important in some applications, yet they remain secondary with respect to the three basic questions: What is the bandwidth, how does it sound, and how much will it cost?

**Rate** is the most objective of the three parameters. It can be expressed in terms of bits per second (possibly averaged) or in terms of AM (amplitude modulation) or FM (frequency modulation) transmission parameters, such as the required normalized bandwidth, typically 1 to 2 Hz/bit/s.

**Quality**, or the inverse distortion, is an important dimension for public acceptance. It has become customary to divide this scale into four broad classes (listed in Table 3.2) together with the typical yardstick used to assess quality in that range.

**Complexity** is a hard dimension to describe. Objective complexity indexes have been proposed, such as the log of the relative count of logic gates [Flanagan, 3.3]. The most useful measure would be the real cost per **coder-decoder pair** (codec) at a given date, but this ordering would be constantly reshuffled by new technologies, higher integration, and large-volume applications. Moreover it is hard to compare off-the-shelf products with those yet to be developed. Sudden jumps in cost also occur whenever the scale of integration reaches the point at which integration of the complete codec into a single chip becomes possible. The total codec complexity is shared between the coder and decoder. Most algorithms tend to require more processing from the coder, which can sometimes go as far as reducing the decoding process to a mere operation of looking in a table. This is not unavoidable by any means. Some techniques exhibit a more balanced repartition than others have [Lim, 3.6] or can be designed to have all the complexity located at the decoder. In some applications, such as broadcast and mobile radio transmission, it might be useful to be able to choose where to place the complexity, whether at the coder or at the decoder.

### 3.1.2 Basic Redundancies in the Speech Signal

Digital coding of speech is an analog-to-digital (A/D) conversion process that includes removal of the redundancies inherent in the nonstationary speech signal. This redundancy comes from constraints upon the manners and means by which speech is produced. Although speech is a highly nonstationary signal, the means

**TABLE 3.2** Broad quality classes for speech transmission systems

| Quality classes | Concern | Typical criteria |
|---|---|---|
| 1. Broadcast quality | Comfort | Bandwidth and signal-to-noise ratios |
| 2. Tool quality | Noise | Signal-to-noise ratio |
| 3. Communication quality | Naturalness | Subjective rating |
| 4. Synthetic quality | Intelligibility | Intelligibility tests |

by which it is produced, namely, the larynx and mouth apparatus, are remarkably universal, as is the hearing equipment!

The presence in speech of these statistical invariants makes efficient speech coding a challenging research area. Coding algorithms make varying attempts to account for that redundancy. At high bit rates, there is no need for sophisticated redundancy removal, and therefore algorithms are modestly aimed at modeling long-time average speech properties. At low bit rates however, coding algorithms must adhere closely to the time-varying properties of speech.

We shall now briefly discuss both amplitude and spectrum characteristics of speech.

### Dynamic Range

The ear exhibits a large dynamic range in its ability to adjust from very loud sounds to very soft whispers. In speech there is an interspeaker difference in volume, which can reach up to 60 dB. For a given speaker there remains 20 to 30 dB of dynamic range between different types of sounds. In particular, it is worth distinguishing between two broad classes of sounds, which differ in the way they are produced. **Voiced sounds** are produced by the vibration of the vocal chords, whereas **unvoiced sounds** are the result of turbulent air passing through the open vocal track. It is not surprising that these two types of sounds have very different characteristics [Fant, 3.7]. Figure 3.1 illustrates the time and frequency domain properties of these sounds.

*Voiced sounds.* Voiced sounds exhibit high-amplitude pseudoperiodic wave-forms. The period, $T$, which is called the pitch period, is determined by vocal chord vibrations. The pitch varies slowly (i.e., with a time constant in tenths or hundredths of a millisecond). The variations, or pitch contour, determine the speech intonation which betrays the mood and disposition of the speaker. The **autocorrelation for voiced sounds**, $R(k)$, exhibits a positive peak for $k = T$. Note also that $R(k)$ is positive for the first few values of $k$; $k = 1, 2$ indicating that adjacent voiced-speech samples are likely to be of the same sign. The log magnitude spectrum of a typical voiced sound is also shown on Fig. 3.1.

The bulk of the energy of voiced sounds is in the low frequency range. In the telephone bandwidth of 4 kHz, voiced speech generally exhibits three or four frequencies, around which the energy is concentrated. These are called **formant frequencies** and are characteristic of the vocal-track resonant cavity. In particular, the lower formant is fairly stable, as it is related to the distance between the throat and the lips of the speaker. The second formant has a wider range of variation. Correct rendering of the second formant is subjectively important as it is intimately linked with the tongue position, which in turn is responsible for differentiating the various vowels of the language. The formant frequencies vary according to the motion of the mouth and tongue. The spectrum of speech may, however, be considered fixed during time intervals below 16 ms or so.

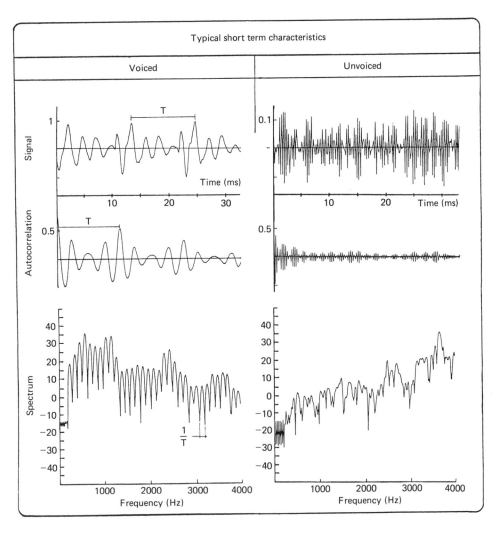

**Figure 3.1** Basic short-term speech characteristics. Typical waveform autocorrelation and log spectrum for both voiced and unvoiced sounds.

A closer look at the voiced log spectrum reveals a periodic structure that corresponds to the harmonics of the fundamental frequency, $1/T$.

*Unvoiced sounds.* Turning our attention to unvoiced sounds, we can say that in general they exhibit low-amplitude noiselike waveforms. The autocorrelation $R(1)$ is generally very negative, indicating that successive samples are likely to be of opposite signs. The log spectrum shows a concentration of energy in the high frequency range.

## 3.2 FUNDAMENTAL CODING FUNCTIONS AND GOALS

### 3.2.1 Quantization: Living With Approximations

We shall review some basic processing which generally takes place in a digital encoding scheme, such as prediction and syllabic adaptation. All these processing operations are reversible in the sense that it is possible to reverse their effect and recover the original signal. There is, however, a point of no return in any digital encoding scheme: This is the quantization stage, sometimes identifiable as the A/D conversion process. At this stage, we must settle for approximations, since the original signal can no longer be recovered exactly. In fact, it is as if an undesirable signal, called the **quantization noise**, were added to the original signal.

In recent years, attention has been focused on the quantizing stage with the implementation of various **vector quantization** (VQ) schemes. VQ is the process of rounding several variables (i.e., the components of a vector) simultaneously. We shall describe VQ in more detail later. For the time being we shall describe the basic **one-dimensional**, or **scalar**, quantizer.

In a (scalar) quantizer, an input variable $v$ is replaced by an approximate value $\hat{v}_i$. The index $i$ ($i = 0, 1, 2 \ldots N-1$) indicates that the value $\hat{v}_i$ is drawn from a finite set of possible rounded values. Whenever $v$ is replaced by $\hat{v}_i$, a quantization error results:

$$e_q = v - \hat{v}_i \tag{3.1}$$

Essentially, the quantizer is a many-to-one mapping, defined by a set of rounded values $\{\hat{v}_i\}$ and a mapping rule typically expressed in terms of minimizing the absolute error or, equivalently, the squared error.

It is often useful to view quantization as a two-step process (whether actually or only conceptually) consisting of a coding part, or A/D conversion part, which finds the index $i$, followed by a decoding part, or digital-to-analog (D/A) conversion part, which is, finally, a simple operation of finding values in a table (see Fig. 3.2).

In many situations, only the coding part of the quantizer is required at the transmitter. This is, for instance, the case with log pulse code modulation (PCM), where the index $i$ is simply transmitted in binary form.

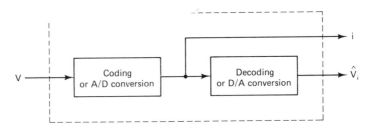

**Figure 3.2** The quantization process viewed as a two-stage operation. The coding stage finds the suitable index $i$ and the decoding stage outputs the rounded-off quantity $\hat{v}_i$. The advantage of viewing the quantization process this way is that it readily carries over to vector quantities.

The log PCM quantizer is of itself a widely used technique for telephone-speech digitization. Log PCM codecs, complete with the accessory filtering circuitry, are available on a single integrated circuit (IC) chip from several manufacturers. We will briefly describe the so-called μ-law PCM quantizer as an illustration of the quantization concept. A detailed description can be found in [Jayant, 3.72].

In order to deal solely with integers, we shall consider that the input variable has the following range:

$$v \in [-8287, +8287] \tag{3.2}$$

With that scale in mind, the standard μ-law PCM encoder uses a set of 256 rounded off values given in the (decoding) Table 3.3.

It is easy to complete this table with a handheld calculator using a simple formula. First note that the table is symmetrical in the sense that each positive rounded value has its negative counterpart, namely,

$$\hat{v}_{128+j} = -\hat{v}_{127-j}; \quad j = 0, 1, 2, \ldots, 127 \tag{3.3}$$

Furthermore, if $j$ is expressed in base 16:

$$j = 16C + S = 0, 1, 2, \ldots, 127 \tag{3.4}$$

Then

$$\hat{v}_{128+j} = \hat{v}_{127-j} = 2^{C+1}(S + 17) - 33 \tag{3.5}$$

where
$C = 0, 1, 2 \ldots 7$ is called the chord number
$S = 0, 1, 2 \ldots 15$ is the step number within a chord

TABLE 3.3  Facsimile of the log PCM quantization table, which can easily be completed using Equation (3.5)

| $i$ | $\hat{v}_i$ |
|---|---|
| 255 | 8159 |
| 254 | 7903 |
| 253 | 7647 |
| 240 | 4319 |
| 239 | 4063 |
| 188 | 431 |
| 187 | 415 |
| 186 | 399 |
| 130 | 5 |
| 129 | 3 |
| 128 | 1 |
| 127 | -1 |
| 126 | -3 |
| 2 | -7647 |
| 1 | -7903 |
| 0 | -8159 |

Suppose, for example, that the input variable $v$ is equal to 420. In the table we find $\hat{v}_{188} = 431$ and $\hat{v}_{187} = 415$. We deduce that the index will be 187, since $\hat{v} - v_{187} = 5$ gives the smallest possible absolute error.

The log scale of the μ-law PCM encoder provides a wide dynamic range for the speech signal over which the signal to quantization noise ratio is practically constant and is larger than 30 dB.

### 3.2.2 Prediction: Removing Blatant Redundancies

A fundamental method that removes the redundancy of a time series is called **predictive coding**. Basically, this technique attempts to remove from the signal prior to transmission anything that can be predicted about it at the receiver. Predictive coding has been used successfully in the encoding process of speech, image, and data sources (see Fig. 3.3). The predictors at the receiver and at the transmitter can be made as "intelligent" as we want, provided that they work in synchronization. This means that they operate deterministically according to the same set of rules and that they have access to the same data base.

The basic encoding idea can be illustrated by a description of the widely used method for transmitting correlated samples called **differential pulse code modulation** (DPCM).

The canonical diagram of DPCM is illustrated in Figs. 3.4 and 3.5. In this diagram, $e(n)$ is the difference between the input sample $x(n)$ and a prediction of $x(n)$ denoted by $\hat{x}(n|n-1, \ldots, n-p)$. The most successful class of predictors has been the class of linear estimators. In this case $\hat{x}(n|n-1, \ldots, n-p)$ is

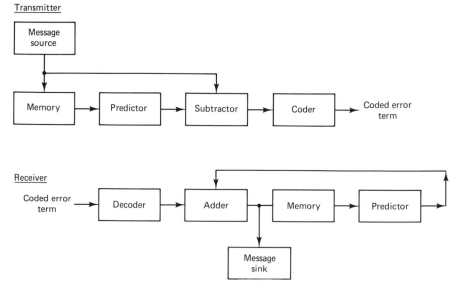

**Figure 3.3** Basic predictive coding paradigm. (Used by permission from the IEEE, [Elias, 3.9]).

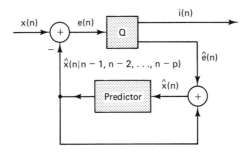

**Figure 3.4** Canonical diagram of differential PCM coder. The predictor computes an estimate $\hat{x}(n|n - 1, n - 2, \ldots, n - p)$ of the input sample $x(n)$ from information based on the $p$ previous reconstructed samples, $\hat{x}(n - k)$. The difference $e(n)$ is then quantized as $\hat{e}(n)$ with corresponding index $i(n)$. Note that since the reconstructed sample $\hat{x}(n)$ is entirely determined by the past sequence of $\hat{e}(n)$, it can be reconstructed in the same manner at the receiver end (providing no transmission error corrupted the sequence of indices).

calculated as a linear combination of the $p$ previously transmitted samples, as follows:

$$\hat{x}(n|n - 1, n - 2, \ldots, n - p) = \sum_{k=1}^{p} a_k \, \hat{x}(n - k) \tag{3.6}$$

where $\hat{x}(n - k)$ stands for the reconstructed sample at time $n - k$ using the quantized difference $\hat{e}(n - k)$.

$$\hat{x}(n - k) = \hat{x}(n - k|n - k - 1, n - k - 2, \ldots, n - k - p) + \hat{e}(n - k)$$

Use of DPCM is predicated on the basis of the fact that the variance of the difference signal $\hat{e}(n)$ is smaller than the variance of the sample $x(n)$. In fact, if we neglect the effect of the quantizer, the contribution of the prediction scheme can be characterized by the energy reduction, expressed in decibels, between the $x(n)$ and the $e(n)$ samples. This energy reduction factor is called the **prediction gain** and is illustrated in Fig. 3.6 as a function of the order $p$. It is generally expressed in decibels. As an example of the type of reasoning which can yield the calculation of the coefficients $a_i$, let us consider briefly the case $p = 1$. Ignoring

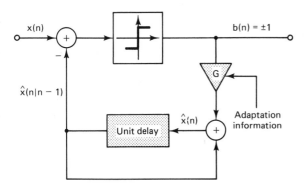

**Figure 3.5** Block diagram of adaptive delta modulators. There exists a wide variety of techniques for generating the adaptation information from the signal's past. In fact, techniques based on almost all combinations of short versus long adaptation-time constant on the one hand and binary ($b_n$) versus analog ($\hat{x}(n|n - 1)$) representation for the signal's past have been proposed in the technical literature.

the quantization effect we suppress the "hats." The prediction algorithm has the form

$$x(n|n - 1) = \alpha x(n - 1) \tag{3.7}$$

We seek a value of $\alpha$ which, in a statistical sense, will minimize the energy of $e(n)$ assuming a stationary input process:

$$\bar{e}^2 = E[e(n)^2] = E[(x(n) - x(n|n - 1))^2] \tag{3.8}$$

$$= E[x^2(n) - 2\alpha x(n)x(n - 1) + \alpha^2 x^2(n - 1)] \tag{3.9}$$

$$= R(0)[1 + \alpha^2] - 2\alpha R(1) \tag{3.10}$$

where $R(k) \stackrel{\Delta}{=} E[x(n)x(n - k)]$ is the autocorrelation function of the input process. Differentiating $\bar{e}^2$ with respect to the parameter $\alpha$, we obtain

$$\frac{\partial \bar{e}^2}{\partial \alpha} = 2\alpha R(0) - 2R(1) \tag{3.11}$$

Since the second derivative (i.e., $2R(0)$) is always positive, equating $\partial \bar{e}^2 / \partial \alpha$ to zero yields the value $\alpha$ that provides the largest prediction gain:

$$\alpha = \frac{R(1)}{R(0)} \tag{3.12}$$

On a short-term basis, $\alpha$ varies between $\pm 1$, in particular, near $+1$ for voiced sounds and near $-1$ for unvoiced sounds. However, on a long-term basis, the voiced sounds dominate and $\alpha$ converges to a value around 0.85. Essentially, with $\alpha = 0.85$, predictive coding removes what the receiver already knows, namely, that low-frequency energies dominate in speech. The predictive scheme is essentially purging the signal from this obvious redundancy.

In terms of signal-to-noise improvement over PCM transmission, a DPCM technique will buy roughly 6 dB, assuming the same coding rate and the same quantizer.

### 3.2.3 Adaptation: The Handling of Nonstationarity in Speech

Because the speech process is fundamentally nonstationary, there is a coding advantage to adapting both coder and decoder to the prevailing speech statistics. Coding schemes can be adaptive in many ways: For instance, a differential PCM scheme can have an adaptive quantizer, or an adaptive predictor, or both. This can be at times a source of confusion.

The basic adaptation principle is to have in the coder—and sometimes in the decoder as well—the calculation of a local estimate for some statistical feature of speech such as energy, spectrum, or pitch. In DPCM-like coders it is the adaptation of the quantizer that provides the most significant improvement in speech quality at a given rate.

In adaptive quantization, a local estimate of the standard deviation of the input signal is calculated, which in turn controls the gain of an amplifier located in front of the (fixed) quantizer that has been optimized for signals with unit variance. Or equivalently, the estimate controls directly the scale of the variable-quantizer step size. A higher gain than 5 dB (around 24 kb/s) in the **signal-to-noise ratio** (SNR) can be achieved with an adaptive quantizer over the fixed quantizer DPCM system [Noll, 3.10].

Typical local estimates include the following.

$$\sigma_n^2 = \frac{1}{N} \sum_{i=1}^{N} x^2(n-i) \qquad \text{running average} \qquad (3.13)$$

$$\sigma_n^2 = \left(1 - \frac{1}{K}\right)\sigma_{n-1}^2 + \frac{1}{K} x^2(n-1) \quad \text{first-order filter} \qquad (3.14)$$

The time constant is equal to $K$ sample intervals.

The terminology to classify the various adaptation characteristics is defined in the following material.

### Adaptation Time Constant: Syllabic Versus Instantaneous Adaptation

Adaptation refers to schemes that estimate the speech characteristic over a duration of several milliseconds (typically 4 to 25 ms) to accommodate changes in phonemes and syllables. On the contrary, *instantaneous* adaptation schemes have very short time constants of a few samples (typically less than 4 ms) like in some **delta modulation** (DM) schemes that adapt to the slope of the waveform [Jayant, 3.2].

DM is a special case of DPCM where the quantizer has only two possible rounded values, resulting in a staircase approximation of the waveform. DM with a fixed quantizer requires a very high sampling rate ($\geq 100$ kb/s) to accommodate the speech dynamics. However when the two rounded quantized values can be made to adapt to the speech energy, the result gives a very simple and inexpensive digitization scheme (see Fig. 3.5). DM chips (such as Motorola's MC3418) are available off the shelf and provide communication quality in the range of 16 to 32 kb/s with a remarkable robustness to channel errors. Channel error rates as high as $10^{-2}$ are acceptable and have no significant effect on the quality of speech.

*DM with Syllabic Adaptation.* The frequently used **continuously variable slope-delta modulation** (CVSD) technique [Seiler, 3.11] has a syllabic adaptation scheme in the form of a first-order digital filter with a time constant greater than 4 ms. It is described by a gain $G_n$ (see Fig. 3.5) of the form

$$G_n = \alpha G_{n-1} + f(b_{n-1}, b_{n-2}, b_{n-3}) \qquad (3.15)$$

where the function $f(\cdot) = 1, 0$ according to whether $b_{n-1}, b_{n-2}, b_{n-3}$ are, or are not, all of the same sign. Basically, $G_n$ is increased whenever the binary stream $b_n$ exhibits too many consecutive bits of the same sign—that is, when the entropy of the bit stream diminishes.

***DM with Instantaneous Adaptation.***   The *one-word-memory adaptation* scheme of Jayant [Jayant, 3.12] is also well known and illustrates an instantaneous adaptation.   The coding strategy is as follows:

$$G_n = G_{n-1}M(b_{n-1}, b_{n-2}) \tag{3.16}$$

The multiplier $M$ takes one of two values according to whether $b_{n-1}$ and $b_{n-2}$ are or are not of the same sign.

### Mode of Adaptation

The parameter(s) to be adapted can vary continuously over a given range or can take a finite number of chosen values.   In this case the method is based on the **switched quantizer** [Dietrich, 3.13], or the **switched predictor** [Adoul, 3.14] techniques.

For instance, adapting the predictor of an ADPCM coder requires the specification of *p predictive coefficients* $a_i$.

Figure 3.6 illustrates the prediction gain in decibels that can be obtained as a function of the prediction order $p$ under six adaptation strategies.

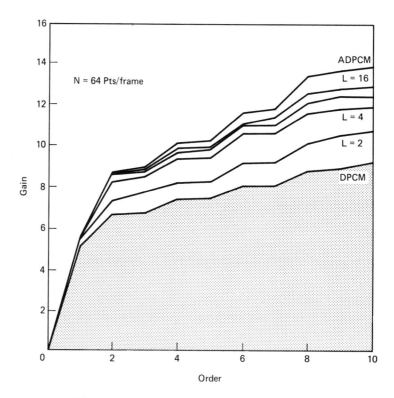

**Figure 3.6**   DPCM prediction gain obtained on a particular sentence as a function of the prediction order for a fixed predictor (DPCM), the best out of $L$ appropriately selected (i.e., fixed) predictors ($L = 2$, 4, 8, 16) and a continuously adaptive predictor (ADPCM).   (Used by permission from the IEEE, [Adoul, 3.14].)

**Information for Adaptation**

The information for adaptation must be identical at the transmitter and at the receiver. There are two philosophies for handling this question (see Fig. 3.7).

In **forward adaptation** the estimation value is calculated from the input samples at the transmitter. The input signal must be buffered and the estimation value (or a coded equivalent) must be transmitted to the receiver. In **backward adaptation** the estimation value is calculated from the transmitted quantized samples independently by the transmitter and the receiver. It follows that no side information about adaptation is required in this case.

For instance, the **nearly instantaneous coding** (NIC) [Duttweiler, 3.15] and the **backward adaptive reencoding** (BAR) techniques [Adoul, 3.16] are two schemes for reducing the bit rate for speech already digitized by standard 8-bit log PCM encoders. Basically, they both accept 8-bit-per-sample log PCM input signals and condense the essential information content into 6-bit words. To achieve this end, both techniques use switched quantizers according to a "maximum-amplitude" estimate. Since the input is already discrete, the quantizers are many-to-one transcoding tables. In the NIC, the estimate, $E_n$, is calculated at the transmitter as the greatest absolute value within a block of input samples. In the BAR system, on the contrary, it is calculated simultaneously at both the transmitter and the receiver using the following recursive formula:

$$E_n = \max\{E_{n-1} - 1, \hat{x}(n-1)\} \tag{3.17}$$

The performance of these two schemes is comparable (30 dB SNR) (see Fig. 3.8). The NIC scheme is simpler to implement; however, the BAR scheme does not require transmission of side (or overhead) information [Adoul, 3.16].

**Type of Processing**

Processing can be **sequential** in the sense that the estimate is updated at each sample, or it can be calculated once per block, as in **block processing**.

To illustrate the difference, let us consider again the adaptive differential

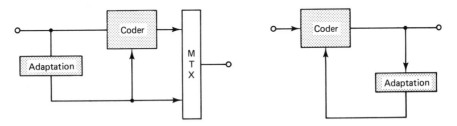

**Figure 3.7** Two approaches for sharing the adaptation information between coder and decoder. In the forward adaptation (a), the information for adaptation is determined unilaterally by the coder from the input signal. However, this side information must be multiplexed with the coded waveform. In backward adaptation (b), the information for adaptation is derived from the coded information, which is indeed available at both transmitter and receiver. Forward adaptation is more efficient at low bit rates and backward adaptation is better at high bit rates.

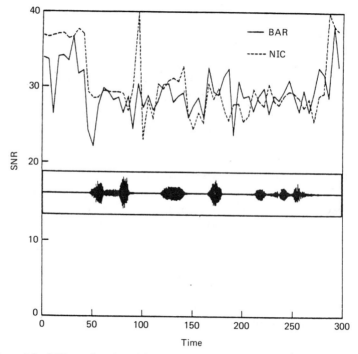

**Figure 3.8** SNR as a function of time for a forward (NIC) and a backward (BAR). bit rate reduction schemes for μ-law PCM companded signals when these schemes are operating at 48 Kb/s (Used by permission from the IEEL [Adoul, 3.16].)

PCM (ADPCM) example with (continuously) adaptive prediction of order $p$. The problem is to select the set of $p$-prediction coefficients that minimizes the expected prediction error:

$$E[(e(n))^2] \stackrel{\Delta}{=} E[(x(n) - \Sigma a_i \hat{x}(n - i))^2]$$

Hats have been used for $x(n - 1)$, indicating that a backward adaptation is implied. Differentiating $e^2$ with respect to the $p$ prediction coefficients yields a set of $p$ equations, which equated to zero give the well-known result [Makhoul, 3.17]

$$
\begin{bmatrix}
R(0) & R(1) & R(2) & \cdots & R(p-1) \\
R(1) & R(0) & R(1) & \cdots & R(p-2) \\
R(2) & R(1) & R(0) & \cdots & R(p-3) \\
\vdots & \vdots & \vdots & & \vdots \\
R(p-1) & R(p-2) & R(p-3) & \cdots & R(0)
\end{bmatrix}
\begin{bmatrix}
a_1 \\
a_2 \\
a_3 \\
\vdots \\
a_p
\end{bmatrix}
= -a_0
\begin{bmatrix}
R(1) \\
R(2) \\
R(3) \\
\vdots \\
R(p-1)
\end{bmatrix}
\tag{3.18}
$$

where $R(k) = E[x(n), x(n - k)]$ is the autocorrelation function. The $a_i$ values can be obtained by inverting the (Toeplitz) matrix using the efficient Levinson-Durbin algorithm [Rabiner, 3.5] or the Le Roux-Gueguen [Le Roux, 3.18] fixed-point algorithm. Hence, in block processing, the autocorrelation functions $R(0)$ through $R(p)$ are estimated from a block of $N$ samples (up to a scaling factor) as

$$R(k) \approx \sum_{i=0}^{N-k} x(n)x(n + k) \tag{3.19}$$

and a new set of prediction coefficients is calculated by matrix inversion. This is the **autocorrelation method**.

There is a *sequential counterpart* to this processing technique, which slightly updates the $a_i$ coefficients at each new sample. This type of processing has been widely experimented on backward adaptive ADPCM schemes [Gibson, 3.19].

The most-popular sequential scheme uses a gradient approach, which can be expressed by the following set of recursive equations:

$$a_i(n) = a_i(n - 1) + \alpha e(n)\hat{x}(n - i) \qquad \text{for } i = 1, 2, \ldots, p \tag{3.20}$$

where $e(n)$ is the (quantized) difference signal (see Fig. 3.4), $\hat{x}(n - i)$ is the received sample at time $n - i$, and $\alpha$ is a positive gain factor which can be constant or, preferably, vary as the inverse of the signal energy. Define

$$\alpha = \frac{B}{C + \sum\limits_{i=1}^{p} [\hat{x}(n - i)]^2} \tag{3.21}$$

where $B$ and $C$ are real constants and the summation reflects the signal energy measured over $p$ samples. The rationale behind the gradient approach is that the expected energy of the difference signal, equation (3.14), is described by a quadratic equation (a paraboloid) with respect to the variables $a_i$ when considering an actual set of sample values (see Fig. 3.9).

$$J \overset{\Delta}{=} E[(e(n))^2] \cong \frac{1}{N} \sum_{n=1}^{N} [x(n) - \sum_{i=1}^{p} a_i \hat{x}(n - i)]^2 = f(a_1, a_1^2, a_2, a_2^2, \ldots) \tag{3.22}$$

It follows that the minimum of $J$ will be obtained at the bottom of the paraboloid. The gradient technique is a procedure for converging progressively toward the optimum set of variables by successive moves in the direction opposite to that of the (projected) gradient. This sequential scheme has been successful in many applications [Widrow, 3.20]. In this technique, the minimization of $E[e^2]$ is replaced by the minimization of $e^2$ itself at each $n$. The component of the (projected) gradient along the $a_i$ dimension becomes:

$$\frac{\partial e^2(n)}{\partial a_i} = \frac{\partial e^2(n)}{\partial e(n)} \cdot \frac{\partial e(n)}{\partial_i} = 2e(n)\frac{\partial e(n)}{\partial a_i} \tag{3.23}$$

$$= -2e(n)\hat{x}(n - i) \tag{3.24}$$

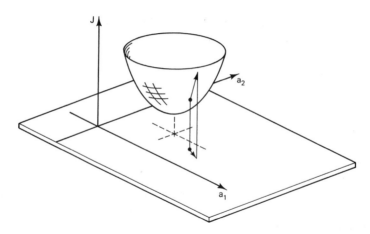

**Figure 3.9** Geometric interpretation of the gradient descent algorithm. The paraboloid gives the energy $J$ of the prediction error (for a particular block of input samples) as a function of the prediction coefficients $a_1, a_2, \ldots, a_p$. The procedure converges to the optimum set of coefficients by successive moves in the direction opposite to that of the (projected) gradient.

### 3.2.4 Readiness and Foresight; Living with the Unexpected

Adaptation helps correct a basic inefficiency of fixed coding schemes, which fail to account for the range of amplitude and spectra that the speech signal can have at any one time. However, due to adaptation requirements, it is more difficult to cope with rapid speech statistics changes. This is particularly true at the onset of voiced spurts. Many adaptation schemes, in particular the backward type, poorly encode the signal for a few milliseconds following an unvoiced-to-voiced transition while estimates are being corrected. Failure to encode voicing attacks properly results in a loss of quality and sometimes a loss of intelligibility.

Adaptive coders benefit from some provision of foresight and/or preparedness. The problem might be illustrated by the intuitive graph of Fig. 3.10, which displays the amount of instantaneous information per sample to be transmitted in the case of adaptive schemes for encoding a particular signal.

It is somewhat easier to circumvent the problem in a forward system by looking to the signal ahead of time or, more precisely stated, by delaying the encoding process in order to accurately estimate the current statistics and foresee future events. In addition, forward-side information can be used to initiate a major change in the encoding scheme employed, such as a switch from one coding scheme to another (e.g., voiced or unvoiced).

Several solutions have been proposed for backward schemes. Consider for instance the pitch compensated quantizer of Cohen and Melsa [Cohn, 3.21] and related techniques [Adoul, 3.22; Bertocci, 3.23]. Whenever either of the outermost quantizer levels occur, the gain is significantly increased. If no further outer levels occur, the gain decays back to a normal value proportional to the previously occupied quantizer level. To attain higher efficiency, these approaches may be associated with variable length encoding (entropy coding) of the quantization levels

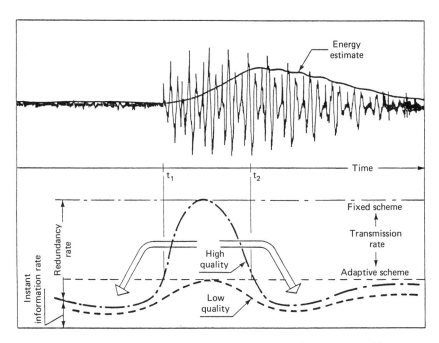

**Figure 3.10** Living through transitions. The upper graph shows a transition at time $t_1$ from an unvoiced to a voiced segment. On the same graph a (backward) energy estimate is also shown. It is only after time $t_2$ that this energy estimate is able to reflect the signal accurately. The lower graph plots various rates. The light horizontal lines exemplify the saving in transmission rate from a fixed to an adaptive scheme. The catch, however, is that with (forward as well as backward) adaptive schemes, there are situations such as during the $t_1$, $t_2$ interval where either a greater rate is required to maintain the nominal quality or else one must momentarily settle for a lower-quality transmission. The dark dotted and mixed lines show an intuitive instantaneous information rate for, respectively, a low- and a high-quality transmission. The solution suggested by the arrows is to spread out this is concentration of instantaneous information over a longer interval.

[Cohn, 3.24]. Other schemes use variable sampling rates [Dubnowski, 3.25] to cope with sudden changes in amplitude. In amplitude-time encoding [Soumagne, 3.26], the quantizer index corresponds to a specific amplitude and sampling time pair (see Fig. 3.11). This technique also has application in companding speech for FM transmission.

### Tree Coding Principle

The most systematic approach to the problem of foresight is addressed by tree coding schemes [Anderson, 3.27]. The principle of tree coding is to resort to delayed coding in an intelligent way. It can be applied usefully to any backward scheme. We briefly illustrate the operative principle of Anderson's *M-L* algorithm [Anderson, 3.27]. Let us suppose, to begin with, a 1-bit-per-sample coding rate. The coding of sample $x(n)$ can take place when future samples up to $x(n + L)$ are available. Let us focus on the decoder. From the state it is in, a decoder would

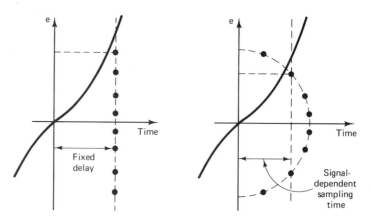

**Figure 3.11** Comparison of DPCM and amplitude-time encoding. In DPCM the (analog) difference signal $e(t)$ is sampled after a fixed delay and quantized by selecting the most appropriate out of the set of rounded values symbolized by black dots. In amplitude-time encoding the index outputed by the quantizer points simultaneously to a rounded amplitude $\hat{e}(t)$ and to a particular sampling time, which are the black-dot coordinates. In both schemes the difference signal can be scaled with respect to some energy estimate prior to quantization.

produce $2^{L+1}$ different output waveforms of the type $y(n), y(n + 1), \ldots, y(n + L)$ in response to the $2^{L+1}$ distinct $(L + 1)$-long binary streams that could be forced as possible inputs.

In fact, we could look for the binary stream (out of the $2^{L+1}$) that yields the closest output sequence, in total **mean-squared error** (MSE) to the sequence $x(n)$, $x(n + 1), \ldots, x(n + L)$ to be coded. This is precisely what the tree coding algorithms attempt to achieve while keeping the calculations to a minimum. Basically, the *M-L* algorithm keeps track of only the $M$ most promising binary sequences, also called *paths*. Traditionally, tree coding is described by discussion involving branches and nodes of a tree. We offer an alternate approach with the hope that it can help understanding and shed light on some of the sequential features of this algorithm.

Figure 3.12 illustrates the operations of the *M-L* algorithm for $M = 4$ and $L = 3$. Prior to the coding of sample $x(n)$, the path memory contains 4 rows, which are the most promising 3-bit paths. The **squared-error memory** (SEM) contains 4 rows, which give the squared errors for each of the paths. For instance, if the first path is forced as input into the decoder, the output waveform is such that $[x(n) - y(n)]^2 = 3$, $[x(n + 1) - y(n + 1)] = 2$ and so on. As soon as sample $x(n + 3)$ becomes available we have all the information necessary to proceed with the computations. We extend each of the 4 rows with both a 0 branch and a 1 branch. We obtain 8 corresponding squared errors (i.e., forced with input path 0100, the decoder will generate $y(n)$ to $y(n + 3)$ such that $[x(n + 3) - y(n + 3)]^2 = 6$; and so on). We then compute the 8 cumulative squared errors and order them. The path with minimum cumulative squared error (i.e., 6) is the extended path 0010. The first bit (i.e., 0) of this extended path is transmitted (see the dotted box in Fig. 3.12). This is the code corresponding to sample $x(n)$. Now that bit

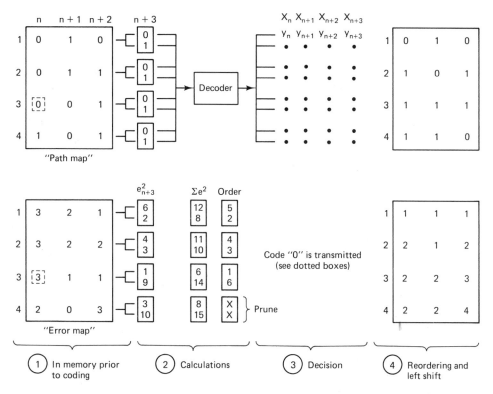

**Figure 3.12** A step-by-step illustration of the (*M-L*) tree encoding procedure (*M* = 4), *L* = 3 in the example). We want to encode sample $x_{(n)}$ at the rate of one bit per sample. Before deciding to send a 0 or a 1, we want to check the future impact of this decision by observing the output waveforms (*L* + 1 = 4 sample long) of the decoder when subjected to the 2*M* = 4 most-promising binary sequences called (extended) paths. The coding algorithm selects the generated waveform closest in total **mean-squared error** MSE to the target input waveform $x_{(n)}$ through $x_{(n+3)}$. The leading bit of the path which generated the winning waveform is the bit transmitted. The M = 4 most promising extended paths (compatible with the decision) are memorized for future use. The lower half of the figure illustrates, from left to right, the four steps involved.

0 has been selected, extended paths beginning with a 1 are no longer of interest and must be discarded. We shift the path and error maps by 1 bit to the left, retaining and reordering the *M* = 4 paths having the minimum cumulative squared error.

### 3.2.5 Interpitch Processing: Tapping the Redundancy of Speech Pseudoperiodicity

One of the most-challenging forms of redundancy to be removed is the pseudo-periodicity of voiced speech. Dependence between successive pitch periods is exhibited in Fig. 3.13, where the successive pitch periods of the French word *oui* have been aligned. This is a stereoscopic picture. The depth can be perceived

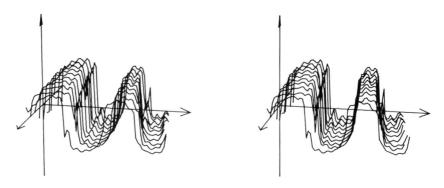

**Figure 3.13** Redundancy in the speech pseudoperiodicity. This is a stereoscopic picture which can be perceived by crossing the eyes (focusing first on a pencil placed midway between the eyes and the page should help the appearance of a third central image. The figure shows ten successive pitch periods of the waveform for the french word *oui* superposed on the same amplitude time referential.

by looking at the figure while crossing the eyes. Several techniques have been developed to exploit pseudoperiodicity for data-compression purposes. All these techniques rely on accurate pitch period extraction. Many pitch-extraction methods have been suggested. Performance comparisons indicate nearly equivalent results, with slightly better performances with increased complexity [Rabiner, 3.28]. One of the simpler methods, which is appropriate to speech coding, is the **amplitude mean difference function** (AMDF) method [Ross, 3.29]. To implement this method, select the integer interval $k$ that minimizes

$$\min_{k} \sum_{n=1}^{N} |x(n) - x(n + k)| \tag{3.25}$$

The speech fundamental frequency, which is the inverse of the pitch period, ranges from 80 to 200 Hz for males, 120 to 300 Hz for females, and up to 350 Hz for children.

### Pitch Prediction

One approach to removing the redundancy inherent in the pseudoperiodicity of speech signals is to use linear prediction across a pitch period in such a way that the difference $e(n) = x(n) - \beta x(n - k)$ is encoded. Here, $k$ is the pitch period in terms of sampling intervals that must be extracted at the coder and then transmitted on a regular basis in a forward manner (e.g., every 16 ms). The real coefficient $\beta$ is calculated in the standard way which leads to the ratio of two autocorrelation coefficients:

$$\beta = \frac{R(k)}{R(0)} \tag{3.26}$$

This time-varying coefficient must also be transmitted periodically. A predictor of order 1 is generally employed when this technique is associated with a standard ADPCM coder. In this case the global scheme is called **adaptive predictive coder**

(APC) [Atal, 3.30]. The total prediction gain is not, however, the sum (in decibels) of the gain achievable by any one prediction scheme alone. The first predictor achieves the bulk of the gain (typically 13 to 14 dB), whereas the second predictor, which operates on a significantly less redundant signal, achieves the balance of roughly 3 dB.

### Time Domain Harmonic Scaling

Another approach to removing the redundant pseudoperiodicity is to transmit one pitch period instead of two (or three). Malah [Malah, 3.31] has invented such technique for providing a smooth half-rate waveform. This technique called **time domain harmonic scaling** (TDHS) has been experimentally used [McClennon, 3.32] with a good measure of success. In fact, TDHS is not in itself a digitization procedure and can consequently be associated with any coding scheme. The procedure achieves a smooth merging of waveforms through the use of complementary windows. The processed half-rate samples are given by the following formula:

$$y\left(\frac{lo}{2} + i\right) = x(lo + i)\, h(i : k) + x(lo + k + i)[1 - h(i : k)] \quad (3.27)$$

For $i = 1, 2, 3, \ldots, k$. The constant $k$ is the pitch period and $lo$ is the time offset. This time offset is incremented by $2k$ prior to processing the next pair of pitch periods. The window $h(i : k)$ can be any function of $i$ with the constraints $h(1 : k) = 1$ and $h(k : k) = 0$ to ensure proper *smoothing*. The most popular window, which performs essentially as well as any other, [Melsa, 3.33] is given by $h(i : k) = (k - i)/(k - 1)$. The value of the pitch period being transmitted once every 10 to 20 ms and the retrieval of a close-to-the-original waveform can be done using the somewhat more involved smoothing procedure:

$$x(lo + i) = y\left(\frac{lo}{2} + i\right)h(i : 2k) + y\left(\frac{lo}{2} + k + i\right)[1 - h(i : 2k)] \quad (3.28)$$

for $i = 1, 2, 3, \ldots, 2k$. The unvoiced sounds, which do not have periodicity, are nevertheless fairly well processed through the TDHS technique. However, transitions between unvoiced and voiced phonema are often poorly rendered (see Fig. 3.14).

A related technique [Pépin, 3.34] consists of sending the actual pitch event instead of the pitch value. In this case the decoder is "synchronized" to the coder and the input signal can be decomposed in two half-rate signals as follows:

$$y(n) = \frac{x(n) + x(n + k)}{2} \quad (3.29)$$

$$z(n) = \frac{x(n) - x(n + k)}{2} \quad (3.30)$$

The signal $y(n)$ is akin to the results of TDHS, whereas the signal $z(n)$ acts as a corrective to $y(n)$. In fact, $z(n)$ has low energy most of the time (typically 13 dB less than $y(n)$), which gives a justification to the TDHS procedure (see Fig. 3.15).

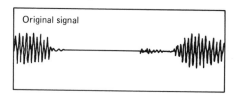

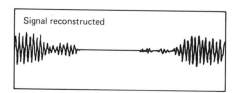

**Figure 3.14** Some defects of the time domain harmonic scaling (TDHS) technique. This scheme basically transmits one pitch period every two intervals. Spurious bursts can appear at transitions when the signal is reconstructed following TDHS processing.

However, $z(n)$ can become comparable to $y(n)$ at transitions, precisely where the TDHS fails. Note that the original signal can be restored entirely from the $y(n)$ and $z(n)$ pair. Vector quantization of $(y(n), z(n))$ provides an encoding procedure that can best take into account the relative energy of these two components and thus best do justice to the original signal.

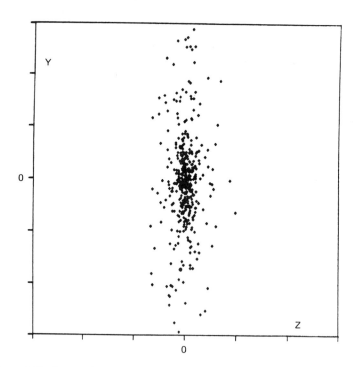

**Figure 3.15** Scatter diagram for the $Y(n)$ and $Z(n)$. These are, respectively, the mean sum and mean difference between pairs of speech samples that are one pitch period apart. The standard deviation of $Y(n)$ is four to five times that of $Z(n)$ most of the time.

### 3.2.6 Noise Control: Paying Attention to Speech Perception

A better understanding of how speech is perceived can help design a better-sounding codec at a given rate; equivalently, a good perception model points to redundancy in the speech signal, which can be removed by an appropriate data-compression strategy.

The theory of auditory masking [Atal, 3.43] indicates that quantization noise is less annoying in frequency regions where the signal energy is itself concentrated, namely, the formants. Hence, several codec design algorithms attempt to distribute noise in the frequency domain in accordance with the signal-power distribution.

#### Subband Coding and Adaptive Transform Coding

The technique of subband coding, introduced by Crochiere [Crochiere, 3.35], consists of splitting the spectrum into nonoverlapping bands, which are then encoded independently by time domain techniques. Galand [Esteban, 3.37] quadrature mirror filters and Rothweiler [Rothweiler, 3.38] polyphase filters are powerful techniques to reduce the computational requirements of nonoverlapping filters. These are digital **passband filter** (BPF) design techniques, which assure that the aliasing error on each band edge is canceled by the aliasing error of the adjacent band.

A more systematic but more complex technique for controlling noise in the frequency domain is the **adaptive transform coding** (ATC) technique introduced by Zellinsky and Noll [Zellinsky, 3.39]. A block of $N$ windowed speech samples is transformed into a corresponding set of coefficients with a frequency-domain interpretation. One efficient transformation is obtained by the so-called cosine transform, which is a real-to-real transform pair (i.e., reversible) defined by

$$y(k) = \sum_{n=0}^{N-1} x(n)g(k)\cos\left[\frac{(2n+1)k\pi}{2N}\right]; \qquad k = 0, 1, \ldots, N-1 \quad (3.31)$$

and its reverse

$$x(n) = \frac{1}{N}\sum_{k=0}^{N-1} y(k)g(k)\cos\left[\frac{(2n+1)k\pi}{2N}\right]; \qquad n = 0, 1, \ldots, N-1 \quad (3.32)$$

where $g(0) = 1$

and
$$g(k \neq 0) = \sqrt{2}$$

Efficient implementations of this transform have been receiving much attention [Bertocci, 3.23]. Several bit-allocation techniques for ATC (as well as for subband coding) have been proposed to assign efficiently the bits available for transmission to the various frequency domain coefficients (or subbands) in order to achieve the goal of controlling the signal to noise ratio in specified frequency bands [Heron, 3.70].

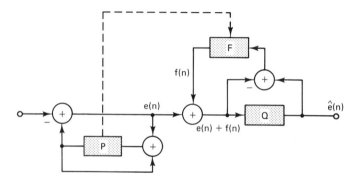

**Figure 3.16** Block diagram of a predictive coding scheme with noise-shaping filter *F*. On the right-hand side is the DPCM structure of Fig. 3.4 except that the quantizer has been taken outside the loop. The difference, or noise, between the input to and the (rounded) output of the quantizer is fed back after being filtered by *F*. This technique is sometimes referred to as *noise feedback coding*.

**Noise Shaping**

A third approach due to Atal and Schroeder [Schroeder, 3.40] controls the quantization noise spectrum of an ADPCM-like coder, forcing it to reflect the signal spectrum. This technique is appropriately called **noise shaping** (NS). The basic diagram is shown in Fig. 3.16.

The quantizer is placed outside of the ADPCM prediction loop:

$$e(n) = x(n) - \hat{x}(n|n-1, \dots, n-p) \qquad (3.33)$$

A filter *F* redistributes the quantization noise power from one frequency band to another, according to the signal spectrum as estimated by the adaptive predictor *P*.

Another characteristic of the perception process, which has not received sufficient attention yet in our judgment, is the **critical band phenomenon**. Basically, the ear does not require the same frequency resolution throughout the spectrum. In fact, it requires a so-called MEL scale, which is a linear frequency scale below 1000 Hz and a logarithmic one (i.e., less resolution) for higher frequencies. Use of the scale has proven to be quite relevant in the speech-recognition area [Davis, 3.41]. Some attempts to use a MEL scale in speech transmission [Imai, 3.42] have been made. The main obstacle is that the simplicity of linear prediction techniques seems difficult to carry into approaches using nonlinear frequency scales.

### 3.2.7 Speech-Production Modeling: When Redundancy Removal Is Not Enough

As the nominal transmission rate decreases, it becomes more difficult to encode the speech waveform itself. The classical approach is to model in part the speech production apparatus. The basic model consists of:

1. An excitation signal typical of the air pressure modulated by the vocal chords;
2. A filter characterizing the vocal track (mouth and nose).

### Parametric Representation of the Time-Varying Filter

To produce speech, the filter modeling the vocal track must be updated at a comparatively low rate (typically 50 times a second) in order to simulate the speed of motion of the mouth and tongue.

The **channel vocoders** model this "vocal-track" filter using a bank of non-overlapping, adjacent BPFs, typically 12–32 filters. Each filter has a separate adjustable gain.

**Linear predictive coding** (LPC) **vocoders** model this vocal-track filter using a single linear all-pole filter. All-pole filters with rather small order $p$, between 6 and 12, are (almost) ideally suited to model a vocal-track transfer function. In effect, they enable the modeling of the 3 to 6 ($=p/2$) resonnant frequencies (i.e., formants), which are characteristic of human speech in the bandwidth of interest (0 to 5 kHz). In addition, the $p$ real parameters that define the vocal track at a particular instant can be extracted efficiently by the linear prediction method from the first $p + 1$ autocorrelations, $R(k)$, of the speech signal to be modeled using fast algorithms or, as will be detailed later, using VQ. When using VQ, the periodic description of the time-varying filter requires merely from 400–500 bits.

### Representation of the Excitation Signal:

Once a particular vocal-track filter has been retained for transmission of a particular speech segment, the ideal excitation can be obtained by inverse filtering of the speech (see Fig. 3.17). By definition, cascading a filter to its inverse filter amounts to a pure delay (i.e., the total impulse response is a delayed unit impulse). It is particularly easy to obtain the inverse filter of an all-pole filter by taking the all-zero filter, which has zeros in place of the poles. The signal that results from inverse filtering of the speech (i.e., the ideal excitation) is called the **residual**. The residual contains a significant redundancy, which led to the development of several encoding schemes. A typical residual signal is shown in Fig. 3.18.

Residuals for voiced sounds are characterized by periodic pulses at the pitch rate, while unvoiced-sound residuals are characterized by white noise. Note that both types have a flat spectrum, albeit with a periodic fine structure in the case of voiced sounds caused by the harmonics of the fundamental frequency. In **pulse-excited linear predictive** (PELP) **vocoders** such as the standard 2400-b/s LPC-10 codec [Atal, 3.43; Tremain, 3.44], the residual is not actually transmitted but is synthetised using an artificial dual source, which is regulated by three parameters (see Fig. 3.19):

**1.** The voiced-unvoiced switch, which determines whether the periodic pulses or the white noise source is activated.

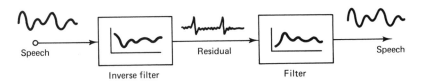

**Figure 3.17** Excitation signal obtained by inverse filtering.

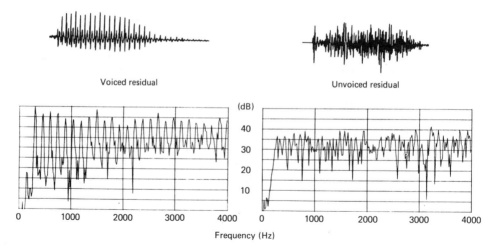

Voiced residual

Unvoiced residual

**Figure 3.18** Typical voiced and unvoiced residual signals with their respective flat spectrum. Note that in the voiced case the spectrum retain its periodic fine structure.

2. In case of the periodic pulse source, a pitch-period parameter defines the spacing between pulses in terms of the number of samples.

3. The overall excitation power is controlled by a gain factor $G$.

In **residual excited** (RELP) **vocoder** [Un, 3.45], the extracted residual is economically encoded and transmitted. It is interesting to note that in RELP vocoders the vocal-track filter acts as a noise-shaping mechanism for any white quantization noise generated in encoding the residual. Several experiments [Wong, 3.46] have shown that transmission of a portion of the residual spectrum (0 to 1000 Hz) is sufficient for obtaining a good communication quality. The procedure is described in Fig. 3.20. In high-frequency regeneration we make use of the fact that the spectrum of the residual is periodic, and hence the missing portion can be best obtained by duplicating the known portion. This duplication can be done either by spectrum folding or spectrum translation [Viswanathan, 3.47].

An application of the perturbation technique notably improves the subjective results by removing some of the unwanted regularity introduced by spectrum duplication.

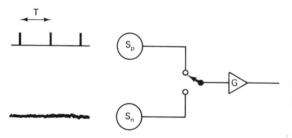

**Figure 3.19** Excitation mechanism in PELP vocoders. It has a voiced-unvoiced switch which selects either a periodic-pulse source or a noise-like source.

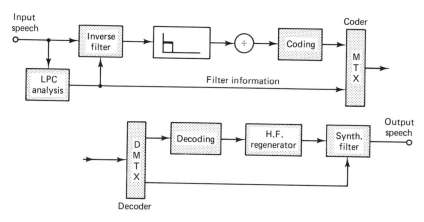

**Figure 3.20** Block diagram of a RELP vocoder. An LPC analysis determines the spectrum of the input speech and, thereby, the inverse filter to be used. The residual is then low-pass filtered, decimated, and time domain encoded. Both residual and filter information are multiplexed.

**Baseband residual** (BBRELP) **vocoders**, operated at a bit rate of 2400 bits per second, have a quality comparable to the currently standard LPC-10 and have a greatly improved background-noise immunity [Adoul, 3.48]. This is made possible by using very efficient VQ schemes applied to the time-domain coding of the baseband residual.

## 3.3 THE QUANTIZATION OF SEVERAL VARIABLES

In this section we shall discuss the principle and design of **vectors quantizers**. Basically, we will consider the simultaneous quantization of several variables. This approach, also called *block quantization*, has recently received renewed interest with the use of a practical design procedure [Linde, 3.50; Adoul, 3.51] called K-means, which we shall describe. VQ has brought spectacular savings in the coding of the vocal-track filter, as it enables the description of the interrelation between formants.

When dealing with a set of random variables $x = \{x_1, x_2, \ldots, x_M\}$, there are a number of more efficient ways to quantize them than using $M$ distinct scalar quantizers. One well-known method uses a linear or orthonormal transformation (the Karhunen–Loève transformation) in order to first obtain $M$ statistically independent variables $y_1, y_2, \ldots, y_M$. Then each variable $y_i$ is scalar quantized individually [Max, 3.49] with its optimum quantizer. The number of rounded values assigned to the quantizer of a particular $y$-component is dependent upon the relative standard deviation of the corresponding random variable. Basically this method involves rotating the coordinate axes to make them coincide with the principal components of the data. This method is well suited to data whose joint distribution $P(x_1, x_2, \ldots, x_M)$ is close to an $M$-dimensional Gaussian distribution.

### 3.3.1 Conditional Quantization

Another, less well known method, which also retains the simplicity of scalar quantization (i.e., threshold logic), is what could appropriately be called **conditional quantization**. The quantization is performed sequentially (i.e., beginning with $x_1$), but the particular quantizer used to encode $x_i$ (i.e., the set of possible rounded values) is dependent upon the result of the previously quantized variables $x_1, x_2, \ldots, x_{i-1}$.

To illustrate this principle we first describe the conditional quantization of two variables (see Fig. 3.21). In the $x_1$, $x_2$ coordinate system we use a cross to represent a particular realization of the two random variables.

The variable $x_1$ is processed first using a scalar quantizer with $N$ ($N = 3$ in the example) rounded values. Variable $x_2$ will be encoded using one of $N$ scalar quantizers. The procedure is described in the caption of Fig. 3.21. The generalization of this scheme is straightforward: The third variable would be conditioned by $i$ only (i.e., $Q_3|i$), and so on.

### 3.3.2 Principle of Vector Quantization

The optimum way of quantizing several variables is the VQ approach. Basically, a vector quantizer operates a (many-to-one) mapping of the input vector $\mathbf{x} = x_1, x_2, \ldots, x_M$ (we will assume $\mathbf{x} \in R^M$) to a finite set of indexed $M$-dimensional rounded vectors.

$$S = \{\mathbf{y}_1, \mathbf{y}_2, \ldots, \mathbf{y}_i, \ldots, \mathbf{y}_k\}; \qquad \mathbf{y}_i \in R^M \tag{3.34}$$

We call these the $K$ prototype vectors of the quantizer, borrowing the terminology from the pattern-recognition literature. The mapping is done according to a minimum distance rule.

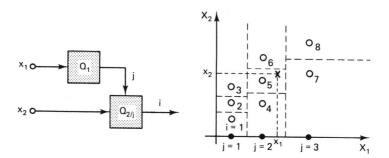

**Figure 3.21** A two-dimensional conditional quantizer. Variable $x_1$ is first quantized with scalar quantizer $Q_1$, which outputs an intermediary index $j$ ($j = 2$ in this particular case). Variable $x_2$ is then processed by a scalar quantizer tailored to the conditional distribution of $x_2$ given $j$. In this example the conditional quantizer $Q_2|j$ has three possible outputs indexed by $i = 4, 5, 6$. Index $i = 5$ is selected in this case. The coordinates of the fifth prototype vector are the rounded values $\hat{x}_1$ and $\hat{x}_2$.

The distance measure to be used depends on the particular problem at hand. The distance measure will be denoted by

$$d(\mathbf{x}, \mathbf{y}_i)$$

It is a real-valued function which reflects the distortion resulting from approximating the (input) vector $\mathbf{x}$ with the (prototype) vector $\mathbf{y}_i$. Note that in using VQ as well as in designing a vector quantizer (i.e., the selection of the set $S$), we use the distance measure only for the purpose of comparing the same input $\mathbf{x}$ with various candidate prototypes $\mathbf{y}_i$. It follows that the distance measure can be fairly general: It does not have to be symmetrical if the first entry is always used for the input $\mathbf{x}$, nor does it have to verify the triangle inequality of a metric. However, a quantized quantity cannot be less distorted than the quantity itself, and hence we must require that

$$d(\mathbf{x}, \mathbf{y}) - d(\mathbf{x}, \mathbf{x}) \le 0 \tag{3.35}$$

and

$$d(\mathbf{x}, \mathbf{y}) - d(\mathbf{y}, \mathbf{y}) \le 0 \tag{3.36}$$

We shall illustrate this principle with a two-dimensional example: consider again an input vector $\mathbf{x} = [x_1, x_2]^T$ in its $x_1, x_2$ coordinates and let us represent with the same coordinates the $K$ prototype vectors, labeling them by their index $i$. The distance is the Euclidean distance defined as follows (in our case $M = 2$):

$$d^2(\mathbf{x}, \mathbf{y}_i) = \sum_{j=1}^{M} (x_j - y_{i,j})^2 \tag{3.37}$$

where $y_{i,j}$ is the $j$th component of the prototype vector $\mathbf{y}_i$ (see Fig. 3.22). From a close inspection of this figure, we conclude that the input vector $\mathbf{x}$ is closer to the fourth prototype vector than to any other. This is indeed the quantized vector, and 4 is the selected index.

The prototypes define a partition of the input space into $K$ cells illustrated by dotted frontiers on the figure. Cell $C_i$ is the subset of the input space that is mapped into the $i$th prototype by the quantizer. It is defined by

$$C_i = \{\mathbf{x} | d(\mathbf{x}, \mathbf{y}_i) < d(\mathbf{x}, \mathbf{y}_k); k < i \quad \text{and} \quad d(\mathbf{x}, \mathbf{y}_i) < d(\mathbf{x}, \mathbf{y}_j); j > i\} \tag{3.38}$$

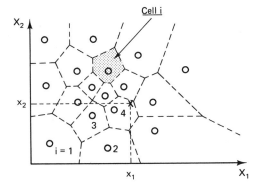

**Figure 3.22** A two-dimensional vector quantizer. The $K$ prototype vectors are symbolized by little circles, which have indices from 1 to $K$. The input vector to be quantized is represented by a cross. If Euclidean distance is assumed, prototype 4 is selected and its coordinates provide the rounded quantities $\hat{x}_1$ and $\hat{x}_2$.

These cells are also called **Voronoi regions**. They are necessarily convex regions and have very complicated frontiers, and hence the implementation of VQ can hardly be brought to threshold decision as in the case of conditional quantization. Actual distances have to be calculated for comparisons.

A fact that is often overlooked is that a VQ offers better (or at least equivalent) performance for scalar quantization even if the components of the vector **x** are statistically independent. For instance, consider the case of two Gaussian variables having the same variance and correlation coefficient ρ. Figure 3.23 illustrates the percentage of error power (MSE) generated by an optimal VQ with $K = N^2 = 2^m$ prototypes as compared to the error power generated by two optimum scalar [Max, 3.49] quantizers with $N$ rounded values. The two approaches are compared in the $x_1, x_2$ coordinates for the case of $N^2 = 16$ or of a 2-bit/dimension quantizer. In this case the error power of the VQ quantizer is 82.7% of that generated by the scalar quantizer (see Fig. 3.24).

If the input vector **x** is a random vector with a known joint (cumulative) distribution $P(\mathbf{x})$, it is possible to introduce two concepts: (1) the mean distance performance, and (2) the notion of centroid of a Voronoi region. Let us begin with the definition of the mean distance performance, as the following expectation:

$$D = E(d) = \int_{\mathbf{x}} \min_{i} d(\mathbf{x}, \mathbf{y}_i) \, dP(\mathbf{x}) \tag{3.39}$$

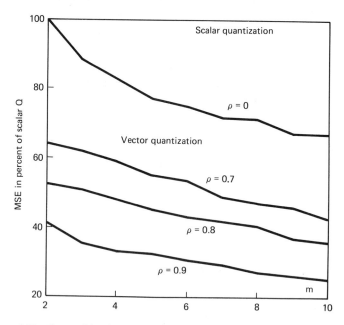

**Figure 3.23** Comparision between scalar and vector quantization for two jointly normal variable with correlation ρ. The plot provides the expected MSE performance obtained by a vector quantizer with $K = 2^m$ prototypes in percent of the MSE for the optimal scalar-quantizer case.

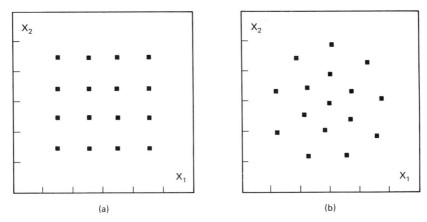

**Figure 3.24** Two-bit-per-dimension quantizers for two independent normal variables. (a) Optimum (max) quantizer. (b) Vector quantizer from $K$-mean design.

If the distance does not depend on $\mathbf{x}$, we can rewrite the expectation as

$$D = \sum_i P(i) \int_{\mathbf{x} \in C_i} d(\mathbf{x}, \mathbf{y}_i) \, d\mathbf{x} \qquad (3.40)$$

where $C_i$ is the Voronoi region for prototype $i$ and $P(i)$ is the probability that $\mathbf{x} \in C_i$. Let us now turn to the centroid of a Voronoi region $C_i$. Basically, it is the vector $\mathbf{z}_i$ which minimizes the distance from the region $C_i$, namely, the quantity:

$$\int_{\mathbf{x} \in C_i} d(\mathbf{x}, \mathbf{z}_i) \, dP(\mathbf{x}) \qquad (3.41)$$

Clearly, if $\mathbf{y}_i \neq \mathbf{z}_i$, the mean distance performance of the quantizer would be improved by taking $\mathbf{z}_i$ as the $i$th prototype (i.e., even if we do not change the Voronoi cell). Of course, replacing $\mathbf{y}_i$ by $\mathbf{z}_i$ will cause some vectors $\mathbf{x}$ to change allegiance from one prototype to another, thus modifying the original partition. However, this moving around can only improve the quantizer performance (i.e., further decrease the quantizer mean distance). As a matter of fact, a vector $\mathbf{x}$ changes prototype only if it can find a closer one.

### 3.3.3 Vector Quantization Design: the K-means Algorithm

The previous discussion suggests a quantizer design algorithm which is an $M$-dimensional version of the Lloyd-Max algorithm [Lloyd, 3.52; Max, 3.49] for minimizing the mean distance. Let us denote $S(m)$ as the set of $K$ prototypes at iteration $m$

$$S(m) = \{y_i(m) | i = 1, 2, \ldots, K\} \qquad (3.42)$$

### Generalized Lloyd-Max Algorithm (P(x) Known)

*Step 1.* Set $m$, the iteration number, to zero ($m = 0$). Select an arbitrary set of $K$ prototype vectors. Call the set $S(0)$.

*Step 2.* Determine the $K$ Voronoi regions for the set $S(m)$.

*Step 3.* Compute the set $S(m + 1)$ as the respective centroid of the Voronoi regions determined in Step 2.

*Step 4.* Calculate $D(m + 1)$, the mean distance for quantizer $S(m + 1)$. If $m \neq 0$ and if $(D(m + 1) - D(m) \leq \varepsilon$ (some preset quantity), increment $m$ by 1 and go back to Step 2. Otherwise terminate.

The Lloyd-Max algorithm, generalized to more than one dimension is very difficult to implement in practice, even for joint distributions that can be expressed analytically. Moreover, most applications deal with distributions that cannot be expressed analytically. Fortunately, good vector quantizers can be designed from a training sequence of $N$ typical *input* vectors, where $N >> K$, using an algorithm analogous to the Lloyd-Max algorithm, which is known in the clustering and pattern-recognition literature as the $K$-means algorithm [Anderberg, 3.53; Diday, 3.54]. Let us define $T = \{\mathbf{x}_j : j = 1, 2, \ldots , N\}$ as the training set, where $S(m)$ is again the set of $K$ prototypes at iteration $m$. Following these definitions we present the following algorithm.

### K-Means Algorithm (Training Set T)

*Step 1.* Set the iteration number, $m$, to zero ($m = 0$). Select an arbitrary set of $K$ prototype vectors. Call the set $S(0)$. A good method consists of taking at random $K$ vectors from the training set $T$.

*Step 2.* Classification: Classify the vectors of $T$ into $K$ classes, $B_i$, according to the minimum distance criterion. Let $N_i$ be the cardinality of class $B_i$.

*Step 3.* Compute the set $S(m + 1)$ as the respective centroid of each class. The centroid for class $i$ is the vector $\mathbf{y}_i(m + 1)$ which minimizes the quantity

$$D_i(m + 1) = \frac{1}{N_i} \sum_{x \in B_i} d(\mathbf{x}, \mathbf{y}_i(m + 1)) \qquad (3.43)$$

*Step 4.* Calculate $D(m + 1)$, the mean distance prior to iteration $m + 1$, by

$$D(m + 1) = \frac{1}{N} \sum_{i=1}^{K} N_i D_i(m + 1)$$

If $m \geq 0$ and if $D(m) - D(m + 1)$ is greater than some preset small quantity, increment $m$ by 1 and go back to Step 2, otherwise, the design is terminated.

There are two domains of application of VQ to speech coding that we consider in this section. The first is the efficient encoding of LPC coefficients and the second is the time domain encoding of waveforms: speech, residuals, and baseband residuals.

### 3.3.4 Application of VQ to the Efficient Encoding of LPC Coefficient

The application of VQ to the encoding of LPC coefficients was pioneered by [Chaffee, 3.55] and in particular by Linde, Buzo, and Gray [Buzo, 3.56; Linde, 3.57]. Buzo and others demonstrated that VQ could reduce by a factor between 2 and 3 the number of bits required to describe the vocal-track filter. Their findings were also confirmed in references [Roucos, 3.58; Mabilleau, 3.59]. This new approach (and a better encoding scheme) has allowed Wong [Wong, 3.60] to demonstrate the standard LPC-10 quality at a third of the rate. VQ of the LPC filter and time domain VQ of the base band residual makes it possible to use nonsynthetic excitation (BBRELP) as low as 2400 b/s.

**Parametric Representations of LPC Spectra and Appropriate Distances**

We briefly review the various parametric representations of the LPC filter (i.e., vocal-track model). We also expand on two interpretations given in [Adoul, 3.14] regarding (1) the $K$-means algorithm, as applied to the so-called likelihood ratio or Itakura distance and (2) the interpretation of the binary (tree search) VQ as a linear pattern classification of the input in the autocorrelation space.

In the sequel we assume that the reader is familiar with the $z$-transform notation [Oppenheim, 3.62]. Let us briefly highlight the salient facts. The $z$-transform of a sequence of real samples $\{x(n)\}$ is a counterpart to the Laplace transform and is defined by:

$$\mathcal{L}\{x(n)\} \overset{\Delta}{=} X(z) = \sum_{n=-\infty}^{+\infty} x(n)^{z-n} \tag{3.44}$$

with the fundamental shift property

$$\mathcal{L}\{x(n)\} = X(z) \leftrightarrow \mathcal{L}\{x(n + n_o)\} = z^{n_o}X(z) \tag{3.45}$$

Let us now turn our attention to a recursive (i.e., all-pole) filter with a time domain difference equation given by

$$y(n) = -\sum_{i-1}^{p} a_i y(n - i) + Gx(n) \tag{3.46}$$

Taking the $z$-transform and using the shift property, we obtain the filter-transfer function $H(z)$.

$$H(z) \overset{\Delta}{=} \frac{Y(z)}{X(z)} = \frac{G}{\displaystyle\sum_{i=0}^{p} a_i z^{-i}} \tag{3.47}$$

where $a_o$ is set equal to 1 and where $X(z)$, $Y(z)$, and $H(z)$ are, respectively, the $z$-transforms of the input sequence $\{x(n)\}$, the output $\{y(n)\}$, and the (causal) filter impulse response $\{h(n)\}$. $H(z)$ is a complex function of the complex variable $z$.

One such typical transfer function is illustrated in Fig. 3.25. Note that we have represented the normalized transfer function with the inherent filter delay removed. The result is an all-pole rational fraction in $z$.

$$\frac{z^{-p}}{G} H(z) = \frac{1}{\sum_{i=0}^{p} a_i z^{p-i}} = \frac{1}{(z - z_1)(z - z_2) \cdots (z - z_p)} \tag{3.48}$$

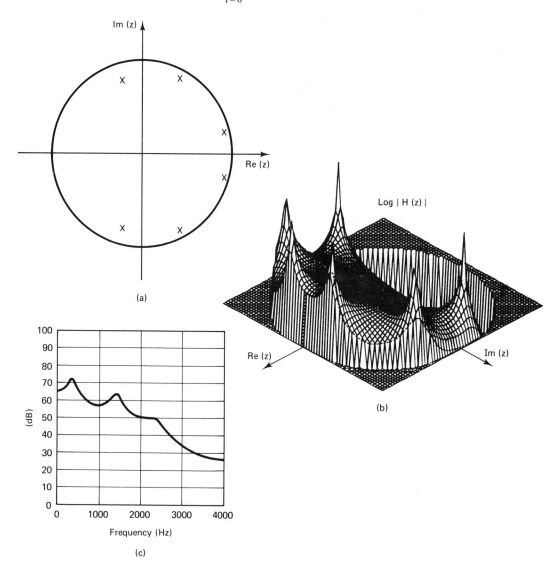

(a)

(b)

(c)

**Figure 3.25** Representations for an all-pole linear filter: Considering an all-pole filter $H(z)$, (a) the location of the poles in conjugate pairs and (b) a three-dimensional representation of the log magnitude of $|H(z)|$ as a function of $z$. The plot is limited to cases of $z$ inside the unit circle. Note that the three-dimensional plot has been rotated to exhibit the first quadrant. The vertical slice along the unit circle indeed gives the log spectrum, also shown on (c).

There are three important facts: (1) The $a_i$'s are real coefficients and the poles are either real or come in complex conjugate pairs. (2) The filter is stable if and only if all poles are inside the unit circle. (3) When evaluating on the unit circle (i.e., $z = e^{i\Theta}$, $\Theta \in [-\pi/2, +\pi/2]$) the z-transform is equal to the discrete Fourier transform and hence the log spectrum of Fig. 3.25 is simply a vertical slice which follows the unit circle of Fig. 3.25(b).

$A(z)$ is the (normalized) inverse filter of $H(z)$:

$$A(z) = \frac{G}{H(z)} = \sum_{i=0}^{p} a_i z^{-i} \tag{3.49}$$

The impulse response of $A(z)$ is simply $a_i$ for $i\epsilon[0, p]$ and zero otherwise. The inverse filter, also called **analysis filter**, is used in LPC to extract the residual from a block of $N$ samples $\{x(n)\}$, which are the windowed and preemphasized speech samples.

Since only finite-energy sequences will be considered in the sequel, we will use lowercase $r$ for autocorrelation to emphasize the context shifting from statistical averages to single finite sequences.

$$r_x(k) = \sum_{n} x(n)\, x(n + k); \qquad -\infty \le n \le +\infty \tag{3.50}$$

From LPC theory, the inverse filter $A(z)$ of order $p$ that minimizes the residual energy, $r_y(0)$, is the filter whose impulse response is the solution to the following set of equations.

$$\sum_{n=1}^{p} a_n r_x(|k - n|) = -a_o r_x(k); \qquad k = 1, 2, \ldots, p \tag{3.51}$$

which can be solved, as already noted, using the **Levinson-Durbin** (L-D) algorithm [Makhoul, 3.17]. As a byproduct to solving the system of equations, the L-D algorithm generates a set of $p$ real parameters, $k_i$, called the *partial correlations*. These coefficients nicely range between $\pm 1$ and are equivalent to the set of $a_i$, which defines the inverse filter $A(z)$.

There are other $p$-parameter sets that uniquely define $A(z)$. The most common are summarized in Fig. 3.26. Along with the parameters already introduced, we find the cepstral and Log-area-ratio coefficients.

All parameter sets have interesting features and interpretations, which are well documented elsewhere [Markel, 3.63; Makhoul, 3.17]. For the purpose of efficiently transmitting the filter characteristics, VQ can be used on any of these equivalent $p$-parameter sets. The problem, however, is one of deciding on the subjectively appropriate and—it is to be hoped—tractable distance to be used. Euclidian distance is a reasonable distance for the sets $k_i$, $L_i$, and $c_i$. Excellent phoneme classification results consistently obtained in the cepstrum space [Miclet, 3.64; Paliwal, 3.65] suggest that this space is well behaved and subjectively meaningful under the Euclidian distance. Figure 3.27 illustrates typical VQ quantizer speech spectra in different parametric representations for $r(1)$, $r(2)$, $k_1 k_2$, $c_1 c_2$ and $L_1 L_2$, respectively. The location of the particular phonemes triangle [Fant, 3.7] is indicated.

**TABLE 3.4** Transformations among LPC parameter sets (the formula's numbers refer to Fig. 3.26)

| # | Input | Calculations | Output |
|---|-------|--------------|--------|
| | | [Levinson-Durbin recursion] | |
| 1 | $\{r(1), r(2), \ldots, r(p)\}$ <br> $r(0) = \alpha$ | (a) *Initialization* <br> $\alpha_0 = r(0)$, $k_1 = \dfrac{r(1)}{r(0)}$ <br> $a_{10} = 1$, $a_{11} = k_1$ <br> $\alpha_1 = \alpha_0(1 - k_1^2)$ <br><br> (b) *Iteration* $m = 2, 3, \ldots, p$ <br> $k_m = -\dfrac{\sum\limits_{j=0}^{m-1} r(m-j)a_{m-1,j}}{a_{m-1}}$ <br> $a_{m0} = 1$ <br> $a_{mj} = a_{m-1,j} + k_m a_{m-1,m-j}$; for $j = 1, 2, \ldots, m-1$ <br> $a_{mm} = k_m$, <br> $\alpha_m = \alpha_{m-1}(1 - k_m^2)$ | $\{k_1, k_2, \ldots, k_p\}$ <br> $r(0) = \alpha$ |
| 2 | $\{r(1), r(2), \ldots, r(p)\}$ <br> $r(0) = \alpha$ | | $\{a_1, a_2, \ldots, a_p\}$ <br> $r(0) = \alpha$ <br> $a_0 = 1$ |
| 3 | $\{k_1, k_2 \ldots, k_p\}$ <br> $r(0) = \alpha$ | (a) *Initialization* <br> $\alpha = r(0)$, $\quad r(1) = -k_1 r(0)$ <br> (b) *Iterations:* $m = 2, 3, \ldots, p$ <br> $a_{m0} = 1$ <br> $a_{mj} = a_{m-1,j} + k_m a_{m-1,m-m}$; for $j = 1, 2, \ldots, m-1$ <br> $a_{mm} = k_m$ <br> $r(m) = -\sum\limits_{j=1}^{m} r(m-j)a_{mj}$ | $\{r(1), r(2), \ldots, r(p)\}$ <br> $r(0) = \alpha$ |

4 $\{a_1, a_2, \ldots, a_p\}$
$r(0) = \alpha$
$a_0 = 1$

5 $\{z_1, z_2, \ldots, z_p\}$
$r(0) = \alpha$

6 $\{c_1, c_2, \ldots, c_p\}$
$c_0$

7 $\{a_1, a_2, \ldots, a_p\}$
$r(0) = \alpha$

8 $\{L_1, L_2, \ldots, L_p\}$
$r(0) = \alpha$

9 $\{k_1, k_2, \ldots, k_p\}$
$r(0) = \alpha$

10 $\{a_1, a_2, \ldots, a_p\}$
$r(0) = \alpha$
$a_0 = 1$

## Root-solving algorithm for polynomial

$\{z_1, z_2, \ldots, z_p\}$

$$(z - a_1)(z - z_2), \ldots, (z - z_p) = \sum_{i=0}^{p} a_i z^{p-i}$$

$\{a_1, a_2, \ldots, a_p\}$
$r(0) = \alpha$
$a_0 = 1$

Identify:

$$\sum_{i=0}^{p} a_i z^{p-i} = (z^2 - 2z \cos \Theta_j + r_j^2)(\ldots)(\ldots)$$

where $z_j = \bar{z}_k = r_j e^{i\theta}_j$ are conjugate pairs

$\{a_1, a_2, \ldots, a_p\}$
$r(0) = \alpha$

$c(0) = \ln[\sqrt{\alpha}]$
$c(1) + a_1 = 0$

$\{c_1, c_2, \ldots, c_p\};$   $c(0)$
for $n \geq p$
Obtain $c_n$ by setting $a_n = 0$

$$n c(n) + n a_n = \sum_{k=1}^{n-1} (n - k) c(n-k) a_k, \qquad \text{for } n = 2, \ldots, p$$

(Explicit $a_n$ or $c_n$.)

$\{k_1, k_2, \ldots, k_p\}$
$r(0) = \alpha$

$$k_j = \frac{(1 - e^{L_j})}{(1 + e^{L_j})}$$

$\{L_1, L_2, \ldots, L_p\}$
$r(0) = \alpha$

$$L_j = \mathrm{Ln}\left[\frac{(1 - k_j)}{(1 + k_j)}\right]$$

$\{k_1, k_2, \ldots, k_p\}$
$r(0) = \alpha$

a) *Initialization*

$a_k = a_{p,k};$   for $k = 1, 2, \ldots, p$

b) *Iteration* (step down) $m = p, p - 1, p - 2, \ldots, 1$

$k_m = a_{m,m}$

$a_{m-1,0} = 1$

$$a_{m-1,j} = \frac{a_{m,j} - k_m a_{m,m-j}}{1 - k^2 m}$$

for $j = 1, 2, \ldots, m - 1$

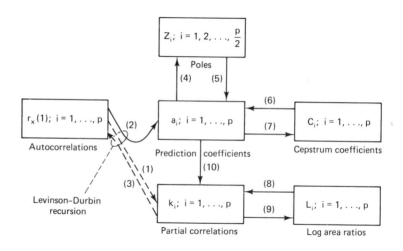

**Figure 3.26** Parametric representations of an all-pole model. The relation between the various equivalent parameter sets is shown. The numbers in parenthesis refer to the transformations compiled in Table 3.4.

### Interpretation of VQ in the Autocorrelation Space

The Buzo [Buzo, 3.56] VQ scheme can be interpreted as a VQ in the autocorrelation space $R \in R^n$. We define the vectors $\mathbf{x}$ and $\mathbf{y}_i$'s as column vectors whose components are the first $p$ autocorrelations $r_x(n)$ or $r_{y_i}(n)$; $n = 1, 2, \ldots, p$) of an energy-normalized sequence, i.e., $r_x(0) = 1$ or $r_{y_i}(0) = 1$.

$$\mathbf{x} = [r_x(1), r_x(2), \ldots, r_x(p)]^T \in R \tag{3.52}$$

$$\mathbf{y}_i = [r_{y_i}(1), r_{y_i}(2), \ldots, r_{y_i}(p)]^T \in R \tag{3.53}$$

In fact, we need not know the actual sequence $\mathbf{x}_i(n)$ in this problem but solely the inverse (analysis) filter, $A_i(z)$, which matches the input sequence—that is, which is a solution to the set of equations (3.51). Figure 3.28 illustrates the $A_i$ filter under two conditions. In the first case the (energy normalized) input $x(n)$ is not matched to the filter. The energy of the residual $e(n)$ is called $\delta_i(x)$ and is expressed as

$$\delta_i(x) = r_e(0) = \sum_n r_{a_i}(n)r_x(n) \tag{3.54}$$

where the sum is over all relevant $n$.

In the second case, the input $y_i(n)$ is matched to the filter; the residual output energy is called $\alpha_i$ and is expressed as

$$\alpha_i = r_e(0) = \sum_n r_{a_i}(n)r_{y_i}(n) \tag{3.55}$$

We know from the derivation of equations (3.52) that the summation in this case is (by construction) minimum and that its value collapses to:

$$\alpha_i = \sum_{n=0}^{p} a_n r_{y_i}(n) \tag{3.56}$$

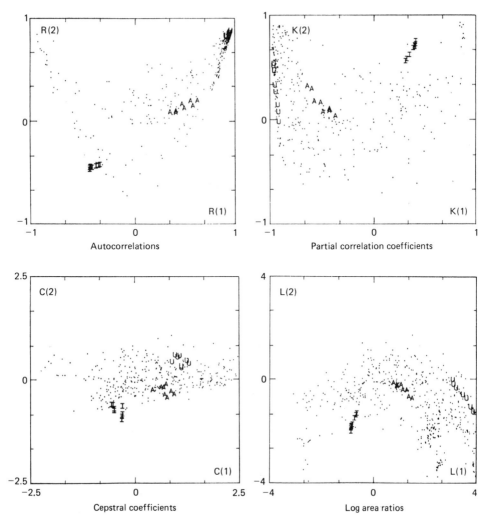

**Figure 3.27** Scatter diagrams of typical speech spectra in different parametric representations. These are: The autocorrelation ($r(1)$, $r(2)$); the partial correlation coefficients ($k_1$, $k_2$); the cepstral coefficients ($c_1$, $c_2$), and the log area ratios ($L_1$, $L_2$). The vowel sounds /a/, /i/ and /u/ are marked in each diagram.

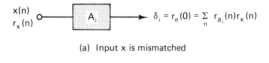

(a) Input x is mismatched

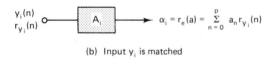

(b) Input $y_i$ is matched

**Figure 3.28** Residual energy. When the input-signal spectrum matches the filter the residual energy, $\alpha$, is minimal. In the general (mismatched) case, the energy is $\delta$.

We now have the necessary parameters to obtain the distance measure for our VQ in the autocorrelation space $R$. The distance is given by

$$d(\mathbf{x}, \mathbf{y}_i) = \delta_i(x) \tag{3.57}$$

This is indeed a legitimate distance measure in the sense that

$$d(\mathbf{x}, \mathbf{y}_i) - d(\mathbf{y}_i, \mathbf{y}_i) \geq 0 \tag{3.58}$$

Recall that the $K$-means algorithm operates on a training set of $N$ ordered vectors, $\mathbf{x}_j$, which in this case are autocorrelation vectors computed from a sequence of speech blocks. The $K$-means algorithm alternates between two basic steps: the classification step and the centroid calculation. In our situation the first step consists of classifying each $\mathbf{x}$ into a class $B_i$ for which $\delta_i$ is less than any other $\delta_{j \neq i}$. In this second step the centroid for class $i$ for iteration $m + 1$ is the $\mathbf{y}_i(m + 1)$ that minimizes

$$D_i(m + 1) = \frac{1}{N_i} \sum_{x \in B_i} d(\mathbf{x}, \mathbf{y}_i(m + 1)) \tag{3.59}$$

$$= \frac{1}{N_i} \sum_{x \in B_i} \delta_i(x) \tag{3.60}$$

But $\sum_x \delta_i(x)$ is the cumulative energy, $r_e(0)$, at the output of a single filter excited by a signal $S_n$. The signal $S_n$ can be thought of as the concatenation of the $N_i$ input signals $x(n)$ that have fallen in class $B_i$ interspaced with $p$ "zero" samples between them (see Fig. 3.29). Based on this interpretation, we can conclude that since $S_n$ is a "real-life" signal, the centroid will yield a stable filter $A_i(m + 1)$.

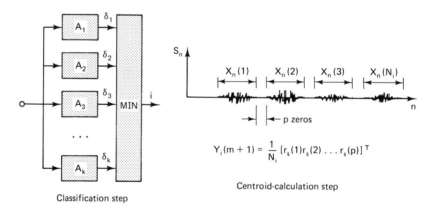

Classification step                Centroid-calculation step

**Figure 3.29** Interpretation of the $K$-Means basic two steps. The VO design algorithm iterates between a classification step in which each vector of the training set is assigned to the class $i$ yielding minimum residual $\delta_i$. In the second step (barycenter calculation), a new improved filter is calculated for each class. It can be interpreted as the filter that matches the signal $S_n$, which is the concatenation of all training signals assigned to that class.

**Illustration with the Case for $p = 2$**

Let us consider a concrete example of the VQ of the LPC vocal-track filter in the case of $p = 2$ and with 8 prototypes. The input signal is defined by its first two normalized autocorrelations. Figure 3.30(a) shows the signal autocorrelation and the corresponding vector $\mathbf{x}$ in the $R$-space of stable filters (Fig. 3.30(b)). Voiced sounds have positively correlated adjacent samples, and therefore their corresponding $\mathbf{x}$ vectors cluster in the upper right corner of the $R$-space. Vectors corresponding to invoiced sounds pretty much cover the rest of the $R$-space. In particular, the input vector $\mathbf{x} = [0, 0]^T$ corresponds to a signal with uncorrelated adjacent samples, an input for which there is no prediction gain possible.

To be able to watch more closely the voiced sound zone, Fig. 3.30(c), 3.30(d), and 3.30(e) have been rotated. Figure 3.30(d) gives an indication of the joint probability distribution of vector $\mathbf{x}$ in speech and emphasizes the great concentration and frequency of occurrence of voiced sounds. Figure 3.30(c) illustrates the minimum residual $\alpha = d(\mathbf{x}, \mathbf{x})$ obtained when the analysis filter is always matched to the input autocorrelation (i.e., no quantization). Note that the maximum $\alpha$ is obtained for the unpredictable input vector $\mathbf{x} = [0, 0]^T$ just discussed, where $\alpha = 1$. Figure 3.30(e) shows for every input vector $\mathbf{x}$ the distance to the closest prototype out of the $K(= 8)$ prototypes selected using the $K$-means algorithm. The VQ mean distance is the expectation of $\min_i \delta_i(x)$ weighed by the joint distribution of $\mathbf{x}$ (i.e., Fig. 3.30(d)). In Fig. 3.30(e) we note that the $\min_i d(\mathbf{x}, \mathbf{y}_i)$ function is made out of eight planes (piecewise linear). Each plane is tangent to the surface of Fig. 3.30(c) at the point $\mathbf{x} = \mathbf{y}_i$. The Voronoi regions have straight-line boundaries in the $R$-space (see dotted line in Fig. 3.30(b)).

**Linear Discriminant Property of the $R$-space under the $\delta_i(\mathbf{x})$ Distance**

When we have to decide which is the closest of two candidate prototypes $\mathbf{y}_i$ and $\mathbf{y}_j$ to an input vector $\mathbf{x}$, we have in effect a linear classification problem in $R$-space [Adoul, 3.14]. We have to determine the sign of the quantity

$$H_{ij}(\mathbf{x}) = \delta_i(\mathbf{x}) - \delta_j(\mathbf{x}) \gtreqless 0 \tag{3.61}$$

where

$$\delta_i(x) = r_e(0) = \sum_k r_x(k)r_a(-k) \tag{3.62}$$

Summing over all relevant $k$ yields

$$\delta_i(x) = r_a(0) + 2 \sum_{k=1}^{p} r_x(k)r_a(k) \tag{3.63}$$

It follows that we can rewrite $H_{ij}(\mathbf{X})$ as a linear equation in terms of the input-signal autocorrelation:

$$H_{ij}(x) = \sum_{k=0}^{p} w_k r_x(k) \tag{3.64}$$

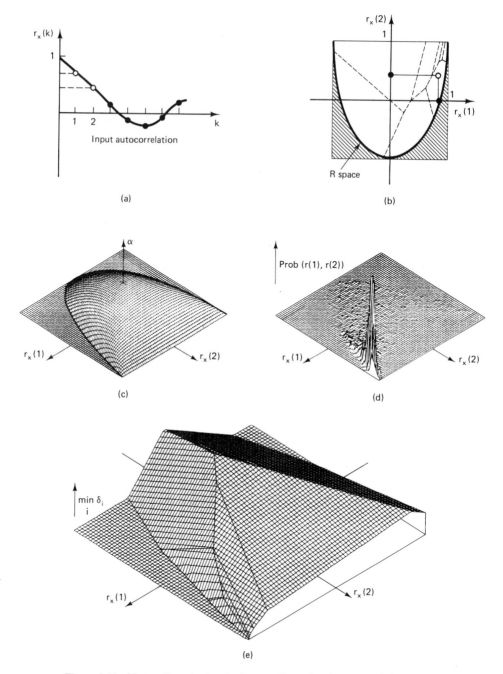

**Figure 3.30** Vector Quantization in the two-dimensional autocorrelation space. The problem is to find the best-matching filter $A_i$ among eight candidates for the input signal $x$ defined through its first two autocorrelation coefficients (a). (d) shows a three-dimensional histogram of the occurrence of $r_x(1), r_x(2)$ pairs in speech (compare with (a)). (c) gives the (minimal) residual energy $\alpha$ and (e) shows the residual energy of the best-matching filter.

where the weight coefficients $w_k$ take on the values

$$w_0 = r_a(0); \quad w_k = 2r_a(k) \quad \text{for} \quad k = 1, 2, \ldots, p \quad (3.65)$$

### Hierarchical VQ

When the calculation of $K$ distances is an issue, VQ can be accomplished more efficiently using a hierarchical approach. Buzo and others [Buzo, 3.56] describe such an approach. The VQ is performed in $l$ successive approximations, which find a path through a tree from depth 1 to depth $l$, where $m_i$ choices (or branches) are available at depth $i$. The binary tree search (i.e., $m_i = 2$ for all $i$) provides the fastest computation, namely,

$$N_c = 2 \log_2 K \quad \text{distance computations}$$

$$N_s = 2(K - 1) \quad \text{vector storage locations}$$

For the binary tree search case, Adoul [Adoul, 3.14] describe a hierarchical technique that further reduces both $N_c$ and $N_s$ by a factor of 2. The two techniques are functionally the same in the sense that the same quantized value is selected. The speed and storage economy of the second technique comes from the fact that it takes advantage of the linear discriminant property described earlier. Distance calculations are replaced by hyperplane calculations and we note that computation and storage of a hyperplane is strictly equivalent to the requirements of a distance computation. Figure 3.31 illustrates the saving using a simple example. To reach the same quantized index, $c$ in this example, the distance approach requires 14 storage locations and computations of distances 1, 2, 3, 4, 9, 10. The hyperplane

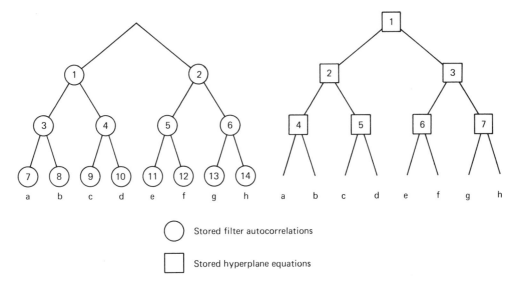

**Figure 3.31** Comparison of computation and storage requirements between distance and hyperplane approaches in binary hierarchical VQ. The second approach requires half the storage and half the computation.

approach requires 7 storage locations and the computation of hyperplanes 1, 2, and 5.

The design of a linear classifier (hyperplane approach) is simple. The design of the hyperplane at a given node is obtained by the $K$-means algorithm for $K = 2$ applied to those subsets of the training set that have "found their ways" to that node.

***Interpretation of VQ in the Prediction-Coefficient Space.*** The VQ technique just described can be interpreted as a quantization of a prediction coefficient vector in the prediction coefficient space $P \in R^{p+1}$ but with a distance measure under a new form. Now we present an additional interpretation of the vectors $\mathbf{x}$ and $\mathbf{y}_i$.

Let the vector $\mathbf{x}$ associated with the input sequence $\{x(n)\}$ be composed of the (prediction) coefficients of the matched analysis filter:

$$\mathbf{x} = [1, a_1, a_2, \dots, a_p]^T \in P \qquad (3.66)$$

Also let

$$\mathbf{y}_i = [1, a_1(i), a_2(i), \dots, a_p(i)]^T \in P \qquad (3.67)$$

be the prototype vectors that are prediction coefficients of some analysis filters.

Finally, we define $R_x$ as the $(p + 1) \times (p + 1)$ Toeplitz matrix formed with the normalized autocorrelations, $r_x(k)$, of the input signal, the $ij$ entry being $r_x(|i - j|)$. Quantization uses the following distance in matrix form:

$$d_1(\mathbf{x}, \mathbf{y}_i) = \mathbf{y}_i^T R_x \mathbf{y}_i = \delta_i(\mathbf{x}) \qquad (3.68)$$

A closely related distance measure is given by

$$d_2(\mathbf{x}, \mathbf{y}_i) = (\mathbf{x} - \mathbf{y}_i)^T R_X(\mathbf{x} - \mathbf{y}_i) \qquad (3.69)$$

It can be shown (see [Tribolet, 3.61]) that this form is equivalent to the quantity $\delta_i(\mathbf{x}) - \alpha(\mathbf{x})$, where $\alpha(\mathbf{x})$ is the minimum residual for input $x$ using its matched analysis filter. Note that the quantizer that minimizes $d_2$ for all sequences of the training set, $T$, is equivalent to the one that minimizes $d_1$, namely:

$$\min_Q \sum_T (\delta_i(\mathbf{x}) - \alpha(\mathbf{x})) = \min_Q \left[ \sum_T \delta_i(\mathbf{x}) \right] - \sum_T \alpha(\mathbf{x}) \qquad (3.70)$$

Applying the $K$-means algorithm with either distance yields the same quantizer (i.e., same classification steps and same steps for calculating the centroid). Distance measure $d_2$ is very attractive in its quadratic form, which makes explicit the difference between the two vectors $(\mathbf{x} - \mathbf{y}_i)$ and is akin to Mahalanobis distance, which is widely used in the context of Gaussian distribution [Anderberg, 3.53]. Note that the set of points $\mathbf{y}_i$ which are at a given $d_2$ distance from an input vector $\mathbf{x}$ form an ellipsoid in $P =$ space (not all $\mathbf{y}_i$ necessarily correspond to stable filters).

Another possible distance is

$$d_3(\mathbf{x}, \mathbf{y}_i) = \frac{(\mathbf{x} - \mathbf{y}_i)^T R_x (\mathbf{x} - \mathbf{y}_i)}{(\mathbf{x}^T R_x \mathbf{x})}$$

$$= \frac{\delta_i(x) - \alpha(x)}{\alpha(x)} \tag{3.71}$$

$$d_3(\mathbf{x}, \mathbf{y}_i) = \frac{\delta_i(x)}{\alpha(x)} - 1 \tag{3.72}$$

The virtue of this third distance is that for small values, $d_3(\mathbf{x}, \mathbf{y}_i)$ becomes equivalent to the **log-likelihood ratio**

$$\log_e\left(\frac{\delta_i(x)}{\alpha(x)}\right) = \log_e(1 + d_3) \cong d_3(\mathbf{x}, \mathbf{y}_i) \tag{3.73}$$

### 3.3.5 VQ Speech Waveforms

Several authors have applied VQ to the quantization of speech waveforms or residuals by considering blocks of $M$ successive samples as $\mathbf{x}$ vectors, which have to be quantized:

$$\mathbf{x} = [x_1, x_2, \ldots, x_M]^T$$

The only distance seriously investigated is Euclidian distance. Abut and others [Abut, 3.66] study VQ up to dimension 8 for autoregressive sources with Gaussian, Laplacian innovations. Mabilleau [Mabilleau, 3.67] studies the VQ of full-band residual signals (4 kHz).

Other interesting variant VQ structures include the following schemes: VQ by stage [Juang, 3.68] where the error vector of a first VQ stage undergoes a second VQ. Differential VQ [Cuperman, 3.69] is a vector version of the ADPCM system. VQ can be associated with Markov chains [Heron, 3.70]. Spherical VQ [Adoul, 3.48] uses properties of regular point lattices in $R^n$ and does not require storage of an actual code book.

## 3.4 CONCLUSION

The presence of intricate statistical invariants has made the design of efficient speech-coding algorithms a challenging research area for more than 20 years and will no doubt continue to do so for many years to come. In fact, this challenge has been made even more attractive by the availability of inexpensive digital processing chips. What used to be a mere exercise in data compression can now be put to work in real time from a few building blocks. Technology seems to be getting ahead of algorithm development. Chips with stored programs make switching from one algorithm to another so easy that no standard algorithm at a given (low) bit rate will be able to withstand the competition from newer and better algorithms for very long.

In the beginning of the seventies, algorithm design focused primarily on doing justice to speech production. Later, this focus widened to encompass concerns for the perception process. This trend is expected to continue. VQ is already being used to respond to both concerns. It is likely that VQ and other techniques will bring about a systematization and merging of some of the fundamental coding functions outlined in this chapter (quantization, prediction etc.). With its codebook approach, VQ paves the way for database-dependent coding algorithms.

## REFERENCES

[3.1] Crochiere, R. E., and J. L. Flanagan. "Current perspectives in digital speech," *IEEE Commun. Mag.*, January, 1983, pp. 32–40.

[3.2] Jayant, N. S. "Digital coding of speech waveforms: PCM, DPCM, and DM quantizers," *Proc. IEEE*, May, 1974, pp. 611–632.

[3.3] Flanagan, J., M. Schroeder, B. Atal, R. Crochiere, N. Jayant and J. Tribolet. "Speech coding," *IEEE Trans. Comm.* April, 1979, pp. 710–737.

[3.4] Gibson, J. D. "Adaptive prediction in speech differential encoding systems," *Proc. IEEE*, Vol. 68, No. 4, April, 1980, pp. 488–525.

[3.5] Rabiner, L. R., and R. W. Schafer. *Digital Processing of Speech Signals*, Prentice-Hall, Englewood Cliffs, N.J., 1978.

[3.6] Lim, J. S., and A. V. Oppenheim. "Reduction of quantization noise in PCM speech coding," *IEEE Trans. ASSP*, Vol. 28, February 1980, pp. 107–110.

[3.7] Fant, G. *Speech Sound and Features*, MIT Press, Cambridge, Mass., 1973.

[3.8] Owen, F. *PCM and Digital Transmission System*, McGraw-Hill, New York, 1982.

[3.9] Elias, P. "Predictive coding—Part I" and "Predictive coding—part II," *IRE Trans. Inform. Theory*, Vol. IT-1, March, 1955, pp. 16–33.

[3.10] Noll, P. "Adaptive quantizing in speech coding systems," *Proc. 1974 Zurich Seminar on Digital Communications*, Zurich, March, 1974.

[3.11] Seiler, N. C., R. Flowers, and J. Friedman. "A monolithic implementation of a CVSD algorithm," *Proc. IEEE International Communication Conf.*, 1976, pp. 31–11/16.

[3.12] Jayant, N. S. "Adaptive quantization with a one-word memory," *Bell Syst. Tech. J.*, Vol. 52, September, 1973. pp. 1119–1144.

[3.13] Dietrich, M. "Coding of speech signals using a switched quantizer," *Proc. IEEE Zurich Seminar on Digital Commun.*, March, 1974, pp. A4(1)–A4(4).

[3.14] Adoul, J.-P., J.-L. Debray, and D. Dalle. "Spectral distance measure applied to the optimum design of DPCM with L predictors," Proc. *the 1980 IEEE International Conf. ASSP*, Denver, Colo., April, 1980, pp. 512–515.

[3.15] Duttweiler, D. L., and D. G. Messerschmitt. "Nearly instantaneous companding for nonuniformly quantized PCM," *IEEE Trans. Commun.*, Vol. COM-24, August, 1976.

[3.16] Adoul, J.-P. "Backward adaptive reencoding: A technique for reducing the bit rate of µ-law PCM transmissions," *IEEE Trans. Commun.*, Vol. COM-30, No. 4, April, 1982, pp. 581–592.

[3.17] Makhoul, J. "Linear prediction: A tutorial review," *Proc. IEEE*, Vol. 63, April, 1975, pp. 561–580.

[3.18] Le Roux J., and C. Gueguen. "A fixed point computation of partial correlation coefficents," *IEEE Trans. ASSP*, Vol. 25, June, 1977, pp. 257–259.

[3.19] Gibson, J. D. "Sequentially adaptive backward prediction in ADPCM speech coders," *IEEE Trans. Commun.*, Vol. 26, -1, January, 1978.

[3.20] Widrow, B., J. M. McCool, M. G. Larimore, and C. R., Johnson, Jr. "Stationary and nonstationary learning characteristics of the LMS adaptive filter," *Proceedings of IEEE*, Vol. 64, August 1976. pp. 1151–1162.

[3.21] Cohn, D. L., and J. L. Melsa. "A pitch compensating quantizer," in Conf. Rec. 1976, *IEEE International Conf. ASSP*, 1976, pp. 258–261.

[3.22] Adoul, J.-P., and M. Boutaleb-Joutei. "Efficient coding of unvoiced sounds in digital Telephony," *Can. Elec. Eng. J.*, Vol. 2, No. 3, 1977, pp. 23–27.

[3.23] Bertocci, G., B. W. Shoenherr and D. A., Messerschmitt. "An approach to the implementation of a discrete cosine transform," *IEEE Trans. Commun.* Vol. 30, April, 1972, pp. 635–641.

[3.24] Cohn, D. L., L. Melsa and A. Arora. "Practical considerations for variable length source coding," *ICASSP-81*, pp. 816–819.

[3.25] Dubnowski, J. J., and R. E. Crochiere. "Variable rate coding of speech," *Bell Syst. Tech. J.*, March, 1979.

[3.26] Soumagne, J., J.-P., Adoul, and S. Morissette. "A new concept for encoding speech: amplitude time quantization," *IEEE, ICASSP-84*, pp. 10.12.1–10.12.4.

[3.27] Anderson, J. B., and J. B. Bodie. "Tree encoding of speech," *IEEE Trans. Inform. Theory*, Vol. IT-21, July, 1975, pp.379–387.

[3.28] Rabiner, L. R., M. J. Cheng, A. E. Rosenberg, and C. A. McGonegal. "A comparative performance study of several pitch detection algorithms," *IEEE Trans. Acoustics, Speech, and Signal Processing*, Vol. ASSP-24, October, 1976, pp. 399–418.

[3.29] Ross, M. et al. "Average magnitude difference function pitch extractor," *IEEE Trans.* Vol. ASCP-22, October, 1974, pp. 353–362.

[3.30] Atal, B. S., and M. R. Schroeder. "Adaptive predictive coding of speech signal," *BSTJ*, Vol. 49, October, 1970, pp. 1973–1986.

[3.31] Malah, D. "Time domain algorithms for harmonic bandwidth reduction and time scaling of speech signals," *IEEE Trans. ASSP*, Vol. 27, No. 2, April, 1979, pp. 21–133.

[3.32] Clennon, S., and A. Leon-Garcia. "TDHS and Subband coding of speech," EHV conf. Montreal, P. Q., Canada, 1981.

[3.33] Melsa, J., and A. Pande. "Mediumband speech encoding using time domain harmonic scaling and adaptive residual coding," *ICASSP*, 1981, pp. 603–606.

[3.34] Pépin, G. "Méthode de correction de l'algorithme TDHS," *CRCS Report*, University of Sherbrooke, Sherbrooke, Qué., Canada, August 1983.

[3.35] Crochiere, R. E. "On the design of sub-band coders for low-bit rate speech communication," *Bell Syst. Tech. J.*, Vol. 56, May–June 1977, pp. 747–770.

[3.36] Gersho, A. "Quantization," *IEEE Commun. Soc. Mag.*, September, 1977, pp. 16–29.

[3.37] Esteban, D., and C. Galand. "Application of quadrature mirror filters to split band voice coding schemes," *Proc. 1977 International Conf. ASSP*, Harford, Conn., May, 1977, pp.191–195.

[3.38] Rothweiler, J. H. "Polyphase quadrature filters—A new subband coding technique," *ICASSP*, Vol. 83, pp. 1280–1283.

[3.39] Zelinski, R. and P. Noll. "Adaptive transform coding of speech signals," *IEEE Trans.* Vol. ASSP–25, August, 1977, pp. 299–309.

[3.40] Schroeder, M., and B. S. Atal. "Predictive coding of speech signals and subjective error criteria," *IEEE-Trans.* ASSP, Vol. 27, No. 3, June, 1979, pp. 247–254.

[3.41] Davis, S. B., and P. Mermelstein. "Comparison of parametric representations for monosyllabic word recognition in continuously spoken sentences," *IEEE Trans. on ASSP*–28, No. 4, August 1980, p. 357.

[3.42] Imai, S. "Cepstral analysis synthesis on the MEL frequency scale," ICASSP-83, Vol. 1, pp. 93–96.

[3.43] Atal, B. S., and M. R. Schroeder. "Predictive coding of speech signals and subjective error criteria," *IEEE Trans. ASSP*, Vol. ASSP-27, June, 1979, pp. 247–254.

[3.44] Tremain, T. E. "The government standard linear predictive coding algorithm: LPC-10," *Speech Technology*, Vol. 1, No. 2, April, 1982, pp. 40–49.

[3.45] Un C. K., and D. T. Magill. "The residual-excited linear prediction vocoder with transmission rate below 9.6 kbits/s," *IEEE* Vol. Com-23, December 1975, pp. 1466–1474.

[3.46] Wong, D. Y. "On understanding the quality problems of LPC speech," *ICASSP*, 1980, pp. 725–728.

[3.47] Viswanathan, V. R., A. L. Higgins, and W. H. Russel. "Design of a robust baseband LPC coder for speech transmission over 9,6 kbits/noisy Channels," *IEEE*, Vol. Com-30, No. 4, April, 1980, pp. 663–673.

[3.48] Adoul, J.-P., C. Lamblin, and A. Leguyader. "Base band speech coding at 2400 bps using spherical vector quantization," *IEEE*, ICASSP-84, pp. 1.12.1–1.12.4.

[3.49] Max, J. "Quantizing for minimum distortion," *IRE Trans. Information Theory*, Vol. 6, March 1960, p. 16–21.

[3.50] Linde, Y., A. Buzo, and R. Gray. "An algorithm for vector quantizer design," *IEEE Trans. Commun.*, Vol. COM-28, January 1980, pp. 84–95.

[3.51] Adoul, J.-P., C. Collin, and D. Dalle. "Block encoding and its application to data compression of PCM speech," in *Proc. Canadian Com. and EHV Conf.* Montreal 1978, pp. 145–148.

[3.52] Lloyd, S. P. "Least squares quantization in PCM," *IEEE-IT*, Vol. IT–28, No. 2, March, 1982, pp. 127–137.

[3.53] Anderberg, *Cluster Analysis for Applications*, Academic Press, New York, 1973.

[3.54] Diday, E. and J. C. Simon. "Clustering analysis" in *Digital Pattern Recognition*, ed. K. S., Fu, Springer-Verlag, New York, 1976.

[3.55] Chaffee, D. L. "Applications of rate distortion theory to the bandwidth compression of speech signals," Ph.D. Thesis, 1975, UCLA.

[3.56] Buzo, A., A. H. Gray, Jr., R. M. Gray, and J. D. Markel. "Speech coding based on vector quantization," *IEEE ASSP*, Vol. 28, No. 5, pp. 562–574.

[3.57] Y. Linde, A. Buzo and R. M. Gray. "An Algorithm for Vector Quantizer Design," *IEEE*, vol. Com.28, No. 1, January 1980, pp. 84–95.

[3.58] Roucos, Makhoul and Schwartz. "A variable order Markov chain for coding of speech spectra." *IEEE-ICASSP*, 1982, pp. 1565–1569.

[3.59] Mabilleau, P. "La quantification vectorielle et son application au codage de la parole," Ph.D. thesis, University of Sherbrooke, 1983.

[3.60] Wong, D. Y., B.-H. Juang, and A. H. Gray Jr. "An 800 bit/s vector quantization LPC vocoder," *IEEE, Trans. ASSP*, Vol. 30, No. 5, October, 1982, pp. 770–780.

[3.61] Tribolet, J. M., L. Rabiner, and M. M. Sondhai. "Statistical properties of an LPC distance measure," *IEEE Trans. ASSP* Vol. 27, No. 5, October, pp. 550–558.

[3.62] Oppenheim, A. V., and R. W. Schaefer. *Digital Signal Processing,"* Englewood Cliffs, N.J. Prentice-Hall, 1975.

[3.63] Markel, J. D., and A. H. Gray, Jr. *Linear Prediction of Speech*, Springer-Verlag, New York, 1976.

[3.64] Miclet, L. and A. Nehme. "Expériences en reconnaissance de la parole par prédiction linéaire," *Conf. Rec. of 2e congrés AFCET-IRIA de Rec. des formes et IA.*, Toulouse France, 1979.

[3.65] Paliwal and Rao. "Evaluation of various LP parametric representation in vowel recognition," *Signal Processing*, Vol. 1, No. 4, 1982.

[3.66] Abut, H. et al. "Vector quantization of speech and speech-like waveforms," *IEEE-ASSP*, Vol. 30, No. 3, June, 1982, pp. 423–435.

[3.67] Mabilleau, P., and J.-P. Adoul. "Medium band speech coding using a dictionary of waveform," *ICASSP*, 1981, pp. 804–807.

[3.68] Juang, B.-H., and A. H. Gray. "Multiple stage vector quantization for speech coding," *ICASSP*, 1982, pp. 597–600.

[3.69] Cuperman, V., and A. Gersho. "Adaptive differential vector quantization of speech," *Conf. Rec. NTC*, 1982.

[3.70] Heron, C. D., R. E. Crochière, and R. V. Cox. "A 32-band sub-band/transform coder incorporating vector quantization for dynamic bit allocation," *IEEE-ICASSP*, 1983, pp. 1276–1279.

[3.71] Gersho, A. "Asymtotically optimal block quantization," *IEEE-IT*-25, July, 1979, pp. 373–380.

[3.72] Jayant, N. S., and P. Noll. *Digital Coding of Waveforms*, Prentice-Hall, Englewood Cliffs, N.J., 1984.

[3.73] Gersho, A., and C. Cuperman. "Vector quantization: a pattern-matching technique for speech coding," *IEEE Commun. Mag.*, December, 1983.

[3.74] Gray, R. M. "Vector quantization," *IEEE Trans. ASSP*, April, 1984.

# 4

# ECHO CANCELLATION IN SPEECH AND DATA TRANSMISSION

**DR. DAVID G. MESSERSCHMITT**

*Professor*
*Department of Electrical Engineering and Computer Sciences*
*University of California*
*Berkeley, California 94720*

*Dr. Messerschmitt is a Professor at the University of California at Berkeley, and was formerly with Bell Laboratories. He has worked and published extensively in digital voice and data transmission and digital signal processing, including both theoretical and implementation aspects.*

## 4.1 INTRODUCTION

Echo cancellation is an application of adaptive filtering technology to the control of echoes in the telephone network. It has been studied in the research laboratory for two decades, and due to the advances in microelectronics technology is now finding widespread practical application. This chapter describes the echo-control problems addressed by echo cancellation for both voice and data transmission, describes and analyzes the echo-cancellation algorithms used in these applications, and discusses the implementation of echo cancelers.

Section 4.2 describes the application of echo cancellation to speech transmission and the principles of echo suppression, and Section 4.3 discusses applications in full-duplex data transmission. Section 4.4 outlines the algorithms used for adaptation of echo cancelers in all these applications and evaluates their performance. Finally, Section 4.5 summarizes advanced techniques in echo cancellation.

## 4.2 ECHO CANCELLATION IN SPEECH TRANSMISSION

The telephone network generates echoes at points within and near the ends of a telephone connection. Echo cancellation is one means used to combat this echo for both speech and data transmission. The requirements for speech and data are quite different, so this section concentrates on speech and Section 4.3 deals with data. The sources of echo are reviewed in Section 4.2 along with the sometimes obscure terminology associated with echoes and hybrids, and the several methods used to combat echo are described in Section 4.2.2. Section 4.2.3 describes several realizations of an adaptive echo canceler for speech.

### 4.2.1 Sources of Echo in the Network

This section describes the sources of echo in the network and the sometimes obscure terminology surrounding echoes and hybrids. It also attempts to explain in tutorial fashion the origin and types of echo problems. Subsequent sections describe potential solutions to these problems. See [Sondhi, 4.1] for another tutorial reference on this topic.

A starting point on echo terminology is given in Fig. 4.1. In Fig. 4.1(a), a simplified telephone connection is shown. This connection is typical in that it contains two-wire segments on the ends (the subscriber loops and possibly some portion of the local network), in which both directions of transmission are carried on a single wire pair. The center of the connection is four-wire, in which the two directions of transmission are segregated on physically different facilities. This is necessary where it is desired to insert carrier terminals, amplifiers, or digital switches (in cases where carrier transmission is used, these are not necessarily four physical wires, but the terminology is still useful and descriptive).

There is a potential feedback loop around the four-wire portion of the connection, and without sufficient loss in this path there is degradation of the transmission or in extreme cases oscillation (called *singing*). The hybrid is a device which provides a large loss around this loop, thereby limiting this impairment. At the same time, the hybrid must not insert significant loss in the two-talker speech paths. The remainder of Fig. 4.1 illustrates more graphically the function of this hybrid. One of the two-talker speech paths is shown in Fig. 4.1(b). In order that this path not have a large attenuation, it is necessary for the hybrid not to have an appreciable attenuation between its two-wire and either four-wire port. There are two distinct echo mechanisms shown in Fig. 4.1(c) and 4.1(d). **Talker echo** results in the talker hearing a delayed version of his or her own speech, while in **listener echo** the listener hears a delayed version of the talker's speech. Both these echo mechanisms are mitigated if the echo has significant loss between its two four-wire ports.

The subjective effect of the echo depends critically on the delay around the loop and the effective transfer function around the loop (which of course incorporates the delay) [Neigh, 4.2]. For short days, the talker echo represents an insignificant impairment if the echo attenuation is reasonable (6 dB or so), since

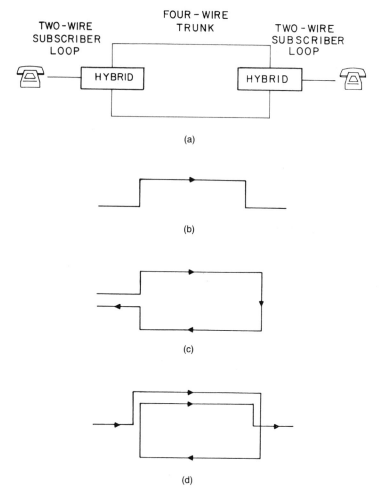

**Figure 4.1** Sources and types of echo in the telephone network. (a) Simplified telephone connection. (b) Talker speech path. (c) Talker echo path. (d) Listener echo path.

the talker echo is indistinguishable from the normal sidetone (a version of the talker speech that is deliberately reproduced in the earpiece, making the telephone sound "live"). For longer delays of 40 ms or so, talker echo represents a serious impairment unless the echo is highly attenuated. In this case the echo is disturbing to the talker, and in fact can make it very difficult to carry on a conversation.

The situation reverses in the case of listener echo. For short delays in the range of 1 ms or so, the listener echo results in an overall transfer function with peaks and valleys due to the destructive and constructive interference at different frequencies. The situation is modeled in Fig. 4.2a. In this figure the echo-transfer function is oversimplified to include just a delay $T$ and attenuation factor $\alpha$. The

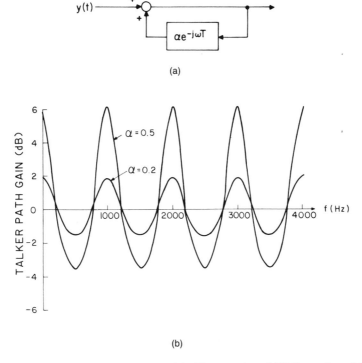

**Figure 4.2** Simplified mathematical model of listener echo. (a) Talker path model for a listener echo path consisting of a flat attenuation and fixed delay. (b) Talker path gain for the model of (a) for a 1-ms delay in the listener echo.

squared magnitude of the effective transfer function of the talker path is given by

$$|H(\omega)|^2 = \frac{1}{1 + \alpha^2 - 2\alpha \cos(\omega T)} \tag{4.1}$$

While the details of the overall effective transfer function depend on the exact attenuation and delay, in general the effect is appreciable when the delay is small, on the order of milliseconds. For example, the argument of the cosine in the denominator goes through one cycle over a frequency band of 1 kHz when the delay is 1 ms, as illustrated in Fig. 4.2(b). This denominator results in frequency-periodic peaks and valleys in the frequency response, which subjectively have an effect similar to talking into a rain barrel.

The basic principle of the hybrid is shown in Fig. 4.3, where Fig. 4.3(a) is a hybrid implemented using two transformers (an older method) and Fig. 4.3(b) is an electronic hybrid (a more modern method). Looking first at Fig. 4.3(a), assume for the moment that the **balance impedance** $Z_B$ is equal to the two-wire **load impedance** $Z_L$, and that the four-wire impedances $R_1$ and $R_2$ are also equal. In this case, the proper choice of turns ratio for the transformer will make port 1

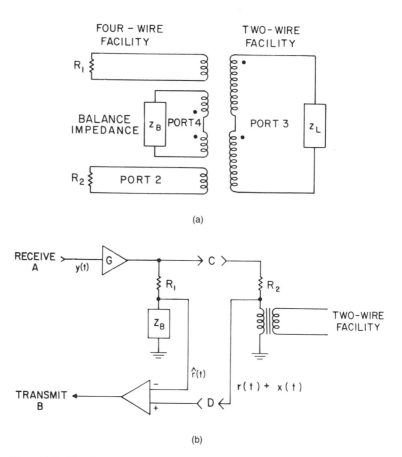

**Figure 4.3** Two implementations of the hybrid function. (a) A two-transformer hybrid. (b) An electronic hybrid.

*conjugate* to port 2 and port 3 conjugate to port 4; that is, if a source is delivering power to port 1 (port 3), a negligible part of that power will flow into port 2 (port 4). If port 1 is the receive four-wire port and port 2 is the transmit four-wire port as shown, then the hybrid provides infinite loss between these ports. The two-talker speech paths are from port 1 to port 3 and from port 3 to port 2. Any power applied to port 1 (port 3) is split equally between port 3 and port 4 (port 1 and port 2), so that ideally there is a 3-dB loss in these talker speech paths. That loss can be compensated by the addition of some amplification.

Since $Z_L$ depends on the details of the subscriber loop (such as gauge, length, and the configuration of bridged taps), it varies from one subscriber loop to another. Any fixed balancing impedance can only be chosen to be a compromise. The compromise balancing impedance is usually chosen to be either a parallel or series resistor-capacitor combination [Neigh, 4.2; Bunkee, 4.3]. The degree to which the far-end talker signal is attenuated depends on the relationship between the two-wire and a balancing impedance.

An electronic version of the hybrid is shown in Fig. 4.3(b), where the transformer shown is not an integral part of the hybrid but is merely used to isolate the electronics from foreign potentials and to serve as a balanced to unbalanced converter. The four-wire receive port includes an amplifier to make up for the loss that the hybrid and transformer would otherwise insert in the talker path (its gain $G$ is typically 7 to 8 dB). Since the near-end and far-end talkers are both superimposed on the transformer windings, the receive signal passes through a voltage divider with resistance $R_2$ in series with the impedance looking into the two-wire line through the voltage divider, which we call $Z_L$. The voltage at the center of this voltage divider is the near-end talker $x(t)$ plus the feedthrough in the voltage divider $r(t)$. To cancel this undesired feedthrough, a second voltage divider consisting of resistor $R_1$ in series with a fixed balancing impedance $Z_B$ generates a replica $\hat{r}(t)$, and a difference amplifier subtracts the replica from the signal across the transformer. The objective in choosing $Z_B$ is to match the transfer functions of the two voltage dividers, which are given by

$$\frac{R_1}{R_1 + Z_B} \approx \frac{R_2}{R_2 + Z_L}$$

When there is exact equality (for example $R_1 = R_2$ and $Z_B = Z_L$), then the replica $\hat{r}(t)$ will be identical to the actual echo signal $r(t)$, and there will be no component of the echo in the transmit signal on port B. On the other hand, the near-end talker signal appears directly at the transmit port B, since the low impedance at the receive amplifier output prevents any of this signal from appearing across the hybrid voltage divider.

In order to quantify the effect of choosing a particular compromise balancing impedance, it is common to define some useful parameters shown in Fig. 4.4. The two balancing impedances $Z_B$ are explicitly shown to emphasize the dependence of the performance of the hybrid on the match of this impedance to the line

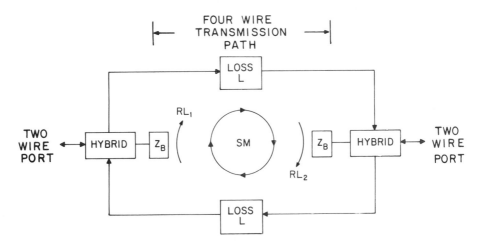

**Figure 4.4** Measures of hybrid balance—transhybrid loss, hybrid return loss, and signing margin (all these quantities can be measured at a single frequency or weighted over the voiceband).

impedance $Z_L$ (which in general will be different on the two ends of the connection). Also shown are two loss pads, one in each direction, with attenuation $L$ decibels. This loss $L$ is defined as the **insertion loss** from one two-wire port to the other, which implies that the hybrids have no internal insertion loss in the talker path (i.e., any loss has been compensated by internal gain). The losses through the hybrids from four-wire receive port to four-wire transmit port are defined as $RL_1$ and $RL_2$, where the letters stand for **return loss**. The **singing margin** $SM$ is the total loss around the four-wire loop. Of course, the return losses and the singing margin are all frequency-dependent since they are determined by frequency-dependent impedances.

The singing margin is a good overall indication of the transmission quality for connections with small delays. It is a particularly good indicator of the deleterious effect of the listener echo. The objective for the spring margin in North America is 10 dB on 95% of connections. When the predominant impairment is listener echo (short delays), there are indications that a smaller singing margin, on the order of 8 dB, is acceptable [Neigh, 4.2; Bunker, 4.3]. These values of $SM$ are defined as the smallest singing margin at any frequency.

Singing margin summarizes the combined return loss of both hybrids in the connection. For long-delay connections where talker echo is dominant, the return loss of the near-end hybrid is of little relevance. Hence, a better indicator of subjective quality on this type of connection is the **echo return loss**, which is simply the loss experienced by the talker echo signal weighted over the frequencies from 500 to 2500 Hz. As the delay of the talker echo increases, a larger echo return loss is required to maintain equivalent subjective quality. This is accomplished by the **via net loss** (VNL) plan described in the next section.

For very long delay connections, the insertion loss that would be necessary to combat talker echo becomes itself an unacceptable impairment. Therefore, on these connections it is necessary to provide other echo control mechanisms, as described in the next section.

### 4.2.2 Echo-Control Methods

This section describes the commonly utilized methods for controlling echo in the telephone network. These include loss plans, echo suppression, and echo cancellation.

#### Controlling Echo with Loss

The singing margin depends directly on the insertion loss in the connection through the equation (see Fig. 4.4).

$$SM = RL_1 + RL_2 + 2L \qquad (4.2)$$

Again we emphasize that each term in this equation is actually frequency-dependent. At any frequency, there is a 2-dB improvement in singing margin for every decibel of additional insertion loss. Both talker and listener echo are attenuated by the insertion loss, the talker echo 2 dB and the listener echo 3 dB for every decibel of insertion loss. The attenuation of the talker speech path represents an impairment, but since both of the echo paths are attenuated more than the talker path, the inclusion of a small connection insertion loss is generally beneficial to the overall subjective quality in the absence of echo suppressors or cancelers.

One approach to controlling echo in the network is thus to introduce loss in the connection in a controlled fashion. The introduction of loss represents a trade-off between its beneficial effects on the subjective effects of echo and its deleterious effects on the talker speech path. The effect of the echo increases as the connection round-trip delay increases. Thus to avoid introducing any more loss than is necessary, the loss should be increasing as the connection round-trip delay increases. This is the principle of VNL, in which a loss is added to a trunk which depends on its length. Without specifying the details of this plan, the objective of the VNL is to achieve a loss between the two local switches in any given connection which approximates

$$DB = 4.0 + 0.4N + 0.102D \tag{4.3}$$

where $DB$ is the overall loss in decibels of the talker speech path, exclusive of the loss of the two subscriber loops at the two ends of the connection. $N$ is the number of trunks in the connection, and $D$ is the echo path round-trip delay in milliseconds [Bell Labs., 4.4].

The use of loss to control echo is an inadequate measure for very long delays, since the loss itself becomes an important impairment. For this reason, the approach of using loss to control echo is used only for terrestrial trunks up to about 1900 km in North America (this number is changing with the introduction of fiber optics systems). For trunks longer than this, the insertion loss is reduced to zero, and either the echo suppressor or the echo canceler described later are used to control echo.

**The Echo Suppressor**

The echo suppressor is a device that inserts in a connection a large loss in one direction or the other. The result is that any echo signal experiences this large loss and is therefore heavily attenuated. The goal is to put the large loss in the direction opposite to the current active speech path, assuming that there is speech in only one direction at a time. This means that there is inevitable clipping during doubletalk (speech in both directions at the same time).

The principle of the echo suppressor is shown in simplified form in Fig. 4.5. A large attenuation pad is normally inserted in one direction or the other. Thus one of the two pads shown is bypassed at any given time. A speech detector is used to determine the direction of the active speech, controlling which pad is bypassed.

The greatest difficulty in the design of the algorithms is in the choice of a strategy during doubletalk. When the propagation delay is large, it becomes difficult to detect and deal with doubletalk adequately, with the result that the subjective performance of an echo suppressor on a long-delay channel is worse than for a channel with the same delay but no echo [Sondhi, 4.1]. The performance of a properly operating echo suppressor is excellent for terrestrial connections but

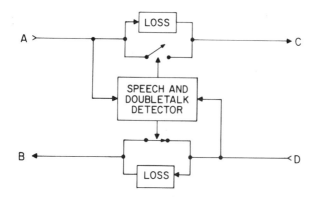

**Figure 4.5** Principle of the echo suppressor.

becomes more marginal for the long delays associated with a satellite connection [Suyderhoud, 4.5]. One approach that has been used to improve the performance of the echo suppressor is to restrict the satellite to a so-called half-hop mode, in which one direction of the four-wire connection was on the satellite and the other direction on a terrestrial circuit (thus limiting the round-trip delay). Many designers of satellite systems have chosen to use the echo canceler in place of the suppressor, as described shortly.

An additional administrative problem with echo suppressors is that two or more suppressors in tandem must be avoided. It is possible for two suppressors to lock out during doubletalk, making further conversation impossible.

### The Echo Canceler

With increasing round-trip delay, the subjective effect of echo becomes more annoying. The introduction of satellite transmission systems, which have round-trip delay of about 550 ms, has introduced a significant new source of large transmission delay. Satellite transmission systems require well-designed echo control subsystems, and these systems were the main motivation for development of the echo canceler. The echo canceler is a sophisticated form of echo control, which can effectively eliminate echo as an impairment even on very long delay channels and without regard to the doubletalk situation.

The echo canceler is based on a similar principle to the electronic hybrid of Fig. 4.3b, which can be generalized as shown in Fig. 4.6. The portion of the four-wire connection near the two-wire interface is shown in this figure, with one direction of transmission between ports A and C and the other direction between ports D and B. The far-end talker is called $y(t)$, the undesired echo is assumed to be $r(t)$, and the near-end talker is $x(t)$. The near-end talker is superimposed with the undesired echo on port D. The far-end talker $y(t)$, which is available to the canceler on port A, and is called the **reference signal** for the echo canceler, is used by the canceler to generate a replica of the echo alone, called $\hat{r}(t)$. This replica is subtracted from the echo plus near-end talker to yield $e(t)$, which ideally contains the near-end talker alone.

In the electronic hybrid, the echo path resulted from the superposition of the two talkers across the transformer; in contrast, in the echo canceler generally the

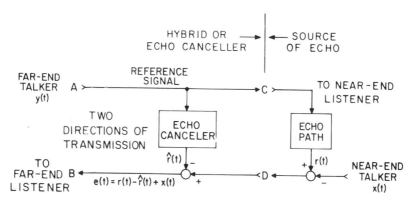

NOTE: IDEALLY $r(t) = \hat{r}(t)$

**Figure 4.6** Principle of the echo canceler based on generating an echo replica and subtracting it from the near-end talker signal plus echo.

echo path will be due to the finite return loss of a hybrid at a point of two- to four-wire conversion at the near end of the connection. In the hybrid the echo replica is generated by applying the reference signal to a simple voltage divider with fixed impedances. In the echo canceler, the echo replica is generated by a circuit, which actually adapts to the unknown echo path and is therefore able to generate a much more accurate replica in the face of uncertainties in the precise echo mechanism encountered in any particular connection.

The echo canceler, shown for only one direction of transmission in Fig. 4.7, generates the echo replica by applying the reference signal to a transversal filter (tapped delay line). The transversal filter coefficients are caused to adapt to the echo transfer function using the circuitry of Fig. 4.8.

For convenience of implementation of the transversal filter, it is assumed in Fig. 4.7 that the interfaces to the four-wire ports are sampled data. Thus the far-end speech signal $y(t)$ is replaced by $y(kT) = y_k$, etc. This sampled data interface would occur naturally at a digital transmission terminal or digital switch. Of course, the interface to analog facilities could also be accomplished with the addition of the usual anti-aliasing and reconstruction **low-pass filters** (LPFs) and samplers. It is also assumed that the sampling rate in the two directions is synchronous. It is also possible to implement an echo canceler when the sampling rate is not synchronous, but the device is somewhat more complicated than shown.

Since the transfer function of the echo path from port C to port D is normally not known in advance, particularly with sufficient accuracy to assure a high degree of cancellation, the canceler adapts the coefficients of the transversal filter in the manner illustrated in Fig. 4.8. This adaption algorithm is derived in Section 4.4.1 and is usually adequate even though there are more sophisticated adaptation algorithms, as discussed in Section 4.5. In Fig. 4.8, the residual cancellation error $e_i$ is used to drive the canceler. The adaptation algorithm infers from this error the appropriate correction to the transversal filter coefficients so as to reduce this error. Specifically, this error is crosscorrelated with successive delays of the far-end talker

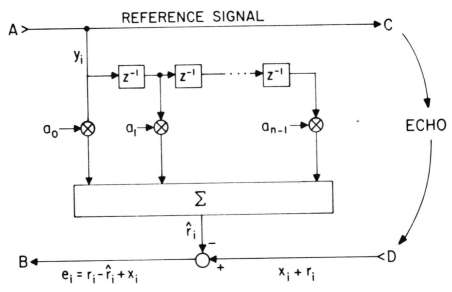

**Figure 4.7** Echo canceler for one direction of transmission.

signal. The summation box is an accumulator, the output of which is the corresponding filter coefficient. The correction to the accumulated value is scaled by a constant $\beta$, which is the "step-size" of the algorithm. This step-size controls the speed of adaptation and the asymptotic error in a manner discussed in Section 4.4.1.

The error signal $e_i$ contains a component of the near-end talker signal $x_i$ in addition to the residual echo-cancellation error. The effect of this on the cancellation is naturally of some concern. As long as $y_i$ is uncorrelated with $x_i$, which should be the case, this near-end talker will not affect the asymptotic mean value of the filter coefficients. However, the asymptotic variation in the filter coefficients will be increased substantially in the presence of a near-end talker due to the introduction of another (large) stochastic component in the adaptation. This effect

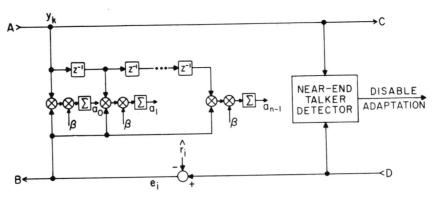

**Figure 4.8** Adaptation mechanism for the echo canceler of Fig. 4.7. This circuit generates the coefficients $a_0, \ldots, a_{n-1}$ used in the transversal filter.

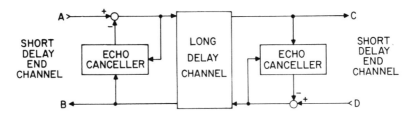

**Figure 4.9** Split echo canceler configuration for two directions with largest delay in the center.

can be compensated by choosing a very slow adaptation (small β), but it is generally preferable to employ a near-end talker detector as shown to disable adaptation when a near-end talker is present. The adaptation then proceeds only when there is no detectable near-end talker, and the adaptation speed can be increased. The near-end talker detector assumes that there is some minimum return loss through the hybrid, that is, between port C and port D, since otherwise the large echo signal would obscure a small near-end talker signal. Only near-end talker signals that are large (relative to the far-end talker echo) can be detected.

In practice it is desirable to cancel the echoes in both directions of a trunk. For this purpose two adaptive cancelers are necessary, as shown in Fig. 4.9, where one cancels the echo from each end of the connection. The near-end talker for one of the cancelers is the far-end talker for the other. In each case, the near-end talker is the closest talker, and the far-end talker is the talker generating the echo which is being canceled. It is desirable to position these two halves of the canceler in a split configuration, as shown in Fig. 4.9, where the bulk of the delay in the four-wire portion of the connection is in the middle. The reason is that the number of coefficients required in the echo-cancellation filter is directly related to the delay of the channel between the location of the echo canceler and the hybrid that generates the echo. In the split configuration the largest delay is not in the echo path of either half of the canceler, and hence the number of coefficients is minimized. Typically cancelers in this configuration require only 128 or 256 coefficients, whereas the number of coefficients required to accommodate a satellite connection in the end link would be impractically large.

### 4.2.3 Implementation of Echo Cancellation

The widespread application of echo cancellation has recently been stimulated by advances in microelectronics, making the computational requirements of the canceler within the reach of an inexpensive single-chip implementation. Because the potential applications of echo cancellation are numerous, there has been considerable activity in the design of devices for echo cancellation.

The first special-purpose chip for speech echo cancellation was developed by Bell Laboratories [Duttweiler, 4.6]. This device implemented a single 128-coefficient canceler using digital techniques, with an interface to the standard μ-255 companding law widely used in the digital transmission network. Advantage was taken of the nearly logarithmic nature of this companding law to simplify the

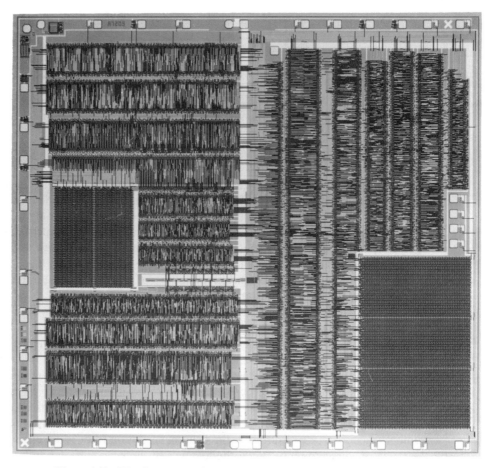

**Figure 4.10** Die photograph of a 128-tap voice echo canceler integrated circuit designed by Bell Laboratories and manufactured by Western Electric. Up to four of these chips can be cascaded to yield a 512-tap echo canceler. (Photo courtesy Bell Laboratories, with permission.)

multiplications required in the implementation of the transversal filter. A later version of this device added the capability to readily cascade chips to achieve up to four times the number of coefficients [Tao, 4.7]. A die photo of this device is shown in Fig. 4.10.

A complete echo-canceler system requires more than a single chip. Shown in Fig. 4.11 are two boards from a Tellabs echo-canceler system, one of which contains 12 single-chip CMOS 256-tap cancelers. Four of these boards are included in a complete 48-channel echo-cancellation system. The other board is required to interface digital transmission facilities.

A complete 48-channel canceler system is the Granger Associates unit shown in Fig. 4.12. An interesting feature is the hand-held "program control and display unit," which enables an operator to program on a per-channel basis various options such as bypass of a canceler, initialization, or disabling of adaptation.

**Figure 4.11**  Two boards, which are a part of the Tellabs 251 48-Channel echo-canceler system with T-1 interfaces.  The 6951 board includes 12 40-pin packages, each one a 256-tap echo canceler (four of these boards are used in the Tellabs 251 48-channel echo canceler system).  The second board is the interface to digital transmission equipment.  (Photo courtesy Tellabs, with permission.)

### 4.2.4  Recent Developments in Echo Control

The telephone network is undergoing a significant transition with respect to echo control.  The primary reasons are the rapid introduction of digital transmission, digital switching, and satellite transmission.

The VNL loss plan was developed and implemented in the context of the analog transmission network.  The telephone network is presently evolving toward a new transmission level and loss plan for the all-digital transmission and switching network [Abate, 4.8].  When a trunk is transmitted in a digitally encoded format, it is not convenient to provision attenuation (which requires a more complicated digital processing), and furthermore attenuation involves an undesirable quantization noise penalty.  An alternative plan has been developed for the digital network, in which a fixed 6-dB loss is provided on any toll connection trunk (trunk between a local switch and toll switch) on a digital facility [Abate, 4.8].  This overall loss is achieved by inserting 3 dB in each of the two toll-connect trunks in

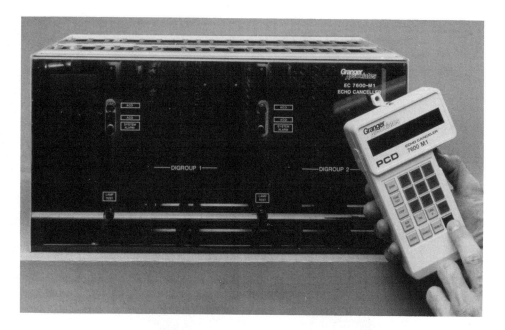

**Figure 4.12** The Granger Associates EC 7600-M1 48-channel echo canceler system. Also shown is a hand-held program and display unit. (Photo courtesy Granger Associates, with permission.)

a connection, with all digital intertoll trunks (between two toll switches) having no loss.

In the interim network with both analog and digital facilities, the VNL plan is retained on all analog links, and the fixed loss plan is used on digital links. The exact connection loss thus depends on the exact configuration of the network in terms of the analog and digital links. It has been shown that adequate subjective quality is maintained during this interim period.

With the advent of digital transmission and digital switching in the local network, the conversion from two-wire to four-wire is moving from the trunk side to the subscriber side of the local switch. There is also a strong desire to retain the same 0-dB insertion loss for a local digital switch as was inherent in the metallic crosspoint analog switches. If this is done, however, two subscribers connected to the same local switch have no insertion loss and the singing margin is degraded. The result is a log of effort to improve the return loss of hybrids at the subscriber interface to a local switch through the use of selective terminations [Bunker, 4.3; Neigh, 4.2] or adapting the termination [Messerschmidt, 4.9; Dotter, 4.10]. An approach to adapting a hybrid is outlined in Section 4.4.3.

### 4.2.5 Different Realizations of the Echo Canceler

Although Fig. 4.7 showed a transversal filter echo canceler, there are other realizations. This section describes, in addition to the transversal filter, a lattice filter and a continuous time realization. All these realizations have the desirable property that there are a small number of filter parameters that can be adapted to realize a wide range of transfer functions.

### Transversal Filter Realization

The transversal filter of Fig. 4.7 is a direct-form realization of a finite impulse response

$$\hat{r}_i = \sum_{m=0}^{n-1} a_m x_{i-m} \tag{4.4}$$

where there are $n$ filter coefficients $a_0, \ldots, a_{n-1}$. These filter coefficients are adapted in the manner described in Section 4.4.1.

### Lattice Filter Realization

There are many realizations of a **finite impulse response** (FIR) filter other than the transversal filter. One alternative of particular importance is the lattice structure shown in Fig. 4.13 [Gray, 4.11; Makhoul, 4.12]. The lattice structure is equivalent to the transversal filter in the sense that any transfer function which can be represented by the transversal structure can also be represented by the lattice structure. While the structure of Fig. 4.13 realizes an FIR filter, lattice filters are also applicable to **infinite impulse response** (IIR) filter realization [Gray, 4.11].

The lattice filter structure consists first of a set of $n - 1$ stages with internal coefficients $k_j$, $1 < j \leq n-1$, which are commonly called the **reflection** or **PARCOR** coefficients. The signals at the output of the $m$th stage, $e_f(1|m)$ and $e_b(i|m)$, are called, respectively, the forward and backward prediction error of order $m$. For purposes of realization of the echo canceler, the salient property of these prediction errors is that when the reflection coefficients are chosen appropriately, the successive backward prediction errors are uncorrelated. The *order update equations* for the prediction errors are given by

$$e_f(i|m) = e_f(i|m-1) - k_m e_b(i|m-1) \tag{4.5}$$

$$e_f(i|0) = y_i \tag{4.6}$$

$$e_b(i|m) = e_b(i-1|m-1) - k_m e_f(i-1|m-1) \tag{4.7}$$

$$e_b(i|0) = y_{i-1} \tag{4.8}$$

The echo replica is generated by forming a weighted linear combination of the backward prediction errors of successive orders with weights $b_1, \ldots, b_n$, given by

$$\hat{r}_i = \sum_{m=1}^{n} b_m e_b(i|m-1) \tag{4.9}$$

One way to think of this structure is as a transversal filter where the delay elements have been replaced by the more complicated lattice stages. The purpose of the lattice stages is to decorrelate the input signal, so that the final replica is formed from a sum of uncorrelated signals. The reflection coefficients are therefore chosen

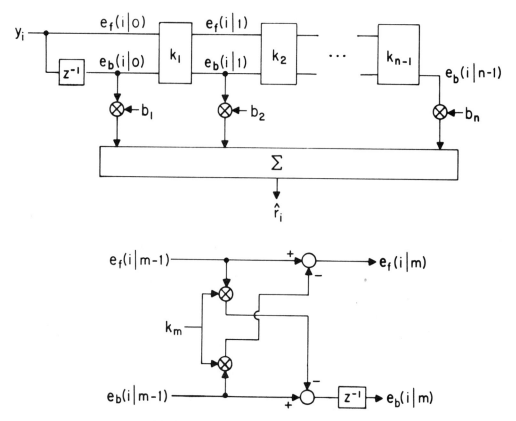

**Figure 4.13**  Echo canceler based on a lattice filter.

irrespective of the desired echo replica on the basis of the reference signal input statistics alone. The transfer function of the overall filter is controlled by the choice of the coefficients $b_1, \ldots, b_n$.

The advantage of the lattice structure arises from its greater speed of adaptation, as discussed in Section 4.5.2.

### The Continuous Time Canceler

An alternative realization for an echo canceler is as a continuous time filter, as was suggested in an early paper on echo cancellation [Makhoul, 4.13]. A practical application for the continuous time version has been suggested [Messerschmitt, 4.9]. This structure, which is illustrated in Fig. 4.14, is applicable to cancellation at the hybrid itself and has been called an **adaptive hybrid**. This approach has been proposed for the interface to a digital switch, where the amount of additional cancellation required relative to a single compromise termination is not great, and the application demands a cost-effective solution.

The adaptive hybrid takes advantage of the limited variability of the input impedance of a two-wire transmission line to which the hybrid is connected to reduce the complexity of the canceler. Since the echo signal $r(t)$ is generated by

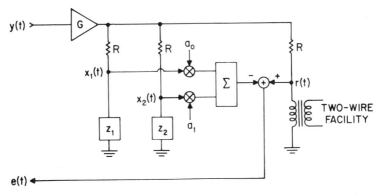

**Figure 4.14** A two-tap continuous-time echo canceler, or adaptive hybrid.

applying the reference input $y(t)$ to a voltage divider, two similar voltage dividers are built using two impedances $Z_1$ and $Z_2$, which are approximations to the two-wire line impedance at its extremes. The outputs of these two voltage dividers are weighted and summed and used to cancel the echo. A method for adapting the two coefficients is described in Section 4.4.3.

In some applications this simple structure with only two coefficients can achieve adequate cancellation, whereas a transversal filter would require many more coefficients. This illustrates that the transversal filter may sometimes have many more degrees of freedom than really necessary for a particular application.

## 4.3 ECHO CANCELLATION IN DATA TRANSMISSION

When data signals are transmitted through the network, some of the same difficulties with echo are encountered as in speech transmission. In transmitted **half-duplex** data (data transmitted in only one direction), echoes present no problem since there is no receiver on the transmitting end to be affected by the echo. In **full-duplex** transmission, where the data signals are transmitted in both directions simultaneously, echoes from the data signal transmitted in one direction interfere with the data signal in the opposite direction.

Most digital transmission is half-duplex. For example, the high-speed trunk digital transmission systems separate the two directions of transmission on physically different wire, coaxial, or fiber optic media. The two directions therefore do not interfere, except perhaps through crosstalk resulting from inductive or capacitive coupling. Similarly, most high-speed voiceband data transmission, in which a **modulator-demodulator (modem)** is used for transmitting computer data over the public telephone network, is half-duplex.

However, full-duplex data transmission over a common media has arisen in two important applications. In both these applications, the need for a common media arises because the network typically only provides a two-wire connection to each customer premise (this is because of the high cost of copper wire and the large percentage of the telephone network investment in this facility).

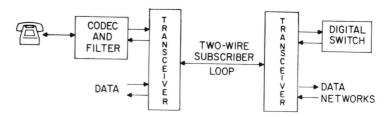

**Figure 4.15**  Application of echo cancellation to subscriber loop digital transmission.

The first application (Fig. 4.15) is digital transmission on the subscriber loop, in which the basic voice service and enhanced data services are provided through the two-wire subscriber loop.  Total bit rates for this application that have been proposed are 80 and 144 kb/s in each direction, where the latter rate includes provision for two voice-data channels at 64 kb/s each plus a data channel at 16 kb/s, and the first alternative allows only a single 64-kb/s voice-data channel.  This digital subscriber loop capability is an important element of the emerging integrated services digital network (ISDN), in which integrated voice and data services will be provided to the customer over a common facility. Voice transmission requires a *codec* (coder-decoder) and anti-aliasing and reconstruction filters to perform the analog-to-digital and digital-to-analog conversion on the customer premises, together with a **transmitter-receiver (transceiver)** for transmitting the full-duplex data stream over the two-wire subscriber loop.  Any data signals to be accommodated are simply connected directly to the transceiver.  The central office end of the loop has another full-duplex transceiver, with connections to the digital central office switch for voice or circuit switched data transmission and to data networks for packet switched data-transport capability.

The second application for full-duplex data transmission is in voiceband data transmission (Fig. 4.16) where the basic customer interface to the network is often the same two-wire subscriber loop.  In this case the transmission link is usually more complicated due to the possible presence of four-wire trunk facilities in the middle of the connection.  The situation can be even more complicated by the presence of two-wire toll switches, allowing intermediate four-two-four wire conversions internal to the network.

The two applications differ substantially in the types of problems which must be overcome.  For the digital subscriber loop, the transmission medium is fairly ideal, consisting of cable pairs with a wide bandwidth capability.  The biggest

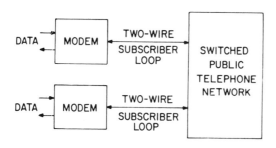

**Figure 4.16**  Application of echo cancellation to full-duplex voiceband data modems.

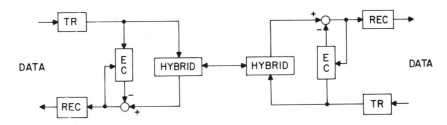

**Figure 4.17** Configuration of echo cancellation for full-duplex data transmission.

complication is the higher bit rate and the presence in some countries of bridged taps—open-circuited wire pairs bridged onto the main line. The voiceband data modem, while requiring a lower speed of transmission, encounters many more impairments. In addition to the severe bandlimiting when carrier facilities are used, there are problems with noise, nonlinearities, and sometimes even frequency offset. Another difference is that the subscriber loop can use baseband transmission, whereas the voiceband data set always uses passband transmission—that is, modulates a carrier with the data stream.

One approach to full-duplex data transmission is the use of an echo canceler (EC) to isolate the two directions of transmission as shown in Fig. 4.17. There are a transmitter (TR) and receiver (REC) on each end of the connection, and a hybrid is used to provide a virtual four-wire connection between the transmitter on each end and the receiver on the opposite end. The problem with this arrangement is the high-level signal partially fed through the hybrid from the transmitter into the local receiver. The hybrid depends on knowledge of the two-wire impedance for complete isolation, and no better than 10 dB or so of attenuation through the hybrid can be guaranteed with a single compromise termination.

In both subscriber loop and voiceband data transmission, the loss of the channel from one transmitter to the receiver on the other end can be as high as 40 to 50 dB. This implies that in the worst case, the undesired transmitter local feedthrough (echo) signal can be 30 to 40 dB higher in level than the data signal from the far end (assuming both transmitters are at the same level). Since signal to interference ratios on the order of 20 dB are required for reliable data transmission, the echo canceler is added to each end to give additional attenuation of the undesired feedthrough signal of up to about 50 to 60 dB.

The echo-cancellation requirements for both applications are discussed in more detail in the following two sections. In each case, alternative methods of providing the full-duplex data transmission are also mentioned.

### 4.3.1 Subscriber Loop Digital Transmission

There are two competing methods of providing full-duplex data capability on the subscriber loop [Ahamed, 4.14]. In spite of this competition, echo cancellation has achieved significant attention because of its ability to provide greater range (distance between subscriber and central office) in some circumstances.

The first competing method is frequency division multiplexing (FDM), in

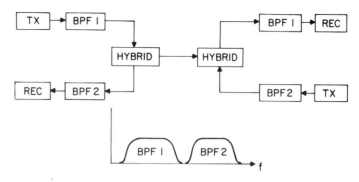

**Figure 4.18** FDM method of full-duplex data transmission.

which the two directions are transmitted in nonoverlapping frequency bands and therefore do not interfere with one another even in the presence of echos (as long as the channel is linear and time-invariant, or nearly so). This is illustrated in Fig. 4.18, where one direction of transmission utilizes the frequency band passed by band-pass filter BPF1, and the other direction uses the nonoverlapping band passed by BPF2. Note that there is no interference between transmitter and receiver on one end, since the two band-pass filters do not overlap.

The second competing method is *ping-pong* or **time compression multiplexing** (TCM), illustrated in Fig. 4.19. The idea is to divide the common medium in time rather than frequency, transmitting first in one direction and then in the other. The switches operate in much the same manner as an echo suppressor, taking advantage of the fact that there is transmission in only one direction at a time to separate the two directions. The "west" transmitter first transmits a burst of bits at a rate somewhat higher than twice the average data rate in one direction, and that burst is received $d$ seconds later, where $d$ is the propagation delay on the cable. The "east" end, which is slaved to the west end, detects this received burst and sends its own transmitted burst shortly after the end of the received burst. This burst arrives at the west end with delay $d$ before the next burst must be transmitted. The west end, being in control, simply transmits its burst of data at regular intervals. Note that the choice of an instantaneous data rate, the length

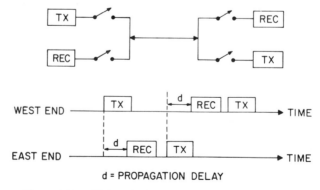

**Figure 4.19** TCM method of full-duplex data transmission.

of a burst, and a desired guard time between bursts puts an upper limit on $d$ and hence on the range of transmission.

Both FDM and TCM are viable alternatives to the use of echo cancellation in the subscriber loop. The primary advantage of the echo-cancellation method is the approximate halving of the transmitted signal bandwidth, which results in less attenuation of the data signal on the medium (it usually has an attenuation which increases with frequency) and reduced crosstalk susceptibility. However, an important consideration in the digital subscriber loop is the crosstalk among many cable pairs which are each carrying a data signal. In particular, the most important crosstalk is **near-end crosstalk** (NEXT), in which other local transmitters are crosstalking into a given local receiver. The importance of this mechanism is due to the high level of the local sources of crosstalk relative to the received far-end data signal. If in TCM the transmitters are all synchronized to transmit their bursts of data at the same time or nearly the same time, NEXT is not a consideration, since none of the local transmitters are active at the same time as the local receivers. Similarly, there is no NEXT interference in an FDM system since the local transmitter and receiver are operating in different frequency bands. In the echo-cancellation method, however, NEXT always results in an impairment.

The comparison among FDM, TCM, and the echo-cancellation method is therefore dependent on the complicated trade-off between the effect of cable attenuation at high frequencies (which is favorable to echo cancellation) and NEXT (which is favorable to FDM or synchronized TCM) and depends on many details, such as the degree of synchronization of the TCM system, line codes, and so on. The comparison among the three methods would usually be done on the basis of range; that is, the permissible distance of transmission. TCM is of course much simpler to implement and therefore is the best choice where range is not a consideration.

The comparison between ranges for these different techniques is still being worked out and depends on the detailed assumption about the degree of echo cancellation that can be obtained. On the basis of information available at this writing, some statements can be made about the comparison. Where the TCM bursts cannot be synchronized, both TCM and echo cancellation suffer near-end crosstalk, and echo cancellation has been shown by theory and experiment to have a range about 30% greater than TCM [Aschrafi, 4.15; Andersson, 4.16]. This comparison assumes that the echo cancellation attenuation is large enough so that other factors (such as NEXT and noise) become the limiting factors in range.

### Implementation of a Digital Subscriber Loop Transceiver

This section describes the complete transceiver that is used to transmit full-duplex data as a baseband signal over a subscriber loop.

The implementation of a data echo canceler differs from a voice echo canceler in several respects. First, the input signal is a data signal, and it is therefore possible if the sampling rate is made synchronous with the data rate to apply a binary-valued signal to the canceler. This is true for binary transmission, has been shown to be true for multilevel transmission [Agazzi, 4.17] and eliminates the need for multiplications. Second, by definition the signals in both directions are present

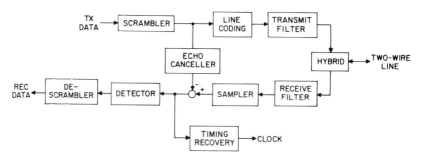

**Figure 4.20** Block diagram of an echo cancellation subscriber loop transceiver.

at all times, making the effect of the far-end signal on adaptation much more significant an issue. Third, the desired degree of cancellation is much larger, making design and implementation much more challenging. Finally, there are significant issues of timing recovery, synchronization, and equalization, which interact with echo cancellation and do not exist in voice cancellation.

A block diagram illustrating the functions of the transceiver is given in Fig. 4.20 [Agazzi, 4.18]. The transmit data is first scrambled to insure that there are sufficient pulses for timing recovery on the other end. Some form of line coding is applied to control the transmitted signal spectrum—for example, to insure that there is no energy at dc. Next is the transmit filter to limit the high frequency components in the signal for **radio-frequency interference** (RFI) and crosstalk purposes. The echo canceler is connected either before or after the line coding. The two constraints are that the signal at the input to the canceler have as few levels as possible (binary is best) for simplicity of implementation, and the echo path between the input and output connections of the echo canceler should be linear. Some forms of line coding are nonlinear, and any such nonlinearities should occur prior to the canceler input.

On the receive side, a receive filter prevents aliasing in the subsequent sampling operation and may also provide equalization of the high-frequency attenuation of the cable. The signal is then sampled, since the echo canceler operates in the sampled data domain. The choice of the sampling rate used in the echo cancellation is a critical design factor, which will be addressed shortly. After echo cancellation, the data is detected, taking into account the line coding and any intersymbol interference present, and descrambled to yield the received data sequence.

This block diagram presumes that the two directions of transmission are synchronous. This is usually imposed by the digital switch in the central office. The synchronization requirements depend on whether the transceiver is at the central office, which serves as the synchronization master clock, or at the subscriber end. At the central office end, the central office clock can serve to drive both the transmitter and receiver, although there are some subtleties relating to the arbitrary phase of the incoming data stream. At the subscriber end, there is no reference clock other than the incoming data stream. Hence it is necessary to perform timing recovery on the incoming data, as shown in Fig. 4.20, and this must be done after the echo canceler so that the recovered timing is not affected by the local transmitted data. The echo canceler is able to remove the local data signal even if the local and remote data signals are asynchronous. Once the timing is recovered, the resulting clock can be used to drive the local transmitter, which will then be synchronized with the remote transmitter.

The choice of sampling rate thus represents a trade-off between the complexity of the echo canceler and the ease of recovering timing. For purposes of data detection, a sampling rate equal to the data symbol rate is adequate, although there are some benefits to doubling this rate and using fractionally spaced equalizers [Ungerboeck, 4.19]. Timing recovery is usually considered to require a sampling rate equal to twice the data symbol rate, since this is the minimum rate at which the signal can be recovered without aliasing (the data signal bandwidth is usually greater than half the symbol rate and less than the symbol rate). There have been attempts at timing recovery with symbol-rate sampling, but they are undesirably dependent on the received pulse shape [Mueller, 4.20].

In conclusion, the sampling rate is usually equal to twice the symbol rate or higher, implying that the echo canceler has different sampling rates at input and output, since the input rate is equal to the symbol rate. The manner in which this is accomplished is outlined in the following section.

### Mathematical Formulation of the Baseband Echo Canceler

The mathematical formulation of the baseband data echo canceler used in a digital subscriber loop transceiver is somewhat more complicated than the formulation of a voice canceler, due to the possibly different sampling rates at the canceler reference input (data stream) and output (echo replica).

Assume that the transmitted signal is

$$s(t) = \sum_m C_m g(t - mT) \tag{4.10}$$

where $C_m$ is the sequence of transmitted data symbols, $g(t)$ is the transmitted pulse shape, and $T$ is the interval between transmitted data symbols (the baud interval). Suppose this signal is passed through a filter with transfer function $H_e(\omega)$ representing the echo response. If we denote the response of this filter to $g(t)$ as $h(t)$ and the echo response as $r(t)$, then the latter becomes

$$r(t) = \sum_m C_m h(t - mT) \tag{4.11}$$

In order to use a transversal filter echo canceler, it is necessary to reconstruct samples of this echo signal. As previously mentioned, it is necessary to sample this signal more often than every $T$ seconds if the objective is to be able to reconstruct $r(t)$ exactly. Since a clock representing the transmitted data signal is available, it is natural to sample the echo signal at a rate which is an integer multiple of the symbol rate, say a multiple $R$. Thus define

$$r_i(l) = r\left(\left(i + \frac{l}{R}\right)T\right), \qquad 0 \le l \le R-1 \tag{4.12}$$

where the index $i$ represents the data symbol epoch and $l$ represents the sample from among $R$ samples uniformly spaced in this epoch. Similarly, define a notation for the samples of the echo pulse response

$$h_i(l) = h\left(\left(i + \frac{l}{R}\right)T\right), \qquad 0 \le l \le R-1 \tag{4.13}$$

From (4.11–4.13),

$$r_i(l) = \sum_m h_m(l)C_{i-m} \tag{4.14}$$

This relation shows that the samples of the echo can be thought of as $R$ independent echo channels, each channel being driven by the same sequence of data symbols. The impulse response of the $l$th echo channel is $h_i(l)$.

Since the $R$ echo channels are independent, the index $l$ can be dropped. The transversal filter has a finite impulse response, so consider building an echo canceler to generate the replica

$$\hat{r}_i = \sum_{m=0}^{n=1} C_{i-m}a_m \tag{4.15}$$

where $a_i$, $0 \le i < n-1$, are the $n$ filter coefficients of one of the $R$ interleaved transversal filters. This transversal filter generates an FIR approximation to the echo response. Note the analogy to (4.4), where the far-end talker samples of the voice canceler have been replaced by the transmitted data symbols at the canceler reference input.

The $R$ independent echo cancelers can be used in the manner of Fig. 4.21 for $R = 4$. Each canceler has the same input data sequence at its reference input. The far-end data plus echo signal from the hybrid is decimated to four independent signals, each at the data symbol rate, and these are independently canceled and recombined into a single sample stream representing the far-end data signal alone. Each of the cancellation error signals is used to drive the adaptation of the corresponding canceler.

Each canceler can be thought of as adapting to the impulse response of the echo channel sampled at a rate equal to the symbol rate but with a particular phase out of $R$ possible phases. These cancelers independently converge, although they do have in common the same input sequence of data symbols. Since the transversal filters all adapt independently, the presence of multiple interleaved canceler filters does not affect the speed of adaptation. Therefore, the choice of an output sampling rate is purely a question of implementation complexity: the adaptation rate and asymptotic error are not affected by the sampling rate.

### 4.3.2 Voiceband Data Transmission

The preceding section describes the application of echo cancellation to full-duplex baseband data transmission (the digital subscriber loop); the present section extends this to the passband case typical of voiceband data transmission modems. Full-duplex voiceband data transmission has been provided for some years at 1.2 kb/s and below (and more recently at 2.4 kb/s) using FDM. However, at 4.8 kb/s and above, frequency separation becomes impractical due to inadequate total band-

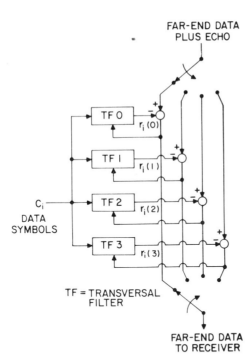

FAR-END DATA
PLUS ECHO

TF 0

$r_i(0)$

TF 1

$r_i(1)$

$C_i$ —
DATA
SYMBOLS

TF 2

$r_i(2)$

TF 3

$r_i(3)$

TF = TRANSVERSAL
FILTER

FAR-END DATA
TO RECEIVER

**Figure 4.21** Interleaved data transmission echo cancelers to achieve higher output sampling rate (R = 4).

width, and the use of echo cancellation must be considered. TCM is not a candidate for voiceband data at any bit rate because of the large propagation delay in the channel that must potentially be accommodated (it can be in the hundreds of milliseconds for a satellite connection). This delay would necessitate impractically long data bursts.

Since voiceband data is transmitted at passband rather than baseband, the details of echo-cancellation provisioning are considerably different from the subscriber loop. For this reason, passband transmission and its effect on echo cancellation are reviewed in the following section.

### Review of Quadrature Amplitude Modulation

The usual method of voiceband data transmission is quadrature amplitude modulation (QAM). Since the echo canceler must operate in the presence of this modulation, this section briefly reviews this technique and establishes a notational framework.

The frequency translation of a baseband data signal up to passband by simply modulating by a carrier is wasteful of bandwidth since both the upper and lower sidebands are transmitted. A more efficient method is **single-sideband** (SSB) or **vestigial sideband** (VSB) transmission. Voiceband data modems typically use QAM, which has the same bandwidth efficiency as SSB modulation and is typically simpler to implement. In QAM, two carriers that are in quadrature to one another are independently modulated by a baseband data signal.

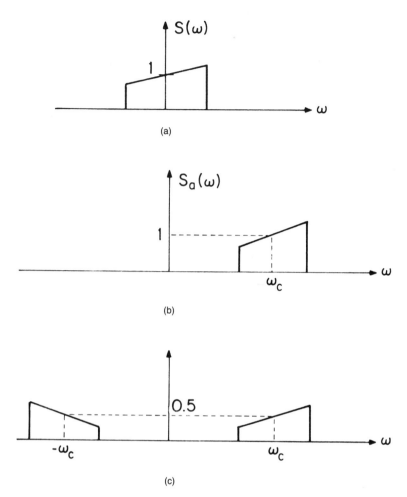

**Figure 4.22** Spectra of signals for QAM. (a) Baseband signal with complex-valued data symbols. (b) After modulation by complex exponential. (c) After taking real part.

In a mathematical formulation of this passband modulation, it is convenient to use complex notation, so consider a baseband data signal identical to (4.10) except that the data symbols are complex.

$$C_i = a_i + jb_i \tag{4.16}$$

and the transmitted pulse $g(t)$ is still real-valued. Since this signal is complex-valued, it has an asymmetrical spectrum, as illustrated in Fig. 4.22a. If this complex data signal is modulated up to passband by multiplying by a complex exponential having frequency $\omega_c$, the carrier frequency, the spectrum is shifted by $\omega_c$ radians, as illustrated in Fig. 4.22b. If the time domain equivalent is called $s_a(t)$, this complex-valued signal is given by

$$s_a(t) = \sum_m C_m g(t - mT)e^{j\omega_c t} \tag{4.17}$$

Finally, to make the transmitted signal real-valued, simply take the real part of $s_a(t)$,

$$s(t) = \text{Re}[s_a(t)] \qquad (4.18)$$

Since the real part can be written alternatively as

$$s(t) = \frac{s_a(t) + s_a^*(t)}{2} \qquad (4.19)$$

and since the Fourier transform of $s_a^*(t)$ is $S_a^*(-\omega)$, the Fourier transform of $s(t)$ is

$$S(\omega) = \frac{S_a(\omega) + S_a^*(-\omega)}{2} \qquad (4.20)$$

as illustrated in Fig. 4.22(c). Note that each portion of the spectrum in Fig. 4.22(c) is half the height of the spectrum in Figs. 4.22(a) and 4.22(b).

For purposes of implementation of the transmitter, it is convenient to expand in terms of (4.16), so that (4.17) becomes

$$s(t) = \sum_m a_m g(t - mT)\cos(\omega_c t) - \sum_m b_m g(t - mT)\sin(\omega_c t) \qquad (4.21)$$

where the amplitude modulation of two independent data signals by quadrature carriers is evident. A method of generating this modulated signal is illustrated in Figure 4.23(a). The input data stream is encoded to yield the $a_n$ and $b_n$ values. There are many ways of doing this, and Fig. 4.23b illustrates one simple way. In this case, the complex-valued data symbol $C_n$ assumes one of 16 distinct values, with the real part $a_n$ having 4 values and the imaginary part $b_n$ having 4 values (each equally spaced). The encoder in this case maps 4 input bits into the complex-valued data symbol, 2 bits into the real part and 2 into the imaginary part. The remainder of the transmitter in Fig. 4.23a modulates a pulse $g(t)$ by the real and imaginary parts, modulates them up to passband by multiplying by quadrature carriers, and sums the results.

The time domain signal with only positive frequency components corresponding to Fig. 4.22b is called the **analytic signal**. In the receiver processing, it is convenient to recover this complex-valued signal prior to further processing. Given the passband QAM-modulated signal, the analytic signal can be recovered in the manner illustrated in Fig. 4.23c utilizing a phase splitter. The imaginary part of the analytic signal is simply the output of a 90° phase shift network (transfer function $-j\,\text{sgn}(\omega)$), also called a **Hilbert transform** filter. To see this, note that the Fourier transform of the complex-valued signal generated in Fig. 4.23(c) is

$$S(\omega) + \text{sgn}(\omega)S(\omega) = \begin{cases} 2S(\omega) & x \geq 0 \\ 0 & x < 0 \end{cases} \qquad (4.22)$$

and this signal has only positive frequency components.

In practice, the phase splitter consists of two filters that are all-pass (have unity magnitude response) and a phase difference of 90°. In addition, it is common to implement the phase splitter in discrete time.

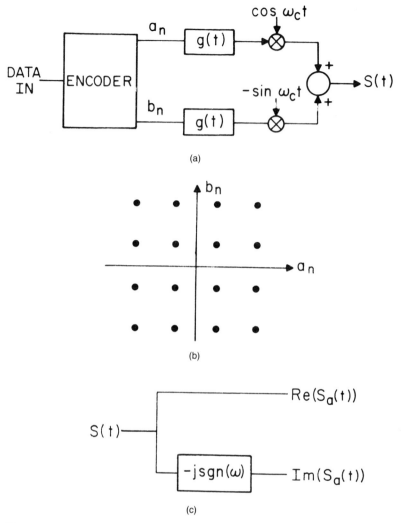

(a)

(b)

(c)

**Figure 4.23** Principle of QAM. (a) QAM modulator. (b) 16-point signal constellation. (c) Phase splitter in receiver.

### Passband Echo Cancellation

The passband echo canceler is quite similar to the baseband case except that a complex-valued data sequence and the complex-valued output of the phase splitter are input to the canceler rather than a real-valued sequence. The mathematical details of this canceler are derived in this section [Weinstein, 4.21].

Consider the response of the echo-transfer function (denoted by $H_e(\omega)$, as in the baseband case) to the transmitted data signal given by (4.18). In fact, since the analytic signal of this echo response is generated by the phase splitter at the receiver input, what is really desired is the analytic signal corresponding to this echo signal. Examining Fig. 4.24, if the response of the echo filter to the trans-

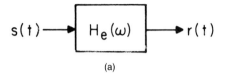

(a)

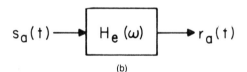

(b)

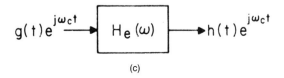

(c)

**Figure 4.24** Response of echo path filter to a single modulated pulse. (a) Transmitted signal and echo. (b) Analytic signals. (c) Isolated pulse response.

mitted signal is denoted by $r(t)$, as shown in Fig. 4.24(a), and the corresponding analytic signal is $r_a(t)$, then—as shown in Fig. 4.24(b)—$r_a(t)$ is the response of the echo filter to the transmitted analytic signal $s_a(t)$. This is because the impulse response of the echo filter is real-valued. Determining the response to the analytic transmitted signal is simple if the response to a single isolated analytic pulse can be determined, as illustrated in Fig. 4.24(c). The output is expressed in terms of another analytic signal, where $h(t)$ is the equivalent baseband pulse which returns through the echo path. Taking the Fourier transform of the input and output of the echo path filter in Fig. 4.24(c),

$$G(\omega - \omega_c)H_e(\omega) = H(\omega - \omega_c) \qquad (4.23)$$

or equivalently,

$$H(\omega) = G(\omega)H_e(\omega + \omega_c) \qquad (4.24)$$

This demonstrates that the baseband output pulse can be obtained by shifting the echo-transfer function in the vicinity of the carrier frequency down to dc, as might be expected. The spectrum of $H(\omega)$ does not have complex-conjugate symmetry about the origin, so $h(t)$ is in general complex-valued.

The analytic echo signal follows from superposition as

$$r_a(t) = \sum_m C_m h(t - mT)e^{j\omega_c t} \qquad (4.25)$$

very much in analogy to (4.11). As in the baseband case, the echo canceler must be implemented in discrete time. In order to be able to reconstruct the echo signal the sampling rate must be an integer $R$ times as high as the data symbol rate, where $R$ is an appropriate integer on the order of 2 or 4. Defining $r_{a,i}(l)$ and $h_i(l)$ as in (4.12) and (4.13), a relation similar to (4.14) is obtained,

$$r_{a,i}(l) = \left( \sum_m C_m h_{i-m}(l) \right) e^{j\omega_c \left( i + \frac{l}{R} \right) T} \qquad (4.26)$$

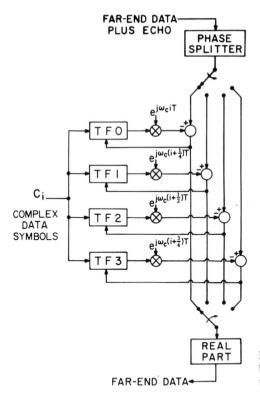

**Figure 4.25** Interleaved complex base-band echo cancelers in a passband data-transmission system (R = 4).

A transversal filter FIR approximation to this echo becomes,

$$\hat{r}_{a,i} = \left( \sum_{m=0}^{n-1} a_m C_{i-m} \right) e^{j\omega_c \left( i + \frac{l}{R} \right) T} \tag{4.27}$$

where $l$ has again been suppressed and the filter coefficients $a_m$ are complex-valued.

An illustration of this approach for $R = 4$ is shown in Fig. 4.25. The same complex data symbols $C_i$ are applied to each of the transversal filters. These transversal filters are operating at baseband and have complex-valued filter coefficients, and the outputs are modulated up to the carrier frequency $\omega_c$ prior to cancellation at passband. The phase of the modulation carriers differs for each of the interleaved cancelers. The output of the hybrid, containing the undesired echo signal, is first converted to the analytic signal with a phase splitter and partitioned into four signals, each at the data symbol rate. Then four interleaved echo cancellations are performed, and the signals are recombined into a single canceler error signal containing the far-end data signal. The real value of this signal is taken to yield a passband representation of the far-end data signal. Typically, this signal would be applied to an interpolation filter and then resampled synchronously with the far-end data, since typically the data signals in the two directions are asynchronous.

It should be noted that many other configurations are possible—for example, deriving two quadrature baseband data signals and individually echo canceling them. Many other practical considerations come into the design of a passband echo canceler, which have not been touched on here, as for example interaction with carrier recovery and compensating for the possible frequency offset of an echo that originates at the far end of a connection.

## 4.4 DESIGN CONSIDERATIONS IN ECHO CANCELLATION

The design of an echo canceler involves many considerations, such as the speed of adaptation, the effect of near- and far-end signals, the impact of signal levels and spectra, and the impact of nonlinearity. Many of these considerations are outlined in this section.

### 4.4.1 Adaptation Algorithm

The algorithm chosen for adaptation is largely a trade-off between complexity of implementation and speed of adaptation. There are a range of alternatives, from the simple stochastic gradient algorithm to several versions of the least-squares algorithms. The range of alternatives is mentioned here, although only the most commonly applied stochastic gradient algorithm is discussed in detail.

There are generally two measures of performance of an adaptive echo canceler: the speed of adaptation and the accuracy of the cancellation after adaptation. Generally there is a trade-off between these two measures: For a particular class of adaptation algorithm, as the speed of adaptation is increased, the accuracy of the transfer function after adaptation gets poorer. This trade-off is fundamental, since a longer averaging time is necessary to increase asymptotic accuracy, but it slows the rate of convergence.

Generally the motivation for adapting an echo canceler is that the transfer function of the echo is not known in advance. It is also probable that the echo transfer function is changing with time, although in most cases the change will be quite slow (say in response to temperature changes of the transmission facilities). Thus in most instances the accuracy of the final cancellation of the echo is the most critical design factor.

Although the ability of the canceler to track a changing echo response rapidly is usually not important, the speed of initial adaptation from an arbitrary initial condition is often important. For a voice echo canceler, where the canceler is not dedicated to a particular subscriber, the adaptation must occur on each new call, and during the adaptation period the voice quality is degraded. For a data echo canceler, the adaptation is part of the initialization sequence before useful data transmission can occur, and there is, therefore, motivation to speed adaptation. However, sufficient time is available for adaptation of the echo canceler that the simple stochastic gradient algorithm gives adequate performance. Nevertheless, some approaches to speeding up the adaptation are discussed in Section 4.5.

The adaptation algorithms for a baseband real-valued echo canceler [Widrow,

4.22; Niek, 4.23; Duttweiler, 4.24] followed by a complex-valued canceler are described here [Weinstein, 4.21]. More detail on adaptation algorithms can be found in [Honig, 4.25].

In order to analyze the convergence rate of an adaptation algorithm, it is necessary to assume that the environment is stationary—that is, that the echo transfer function is not changing and neither is the spectrum of the reference signal. As long as the speed of adaptation is significantly greater than the rate at which these parameters are changing, this is a valid assumption. As before, write the $m$th filter coefficient as $a_m$ and the output error as at time $i$ as $e_i$. It is also convenient to define a vector notation for the vector of $n$ filter coefficients

$$\mathbf{a}^T \equiv [a_0, a_1, \ldots, a_{n-1}] \tag{4.28}$$

where $T$ denotes transpose, and $n$ is the number of coefficients in the transversal filter. Also define a vector of the current and $n-1$ past input samples

$$\mathbf{y}_i^T \equiv [y_i, y_{i-1}, \ldots, y_{i-n+1}] \tag{4.29}$$

If the impulse response of the echo channel is $h_i$, $0 \le k < \infty$, then it is also convenient to define a vector of the first $n$ of these impulse response samples,

$$\mathbf{h}^T = [h_0, h_1, \ldots, h_{n-1}] \tag{4.30}$$

With this notation in hand, the error signal for a voice echo canceler as shown in Fig. 4.7 can be written as

$$
\begin{aligned}
e_k &= \sum_{m=0}^{\infty} h_m y_{k-m} - \sum_{m=0}^{n-1} a_m y_{k-m} + x_k \\
&= (\mathbf{h} - \mathbf{a})\mathbf{y}_k + v_k
\end{aligned}
\tag{4.31}
$$

where

$$v_k = \sum_{m=n}^{\infty} h_m y_{k-m} + x_k \tag{4.32}$$

is the residual uncancelable echo corresponding to echo delays that exceed the number of coefficients in the transversal filter plus the near-end talker signal. For the data echo canceler (4.31) is still valid, where the reference data symbols $C_{i-m}$ replace $y_{i-m}$. It should be noted that the two terms in (4.31) are correlated when the reference signal samples are correlated. In subsequent sections, several ways of minimizing this cancellation error will be defined.

### Least-Mean-Square (LMS) Solution

In this section it is assumed that the signal input $y_k$ is a wide-sense stationary discrete time random process. The unrealistic assumption will be made that the autocorrelation and power spectrum of the reference are known, as well as the echo-transfer function. The mean-square echo residual signal will be minimized for this case. The solution for this case will serve as motivation for the subsequent techniques, which are oriented toward the unknown statistics or nonstationary case.

The criterion that will be used to minimize the error for this case is the mean-square error $E(e_i^2)$, where $E$ denotes expectation. Define a vector whose components are the crosscorrelations between the $v_i$ and the current and past $n-1$ inputs $y_{i-j}$ as

$$\mathbf{p} = E[v_i \mathbf{y}_i] \tag{4.33}$$

where this quantity is assumed to be independent of $i$ due to the stationarity assumption. Similarly, define an autocorrelation matrix for the $y_i$ input process as

$$\Phi = E(\mathbf{y}_i \mathbf{y}_i^T) = \begin{bmatrix} R_0 & R_1 & \cdot & \cdot & R_{n-1} \\ R_1 & \cdot & & & \cdot \\ \cdot & & \cdot & & \cdot \\ \cdot & & & \cdot & \cdot \\ \cdot & & & & \\ R_{n-1} & \cdot & \cdot & \cdot & R_0 \end{bmatrix} \tag{4.34}$$

where

$$R_j = E[y_i y_{i+j}] \tag{4.35}$$

are the autocorrelation coefficients of the reference process. Explicitly evaluating the mean-squared error (MSE),

$$E[e_i^2] = (\mathbf{H} - \mathbf{a})^T \, \Phi(\mathbf{h} - \mathbf{a}) + 2(\mathbf{h} - \mathbf{a})^T \, \mathbf{p} + \sigma_v^2 \tag{4.36}$$

where

$$\sigma_v^2 = E[v_i^2] \tag{4.37}$$

This is a quadratic form in the tap weight vector $\mathbf{a}$, and hence there is a unique minimum, which can be obtained by completing the square or setting the gradient equal to zero. In particular, by completing the square (4.36) becomes

$$E[e_i^2] = \xi_{\min} + (\mathbf{a} - \mathbf{a}_{\text{opt}})^T \, \Phi(\mathbf{a} - \mathbf{a}_{\text{opt}}) \tag{4.38}$$

where

$$\mathbf{a}_{\text{opt}} = \mathbf{h} + \Phi^{-1}\mathbf{p} \tag{4.39}$$

$$\xi_{\min} = \sigma_v^2 - \mathbf{p}^T \Phi^{-1} \mathbf{p} \tag{4.40}$$

The $\Phi$ matrix has several important properties. First, it is symmetric and the $i, j$ element is a function of $(i - j)$. A matrix with this property is known as a **Toeplitz** matrix. Second, since the matrix is an autocorrelation matrix, it is nonnegative definite, and in fact for practical purposes can be assumed to be positive definite. In this case it has positive real-valued eigenvalues and is invertible. This inverse $\Phi^{-1}$ is also a symmetric matrix.

Since $\Phi$ is positive definite, the second term in (4.38) is nonnegative and can be minimized by the choice

$$\mathbf{a} = \mathbf{a}_{\text{opt}} \tag{4.41}$$

This choice also minimizes the MSE, which has a resultant minimum value

$$E[e_i^2] = \xi_{min} \qquad (4.42)$$

Note that in general the optimum tap weight vector is not equal to the first $n$ samples of the echo impulse response. However, if we assume that the reference samples are mutually uncorrelated, and that $x_k$ is uncorrelated with the reference samples, then we get

$$\mathbf{p} = 0 \qquad (4.43)$$

$$\mathbf{\Phi} = R_0 \mathbf{I} \qquad (4.44)$$

where $\mathbf{I}$ is the identity matrix. In this case the optimum tap weight vector is equal to the echo impulse response

$$\mathbf{a}_{opt} = \mathbf{h} \qquad (4.45)$$

and the resultant MSE is equal to the variance of the uncancelable echo.

$$\xi_{min} = \sigma_v^2 \qquad (4.46)$$

Another condition under which this is true is when the echo impulse response is truncated to $n$ coefficients or less. In practice, $n$ will be chosen sufficiently large for this latter condition to be nearly valid (approximately 128 for a voice echo canceler, 8 or 16 for a baseband data echo canceler).

### LMS Gradient Algorithm

The LMS solution of (4.39) requires in the general case the solution to a system of linear equations. As a starting point in the quest for a practical adaptive filter algorithm, an impractical algorithm called the LMS gradient algorithm is derived. This algorithm is impractical because it requires knowledge of the ensemble average of the input signals $y_i$ and $d_i$, but it is a step in the right direction because it substitutes an iterative algorithm for the matrix inversion of the LMS algorithm.

The LMS gradient algorithm is iterative; that is, the filter coefficient vector is iteratively updated and approaches the optimum vector only asymptotically. Since the coefficient vector is therefore changing, an expanded notation is required. Call the coefficient vector $\mathbf{a}_j$, where the current iteration is labeled $j$. Given the present tap vector $\mathbf{a}_j$, by subtracting a term proportional to the error gradient, $\nabla_a\{E[e^{2i}]\}$, the resultant tap vector should be closer to $\mathbf{a}_{opt}$. This is because the gradient of the error is a vector in the direction of maximum increase of the error. Moving a short distance in the opposite (negative) direction to the gradient should therefore reduce the error. On the other hand, moving too far in that direction might actually overshoot the minimum and result in instability.

The LMS gradient algorithm is explicitly

$$\mathbf{a}_{j+1} = \mathbf{a}_j - \frac{\beta}{2} \nabla_{a_j}\{E[e^{2i}]\} \qquad (4.47)$$

where $\beta$ is a small adaptation constant that controls the step size or change in $\mathbf{a}_j$

at each update. The division by 2 is included to avoid a factor of 2 in the subsequent adaptation algorithm. Referring back to (4.36) this algorithm becomes

$$\mathbf{a}_{j+1} = \mathbf{a}_j + \beta(\mathbf{p} - \Phi\mathbf{a}_j + \Phi\mathbf{h})$$
$$= (\mathbf{I} - \beta\Phi)\mathbf{a}_j + \beta(\mathbf{p} + \Phi\mathbf{h}) \tag{4.48}$$

where $\mathbf{I}$ is the identity matrix. It is to be hoped that if this algorithm is simply iterated from some arbitrary initial guess $\mathbf{a}_0$, it will converge to $\mathbf{a}_{\text{opt}}$ of (4.39).

Fortunately it is easy to investigate this convergence behavior. If $\mathbf{a}_{\text{opt}}$ from (4.39) is subtracted from both sides of (4.48) and if

$$\varepsilon_j = \mathbf{a}_j - \mathbf{h} - \Phi^{-1}\mathbf{p} \tag{4.49}$$

is defined as the error between the actual and optimal coefficient vector, then an iterative equation for the error results,

$$\varepsilon_{j+1} = (\mathbf{I} - \beta\Phi)\varepsilon_j \tag{4.50}$$

Iterating this equation,

$$\varepsilon_j = (\mathbf{I} - \beta\Phi)^j\varepsilon_0 \tag{4.51}$$

The question becomes whether this error converges to zero.

Since $\Phi$ is a symmetric matrix, it can be written in the form

$$\Phi = \mathbf{V}\Lambda\mathbf{V}^T \tag{4.52}$$

where $\Lambda$ is a diagonal matrix of eigenvalues of $\Phi$

$$\Lambda = \text{diag}[\lambda_1, \ldots, \lambda_n] \tag{4.53}$$

and $\mathbf{V}$ is an orthonormal matrix,

$$\mathbf{V}\mathbf{V}^T = \mathbf{I} \tag{4.54}$$

whose $j$th column is the eigenvector of $\Phi$ associated with the $j$th eigenvalue. Since $\Phi$ is assumed to be positive definite, the eigenvalues are positive real-valued.

The matrix in (4.51) can be written in the form

$$(\mathbf{I} - \beta\Phi)^j = (\mathbf{V}\mathbf{V}^T - \beta\mathbf{V}\Lambda\mathbf{V}^T)^j$$
$$= (\mathbf{V}(\mathbf{I} - \beta\Lambda)\mathbf{V}^T)^j \tag{4.55}$$
$$= \mathbf{V}(\mathbf{I} - \beta\Lambda)^j\mathbf{V}^T$$

Thus the vector $\varepsilon_j$ obeys a trajectory that is the sum of $n$ modes, the $i$th of which is proportional to $(1 - \beta\lambda_i)^j$. Assume for simplicity that all the eigenvalues are distinct, and order them from smallest $\lambda_{\text{min}}$ to largest $\lambda_{\text{max}}$. It follows that $\varepsilon_j$ decays exponentially to zero as long as

$$0 < \beta < \frac{2}{\lambda_{\text{max}}} \tag{4.56}$$

which is the condition for each of the factors to be less than unity. Convergence of the gradient algorithm is thus assured if $\beta$ is sufficiently small. Conversely, an excessively large value for $\beta$ leads to instability in the form of an exponentially growing error.

For a fixed $\beta$, the speed of convergence of the algorithm is determined by the slowest converging mode, which is proportional to the largest value of $|1 - \beta\lambda_i|$. This largest value is minimized, resulting in the fastest convergence, by choosing a step size of

$$\beta = \frac{2}{\lambda_{min} + \lambda_{max}} \tag{4.57}$$

in which case the slowest converging mode is proportional to

$$\left( \frac{\dfrac{\lambda_{min}}{\lambda_{max}} - 1}{\dfrac{\lambda_{min}}{\lambda_{max}} + 1} \right)^j \tag{4.58}$$

The ratio of largest to smallest eigenvalue is thus seen to be of fundamental importance; it is called the **eigenvalue spread**. The eigenvalue spread has a minimum value of unity and can be arbitrarily large. The larger the eigenvalue spread of the autocorrelation matrix, the slower the convergence of the gradient algorithm.

It is instructive to relate the eigenvalue spread to the power spectral density of the wide-sense stationary reference random process. It is a classical result of Toeplitz form theory [Grenander, 4.26] that

$$\min_{\omega} S(\omega) < \lambda_i < \max_{\omega} S(\omega) \tag{4.59}$$

where $S(\omega)$ is the power spectral density. While the eigenvalues depend on the order of the matrix, $n$, as $n \to \infty$ the maximum eigenvalue

$$\lambda_{max} \to \max_{\omega} S(\omega) \tag{4.60}$$

and the minimum eigenvalue

$$\lambda_{min} \to \min_{\omega} S(\omega) \tag{4.61}$$

It follows that the spectra which result in slow convergence of the gradient algorithm are those for which the ratio of the maximum to minimum spectrum is large, and spectra which are almost flat (have an eigenvalue spread near unity) result in fast convergence.

Intuitively, the reason for slow convergence of the gradient algorithm for some spectra is the interaction among the adaptation of the different coefficients, which becomes more pronounced as the successive reference samples become more correlated. This showing of convergence is often more of a problem for a voice canceler than a data canceler, since voice samples are highly correlated and successive data symbols are frequently uncorrelated.

Since the modes of convergence of the gradient algorithm are all the form of

$\gamma^j$ where $\gamma$ is a positive real number less than unity and $j$ is the iteration number, the error in decibels can be determined by taking the logarithm of the square.

$$10 \log_{10}(\gamma^{2j}) = \zeta j \qquad (4.62)$$

where $\zeta$ is a constant, and thus the error expressed in decibels decreases linearly with iteration number (the constant factors multiplying these exponentially decaying terms give a constant factor in decibels). The convergence of a gradient algorithm is thus often expressed in units of *decibel/iteration*, which is the number of decibels of decrease in the error power per iteration. This convergence rate can be made larger by increasing the step size $\beta$, within the limits of stability criterion (4.56). There is a limit on the maximum convergence rate imposed by the eigenvalue spread of the autocorrelation matrix.

### The LMS Stochastic Gradient Algorithm

The most widely used practical algorithm for adaptation of an echo canceler is the LMS stochastic gradient algorithm.

The **LMS stochastic gradient algorithm** derives its name from the origin of the adaptation algorithm. Specifically, it is derived starting from the LMS gradient algorithm described earlier. In that algorithm the quantity $\nabla_a\{E[e^2(i)]\}$ is usually unavailable since it requires knowledge of the ensemble statistics. The approach taken here is to in effect substitute a time average for the ensemble average.

Since the gradient and expectation in (4.47) can be interchanged, an alternative expression for the gradient is

$$\nabla_a\{E[e_i^2]\} = -2\, E[e_i \mathbf{y}_i] \qquad (4.63)$$

The troublesome part of this expression is the expectation operator. The stochastic gradient algorithm ignores the expectation operator. The quantity left, although random, is an unbiased estimate of the gradient. This "noisy" or "stochastic" gradient is substituted for the actual gradient in the algorithm of (4.47) resulting in

$$\mathbf{a}_i = \mathbf{a}_{i-1} - \frac{\beta}{2}\, \nabla_a[e_i^2] \qquad (4.64)$$

or

$$\mathbf{a}_i = \mathbf{a}_{i-1} + \beta e_i \mathbf{y}_i. \qquad (4.65)$$
$$= (\mathbf{I} = \beta \mathbf{y}_i \mathbf{y}_i^T)\, \mathbf{a}_i + \beta\, (\mathbf{y}_i \mathbf{y}_i^T \mathbf{h} + v_i \mathbf{y}_i)$$

The first form of this expression is suitable for implementation, since all the quantities are known without knowledge of the echo impulse response. The second form requires knowledge of the echo impulse response and is therefore useful only for analysis purposes.

There is another subtle but important difference between this stochastic gradient algorithm and the gradient algorithm of (4.48). In (4.48), the index $j$ corresponds to the iteration number for the iterative algorithm for inverting a matrix. In (4.65), the iteration number $i$ corresponds to the index of the current sample.

Thus each iteration corresponds to a new given data sample. The algorithm in effect performs a time average in order to determine the gradient.

Note the similarity between the second form of (4.65) and (4.48). The former substitutes the stochastic matrix $\mathbf{y}_i \mathbf{y}_i^T$ for $\Phi$ and the stochastic vector $v_i \mathbf{y}_i$ for the vector $\mathbf{p}$. In each case the deterministic matrix or vector corresponds to the ensemble average of the stochastic matrix or vector for the stationary case.

The nature of the adaptation algorithm (4.65) is illustrated in Fig. 4.8. The first form of (4.65) is assumed, as this is a simpler form for implementation. Specifically, $y_{i-j}$ is taken from the output of the $j$th unit delay and multiplied by the $j$th coefficient at time $i$, which is denoted by $[\mathbf{a}_i]_j$. The resultant value contributes to the summation, which is subtracted from the second input to obtain the canceler output error $e_i$. Using this notation for the $j$th coefficient, the adaptation algorithm of (4.65) can be rewritten for the $j$th component as

$$[\mathbf{a}_i]_j = [\mathbf{a}_{i-1}]_j + \beta e_i y_{i-j} \tag{4.66}$$

In accordance with this equation, $e_i$ is obtained by crosscorrelating (using a time average) the estimation error $e_i$ with the delayed input $y_{i-j}$. This crosscorrelation consists of taking the product of $e_i$ with $y_{i-j}$ and step size $\beta$ and accumulating the result.

Another important question is the nature of the adaptation of the LMS stochastic gradient algorithm. A difference between stochastic gradient algorithm of (4.65) and the gradient algorithm of (4.48) is that in (4.48) the coefficient vector follows a deterministic and predictable trajectory, whereas the trajectory in (4.65) is random or stochastic. The cause of this random fluctuation is the use of a time average in place of ensemble average, or, alternatively, the use of reference samples in place of ensemble averages.

A common and useful method for analyzing the convergence behavior of an echo canceler is to assume (often unrealistically) that the reference input can be modeled as a wide-sense stationary random process. The coefficient vector can be expected to converge (in some sense to be determined) to the value corresponding to the LMS solution for the known statistics. The speed of this convergence, although not directly applicable to a case where the input statistics are unknown or changing with time, is a good indication of the convergence performance of the algorithm.

If the assumption is made that the reference signal is a wide-sense stationary random process, then the expectation of the coefficient vector in (4.65) can be taken,

$$E[\mathbf{a}_i] = E[(\mathbf{I} - \beta \mathbf{y}_i \mathbf{y}_i^T) \mathbf{a}_{i-1}] + \beta E[\mathbf{y}_i \mathbf{y}_i^T \mathbf{h} + v_i \mathbf{y}_i]$$
$$\approx (\mathbf{I} - \beta \Phi) E\mathbf{a}_{i-1} + \beta(\Phi \mathbf{h} + \mathbf{p}) \tag{4.67}$$

The key approximation that has been made is that the reference signal is uncorrelated with the filter coefficient vector. This approximation, which is necessary to make tractable the stochastic analysis of the gradient algorithm, is fortunately valid for small $\beta$ because of the resultant slowly varying trajectory of $\mathbf{a}_i$. Comparing (4.67) with (4.48), this approximate average trajectory precisely obeys the earlier deterministic algorithm (4.48). Therefore, within the accuracy of this approximation the average trajectory of the stochastic gradient converges to the optimum coefficient vector under the same condition as guarantees convergence of (4.48). The nature of the average trajectory convergence is identical to that discussed earlier.

This does not imply that any particular coefficient vector trajectory itself converges to the optimum but only that the average of all trajectories converges to the optimum. In fact, the coefficient vector does not converge to the optimum. To see this, observe that even after convergence of the coefficient vector in the mean-value sense, difference equation (4.65) still has a stochastic driving term, and therefore the coefficient vector continues to fluctuate about the optimum coefficient vector randomly. The larger the value of the step size $\beta$, the larger this fluctuation.

Choosing a larger $\beta$ results in faster convergence of the average trajectory as seen earlier. Aside from the faster initial convergence, in a nonstationary environment this faster convergence also enables the algorithm to track more rapid variations in the statistics. The price that is paid is the larger fluctuation of the coefficient vector about its optimum value after convergence of the average coefficient vector.

It is appropriate to study the fluctuation of the tap weight vector about its mean after convergence. This can be done for the general case but is cumbersome. Therefore, to simplify this analysis it will be assumed here that the reference signal samples are uncorrelated (the reference signal is white). This assumption is not valid for speech signals, but the analysis nevertheless gives useful insight. This assumption is fortunately quite valid for a data echo canceler.

Calculating the norm of the difference between the tap weight vector and the optimum vector (which is $\mathbf{h}$ for the white reference signal case), from (4.65) we have

$$
\begin{aligned}
|\mathbf{a}_i - \mathbf{h}|^2 &= (\mathbf{a}_{i-1} - \mathbf{h})^T (\mathbf{I} - \beta \mathbf{y}_i \mathbf{y}_i^T)^2 (\mathbf{a}_i - \mathbf{h}) \\
&+ \beta^2 v_i^2 \mathbf{y}_i^T \mathbf{y}_i \\
&+ 2\beta v_i \mathbf{y}_i^T (\mathbf{I} - \beta \mathbf{y}_i \mathbf{y}_i^T)(\mathbf{a}_{i-1} - \mathbf{h})
\end{aligned}
\tag{4.68}
$$

The third term disappears when the expected value is taken and $E[v_i]$ is assumed zero. Invoking the white reference signal assumption and again making the assumption that the reference signal and the tap weight vector are uncorrelated,

$$
E|\mathbf{a}_i - \mathbf{h}|^2 = E[(\mathbf{a}_{i-1} - \mathbf{h})^T E(\mathbf{I} - \beta \mathbf{y}_i \mathbf{y}_i^T)^2 (\mathbf{a}_{i-1} - \mathbf{h})] + \beta^2 \sigma_v^2 n R_0 \tag{4.69}
$$

The final quantity to calculate,

$$
E(\mathbf{I} - \beta \mathbf{y}_i \mathbf{y}_i^T)^2 = \mathbf{I} - 2\beta R_0 \mathbf{I} + \beta^2 E[(\mathbf{y}_i^T \mathbf{y}_i)(\mathbf{y}_i \mathbf{y}_i^T)] \tag{4.70}
$$

depends on fourth-order statistics of the reference signal.

For the speech canceler, it is tractable and perhaps somewhat valid to assume that the reference signal is white and Gaussian. For this case, the fourth moment in (4.70) can be shown to be

$$
E[(\mathbf{y}_i^T \mathbf{y}_i)(\mathbf{y}_i \mathbf{y}_i^T)] = (n + 2) R_0^2 \mathbf{I} \tag{4.71}
$$

For a baseband data canceler, if the reference signal $y_i$ assumes the values $+1$ and $-1$ independently and with equal probability, then

$$\mathbf{y}_i^T \mathbf{y}_i = nR_0 \qquad (4.72)$$

deterministically, and thus

$$E[(\mathbf{y}_i^T \mathbf{y})(\mathbf{y}_i \mathbf{y}_i^T)] = nR_0^2 \mathbf{I} \qquad (4.73)$$

Since these answers are so close, particularly for large $n$, subsequently the version of (4.73) is used.

Substituting (4.73) back into (4.69),

$$E|\mathbf{a}_i - \mathbf{h}|^2 = (1 - 2\beta R_0 + \beta^2 nR_0^2)E|\mathbf{a}_{i-1} - \mathbf{h}|^2 + \beta^2 nR_0\sigma_v^2 \qquad (4.74)$$

This simple recursion shows that for small $\beta$ the mean-square tap weight error approaches an asymptotic value of

$$E|\mathbf{a}_i - \mathbf{h}|^2 \to \frac{\beta n}{2 - \beta nR_0} \sigma_v^2 \qquad (4.75)$$

or the rms error per tap is

$$\frac{E|\mathbf{a}_i - \mathbf{h}|^2}{n} \approx \frac{\beta \sigma_v^2}{2} \qquad (4.76)$$

This illustrates that the tap weight vector can be made more accurate asymptotically by reducing $\beta$, which also slows adaptation, or by keeping $\sigma_v^2$ small during periods of adaptation. The latter is accomplished for a voice canceler by disabling adaptation during periods of significant near-end talker energy.

The error approaches the asymptote of (4.75) as

$$(1 - 2\beta R_0 + \beta^2 nR_0^2)^i \qquad (4.77)$$

and hence only approaches this asymptote if the quantity in parenthesis is less than unity in magnitude. This in turn requires that

$$0 < \beta < \frac{2}{nR_0} \qquad (4.78)$$

It is instructive to define the time constant $\tau$ of the convergence of the mean-square tap weight vector error as the value which satisfies the equation

$$(1 - 2\beta R_0 + \beta^2 nR_0^2)^\tau = \frac{1}{e} \qquad (4.79)$$

When $\beta$ is very small, this can be solved for $\tau$ as

$$\tau \approx \frac{1}{2\beta R_0} \qquad (4.80)$$

iterations.

In comparing (4.78) with (4.56), note that for the simpler white reference signal case considered here, all the eigenvalues of $\Phi$ are equal to $R_0$. Hence, $\beta$

in (4.78) is a factor of $n$ more stringent than (4.56), which will make quite a difference for large $n$. Further, (4.77) has an optimum $\beta$ for fastest convergence given by

$$\beta_{\text{opt}} = \frac{1}{nR_0} \tag{4.81}$$

which is also a factor of $n$ smaller than the optimum step size of (4.57). The resultant convergence is

$$\left(1 - \frac{1}{n}\right)^i \tag{4.82}$$

or a time constant of

$$\tau \approx \frac{n}{2} \tag{4.83}$$

The preceding gives the MSE in the tap vector. This quantity can also be related to the mean-square echo residual. The latter is given by (4.36) for any particular tap weight vector $\mathbf{a}$. If $\mathbf{a}_i$ is substituted for $\mathbf{a}$ in this relation, assuming that $\mathbf{p} = 0$ and $\Phi = R_0\mathbf{I}$, the resulting MSE is again stochastic and must be averaged over the ensemble of tap weight vectors. Hence, the MSE becomes

$$E[e_i^2] = R_0 E|\mathbf{a}_i - \mathbf{h}|^2 + \sigma_v^2 \tag{4.84}$$

The second term is the minimum MSE that would be possible if the correct constant tap weight vector was used. The first term therefore represents an *excess MSE*, which is due to the adaptation algorithm. When the step size of (4.81) that results in maximum convergence rate is used, (4.84), becomes

$$E[e_i^2] = 2\sigma_v^2 \tag{4.85}$$

and the excess MSE is equal to the variance of the uncancelable signal. For the case of the data echo canceler, the uncancelable signal $v_i$ is in fact the far-end data signal. The choice of step size (4.81) therefore results in an excess MSE equal to the variance of the far-end data signal. This will result in an unacceptable error rate in the detection of that far-end error signal. Therefore, a step size smaller than (4.81) will be necessary, somewhat slowing the convergence from the optimum.

Several properties of this convergence should be noted. First, the $\beta$ that should be chosen to keep the excess MSE small is considerably smaller than would be predicted by (4.57), which relates only to the convergence of the mean value of the tap weight vector. Second, the fastest convergence of the MSE depends on the number of coefficients in the echo canceler: as the number of coefficients $n$ increases, the fastest convergence slows down. Third, if it should happen that $\sigma_v^2 = 0$, then there is no excess MSE, due to the fact that the error approaches zero and there is no further adaptation. Finally, the optimum step size for fastest convergence in (4.81) is inversely proportional to the reference signal power, $R_0$. Thus to choose $\beta$ properly, $R_0$ should be known, and if $R_0$ is actually varying slowly with time (as in speech cancelers), best performance requires that $\beta$ track this variation. This is considered in Section 4.4.3.

### 4.4.2 Deterministic Theory of Convergence

There are several shortcomings of the convergence analysis just performed. The assumption of wide-sense stationarity is invalid for speech, and the approximation of a coefficient vector that is uncorrelated with the reference signal is of doubtful validity as $\beta$ gets large.

There is also a fundamental question of what happens when the input signal does not cover the entire frequency band up to half the sampling rate (an extreme case would occur when the reference signal is a sinusoid). Then from (4.61) the minimum eigenvalue approaches zero and the eigenvalue spread approaches infinity. Does this imply that the average coefficient trajectory does not approach the echo impulse response? Indeed it does, since there are many coefficient vectors which will result in complete cancellation under this condition, each of these vectors having a different transfer function in the frequency band where there is no reference signal power. Fortunately, under these conditions the cancellation error still approaches zero on average, although there are practical problems due to the very large coefficient values that can result.

There is an important deterministic theory of canceler adaptation which is able to rigorously derive upper and lower bounds on the coefficient vector error for a given input reference signal waveform [Weiss, 4.27; Sondhi, 4.1]. This theory gives qualitatively similar results to the stochastic analysis previously described, but it is very comforting to be able to state rigorous conditions under which convergence is guaranteed. In particular, this theory assumes a "mixing condition" to be satisfied in order for convergence of the coefficient vector to the region of the actual echo impulse response to occur. This mixing condition is analogous to the reference signal having power over the entire frequency band up to half the sampling rate.

### 4.4.3 Modifications to the Adaptation Algorithm

There are several useful modifications that can be made to the adaptation algorithm to improve performance or simplify implementation. These modifications are described in this section.

#### Nonlinear Correlation Multipliers

An interesting modification often made to the algorithm is to replace the error signal $e_i$ by the sign of this signal, $\text{sgn}(e_i)$. The motivation for doing this is simplification of the hardware: The adaptation equation of (4.66) becomes

$$[\mathbf{a}_i]_j = [\mathbf{a}_{i-1}]_j + \beta \, \text{sgn}(e_i)y_{i-j} \tag{4.86}$$

where the only multiplication is by $\beta$ or $-\beta$. This limited multiplication is particularly simple to implement if $\beta$ is chosen to be a power of $2^{-1}$ in a digital implementation.

Intuitively, the value of the crosscorrelation would not be affected in a major way by using the sign of the error. It has been shown [Duttweiler, 4.28] that the convergence of the algorithm is not compromised by this simplification, but that the speed of adaptation is reduced somewhat. In fact, it is shown more generally that any nonlinearity in the multiplication adversely affects convergence to some extent.

In spite of the simplification in using the sign of the error, in a voice echo canceler multiplications are still inevitable in implementing the transversal filter itself. In the data echo canceler, on the other hand, the transversal filter does not require multiplications for binary valued data symbols (and as will be shown later for multilevel data as well), and the adaptation algorithm does not require multiplications even for the original algorithm of (4.66). Hence in the data canceler there is little motivation for use of the nonlinear correlation multiplier.

**Normalization of Step Size**

The stochastic gradient algorithm displays an undesirable dependence of speed of convergence on input signal power. This can be seen from (4.80), where the time constant of adaptation is inversely proportional to $R_0$. Further, if $\beta$ is kept constant and the signal power is increased, from (4.78) the algorithm eventually becomes unstable. These are particularly problems in applications such as speech where the input signal power varies considerably.

A frequent solution to this difficulty is to normalize the step size of the algorithm. From (4.81), it is desirable to normalize $\beta$ by the reference signal power, or an estimate of same. The step size could be chosen as

$$\beta_i = \frac{a}{\sigma_i^2 + b} \tag{4.87}$$

where $\beta_i$ is the step size at sample $i$, $a$ and $b$ are some appropriately chosen constants, and $\sigma_i^2$ is an estimate of the input signal power at time $i$. The purpose of the $b$ in the denominator is to prevent $\beta_i$ from getting too large (causing instability) when the input signal power gets very small.

As an example of how the estimate of input signal power can be developed, an exponentially weighted time average of the input signal power can be used.

$$\sigma_i^2 = (1 - \alpha) \sum_{j=0}^{\infty} \alpha_j \, y_{i-j}^2 \tag{4.88}$$

where $\alpha$ is an appropriately chosen constant and the $(1 - \alpha)$ factor normalizes the estimate to be an unbiased estimate of the input signal power. The reason for choosing this estimate is that it can be written recursively as

$$\sigma_i^2 = \alpha \sigma_{i-1}^2 + (1 - \alpha)y_i^2 \tag{4.89}$$

Normalization of the step size is usually employed for voice echo cancelers due to the wide variability in talker power. For a data canceler, however, the signal levels are usually better controlled and normalization is not as important.

### Adaptation of a Complex Canceler

The adaptation of a complex data canceler is similar to the real-valued case. It is readily shown [Weinstein, 4.21] that the algorithm analogous to (4.65) is

$$\mathbf{a}_i = \mathbf{a}_{i-1} + \beta e_i \mathbf{C}_i^* \, e^{-j\omega_c(i + l/R)T} \tag{4.90}$$

where $e_i$ is the analytic cancellation error and $\mathbf{C}_i^*$ is the conjugate of a vector of complex data symbols.

### Adaptation of Continuous Time Canceler

The continuous time echo canceler with limited degrees of freedom of Fig. 4.14 gives an opportunity to show how a stochastic gradient algorithm could be derived in a continuous time context. For this structure, the error signal at the output is

$$e(t) = r(t) - (a_0 x_1(t) + a_1 x_2(t)) \tag{4.91}$$

where due to the special application only two coefficients are required. In fact, the number of coefficients can be further reduced by limiting the range of adaptation. In particular, take $a_0 = \theta$ and $a_1 = (1 - \theta)$, where $0 \le \theta \le 1$. As $\theta$ is varied, in a sense the termination is adapting to a point intermediate to the two terminations $Z_1$ and $Z_2$. Taking the derivative of $e(t)$ with respect to the single coefficient $\theta$,

$$\frac{\partial}{\partial \theta} e^2(t) = 2e(t)(x_2(t) - x_1(t)) \tag{4.92}$$

The continuous time analog of incrementing the coefficient by the gradient would be to make an adjustment $d\theta$ to $\theta$ proportional to the partial derivative of (4.92) times a small increment of time $dt$. Doing this,

$$d\theta(t) = \beta e(t)[x_2(t) - x_1(t)] \, dt \tag{4.93}$$

and integrating both sides of (4.93).

$$\theta(t) = \theta(0) + \beta \int_0^t e(\tau)[x_2(\tau) - x_1(\tau)] \, d\tau \tag{4.94}$$

The implementation of resulting adaptation algorithm is illustrated in Fig. 4.26. It is analogous to the algorithm of Fig. 4.8 except that the accumulator is replaced by the continuous time integrator. As in the discrete time case, the step size $\beta$

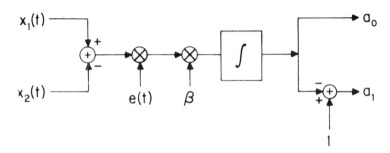

**Figure 4.26**  Adaptation circuit for continuous-time echo canceler of Fig. 4.14

cannot be chosen too large or instability results, and the choice of β affects the trade-off between speed of adaptation and asymptotic fluctuation of the coefficient θ(*t*) [Messerschmitt, 4.9].

## 4.5 RECENT ADVANCES IN ECHO CANCELLATION

The major portion of recent work in echo cancellation has focused on three areas: the extension to nonlinear echo models, speedup of the adaptation, and implementation. This section reviews this recent activity.

### 4.5.1 Nonlinear Echo Cancellation

The basic echo-cancellation model presented earlier is capable, within the constraints of a finite impulse response, of exactly canceling only an echo which is a linear function of the reference signal. Some older work in speech cancelers and some more recent work in data cancelers have extended the adaptive echo-canceler technique to nonlinear echo-generation phenomena [Thomas, 4.29; Agazzi, 4.17]. This extension is not critical to speech echo cancelers because the actual echo-generation mechanisms, although not necessarily precisely linear, are close enough that a linear canceler is able to meet the cancellation objectives. In data transmission, on the other hand, the objectives for degree of cancellation are sufficiently ambitious that nonlinear echo generation phenomena are of importance.

In speech echo cancellation, the sources of nonlinearity would be primarily the data converters in an analog-to-digital-to-analog conversion which may reside in the echo path or at the interface to the canceler, saturation of transformer magnetic media, and amplifiers in analog transmission systems. None of these sources of nonlinearity are of serious concern in voice echo cancellation.

In data echo cancellation, the primary sources of nonlinearity are the data converters in the implementation of the canceler itself as well as transmitted pulse asymmetry and saturation of transformer magnetic media. Although these mechanisms result in a very small degree of nonlinearity, they are nevertheless of importance when the objective is 50 to 60 dB of cancellation.

The method of extending echo cancellation to nonlinear echo mechanisms that has been proposed uses the Volterra expansion. This expansion is capable of representing any echo mechanism that is time-variant. That is, the nonlinearity can have memory, such as in the hysteresis in a magnetic medium, but the nature of the nonlinearity cannot change with time. Of course, since the canceler is adaptive, the nonlinearity can change slowly in time, as long as the adaptation can keep up with it.

In the speech canceler, the Volterra expansion requires an infinite number of "coefficients," or adaptation parameters, in the general case [Thomas, 4.29]. For a data canceler, however, the number of coefficients is finite because of the finite number of possible transmitted signals [Agazzi, 4.17]. This assumes as in the linear case that the echo can be represented as a function of a finite number of past transmitted data symbols. In both speech and data cancelers, practical constraints dictate that the Volterra expansion be truncated.

### 4.5.2 Speedup of Adaptation

In Section 4.4.1 we describe a basic trade-off between speed of adaptation and asymptotic excess mean-square error. The nature of this trade-off is that in order to achieve great asymptotic accuracy, speed of adaptation must be sacrificed. The presence of an uncancelable signal, such as the near-end talker for a voice canceler or the far-end data signal for a data canceler, also has an adverse impact on the speed of adaptation, since the adaptation must be slowed to maintain sufficient asymptotic accuracy in the presence of such a signal. This is evident from (4.86), where the rms tap error is directly proportional to $\sigma_v^2$.

The importance of speed of adaptation also depends on the application. For voice echo cancellation, the canceler must generally readapt at the beginning of each telephone call. Since the degree of cancellation is likely to be inadequate during this period of adaptation, there is motivation to improve the adapation speed. Fortunately, for this case a relatively high speed of adapation is achievable because of the ability to suspend adaptation during a significant near-end talker signal.

For voiceband data cancellation, adaptation occurs during the training period at the initialization of the connection. In addition to adapting the canceler, this time is also used to adapt the equalizer, acquire timing and carrier, and so on. During the training period there is no actual data transmission occurring, and therefore it is reasonable to allocate sufficient time for convergence of the echo canceler. Thus adaptation speed is of concern but not of critical concern.

Finally, on the digital subscriber loop, the echo canceler is dedicated to one subscriber loop. The echo-transfer function only changes by small perturbations between calls, those perturbations due mainly to temperature changes. If the coefficients of the canceler can be held between calls, which is practical for a digitally implemented adaptation algorithm, the only adaptation required is to adjust to small perturbations. The only major adaptation is on first installation or after a power failure or rearrangement of facilities. Speed of adaptation is therefore of little concern for this application.

In summary, for most applications, the stochastic gradient method of adaptation described earlier is adequate. For those applications that demand faster adaptation, the following subsections describe several techniques which can be used.

#### Gear-Shift Algorithms

In almost all echo-cancellation applications, high-speed of convergence is desirable for initial acquisition but is not necessary for tracking following acquisition, since the echo-transfer function changes very slowly. Thus a large constant of adaptation $\beta$ is desirable during acquisition but is not necessary following reasonable convergence. A natural algorithm to consider is to use a relatively large $\beta$ initially and then "gear-shift" to a smaller $\beta$ after convergence is achieved.

This algorithm is straightforward and effective as long as the need for a readaptation is known. On a voice echo canceler, it is difficult to detect a new call, so that this algorithm would require making the signaling information available to the canceler. For a data canceler, the need for readaptation is generally known since it is at the initialization of the call controlled by the modem, so this algorithm becomes more practical.

**Lattice Filter Echo Cancelers**

Under some circumstances the lattice filter can be effective in speeding up convergence of an echo canceler. As detailed in Section 4.4.1, the transversal filter algorithm convergence suffers when the reference samples are highly correlated. In effect, the lattice filter does an adaptive prefiltering of the reference signal prior to the adaptation to the echo transfer function. This prefilter serves the purpose of whitening the reference signal, thereby speeding the convergence of the canceler.

The extension of the stochastic gradient algorithm to the lattice filter is straightforward. The simplest, but not necessarily best performing approach is described next (see [Makhoul, 4.30; Satorius, 4.31; Honig, 4.25] for more details). Referring to (4.9), if the derivative of $e_i^2$ with respect to the tap weight $b_m$ is taken, a stochastic gradient algorithm becomes

$$[\mathbf{b}_i]_m = [\mathbf{b}_{i-1}]_m + \beta e_i e_b(i|m-1) \qquad (4.95)$$

which is identical to the transversal filter algorithm of (4.67) except for the substitution of $e_b(i|m-1)$ for the reference signal $y_{i-m}$. The adaptation of the reflection coefficients can proceed by taking the derivative of $e_f^2(i|m)$ with respect to $k_m$, yielding the stochastic gradient algorithm

$$k_m(i) = (1 - \beta e_b^2(i|m-1))k_m(i-1) + \beta e_f(i|m-1)e_b(i|m) \qquad (4.96)$$

Note that the adaptation of each lattice stage is based on the input signals to that stage. Therefore, the algorithm converges sequentially by stage: the adaptation of the first stage is necessary before the adaptation of the second stage, and so on. Since each stage requires only a scalar adaptation, the speed of adaptation is essentially independent of eigenvalue spread. The adaptation serves to minimize the forward prediction error, which it turns out is equivalent to making the successive orders of backward prediction error uncorrelated. The latter, in turn, speeds the adaptation of the $b_m$ coefficients, since it eliminates the interaction among those coefficient adaptations. Experience has shown that for input signals with a large eigenvalue spread, the overall adaptation of the lattice filter is considerably faster than the adaptation of the transversal filter.

For the data canceler, the data symbols are usually approximately uncorrelated. This is particularly true in the presence of scrambling, which "breaks up" probable highly correlated patterns, such as marking sequences. While the primary purpose of scrambling is to aid timing recovery, it is also of benefit to canceler adaptation. Thus for data cancelers, there is little benefit to using a lattice filter.

For speech cancelers, on the other hand, the signal samples are usually highly correlated. Unfortunately, however, the exact nature of the correlation is changing fairly rapidly with time. The reflection coefficients of the lattice filter therefore have to readapt continually to the spectrum of the reference speech signal. As the reflection coefficients adapt, the $b_m$ coefficients also have to readapt to maintain the same overall transfer function. This illustrates a disadvantage of the lattice filter for echo-cancellation applications: although the transversal filter coefficients are largely independent of the reference signal statistics and depend mostly on the impulse response of the echo channel, the lattice filter coefficients depend strongly on the reference signal statistics and must continually readapt as those statistics change.

### Least-Squares Algorithms

The preceding sections emphasized the widely used stochastic gradient algorithm for adaptation. Another class of adapation algorithms called the **least-squares (LS) algorithms** [Honig, 4.25] is closely related to the Kalman filter in control theory. It is based on the minimization of the LS cancellation error over the choice of adaptive filter parameters, with a weighting function decreasing exponentially into the past to give the algorithm finite memory. This class of algorithms can be illustrated by a simple example. Define a squared error at time $i$ as

$$\varepsilon_i = \sum_{j=0}^{i} \alpha^j \, (\mathbf{r}_{i-j} - \mathbf{a}_i^T \mathbf{y}_{i-j})^2 \tag{4.97}$$

The cancellation error is squared and summed with a weighting function, which decreases exponentially into the past. Note that the tap weight vector at time $i$, $\mathbf{a}_i$, is used in forming the error for the past. Defining

$$\mathbf{p}_i = \sum_{j=0}^{i} \alpha^j r_{i-j} \mathbf{y}_{i-j} \tag{4.98}$$

and

$$\Phi_i = \sum_{j=0}^{i} \alpha^j \mathbf{y}_{i-j} \mathbf{y}_{i-j}^T \tag{4.99}$$

then the squared error at time $i$ can be expressed as

$$\varepsilon_j = \sum_{j=0}^{i} \alpha^j r_{i-j} - 2\mathbf{a}_i^T \mathbf{p}_i + \mathbf{a}_i^T \Phi_i \mathbf{a}_i \tag{4.100}$$

which is minimized by the choice

$$\mathbf{a}_i = \Phi_i^{-1} \mathbf{p}_i \tag{4.101}$$

Further, there are simple recursive relations for $\mathbf{p}_i$ and $\Phi_i$,

$$\mathbf{p}_i = \alpha \mathbf{p}_{i-1} + r_i \mathbf{y}_i \tag{4.102}$$

$$\Phi_i = \alpha \Phi_{i-1} + \mathbf{y}_i \mathbf{y}_i^T \tag{4.103}$$

The stochastic gradient algorithm is an approximation to an iterative approach to solving the LS problem. The LS algorithm of (4.101) on the other hand, represents an exact theoretical solution to a well-posed minimization problem. The adaptive nature of the solution is enabled by the exponentially decreasing weighting into the past in (4.97), resulting in greater weight for more recent reference samples.

The LS algorithm is more similar to the stochastic gradient algorithm than might first appear. Substituting from (4.102) into (4.101) and defining the error signal

$$e_i = r_i - \mathbf{a}_{i-1}^T \mathbf{y}_i \qquad (4.104)$$

a recursive update equation for the tap weight vector results,

$$\mathbf{a}_i = \mathbf{a}_{i-1} + \Phi_i^{-1} e_i \mathbf{y}_i \qquad (4.105)$$

Comparing this with the stochastic gradient algorithm of (4.65), the only difference is the replacement of $\beta$ with $\Phi_i^{-1}$. This replacement, which substantially increases the complexity of the algorithm, also speeds up adaptation substantially for reference signals with a large eigenvalue spread.

To see the effect on adaptation, note that

$$E(\Phi_i) = \frac{1 - \alpha^{i+1}}{1 - \alpha} \Phi \rightarrow \frac{1}{1 - \alpha} \Phi \qquad (4.106)$$

where $\Phi$ is given by (4.34). Further, because of the averaging, the variance of $\Phi_i$ will be asymptotically small. Therefore, it is reasonable to replace $\Phi_i$ by $\Phi$ in (4.101), resulting in the algorithm

$$\mathbf{a}_i = (\mathbf{I} - \beta\Phi^{-1}\mathbf{y}_i\mathbf{y}_i^T)\mathbf{a}_{i-1} + \beta\Phi^{-1}(\mathbf{y}_i\mathbf{y}_i^T + v_i\mathbf{y}_i) \qquad (4.107)$$

where $\beta = 1 - \alpha$. This modification to the gradient algorithm of (4.65) results in a mean value for the tap weight vector modified from (4.67) to

$$E[\mathbf{a}_i] = (1 - \beta)E[\mathbf{a}_{i-1}] + \beta(\mathbf{h} + \Phi^{-1}\mathbf{p}) \qquad (4.108)$$

which approaches the $\mathbf{a}_{\text{opt}}$ of (4.39) as $(1 - \beta)^i$ virtually independently of the eigenvalue spread.

There are many versions of the LS algorithms, including those based on both transversal filter [Falconer, 4.32] and lattice filter realizations [Friedlander, 4.33]. Neglecting finite precision effects, the performance of these algorithms are identical since the same quantity is minimized in each case. There are also versions of the LS algorithms that have much reduced computational requirements relative to the simple algorithm just derived [Friedlander, 4.33].

The LS algorithms, though usually considerably more complicated to implement than the stochastic gradient algorithm, can yield an improvement in the speed of adaptation. This is particularly true for the data canceler case, where the data symbols can be chosen during a training period to assist in the canceler adaptation. In this case, the reference signal algebraic properties become much more important than the stochastic properties, and it has been shown that the mean values of the filter coefficients of a canceler based on LS can converge in $n$ data symbols for an $n$-tap canceler [Mueller, 4.34]. Furthermore, it has been shown that the LS algorithm can be virtually as simple as the stochastic gradient algorithm for a reference signal which is chosen to be a pseudorandom sequence [Salz, 4.35]. This sequence is also particularly simple to generate during a training period.

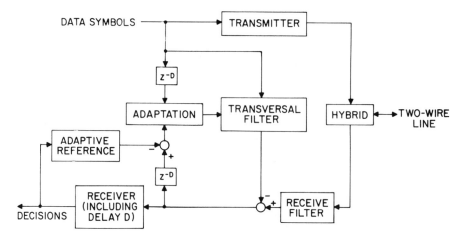

**Figure 4.27**  Baseband adaptative reference echo canceler

### Adaptive Reference Echo Canceler

In data echo cancellation, a significant factor slowing adaptation is the far-end data signal. This suggests another means of speeding adaptation, in which the data signal is adaptively removed from the cancellation error in a decision-directed fashion [Falconer, 4.36] in an approach called an **adaptive reference** canceler. This is illustrated in Fig. 4.27, where the adaptive reference is attached to the output of the receiver. It forms a linear combination of the current and past receiver decisions to form a replica of the far-end data signal appearing in the cancellation error signal. This replica is subtracted to yield a new error signal, which drives the echo-canceler adaptation algorithm. Because the far-end data signal has been removed from the error, $\sigma_v^2$ is made smaller and the adaptation constant $\beta$ can be chosen to be larger for a given excess mean-square cancellation error. This in turn speeds adaptation.

Because the receiver decision making typically includes delay, a compensating delay is inserted in the other inputs to the canceler adaptation algorithm. In effect the entire adaptation operates on a delayed basis.

Of course there is a problem getting started, when the cancellation may not be adequate to support a reasonable receiver error rate. This problem is solved by either starting out with a known training sequence (not requiring receiver decisions), or by disabling the adaptive reference and using a larger $\beta$ initially until the echo is canceled sufficiently for a reasonable error rate.

### 4.5.3 Implementation of Data Echo Cancelers

The implementation of a data echo canceler in monolithic form represents special challenges [Niek, 4.23; Agazzi, 4.18]. The major difficulty is achieving the required accuracy without incorporating expensive off-chip precision components or trimming during manufacture. The source of the problem is that the adaptation must be implemented digitally to achieve the long-time constants required for high

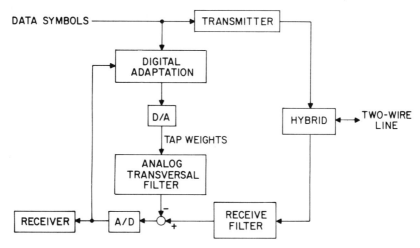

**Figure 4.28** Implementation of baseband data echo canceler insensitive to data converter nonlinearity.

asymptotic accuracy, but the interface to the transmission medium is inherently analog. This implies the need for high-speed A/D conversion somewhere between the medium and the adaptation, and this conversion must have an accuracy on the order of 12 bits with almost perfect linearity. While this accuracy can be achieved with trimming, without trimming special measures must be taken to overcome the nonlinearity of the data converters.

Two solutions have been proposed to this problem. One solution [Agazzi, 4.18] is shown in Fig. 4.28, in which the transversal filter is implemented in analog circuitry where good linearity can be insured. The adaptation circuitry is digital, ensuring that the long time constants which are necessary can be achieved. The D/A conversion occurs at the output of the adaptation circuitry, where the filter

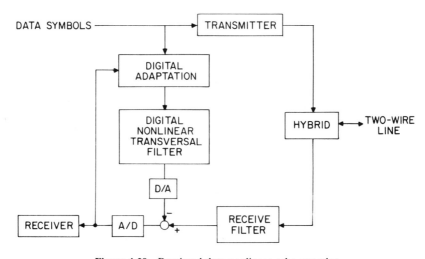

**Figure 4.29** Baseband data nonlinear echo canceler.

coefficients are converted to analog. In this case, a mild nonlinearity can be compensated by the adaptation algorithm as long as it is monotonic.

The second solution [Agazzi, 4.17] is to use the nonlinear cancellation algorithm described earlier. This approach is illustrated in Fig. 4.29, where the transversal filter is implemented digitally. The nonlinearity of the D/A converter in this case is compensated by extra nonlinear coefficients in the canceler. This method has the advantage that sources of nonlinearity external to the canceler, such as transmitted pulse asymmetry and transformer saturation, can also be compensated. Finally, the two solutions can be combined, extra nonlinear coefficients can be added to the analog transversal filter in Fig. 4.28 for the compensation of any external nonlinearities.

## 4.6 CONCLUSIONS

This chapter describes the principles, applications, algorithms, and technology of echo cancellation. Due to advances in microelectronics, which make echo cancellation more economical, broad application of these techniques can be expected. New applications of echo cancellation, such as the elimination of acoustic reverberation, can also be expected.

## 4.7 ACKNOWLEDGMENTS

The author is indebted to many colleagues who have contributed to his understanding of echo cancellation, particularly Drs. D. Duttweiler, O. Agazzi, D. Falconer, S. Weinstein, and D. Hodges. Dr. K. Feher and M. R. Aaron carefully read this chapter and provided many valuable comments. Portions of the reported work were supported by grants from the National Science Foundation, the University of California Microelectronics and Computer Research Opportunities Program, Racal-Vadic, National Semiconductor, Harris, Fairchild, and Advanced Micro Devices.

## REFERENCES

[4.1] Sondhi, M., and D. A., Berkley. "Silencing echos on the telephone network," *Proc. IEEE*, Vol. 8, August, 1980.

[4.2] Neigh, J. L. "Transmission planning for an evolving local switched digital network," *IEEE Trans. Commun.* COM-27, July, 1979, p. 1019.

[4.3] Bunker, R. et. al. "Zero loss considerations in digital class 5 offices," *IEEE Trans. Commun.* Vol. COM-27, July, 1979, p. 1013.

[4.4] Bell Laboratories Members of Technical Staff, *Transmission Systems for Commun.*, Western Electric Co., Winston-Salem, N.C., 1970.

[4.5] Suyderhoud, H. G., S. J. Campanella, and M. Onufry. "Results and analysis of a worldwide echo cancellation field trial," *COMSAT Technical Review* Vol. 5 Fall, 1975, p. 253.

[4.6] Duttweiler, D. L., and Y. S. Chen. "A single-chip VLSI echo canceler," *Bell Syst. Tech J.*, Vol. 59, No. 2, February, 1980, pp. 149–60.

[4.7] Tao, Y. G., K. D. Kolwicz, C. W. K. Gritton, and D. L. Duttweiler. "A cascadable VLSI echo canceller," *IEEE Trans. Commun.* Vol. COM-32, February 1984.

[4.8] Abate, J. et. al. "The switched digital network plan," *Bell Syst. J.* September, 1977, p. 1297.

[4.9] Messerschmitt, D. G. "An electronic hybrid with adaptive balancing for telephony," *IEEE Trans. Commun.* Vol. COM-28 August, 1980, p. 1399.

[4.10] Dotter, B. et. al. "Implementation of an adaptive balancing hybrid." *IEEE Trans. Commun.* Vol. COM-28, August, 1980, p. 1408.

[4.11] Gray, A H., and J. D., Markel. "Digital lattice and ladder filter synthesis," *IEEE Trans. Audio Electroacoustics*, AU-21, 1973 p. 491.

[4.12] Makhoul, J. "A class of all-zero lattice digital filters," *IEEE Trans. ASSP*, Vol. ASSP-26, August, 1978, p. 304.

[4.13] M. M., Sondhi and A. J. Presti. "A self-adaptive echo canceller," *Bell Syst. Tech. J.*, Vol. 45, 1966, pp. 1851–1850.

[4.14] Ahamed, S. V., P. P. Bohn, and N. L. Gottfried. "A tutorial on two-wire digital transmission in the loop plant," *IEEE Trans. Commun.* COM-29, November, 1981.

[4.15] Aschrafi, B., P. Meschkat, and K. Szechenyi. "Field trial of a comparison of time separation, echo cancellation, and four-wire digital subscriber loops," *Proceedings of the International Symp. on Subscriber Loops and Services.* September, 1982.

[4.16] Andersson, J-O, B. Carlqvist, and G. Nilsson. "A field trial with three methods for digital two-wire transmission," *Proceedings of the International Symp. on Subscriber Loops and Services*, September, 1982.

[4.17] Agazzi, O., D. G. Messerschmitt, and D. A. Hodges. "Nonlinear echo cancellation of data signals," *IEEE Trans. Commun.* Vol. COM-30, November, 1982, p. 2421.

[4.18] Agazzi, O., D. A. Hodges and D. G. Messerschmitt. "Large-scale integration of hybrid-method digital subscriber loops," *IEEE Trans. Commun.* Vol. COM-30, September, 1982, p. 2095.

[4.19] Ungerboeck, G. "Fractional tap-spacing equalizer and consequences for clock recovery in data modems," *IEEE Trans. Commun.* Vol. COM-24, August, 1976, p. 856.

[4.20] Mueller, K. H., and M. Muller. "Timing recovery in digital synchronous data receivers," *IEEE Trans. on Commun.* Vol. COM-24, May, 1976, p. 516.

[4.21] Weinstein, S. B. "A passband data-driven echo canceller for full-duplex transmission on two-wire circuits," *IEEE Trans. on Commun.* Vol. COM-25, No. 7, July, 1977.

[4.22] Widrow, B. et. al. "Stationary and nonstationary learning characteristics of the LMS adaptive filter," *Proc. IEEE*, Vol. 64, No. 8, August, 1976.

[4.23] Niek, A. M., Verhoeckx et. al. "Digital echo cancellation for baseband data transmission," *IEEE Trans. ASSP*, Vol. ASSP-27 No. 6, December, 1979.

[4.24] Duttweiler, D. L. "A twelve-channel digital echo canceler," *IEEE Trans. on Commun.* Vol. COM-26, No. 5, May, 1978.

[4.25] Honig, M., and D. G. Messerschmitt. *Adaptive Filters: Structures, Algorithms, and Applications*, Kluwer Academic Press, Hingham, Mass, 1984.

[4.26] Grenander, U., and G. Szego. *Toeplitz Forms and Their Applications*, University of California Press, 1958.

[4.27] Weiss, A., and D. Mitra. "Digital adaptive filters: conditions for convergence, rates of convergence, effects of noise and errors arising from the implementation," *IEEE Trans. on Information Theory*, Vol. IT-25, No. 6, November, 1979, pp. 637–652.

[4.28] D. L., Duttweiler. "Adaptive filter performance with nonlinearities in the correlation multiplier," *IEEE Trans. ASSP*, Vol. 30, August 1982, p. 578.

[4.29] Thomas, E. J. "An adaptive echo canceller in a nonideal environment (nonlinear or time variant)," *Bell Syst. Tech. J.*, Vol. 50, No. 8, October, 1971, pp. 2779–2795.

[4.30] Makhoul, J. "Stable and efficient lattice methods for linear prediction," *IEEE Trans. ASSP*, Vol. ASSP-25, October, 1977, p. 423.

[4.31] Satorius, E. H., and S. T. Alexander. "Channel equalization using adaptive lattice algorithms," *IEEE Trans. Commun.* Vol. COM-27, June, 1979, p. 899.

[4.32] Falconer, D. D., and L. Ljung. "Application of fast kalman estimation to adaptive equalization," *IEEE Trans. Commun.* Vol. COM-26, October, 1976, p. 1439.

[4.33] Friedlander, B. "Lattice filters for adaptive processing," *Proc. IEEE*, Vol. 70, August, 1982, p. 829.

[4.34] Mueller, M. S. "On the rapid initial convergence of least-squares equalizer adjustment algorithms," *Bell Syst. Tech. J.*, Vol. 60, No. 10, December, 1981, pp. 2345–2358.

[4.35] Salz, J. "On the start-up problem in digital echo cancelers," *Bell Syst. Tech. J.* Vol. 62, July–August, 1983, p. 1353.

[4.36] Falconer, D. D. "Adaptive reference echo cancellation," *IEEE Trans. Commun.* Vol. COM-30, September, 1982, p. 2083.

# 5

# DIGITAL SPEECH INTERPOLATION SYSTEMS

**DR. S. J. CAMPANELLA**

*Vice-President, COMSAT Laboratories*
*Clarksburg, Maryland 20734*

## 5.1 INTRODUCTION

Speech signals occurring on telecommunications links are the product of two-way conversations. It is customary for one talker to pause while the other speaks; thus an active speech signal is present on a transmission channel for only a fraction of the time. In addition, even when only one talker is speaking, pauses occur between utterances and there are times when the circuit is simply idle. Therefore, it can be expected that, on the average, speech is present for considerably less than 50% of the time. Measurements show that speech is present on a telephone channel approximately 40% of the time [Brady, 5.1], averaged over a large number of trunks. This chapter discusses digital time-assigned speech interpolation (TASI) [Daymonett, 5.2; Hashimoto, 5.3; Tygohounis, 5.4; Poretti, 5.5] without and with overload channels, variable rate delta modulation used with speech interpolation, and speech predictive encoded communications (SPEC) [Campanella, 5.6; Sevilli, 5.7; Suyderhoud, 5.8] techniques, which exploit these activity properties to reduce the information rate needed to handle a multiplicity of speech-carrying telephone channels.

Digital TASI and SPEC have been implemented using the customary 8-bit-per-sample pulse code modulation (PCM) format for 64-kb/s digital telephone transmission with either μ-law companding for T-carrier systems used in North America and Japan [5.9] or A-law companding for CEPT-32 used in Europe and most other countries outside of North America and Japan [CCITT, 5.9]. With such systems, the number of channels carried on a transmission facility can be multiplied by between 2 : 1 and 2.5 : 1 depending on the number of trunks interpolated and the mean voice-spurt activity on them. Interpolation techniques have also been extended to other methods of digital source coding such as delta modulation (DM) [Steele, 5.10], adaptive differential PCM (ADPCM) [Fukasawd, 5.11; Jayant, 5.12; Goodman, 5.13] and nearly instantaneous companding (NIC] [Dutweiller, 5.14; Dutweiller, 5.15]. At the time of this writing, such source coders are easily capable of providing telephone-quality speech reproduction at a rate of 32 kb/s and, when combined with speech interpolation, will yield a net channel multiplication of 4 : 1 to 5 : 1 compared with conventional noninterpolation PCM telephony. It is further expected that by 1990, 16-kb/s source coders [Osborne, 5.16; Cheung, 5.17; Crochiere, 5.18] will be perfected and, when combined with interpolation, these will yield a net channel-multiplication ratio of between 8 : 1 and 10 : 1.

The original application of speech interpolation was on analog telephone circuits and was intended principally for use on transoceanic telephone cable circuits. Such analog implementation, based on TASI, was subject to impairments caused by the difficulty to realize ideal, click-free analog switches and differences in the quality of the interconnecting transmission channels themselves. TASI-type systems will randomly seize any one of the available transmission channels on successive speech spurts, and variations due to gain and noise differences on the transmission links can become noticeable. Digital implementation of speech-interpolation techniques applied to digital speech transmission channels eliminates both of these impairments. When interpolation is implemented by digital means on digital transmission facilities, the method is referred to as **digital speech interpolation**. The TASI technique, when implemented by digital techniques, is frequently referred to as **digital TASI** (TASI-D). An alternative digital implementation using sample-by-sample prediction of PCM speech has been called **speech predictive encoding**.

## 5.2 TASI

TASI [Bullington, 5.19; Frazer, 5.20; Midema, 5.21] is a technique in which the idle time between calls and also that punctuating conversation during calls is used to interpolate additional calls. With a sufficiently large number of channels involved in the interpolation, most of the idle time can be filled on the transmission link so that the transmission capacity is enhanced by a factor greater than two. Each period of time occupied by a caller's speech is called a **speech spurt**, and each speech spurt may be carried on a separate transmission channel. If all connections supplied to a TASI terminal are busy, it can be expected that the speech spurt activity will average 40%. If, on the other hand, all circuits are not busy, the average speech activity experienced by the TASI terminal will be proportionately decreased. The percentage of busy circuits is called the **incoming channel activity** and the percentage of time that speech spurts occupy a channel is the **speech spurt activity**, or simply speech activity. The resultant ensemble speech spurt activity is the product of the two. Thus if the incoming channel activity is 85% and the speech spurt activity on an active channel is 40%, the resultant ensemble speech spurt activity will be 34%.

## 5.3 QUALITY ASPECTS OF TASI

### 5.3.1 Competitive Clipping and Freeze-Out

TASI exploits low speech spurt activity by assigning transmission channels only when a speech spurt is present. It is intuitively evident that efficient application of the process requires a reasonably large number of channels. When a large number of independent conversations compete for some smaller number of transmission channels, there is always some finite probability that the number of conversations demanding service will exceed the number of available transmission channels. The competing input channels are placed in a queue to await assignment. This competition manifests itself in clipping of the initial portion of a speech spurt while waiting in the queue and is called **competitive clipping**, or **freeze-out**. Provided that the population of incoming channels processed is sufficiently large and the ratio of the number of incoming channels to the number of channels available for transmission* is sufficiently small, the fraction of speech lost to freeze-out can be rendered acceptably small.

Freeze-outs consist principally of short clips of the initial portions of speech spurts, ranging in duration from zero to a few hundred milliseconds. Speech clips longer than 50 ms cause perceptible mutilation of initial plosive, fricative, and nasal consonants. To maintain high quality, the frequency of occurrence of speech clips longer than 50 ms must be kept at a very low level [Ahmed, 5.22]. In the following, the performance of TASI is analyzed in terms of the criterion that the percentage of clips longer than 50 ms is less than 2%.

### 5.3.2 Connect Clipping

At each TASI terminal the presence of speech on a telephone channel is sensed by a speech detector, which initiates a request for a transmission channel. The common control equipment assigns an idle transmission channel to the incoming channel in response to the request. It also sends a connect signal to the TASI terminal at the far end, specifying the outgoing channel to which the transmission channel is to be connected. During the time required to make the channel assignment and connect the listener and talker, the speech can be clipped. This is called a **connect clip**, as opposed to the competitive clip previously discussed. The duration of this type of clip is kept very small by providing adequate assignment-message-signaling capacity to minimize subjective degradation.

### 5.3.3 Speech Detector Clipping

In addition to competitive and connect clipping, processing time is required to accomplish speech detection, and a detector clip can also occur. However, it is possible virtually to eliminate such detector clipping by introducing a small fixed

---

*This ratio is the interpolation advantage and is frequently referred to as the *channel multiplication ratio*. The number of channels carried on the transmission facility is the product of the number of transmission channels and the channel multiplication ratio.

delay in the speech path of 5 to 10 ms to compensate for processing time in the detector. Hence, for all practical purposes impairments due to detector clipping can be eliminated.

## 5.4 DIGITAL TASI IMPLEMENTATION

The TASI form of digital speech interpolation, shown functionally in Fig. 5.1, is based on the TASI principle described in the preceding section. The incoming telephone channels are digitized into conventional 8-bit-per-sample, 8000-sample-per-second, PCM, time division multiplexed (TDM) format as part of the existing telephone plant or specifically to interface with the DSI system. The digitized signals are then processed by the transmit assignment processor, which is sequentially shared among all the incoming telephone channels.

The transmission channels consist of time slots arrayed consecutively in a TDM frame. The PCM samples selected for transmission by the action of the assignment processor are assigned to the slots that next become available when the assignment processor demands a transmission channel to service a particular incoming channel. The control channel needed to carry the channel assignment to the far end is included in the same TDM frame.

Referring to Fig. 5.1, channel time slots 1, 2, and 4 in the incoming TDM frame at the transmitter have been determined to contain active speech spurts by the transmit assignment processor, and in the example shown these have been randomly assigned to time slots 2, 1, and 3, respectively, in the transmitted frame. These are called **service channels** (SCs). Messages indicating these assignments are transmitted to the receiver in the assignment channel. The assignment of active speech spurts to SCs may be performed on a next-available, next-assigned basis or on the basis of some other rule, such as assignment to the lowest numerically ordered next-available service channel. The latter procedure will tend to group the assignments toward the beginning of the transmit frame and is referred to as *herding*. Such herding can be performed elsewhere in the frame if desired. Herding is valuable in preparing the structure of a transmission frame for a change in its duration associated with demand assignment in time division multiple access (TDMA) systems. At the receiver, the received SCs are reformatted into their appropriate outgoing TDM frame positions. Thus time slots 1, 2, and 3 appearing in the received frame are assigned time slots 2, 1, and 4, respectively. This reformatting is done by the receive assignment processor, which uses the received assignment messages to direct the reformatting of the time slot locations. In the case shown here, the incoming channels are returned to their original incoming TDM frame positions in the outgoing TDM frame. This is not necessarily always the case. In multidestinational-multiorigin satellite service, it is most likely that the positions assigned in the outgoing TDM frame will not be the same as those in the incoming because of the necessity to reformat channels received from several originating TASI terminals.

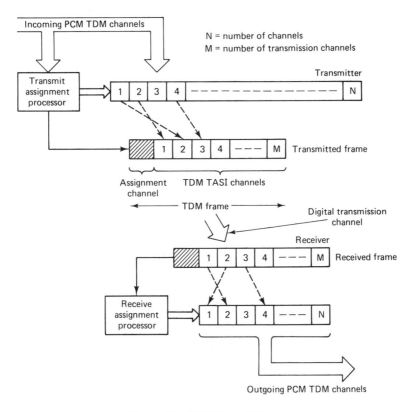

**Figure 5.1** TASI-Type DSI system.

## 5.5 DIGITAL SPEECH INTERPRETATION GAIN AND CHANNEL MULTIPLICATION

Speech interpolation refers in general to the process of time sharing a multichannel transmission facility among a number of telephone conversations. Individual segments of conversations, called *talk spurts*, are given access to a transmission channel as soon as possible after their onset. In this process, $N$ conversations are carried on $M$ transmission channels, where $M$ is less than $N$; the ratio of $N/M$ is referred to as the **channel-multiplication** (CM) ratio. For an accurate measure of the CM ratio, the value of $M$ should also contain the channel capacity for the assignment information needed by the system.

Speech interpolation implies that the talk spurts of different conversations are transmitted in sequence. Thus during a sufficiently long time interval, any transmission channel will carry talk spurts from a number of different conversations. The achievable CM ratio is a function of the activity, $a$, occurring on the lines that arrive at the speech-interpolation processor. Activity is defined as the ratio of time that speech is present to the total elapsed time. For an ensemble of $N$ lines, activity also represents the average number of lines that simultaneously carry talk spurts. With the parameters of average talk spurt length, $L$, and average pause between talk spurts, $P$, activity is expressed as

$$a = \frac{L}{L + P} \tag{5.1}$$

Measurements of talk spurt activity during typical two-way telephone conversations indicate that average activity rarely exceeds 40%. This low value is to be expected because separate channels carry the forward and return conversation paths, and usually one party listens while the other talks. In addition, short pauses by both parties occur frequently.

Talk spurt activity is the principal factor in determining the CM ratio of speech-interpolation systems. In general, as the number of busy lines, $N$, involved in the interpolation process becomes very large, the CM ratio asymptotically approaches the reciprocal of the activity (excluding the capacity needed for assignment control). Thus for $a = 0.4$, the CM ratio will approach 2.5 when $N$ becomes large. For smaller values of $N$, the CM ratio is smaller. The actual value is determined by limiting the degradation caused by competition for assignment of a talk spurt when the demand for transmission space exceeds that available. This is a classical problem in the theory of queues. The smaller the value of $N$, the greater is the variation in demand for transmission space; therefore, the CM ratio that can be achieved without exceeding some preassigned degradation limit decreases.

TASI-D has a number of advantages over analog TASI. For example, digital voice detectors perform better than their analog counterparts.* In addition, more precise and efficient switching of digital speech samples among channel slots of the TDM time frame is inherent to digital techniques. An all-digital approach is also compatible with the use of a digital channel-assignment processor for recording the channel assignments at any instant and communicating this information to a companion digital channel-assignment processor at the receiver.

An additional advantage of the digital implementation of TASI is the ability to expand the number of transmission channels to avoid freeze-out during instants of overload by reappropriating the least significant bits of the digital transmission time slots. This technique, known as **channel augmentation by bit reduction**, can be invoked during overload conditions to avoid excessive competitive speech clipping. The channels so generated are called **overload channels**. Although reducing the number of quantizing levels per slot from 256 (8 bits) to 128 (7 bits) produces a 6- to 8-dB increase in quantization noise, the fraction of time that bit reduction is required is very low. Hence, its presence is not apparent.

To summarize, relative to analog TASI, TASI-D has the following advantages deriving from the use of digital implementation:

1. A digital speech detector, which performs better than its analog counterpart.
2. Click-free digital switching.

*In a digital speech-interpolation (DSI) system, the digital voice detector is part of a central assignment processor that is time-shared among all the telephone channels. Therefore, a more sophisticated design can be incorporated. An equivalent analog design may be much more costly since it must essentially be repeated in every detail for each incoming telephone channel.

**3.** Precise and efficient assignment of PCM speech samples to TDM transmission slots using a digital assignment processor.

**4.** Generation of overload channels by means of the bit-reduction method.

Because of the superior performance of digitally implemented TASI, it is used even on analog circuit applications. This is accomplished by first converting the analog speech signals to the digital TDM/PCM form (either $T$-carrier or CEPT-32), processing them in DSI equipment, and converting the interpolated signal-output channels again to the analog format. In this way the same DSI equipment can be used for both digital and analog TASI applications. However, such implementation does not overcome the impairments to TASI operation on analog circuits caused by the variable quality of those circuits, nor does it provide means for overload channel generation.

## 5.6 ANALYSIS OF TASI-D PERFORMANCE

### 5.6.1 Statistics of Competitive Clipping

As discussed in previous sections, a principal factor governing TASI performance from the subjective point of view is the probability of occurrence of speech spurt clips of 50 ms or longer. A 2% probability of occurrence of clips equal to or longer than 50 ms is used as a threshold of acceptability in the following analysis. This means that a clip of this type will occur once every 1.5 min in a telephone conversation to one of the subscribers, assuming an average voice spurt duration of 1.5 s and an activity (average percentage of time during which the speech is present on a channel) of 40%.

At any particular instant, the probability that the number of simultaneous talkers on $N$ incoming channels with activity $a$ will equal or exceed $M$ (where $M$ is the number of transmission channels) is given by the binomial distribution

$$B_{M,N,a} = \sum_{x=M}^{N} \frac{N!}{x!(N-x)!} a^x (1-a)^{N-x} \tag{5.2}$$

If it is further assumed that the speech spurts have durations exponentially distributed with mean $L$, then the probability that a spurt is frozen out (i.e., clipped) for longer than $T$ is given by $B_{M,N,\theta}$, where

$$\theta = a\varepsilon^{-T/L} \tag{5.3}$$

This results from the assumption of an exponential distribution of speech spurt length. A fraction $\varepsilon^{-T/L}$ of the speech spurts remain unchanged after $T$ seconds have elapsed to yield an average activity of $a\varepsilon^{-T/L}$. Using this latter activity in equation (5.2) expresses the probability of encountering $M$ or more simultaneous talk spurts after $T$ seconds, which is the same as the probability of a talk spurt being clipped for $T$ seconds or more.

From these expressions, the probability of occurrence of clips with $T > 50$ ms for $L = 1.5$ s as a function of the number of transmission channels, $M$, for $N$

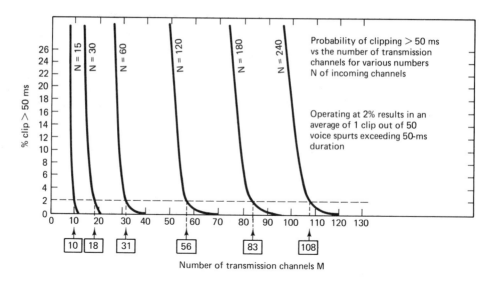

**Figure 5.2** TASI performance in terms of competitive clipping.

= 15, 30, 60, 120, 180, and 240 incoming channels has been calculated. The results are shown in Fig. 5.2. Figure 5.2 also shows the 2% boundary and the number of transmission channels needed to accommodate the various numbers of incoming channels, $N$. The values of $M$ indicated correspond to the 2% threshold. The probability of a competitive clip exceeding 50-ms duration increases abruptly for values of $M$ less than this threshold. The ratio $N/M$ is the DSI gain or channel multiplication ratio and is plotted in Fig. 5.3 as a function of $N$. It is seen that the TASI channel multiplication ratio reaches a value of 2.2 for 240 incoming channels and exceeds 2 for all cases in which the number of available transmission channels exceeds 38.

The distribution of competitive clip durations is a function of the number of incoming channels, as shown in Fig. 5.4, which indicates the distributions of competitive clip duration for $N = 60$ and 240. It is evident that longer-duration clips are more likely to occur for $N = 60$ than for $N = 240$. However, at the 2% threshold level, the clip duration exceeded is the same for both.

### 5.6.2 TASI with PCM Bit Reduction for Overload Channels

The sudden onset of speech spurt clipping can be prevented by providing access to additional channels, called **overload channels**. In a PCM-D system, overload channels can be obtained by appropriating the **least significant bit** (LSB) or bits of existing transmission channels. Thus if $M$ 8-bit PCM transmission channels are provided, then $\lfloor M/7 \rfloor$* additional 7-bit overload channels can be generated by appropriating one LSB; an additional $M/6$ channel can be generated by appropriating two LSBs, and so on. This bit-reduction technique shown in Fig. 5.5 gen-

---

* Topless bracket indicates that the value within is rounded to the next lowest integer.

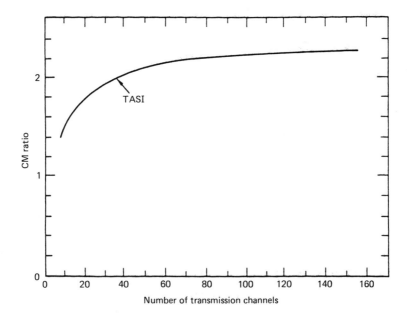

**Figure 5.3** Channel multiplication ratio for TASI DSI.

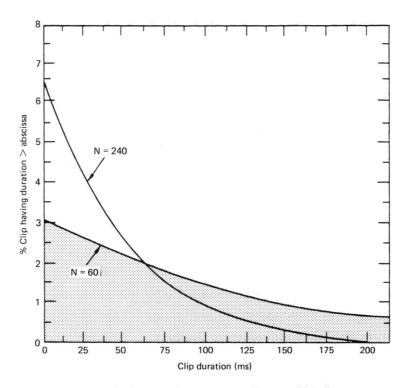

**Figure 5.4** Distributions of durations of competitive clip.

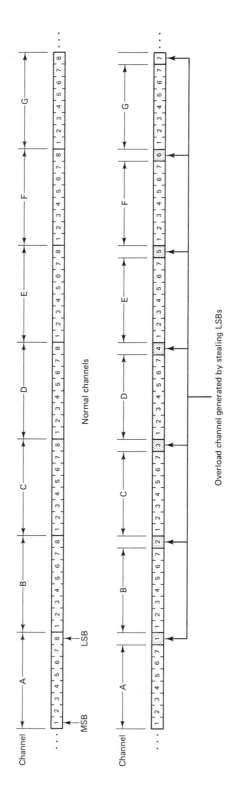

**Figure 5.5** Overload channel generation by bit reduction.

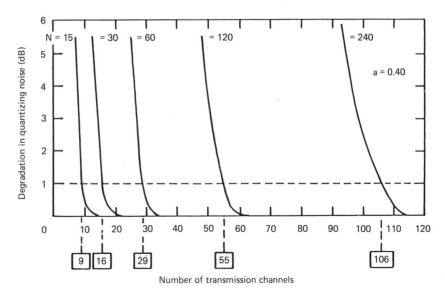

**Figure 5.6** TASI performance with overload channels generated by bit reduction.

erates overload channels,* which prevent speech spurt clipping degradation but increase quantizing noise. The resulting degradation in signal-to-quantizing noise can easily be calculated by using the probabilities of exceeding $M$, $M + M/7$, $M + M/7 + M/6$, and so on, obtained from equation (5.2) to determine the probability of using bit rate and by attaching a 6- to 8-dB degradation to the loss of each bit. This degradation occurs not only on the overload channel generated but also on the normal channels from which bits are stolen to generate the overload channel.

When the calculation is performed, curves of quantizing noise degradation versus the number of 8-bit transmission channels shown in Fig. 5.6 result for the various number of input lines processed. These curves, which reveal a sudden increase in quantizing noise below a certain number of transmission channels, appear similar to the curves for speech spurt clipping in Fig. 5.2. The difference is that quantizing noise degradation is significantly less damaging to speech quality than excessive speech spurt clipping. If a 1-dB degradation threshold of quantizing noise is accepted as tolerable, then the number of transmission channels indicated in Fig. 5.6 is needed to accommodate the various input lines. Analysis of Figs. 5.2 and 5.6 indicates that only a small reduction in the number of transmission channels causes rapidly increasing degradation; however, the soft degradation of the bit-reduction method is preferable to talk spurt clipping encountered without it. The resulting bit-reduction TASI DSI channel multiplication as a function of the number of busy input lines processed is plotted in Fig. 5.7.

In the foregoing discussion, the DSI CM ratio is increased until the degradation due to increased quantization distortion reaches 1 dB on the channels involved. This results in a slight increase in the realizable CM ratio compared to that achieved without overload channel degradation. However, another way to

---

* An overload channel is an extra channel generated during short term traffic peaks.

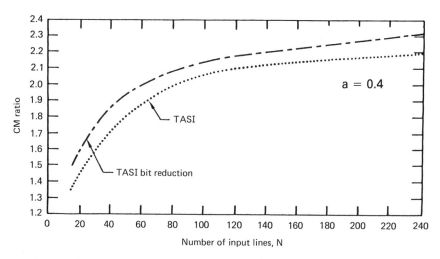

**Figure 5.7** Channel multiplication ratio for TASI DSI with overload channels.

view the use of the overload channels is simply an increased number of transmission channels. If this is done without increasing the input loading (number of input trunks) over that determined as suitable for operation without overload channel generation, then the probability of occurrence of speech clips of less than 50-ms duration will decrease significantly. For 1 bit of bit reduction this probability can be easily determined by substituting $(M + M/7)$ for $M$ in equation (5.2). The resulting probabilities of occurrence of clip duration greater than 50 ms for various values of $N$ are given in Table 5.1. The values tabulated show that bit reduction to the level of only 1 bit greatly reduces the probability of competitive clipping in TASI.

Of course the channels involved in the bit reduction, both the channels generated and the channels used to generate them, suffer an increase in quantizing distortion of 6 to 8 dB due to the truncation from 8 to 7 bits. For CM ratios based on loading such that the probability of a clip of duration greater than 50 ms is <2%, the probability of use of the first overload channel is 0.69%, of the second is 0.53%, and of the third is 0.38%. These values are determined, respectively, from the relations

**TABLE 5.1** Probability of clip duration greater than 50 ms with and without bit reduction ($\alpha = 40\%$)

| Number of incoming channels, $N$ | Probability of clip duration >50 ms (%) | |
|---|---|---|
| | With bit reduction | Without bit reduction |
| 60 | 0.13 | 2 |
| 120 | 0.07 | 2 |
| 180 | 0.01 | 2 |
| 240 | 0.004 | 2 |

Prob. of occurrence of 1st OL channel $= B_{M,N,a} - B_{M+1,N,a}$

Prob. of occurrence of 2nd OL channel $= B_{M+1,N,a} - B_{M+2,N,a}$        (5.4)

Prob. of occurrence of $n$th OL channel $= B_{M+n-1,N,a} - B_{M+n,N,a}$

These probabilities are quite low and represent an insignificant increase in perceived quality degradation to the talkers carried on the service.

Yet another way to use overload channels is as if they are normal transmission channels, so that the probability of clip duration less than 50 ms is kept at 2% when they are included. Thus in the computation of the probability of clip duration, the value $M' + M'/7$ is used in place of $M$, where $M'$ is selected such that

$$M' + \frac{M'}{7} = M \tag{5.5}$$

and the CM ratio is assumed to be $N/M'$.

Thus when a 240-channel system operating with $M = 108$ normal interpolated channels is augmented by overload channels derived by 1-bit of bit reduction, $M' = 95$ and $M'/7 = 13$. Thus the CM ratio increases from $240/108 = 2.2$ to $240/95 = 2.5$. However, this is accomplished at a significant increase in the fraction of time that the overload channels are invoked. For the $M' = 95$ case the probability that one or more overload channels are used is given by $B_{M,N,a}$ with $M = 95$, $N = 240$, and $a = 0.4$, which is 56%. For normal operation at $M = 108$ with overload channels, the probability of overload channel use is only 6%. Thus it is apparent that loading the overload channels to the extent that 50 ms or greater clips occur 2% of the time will significantly increase the degradation due to quantization distortion and should not be used except in exceptional circumstances. It is better to adopt the intermediate advantage of limiting the mean degradation to 1-dB increase in mean quantizing distortion.

## 5.7 TASI WITH DELTA MODULATION BIT REDUCTION

Another means of creating overload channels for TASI is achieved by use of speech interpolation with delta modulation. In this case the sampling rate is reduced to generate space for the overload channels. Delta modulation is particularly well suited in this application because of its inherent sampling rate flexibility. This combination of delta modulation with speech interpolation has been called **DELSI** by the author.

Figure 5.8 illustrates the operation of the DELSI technique. Here, $N$ telephone trunks are supplied as the input. The speech spurts occurring on each trunk are detected by a voice detector. When a speech spurt is present, the voice detector signals a DSI TDM frame buffer to assign the active trunk to a time slot location in the transmission TDM frame. The system is designed to accommodate a minimum of $M$ transmission channels at a transmission bit rate of $R_o$ bits per second. Thus the minimum bit rate on the transmission link is $R_o M$, and if the TDM frame has a duration $T_F$, then the number of bits per channel is $R_o T_F$ and the number

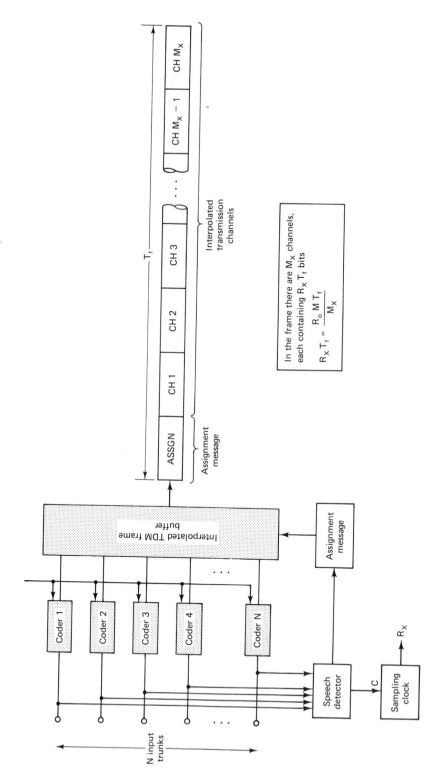

**Figure 5.8** Delta modulated speech interpolation with overload channels.

per frame is $MR_oT_F$. Whenever the number of voice spurts to be assigned is less than $M$, the arrangement described above is not modified (i.e., $R_o$ remains unchanged), and some of the time slots may remain unassigned. This is done to set an upper bound on channel quality. However, if the number of voice spurts exceeds $M$, then to produce overload channels the delta mod sampling rate can be reduced to a value $R_x$, which results in a total of $M_x = MR_o/R_x$ transmission channels, an increase of $M(R_o - 1)/R_x$ channels. The value of $R_x$ must be constrained such that the number of bits per channel in the TDM frame is an integer value and the product $M_xR_xT_M \leq MR_oT_M$. This requirement is imposed by the objective of not exceeding the number of bits in a TDM frame initially established for the delta modulation (mod) rate $R_o$. To satisfy this requirement,

$$R_xT_f = \left\lfloor \frac{MR_oT_f}{M_x} \right\rfloor \tag{5.6}$$

where the topless bracket indicates that the result is rounded to the next lowest integer.

If the value of $R_x$ is to be further constrained such that the TDM frame containing $MR_oT_F$ bits is divided precisely into $M_x$ channels, each containing an integer number of bits with no excess bits left over, then $R_x$ must be chosen such that

$$R_xT_f = \frac{R_oMT_f}{M_x} - \text{an integer} \tag{5.7}$$

This is a more-constraining condition but it greatly simplifies the equipment design.

To clarify further the operation of DELSI, consider the following example. Let the number of incoming interface trunks be $N = 48$ and the number of normal transmission channels be $M = 24$. This corresponds to a channel multiplication ratio of 2. Furthermore, assume that the maximum delta mod sampling rate and, hence, bit rate per channel is $R_o = 32$ kb/s and the TDM frame period is $T_f = 1$ ms. In this case, each normal transmission channel time slot would contain $R_oT_f = 32$ bits and each TDM frame would contain $MR_oT_f = 768$ bits for accommodating 24 transmission channels. This condition would prevail as long as the number of speech spurts to be assigned among the 48 incoming trunks is less than 24. However, due to the random nature of the voice-spurt activity, occasionally the number of voice spurts to be assigned may exceed 24. If the number is 25, then from equation (5.6) the delta mod sampling rate is constrained to be 30 kb/s. This results in the generation of 25 transmission channels each containing 30 bits or a total of 750 bits, thus leaving $768 - 750 = 18$ unassigned bits in the frame. If two overload channels are needed, the channel rate is reduced to 29 kb/s, resulting in 26 transmission channels, each containing 29 bits for a total of 754 bits per TDM frame, thus leaving $768 - 754 = 14$ unassigned bits. If the condition of equation (5.7) is used, the next closest value of $M_x$ less than $M$ that will comply is 32 (the integer value is 8 in this case), yielding a bit rate of 24 kb/s per channel. This results in the generation of 36 channels, each containing 24 bits, yielding a TDM frame of 768 bits, which leaves no unassigned bits. This latter choice is considerably

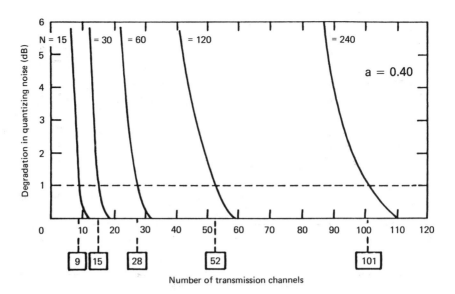

**Figure 5.9** Performance of delta modulation with speech interpolation (DELSI).

simpler to implement but does exact a greater level of degradation due to increased quantizing distortion on all transmission channels.

If a second-order delta modulator is used, the delta modulation quantizing noise power (including slope overload noise) can be assumed to increase by 50 $\log(R_0/R_x)$ [Steele, 5.10]. The result of performing these calculations for various numbers of input lines to the processor in terms of signal-to-quantizing noise degradation is shown in Fig. 5.9. With 1-dB degradation as the threshold of acceptability, the number of transmission channels needed at the sampling rate $R_0$ is indicated in Fig. 5.10. If it is assumed that the equivalent of one addition channel

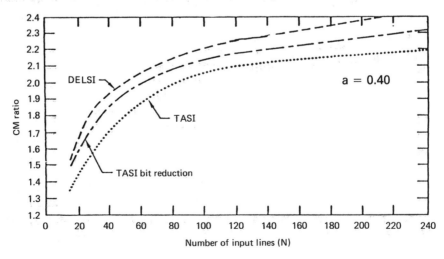

**Figure 5.10** CM ratio with DELSI (excluding multiplication of 2:1 due to reduced bit rate).

at rate $R_0$ is used for talk spurt assignment information, the DSI gain characteristic shown in Fig. 5.5 is obtained. The DSI gain is slightly better than that for PCM TASI or PCM TASI with bit reduction. Since delta modulation can operate with high quality at a bit rate of 32 kb/s, which is half the standard PCM bit rate, and a DSI gain of 2 is easily obtained, combining the two will yield an overall channel multiplication of 4 : 1.

## 5.8 DSI WITH OTHER LOW-RATE SPEECH ENCODERS

The DELSI method described previously combines delta modulation and speech interpolation to achieve high CM ratios. This can equally well be done with other low-rate speech encoders such as adaptive differential PCM ADPCM [Fukasawa, 5.11; Jayant, 5.12; Goodman, 5.13], NIC [Dutweiller, 5.14; Dutweiller, 5.15], and **subband coding** (SBC) [Osborne, 5.16; Cheung, 5.17; Crochiere, 5.18]. Such combinations will appear extensively in both terrestrial and satellite telephone communications systems in the latter half of the 1980s. The TAT-8 fiber optic cable will incorporate digital speech interpolation with 32-kb/s ADPCM to achieve a CM of greater than 4 : 1. The International Satellite Communications Organization, INTELSAT, system will probably incorporate similar equipment. It is likely that by the 1990s, 16-kb/s **low-rate encoding** (LRE) speech will be used with interpolation to yield channel multiplication as great as 8 : 1.

## 5.9 DSI ASSIGNMENT CHANNEL PROTOCOL

### 5.9.1 Interface Channels (IC) and Service Channels (SC)

Speech spurts are assigned to transmission SC locations as soon as possible after they appear at an **interface channel** (IC) to the DSI equipment. As described previously, by proper choice of the number of interpolated channels $M$ in the transmission frame for a given number of input trunks $N$, the waiting time before assignment can be made sufficiently small to result in insignificant degradation. When a speech spurt appears on an IC to the DSI equipment it is detected and assigned to an available (unoccupied) SC. Since the appearance of voice spurts is a purely random process, so will the associations between SCs and ICs. Under heavily loaded conditions, during a call each voice spurt of the subscriber involved may be assigned to a different SC each time it occurs. Literally, the SC locations of calls will vary randomly during the call. Under lightly loaded conditions, when contention for reassignment of an SC does not exist, an IC may be continuously connected to the same SC for the entire duration of a call.

### 5.9.2 IC/IC and SC/IC Associations

At the receiver, the receive-side demultiplexer must, of course, be able to keep track of the speech spurts associated with the call and reassemble them onto the same outgoing trunk. To accomplish this, it must have two pieces of information about each voice spurt, the first being the SC/IC association of each speech spurt assignment made at the transmit side and the second the IC/IC association between the transmit interface and the receive interface. These associations are illustrated in Fig. 5.11. In this figure, DSI modules residing at three terminals A, B, and C are shown. A fourth terminal D is also referenced but not shown in the figure. Each terminal is assumed to have 16 ICs and communicates in a multidestinational manner to all others. Such a multidestinational configuration is encountered in satellite communications systems, where multidestinational operation on the transmit side and multiorigin operation on the receive side are characteristically encountered. Figure 5.11 also shows how the receive side of terminal B is configured to receive DSI channels from terminals A, C, and D.

The transmit ICs at terminal A are identified in terms of groups destined to terminals B, C, and D and at terminal C in terms of groups destined to A, B, and D. Thus at terminal A, ICs 1 through 6 are destined to B, 7 through 12 to C, and 13 through 16 to D, whereas at terminal C, ICs 1 through 4 are destined to B, 5 through 10 to A, and 11 through 16 to D. The transmit sides of all other terminals, not shown, would be similarly arranged. Such arrangements of the interfaces among the terminals can be prearranged or scheduled by previous long-term agreement, in which case they may be referred to as being **preassigned**, or they can be scheduled by short-term signaling based on demand on arrival, in which case they may be referred to as **demand assigned**.

### 5.9.3 DSI Transmit and Assignment of Receive Side Assignment Channel

It is the objective here to describe how the receive-side DSI processing equipment would operate to assure that the calls arriving from corresponding originating terminals are properly routed to the receive terminal's output interfaces. Each interface, of course, has a transmit and receive port and processing in both directions would be the same.

Consider now the receive-side situation at terminal B described earlier and illustrated in Fig. 5.11. The IC/IC assignments existing at terminal B are those shown at the lower right corner of figure. Note that each IC in the system is uniquely identified in terms of its terminal identification, namely, A, B, C, . . . , and the port number on that terminal, namely, 1, 2, 3, . . . . Thus at the receive side of terminal B all signals originating from terminal A on ports 1, 2, 3, 4, 5, and 6 are routed to terminal B on ports 1, 2, 3, 4, 5, and 6; all signals originating from terminal C on ports 1, 2, 3, and 4 are routed to terminal B on ports 7, 8, 9, and 10, and all signals originating from terminal D on ports 1, 2, 3, 4, 5, and 6 are routed to terminal B on ports 11, 12, 13, 14, 15, and 16. Such IC/IC assignments exist at each terminal of a multidestinational network and are stored in each terminal as the *IC/IC assignment map*.

Consider next the IC/IC assignments. These are made in response to the detection of speech spurts on the ICs incoming to the transmit side of each DSI terminal. Assume that a speech spurt is detected on IC *i* by the speech detector. This event is identified to the channel assignment processor of the transmit side

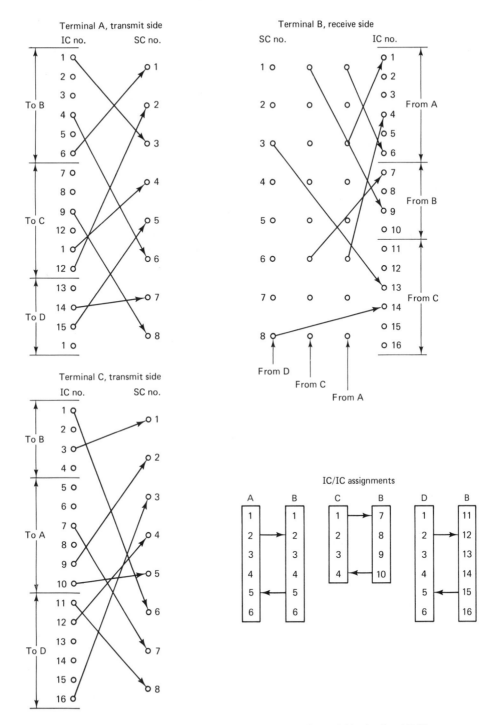

**Figure 5.11** IC/IC and SC/IC assignment maps for multidestinational DSI.

interpolation buffer and it then selects an available SC $j$. This generates an SC $j$ to IC $i$ assignment message that must be sent to the receiver to form an *SC/IC assignment map*.

### 5.9.4 DSI TDM Frame Structure

Assignment messages are sent in a TDM frame along with the SCs, which carry the speech spurts. An assignment message comprising identification of the numbers $i$ and $j$ is formatted and sent in the assignment message channel time slot of the next TDM frame. The first appearance of the actual SC $j$ to IC $i$ assignment occurs in the TDM frame following the one carrying the assignment message. The structure of a typical interpolated TDM frame used for the INTELSAT TDMA/DSI system is shown in Fig. 5.12. In this case the TDM frame period is 2 ms. Depending on design consideration, it may range from as short as 125 μs to as long as several tens of ms. Since the duration of the TDM frame causes a lag in the time between speech spurt detection and the time of actual assignment (this is in addition to the lag encountered in speech spurt detection and waiting for assignment), the design for long-duration frames may require the first appearance of the SC $i$ to IC $j$ connection to occur in the same frame as the assignment message itself. If this is done, the delay between SC, IC assignment message reception and actual connection is at most one frame period. However, the implementation for such operation is more complex and need be considered only if a long frame period (greater than 16 ms) is to be used.

## 5.10 ASSIGNMENT MESSAGE STRUCTURE

The convention described here is that intended for use in the international satellite TDMA/DSI application. It is presented as an example of current operational practice adopted for multidestinational satellite communications. Arrangements adopted for TASI-D and TASI-E for operation on digital cable would be similar.

### 5.10.1 IC and SC Numbering

Each DSI unit serves up to 240 telephone trunks. These are referred to as ICs* and are numbered from 1 to 240 using an 8-bit binary code. Thus IC 1 is designated 00000001, IC 2 as 00000010, and so on. Identification of the DSI terminal itself is determined by its location in the 2-ms TDMA frame.

Interpolated SCs are also identified by an 8-bit number. There are normal SCs and overload SCs. The normal SCs are numbered from 1 through 128, SC 1 is 00000001, 2 is 00000010, . . . , and 128 is 01111111. **Overload SCs** (OL SCs) are numbered from 255 to 240 (16 OL SCs). Thus the first OL SC is numbered 11111111, the second 11111110 and so on. As illustrated in Fig. 5.12, the first OL

---

* In the INTELSAT TDMA/DSI system specification, the IC is referred to as an international channel and the SC, as a satellite channel.

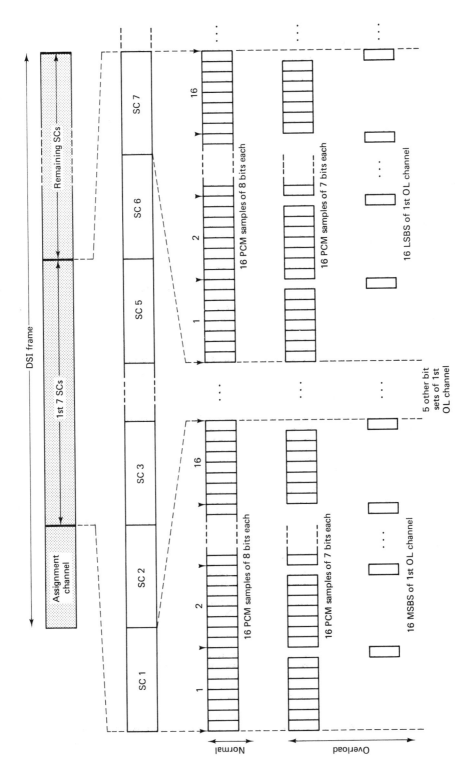

**Figure 5.12** Typical DSI TDM transmission frame.

SC is constructed from the LSBs of normal SCs 1 through 7, the second from SCs 8 through 14, and so on to the sixteenth, which is constructed from normal SCs 106 through 112.

### 5.10.2 Assignment Messages and Channels

The assignment message and channel structure discussed here assumes a 2-ms TDM frame period such as that specified for the INTELSAT system. Each assignment message contains a total of 16 bits, comprising 8 bits to designate the SC and 8 bits to designate the associated IC. Three of these SC/IC messages are carried at the beginning of each interpolated TDM frame in 128 bits allocated to the **DSI assignment channel** (DSI-AC). As can be expected, errors in the assignment messages can cause serious disruption of DSI transmission. Not only is the desired channel misdirected, but the channel to which this channel is misdirected by the mutated bits is also interferred with. For this reason a [Phiel, 5.24; Jayant, 5.12] rate one-half double-error correction, triple-error detection code is used to carry the assignment messages. The (24, 12) code is obtained by adding a dummy bit to a 23, 12 Golay code. The generator polynomial for the 23, 12 code is

$$g(x) = x^{11} + x^9 + x^7 + x^6 + x^5 + x + 1$$

The result of the Golay code is to generate blocks of 24 bits, each carrying 12 information bits. Four of these blocks are used, as shown in Fig. 5.13, to carry the three 16-bit assignment messages; an additional 32 dummy bits must be added to generate the 128 bits needed for defining the DSI-AC. The requirement for 128 bits stems from a fundamental unit of capacity of 128 bits per channel, which represents a rate of 64 kbit/s for the 2-ms frame duration.

### 5.10.3 Assignment Message Freeze-Out

In the version of DSI for international satellite TDMA use, assignment messages are transmitted by threes every 2 ms as part of the DSI subburst. Each message consists of 12 bits of assignment information which is rate one-half coded for transmission error protection and can accommodate one connect operation. Only one such message needs to be sent to accommodate each voice spurt.

The probability that the assignment messages will be frozen out can be analyzed in terms of the binomial probability expression $B_{M,N,q}$ using equation (5.2). Since there are three assignment transmission channels, $M$ is equal to three. These channels must serve the randomly occurring assignment messages arriving on the $N$ parallel incoming channels. The activity, $q$, is the ratio $aT_f/L$, where $T_f$ is the length of the TDM frame and $L$ is the average speech spurt length. For $T_f = 2$ ms, $L = 1.5$ s, and $a = 0.40$, $q = 5 \times 10^{-4}$. The freeze-out probability of an assignment message for $N = 240$ channels, where $M = 3$ and $q = 5 \times 10^{-4}$, is $B_{M,N,q} = 10^{-1}$. This is the probability of having to wait more than one TDMA frame period, i.e., 2 ms, before an assignment message is transmitted. This situation will occur once every 10 voice spurts or once every 38 s in each direction of conversation. If statistical independence is assumed, the probability of having

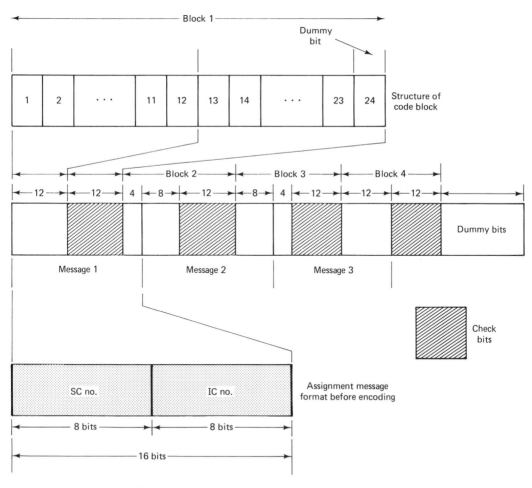

**Figure 5.13** Assignment message structure for 2-ms frame period.

to wait $s$ TDM frames for assignment is simply $B^s_{M,N,q}$. For $s = 4$, the probability of having to wait longer than 8 ms is $10^{-4}$ and the time between occurrences is 10 h. Hence, it is obvious that speech clips caused by assignment message channel congestion are of negligible consequence for the case described here.

## 5.11 TASI/TDMA IMPLEMENTATION

### 5.11.1 Transmit-Side Equipment

The transmit terminal of a typical digital TASI system for use in TDMA equipment is shown in Fig. 5.14. The terminal is configured to accommodate 240 ICs. Analog terrestrial input trunks in groups of 30* are supplied to CEPT-32 PCM A/D con-

---

*The CEPT-32 primary multiplex provides 32 eight-bit PCM channels per frame. Thirty are used for voice channels and 2 are used for signaling and supervision.

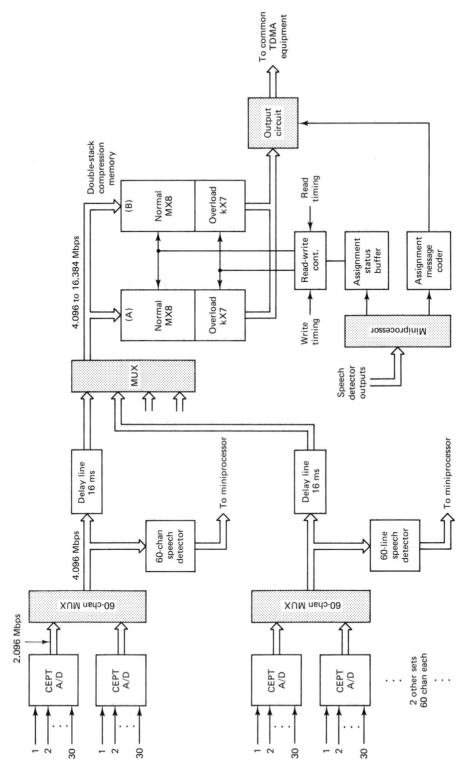

**Figure 5.14**  TASI/TDMA transmit terminal.

verters. Each of these converters delivers a PCM/TDM digital stream at a bit rate of 2.048 Mb/s. These outputs are then multiplexed by a 60-channel multiplexer into a single stream at a bit rate of 4.096 Mb/s. Modular grouping into 60 channels is used as the basic building block of the system. The system can handle additional channels in groups of 60 by adding more building blocks of this type in parallel.

The digital output of each 60-channel multiplexer is next fed to a digital delay line having a delay of approximately 5 ms to offset any delay in the speech spurt detection circuits. The same digital stream is also fed to a speech detector, which simultaneously processes 60 channels in digital form. The speech detector decisions are supplied as input to the microprocessor section of the DSI equipment. The microprocessor assigns active speech spurts to SCs as they become available and formulates assignment messages relating the originating interface connection to satellite channel connection. Under microprocessor control, each IC on which a speech spurt is detected is assigned to an available SC location in the compression buffer. The TDMA compression buffer is a double-stack design, in which one stack is filled while the other is emptied to prevent conflict due to read/write function overlapping. Each section can hold up to 128 PCM channels, which is the maximum number required for 60 to 240 interface channels. In addition, each section has an overload memory to accommodate overload channels generated by bit reduction. The bit-reduction strategy essentially increases the number of channels available in the DSI transmit frame to $(8/7)M$ rounded to the next lower integer. Thus if the system is operating with 128 8-bit time slots, the capacity is increased to 144 7-bit time slots when bit reduction is used. The microprocessor controls the onset of the bit-reduction mode and establishes the necessary signaling to inform the receiver that bit reduction is being used.

### 5.11.2 Receive-Side Equipment

The receive side of the DSI subsystem, shown in Fig. 5.15, must be capable of accepting bursts from multiple sources and appropriately distributing the information contained in these bursts to the terrestrial channels. Assignment messages from all sources are stored in the microprocessor in the form of maps, which associate SC slots in the various received DSI subbursts with outgoing terrestrial channels. The expansion buffer is a dual-stack structure. Each stack is capable of storing the contents of all DSI frames destined to the receive terminal and stores as many PCM channels as there are outgoing terrestrial channels. Under control of the assignment status buffer at the receive terminal, the samples stored in the expansion buffer are read out to the appropriate output channels.

To accomplish its function, the receive-side equipment must know the locations of all of the SCs in all of the DSI frames which carry traffic destined to it. This information is contained in the *assignment status buffer*. It gets updated each TDMA frame by the DSI-AC assignment messages described previously, which it receives from its corresponding originating terminals. This formats a complete SC/IC map, which it uses to direct the appropriate SCs to specific locations in the *expansion memories*. Then, using the information contained in its originating-IC/ receive-IC map, it directs SCs stored in the expansion memories to the proper

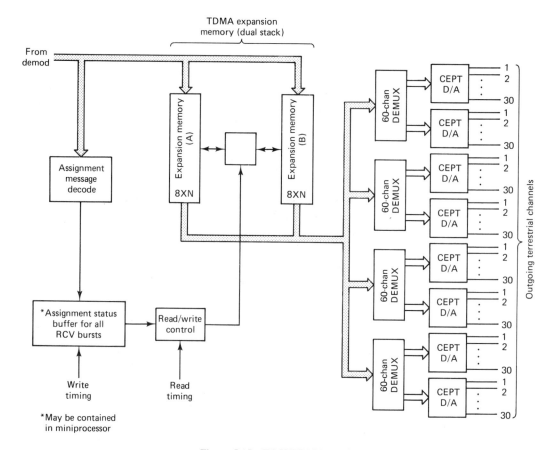

**Figure 5.15** TASI/TDMA receive terminal.

output IC ports. Outgoing channels are configured into the appropriate primary multiplex format, which, in the case shown, is CEPT-32.

## 5.12 DSI OPERATION

### 5.12.1 Operation Using Normal SCs

1. When a signal to be transmitted appears on any originating IC, a voice detector recognizes it as active and assigns it to an available normal SC.

2. An assignment message is generated and transmitted in the DSI-AC to inform the receiving end of the service channel to interface channel (SC to IC) association. The assignment message for a newly active channel is called a *new assignment message*.

3. At the receiving end, the normal service channel is connected to the appropriate output channel according to the SC to IC association defined in the new assignment message on an IC determined from the stored IC/IC map.

4. After an originating IC has been assigned to a normal SC, the connection is not released as long as the originating IC is active, unless a reassignment request to accommodate traffic rearrangements occurs. When the originating IC becomes inactive, it shall not be disconnected unless a new connection or reassignment request for the SC to which it is connected occurs.

5. During operation in nonoverload conditions, reassignments from one SC to another SC (normal or overload) may be used during traffic rearrangements.

6. When an IC on the receive side is not connected to an SC, it is recommended that a pseudorandom bit sequence to simulate random noise of $-65$ dBm0 be transmitted on the output.

### 5.12.2 Operation Using Overload SCs

1. When activity is detected on an originating IC that is not connected to an SC and there is no normal SC available, an overload SC is used.

2. A distant terminal is informed that overload SCs are invoked by the occurrence of SC numbers exceeding 239. On the receive side, the least significant bit of a normal SC need not be masked* when an overload SC is allocated.

3. An active IC connected to an overload SC is reassigned to an available normal SC when the IC assigned to the latter becomes inactive and there is no new assignment to be made. Reassignments may also be used for traffic rearrangement.

4. When no SC (normal or overload) is available, the connection shall be delayed until an SC becomes available. During this time the IC is said to be frozen out.

5. When an SC (normal or overload) becomes available and more than one IC is frozen out, the IC/SC assignment shall be made to the IC that first became active.

6. An active IC connected to an overload SC shall be eliminated by an assignment message, which reassigns it to IC 0.**

## 5.13 DSI CONNECTION PROCEDURES

Channel connections are established by allocating an SC to an active IC and by informing the distant receive DSI module of the association by the use of an assignment message. When a multidestinational transmit DSI module is corresponding with more than one receive DSI module, the SCs are shared in a common pool. Consequently, the assignment messages transmitted by one station shall be received by all corresponding stations. Originating ICs are preassigned to destination ICs. Such preassignments may be changed during traffic reconfigurations.

*Experiments have shown that even though the LSB of an 8-bit PCM channel is carrying the signal of some other channel, the resulting interference is nonintelligible. Hence masking (i.e., clamping to 0 or 1) is unnecessary.

**IC 0 is a nonexistent terrestrial channel reserved for this disconnection function.

### 5.13.1 New Assignment Procedure

The procedure for assigning a newly active IC $i$ to a normal or overload SC $j$ consists of the following message:

$$(\text{SC number } j, \text{ IC number } i)$$

A new assignment message deletes existing connections to SC number $j$ and the receive IC corresponding to the originating IC number $i$ at any corresponding destination DSI module. A new assignment shall be made to an available SC. If the assignment message is transmitted in TDM frame $n$, the first transmission of the related signal shall occur in TDM frame $n + 1$. The new assignment procedure has priority over all other assignment procedures.

### 5.13.2 Reassignment Procedure

This procedure is used to transfer active ICs from overload channels to normal channels. The procedure for reassigning IC $i$ connected to SC $k$ to SC $j$ consists of the following message:

$$(\text{SC number } j, \text{ IC number } i)$$

A reassignment message deletes existing connections to SC number $j$ and the receive IC corresponding to originating IC number $i$ at any corresponding destination DSI module. There shall be no loss of signal information in the process. A reassignment is made to an available normal SC whenever the currently connected originating IC becomes inactive and there is no new assignment. If the message is transmitted in TDM frame $n$, the reassignment shall be executed in frame $n + 1$.

The reassignment procedure may also be used to reassign an IC from any SC to any other SC, overload or normal.

The reassignment procedure has less priority than the new assignment procedure but greater priority than the disconnection or refreshment procedures.

### 5.13.3 Disconnection Procedure

This procedure is used to disconnect an inactive IC from an overload SC. The procedure disconnects SC number $j$ at any corresponding destination DSI module by the following message:

$$(\text{SC number } j, \text{ IC number } 0)$$

If the disconnection message is transmitted in TDM frame $n$, the disconnection will be effective from TDM frame $n + 1$. The disconnection procedure has less priority than the new assignment or reassignment procedure but a greater priority than the refreshment procedure.

For traffic rearrangement, the disconnection procedure may be used to disconnect normal SCs which are to be eliminated.

### 5.13.4 Refreshment Procedure

The refreshment procedure cyclically repeats the connections of all satellite channels in the DSI pool. It is used only when no other assignment functions are required. It consists of the following message:

$$(\text{Sc number } j, \text{ IC number I}) \quad \text{or} \quad (\text{SC number } j, \text{ IC number 0})$$

A refreshment message deletes existing connections to SC number $j$ and the receive IC corresponding to originating IC number $i$ and connects SC number $j$ to the receive IC corresponding to originating IC number $i$ at any corresponding destination DSI module. It is transmitted cyclically following the SC numbering sequence for normal and overload satellite channels. When such a message is transmitted in TDM frame $n$, refreshment of the connection occurs in TDM frame $n + 1$. This procedure has lowest priority.

## 5.14 SPEC

SPEC is a form of digital speech interpolation that differs significantly from digital TASI. One of its principal merits is total avoidance of the competitive clip problem experienced by TASI. Its **sample assignment word** (SAW), which is refreshed in each frame avoids record keeping for connections from interface channels to service channels. Hence, separate channel-assignment memory and assignment message channel implementation is unnecessary. Its adaptive processing method results in only a slight increase in quantizing noise when confronted with overload. CM ratios achieved are competitive with those of TASI-D for PCM telephony.

## 5.15 QUALITY ASPECTS OF SPEC

### 5.15.1 Predictor Distortion

SPEC requires the signal to pass a speech detector before samples are admitted to the predictor. The signals passed by the speech detector exhibit an activity similar to that experienced with TASI. The predictor algorithm reduces this activity by eliminating predictable PCM samples in the instantaneous speech waveform and in the short intersyllabic pauses not sensed by the voice detector. Under average load conditions the SPEC predictor removes approximately 6% of the PCM samples occurring during the speech spurts. This results in an increase in **quantization** distortion. The **ratio** $(S/D)$ decreases only 0.5 dB compared to that of conventional 8-bit-per-sample PCM. The achievable CM ratio is shown in Fig. 5.16 as a function of the number of incoming channels. For comparison, this figure also shows the performance of TASI without overload channels.

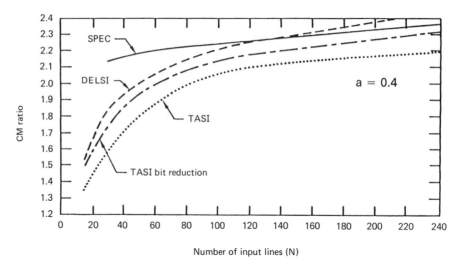

**Figure 5.16** CM ratio for SPEC compared to other methods.

### 5.15.2 Competitive Clipping

SPEC cannot cause clips such as those encountered in TASI. PCM samples would have to be frozen out for a succession of 80 SPEC frames, which occur at a rate of 8000 Hz to produce a 10-ms clip. The probability of occurrence of such an event is vanishingly small.

### 5.15.3 Speech Detector Clipping

SPEC incorporates a speech detector and, as in the case of digital TASI, the inclusion of adaptive threshold features in the design results in negligible speech detector–induced clipping.

SPEC transmits sets of unpredictable PCM samples in a TDM frame of 125-μs duration. The samples are destination directed by a SAW, which accompanies each frame. SAW structure is discussed in detail later.

### 5.15.4 Connect Clipping

There is 1 bit in the SAW for each incoming channel. This bit informs the receiver whether or not the frame contains a sample for the channel designated by the bit position in the SAW. The SPEC SAW allows fully flexible connectivity for all incoming channels. Hence, clipping caused by waiting for channel assignment cannot occur.

## 5.16 SPEC IMPLEMENTATION

In the SPEC system, shown functionally in Fig. 5.17, the speech information contained on $N$ incoming 8-bit-per-sample PCM telephone channels is transmitted in the space of $(N/8) + c$ 8-bit per-sample transmission channels; $N/8$ is the number of 8-bit channels needed for the SAW, and $c$ is the number of PCM sample channels used for transmission and is typically equal to slightly more than $N/3$. SPEC processes all incoming trunks and transmits a frame once per period of the PCM Nyquist rate, which is the 125-µs sampling interval in typical commercial digital telephone usage. Several SPEC frames can be concatenated for transmission over TDMA systems having longer frame periods.

### 5.16.1 PCM Sample Prediction

SPEC operation is now described in terms of an experimental implementation [Campanella, 5.26], for accommodating 4 incoming telephone channels (corresponding to a bit rate of 4.096 Mb/s) in the transmission space normally allotted to 32 PCM channels (corresponding to a bit rate of 2.048 Mb/s). Referring to Fig. 5.17, the PCM samples derived during each sample period from the 64 incoming channels are compared with the samples previously sent to the receiver and stored

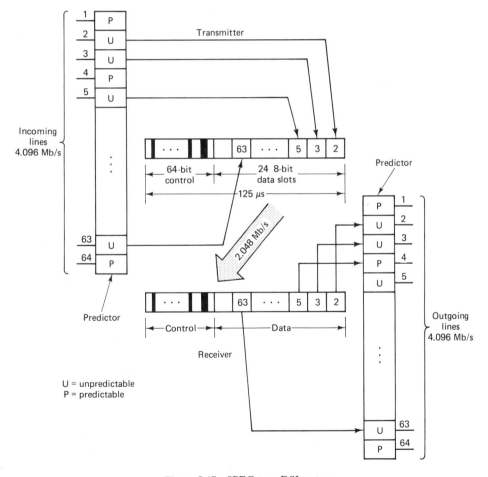

**Figure 5.17** SPEC-type DSI system.

in a memory at the transmitter. Any that differ by an amount equal to or less than a given number of quantizing steps, called the *aperture*, are discarded and not sent to the receiver. These are referred to as **predictable** (P) samples. The remaining **unpredictable** (U) samples are transmitted to the receiver and replace the values formerly stored in memories at both the transmitter and receiver. The aperture is automatically adjusted as a function of activity observed over the 64 incoming channels on each frame so that the number of samples transmitted is less than 24.

### 5.16.2 SAW

The SPEC transmission frame is composed of an initial SAW followed by a number of 8-bit time slots that carry the individual PCM samples judged unpredictable by the transmitter's prediction algorithm. The SAW contains 1 bit for each of the incoming telephone channels. Thus for the 64-terrestrial-channel system, it contains 64 bits. The bit corresponding to a given channel is a 1 if the frame contains a sample for that particular channel and a 0 if it does not. Thus the SAW contains all of the information needed to distribute the samples among the 64 outgoing channels at the receive end.

At the receiver the unpredictable samples received in the SPEC frame replace previously stored samples in the receiver's 64-channel memory as directed by the SAW. The samples in the memory are demultiplexed into the outgoing channels at the 8000-Hz rate in the form of a conventional PCM/TDM frame. The most recent frame thus contains new samples on the channels that have been updated by the most recently received SPEC frame and repetitions of the samples that have not been updated.

## 5.17 ANALYSIS OF SPEC PERFORMANCE

### 5.17.1 Activity Reduction by Prediction

The PCM samples transmitted by SPEC are determined by selecting that set in which the differences between the samples previously sent and those currently being sent are equal to or greater than some number of quantizing steps called the **aperture**, $l$. The value of the aperture is adjusted so that the number of samples transmitted does not exceed the number of available transmission channel slots.

If the incoming channel activity for a given frame is such that all new samples can be transmitted with zero aperture, that is, if the number of samples to be transmitted with differences greater than zero is equal to or less than the number of transmission slots, no error is made in the transmission. If the incoming channel activity for a given frame is such that the aperture $l = 1$, samples differing from the previously transmitted samples by a number of quantizing steps greater than one are transmitted. Hence, the number of samples requiring transmission is reduced by culling out those with differences less than or equal to 1. The activity resulting from this action is

$$a_1 = a[1 - P(\Delta = 0) - P(\Delta = 1)] \tag{5.8}$$

where $\quad a =$ activity experienced on the ensemble of incoming
terrestrial channels with no prediction
$a_1 =$ activity after culling out samples with differences
less than or equal to 1
$P(\Delta = k) =$ probability of samples differing from those
previously sampled by $k$ quantizing steps

In this case, errors of magnitude of one quantizing step will be made for a fraction of the samples equal to $P(\Delta = 0)$.

In the general case, if a set of SPEC processed samples is transmitted with an aperture of value $l$, then the resulting activity for that set is

$$a_l = a\left[1 - \sum_{k=0}^{l} P(\Delta = k)\right] = a(1 - P_l) \tag{5.9}$$

The amount given by the summation and designated as $P_l$ is called sample predictability for aperture $l$. On the average, the value of incoming channel activity, $a$, is approximately 40%.

SPEC is designed to operate at an average transmit channel activity (given by the ratio of the number of transmission channels available for samples, $c$, to the number of incoming channels, $N$, of $c/N = 0.3$. The aperture must be selected such that

$$a_l \leq \frac{c}{N} \tag{5.10}$$

and the predictability must be such that

$$P_l \geq 1 - \frac{c}{Na} \tag{5.11}$$

These rules establish the average aperture at which SPEC operates.

### 5.17.2 Distribution of Prediction

For a given incoming channel activity, $a$, the number of incoming channels, $n$, requiring transmission due to the presence of voice spurts is a random variable. The probability that the number of incoming channels requiring service at a given instant will exceed the number of transmission channels is given by the binomial distribution

$$P(n > c) = B_{c,n,a} = \sum_{x=c}^{n} \frac{n!}{x!(n-x)!} a^x (1 - a)^{n-x} \tag{5.12}$$

The system is constrained so that the number of samples needing transmission never exceeds $c$. This is accomplished by using prediction to reduce the activity in the entire set of incoming channels to a new activity, $a_l$, as described previously. The average number of channels requiring service under these circumstances is then

$$n_l = a_l N \qquad (5.13)$$

Substituting the new value of activity into the binomial distribution yields

$$P(n > c) = B_{c,n,al} \qquad (5.14)$$

where $a_l = a(1 - P_l)$, which expresses the modification in activity resulting from the introduction of the predictor. Since the average number of transmission channels used with prediction is

$$c = Na(1 - P_l) \qquad (5.15)$$

the binomial distribution can be redesignated as

$$P[n > Na(1 - P_l)] = B_{c,n,al} \qquad (5.16)$$

which can be further modified as follows:

$$P\,a(1 - P_l) < \frac{n}{N} = Pa < \frac{n}{N(1 - P_l)}$$

$$= PP_l > 1 - \frac{n}{Na}$$

$$= P\,P_l > 1 - \frac{a_l}{a}$$

$$= B_{c,n,al} \qquad (5.17)$$

Equation (5.17), which gives the probability that the prediction will exceed the value $(1 - a_l)/a$, is very useful in assessing the level of degradation introduced by the predictor.

Figure 5.18 shows the performance of SPEC in terms of the predictability $P_l$ = 0.25, calculated from equation (5.16) for $a = 0.40$ and $N = 30, 60, 120, 180,$ and 240 incoming channels plotted against the number of transmission channels, $c$. The ordinate is the probability that the indicated value of $P_l$ is exceeded for the number of transmission channels on the abscissa, which includes those channels needed to accomodate the SAW. Each transmission channel is assumed to be 8 bits wide. A predictability of $P_l = 0.25$ means that, on the average, one sample out of four is predicted during a voice spurt. In the case of $N = 120$, it can be seen that for $c = 54$ transmission channels, the probability that the predictability will exceed $P_l = 0.25$ is 25%. This can also be interpreted to mean that for 75% of the time the predictability will be such that less than one in four samples will be predicted during a voice spurt when 120 incoming channels are carried in the space of 54 transmission channels. In this case, the SPEC DSI channel multiplication is $120/54 = 2.22$.

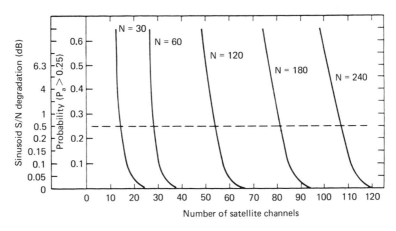

**Figure 5.18** SPEC predictor performance.

### 5.17.3 Signal-to-Distortion Ratio

It is possible to relate the probability that $P_l > 0.25$ to degradation in the signal-to-distortion ratio $S/D$ over SPEC channels. This is done by using experimental data regarding the $S/D$ value observed for a 0-dBm0 sinusoidal signal passed through one channel of an experimental SPEC terminal built to accommodate 64 incoming channels in 32 PCM transmission channels plus 64 SAW bits. The experimental data given in Fig. 5.19 show the degradation of the sinusoidal $S/D$ as the average activity on the 64 incoming channels varies over a wide range of values. The overall degradation is due to a combination of PCM quantization noise and prediction predistortion. As the incoming channel activity increases, the predictor must handle a greater number of samples by increasing its aperture, thus causing the total distortion to increase. Figure 5.19 also shows the signal-to-Gaussian-noise ratio ($S/N$) for a 0-dBm0 sinusoidal signal on a Gaussian-noise-perturbed

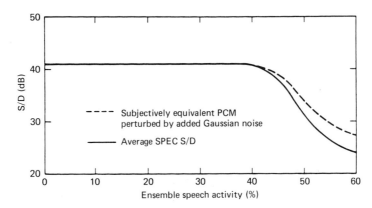

**Figure 5.19** Measured and subjectively equivalent $S/D$ as functions of ensemble speech activity.

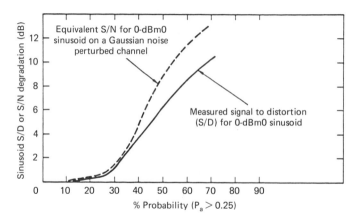

**Figure 5.20** Sinusoid $S/D$ and Equivalent $S/N$ versus $P(P_l > 0.25)$.

PCM channel with speech quality equivalent to that experienced with prediction distortion. This curve indicates that a given level of prediction distortion corresponds to a lower level of Gaussian noise.

The data presented in Fig. 5.19 can be directly related to predictability by using equation (5.17). The results are found in Fig. 5.20, which shows the dB degradation in sinusoidal $S/D$ incurred by prediction as a function of the probability that $P_l > 0.25$.

Also shown in Fig. 5.20 is a curve giving the equivalent performance of the PCM channel perturbed by additive Gaussian noise. This curve indicates that, for conditions such that $P(P_l > 0.25) = 25\%$, the decibel degradation in the sinusoidal $S/D$ is only 0.75 dB. The decibel degradation in the sinusoidal $S/N$ on a subjectively equivalent channel perturbed by Gaussian noise is only 0.5 dB. This indicates that $P(P_l > 0.25) = 25\%$ is an acceptable threshold for determining the advantage offered by the SPEC system. The ordinate of Fig. 5.18 also shows a scale giving the decibel degradation in the subjectively equivalent $S/N$ obtained by using the data provided in Fig. 5.20.

The number of transmission channels needed to accommodate each number of incoming channels is determined from the intersection of each curve shown in Fig. 5.20 with the 0.5-dB degradation boundary. The plot of SPEC CM versus the number of transmission channels (including SAW bits) shown in Fig. 5.16 has been obtained from this result.

## 5.18 SPEC TDMA IMPLEMENTATION

### 5.18.1 Transmit Terminal

Figure 5.21 is a block diagram of a SPEC terminal configured for 240 incoming terrestrial channels handled in modules of 60 channels each. Each group of 60 channels is processed through a pair of CEPT-32 A/D converters to convert them

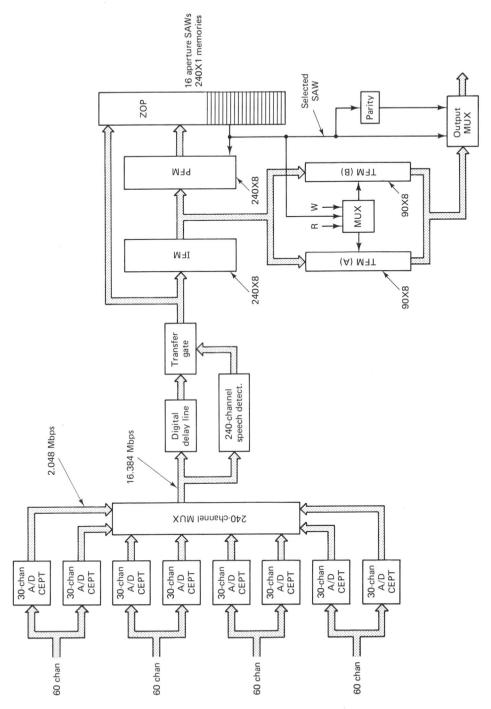

**Figure 5.21** SPEC transmit terminal.

to 2.048-Mb/s primary multiplex PCM digital form. The outputs of all CEPT units are multiplexed in a single multiplexer to a bit rate of 16.384 Mb/s. The bit stream is then supplied to two devices: a 5-ms digital delay line and the 240-channel speech detector. The speech detector detects the presence of voice spurts on each of the 240 incoming terrestrial channels. Whenever a voice spurt is present, it permits the PCM samples to pass to the **intermediate frame memory** (IFM) and the **zero-order predictor** (ZOP).

The IFM is capable of storing 240 8-bit PCM samples. Depending on decisions made by the ZOP, the samples stored in the IFM will be transferred to the **predictor frame memory** (PFM). Specifically, the ZOP calculates the difference between values stored in the PFM and the most recent set of values that has been supplied to the IFM. For those PCM values whose difference is greater than the aperture, the values in the IFM are transferred to the PFM to replace the old values and are stored in the **transmit frame memory** (TFM) for transmission to the appropriate destinations.

Within the predictor there are a number of 240-bit-long storage units in which SAWs are stored for several aperture values. The SAW used in a given frame to control the transfer of values from the IFM to PFM and TFM is the one corresponding to an aperture such that the number of samples transmitted is just less than the number of sample slots available in the TFM. For a 240-channel system, the number of sample slots available in the TFM is 90. With low activity on the incoming channels, it will be found that all samples transmitted are the result of applying a small aperture value, which yields a low quantization noise. If, on the other hand, the activity on the incoming channels is high, the aperture needed to reduce the population to a number below 90 will be larger and the quantization noise will be correspondingly greater. Thus this implementation of the SPEC system permits the aperture occurring from each SPEC frame to vary over a large range to accommodate wide fluctuations in activity.

All values that are updated in the PFM each time the ZOP executes its function are called unpredictable values and are transferred to the TFM, where they await transmission in the next frame. The TFM is a dual-stack buffer configuration consisting of two 90 by 8 sections, each able to store an entire TFM. While one stack is being filled, the other stack is being emptied, thus permitting a continuous flow of signal information to the outgoing communications link. An output multiplexer combines the samples of the TFM with the appropriate SAW to constitute the transmission frame. The frame duration upon which the SPEC transmitter operates is 125 $\mu$s; it is synchronized to the Nyquist sampling frame used in the A/D PCM converters at the input.

In a TDMA application, frame periods of several milliseconds may be used. For the SPEC system implementation to be used with a 2-ms TDMA frame, the 125-$\mu$s SPEC frames must be accumulated in a TDMA burst-compression buffer. This compression buffer would consist of a dual-stack configuration, which stores 16 SPEC frames and outputs them at the burst rate of the TDMA terminal equipment.

### 5.18.2 Receive Terminal

A SPEC receive terminal must select from each received SPEC frame only those samples destined to its outgoing interface channels by examining each source's SAW and selecting and storing only the corresponding samples contained in the PCM sample part of each SPEC DSI frame. A possible implementation is shown in Fig. 5.22. Each SPEC frame obtained from the TDMA demodulator is supplied to a demultiplexer that separates the SAW and PCM sample parts. The SAW is supplied to a **parity check** and a **saw subset select** unit. If the parity check is successful, that subset of SAW bits [designated as (SAW)] identifying the samples destined to the particular terminal of interest is passed on to the SAW memory. The SAW subset selector then admits to the TFM only the appropriate subset of PCM samples. For any SPEC frame on which the SAW parity check fails, none of the received samples are stored in the TFM, and all samples for that SPEC frame are treated as predictable. The SAW and corresponding TFM samples are passed into the PFM, which reconstitutes the conventional PCM/TDM format with all predicted samples appropriately filled in. The conventional PCM/TDM stream is then supplied to the D/A sections of CEPT-32 PCM units and hence to the outgoing trunk.

## 5.19 COMPARISON OF TASI AND SPEC IMPLEMENTATIONS

The TASI and SPEC DSI implementation schemes described have certain common features. In particular, each implementation requires the same kinds of A/D and D/A conversion interfaces. Each also requires speech detectors and digital delay lines to implement these detectors appropriately. There are essentially no differences in the voice-detector performance requirements of the two systems.

There are differences in the manner in which the samples are carried on the transmission facility. In digital TASI, incoming channels are connected to and disconnected from service channels, and the service channels are connected to and disconnected from destination interface channels on the basis of voice spurt demand. PCM samples on active channels are carried in a TDM form and a separate assignment message channel is used to signal the appropriate connects and disconnects at various destinations.

By comparison, in SPEC the individual PCM samples occurring at the Nyquist rate on each incoming channel are compared in a ZOP, and those determined to be unpredictable are sent to appropriate destinations on a sample-by-sample demand basis. The unpredictable samples are carried in TDM form. Unlike the TASI system, however, the SPEC system does not use a separate assignment message channel for connect and disconnect information but instead incorporates a set of bits called a SAW, which is included as part of each SPEC frame, to direct individual samples to outgoing channels at their destination. TASI hardware also

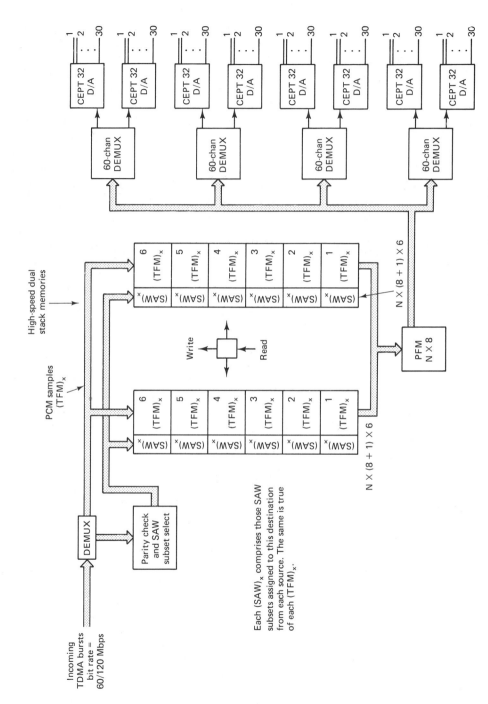

**Figure 5.22** SPEC receive terminal for TDMA operation.

$N$ = number of outgoing terrestrial channels

Each $(SAW)_x$ comprises those SAW subsets assigned to this destination from each source. The same is true of each $(TFM)_x$.

**TABLE 5.2**  Comparison of SPEC and TDMA techniques

| Feature | SPEC | TASI |
|---|---|---|
| DSI advantage for 240 incoming channels* | 2.2 | 2.2 |
| DSI advantage for 60 incoming channels* | 2.1 | 2.0 |
| Susceptible to speech spurt clipping without special precautions | No | Yes |
| Requires changes in channel-assignment protocol during overload | No | Yes |
| Uses miniprocessor for channel-assignment control connectivity map | No | Yes |
| Susceptible to connect message channel overload | No | Yes |
| Relative hardware complexity | Lower | Higher |

* Based on an assumed voice spurt activity of 40%.

requires a memory to retain maps, giving all the connectivities needed to steer speech spurts to their proper interface channels at the destinations. Storage of these maps is not required in SPEC, since all the destination information needed to distribute the PCM samples for each frame is transmitted in the SAW.

SPEC and TASI also differ with respect to operation during high peaks of activity. In SPEC, because of its operating characteristics, the predictor works harder, removing a greater fraction of the incoming speech samples by increasing its aperture value. Although this action does increase the quantization noise, this increase is not subjectively significant under conditions that yield channel multiplication ratios of 2 : 1. In TASI, unless special precautions are taken, high peaks of activity will produce perceptible ($<50$ ms) clips with a frequency that may be unacceptable when operating at a CM 2 : 1. This deficiency of TASI is overcome by adopting the bit-reduction strategy, which has been shown to be very effective. However, introduction of the bit-reduction strategy complicates voice spurt channel assignment protocol and raises the implementation cost.

The salient features of the SPEC and TDMA techniques are compared in Table 5.2.

## 5.20 DSI TRAFFIC REDISTRIBUTION AMONG DESTINATIONS

In satellite communications applications, DSI modules can be used to serve multiple destinations. In such usage, a DSI terminal naturally redistributes its transmission capacities among the destinations it serves to accommodate changes in traffic load. It does this by exploiting the speech spurt assignment capabilities inherent to its operation. The ability of the DSI module inherently to accomplish traffic redistribution eliminates the necessity for a centralized demand assignment network controller, while fully maintaining the benefits of demand assignment within a network of users. This inherent demand assignment accommodation results from the fact that a DSI terminal assigns SCs only when demanded by the occurrence of voice spurts at the ICs. The destination of the voice spurts is governed by the assignment messages, and hence individual SCs can be directed to any of the destinations. All that is required is that a sufficient number of ICs be required to handle the peak load to each destination.

The self-regulating capabilities of the DSI terminal are best described through the following example. Consider a DSI terminal carrying traffic to three different destinations ($A$, $B$, and $C$). If the traffic load ($T_x$) to each destination ($x$) is plotted as a function of time and the peak loads are coincident as shown in Fig. 5.23, then the DSI terminal must be configured to accommodate a worst-case composite load, which is the sum of the individual loads $T_A + T_B + T_C$.

A total of $(T_A + T_B + T_C)/$(DSI gain) SCs would be required to accommodate this worst-case load. In this case, the DSI does not benefit from its natural redistribution processes because the load peaks are coincident. In fact no other form of demand assignment would benefit either. However, if the individual load peaks are separated in time, as shown in Fig. 5.24, then the worst-case composite load has a peak that is less than the sum of the individual load peaks, that is, less than $T_A + T_B + T_C$. Under these conditions, the DSI module will benefit from natural traffic redistribution.

As an example, consider that 100 ICs are directed towards each of destinations $A$, $B$, and $C$, yielding a total of 300 ICs, and that their peaks are separated in time to result in a composite load which peaks at 150 channels, as shown in Fig. 5.24. Due to the time separation of the individual load peaks, the DSI terminal can accommodate the 300 ICs with 150/(DSI gain) SCs. If the DSI gain is 2, only 75 SCs are required to serve the 300 ICs. The DSI terminal has appropriately re-

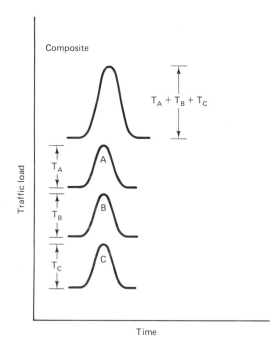

**Figure 5.23** Traffic load versus time (coincident peaks).

Digital Speech Interpolation Systems     Chap. 5

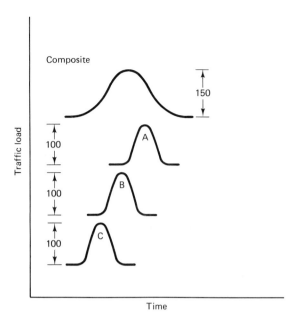

Traffic load

Composite

150

100

A

100

B

100

C

Time

**Figure 5.24** Traffic load versus time (separated peaks).

distributed its SCs to serve the shift in destinations by its inherent demand assignment property. In this example, this has resulted in an apparent channel multiplication ratio of 4 : 1. To protect the DSI module from chance overload, a dynamic load-control signal must be sent to the switching center.

## 5.21 FUTURE DIRECTIONS IN DSI

As digital technologies permeate the world's communications networks, both terrestrial and satellite, the forces of competition will demand application of the most-efficient, least-cost voice-transmission methods. DSI is very efficient and low in cost. The next generation of DSI equipment is already taking form. One example is the recent development of ADPCM exhibiting toll-quality telephone service at a bit rate of 32 kb/s; the CCITT has prepared Rec. G.721 detailing an ADPCM algorithm to be used to achieve worldwide equipment compatability. Integration of such 32-kb/s ADPCM with DSI will provide CMs of 4 : 1; to 5 : 1 compared to the conventional 64-kb/s PCM channel. The terminology **low-rate encoding with DSI** (LRE/DSI) has been used to identify such equipment. The integration of DSI and ADPCM poses a new problem, which stems from the inherent differential coded nature of ADPCM. At any instant, the present value at the output of an ADPCM decoder depends on the present sample as well as those of preceding samples. Consequently, an already-coded ADPCM signal cannot be arbitrarily gated in time without interrupting sample history and introducing distortion. This significantly increases the degradation impact of any clipping and chopping resulting from the DSI implementation, and precautions to avoid such effects must be exerted in the design of LRE/DSI using ADPCM.

Also, a concept called *multiclique* for operation of DSI is being studied. This concept avoids a problem of inefficient backhaul of traffic on terrestrial return links that occurs if the DSI equipment is located remotely from a multidestinational satellite earth terminal. The problem has its origin in the fact that each station in a multidestinational satellite network assigns channels to a single pool that goes to all destinations. Viewed from the receive side, a DSI unit must receive all such multidestinational traffic pools and cull out only the traffic it uses. When the DSI is located remotely from the earth station, such operation requires inefficient backhaul of the traffic from all originating stations over terrestrial links. The multiclique method is designed to avoid this inefficiency by dividing the pool of channels at each originating DSI into destination directed subsets. By this means only those subsets destined to a DSI need be backhauled.

## REFERENCES

[5.1] Brady, P. T. "A statistical analysis of on-off patterns in 16 conversations," *Bell Syst. Tech. J.*, January, 1968.

[5.2] Daynonnet, F. D., A. Jowsset, and A. Profit. "LeCELTIC: cencentrateur exploitant les temps d'Inactivete des circuits," *L'Onde Electrique*, Vol. 42, No. 426, September, 1962.

[5.3] Hashimoto, M. et al. "An application of the digital speech interpolation technique to a PCM-TDMA demand assignment system," International Conference on Space and Communications, Paris, France, March, 1971.

[5.4] Lyghounis, E. "Il sistema A.T.I.C.," *Telecommunicazioni*, No. 26, March, 1968.

[5.5] Poretti, I., G. Monty, and A. Bagnoli. "Speech interpolation systems and their applications in TDM/TDMA systems," *International Conference on Digital Satellite Commun.* Paris, France, 1972.

[5.6] Campanella, S. J. "Digital speech interpolation," *COMSAT Tech. Rev.*, Vol. 6, No. 1, Spring, 1976, pp. 127–158.

[5.7] Sciulli, J., and S. J. Campanella. "A speech predictive encoding communication system for multichannel telephone," *IEEE Trans. Commun.*, Vol. COM–21, No. 7, July, 1973.

[5.8] Suyderhoud, H., J. A. Jankowski, and R. P. Riddings. "Results and analysis of the speech predictive encoding communications system field trial," *COMSAT Tech. Rev.*, Vol. 2, 1974.

[5.9] CCITT, G.711. "Pulse code modulation (PCM) of voice frequencies," CCITT Recommendation G.711, Yellow Book, Geneva, Switzerland, November, 1980.

[5.10] Steele, R. *Delta Modulation Systems*, John Wiley, New York, 1975.

[5.11] Fukasawa, A., K. Hosoda, and R. Miyamoto. "A 32 Kbps ADPCM Codec Based on a new algorithm," Research Laboratory and Hideo Sugihara, Transmission Engineering Department, OKI Electric Industry Company, Ltd., Tokyo, Japan.

[5.12] Jayant, N. S. "Adaptive quantization with a one-word memory," *Bell Syst. Tech. J.*, Vol. 52, No. 7, September, 1973, pp. 1119–1144.

[5.13] Goodman, D. J., and R. M. Wilkinson. "A robust adaptive quantizer," *IEEE Trans. Commun.*, November, 1975, pp. 1362–1365.

[5.14] Dutweiler, D. L., and D. G. Messerschmitt. "Nearly instantaneous companding for nonuniformly quantized PCM," *IEEE Trans. Commun.*, Vol. COM-24, No. 8, August, 1976, pp. 843–864.

[5.15] Dutweiler, D. L., and D. G. Messerschmitt. "Nearly instantaneous companding and time diversity as applied to mobile radio transmission," *Proceedings of International Conference on Communications*, June, 1975, pp. 40(12)–40(15).

[5.16] Osborne, D. W. "Digital sound signals: Further investigation of instantaneous and other rapid companding systems," *BBC Research Department Report*, 1972, p. 3131.

[5.17] Cheung, R. S., and R. L. Winslow. "High quality 16-Kb/s voice transmission: The subband coder approach," *Proceedings of International Conference on Acoustics, Speech, and Signal Processing*, 1980, pp. 319–322.

[5.18] Crochiere, R. E. "On the design of subband coders for low bit rate speech communications," *Bell Syst. Tech. J.*, Vol. 56, May, 1977, pp. 747–770. B. Esteban and C. Galand, "Application of Quadrature Mirror Filters to Split Band Voice Coding Schemes," *Proceedings of International Conference on Acoustics, Speech, and Signal Processing*, Hartford, Connecticut, May, 1977, pp. 191–195.

[5.19] Bullington, K., and J. M. Fraser. "Engineering aspects of TASI," *Bell Syst. Tech. J.*, Vol. 38, No. 2, March, 1959.

[5.20] Fraser, J. M., D. B. Bullock, and H. G. Long. "Overall characteristics of a TASI system," *Bell Syst. Tech. J.*, Vol. 41, No. 4, July, 1962.

[5.21] Midema, H., and M. G. Schachtman. "TASI quality-effect of speech detectors and interpolation," *Bell Syst. Tech. J.*, Vol. 41, No. 4, July, 1962.

[5.22] Ahmend, R., and R. Fatechand. "Effects of sample duration on the articulation of sounds in normal and clipped speech," *J.of the Acoustical Society of America*, Vol. 31, No. 7, July, 1959.

[5.23] Feher, K. *Digital Communications: Satellite/Earth Station Engineering*, Prentice-Hall, Englewood Cliffs, N.J., 1983.

[5.24] Phiel, J. F., and R. C. Thorne. "INTELSAT TDMA system monitor", *Sixth International Conference on Digital Satellite Communications*, Ariz., 1983.

[5.25] Pontano, B. A., G. Forcina, J. L. Dicks, and J. Phiel. "A description of the INTELSAT TDMA/DSI System", *Fifth International Conference on Digital Satellite Communications*," Ariz., 1983.

[5.26] Campanella, S. J., and J. A. Sciulli. "Speech predictive encoded communications," *International Conference on Digital Satellite Communications*, Paris, France, 1972, p. 342.

# 6

# DIGITAL TELEVISION-PROCESSING TECHNIQUES

**DR. HISASHI KANEKO**

*Vice President, Nippon Electric Company (NEC)*
*Kawasaki City, Japan*

**T. ISHIGURO**

*Manager, Digital Television Laboratory*
*Nippon Electric Company (NEC)*
*Kawasaki City, Japan*

## 6.1 INTRODUCTION

Digital transmission and data-compression coding have long been viewed as promising and powerful means to achieve efficient TV transmission. Recent progress in **large-scale integration** (LSI) and digital technologies have made complicated signal processing a technically feasible reality and have led to progress in digital TV encoding, particularly in interframe coding by which the transmission bit rate can greatly be reduced. At the same time, progress in frequency spectrum usage by multiphase modulation or multilevel quadratic amplitude modulated (QAM) has allowed the use of digital transmission format in terrestrial microwave and satellite links. Also, digital fiber optic transmission is being introduced rapidly in communication networks. These trends have motivated serious efforts to develop and use digital approaches in actual TV transmission.

In addition, video-conferencing services are growing. Visual communication is a key to teleconferencing, because a full-motion video signal is very helpful in accomplishing interactive communications. However, a full-motion TV signal requires a bandwidth a thousand times wider than a voice telephone channel. Bandwidth compression is, therefore, a powerful means to provide economical teleconferencing systems with compression ratio of 1 : 40 or even less. Such high data compression can be achieved only by digital video-processing techniques.

282

Many investigations have been made to realize efficient digital encoding (Timb, 6.1; Netravali, 6.2; Pratt, 6.3; IEEE, 6.4]. Digital TV-encoding schemes are generally categorized into three classes: (1) conventional PCM, (2) intraframe coding, and (3) interframe coding, as shown in Fig. 6.1. For **National Television Standard Committee** (NTSC) color TV signals, conventional pulse code modulation (PCM) or straight analog-to-digital (A/D) conversion provides high-quality encoding with 7- or 8-bits/sample at about a 10.7-MHz sampling, resulting in about 80-Mb/s transmission rate. Intraframe coding is a technique to reduce the transmission bit rate by intraframe processing such as differential PCM or orthogonal transform coding. By these intraframe coding methods, the transmission bit rate can be reduced to about 30 to 60 Mb/s, depending on quality requirements and technique employed. There are also certain trade-offs between picture quality, bit rate, and hardware complexity.

Much greater reduction in transmission bit rate can be achieved by use of interframe coding [Kaneko, 6.5; Ishiguro, 6.6]. The general concept of bit rate reduction by interframe coding is to transmit the difference information of the two successive frames instead of transmitting the entire frame information. The information to be transmitted is dependent on picture object movement: The more active the movement, the greater the information becomes. As will be stated later, network TV signals, where much more active motion is encountered, can be transmitted at a bit rate around 6 to 30 Mb/s. Relatively still pictures such as those encountered in conference room scenes can be transmitted at 1.5 Mb/s or at an even lower bit rate.

In the 1970s, the data-compression coding technologies have made great progress owing to the extensive studies on coding algorithms and developments of codec hardware. Sophisticated signal processing has become a reality, leading to a theoretically optimized algorithm implemented with practical hardware complexity.

This chapter first describes fundamentals of TV signal encoding and then discusses typical examples of video codecs that are commercially available in North

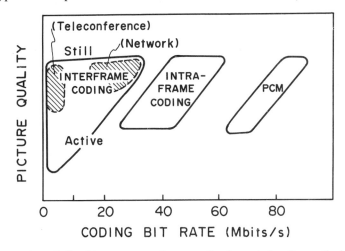

**Figure 6.1** Coding bit rate versus picture quality for typical coding methods.

America. Although there are many other systems developed in the world, discussion will focus mostly on NTSC signals cases.

## 6.2 FUNDAMENTALS OF TV-SIGNAL CODING

### 6.2.1 NTSC Color TV Signals

First let us look at the NTSC color TV signal. The NTSC television signal has a format with 525 lines/frame and 60 fields/s 2 : 1 interlaced scan. The composite NTSC color television signal is basically comprised of three components, $Y$, $I$, and $Q$, where $Y$ is luminance signal and $I$ and $Q$ are color difference signals. The $Y$, $I$, and $Q$ components are related with $R$ (red), $G$ (green), and $B$ (blue) signals, and $Y$, $I$, $Q$s and R, G, Bs are mutually convertible with a linear transformation.

$$\begin{bmatrix} Y \\ I \\ Q \end{bmatrix} = \begin{bmatrix} 0.30 & 0.59 & 0.11 \\ 0.60 & -0.27 & -0.32 \\ 0.21 & -0.52 & 0.31 \end{bmatrix} \cdot \begin{bmatrix} R \\ G \\ B \end{bmatrix} \tag{6.1}$$

The composite NTSC signal is converted from the $Y$, $I$, and $Q$ components through a color-encoding process, as shown in Fig. 6.1. The luminance signal $Y$ is filtered with bandwidth of about 4.0 MHz. The chrominance signals $I$ and $Q$ are band-limited to 1.5 and 0.5 MHz, respectively. Two subcarrier signals, which are 90° out of phase with respect to each other, are amplitude-modulated (AM) with the bandlimited $I$ and $Q$ signals. The modulated subcarrier signals are then band-pass filtered and summed with the luminance signal. The composite NTSC signal is composed by adding the sync signal. The frequency spectrum is shown in Fig. 6.2.

The color-decoding process, obtaining $Y$, $I$, and $Q$ signals from the composite NTSC signal, is an inverse operation of the color-encoding process. The reconstructed $Y$, $I$, and $Q$ components are not exactly identical with the original $Y$, $I$, and $Q$ signals because of the band limitation and frequency overlapping of the components in the composite signal.

### 6.2.2 Pulse Code Modulation Coding

Let us first consider the straight A/D conversion of a TV signal with uniform quantization. This is the basis of digital TV and is often simply referred to as pulse code modulation (PCM) coding. Basic coding parameters in PCM are sam-

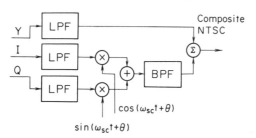

**Figure 6.2** A functional block diagram for composite NTSC color TV signal encoder.

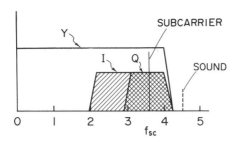

**Figure 6.3** Frequency spectrum of composite NTSC color TV signal.

pling frequency and number of bits per sample. Since the composite bandwidth of the color-video signal is limited to 4.2 MHz, the sampling frequency must be higher than about 9 MHz. Generally, the sampling frequency is chosen to be three or four times the color subcarrier frequency of 3.58 MHz for ease in signal processing.

The number of bits per sample is determined from loading condition and signal-to-noise ratio (SNR) requirement. The NTSC composite color TV signal has a waveform as shown in Fig. 6.3 for a 100% color bar signal. Since black-to-white signal amplitude is defined as 0 to 100 IRE units, sync pulse amplitude is $-40$ IRE units, and the upper peak of the composite signal is 133 IRE units. The amplitude range of A/D conversion should, therefore, cover the range of $-40$ to $+133$ IRE units. For an 8-bit PCM, for example, allowing 5% operation margin, the peak-to-peak amplitude (173 IRE units) is allocated to 242 levels out of 256 quantization steps. Under this appropriate signal loading, the 100 IRE loading level corresponds to 141, or 5.2 dB below the peak-to-peak range of A/D conversion.

The rms quantizing noise of PCM coding is known to be $\Delta/\sqrt{12}$, where $\Delta$ is a unit step size of quantization. The full range of $n$-bit PCM coding is $2^n$ steps, and hence the peak-to-peak signal to rms noise ratio is given by

$$\frac{S}{N} = \frac{2^n \cdot \Delta}{\Delta/\sqrt{12}} = 2^n \sqrt{12}$$

$$= 6 \cdot n + 10.8 \text{ dB}$$

Considering the loading level of $-5.2$ dB stated earlier, $S/N$ is given by

$$\frac{S}{N} = 6n + 5.6 \text{ dB} \qquad \text{(unweighted)} \qquad (6.2)$$

Weighted noise power is measured through a weighting filter with the amplitude characteristic shown in Fig. 6.4 [EIA, 6.7]. When the noise spectrum is assumed to be white, the weighted SNR increases about 7 dB from the value given by equation (6.2).

The theoretical SNR of 8-bit PCM is thus calculated as 53.6 dB unweighted or about 60 dB weighted. For television transmission systems, the EIA RS-250B specification requires 54 dB weighted for long haul, and 56 dB weighted for satellite links, and these specifications are met by 8-bit PCM. In order to satisfy the short-haul or medium-haul specification requiring 60 dB or 67 dB weighted, respectively,

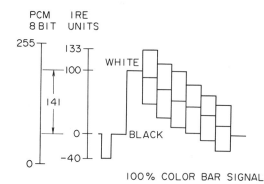

Figure 6.4  An example of PCM coding for composite NTSC color TV signals.

9- or 10-bit PCM should be used. In studio equipment such as a frame synchronizer, time-base corrector, or video switcher, the number should be around 10 bits to allow for multiple digital processing.

Since a straight PCM codec terminal is comparatively economical, it is used for relatively short-haul links or the transmission links, where the bit rate economy is not mandatory. Also, the straight PCM codec is an important gateway component to further sophisticated digital signal processing to be discussed in the next sections.

### 6.2.3 Data Compression

In order to attain further reduction of the transmission bit rate, the PCM-encoded signal has to be manipulated by digital signal processing. This processing is called **data compression**. Data compression is generally achieved by redundancy removal and trade-offs with picture quality.

As is well known, the TV signal has strong correlation between picture elements. For example, neighboring picture elements have nearly equal amplitude in most cases, not only in horizontal but also in vertical directions. The strong correlation also exists betgween picture elements of successive picture frames. If we can decorrelate the input signal, the redundancy contained in the TV signal can be removed, and data compression will be achieved. Typical methods of redundancy removal are predictive encoding and orthogonal transform coding.

Further data compression is achieved by trade-offs with picture quality. The technology for trade-offs should be such that the maximum data compression is achievable with minimum impairment of picture quality considering human perception of TV pictures. Such techniques as nonlinear quantization, noise filtering, moving area segmentation, subsampling, subline, subframe, and various adaptation schemes are appropriately combined with the redundancy removal, resulting in various types of data-compression encoding.

#### Predictive Coding

One widely used method of redundancy removal is **predictive encoding**. Suppose that prediction of an input signal is performed based on the past input signals. Since the input signal varies statistically, the predicted signal is not necessarily the same as the input signal, although it will be very close to it. If we transmit only the difference between the input and predicted signals, the amount of information transmitted can be reduced significantly. In other words, in predictive encoding, predictable redundancy is removed from the input signal, allowing data compression.

In predictive encoding, the present sample, $x(k)$, is estimated from the past samples, $x(k - i)$s. In linear predictive encoding, the predicted signal, $\hat{x}(k)$, is given by a linear combination of the past sample values and is

$$\hat{x}(k) = \sum_{i=1}^{N} a_i \cdot x(k - i) \tag{6.3}$$

where $a_i$'s are constant coefficients and $x(k - i)$ is the $i$th past sample value. Employing the $z$-transform representation

$$\hat{x}(z) = P(z) \cdot x(z), \tag{6.4}$$

where

$$P(z) = \sum_{i=1}^{N} a_i \cdot z^{-1} \tag{6.5}$$

The prediction error is given by

$$\begin{aligned} e(z) &= x(z) - \hat{x}(z) \\ &= \{1 - P(z)\} \cdot x(z) \end{aligned} \tag{6.6}$$

This prediction error, $e(z)$, is encoded and transmitted to the receiving side. At the receiving side, the original input signal, $x(z)$, is reconstructed from the received signal, $e(z)$. In general, due to the strong correlation of the input television signal, the prediction error, $e(z)$, is relatively small compared to the input signal, $x(z)$. Therefore, $e(z)$ is encoded with much smaller number of bits than input, $x(z)$.

The principle of this processing is conceptually illustrated in Fig. 6.5. The process of prediction is considered to be a linear filtering with transfer function given by (6.5), and it is implemented by a nonrecursive digital filter, as shown in Fig. 6.6. As a result, the encoder transfer function is represented as $\{1 - P(z)\}$. At the receiving side, the original signal $x(z)$ is reconstructed by a circuit, shown in Fig. 6.5, with local predictor $P(z)$ in the feedback path. The entire decoder is simply a digital filter with transfer function $1/[1 - P(z)]$. For example, in a simple case where the prediction is made by using only one previous sample, the encoder transfer function is $(1 - z^{-1})$, a differentiator, and the decoder transfer function is $1/(1 - z^{-1})$, an integrator.

Usually, predictive encoding is implemented in a feedback loop configuration, as shown in Fig. 6.7. With respect to the performance for the input signal, the feedback configuration of Fig. 6.7 is exactly the same as the forward configuration in Fig. 6.5. When quantization is considered for prediction error $e(z)$ in Fig. 6.5, the quantization error is integrated through the decoder, the SNR of the decoded signal being decreased greatly. The feedback configuration can avoid this, since the quantization error, $q(z)$, caused at the quantizer, is fed back through the

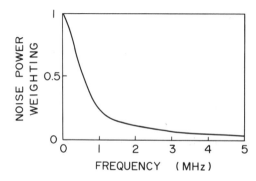

**Figure 6.5** Weighting function for noise power measurement.

predictor so that the quantization error becomes identical with $q(z)$ at the decoder output.

An optimum design of the predictor is given by minimizing the mean-squared error of the prediction. From equations (6.5) and (6.6),

$$E[e^2(k)] = E\left[\left\{x(k) - \sum_{i=1}^{N} a_i \cdot x(k - j)\right\}^2\right] \qquad (6.7)$$

where $E[\cdot]$ depicts the statistical average. Differentiating by $a_i$ and equating the result to zero, we get

$$\sum_{i=1}^{N} a_i E[x(k - i) \cdot x(k - j)] = E[x(k)x(k - j)], \qquad j = 1, \ldots, N \quad (6.8)$$

Assuming $e[x(k)] = 0$, this equation is rewritten in a matrix representation as

$$\begin{bmatrix} R_0 & R_1 & \cdots & R_{N-1} \\ R_1 & R_0 & \cdots & R_{N-2} \\ \cdot & \cdot & \cdot & \cdot \\ \cdot & \cdot & \cdot & \cdot \\ \cdot & \cdot & \cdot & \cdot \\ R_{N-1} & R_{N-2} & \cdots & R_0 \end{bmatrix} \cdot \begin{bmatrix} a_1 \\ a_2 \\ \cdot \\ \cdot \\ \cdot \\ a_N \end{bmatrix} = \begin{bmatrix} R_1 \\ R_2 \\ \cdot \\ \cdot \\ \cdot \\ R_N \end{bmatrix} \qquad (6.9)$$

where the $R$'s are autocovariance $e[x(k)x(k - i)]$. From the inverse transformation, an optimal set of predictor coefficients can be determined. The minimized prediction error power is given by

$$E[e^2(k)]_{\min} = 1 - \sum_{i=1}^{N} a_i \left(\frac{R_1}{R_0}\right) \qquad (6.10)$$

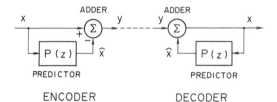

**Figure 6.6** Predictive coding conceptual block diagram.

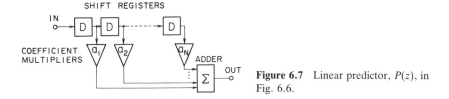

SHIFT REGISTERS

**Figure 6.7** Linear predictor, $P(z)$, in Fig. 6.6.

An example is shown for a simplified Markov signal model simulating a monochrome TV signal. Assume that the $R_i$'s are given by

$$R_i = R_0 \exp(-\alpha \cdot i) \tag{6.11}$$

where $\alpha$ is a constant. In this special case, the optimum predictor is a first-order filter and its coefficient is given by

$$a_i = \begin{cases} 1 - \left(\dfrac{R_1}{R_0}\right) & for\ i = 1 \\ 0 & for\ i \geq 2 \end{cases} \tag{6.12}$$

and the prediction error power is

$$P_{e,\min} = \left(1 - \left\{\frac{R_1}{R_0}\right\}\right)\overline{X^2} \tag{6.13}$$

In this case, $P(z) = a_1 z^{-1}$, and it is simply implemented by a coefficient multiplier with unit delay.

In two-dimensional prediction, the samples in the previous lines are used as well as the previous samples, and optimum predictor coefficients are determined in the similar fashion. The prediction performance is dominantly contributed by the neighboring samples, especially $S(1)$ and $S(-1, 1)$ or $S(1)$ and $S(0, 1)$ in Fig. 6.8. Further inclusion of additional samples for prediction will not significantly improve the effect of prediction.

In practical applications, since the statistical nature of actual picture is nonstationary and spacially nonuniform, it is desirable to vary the prediction function adaptively depending on the local nature of picture. A typical way of adaptation is to select an optimum predictor out of the many prediction functions, as shown in Fig. 6.9. The selection of the optimum prediction function is determined, in general, by observing past neighboring picture elements (pels). This operation

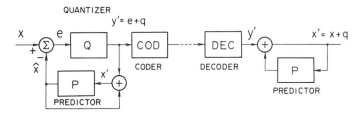

**Figure 6.8** Predictive encoder and decoder block diagram.

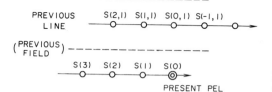

PREVIOUS LINE    S(2,1)  S(1,1)  S(0,1)  S(-1,1)

$\left(\begin{array}{c}\text{PREVIOUS}\\\text{FIELD}\end{array}\right)$

S(3)  S(2)  S(1)  S(0)

PRESENT PEL

**Figure 6.9** The samples used for prediction.

can be considered as a nonlinear prediction in which the prediction function varies with the local structure of a picture. The adaptive prediction is effective in decreasing the large prediction error caused at sharp brightness change edges. By such adaptive prediction, the prediction error entropy is reduced by about $\frac{1}{2}$ bit/sample.

The quantization is made by using a nonuniform quantizing characteristic to reduce the coding bit rate as much as possible. An algorithm for designing an $N$-level quantizer is known as the **Max quantizer**; it minimizes the mean-square quantizing error for a given number of quantizing levels [Max, 6.8].

The mean-square quantizing error is expressed by

$$\sigma_q^2 = \sum_{k=1}^{N} \int_{d_k}^{d_{k+1}} (e - e_k)^2 \, P_b(e) \, de \tag{6.14}$$

where $P_b(e)$ is the probability density function of prediction error $e$, $d_k$ is the $k$th quantizer decision level, and $e_k$ is the $k$th quantizing output level for $d_{k-1} \leq e \leq d_k$. For minimizing $\sigma_q^2$, the following conditions must be satisfied.

$$\int_{d_k}^{d_{k+1}} (e - e_k) \, P_b(e) = 0 \tag{6.15}$$

for $k = 1, \ldots, N$, and

$$d_k = \tfrac{1}{2} (e_{k-1} + e_k), \tag{6.16}$$

for $k = 2, 3, \ldots, N$. Equations (6.15) and (6.16) are solved by an iterative procedure.

Another method of quantizer design is to decide on the quantizer characteristic by taking into consideration human visual perception. Measurement of the visibility threshold for quantizing error has shown that for larger luminance change

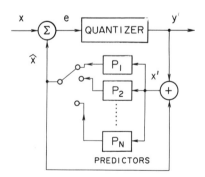

**Figure 6.10** An adaptive predictive encoder.

at an image edge, the human eye is less sensitive to additive noise. Figure 6.10 shows an example of a nonuniform quantization characteristic derived from this visibility threshold criteria (Thoma, 6.9]. This approach generally gives better quantization than that of the Max criterion.

The number of bits, $n$, for quanization ($N = 2^n$) is generally chosen to be 3 to 5 bits, depending on the picture-quality requirements, permitting about 3 bit/pel data reduction compared to PCM. The encoded digital bit stream still contains some redundancy. Code words corresponding to small prediction errors occur more frequently than those for large errors. If we assigns a shorter-length code word to more-frequently occurring levels and longer codes for less-frequently occurring levels, the average number of bits is further reduced. The reduction by this factor is 0.5 to 1 bit per sample. This technique is called **variable word length coding**, or **entropy coding**, and is effective not only for a predictive coding but also for any other data-compression technique.

### Transform Coding

Another widely used data-compression methodology is transform coding [Pratt, 6.10]. The scanned image samples are divided into blocks of length $N$, as shown in Fig. 6.11. Each block of input samples, $x_1, x_2, \ldots, x_N$, is transformed by a linear transformation into transform components $y_1, y_2, \ldots, y_N$, given by

$$
\begin{bmatrix} y_1 \\ y_2 \\ \cdot \\ \cdot \\ \cdot \\ y_N \end{bmatrix} = \begin{bmatrix} h_{11} & h_{12} & \cdots & h_{1N} \\ h_{21} & h_{22} & \cdots & h_{2N} \\ \cdot & \cdot & & \\ \cdot & & \cdot & \\ \cdot & & & \cdot \\ h_{N1} & h_{N2} & \cdots & h_{NN} \end{bmatrix} \cdot \begin{bmatrix} x_1 \\ x_2 \\ \cdot \\ \cdot \\ \cdot \\ x_N \end{bmatrix}
\tag{6.17}
$$

The components, $y_i$, are then quantized and coded for transmission.

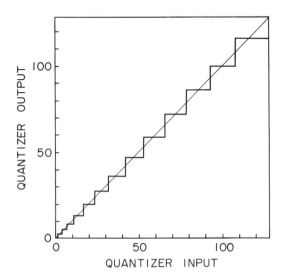

**Figure 6.11** A nonuniform quantization characteristic derived from the visibility threshold criteria.

**Figure 6.12** Pel block for transform coding.

Well-known transformations for this purpose are Hadamard transformation, slant transformation, cosine transformation, and Karhunen-Loeve transformation. Some transformation examples are shown in Fig. 6.12. These transformations are kinds of spatial filtering on the input signal, in which some components will be emphasized and others will be deemphasized, depending on the nature of the input signal. For most images, many of the transform components are of relatively low amplitude. These components can be represented with a smaller number of bits than required for straight PCM and are often discarded with minor effect of image distortion. Thus, on the average, the total number of bits for representing the components can be reduced. The transform coding is easily extended to the two-dimensional picture block case.

Figure 6.13 shows a performance comparison of the various transform coding schemes [Pratt, 6.10]. As the block size increases, the mean-square coding error

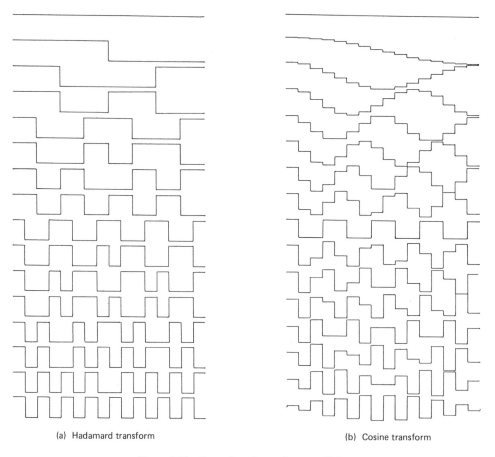

(a) Hadamard transform     (b) Cosine transform

**Figure 6.13** Examples of transform coefficients.

decreases monotonically and almost saturates beyond the block size of 16 × 16. Recent development of LSI has made it possible to implement a real-number coefficient multiplication with reasonable hardware size and to build a transform-calculation circuit for large block size, such as 16 × 16 pels.

The transform component's coding is accomplished by assigning different number of bits for each component. The bit assignment is made according to the statistics of component's amplitudes. An example is shown in Fig. 6.14 [Pratt, 6.10]. In this case, 16 × 16 pel block components are coded at an average rate of about 1.5 bits/pel. The number of assigned bits ranges from 8 to 1 bits and more than half of the components are discarded (zero assignment). The discarded components correspond to two-dimensional high-frequency components, causing only negligible distortion for most images.

Further improvement can be achieved by adaptively varying the bit assignment depending on the picture characteristics. Only components having magnitude greater than a given threshold are coded. This method selects more-important components to represent a block of image, decreasing distortion due to discarding components. In this case, however, it is necessary to code the position of each significant component. A simple way is to use the run-length coding scheme for the position coding. Since the amount of coded data varies with blocks, a buffer memory is necessary to smooth out the data rate variation.

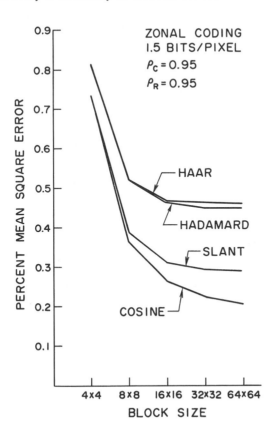

**Figure 6.14** MSE performance comparison of various transform coding as a function of block size. [Pratt, 6.10].

## Interframe Coding

Much-greater reduction in transmission bit rate can be achieved by use of interframe coding, in which the difference signal between two successive frames is encoded and transmitted. The general concept of bit rate reduction by interframe coding can be better illustrated by a simplified model, as shown in Fig. 6.15. Suppose that a soccer ball is crossing a TV screen. On looking at two successive TV frames, the soccer ball in the present frame is positioned slightly differently from that of previous frame, and the difference corresponds to the movement of the soccer ball during the time period for one frame. Instead of transmitting the entire information, if the difference information of the two successive frames is transmitted, the amount of information to be transmitted can be greatly reduced. In fact, if the movement is zero, theoretically, no information needs to be transmitted. The information to be transmitted is dependent on picture object movement: The more active the movement, the greater the information to be transmitted becomes.

The basic configuration of the interframe coder is shown in Fig. 6.16. The digitized TV signal, converted by an A/D converter, is encoded by an interframe coder in which essentially the difference signal between the present and the previous frame is encoded. The previous frame signal is obtained from frame storage built in the interframe coder. The output of the interframe coder is again processed through a variable-length coder to remove the redundancy contained in the frame differential signal. As mentioned earlier, since the information rate is directly dependent on the movement of picture objects, it should be smoothed out to obtain a constant transmission bit rate. This function is accomplished through buffer memory and an adaptive feedback control to the interframe coder to suppress the excessive generation of information. Since the control of the information-generation rate is made by changing the quantization step size, the SNR is decreased as the amount of motion increases. When the information-generation rate exceeds the transmission rate, picture quality starts to degrade. The encoded picture quality thus varies according to the video source materials.

```
8 8 8 7 7 7 5 5 4 4 4 4 4 4 4 4
8 8 7 6 5 5 3 3 3 3 3 2 2 2 2 2
8 7 6 4 4 3 3 2 2 2 2 2 2 2 2 2
7 6 4 3 2 2 2 2 1 1 1 1 . . . .
7 5 4 2 2 2 2 1 1 1 . . . . . .
7 5 4 2 2 2 1 1 . . . . . . . .
5 3 3 2 2 1 1 . . . . . . . . .
5 3 3 2 1 1 . . . . . . . . . .
4 3 2 1 1 . . . . . . . . . . .
4 3 2 1 1 . . . . . . . . . . .
4 3 2 1 . . . . . . . . . . . .
4 3 2 1 . . . . . . . . . . . .
4 2 2 . . . . . . . . . . . . .
4 2 2 . . . . . . . . . . . . .
4 2 2 . . . . . . . . . . . . .
4 2 2 . . . . . . . . . . . . .
```

**Figure 6.15** Typical bit assignments for transform coding in 16 × 16 pel blocks at a rate of 1.5 bits/pel [Pratt, 6.10].

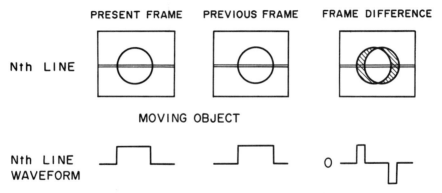

PRESENT FRAME     PREVIOUS FRAME     FRAME DIFFERENCE

Nth LINE

MOVING OBJECT

Nth LINE
WAVEFORM

**Figure 6.16**  Principle of basic interframe coding.

More precisely, the interframe coder in Fig. 6.16 is illustrated in Fig. 6.17. The interframe coder is considered to be a predictive coder with respect to frames. The prediction signal here is the previous frame signal and is given by an output of one frame delay memory.  The special feature of the function lies in a significant pel detection placed in the prediction error signal path [Mounts, 6.12; Candy, 6.13]. The determination of significance is made by comparing the prediction error amplitude with a given threshold, in which nonsignificant change of pels is discarded. Only the significant error signal and the position of significant pels are transmitted. The amount of information to be transmitted is proportional to the significant pels and, hence, to the area of moving part of picture.  Most TV scenes are relatively still, and motion occurs less frequently.  Therefore, as an average, the data rate is greatly reduced.

Motion-compensated interframe coding is a further sophisticated adaptive interframe prediction [Rocca, 6.14; Haskell, 6.15].  The concept of motion compensation itself is very simple.  When the object in the picture moves to some extent between the frames, the pel on the same position in the previous frame is no longer a good sample for prediction.  If it is assumed that the movement of the picture is only a shift of object position and the displacement of the object can be known, an exact interframe prediction is, in principle, possible even for rapidly moving pictures.

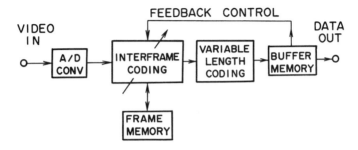

**Figure 6.17**  A basic configuration of interframe encoder.

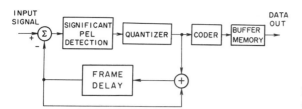

**Figure 6.18** Conditional replenishment coding block diagram.

The principle of motion compensation is illustrated in Fig. 6.18. The pel $S_n(i, j)$ is the sample to be transmitted. The subscript $n$ represents the frame number, and $i$ and $j$ indicate the horizontal and vertical sample positions in the frame. Let us assume that the object displacement has taken place with an amount represented by a vector $V(x_c\, y)$. Obviously, in order to predict the present sample $S_n(i, j)$, the $S_{n-1}(i - x, j - y)$ is a much better sample than $S_{n-1}(i, j)$. Therefore, estimation of the motion vector is an important key in the motion-compensation algorithm.

Motion-compensated interframe predictive coding is realized primarily by the functions shown in Fig. 6.19 [Koga, 6.16; Robbins, 6.17]. The predictor is a variable-frame delay instead of the fixed-frame delay used in simple interframe coding. The variable delay is controlled by a motion-vector-detection circuit. As shown in Fig. 6.19(a), the motion vector is estimated from the input TV signal in a forward-acting manner. The present frame is divided into small rectangular blocks, and each block is compared with the picture of the previous frame to find out the best-matched position. Thus the motion vector for each block is determined as a vector from the block position in the present frame toward the best-matched position in the previous frame. By varying the variable delay line in the interframe coder by an amount defined by the motion vector, the motion of the picture object is actually compensated.

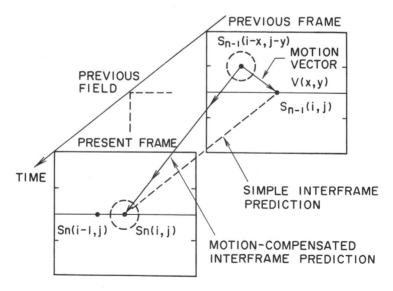

**Figure 6.19** Principle of motion compensation.

There is an alternative method of motion compensation, as shown in Fig. 6.19b, in which the motion vector is estimated in backward manner from the locally decoded signal. In this case, utilizing the relation between the spatial differential and temporal differential signals, the vector estimation is performed by a method called *pel recursive*, in which the motion-estimate error is recursively minimized on the basis of the steepest descent algorithm.

Figure 6.20 shows the picture of significant pels measured at the quantizer input, which are displayed with white dots [Ishiguro, 6.6]. It is observed that the number of significant picture elements in motion-compensated coder is much reduced as compared with the simple interframe coder. The prediction error entropy reduction by this technology is about one-half, as shown in Fig. 6.21 [Koga, 6.16].

### Data Compression for Color TV Signals

Although the data-compression technologies discussed in the previous sections can be applied to either monochrome or color TV signals, more detailed discussion will be made for the handling of color TV signals. Color TV-signal encoding is categorized into **component coding** and **composite coding**. In the component coding, each component of luminance and chrominance signals is separately encoded. Whereas in the composite coding the entire composite color TV signal is directly encoded.

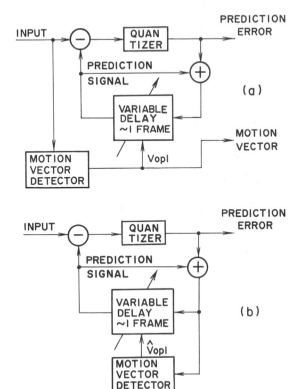

**Figure 6.20** Basic configuration of motion-compensated interframe coder. (a) Forward-acting motion vector detection type. (b) Backward-acting type.

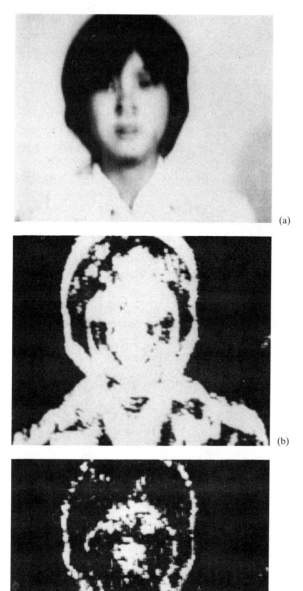

(a)

(b)

(c)

**Figure 6.21** Frame difference comparison between simple interframe coding and motion-compensated interframe coding [Ishiguro, 6.6]. (a) Original picture where a girl is standing up. (b) Significant pels for simple interframe coding (c) Significant pels for motion-compensated coding.

A basic configuration of component coding is shown in Fig. 6.22. A composite NTSC color TV signal is color-decoded into luminance component $Y$ and chrominance components $C_1$ and $C_2$. Each component is bandlimited to an appropriate frequency bandwidth and then encoded separately. Sometimes, two color components are alternately transmitted, discarded every other line to reduce the bit rate. Since component coding gives a great deal of freedom of choice of design parameters, it is widely applied when high data compression is desired.

In contrast, composite coding is used in applications where more fidelity is required. The spectrum of the composite NTSC color video signal has peaks at zero frequency for luminance signal and at the color subcarrier frequency for chrominance signals. The optimum predictor design described on pages 286–297 means that the amplitude characteristic of predictive encoder transfer function $[1 - P(z)]$ should be an inverse of the input signal frequency spectrum. Therefore, the function $[1 - P(z)]$ for composite coding is designed to have zeros in its amplitude characteristic at the zero and the subcarrier frequencies in order to obtain efficient prediction. To satisfy this, the prediction function $P(z)$ should be a higher-order function. In the one-dimensional prediction case, the following expression is derived [Limb, 6.18]:

$$1 - P(z) = (1 - z^{-1})(1 + \beta \cdot z^{-1} + z^{-2})A(z) \qquad (6.18)$$

The first term, $1 - z^{-1}$, has a zero at zero frequency, providing prediction for luminance component. The second term, $1 + \beta \cdot z^{-1} + z^{-2}$, gives zero at the color subcarrier chrominance signals. This means that for the composite predictive coding, at least a third-order function is needed. The last term, $A(z)$, is used to

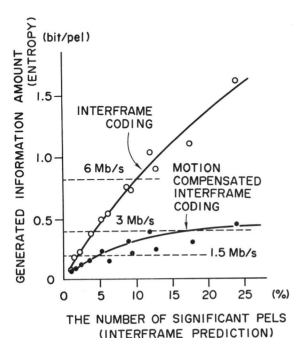

Figure 6.22 Data-compression factor improvement by motion compensation.

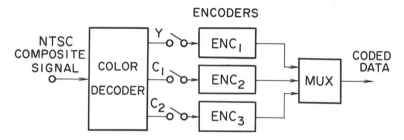

**Figure 6.23** Component-coding encoder block diagram.

optimize the overall prediction efficiency. It is noted that the sampling frequency can be arbitrarily chosen, providing some design freedom advantage.

Since equation (6.18) can be written as

$$1 - P(z) = [1 - P_C(z)][1 - P_L(z)] \qquad (6.19)$$

the higher-order predictive coding can be considered as a cascade prediction of $P_C(z)$ for chrominance and $P_L(z)$ for luminance signals, as composed by a double-loop configuration shown in Fig. 6.23. An example of the prediction characteristics in comparison with a monochrome signal prediction is shown in Fig. 6.24.

This concept of cascaded prediction can be expanded to the two-dimensional case [Iijima, 6.19]. The second term in equation 6.18 can be replaced with a function $1 + z^{-H}$. This means that the previous line sample is used for the chrominance prediction with prediction coefficient of $-1$, since the subcarrier phase differs 180° between horizontal lines. The sampling frequency for this case should be an integral multiple of the horizontal scan frequency. The coding performance improvement compared with the one-dimensional case is about 0.2 to 0.5 bit/sample measured in prediction error signal entropy. This corresponds to SNR improvement of 1 to 3 dB.

## 6.3 CODEC TERMINAL IMPLEMENTATION

Video signal codec terminals have been extensively developed since 1970s, and great progress has been made in hardware implementation as well as coding algorithms. Rapid progress in LSI technologies and cost reductions for memory

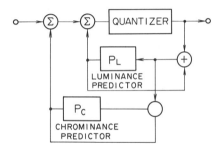

**Figure 6.24** Composite coding: a double loop configuration for higher-order predictive encoder.

integrated circuits have accelerated this tendency. At present, a variety of codec terminals have become available throughout the world. In Table 6.1, presently known codec terminals are listed. Most of codecs are available on the market, but the list includes some of the historical milestones. The list ignores the straight PCM codec.

It is observed from this table that intraframe coding is mostly used for transmission of CATV or dedicated television at a bit rate around 30 to 45 Mb/s. Interframe coding is mostly used for teleconferencing or similar applications at a bit rate of 1.5 Mb/s. It is also applicable to the network television transmission at a bit rate around 20 Mb/s. Another recent interest is in 56 kb/s codec, in which a moving scene of limited size is transmitted by a 56-kb/s data link.

Since it is difficult to describe all these codec implementations in detail, some of the typical examples will be explained later. For other implementations, the reader should refer to the cited papers.

### 6.3.1 Intraframe Coding

For broadcast program network use, the number of bits of PCM coding must be 8 bits/sample or higher, as is stated in Section 6.2.2. The transmission bit rate with PCM coding, therefore, ranges from 90 to 140 Mb/s, depending on the choice of sampling frequency and number of bits per sample. Such a PCM-encoded signal can be carried through fiber optic transmission or digital radio transmission systems with a transmission bit rate of 90, 135, or 140 Mb/s. The PCM video links provide a transparent high-quality transmission media compatible with conventional analog links.

For applications in which the picture-quality requirement is not so severe as for broadcast network use, a relatively simple codec with a lower bit rate can be designed. A typical example is a DPCM codec operating at 45 Mb/s. Figure 6.25 shows a functional block diagram. An input analog TV signal is first encoded into an 8-bit PCM signal by an A/O converter and then coded into 5-bit differential PCM (DPCM) code. The sampling frequency is chosen to be 8.9 MHz, one-fifth of the T3 transmission bit rate. A one-dimensional fifth-order prediction function is used to provide efficient prediction for composite NTSC color TV signals. The quantizer has a 31-level nonlinear quantization characteristic, which is similar to that shown in Fig. 6.10. The minimum quantization step size is 1 (8-bit PCM quantizing step) for small amplitude. Thus high SNR corresponding to 8-bit PCM is obtained for flat parts of picture. The dynamic range of the quantizer is as wide as 104, providing fast transient response for sharp brightness change edges. The sampling frequency is not locked to the input video signal scan frequency and the color subcarrier frequency. The codec can handle different scan-rate video signals, such as NTSC color and monochrome TV signals with $H$ sync frequency of 15.734 kHz and 15.750 kHz, respectively. An audio signal with 10 kHz bandwidth is encoded with 30 kHz sampling rate 12-bit linear PCM and is multiplexed with the video data. Fig. 6.26 shows a photograph of such a codec, the HO-DPCM 45B. The video signal encoder and decoder are built in the unit, the second left and the second right, respectively.

**TABLE 6.1** List of codec terminals

| Codec | Bit rate (Mbps) | Broadcast | Dedicated | Teleconf | Intraframe | Interframe | Predictive | Transform | Component | Composite | Author | Institute | Reference Number |
|---|---|---|---|---|---|---|---|---|---|---|---|---|---|
| DITEC | 30 | | □ | | □ | | □ | | | □ | L.S. Golding | COMSAT | [6.21] |
| (DPCM) | 32 | | □ | | □ | | □ | | | □ | K. Sawada et al. | NTT | [6.22] |
| (H-TRANS) | 32/22 | □ | | | □ | | | □ | | □ | T. Ohira et al. | OKI | [6.23] |
| (DPCM) | 32 | □ | | | □ | | □ | | | □ | J. Yamagata et al. | NTT | [6.24] |
| (DPCM) | 45 | | □ | | □ | | □ | | | □ | B.D. Buschman | DCC | [6.25] |
| HO-DPCM 45B | 45 | | □ | | □ | | □ | | | □ | | NEC | |
| HO-DPCM 45A | 45 | | | | F | | □ | | | □ | N. Suzuki et al. | NEC | [6.26] |
| NETEC-22H | 30–20 | | □ | | △ | □ | □ | | □ | | T. Ishiguro et al. | NEC | [6.27] |
| (INTERFIELD) | 30 | | | | F | □ | □ | △ | □ | | H. Yamamoto et al. | KDD | [6.28] |
| (INTERFRAME) | 30/15 | | □ | | △ | MC | □ | | □ | | H. Murakami et al. | KDD | [6.29] |
| NETEC-6/3 | 6/3 | | | □ | △ | □ | □ | | □ | | T. Ishiguro et al. | NEC | [6.30] |
| NETEC-Z | 6 | | | □ | △ | MC | □ | | □ | | | NEC | |
| TRIDEC | 6 | | | □ | △ | □ | □ | | □ | | H. Kuroda et al. | NTT | [6.31] |
| NETEC-X1MC | 1.5 | | | □ | △ | MC | □ | | □ | | K. Iinuma et al. | NEC | [6.32] |
| VTS-1.5 | 1.5 | | | □ | △ | □ | △ | □ | □ | | W. Chen | CLI | [6.33] |
| COST-211 | 2/1.5 | | | □ | □ | □ | □ | □ | □ | | R.C. Nicol et al. | BT | [6.34] |
| TRIDEC-1.5 | 1.5 | | | □ | △ | MC | △ | | □ | | | NTT | |
| (COS-TRANS) | 56 kb | | | □ | △ | △ | □ | □ | □ | | | WIDCOM | |
| NETEC-XD | 56 kb | | | □ | △ | △ | △ | | | □ | | NEC | |

F = interfield
MC = motion compensation

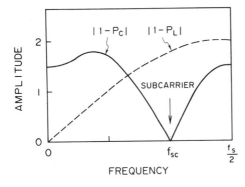

**Figure 6.25** An example of frequency characteristics of the chrominance prediction and the luminance prediction in Fig. 6.24.

Pictures of much-better quality can also be transmitted with the same bit rate, 45 Mb/s, by employing more-sophisticated coding algorithms [Suzuki, 6.26]. Figure 6.27 shows a functional block diagram of such a codec. In order to improve prediction efficiency, an adaptive prediction is used, in which three kinds of predictors—one-dimensional prediction, previous line prediction, and previous field prediction—are adaptively switched depending on local characteristics of pictures. The bit rate reduction is made by a variable-word-length coding (entropy coding). The coded data is once stored in a buffer memory and then transmitted at a constant rate. The buffer occupancy increases if the input signal entropy exceeds the transmission rate. When the buffer becomes full, the information generation rate is decreased by discarding the least-significant bit of the input PCM signal. This reduces the data rate by 1 bit/sample on an average at a sacrifice of 6 dB SNR.

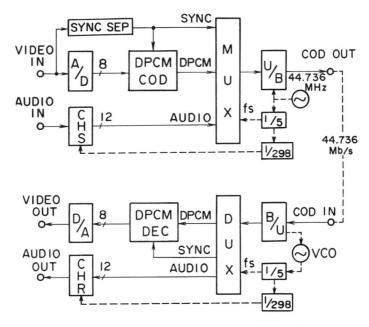

**Figure 6.26** Block diagram of a simple DPCM codec.

**Figure 6.27** Photograph of HO-DPCM 45B codec.

The coding performance is measured from the broadcast signal entropy statistics. The measured data shown in Fig. 6.28 [Suzuki, 6.26] indicate that the TV signal entropy with 8-bit precision is less than 60 Mb/s. In 45-Mb/s transmission, 75% of TV program signals can be encoded with 8-bit precision, although 25% of such signals are truncated to 7-bit coding. An example of this type of codec, HO-DPCM-45A, is shown in Fig. 6.29. The hardware size for the video signal processing is several times larger compared with the simple codec shown in Fig. 6.26, because of three-dimensional adaptive prediction, variable-word-length coding and large-capacity buffer memory.

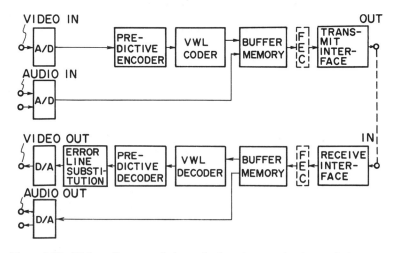

**Figure 6.28** High-quality transmission codec based on an adaptive predictive coding.

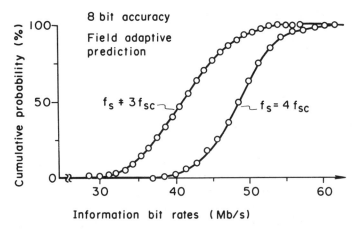

Figure 6.29  Cumulative probability of information bit rates measured from 260 TV pictures [Suzubi, 6.26].

### 6.3.2 Interframe Coding for Network-Quality TV

Network quality TV signals can be transmitted at the 20 to 30-Mb/s rate by using interframe coding.  As an example, encoder/decoder arrangements of NETEC-22H is shown in Fig. 6.30 [Ishiguro, 6.27].  An input composite NTSC color TV signal is first converted by an A/D converter.  In an NTSC color TV signal, the color subcarrier phase is different by 180° between successive frames.  This is undesirable because it generates a large frame difference even when the picture is perfectly still.  To solve this, phase inversion is made by a preprocessing, in which

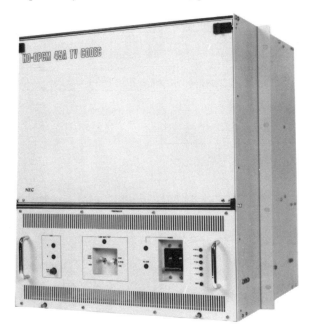

Figure 6.30  A photograph of HO-DPCM 45A codec.

the composite signal is converted into luminance and chrominance components by reversible transformation, and the chrominance component polarity is inverted frame by frame. After the phase correction, the signal is encoded by an adaptive predictive coding, in which either an intraframe prediction or interframe prediction is selected. The quantized prediction error is then coded with a variable-length coder to reduce the average bit rate. The compressed data are stored in a buffer memory with a capacity of about 1 Mb and is then transmitted to a transmission line. **Buffer memory occupancy** (BMO) value is fed back to the adaptive quantizer to control the encoded data generation rate in order to prevent buffer overflow.

The performance of the interframe coder is better illustrated by way of example in Fig. 6.31. The example is taken from the scene of Superbowl 1979, a most-active picture example, for a period of 2 min. The waveform shows the information generated from the interframe coder without feedback control. The amount of information here indicates more or less original source information. As is seen, the amount of information varies tremendously according to the source scenes. Then, by adding feedback control, the system operates to allow coarse quantization for the portions of excessive source information rate, and as a result, keeps the output bit rate constant. This is the case for active motion, although generally television programs yield less source information. Figure 6.32 shows the statistical data of source information for various source scenes. The dotted line shows the probability of source information rate for various source materials taken from broadcast TV programs for 36 h [Ishiguro, 6.27]. It is seen that the average rate is around 15 Mb/s and, as is seen on the solid line for the cumulative probability, 93% of the scenes are handled at 20 Mb/s and 99% at 30 Mb/s without appreciable noise impairments.

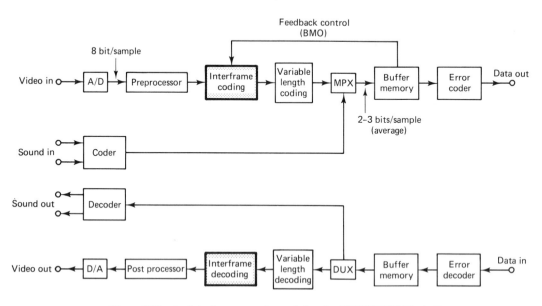

**Figure 6.31** An interframe encoder and decoder NETEC-22H block diagram.

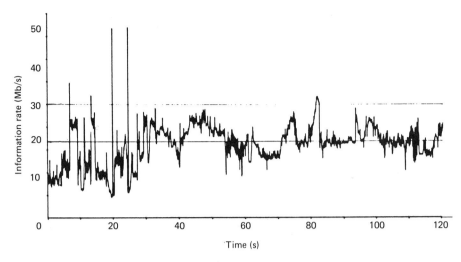

**Figure 6.32** Variation on interframe-encoded data rate with time signal source is "Superbowl '79," a very active picture example.

From what we have observed through the statistics, excellent picture quality can be obtained by 20-Mb/s coding most of the time. Buffer fill occurs with very small probability. According to actual measurements, buffer fill probability decreases very rapidly as the transmission bit rate is increased. The concept of sharing the transmission bit rate among plural TV channels becomes quite effective in reducing the probabilities of buffer fill and in improving picture quality for extremely active motion. Adaptive bit sharing is a concept quite similar to that of time-assigned speech interpolation (TASI), using the advantage of statistical difference among plural channels. When a channel is transmitting a rapidly moving picture, other channels may be transmitting relatively quiet pictures, because the probability of rapid picture motion occurring is generally very small. Therefore, we can assign a larger bit rate to the rapidly moving channel and a smaller bit rate to the other relatively quiet channels, keeping the total bit rate constant. In fact, for three-channel transmission with total 60-Mb/s rate, the bit rate for each channel can be adaptively varied within the 17 to 27 Mb/s range, with an average of 20 Mb/s. Interframe coding performance improvement by adaptive bit sharing is measured [Ishiguro, 6.27] by using a real hardware system shown in Fig. 6.33. Three different broadcast TV program signals are supplied to the three encoders, and the adaptive bit rate assignment is performed by an **adaptive bit-sharing multiplexer** (ABS-MUX) with the total bit rate kept constant at 60 Mb/s. Figure 6.34 shows a photograph of the multichannel transmission system.

In interframe coding, transmission bit errors cause a large effect on the decoded pictures. A forward acting error control circuit with 239/255 double error-correcting BCH code, which is used in this system, has excellent error-correction capability, as shown in Fig. 6.35 [Ishiguro, 6.27]. The calculated mean error free time is longer than 1 h at a line bit error rate of $10^{-5}$ and is about 5 s even at $10^{-4}$, which is generally considered the digital link threshold. The actual observed mean

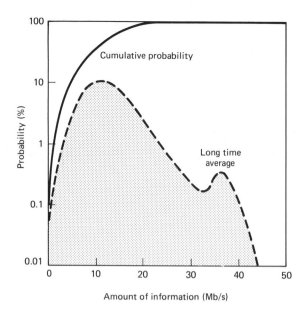

**Figure 6.33** Statistics of the encoded data rate for actual broadcast TV signals for 36 h in total.

error-free time, obtained from a satellite transmission experiment, is plotted in the figure. These data coincide very well with calculated error-control performance. In normal satellite conditions, the bit error rate (BER) performance is generally better than $10^{-7}$, and the mean error-free time gets to be much longer than a year. In addition to forward-acting error control, an erroneous line is automatically replaced by the previous line in the decoder, making the error less observable to the human eye.

### 6.3.3 Interframe Coding for Video Teleconferencing

The advantage of interframe coding is more significant in teleconferencing applications where the transmission bit rate is required to be 1.5 Mb/s, or less. Basic principles of interframe coding are the same as those for network TV, but there exist differences in technologies related to achieve high data compression under conditions of slower picture movement. For this application, transparency is not as important as in broadcast network use. Algorithm design is therefore directed to a best compromise between the transmission bit rate and picture quality.

A functional block diagram of an interframe encoder and decoder terminal is shown in Fig. 6.36 [Iinuma, 6.33]. The terminal encodes color video and audio signals into a digital stream and multiplexes them into a bit stream of 1.5 Mb/s. Video signal input and output are NTSC color TV signals, which are widely used in conventional video equipment. An audio signal with 7-kHz bandwidth is encoded at a bit rate of 128 kb/s.

The composite NTSC signal is first A/D converted into a PCM signal and then transformed by a **color signal processor** (CSP-S) into a time division multiplexed (TDM) signal in which a time-compressed chrominance signal is inserted

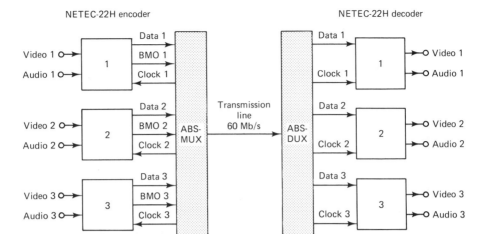

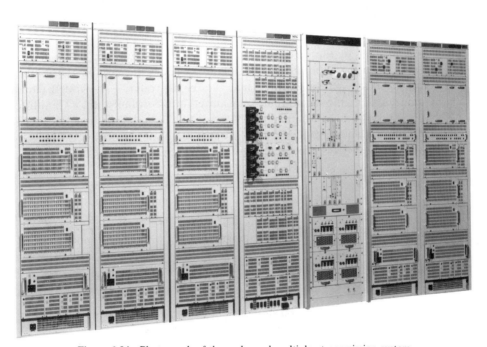

**Figure 6.34** Photograph of three-channel multiplex transmission system.

into the horizontal blanking interval, as shown in Fig. 6.37. The TDM signal is fed to a movement detector and an interframe coder.

In the movement detector, a TV picture field is subdivided into subblocks of 7 pels times 7 lines. The motion vector is calculated for each block by a movement-detection circuit, which consists of high-speed parallel processing elements. The interframe coding is a frame-to-frame differential coding, where the frame delay

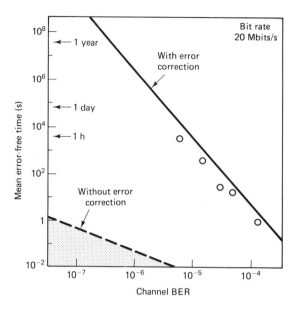

**Figure 6.35** Error correction performance of 239/255 BCH code. The solid line is theoretical and the circles are the satellite transmission test results. [Kaneko, 6.5].

is controlled by the motion vector [Ishiguro, 6.31; Kuroda, 6.32]. The prediction error signal and the motion vector are coded with variable-word-length coding, which utilizes Huffman codes and run-length codes. The coded data are stored in a buffer memory and then transmitted at a constant rate. The buffer memory occupancy is used for feedback control to the coded data-generation rate. The coding parameter-mode control to prevent buffer overflow is made by adaptive quantization, field repeating, subsampling, and frame freezing.

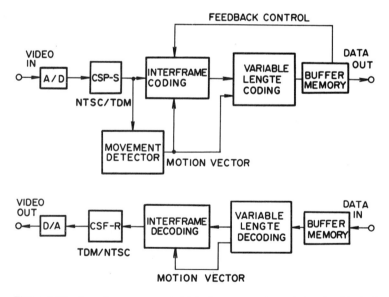

**Figure 6.36** A motion-compensated interframe encoder and decoder block diagram.

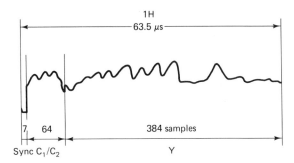

**Figure 6.37** TDM color TV signal format.

The picture quality of the interframe coding is described in relation to the coding parameter-mode control as shown in Fig. 6.38. When buffer occupancy is in the lowest level, the encoding is made with the finest quantization step size to provide the highest value of $(S/N)$ $q$. As buffer occupancy increases, the quantization step size is made coarser. The data-generation rate reduces as the step size increases. When the occupancy is further raised and exceeds a certain level, a field-repeating mode is applied, in which only the information of every other field, either odd or even, is transmitted. The deleted field is interpolated from the adjacent fields at the receiving side. By the field-repeat mode, the information-generation rate can be halved. For further occupancy increase, a subsampling mode is added to the field repeating. The subsampling is to transmit every other sample. If the buffer becomes full in spite of the controls described above, the

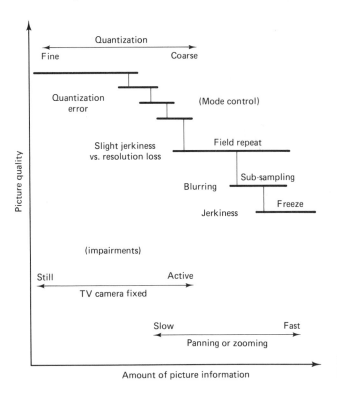

**Figure 6.38** Interframe-coding performance features.

encoding is stopped until the buffer occupancy decreases to a low level. At the receiving side, the same picture is repeated until a new frame picture is received.

It is generally the common nature of interframe coding having the buffer occupancy feedback that the codec terminal is adaptive to any bit rate, which is commanded by an external clock signal. By reducing the external clock rate below 1.5 Mb/s, such as 768 kb/s, the codec is performing even at that rate, although incurring picture quality trade-offs by its own algorithm. Figure 6.39 shows a photograph of NETEC-X1MC terminal in which a motion-compensated interframe coding is implemented with hardware of practical size.

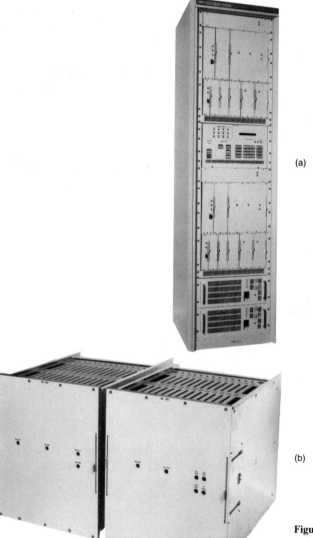

(a)

(b)

**Figure 6.39** A photograph of 1.5 Mb/s codec. NETEC-X1MC.
(a) Bay and (b) motion-compensated interframe coding and decoding units.

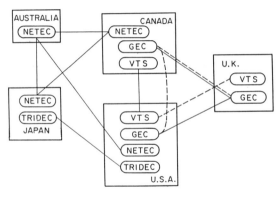

NETEC : NETEC × I (MC)      TRIDEC : TRIDEC 1.5
VTS : VTS I.5E (CLI)        GEC : GEC-Jerrold
                                   (CEPT-BT)

**Figure 6.40**  1.5 Mb/s video codes used in ITS (International Teleconference Symposium) held in April 1984.

## 6.4 CONCLUSION

In accordance with the growth of digital transmission links through satellite, digital coaxial cable, digital radio links, or digital fiber optics, digital television transmission has been introduced in real applications. Teleconferencing has already started use of digital codec technologies and its system effectiveness has been well recognized. In April 1984, there was a memorable event: The International Teleconference Symposium was held simultaneously at 5 locations in the world. The symposium sites in Philadelphia, Toronto, Tokyo, Sydney, and London were interconnected by digital satellite and terrestrial links. Different digital television codecs were used to connect these locations as shown in Fig. 6.40. As a result of this event, the advantage of worldwide teleconferencing was recognized. This event has drawn people's attention to the problem of worldwide compatibility and standardization of codec terminals.

Looking at the future, there will continue to be innovation in digital television technologies. LSI and very large scale integration (VLSI) for high-speed logic and memories will contribute to reducing the size and cost of hardware. Also, VLSI technologies will make it possible to use much more sophisticated signal-cessing techniques such as vector quantization, three- or four-dimensional processing, and image analysis and synthesis. These will further enhance the advantage of digital technologies and open many new application areas.

## REFERENCES

[6.1] Limb, J. O., C. B. Rubinstein, and J. E. Thompson. "Digital coding of color video signals—A review," *IEEE Trans. Commun.*, Vol. COM–25, November, 1977, pp. 1349–1385.

[6.2] Netravali, A. N., and J. O. Limb. "Picture coding: A review," *Proc. IEEE*, Vol. 68, March, 1980, pp. 366–406.

[6.3] Pratt, W. K. *Image Transmission Techniques*, Academic Press, New York, 1979.

[6.4] Special Issue on picture communication systems, *IEEE Trans. Commun.*, Vol. COM–29, No. 12, December, 1981.

[6.5] Kaneko, H., and T. Ishiguro. "Digital television transmission using bandwidth compression techniques," *IEEE Commun. Mag.*, Vol. 18, No. 4, July, 1980, pp. 14–22.

[6.6] Ishiguro, T., and K. Iinuma. "Television bandwidth compression transmission by motion-compensated interframe coding," *IEEE Commun. Mag.*, Vol. 20, No. 6, November, 1982, pp. 24–30.

[6.7] EIA Standard RS-250B, September, 1976.

[6.8] Max, J. *IRE Trans. Inf. Theory*, Vol. IT–6, 1960, pp. 7–12.

[6.9] Thoma, W. *Proc. Int. Zurich Semin. Digital Commun.* 3rd, 1974, pp. C3(1)–C3(7).

[6.10] Pratt, W. K. *Digital Image Processing*, John Wiley, New York 1978.

[6.11] Wintz, P. A. *Proc. IEEE*, Vol. 60, No. 7, 1972, pp. 809–820.

[6.12] Mounts, F. W. "A video encoding system using conditional picture replenishment," Bell Syst. Tech. J., Vol. 48, September, 1969, pp. 2545–2554.

[6.13] Candy, J. C., M. A. Franke, B. G. Haskell, and F. W. Mounts. "Transmitting television as clusters of frame-to-frame differences," *Bell Syst. Tech. J.*, Vol. 50, July–August, 1971, pp. 1889–1919.

[6.14] Rocca, F. "Television bandwidth compression utilizing frame-to-frame correlation and movement compensation," *Proceedings Symp. Picture Bandwidth Compression*, April, 1969, Gordan and Breach, N. Y., 1972.

[6.15] Haskell, B. G. "Entropy measurements for nonadaptive and adaptive, frame-to-frame, linear predictive coding of video telephone signals," *Bell Syst. Tech. J.*, Vol. 54, August, 1975, pp. 1155–1174.

[6.16] Koga, T., K. Iinuma, H. Hirano, Y. Iijima, and T. Ishiguro. "Motion compensated interframe coding for teleconferencing," *Proc. National Telecommun. Conf.*, Vol. 4, G5.3, November, 1981.

[6.17] Robbins, J. D., and A. N. Netravali. "Interframe coding using movement compensation," *IEEE International Conf. on Commun., Conf. Rec.*, Vol. 23.4, June, 1979.

[6.18] Limb, J. O., C. B. Rubinstein, and K. A. Walsh. "Digital coding of color picture-phone signals by element-differential quantization," *IEEE Trans. Commun.*, Vol. COM–19, No. 6, December, 1971, pp. 992–1006.

[6.19] Iijima, Y., and N. Suzuki. "Experiments on higher-order DPCM for NTSC color TV signals, (in Japanese), *Tech. Group Commun. Syst., Monograph*, Vol. CS74–63, August, 1974.

[6.20] Hatori, Y., and H. Yamamoto. "Predictive coding for NTSC composite color television signals based on comb-filter integration method," *IECEJ Trans.*, Vol. E62, April, 1979.

[6.21] Golding, L. S., "DITEC-A digital television communications System for satellite links," *Second International Conf. Digital Sattelite Commun.*, Paris, France, November, 1972.

[6.22] Sawada, K., and H. Kotera, "A 32 Mbit/s component separation DPCM coding system for NTSC color TV," *IEEE Trans. Commun.*, Vol. COM–26, April, 1978, pp. 458–465.

[6.23] Ohira, T., M. Hayakawa, and K. Matsumoto. "Orthogonal transform coding system for NTSC color television signals," *IEEE Trans. Commun.*, Vol. COM–26, No. 10, October, 1978, pp. 1454–1463.

[6.24] Yamagata, J., H. Takashima, N. Bando, and T. Doi. "Asynchronous intra-frame coding with one-dimensional prediction," *IEEE Int. Conf. Commun. Rec.*, 62.1.1, June, 1981.

[6.25] Buschman, B. D., "A 45 Mbps digital television codec," *IEE National Telecommun. Conf. Rec.*, C9.5, November, 1981.

[6.26] Suzuki, N., K. Iinuma, and T. Ishiguro. "Information preserving coding for broadcast television signals," *IEEE Globecom 82, Conf. Rec.* B6.7, November, 1982.

[6.27] Ishiguro, T., K. Iinuma, Y. Iijima, T. Koga, S. Azami, and T. Mune. Composite interframe coding of NTSC color television signals," *Proc. National Telecommun. Conf.*, Dallas, TX, November, 1976, pp. 6.4.1–6.4.5.

[6.28] Koga, T., Y. Iijima, K. Iinma, and T. Ishiguro, "Statistical performance analysis of an interframe encoder for broadcast television signals," *IEEE Trans. on Commun.*, Vol. COM–29, No. 12, December, 1981.

[6.29] Yamamoto, H., Y. Hatori, and H. Murakami. "30-Mbit/s codec on the NTSC color TV signal using interfield-intrafield adaptive prediction," *IEEE Trans. Commun.*, Vol. COM–29, No. 12, December, 1981, pp. 859–867.

[6.30] Murakami, H., S. Matsumoto, Y. Hatori, H. Yamamoto, T. Ohshima, and Y. Sano. "A 15-Mbit/s universal codec for TV signals using a median adaptive prediction coding method," *Sixth International Conf. on Digital Satellite Communications*, Phoenix, Ariz., September, 1983.

[6.31] Ishiguro, T., K. Iinuma, Y. Iijima, T. Koga, and H. Kaneko. "NETEC system: Interframe encoder for NTSC color television signals," *Third International Conf. on Digital Satellite Commun.*, Kyoto, Japan, November, 1975.

[6.32] Kuroda, H., F. Kanaya, and H. Yasuda. "TRIDEC system configuration," *Rev. Elec. Commun. Labs.*, Vol. 25, November–December, 1977, pp. 1347–1351.

[6.33] Iinuma, K., T. Koga, Y. Iijima, and A. Tomozawa. "A 1.5 Mb/s full motion video teleconference system," *The Sixth International Conf. on Digital Satellite Commun.*, September, 1983.

[6.34] Chen, W. "Scene adaptive coder, "*IEEE ICC Conf. Rec.*, 1981, pp. 22.5.1–22.5.6.

[6.35] Nicol, R. C., and T. S. Duffy. "A codec system for worldwide videoconferenceing," *Sixth International Conf. Digital Satellite Commun., Conf. Rec.*, September, 1983, pp. VII.A.9–VII.A.16.

# 7

# DIGITAL MODEM (MODULATION-DEMODULATION) TECHNIQUES

**DR. KAMILO FEHER**

*Professor, Electrical Engineering*
*University of California, Davis,*
*Davis, California 95616*
*and*
*Director, Consulting Group, DIGCOM, Inc.*

## 7.1 INTRODUCTION—LIST OF MODEM TECHNIQUES

A review of the principles of operation of frequently used digital modem techniques is followed by a description of recently developed power-efficient and spectrally efficient modems. For a particular terrestrial microwave, satellite, cable, wire, or other system application, the experienced system designer has a large number of candidate modem techniques at his or her disposal (See Table 7.1), which is an asset. The appropriate choice of a particular modulation technique will lead to the best system performance and to a lower-cost, more-competitive transmission subsystem. However, the less-experienced engineer could consider the large number of modem techniques, listed in Table 7.1, as a strange alphabet soup without significant meaning for his or her decision and system implementation.

The main objective of Section 7.2 is to present the principle of operation of modems, starting with the simplest two-state PSK and four-state QPSK modems, thus leading to the comprehension of complex, 256-state QAM modems. The nomenclature used in this chapter is also defined in this review section.

**TABLE 7.1(a)**   Classical and recently developed power and spectrally efficient digital modem techniques

|  | Abbreviation | Alternate abbreviation | Descriptive name of modem technique |
|---|---|---|---|
| 1 | DSB-SC-AM | DSB-SC | Double-Sideband-Suppressed Carrier—Amplitude Modulation |
| 2 | PSK | BPSK | Phase-Shift Keying; Binary-PSK |
| 3 | DPSK | DBPSK | Differential PSK; Differential Binary PSK (*no* carrier recovery) |
| 4 | DEPSK | DEBPSK | Differentially Encoded PSK (*with* carrier recovery) |
| 5 | QPSK | CQPSK | Quadrature (quaternary) PSK; Coherent QPSK |
| 6 | OQPSK | OKQPSK or SQPSK | Offset QPSK; Staggered QPSK |
| 7 | DQPSK | | Differential QPSK (*no* carrier recovery) |
| 8 | DEQPSK | | Differentially encoded QPSK (*with* carrier recovery) |
| 9 | MSK | FFSK | Minimum-shift keying; Fast-frequency-shift keying |
| 10 | DMSK | | Differential MSK (minimum-shift keying) |
| 11 | GMSK | | Generalized or Gaussian MSK |
| 12 | TFM | | Tamed frequency modulation |
| 13 | Multi-h FM | Correlative FM | Multi-index; correlative; duobinary FM |
| 14 | IJF-OQPSK | NLF-OQPSK | Intersymbol-jitter-free OQPSK; Non-Linearly Filtered OQPSK |
| 15 | SQAM | | Superposed quadrature amplitude modulation (QAM) |
| 16 | TSI-OQPSK | | Two-symbol-interval OQPSK |
| 17 | X-PSK | | Crosscorrelated PSK |
| 18 | DCTPSK | | Differentially continuous transition PSK |
| 19 | CPFSK | | Continuous-phase frequency-shift keying |
| 20 | SFSK | | Sinusoidal frequency-shift keying |
| 21 | QORC | | Quadrature overlapped raised cosine |
| 19 | QAM | M-ary QAM | Quadrature amplitude modulation |
| 20 | APK | | Amplitude Phase Keying |
| 21 | QPRS | | Quadrature Partial-Response System |
| 22 | SSB | | Single sideband Amplitude Modulation |

**TABLE 7.1(b)**   Power-efficient modem $P_e$-performance requirement as a function of $E_b/N_O$*

| $P_e$ | $E_b/N_O$ (dB) (theoretical) | $E_b/N_O$ (dB) (practical) |
|---|---|---|
| $10^{-2}$ | 4.3 | 5.3 |
| $10^{-4}$ | 8.4 | 10.8 |
| $10^{-6}$ | 10.4 | 13 |
| $10^{-8}$ | 12 | 15 |

*Coherent demodulators are assumed.

Sec. 7.1   Introduction—List of Modem Techniques

Recently discovered coherent and differentially coherent *power-efficient* modulation techniques for satellite and other nonlinear radio applications are described in Sections 7.3 through 7.7. The spectral efficiency of *power-efficient* modems is *less than* 2 bits/s/Hz. Most currently operational satellite systems and digital mobile radio systems are *power-limited*; that is, the available carrier-to-noise ratio, $C/N$, is insufficient to enable the utilization of *spectrally efficient* modems, that is, modems that have a spectral efficiency of more than 2 bits/s/Hz.

For terrestrial line-of-sight (LOS) microwave systems, numerous cable systems, and voiceband modem applications, a much-higher spectral efficiency is required. The newest generation of modems has a practical spectral efficiency of more than 6 bits/s/Hz. Such QAM modems and also QPRS modems, which can be operated above the Nyquist rate, are described in Sections 7.8 through 7.11. Finally, in Section 7.13 synchronization problems, and in particular degradations due to carrier recovery and symbol-timing recovery are investigated.

In Sections 7.6 through 7.12 the research and development projects of the Digital Communications Research Group undertaken at our University and a number of industrial collaborating laboratories are highlighted. We present some of our most-important modem techniques, particularly modems which have already found applications in satellite and microwave systems. In this chapter, we concentrate on the description of a limited number of recently discovered modem techniques. Modulation techniques discovered by other research teams are described in Chapters 8 and 10, respectively. In these chapters, power-efficient modem techniques are described. Chapter 9 presents a study of spectrally efficient modems which operate in an interference-controlled environment. Finally, Chapter 12 describes theory and techniques of adaptive equalizers, which are an integral part of a large class of advanced modems.

## 7.2 PRINCIPLES OF MODEM TECHNIQUES

The principles of operation of DSB-SC modems are explained with the aid of Fig. 7.1. A large number of digitally modulated signals, such as QPSK, MSK, QAM, and QPRS may be generated by baseband-processed quadrature modulated DSB-SC subsystems. To simplify our initial discussion, we assume that the premodulation low-pass filters (LPFs) and also the postmodulation band-pass filter (BPF) are bypassed; in other words, there are no bandlimiting filters in the modulator.

The $f_b$-rate binary baseband source is commuted into two binary symbol streams, each having a rate of $f_b/2$. In the design of QAM modulators, the following 2-to-L-level baseband converter converts these $f_b/2$-rate data streams into L-level, pulse amplitude modulated (PAM) baseband signals having a symbol rate of

$$f_s = \left(\frac{f_b}{2}\right) \div (\log_2 L) \qquad \text{symbols/second} \qquad (7.1)$$

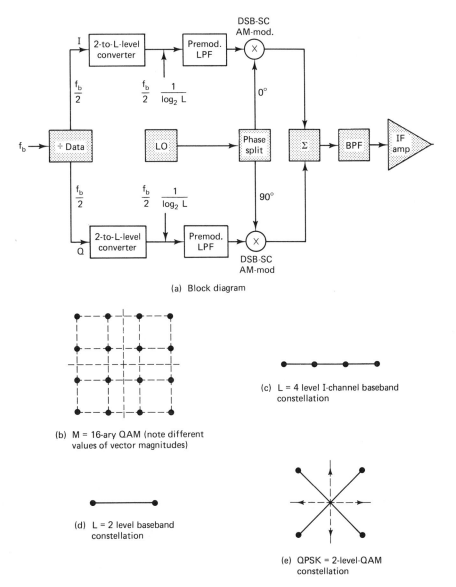

(a) Block diagram

(b) M = 16-ary QAM (note different values of vector magnitudes)

(c) L = 4 level I-channel baseband constellation

(d) L = 2 level baseband constellation

(e) QPSK = 2-level-QAM constellation

**Figure 7.1** Block and constellation diagram of a QAM modulator. The same generic block diagram may be used in the design of QPSK and QPRS modulated signals. (With permission of Prentice-Hall, [Feher, 7.3].)

For example, if the source bit rate is $f_b$ = 6 Mb/s, then the commuted "in-phase" and "quadrature" channel binary baseband streams have a rate of $f_b/2$ = 3 Mb/s. If an $M$ = 16-ary QAM-modulated signal is desired, these commuted binary streams are converted into $L$ = 4-level PAM baseband streams, having a symbol rate of 3 Mb/s : $\log_2 4$ = 1.5 million symbols/s. Four PAM levels of the baseband **in-phase** (I) and of the baseband **quadrature** (Q) channels drive the DSB-SC-AM modulator (mixer) baseband ports. The **local oscillator** (LO) provides the required quadrature (90° phase shifted) intermediate frequency (IF) or radio frequency (RF) carrier signals. As the I- and Q-channels are independent, it is possible to generate any of the $4 \times 4 = 16$ solid dots indicated in Fig. 7.1(b). Let us assume that we terminate the Q-baseband channel. In this case, the four-level I-channel PAM baseband signal, after DSB-SC-AM modulation, generates a vector state space diagram, also known as a **constellation diagram**, such as illustrated in Fig. 7.1(c). Thus in this case we have a one-dimensional, four-level 4-DSB-SC-AM system. Now, let us further simplify our modulator. In addition to terminating the Q-channel baseband input, we decide to bypass the 2-to-L-level PAM converter in the I-baseband signal path. In this case, we have the simplest digital modulator, namely, a two-level DSB-SC-AM modulator. The corresponding constellation diagram contains only two dots (see Fig. 7.1(d)); namely, $+A \cos w_c t$ and $-A \cos w_c t$. This can be noticed from the fact that a balanced **non–return-to-zero** (NRZ) baseband signal has only two possible states, namely, $+1$ or $-1$, and that the DSB-SC-AM modulator is a "product-modulator," which multiplies (time domain multiplication) the $+1$ or $-1$ baseband input by the LO signal $A \cos w_c t$. Multiplication in the time domain corresponds to convolution in the frequency domain. Convolution of the baseband data spectrum with the unmodulated carrier wave is equivalent to frequency translation. Thus the NRZ spectrum is translated around the carrier frequency $f_c$. Our discussion of the modulator having a constellation diagram of Fig. 7.1(d) leads to the following statement:

$$\boxed{\text{BPSK} = \text{DSB-SC-AM}} \qquad (7.2)$$

namely, a binary-phase-shift-keyed modulator is identical (at least in theory) with a double-sideband-suppressed carrier-amplitude modulator *if the baseband drive signal is an unfiltered balanced NRZ data stream*. In other words, multiplication by the $+1$ NRZ state corresponds to a 0° shift of the unmodulated carrier, whereas multiplication by the $-1$ NRZ state corresponds to a 180° phase-shifted carrier. Now, let us connect *both* the I- and Q-channels to the mixer inputs and bypass the 2-to-L-level PAM converters. In this case, the baseband drive signals to both mixers will be binary, balanced NRZ signals. The bit rate of the individual channels is $f_b/2$. The I-mixer would lead to a one-dimensional constellation, as illustrated in Fig. 7.1(d), whereas the Q-mixer would lead to the same type of constellation but rotated by 90°. At all instants of time, both mixers are driven, leading to a resultant two-dimensional constellation shown in Fig. 7.1(e). We just generated a four-phase state constellation. The phase states may change instantly by integer multiples of 90°; thus we have a QPSK system. However, a somewhat different view of the same modulator reveals that we have two DSB-SC-AM modulators in quadrature. The resultant modulated signal is a QAM signal. Thus we can state that

$$\boxed{\text{QPSK} = 4 \text{ QAM}} \qquad (7.3)$$

*if the baseband drive signal of both of the I- and Q-channel modulators is an unfiltered balanced binary data stream.* Even though the block diagrams of *M*-ary QAM and of *M*-phase PSK systems are similar, the performance of higher-state (more than four-state) QAM systems and PSK systems is not identical. For example, the constellation diagram and the performance of a 16-QAM system is not identical to that of a 16-PSK system [Feher, 7.2]. The block diagram used for the generation of *M*-ary QAM and *M*-phase PSK systems can be used for the generation of *N*-ary QPRS. For example, partial-response "cosine-shaped" LPFs, in the I- and Q-channels of Fig. 7.1 convert two-level binary baseband signals into three-level partial-response signals, having a controlled amount of intersymbol-interference (ISI). In this case an $N = 3 \times 3 = 9$ QPRS-modulated signal is obtained. An advantage of partial-response baseband and of QPRS-modulated signals is that they may be transmitted above the Nyquist rate, that is, in an RF bandwidth less than the symbol rate [Lender, 7.6; Wu, 7.32]. This property has been exploited in a number of practical systems. For example, one of the longest operational microwave systems, developed by Bell-Northern Research, employs a QPRS modulation technique. This Canadian coast-to-coast (Atlantic to Pacific) approximately 8000-km microwave system has a capacity of 90 Mb/s in a 40-MHz RF bandwidth. Thus in this case, the spectral efficiency of this radio system is 90 Mb/s : 40 MHz = 2.25 bits/s/Hz. This efficiency is 2.25 : 2 = 1.125, which is 12.5% higher than the spectral efficiency of theoretical QPSK systems [Lender, 7.6] operated at the Nyquist rate. In general, pulse-shaping bandlimitation in a modem is achieved by LPFs and/or BPFs (see Fig. 7.1). Note that Q-double-sideband modulators have the same RF spectral efficiency as I- or Q-baseband systems. Also, the spectral efficiency and the performance of the double-sideband QAM system is identical to that of a single-sideband (SSB) QAM digital system. This is so because in QAM systems there are two sidebands and also two independent baseband channels. In SSB systems there is only one baseband stream (in the Q-SSB implementation the Q-baseband stream does not contain independent information). The required modulated carrier-power-to-noise-power ratio, *C/N*, measured in the *Nyquist* IF or RF *bandwidth* is equivalent with the baseband *S/N* requirement.

Nyquist filters and an overall Nyquist channel response leads to ISI-free transmission systems. The most-frequently approximated Nyquist channel response is the **raised-cosine filter**, given by

$$H(jw) = \begin{cases} \dfrac{\omega T_s/2}{\sin(\omega T_s/2)} & 0 \leq \omega \leq \dfrac{\pi}{T_s}(1-\alpha) \\[3ex] \dfrac{\omega T_s/2}{\sin(\omega T_s/2)}\cos^2\left\{\dfrac{T_s}{4\alpha}\left[\omega - \dfrac{\pi(1-\alpha)}{T_s}\right]\right\} & \dfrac{\pi}{T_s}(1-\alpha) \leq \omega \leq \dfrac{\pi}{T_s}(1+\alpha) \\[3ex] 0 & \omega > \dfrac{\pi}{T_s}(1+\alpha) \end{cases}$$

$$(7.4)$$

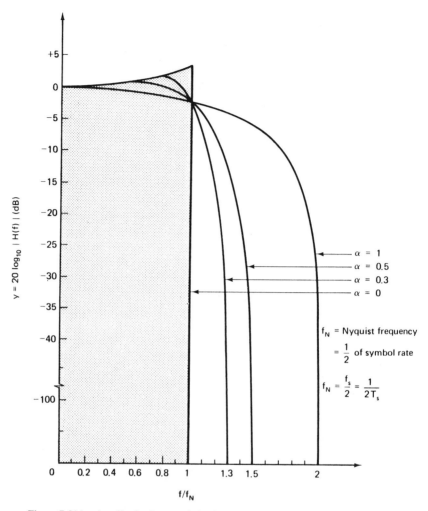

**Figure 7.2(a)** Amplitude characteristics (plotted in decibels) of the Nyquist raised-cosine channel in cascade with an ($x$/sin $x$)-shaped aperture equalizer required for rectangular pulse transmission. (With permission of Prentice-Hall, Inc. [Feher, 7.2; 7.3].)

where $\alpha$ is the roll-off factor. For $\alpha = 0$ an unrealizable minimum-bandwidth filter having a bandwidth equal to $f_N = 1/2T_s$ is obtained. For $\alpha = +0.5$, a 50% excess bandwidth is used, whereas for $\alpha = 1$, the transmission bandwidth is double the theoretical minimum bandwidth. The amplitude characteristics for various values of the bandwidth parameter are shown in Fig. 7.2(a). Theoretically, at the frequency $f = (1 + \alpha)f_N$, the attenuation has an infinite value. For practical realizations an attenuation of 20 to 50 dB is specified depending on the adjacent channel-interference allowance.

The ideal phase of the Nyquist channel is linear up to the infinite attenuation point, e.g., $1.3f_N$ for a 30% roll-off factor channel. Practical channels are phase equalized up to the 10- to 15-dB attenuation point. The ($x$/sin $x$)-shaped aperture

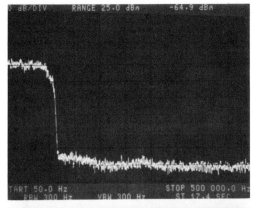

Measured power spectrum

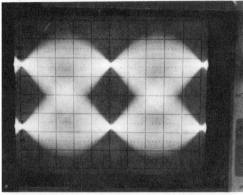

Two-level eye diagram

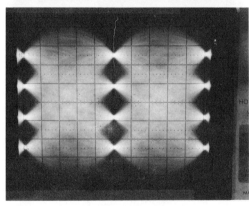

Four-level eye diagram

**Figure 7.2(b)**  Measured power spectrum and eye diagrams of $\alpha = 0.2$ raised-cosine-filtered and $(x/\sin x)$-aperture equalized equiprobable-random 200,000 symbol/s rate binary and four-level data.  In this experiment, performed at the University of Ottawa, filters designed by Karkar Electronics, Inc. of San Francisco are used.  These filters have 55 dB attenuation at $1.2 f_N$, that is, at 120 kHz.

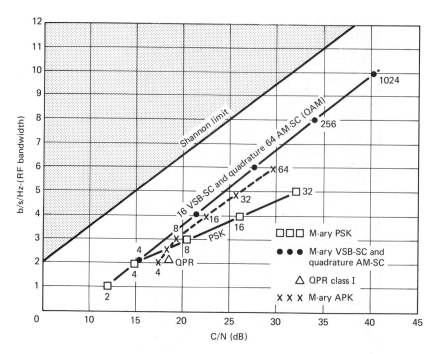

**Figure 7.3** Bit rate efficiency (in bits per second per hertz) of *M*-ary coherent PSK, VSB-SC, quadrature AM-SC, APK, and QPR systems as a function of the available *C/N* at $P(e) = 10^{-8}$. The average *C/N* is specified in the double-sided Nyquist bandwidth, which equals the symbol rate. Ideal $\alpha = 0$ filtering has been assumed. [Feher, 7.2; 7.3].

equalizer ($x = wT_s/2$) is required for binary and multilevel PAM pulse transmission [Feher, 7.2; 7.3].

The bit rate efficiency expressed in bits per second per hertz of RF bandwidth is summarized in Fig. 7.3. In this figure, ideal minimum-bandwidth ($\alpha = 0$) Nyquist channel filtering is assumed. For example, a 64-QAM system has a theoretical spectral efficiency of 6 bits/s/Hz, assuming $\alpha = 0$ filters. With $\alpha = 0.33$, the spectral efficiency of this system is 4.5 bits/s/Hz. A number of 90-Mb/s rate microwave systems use 64 QAM for the authorized 20-MHz RF bandwidth. As noticed from Fig. 7.3, a higher spectral efficiency modem requires an increased value *C/N*.

In Fig. 7.2(b) an illustrative power spectral density measurement result and the corresponding eye diagrams of raised-cosine filtered binary baseband (4-state QAM or QPSK) and of 4-level baseband (16-state QAM) are illustrated [Feher, 7.2; 7.3]. In the experiments raised-cosine $\alpha = 0.2$ filters followed by ($x/\sin x$)-shaped aperture equalizers are used. These filters, designed by Karkar Electronics Inc., have an out-of-band attenuation of 55 dB at $(1 + \alpha)f_N$, that is, at $(1 + 0.2)100$ kHz $= 120$ kHz, as $f_N$ equals 100 kHz for a 200-kbaud (200,000-symbols/s) transmission rate. Note that the resulting eye diagrams of these phase equalized filters are almost completely open.

*Demodulation* is accomplished by multiplication of the received signal by a *coherent* recovered carrier and a 90° shifted (Q) carrier. A coherent demodulator block diagram is illustrated in Fig. 7.4. The I- and Q-multipliers, followed by LPFs, are essentially amplitude and phase detectors. At the LPF outputs, the baseband eye diagrams are measured. The carrier recovery circuit generates an unmodulated carrier frequency from the received modulated data signal. As the transmitted spectrum does not contain a discrete tone, which could be directly related to the carrier RF phase, nonlinear processing such as quadrupling, remodulation, or Costas loops are required in the implementation of carrier recovery circuits [Feher, 7.2]. Carrier recovery circuit oscillators tend to introduce **phase noise**. In general, phase noise may degrade the system performance and is con-

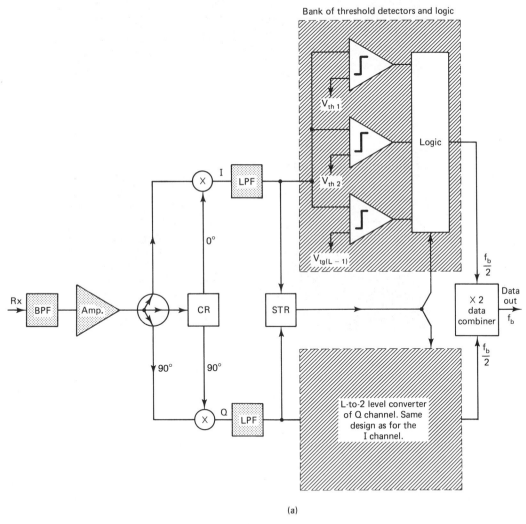

(a)

**Figure 7.4(a)** *M*-ary QAM or PSK demodulator block diagram. For synchronous demodulation the CR circuit generates an unmodulated carrier which is coherent (frequency and phase) with the transmitted carrier. (With permission of Prentice-Hall, Inc. [Feher, 7.3].)

Sec. 7.2    Principles of Modem Techniques

**Figure 7.4(b)** Measurement setup, constellation, and eye diagram display of a 16-ary QAM modem designed at the University of Ottawa (upper photograph). Doctoral and masters students, jointly with an industrial team of research engineers and postdoctoral fellows, design new improved QAM modems in Dr. Feher's laboratory (lower photograph).

sidered to be a major contributor to system-performance degradation in many satellite, LOS microwave and mobile radio systems. Particularly in relatively low bit rate systems, operated at a rate of less than 64 kb/s, phase noise may cause an "error floor" in the measured bit error rate (BER) curve. For such systems applications, differentially coherent demodulators—that is, demodulators that do not have a carrier recovery circuit with its additional oscillators—may be required.

The demodulated baseband signals, corrupted by additive white Gaussian noise (AWGN), phase noise, ISI, and possible cochannel or adjacent channel interference are fed to threshold detectors. The binary outputs of the threshold detector outputs are sampled at the symbol rate and fed to logic circuits and to a parallel-to-serial data converter (combiner).

### $P_e = f(C/N)$ and $P_e = f(E_b/N_o)$ Performance-Ideal Channel

The probability of error performance, $P_e$, of $M$-ary QAM, $M$-ary PSK, and $N$-ary QPRS systems is derived in [Feher, 7.3; Proakis, 7.5]. The final results are illustrated in the $P_e = f(C/N)$ curves shown in Fig. 7.5. Note that the mean value of $C/N$ is specified in the **double-sided Nyquist bandwidth**, that is, the symbol-rate bandwidth, and it is assumed that the probability density of the additive channel noise is Gaussian. For example, the double-sided Nyquist bandwidth for a 90-Mb/s 8-PSK radio system corresponds to 30 MHz, whereas in the case of a 16-QAM system, which also has a capacity of 90 Mb/s, this bandwidth corresponds to 22.5 MHz. The measured double-sided Nyquist bandwidth is independent of the roll-off factor ($\alpha$) of the raised-cosine filter and of the noise probability-density function. In $C/N$ measurements, particular attention should be given to the type of instrumentation used. A *true* power meter (or *rms* voltmeter) must be used to avoid the possibility of making measurement errors. A number of voltmeters, power meters, and spectrum analyzers have built-in peak detectors. With sinu-soidal waveforms, which have a 3-dB difference between the peak and rms voltage, correct results are given. However, modulated and bandlimited QPSK, QAM, and QPRS systems may have a peak-to-rms voltage ratio in the range of 3 to 15 dB. Furthermore, the noise "crest factor" of typical sources is frequently more than 15 dB. For such modulated signals and noise sources, conventional peak-detector-equipped instruments could lead to unacceptably high measurement errors.

### Relationship Between $E_b/N_O$ and $C/N$

In many references, the $P_e = f(E_b/N_O)$ requirement is derived and specified instead of the $P_e = f(C/N)$ specification, where

$$E_b = C \cdot T_b = \text{average energy of a modulated bit}$$

and

$$N_O = \text{noise power spectral density (noise in 1-Hz bandwidth)}$$

Conventional power meters measure average carrier power, $C$, and average noise power, $N$. The following simple relations are useful for the $E_b/N_O$ and $C/N$ transformations:

$$E_b = CT_b = C\left(\frac{1}{f_b}\right) \tag{7.5}$$

$$N_O = \frac{N}{B_w} \tag{7.6}$$

$$\frac{E_b}{N_O} = \frac{CT_b}{N/B_w} = \frac{C/f_b}{N/B_w} = \frac{CB_w}{Nf_b} \tag{7.7}$$

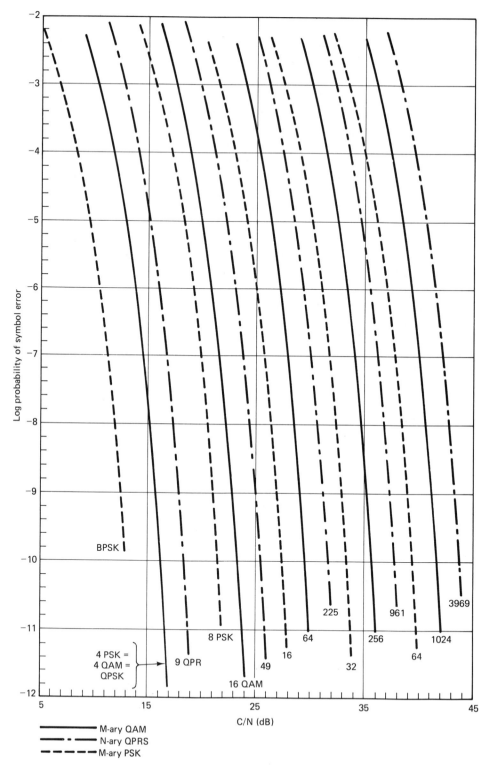

**Figure 7.5** Probability of error performance curves of *M*-ary PSK, *M*-ary QAM, and *N*-ary QPR modulation systems versus the carrier-to-thermal-noise ratio. White Gaussian noise in the double-sided Nyquist bandwidth is specified [Kucar, Feher 7.23].

$$\boxed{\frac{E_b}{N_O} = \frac{C}{N} \cdot \frac{B_w}{f_b}}\qquad(7.8)$$

The ratio $E_b/N_O$ equals the product of the average ratio $C/N$ and of the receiver noise bandwidth-to-bit-rate ratio, $B_w/f_b$. It should be noted, of course, that any instrument measuring $C/N$ can be recalibrated to read $E_b/N_O$ directly, if required [Feher, 7.2].

Figure 7.6 illustrates the $P_e = f(E_b/N_O)$ performance of BPSK, QPSK,

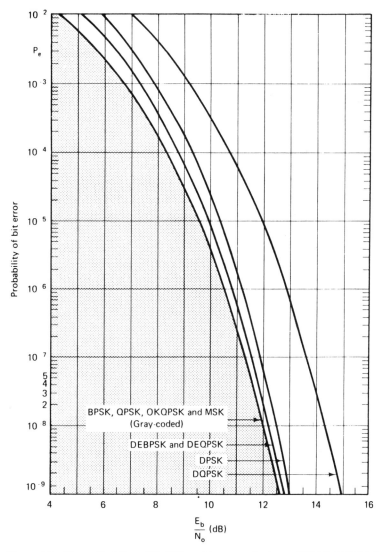

**Figure 7.6** Theoretical $P_e = f(E_b/N_o)$ performance of coherent BPSK, DEBPSK, coherent QPSK, and DQPSK modems (Gray encoded). AWGN and ISI-free model. (With permission of Prentice-Hall, Inc. [Feher, 7.2].)

OKQPSK, MSK, DEBPSK, DEQPSK, DPSK, and DQPSK modems [Feher, 7.2, Prentice-Hall, Inc., 1983]. In this figure, coherent BPSK and QPSK systems are shown to have the same $P_e = f(E_b/N_O)$ performance curve, whereas in Fig. 7.5, it is indicated that the QPSK system requires a 3-dB higher $C/N$ than the BPSK system. These two results are consistent, and from equation (7.28) we note the 3-dB differential between $C/N$ and $E_b/N_O$ because $B_w/f_b = 0.5$. Similarly, for 64-QAM systems, the $C/N$ requirement is 10 log 6 = 7.8 dB higher than the corresponding $E_b/N_O$ requirement.

> Why more and more frequently do we have the $P_e = f(E_b/N_O)$ specification instead of the $P_e = f(C/N)$ requirement?

The $P_e = f(C/N)$ measurement and $C/N$ specification is *meaningless* unless the noise bandwidth of the receiver is carefully specified. The noise bandwidth of many practical receivers is different from the double-sideband Nyquist bandwidth. Furthermore, postmodulation receive LPFs may change the effective receiver noise bandwidth, particularly if it is measured at IF or RF after a receive BPF. To enable a comparison with theoretical modem performance and also to compare modems manufactured by different suppliers, it has been found that the $P_e = f(E_b/N_O)$ specification leads to more-accurate system measurements. Note that $E_b/N_O$ is a normalized quantity, independent of the bandwidth of the receiver. For power-efficient modems, the $E_b/N_O$ parameter is used, whereas for spectrally efficient modems, the $C/N$ parameter is used in the literature as well as in later sections of this chapter.

Before the completion of our review of the principles of modulation techniques, we wish to *highlight the limitations* of a frequently taught and described *optimal demodulator structure*, namely, the correlation receiver or matched filter receiver, by addressing the following question:

> Is the matched filter (correlation receiver) optimal for the reception of synchronous bandlimited data?

The derivation of the matched filter receiver presents a mathematical insight into both the $P_e$ performance and the concepts of optimal binary systems. This receiver is optimal for single-shot transmission, that is, for systems in which only one symbol is transmitted, and in wideband systems where each pulse is confined to its bit interval (so ISI is negligible). For optimal spectral and power efficiency (i.e., a system that has a bandwidth narrower than the bit rate), the Nyquist and the matched receiver criteria must be simultaneously satisfied.

Thus the answer to our question is no; the matched filter (correlation) receiver, in itself, is insufficient for optimal efficiency in the reception of synchronous bandlimited data.

A detailed description of matched filter-correlation receivers and a comparison with bandlimited ISI-free matched Nyquist channel receivers is given in [Feher, 7.2].

**Figure 7.7** Transmit/receive 14/12-GHz Earth station antenna subsystem of Dr. Feher's Digital Communications Research Group at the University of Ottawa. Research and development projects are undertaken in cooperation with Canadian, U.S., and other overseas industries and governments.

## 7.3 POWER-EFFICIENT MODEM TECHNIQUES FOR NONLINEAR SATELLITE AND TERRESTRIAL RADIO SYSTEMS

A modem is considered to be **power efficient** if it can operate in a relatively low $E_b/N_O$ environment, defined in Table 7.1(a). From this table, we note that an ideal uncoded modem requires $E_b/N_O = 8.4$ dB for a $10^{-4}$ performance while a practical systems requirement could be about 10.8 dB. From Figs. 7.5 and 7.6, we conclude that BPSK and QPSK type of modems satisfy this low $E_b/N_O$ requirement, whereas the more spectrally efficient modems like 16-QAM, 64-QAM, . . . , require a higher $E_b/N_O$. For this reason, the theoretical spectral efficiency of uncoded power-efficient modems is limited to 2 bits/s/Hz.

So far we have considered linearly amplified modulated systems; that is, the radio transmitter has been assumed to operate in a linear mode. This implies a large output backoff and a significant loss in the power efficiency of the overall transmission system. A lower power efficiency leads to larger antenna requirements and a more-expensive transmission system.

In this section, following a literature review of power-efficient modems developed by various **research and development** (R&D) groups throughout the world, a more-detailed description of modems developed in our research laboratories is given. These modems do not require a linearly amplified channel; that is, they can utilize fully saturated power-efficient amplifiers. After performing theoretical and computer-simulation studies, we have been developing the prototype hardware of these modems. Field tests are performed over our transmit-receive Earth station and Telesat's ANIK satellites (see Fig. 7.7).

## 7.4 WHAT IS WRONG WITH QPSK? ARE THERE BETTER MODEMS?

During the 1970s and early 1980s, QPSK modulation techniques have been extensively used in high- and medium-speed time division multiple access (TDMA) and **single channel per carrier** (SCPC)—thin route—digital satellite systems. For example, the Intelsat standard SCPC system, operational in many countries throughout the globe, carries 64 kb/s QPSK data. The SBS system is equipped with QPSK modems and the Intelsat V system carries 120-Mb/s QPSK traffic. Earlier generations of terrestrial radio systems such as mobile radio and LOS microwave systems also used QPSK modems.

In analog and in digital radio systems, power and spectral efficiency are among the most-important system parameters. Since power spectra of QPSK signals exhibit sidelobes that may interfere with adjacent channels, a certain amount of filtering is necessary at the transmitter. However, this filtering results in an increased amount of envelope fluctuation in the signal, which leads to a considerable spectrum spreading due to the AM/AM and AM/PM nonlinear effects of the transmit high-power amplifiers (HPAs). These nonlinearities tend to restore the spectral sidelobes that have been previously removed by filtering. To operate the radio transmitter's HPA in a power-efficient mode (power efficiency and low cost are closely related terms), the HPA must be operated in saturation (or as close as possible). Zero-decibel output backoff indicates a saturated mode of operation, whereas for an approximately linear mode of operation of a QPSK modem, a 5-dB output-backoff (10-dB input-backoff) would be required.

Alternatively, BPFs located after the saturated HPA could reduce the QPSK sidelobes and with them the interference into adjacent channels. However, in many applications it *is not practical to design BPFs following the HPA*. For example, assume that for a speech-compressed 28-kb/s data-rate requirement, the transmit Earth station must operate at an uplink frequency of 14 GHz. The minimum bandwidth required for this QPSK system is 14 kHz. Evidently it would not be practical to design a Nyquist type of linear-phase BPF having a center frequency of 14 GHz and a bandwidth of 14 kHz. (Can you imagine a practical filter with a Q-factor of 1,000,000?!) Also, for high-speed 120-Mb/s TDMA systems, requiring frequency hopping (such as Intelsat V), stringent post-HPA filtering is not a desirable design approach. For this reason, in Intelsat V TDMA earth stations, the HPA will operate at approximately a 3- to 5-dB output-backoff and/ or fairly expensive HPA linearizers will have to be employed.

Thus for many power-efficient (saturated HPA) and spectral efficient applications, *QPSK is not a very attractive modulation technique.* One of the solutions to this problem is finding applicable modulation techniques with restored sidelobes due to nonlinear effects that are as low as possible. Several candidate techniques are examined in the following sections.

## 7.5 LITERATURE REVIEW OF POWER-EFFICIENT MODEM TECHNIQUES

This literature review, based on recent original research [Le-Ngoc, 7.10], is presented in historical progression from the widely used QPSK, OQPSK, and MSK modulation schemes, to the recently developed digital modulation schemes. This historical review gives us some insight into understanding the evolution of power-efficient modulation schemes applied to nonlinear channels.

### 7.5.1 QPSK, OQPSK, and MSK Modulation Schemes

The spectral properties and error-probability performance of QPSK, OQPSK, and MSK modulation schemes, in both linear and nonlinear channels, have been investigated in considerable depth by many authors. A detailed description of these modems and an extensive list of references is given in [Feher, 7.2]. The most-important properties of these modulation schemes are summarized in the following paragraphs, with an emphasis on parameters that yield the difference in spectral spreading caused by nonlinear amplification of the QPSK, OQPSK, and MSK signals.

QPSK is frequently used because of its 2-bit/s/Hz theoretical (1.5- to 1.8-bit/s/Hz practical) bandwidth efficiency, low $E_b/N_O$ requirement for good error probability performance, and relatively simple hardware design. Since the power spectrum of a QPSK signal exhibits side lobes that may interfere with adjacent channels, a certain amount of filtering is necessary at the transmitter. However, this filtering leads to a great amount of signal envelope fluctuation and hence a considerable spectral spreading due to the AM/AM and AM/PM nonlinear effects of the transmit HPA.

OQPSK, also known as SQPSK (staggered-QPSK), has relatively low spectral side lobes after undergoing nonlinear amplification [Feher, 7.2]. The main difference between conventional and OQPSK schemes is in the alignment of the I- and Q-baseband components of the modulated signals. This difference in time alignment in the baseband components does not change the power spectral density of the modulated signal, and hence in linear channels both QPSK and OQPSK spectra have the same shape. However, the two modulated signals respond differently when they undergo bandlimiting and then nonlinear amplification. Because of the coincident alignment of the two baseband components in QPSK modulation, an instantaneous phase transition of 180° can take place; hence the bandlimited QPSK signal exhibits 100% envelope fluctuation. In OQPSK, the two baseband signals cannot change their states simultaneously. One of the baseband signals has data transitions in the middle of the symbol interval of the other baseband signal, eliminating the possibility of instantaneous phase transitions of 180°. Thus the bandlimited OQPSK signal has a much smaller envelope fluctuation than the QPSK signal. Consequently, the absence of fast phase transitions (180°) means that nonlinear (soft or hardlimited) amplification will not regenerate the undesired high-frequency components originally removed by the bandlimiting filter.

Spectral advantages of OQPSK stem mainly from the fact that OQPSK avoids the large phase transitions of 180° associated with the QPSK format. This suggests that further suppression of spectral spreading in bandlimited nonlinear applications can be obtained if the OQPSK signal format can be modified to avoid step-phase transitions altogether. (Step functions are discontinuities characterized by wide-bandwidth frequency components.) This can be thought of as an obvious motivation for designing constant-envelope modulation schemes with continuous phase. MSK, intoduced by Doelz and Heald [Doelz, 7.46] is one such scheme.

MSK may be considered as a form of OQPSK in which the symbol pulse shape is a half-cycle sinusoid rather than the usual rectangular form. It can also be viewed as a special case of CPFSK [Pelchat, 7.47; Osborne, 7.48] with the frequency deviation ratio equal to 0.5. In this view, it is also known as FFSK because it can transmit faster pulse trains than any other ordinary FSK or BPSK of equal bandwidth and SNR. Owing to its constant envelope and continuous phase, an unfiltered MSK signal, when nonlinearly amplified (by a soft or hard limiting amplifier), suffers no significant spectral spreading into adjacent channels and provides faster spectral roll-off and lower high-order spectral side lobes than the filtered OQPSK signal. However, its spectral main lobe is 50% wider than that of unfiltered QPSK and OQPSK signals. Therefore, it causes some difficulties to determine whether or not the MSK scheme provides better power and bandwidth efficiencies than the QPSK and OQPSK schemes. Attempting to answer this question, many authors such as [Fang, 7.49; Jones, 7.50; Amoroso, 7.51] have investigated the spectral spreading and error-probability performance of QPSK, OQPSK, and MSK systems in various nonlinear channel models, including **adjacent-channel interference** (ACI) effects. Most of these investigations have been performed with the aid of computer simulations. The comparative results are usually expressed in terms of $E_b/N_O$ requirements for given error-probability performance and channel spacings. The results also depend on the parameters used in the simulation models, such as the transmit and receive channel filters and nonlinear devices. In general, it has been demonstrated that MSK provides better performance for wide channel spacings and worse performance for narrow channel spacings, relative to QPSK and OQPSK.

So far it has been shown that the baseband pulse shape, phase transition, and envelope fluctuation of a modulated signal are important parameters, which affect the spectral properties of the modulated, filtered, and amplified signals. For this reason, efforts have been expanded to search for different pulse shapes, phase transition, and envelope fluctuation characteristics that yield power and bandwidth efficient modulation techniques. Based on envelope fluctuation, these modulation techniques are classified in two categories:

1. Constant-envelope modulation techniques
2. Nonconstant-envelope modulation techniques

### 7.5.2 Constant-Envelope Modulation Techniques

A number of research engineers have studied and introduced techniques to generate an entire class of constant-envelope MSK-type modulated carriers whose spectral properties are more desirable in some applications than those of MSK or OQPSK. These studies are classified in two main categories: one considers MSK as a special case of OQPSK and the other as a special form of CPFSK.

**Constant-Envelope QAM—A Special Case of OQPSK**

An MSK-modulated signal can be generated by two antipodal pulse streams, which modulate I- and Q-carrier components, respectively (see Fig. 7.7) The modulated carrier is given by

$$y(t) = x_I(t)\cos \omega_c t + x_Q \left(\frac{t - T_s}{2}\right)\sin \omega_c t \qquad (7.9)$$

where
$$x_I(t) = \sum_{n=-\infty}^{+\infty} a_n p(t)$$
$$x_Q(t) = \sum_{m=-\infty}^{+\infty} b_m p(t)$$

are the I and Q pulse streams, respectively. Also, $a_n$, $b_m = \pm 1$ with probability of $\frac{1}{2}$; $p(t)$ is a single-interval pulse, that is, $p(t) = 0$ for $|t| > T_s/2$; and $T_s$ is the symbol interval.

The **constant-envelope property** of the modulated carrier is retained if $p(t)$ satisfies the following contraints:

$$p(t) = p(-t) \qquad (7.10a)$$

$$p^2(t) + p^2\left(\frac{T_s}{2 - t}\right) = 1 \qquad \text{for } 0 \le |t| \le \frac{T_s}{4} \qquad (7.10b)$$

For the well-known MSK-modulated carrier, the pulse $p(t)$ is

$$p(t) = \begin{cases} \cos \dfrac{\pi t}{T_s}, & |t| \le \dfrac{T_s}{2} \\ 0 & |t| > \dfrac{T_s}{2} \end{cases} \qquad (7.11)$$

The power spectrum of the MSK-type modulated carrier has the same shape as that of the pulse, $p(t)$. Thus the spectral properties of the MSK-type modulated signal can be improved by using bandwidth-efficient pulse shapes, $p(t)$. With this purpose in mind, a number of authors [Amoroso, 7.52; Simon, 7.53; Rabzel, 7.54], to name only some of the contributors, have proposed different pulse shapes, $p(t)$, to improve the spectral properties of the modulated signal $y(t)$ in retaining its envelope constancy (i.e., satisfying equations (7.10a) and (7.10b)).

Amoroso [Amoroso, 7.52] proposed a class of pulse shapes represented by

$$p(t) = \begin{cases} \cos\left[\dfrac{\pi t}{T_s} - U \sin\dfrac{4\pi t}{T_s}\right], & |t| \leq \dfrac{T_s}{2} \\ 0, & |t| > \dfrac{T_s}{2} \end{cases} \tag{7.12}$$

where $U$ is constant in the range $(0, \frac{1}{2})$. Note that for $U = 0$ the MSK signal is obtained.

The resulting modulation technique is called SFSK by the fact that the modulator can be synthesized by applying a keyed sine wave to a linear integrator followed by a linear frequency modulator (i.e., CPFSK modulator structure, as shown in Fig. 7.8).

Following in the same path, Simon [Simon, 7.53] derived a set of conditions on the input pulse shapes and derived the general power spectral density of these pulse shapes. Rabel and Pasupathy [Rabzel, 7.54] and Bazin [Bazin, 7.55] presented classes of pulse shapes having sharp spectral roll-off by applying the properties of pulse with many continuous derivatives, that is, searching for pulse shapes $p(t)$ that

1. Satisfy (7.10a) and (7.10b), that is, the constant-envelope conditions;
2. Have $N$ continuous derivatives.

This approach is based on the following well-known theorem: If $p(t)$ has $N$ continuous derivatives, then its Fourier transform $p(f)$ decays asymptotically as $|f|^{-(N+2)}$, and its power spectrum $|P(f)|^2$, as $|f|^{-2(N+2)}$.

Boutin and others [Boutin, 7.56] and Deshpande and Wittke [Deshpande, 7.57] presented classes of pulse shapes that attain a low out-of-band power fraction using prolate spheroidal wave functions, that is, searching for pulse shapes $p(t)$ that

1. Satisfy (7.10a) and (7.10b);
2. Have maximum in-band-to-total energy ratio

$$r = \frac{E_B}{E_T}$$

where the in-band energy is

$$E_B = \int_{-B}^{+B} |P(f)|^2 \, df \tag{7.13}$$

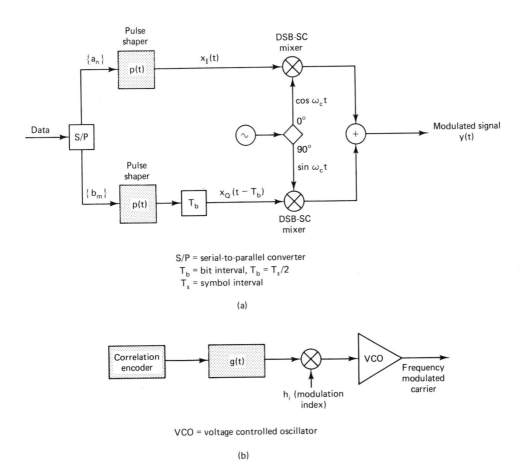

(a)

Correlation encoder → g(t) → ⊗ → VCO → Frequency modulated carrier

$h_i$ (modulation index)

VCO = voltage controlled oscillator

(b)

**Figure 7.8** (a) Block diagram of a quadrature amplitude modulator. (b) Block diagram of CPFSK modulator

and the total energy is

$$E_T = \int_{-T_s/2}^{+T_s/2} |p(t)|^2 \, dt \qquad (7.14)$$

$P(f)$ is the Fourier transform of $p(t)$ and $(-B, +B)$ is a chosen bandwidth. The results of the aforementioned studies indicate that the spectra of the obtained constant-envelope MSK-type signals have lower side lobes but wider main lobes than that of QPSK (or OQPSK) signals: The spectral side lobes of these constant-envelope modulated signals are further reduced relative to those of the MSK signal, but their spectral main lobes are still similar to that of the MSK signal (i.e., in the range of 1.5 times the symbol rate). Hence they are not attractive for narrow channel-spaced systems (i.e., systems in which the difference of the center frequencies of adjacent channels is smaller than the bit rate frequency), such as INTELSAT satellite systems.

### CPFSK Modulation Schemes

Figure 7.8(b) shows the block diagram of a binary CPFSK modulator. The binary CPFSK signal [Aulin, 7.58; 7.59] and [Bhargava, 7.9] is defined by

$$y(t) = \cos [\omega_c t + \psi(t)] \qquad (7.15)$$

The phase $\psi(t)$ is

$$\psi(t) = \sum_{i=-\infty}^{+\infty} 2\pi h_i c_i \int_{-\infty}^{t} g(x - iT_b) \, dx \qquad (7.16)$$

where
$c_i$ = the $i$th bit
$c_i$ = $\pm 1$ with equal probability of $\frac{1}{2}$
$h_i$ = the modulation index during $i$th bit interval
$T_b$ = the bit interval (i.e., $T_b = \frac{1}{2} T_s$)
$g(t)$ = the baseband pulse shape

The amplitude of $g(t)$ is chosen so that the maximum phase change over the $i$th bit interval is $c_i h_i \pi$ radians. To obtain a CPFSK-modulated carrier, the phase $\psi(t)$ should be a continuous function of time $t$, which implies that the baseband pulse shape $g(t)$ should not contain any impulses.

Efforts have been recently expended to search for various baseband pulse shapes $g(t)$, modulation index $h_i$, and correlation between symbols [Muilwijk, 7.61] to improve the system performance. CPFSK systems can be classified in the following manner.

*Full-Response CPFSK Systems.* Full-response CPFSK systems are those with the baseband pulses $g(t)$ limited within one bit interval, $T_b$.

1. *Linear-phase, constant h*: In this case the baseband pulse shape $g(t)$ is a single-interval rectangular pulse defined as

$$g(t) = \begin{cases} \dfrac{1}{2T_b} & |t| \leq \dfrac{T_b}{2} \\ 0 & |t| > \dfrac{T_b}{2} \end{cases} \qquad (7.17)$$

The well-known MSK scheme can be considered as a special case of the linear-phase full-response CPFSK scheme with modulation index $h = 0.5$.

2. *Linear phase, multi-h*: The error-probability performance or spectral properties of the previous systems can be further improved by considering schemes with different modulation index $h_i$ for each symbol interval. In [Anderson, 7.60] schemes with a set of cyclic values of the modulation index $h_i$ used in the modulator were considered.

**3.** *Smooth baseband pulse shape* $g(t)$: Another method of improving the power spectra of the linear phase system is to use continuous baseband pulse shapes $g(t)$ or, better still, pulse shapes with one or more continuous derivatives. Only constant-$h$ schemes have been considered with smooth pulses. An SFSK modulation scheme [Amoroso, 7.52] is an example of this type with $g(t)$ having one continuous derivative.

Spectral properties of full-response CPFSK signals can be summarized as follows. As the baseband pulse $g(t)$ is smoother (i.e., has a higher number of continuous derivatives), the power spectrum of the full-response CPFSK signal rolls off faster and its main lobe becomes wider. A higher average frequency deviation, $h$, gives wider spectral main lobe. The spectral main lobes of these full-response CPFSK signals are still wider than that of QPSK (or OQPSK) signals. For this reason, the full-response CPFSK signals are not attractive in systems that require a narrow adjacent channel spacing.

Bounds on the error probability performance of full-response CPFSK systems in linear channels have been computed by using their Euclidean distance properties [Aulin, 7.58]. It has been shown that these systems with $h = 0.5$ have the same error performance as MSK and QPSK. However, the performance of these systems in a nonlinear multichannel environment has not been documented in the readily available literature.

*Correlated* **(***Partial-Response***)** *CPFSK Systems*. Correlated, or partial-response, CPFSK systems are those with the baseband pulses $f(t)$ wider than one bit interval, $T_b$. Only constant-h schemes have been considered for partial-response CPFSK systems [Aulin, 7.59]. A detailed description of these systems is given in Chapter 8.

**1.** *Linear phase*: Partial-response CPFSK systems with linear phase are considered in [Muilwijk, 7.61]. For these systems the baseband pulses $g(t)$ are piecewise constant, providing piecewise-linear phase trajectories.

**2.** *Smooth pulses*: Partial-response CPFSK schemes based on smooth multiple-interval baseband pulses $g(t)$ are considered in [Muilwijk, 7.61]. The **tamed-frequency modulation** (TFM), introduced by deJager and Dekker [DeJager, 7.18], can be considered as a case of partial-response CPFSK with a smoothed pulse shape $g(t)$.

In principle, the power spectrum of the TFM signal is improved by correlating the phase shifts so that the phase shift over a bit interval is a function of one or more previous bits and the present bit. In other words, the TFM scheme makes use of the correlative level coding techniques [Lender, 7.64] in the phase shifts of the modulated signal. Rhodes [Rhodes, 7.62] extended this technique to a scheme called **frequency-shift offset quadrature** (FSOQ) modulation. Muilwijk [Muilwijk, 7.61] generalized further this correlative encoded FM technique to a class of **correlative phase-shift keying** (CORPSK) modulation techniques.

With smooth, multiple-interval baseband pulses, $g(t)$, partial-response CPFSK modulation schemes provide power spectra having both narrow main lobes and fast roll-off. For $g(t)$ occupying more than four bit intervals, the spectral main lobes of the corresponding CPFSK signals are compatible with that of QPSK, and their spectral roll-off is steeper. However, to obtain the best error probability performance signal detectors must be coherent. Reference signals (i.e., carrier and clock) must then be generated in the receiver. For CPFSK modulation techniques, the extraction of these reference signals is rather complicated in most cases. Furthermore, phase variations are small due to multilevel correlation, demanding careful carrier and clock synchronization. In summary, improvement in CPFSK schemes is achieved by increased hardware complexity and synchronization problems.

### 7.5.3 Nonconstant-Envelope PSK-Type Modulation Techniques

A constant-envelope modulated carrier would not undergo major spectral spreading effects caused by nonlinear amplification. Alternately, nonconstant-envelope bandwidth-efficient modulation techniques can be attractive in nonlinear channel applications if they can provide low spectral spreading caused by nonlinear amplification and have a good $P_e = f(E_b/N_O)$ performance and a simpler hardware structure.

Prabhu [Prabhu, 7.17] proposed bandwidth-efficient PSK-type modulation schemes with overlapping baseband pulses. The modulator block diagram for this type of modulation is similar to that shown in Fig. 7.7, but the baseband pulse $p(t)$ is now wider than one symbol interval. For equiprobable binary data the power spectrum of the modulated signal has the same shape as that of the overlapping baseband pulse $p(t)$ [Feher, 7.2]. Hence the spectral advantages are obtained by the fact that overlapping baseband pulses have a better bandwidth efficiency than nonoverlapping pulses of the same form [Greenstein, 7.63].

As an extension of Prabhu's work, Austin and Chang [Austin, 7.19] studied the spectral properties and error-probability performance of the *quadrature-overlapped raised-cosine* (QORC) and staggered QORC (SQORC) schemes in linear and nonlinear single-channel environments. The block diagram of the SQORC modulator is similar to that shown in Fig. 7.7 with the double-interval raised-cosine pulse defined by

$$p(t) = \begin{cases} \dfrac{1}{2}\left(1 + \cos\dfrac{\pi t}{T_s}\right) & |t| \leq T_s \\ 0 & |t| > T_s \end{cases} \tag{7.18}$$

where $T_s$ = the symbol interval

The results presented in [Pelchat, 47] show that the SQORC scheme has lower (better) spectral spreading, due to a saturated HPA, than OQPSK and MSK schemes.

Amoroso [Amoroso, 7.20] presented a class of quasi-bandlimited double-interval pulses $p(t)$, which can be used in PSK-type modulation schemes. This class of pulses is defined by

$$p(t) = \begin{cases} \left[ \dfrac{\sin (\pi t/T_s)}{\pi t/T_s} \right]^n & -T_s \leq t \leq T_s \\ 0 & \text{otherwise} \end{cases} \qquad (7.19)$$

where $n = 1, 2, \ldots$

These pulses are more and more bandlimited with increasing $n$ because the truncated energy outside $[-T_s, +T_s]$ is a rapidly decreasing function of $n$. This can be seen by noting that the unlimited pulse $(\sin \pi t/T_s)/\pi t/T_s$ with $-\infty < t < +\infty$ is strictly bandlimited.

Amoroso also proposed a method of generating a constant envelope modulated carrier by introducing an ideal hardlimiter after the modulator. This ideal hardlimiter removes the carrier envelope fluctuations but also introduces impairments due to signal distortions.

In the studies mentioned earlier, it has been shown that pulse-overlapping PSK-type signals provide low spectral spreading caused by nonlinear amplification. They have spectral main lobes comparable to that of QPSK and a faster spectral roll-off of the higher-order lobes.

### 7.5.4 Combined Modulated and Coded Systems

Recent research of combined modulated and coded systems led to the development of *power-efficient systems that have a spectral efficiency of more than 2 bits/s/Hz*. Ungerboeck [Ungerboeck, 7.11] introduced channel coding with multilevel phase signals. Ungerboeck's theme was to transmit $n$ bits per signaling interval with a $2^{n+1}$-ary QAM constellation, the modulator symbols being determined by a short constraint length convolutional encoder. By using a modulator constellation twice as large as that necessary for uncoded transmission and with the same dimensionality, Ungerboeck was able to produce codes without bandwidth expansion and with several decibels of power gain for surprisingly small trellises. Expansion by a factor of two is practically convenient and affords essentially as much improvement as larger signal sets would provide.

In [Ungerboeck, 7.11; 7.12], a range of spectral efficiencies (up to 5 bits/interval) with expanded signal sets was analyzed. Lebowitz and Rhodes [Lebowitz, 7.13] have studied a related special case, coded 8-PSK, for application in bandlimited, nonlinear satellite channels and have found the scheme robust relative to its uncoded counterpart (QPSK) under typical channel distortion. Taylor and Chan [Taylor, 7.14] have also provided simulation results for rate $\frac{3}{4}$-coded 8-PSK for various channel bandwidths and (TWT) operating points.

Wilson and others [Wilson, 7.7] achieve the spectral efficiency of 8-PSK—that is, 50% higher than QPSK—yet without sacrificing the usual $E_b/N_O$ penalty of 8-PSK as compared to QPSK. This power-efficiency improvement is achieved by combined modulation and coding. A $\frac{3}{4}$-rate convolutional encoder and a 16-PSK modem are combined in their investigation.

Combined modulated and coded systems are currently being investigated by a number of research teams. To limit the length of this chapter, we will not describe the new, exciting (but unfortunately somewhat complex) developments. Current issues of the IEEE Transactions on Communications and Conferences contain a detailed description of new developments of combined modulated and coded systems. A comprehensive tutorial on combined block-coded and trellis-coded modulation systems is given in [Forney, 7.68].

### 7.5.5 Summary of Literature Review

So far, major studies and contributions in the field of power-efficient digital modulation techniques have been reviewed. In most of these studies, the spectral properties and/or error-probability performance in a single-channel environment were investigated. However, the error-probability performance in a complex interference (e.g., adjacent-channel, cochannel interference) environment, synchronization aspects, and hardware feasibility of proposed modulation schemes have not been described in detail in the readily available literature.

In the following section, we describe the performance of a new family of power-efficient modem techniques and digital transmission systems that use these modems. In addition to AWGN, practical ACI and **cochannel interference** (CCI) constraints will be taken into account. Modulation techniques developed at the University of Ottawa research laboratory in collaboration with Canadian, U.S., Japanese, and European companies will be highlighted and compared to other power-efficient modulation techniques.

## 7.6 A NEW FAMILY OF POWER-EFFICIENT COHERENT MODEMS

A number of modems that are suitable for nonlinearly amplified radio systems and coherent demodulation are described. All these modems have different signal processors, spectral and $P_e = f(E_b/N_O)$ characteristics. We consider them all within one family because *prior* to a hard-limiter, the envelope of the modulated signals is not constant and a hard-limiter inserted into the transmission channel does not significantly spread the processed signal. Thus these modems are suitable for power-efficient applications requiring saturated high power amplifiers (HPAs). For most practical systems applications, a low-power hard-limiter is inserted into the IF or RF chain. Readily available low-power hard-limiters have at least ten times better (smaller) AM/PM coefficients than HPAs. The modulated, low-power, hard-limited signal has no envelope fluctuations. Thus at the output of the hard-limiter, the signal is a constant-envelope signal. This bandlimited and hard-limited signal does not suffer degradations once transmitted through the HPA of the radio transmitter.

### 7.6.1 IJF and TSI: Intersymbol-Interference/Jitter-Free and Two-Symbol-Interval (TSI) Hardlimited OQPSK Modems

An illustrative IJF baseband signal, a conceptual hardware diagram, the corresponding eye diagram, and a power spectrum are shown in Figs. 7.9 and 7.10,

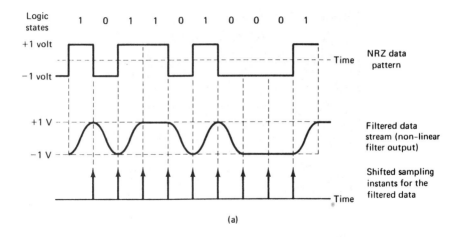

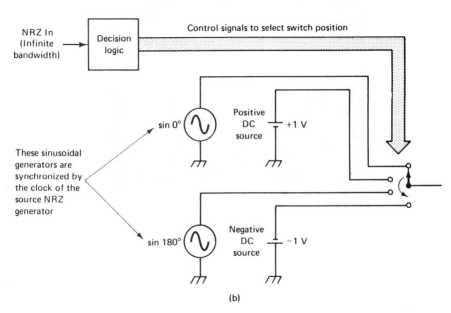

**Figure 7.9** Principle of operation of bandwidth-efficient, IJF nonlinear filters patented by Feher. (a) Block diagram. This conceptual diagram is suitable for the generation of relatively low bit rate signals (up to 200 kb/s). For higher transmission rates (up to several hundred megabits per second), trans-versal filter, two-symbol, interval-overlapped pulse transmission-implementations are used. [Feher, 7.2; 7.3; Sewerinson, 7.31].

respectively. Such IJF signals can be generated by generalized nonlinear switching filters (NLSF) patented by Feher or by TSI overlapped transversal filters [Feher, 7.2; 7.3; 7.31].

In this section, the properties of a class of hard-limited IJF Q-modulated signals are presented. The I- and Q-crosstalk effects due to hard-limiting are analytically investigated. A finite-state Markov chain model is used to calculate

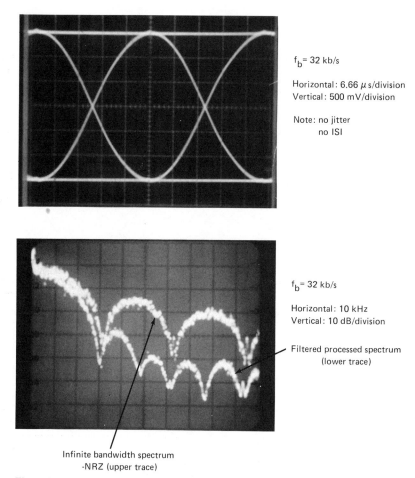

$f_b$= 32 kb/s

Horizontal: 6.66 $\mu$s/division
Vertical: 500 mV/division

Note: no jitter
no ISI

$f_b$= 32 kb/s

Horizontal: 10 kHz
Vertical: 10 dB/division

Filtered processed spectrum
(lower trace)

Infinite bandwidth spectrum
-NRZ (upper trace)

**Figure 7.10** ISI- and jitter-free eye diagrams and power spectrum of one of Feher's nonlinear IJF filters.

the power spectral density of hard-limited IJF Q-modulated signals. The model provides insights into the spectrum spreading action of the ideal hard-limiter on the IJF quadrature modulated signals with different offset values $d$ between the I- and Q-components. The analytical results show that the spectral regrowth of the hard-limited IJF Q-modulated signals is minimum at $d = T_s/2$, where $T_s$ is the symbol interval. Following our theoretical study, applications of IJF-OQPSK satellite Earth station modems operated over $\frac{14}{12}$-GHz links are presented. The ACI environment and its impact on system performance are highlighted.

### Derivation of the Power Spectral Density Function of Hard-Limited IJF-OQPSK Modulated Signals

Figure 7.11 shows the block diagram of an IJF-OQPSK-modulator [Le-Ngoc, 7.78], where $T_s$ is the symbol interval, $0 < d < T_s$ represents the relative time

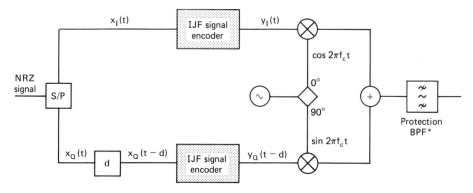

**Figure 7.11** Block diagram of an IJF-OQPSK modulator. Protection BPF removes higher-order intermodulation products. The same block diagram is used for the generation of TSI-OQPSK modems by replacement of IJF encoders with TSI processors.

offset between the I and Q NRZ signals and $x_I(t)$ is given by

$$x_I(t) = \sum_{n=-\infty}^{+\infty} x_n g(t - nT_s)$$

where

$$g(t - nT_s) = \begin{cases} 1, & |t - nT_s| \le T_s/2 \\ 0, & \text{elsewhere} \end{cases} \tag{7.20}$$

and

$$x_n = +1, \text{ or } -1 \text{ with probability of } \tfrac{1}{2}$$

The NRZ signals $x_I(t)$ and $x_Q(t - d)$ are then encoded into IJF baseband signals $y_I(t)$ and $y_Q(t - d)$, respectively:

$$y(t) = \sum_{n=-\infty}^{+\infty} y_n(t) \tag{7.21a}$$

where

$$y_n(t) = \begin{cases} s_1(t - nT_s) = s_e(t - nT_s) & \text{if } x_n = x_{n-1} = +1 \\ s_2(t - nT_s) = -s_e(t - nT_s) & \text{if } x_{n-1} = -1 \\ s_3(t - nT_s) = s_o(t - nT_s) & \text{if } x_n = +1, x_{n-1} = -1 \\ s_4(t - nT_s) = -s_o(t - nT_s) & \text{if } x_n = -1, x_{n-1} = +1 \end{cases} \tag{7.21b}$$

and $s_e(t - nT_s)$ and $s_o(t - nT_s)$ are even and odd single interval pulses satisfying

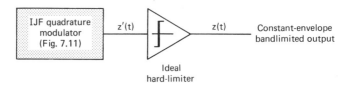

**Figure 7.12** Block diagram of an IJF Q modulator followed by an ideal hardlimiter. In a number of practical systems applications, a low-power hard-limiter precedes the system HPA. The low-power hard-limiter has a very low AM/PM (compared to the HPA) and assures a constant-envelope signal constellation.

the IJF conditions [Le-Ngoc, 7.10]

$$s_e(t - nT_s) = s_e(-t + nT_s) \qquad \text{for } |t - nT_s| < T_s/2$$

$$s_o(t - nT_s) = -s_o(-t + nT_s) \qquad \text{for } |t - nT_s| < T_s/2$$

$$s_e(t - nT_s) = s_o(t - nT_s) = 0 \qquad \text{for } |t - nT_s| \geq T_s/2 \tag{7.22}$$

$$s_e\left(\frac{T_s}{2}\right) = s_o\left(\frac{T_s}{2}\right) \neq 0$$

The **power spectral density** (PSD) function of the encoded signal $y(t)$ is

$$Y(f) = \frac{1}{T_s} [s_e(F) \cos \pi FT_s - js_o(F) \sin \pi FT_s]^2 \tag{7.23}$$

where $S_e(f)$ and $S_O(f)$ are Fourier transforms of $s_e(t)$ and $s_o(t)$, respectively.

It can be shown that the PSD of the IJF Q-modulated signal

$$z(t) = (y_I(t) + jy_Q(t - d))e^{j2\pi f_c t} \tag{7.24}$$

has the same shape as that of the baseband signal $y_I(t)$ or $y_Q(t - d)$.

Consider the IJF-Q modulator followed by an ideal hard-limiter shown in Fig. 7.12.

The hard-limited IJF quadrature modulated signal $z'(t)$ can be represented as

$$z'(t) = [y_I'(t) + jy_Q'(t - d)]e^{j2\pi f_c t} \tag{7.25a}$$

where

$$y_I'(t) = C \frac{y_I(t)}{\sqrt{y_I^2(t) + y_Q^2(t - d)}} \tag{7.25b}$$

$$y_Q'(t) = C \frac{y_Q(t)}{\sqrt{y_I^2(t) + y_Q^2(t - d)}} \tag{7.25c}$$

$$C = \text{a constant}$$

Equations (7.25b) and (7.25c) indicate that for any change in $y_I(t)$ (or $y_Q(t)$), $y_Q'(t)$ and $y_I'(t)$ are affected. In other words, hard-limiting introduces crosstalk between the I and Q equivalent baseband components.

The PSD function of the *hard-limited* IJF Q-modulated signal is derived as follows.

Let the baseband component $Y'_I(t)$ (or $y'_Q(t - d)$) be represented as the sum of an infinite number of single-interval pulse shapes $q(t_n)$ defined in one symbol interval $(n - \frac{1}{2}T_s, (n + \frac{1}{2})T_s)$, that is,

$$y'_I(t) = \sum_{n=-\infty}^{+\infty} q(t_n), \quad t_n = t - nT_s. \tag{7.26}$$

From the above equations, it can be shown that $q(t_n)$ belongs to the set $\{v_i(t_n), i = 1, \ldots, 16\}$ where $v_i(t_n)$ are represented by

$$v_j(t_n) = -v_{17-j}(t_n) = C \frac{s_e(t_n)}{\sqrt{s_e^2(t_n) + s'^2_j(t_n)}} \tag{7.27a}$$

$$v_{j+4}(t_n) = -v_{13-j}(t_n) = C \frac{s_o(t_n)}{\sqrt{s_o^2(t_n) + s'^2_j(t_n)}} \tag{7.27b}$$

where $j = 1, 2, 3, 4$ and pulse shapes $s'_j(t_n)$, $j = 1, 2, 3, 4$, are

$$s'_1(t_n) = \begin{cases} s_e(-t_n - d) & -\dfrac{T_s}{2} \leqslant t_n < d - \dfrac{T_s}{2} \\[2ex] s_e(t_n - d)d & -\dfrac{T_s}{2} \leqslant t_n < \dfrac{T_s}{2} \end{cases}$$

$$s'_2(t_n) = \begin{cases} s_o(-t_n - d) & -\dfrac{T_s}{2} \leqslant t_n < d - \dfrac{T_s}{2} \\[2ex] -s_o(t_n - d)d & -\dfrac{T_s}{2} \leqslant t_n < \dfrac{T_s}{2} \end{cases} \tag{7.28}$$

$$s'_3(t_n) = \begin{cases} s_o(-t_n - d) & -\dfrac{T_s}{2} \leqslant t_n < d - \dfrac{T_s}{2} \\[2ex] s_e(t_n - d)d & -\dfrac{T_s}{2} \leqslant t_n < \dfrac{T_s}{2} \end{cases}$$

$$s'_4(t_n) = \begin{cases} s_e(-t_n - d) & -\dfrac{T_s}{2} \leqslant t_n < d - \dfrac{T_s}{2} \\[2ex] -s_o(t_n - d)d & -\dfrac{T_s}{2} \leqslant t_n < \dfrac{T_s}{2} \end{cases}$$

In each symbol interval, $q(t_n)$ is chosen from the set $\{v_i(t_n), i = 1, 2, \ldots, 16\}$. The waveshape of $q(t_n)$ is dependent on the waveshape of the previous $q(t_{n-1})$, $y_I(t_n)$, and $y_Q(t_n)$. The stationary probabilities

$$p_i = P\{q(t_n) = v_i(t_n)\}, \quad i = 1, 2, \ldots, 16$$

and transition probabilities

$$p_{ik} = P\{q(t_n) = v_k(t_n)|q(t_{n-1}) = v_i(t_{n-1})\},$$

$$i, k = 1, 2, \ldots, 16$$

can be obtained by investigating the encoding law of $y_I(t)$, $y_Q(t)$ and construction of $v_i(t_n)$, $i = 1, 2, \ldots, 16$, that is,

$$p_i = \tfrac{1}{16}, \qquad i = 1, 2, \ldots, 16$$

and $p_{ik}$'s are arranged as entries in the $16 \times 16$ transition matrix

$$P = [p_{ik}, \quad i, k = 1, 2, \ldots, 16] = \frac{1}{4} \begin{bmatrix} A & 0 & A & 0 \\ A & 0 & A & 0 \\ 0 & B & 0 & B \\ 0 & B & 0 & B \end{bmatrix}$$

where $A$ and $B$ are $4 \times 4$ matrices

$$A = \begin{bmatrix} 1 & 0 & 0 & 1 \\ 0 & 1 & 1 & 0 \\ 1 & 0 & 0 & 1 \\ 0 & 1 & 1 & 0 \end{bmatrix}, \qquad B = \begin{bmatrix} 0 & 1 & 1 & 0 \\ 1 & 0 & 0 & 1 \\ 0 & 1 & 1 & 0 \\ 1 & 0 & 0 & 1 \end{bmatrix} \qquad (7.29)$$

and 0 is a $4 \times 4$ null matrix.
The PSD function of $y_I(t)$ is given by [Lindsey, 7.16]

$$Y(f) = \frac{1}{T_s^2} \sum_{n=-\infty}^{+\infty} \left| \sum_{i=1}^{16} p_i V_i\left(\frac{n}{T_s}\right) \right|^2 \delta\left(f - \frac{n}{T_s}\right)$$

$$+ \frac{1}{T_s} \sum_{i=1}^{16} p_i |V_i(f)|^2 \qquad (7.30)$$

$$+ \frac{2}{T_s} \text{Re}\left\{ \sum_{i=1}^{16} \sum_{k=1}^{16} p_i V_i^*(f) V_k(f) V_{ik} \right\}$$

where

$$V_i(f) = \text{the Fourier transform of } v_i(t), \, i = 1, \cdots, 16$$

$$V_i^*(f) = \text{the complex conjugate of } V_i(f)$$

$$V_{ik} = \sum_{n=1}^{\infty} p_{ik}^{(n)} e^{-j2\pi f T_s}$$

$$P_{ik}^{(n)} = \text{the } ik\text{th entry in the matrix } P^n$$

$$P^n = \underbrace{P \times P \times \cdots \times P}_{n \text{ times}}$$

and

$$\text{Re} \{\cdots\} \text{ is the real part of } \{\cdots\}$$

Substituting values of $V_i(f)$, $V_{ik}$, and $p_i$ into (7.30) yields

$$Y(f) = \frac{1}{8T_s} \left\{ 2 \sum_{i=1}^{8} |V_i(f)|^2 + \text{Re} \{([V_1(f) + V_4(f) \right.$$

$$- V_5(f) - V_8(f)]^2 + [V_2(f) + V_3(f) \qquad (7.31)$$

$$\left. - V_6(f) - V_7(f)]^2)e^{-j2\pi fT_s} \right\}$$

It can be shown that the hard-limited IJF Q-modulated signal $z'(t)$ has the same spectral shape as the baseband component $y'_I(t)$ (or $y'_Q(t)$), that is, its normalized PSD function can be represented by

$$Z_h(x) = Z_0 \left\{ 2 \sum_{i=1}^{8} |V_i(x)|^2 \right.$$

$$+ \text{Re} \{([V_1(x) + V_4(x) - V_5(x) - V_8(x)]^2 \qquad (7.32)$$

$$\left. + [V_2(x) + V_3(x) - V_6(x) - V_7(x)]^2) \cdot e^{-j2\pi x} \right\}$$

where $Z_0$ is a normalizing coefficient chosen such that $Z_h(0) = 1$ and

$$x = (f - f_c)T_s$$

Equation (7.32) represents the general PSD function of hard-limited IJF Q-modulated signals with a given offset value $d$. Equation (7.28) indicates that different values of $d$ result in different sets of pulse shapes $s'_i(t_n)$, $i = 1, 2, 3, 4$ and, hence, for the same IJF Q-modulated signal, $z(t)$, the spectral regrowth of the hard-limited IJF Q-modulated signal $z'(t)$ depends on the offset value $d$ between $I$ and $Q$ baseband components.

### An Illustrative IJF Double-Interval Pulse

As an illustrative example, the spectral regrowth of the hard-limited IJF-OQPSK modulated signal having the IJF double-interval pulse

$$s(t) = \frac{1}{2} \left[ 1 + \cos \left( \frac{\pi t}{T_s} \right) \right] \qquad (7.33)$$

is computed, based on equation (7.32).

This double-interval pulse is an element of a signal pattern and an eye diagram such as the one shown in Figs. 7.9 and 7.10. The results shown in Fig. 7.13 indicate that the minimum spectral regrowth of the hard-limited IJF Q-modulated signal is obtained at $d = T_s/2$. The IJF Q-modulation scheme with $d = T_s/2$ is called **IJF-OQPSK**.

Figure 7.14 shows the computed eye diagrams of the equivalent baseband components of the hard-limited IJF Q-modulated signal with different values of $d$ using (7.25b) and (7.25c). Note the IJF conditions apply to the transmitted base-

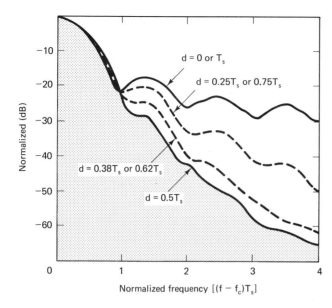

**Figure 7.13** Power spectra of hard-limited IJF Q-modulated signals with different values $d$ (for $s(t) = \frac{1}{2}[1 + \cos(\pi t/T_s)]$). [Feher 7.40; Le-Ngoc 7.41; 7.42; 7.10]

band eye diagrams. However, after Q-modulation, nonlinear amplification, and demodulation, the signals contain ISI and jitter. The QORC scheme is a subclass of the IJF Q-modulation techniques for $s(t) = \frac{1}{2}[1 + \cos(\pi t/T_s)]$. Figure 7.14 shows that the ISI introduced by the $I$ and $Q$ crosstalk effect increases with $d$. It is maximum (3 dB peak-to-peak) at $d = T_s/2$ and zero at $d = 0$.

Both spectral spreading and $I$ and $Q$ crosstalk affect the $P_e$ performance of the IJF Q-modulated signals.

In order to verify that an ideal hard-limiter is a good first approximation for spectral spreading (regrowth) of modulated and bandlimited IJF-OQPSK type of signals, we performed a number of computer simulations and measurements. In these measurements we replaced the ideal hard-limiter with standard high-power amplifiers. From the results of Fig. 7.15 we note that the HPA with a 0-dB **backoff** (BO) led practically to the same spectral restoration as the low-power hard-limiter.

### TSI Modems

A new class of TSI pulse shapes is introduced in this section. These pulse shapes are an extension and have additional desirable properties to our previously described class of IJF signals [Le-Ngoc, 7.10].

So far we have highlighted the properties of the pulse shapes for $N = 1$ and $m_o = 1$, that is,

$$
S_1(t) = \begin{cases} \frac{1}{2}\left[1 + \cos\dfrac{\pi t}{T_s}\right] & |t| \le T_s \\[2em] 0 & \text{elsewhere} \end{cases}
$$

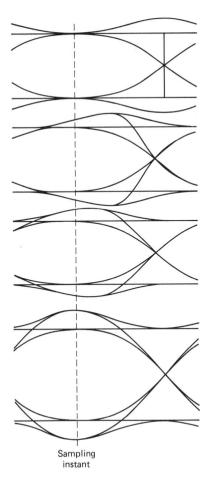

**Figure 7.14** Computed eye diagrams of the demodulated baseband components of hard-limited IJF Q-modulated signals. (a) $d = 0$ (zero ISI). (b) $d = 0.25T_s$ (c) $d = 0.38T_s$ (d) $d = 0.5T_s$ (maximum 3-dB peak-to-peak ISI).

Sampling instant

Let us define our *new class* of TSI pulse shapes by

$$s_n(t) = \begin{cases} \frac{1}{2}\left\{1 - \dfrac{\sin\left[\pi/nT_s(|t| - T_s/2)\right]}{\sin(\pi/2n)}\right\} & |t| \leq T_s, \, n = 1, 2, \ldots \\ 0 & \text{elsewhere} \end{cases} \tag{7.34}$$

The maximal value of $s_n(t)$ equals 1 at $t = 0$, and the minimal value equals zero at $t = \pm T_s$. These values must hold in order to have an ISI-free condition. The amplitudes of $s_n(t)$ at $t = \pm T_s/2$ must be equal to 0.5 in order to have a jitter-free condition.

The pulse shapes $s_n(t)$, illustrated in Figs. 7.16 and 7.17, can be separated

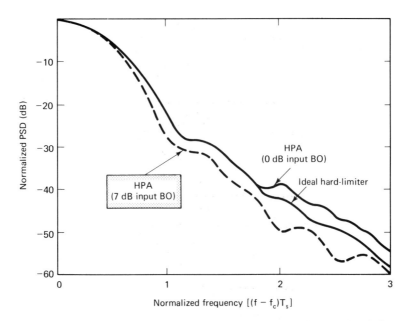

**Figure 7.15** Comparison of the spectral spreading of an IJF-OQPSK signal after an ideal hard-limiter and after a high-power amplifier [Le-Ngoc 7.10; 7.42]. Note that in the $0 \leq (f - f_c) T_s \leq 2$ normalized frequency region, the ideally-hard-limited and high-power amplified (a real HPA amplifier is used here) power spectral density is, by all practical means, the same.

into a pair of even and odd single-interval pulses $s_{en}(t)$ and $s_{on}(t)$ given by

$$
s_{en}(t) = \begin{cases} s_n\left(\dfrac{t - T_s}{2}\right) + s_n\left(\dfrac{t + T_s}{2}\right) & |t| \leq \dfrac{T_s}{2}, \text{ even pulse} \\ \\ 0 & \text{elsewhere} \end{cases}
$$

$$
s_{on}(t) = \begin{cases} s_n\left(\dfrac{t + T_s}{2}\right) - s_n\left(\dfrac{t - T_s}{2}\right) & |t| \leq \dfrac{T_s}{2}, \text{ odd pulse} \\ \\ 0 & \text{elsewhere} \end{cases}
$$

Hence, the even and odd functions of $s_n(t)$ given in equation (7.34) can be found as:

$$
s_{en}(t) = \begin{cases} 1 & |t| \leq \dfrac{T_s}{2} \\ \\ 0 & \text{elsewhere} \end{cases} \tag{7.35}
$$

$$
s_{on}(t) = \begin{cases} \dfrac{\sin(\pi t)/nT_s}{\sin(\pi/2n)} & |t| \leq \dfrac{T_s}{2} \\ \\ 0 & \cdot \text{ elsewhere} \end{cases} \tag{7.36}
$$

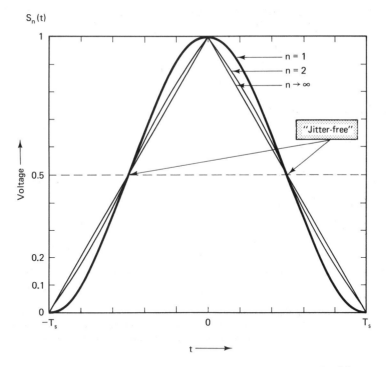

**Figure 7.16** TSI pulse shapes $s_n(t)$. These pulse shapes transmitted in a synchronous sequence for the $Y_{in}(t)$ and $Y_{on}(t)$ data streams. Note that these pulse shapes are also IJF. [Pham Van, 7.36].

The amplitudes of $s_{on}(t)$ are always equal to either $+1$ or $-1$ at the transition points ($t = \pm T_s/2$) to eliminate any discontinuity that may occur when the waveforms are generated. For all values of $n$, the values of $S_{en}(t)$ are the same and are equal to 1.

By replacement of the IJF processors described in Section 7.6.1 with TSI processors (see Fig. 7.11), TSI-OQPSK modulated signal is generated. Eye dia-

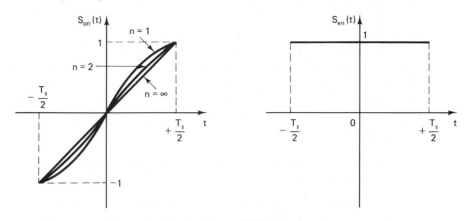

**Figure 7.17** Even and odd pulse shapes of TSI pulse shapes.

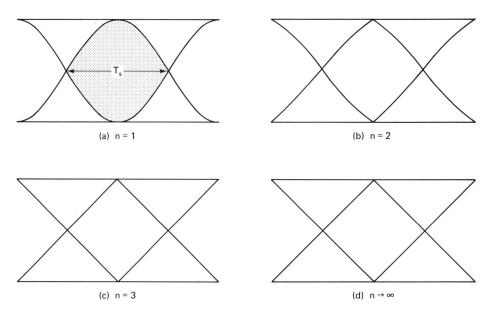

**Figure 7.18** (a), (b) Signal eye diagrams generated by TSI encoders. (a) corresponds to the IJF-OQPSK signaling element described in the previous section (see (7.33)). (c), (d) Signal eye diagrams generated by TSI encoders [Pham Van, 7.36].

grams obtained at the output of TSI processors for $n = 1, 2, \ldots$ are illustrated in Fig. 7.18. The TSI IJF eye diagrams at the modulator (mixer) inputs are illustrated.

The power spectrum derivation of TSI-OQPSK modems is similar to the one presented in the previous section and for this reason is not repeated here. The resultant baseband PSD (prior to modulation and nonlinear amplification) and the modulated and nonlinearly amplified (saturated HPA with O-dB backoff) spectrum of illustrative TSI signals are shown in Fig. 7.19 and 7.20. Note that the TSI signal, having an $n = 1$ parameter, is identical to the IJF signal described in the previous section. TSI signals with n = 2, 3, . . . have a narrower main lobe than the one with $n = 1$.

A major *advantage* of IJF and TSI-OQPSK modems, as compared to conventional filtered QPSK and OQPSK modems, may be observed from the nonlinearly amplified PSDs, which are 38 dB below the maximal density centered around the carrier frequency $f_c$. The spectral components that are the cause of significant ACI are 10 dB lower with our patented signal-processed modems. We also note that in the $O < | (f - f_c)T_s| < 1$ region, the N = 2 parameter TSI processor has a "better" spectrum than the IJF processed (N = 1) modem.

### Performance Advantage of IJF and TSI-OQPSK Modems Operated in an ACI Environment

ACI combined with additive Gaussian noise, phase noise, and ISI is among the most-frequent causes of system performance degradations. Practical receivers

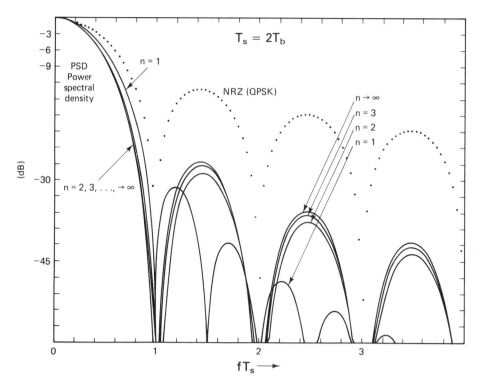

**Figure 7.19** Baseband power spectral density of TSI signals prior to modulation and nonlinear amplification [Pham Van 7.36].

have protection BPFs and also post-demodulation LPFs. The baseband equivalent of the cascade of these filters [Feher, 7.2] removes the ACI components that do not fall into the bandwidth of the desired receiver. Thus the total ACI power of the adjacent channel signals $P_U$ and $P_L$ (see Fig. 7.21) and their respective spectral densities and integrated out-of-band power (which fall within the desired receiver bandwidth) are among the most-critical system parameters. Here an analogy with our daily life may be appropriate. The noise interference caused by our first neighbors is more annoying than the interference caused by the second, third, and farther-away neighbors, assuming that they transmit the same interference power.

In Fig. 7.21 a desired nonlinearly amplified modulator and two interfering modulators (upper and lower ACI) are illustrated. For simplicity, let us assume that the desired demodulator is preceded by an ideal brick-wall BPF. The purpose of this BPF is to minimize the ACI power falling into the band of the desired receiver. We also assume that the integrated power $P_D$ of the desired modulated carrier is lower than the power of the adjacent interfering carriers, $P_L$ and $P_U$. The depicted power differential corresponds to a very realistic, frequently occurring system scenario. For example, assume that the desired signal is transmitted from a smaller-diameter antenna and lower-transmit-power Earth station than the adjacent signals. Other causes of the power differential may be a fade of the desired uplink signal or simply a pointing error of the transmit Earth station antenna.

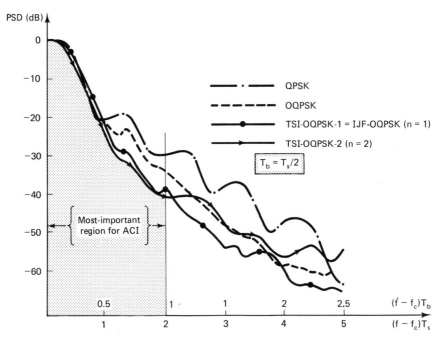

**Figure 7.20** Nonlinearly amplified (saturated HPA with O-dB input BO) PSD of TSI- and IJF-OQPSK signal [Pham Van 7.36].

Uplink fade depth differentials are more critical in $\frac{14}{12}$-GHz satellite systems than in 6/4-GHz systems. In a number of high-speed (120 Mb/s) experiments on 14/12-GHz satellites, fades of up to 14 dB have been measured. In mobile radio systems, fade-depth differentials of more than 40 dB have been observed. As the main (desired) channel is more attenuated, the ACI becomes more and more predominant, and for this reason the modulation technique that creates less ACI leads to better performance.

In Fig. 7.22, two examples are given that highlight the impact of ACI on the $P_e = f(E_b/N_o)$ degradation of IJF and TSI-OQPSK modems. As a reference, the performance degradation curves of conventional QPSK and OQPSK systems are also included. Here we assume that the channel spacing is at 92% and 77% of the bit rate. For example, in 64-kb/s SCPC satellite systems, adjacent carriers would be spaced 58.88 kHz and 49.28 kHz, respectively. This spacing is less efficient than is used in the currently operational, linearly amplified INTELSAT SCPC systems, in which a spacing of 45 kHz is used. However, we wish to reiterate that the major advantage of these IJF and TSI systems is in their power efficiency— namely, that fully saturated HPSs can be used. For a channel spacing equaling 77% of the bit rate, an uplink fade depth of 9 dB and a specified $P_e = 10^{-4}$, we note from Fig. 7.22(b) that the TSI-OQPSK ($n = 2$) system suffers a 2.3-dB degradation, the IJF-OQPSK ($n = 1$) system has a 3-dB degradation, and the conventional QPSK system has a 6-dB degradation from the theoretical

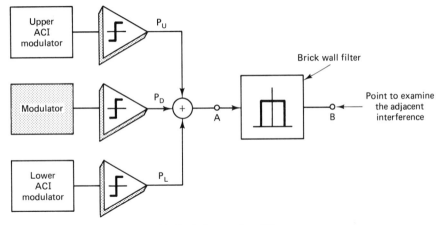

(a) Block diagram of an ACI.

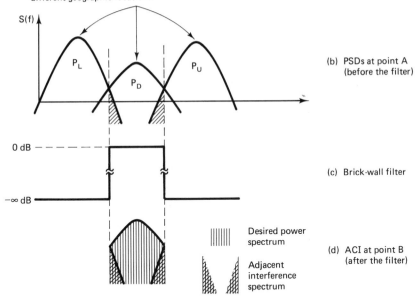

The power of individual modulated signals is different. This variation is due to the different EIRP of the transmit earth stations and the different free-space losses from different geographic locations.

(b) PSDs at point A (before the filter)

(c) Brick-wall filter

Desired power spectrum

Adjacent interference spectrum

(d) ACI at point B (after the filter)

**Figure 7.21**  An example of an ACI environment.

$P_e = f(E_b/N_o)$ performance [R-44].   That is, the hard-limited TSI-OQPSK system has 4 dB less degradation than the conventional nonlinearly amplified QPSK system.   AT $P_e = 10^{-8}$, this differential amounts to almost 6 dB.   A 6-dB differential in the $E_b/N_0$ requirements could enable up *to 50% reduction of the satellite Earth station antenna diameter.*

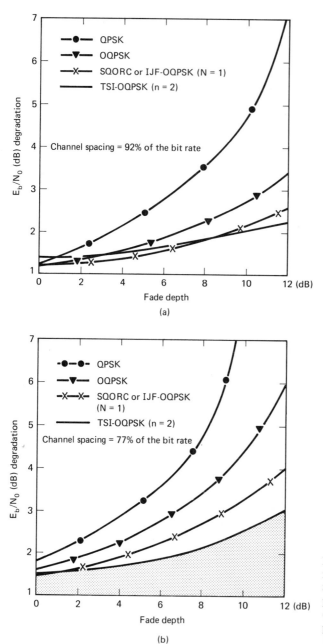

**Figure 7.22** $E_b/N_O$ degradation versus fade depth of the desired channel. (a) Spacing of 92% of the bit rate, compared to $E_b/N_O = 8.4$ dB for $P_e = 10^{-4}$ (b) Spacing of 77% of the bit rate, compared to $E_b/N_o = 8.4$ dB for $P_e = 10^{-4}$. [Pham Van, 7.36]

### 7.6.2 SQAM: Superposed-QAM Technique

In our search for power-efficient modems we found that SQAM offers an attractive alternative to IJF and TSI-OQPSK modems described in the previous Section [See 7.33, 7.35]. The functional diagram of an SQAM modulator is similar to the one

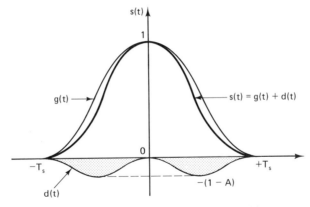

(a) An example of an SQAM double-interval pulse s(t), which is a superposition of g(t) and d(t).

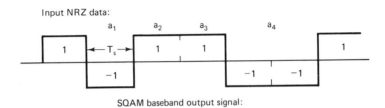

SQAM baseband output signal:

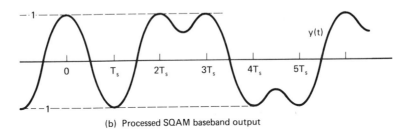

(b) Processed SQAM baseband output

**Figure 7.23** SQAM baseband signal shaping process using the double-interval pulse overlapping concept. (*Note:* $T_s = 2T_b$ is the symbol interval, where $T_b$ is the bit interval) [Seo 7.33; 7.35].

given in Fig. 7.8(a). A conventional Q-modulator with two DSB-SC mixers is used as the modulator. In baseband, an offset delay having a value of $T_b = T_S/2$ is used. The baseband pulse shaper circuits may be IJF, TSI or superposed pulse shape processors.

SQAM baseband signals can be generated by using either the pulse overlapping concept or the NLSF technique [Feher, 7.2]. Figure 7.23 illustrates the SQAM baseband signal-shaping process using the double-interval pulse-overlapping concept [Prabhu, 7.17; Austin, 7.19]. The SQAM baseband signal processor has an impulse response $s(t)$, which is a double-interval $(2T_s)$ raised-cosine pulse super-posed with weighted single-interval $(T_s)$ raised-cosine pulses; that is,

$$s(t) = \tfrac{1}{2}\left(1 + \cos\frac{\pi t}{T_s}\right) + d(t) \tag{7.37}$$

where

$$d(t) = -\frac{1-A}{2}\left(1 - \cos\frac{2\pi t}{T_s}\right) \qquad 0.5 \leq A \leq 1.5, \ -T_s \leq t \leq T_s \tag{7.38}$$

and $A$ is an amplitude parameter of the SQAM signal.

Since the input signal duration is $T_s$ seconds and $s(t)$ has a $2T_s$-second duration, the processed SQAM output signal is obtained by overlapping the double-interval pulses $s(t - nT_s)$ and $s[t - (n + 1)T_s]$. Therefore, the SQAM baseband output signals $y_1(t)$ to $y_4(t)$ for the consecutive two data inputs of $a_n = a_{n+1} = -1, a_n = 1; a_{n+1} = -1, a_n = -1;$ and $a_{n+1} = 1,$ and $a_n = a_{n+1} = 1$ are defined, respectively, as follows:

$$y_1(t) = -A - (1 - A)\cos\left(\frac{2\pi t}{T_s}\right)$$

$$y_2(t) = -\cos\left(\frac{\pi t}{T_s}\right)$$

$$y_3(t) = \cos\left(\frac{\pi t}{T_s}\right) \tag{7.39}$$

$$y_4(t) = A + (1 - A)\cos\left(\frac{2\pi t}{T_s}\right)$$

where
$$0.5 \leq A \leq 1.5$$
$$0 \leq t \leq T_s$$

In 7.41 the value of the amplitude parameter $A$, in the $I$- and $Q$-channel signals respectively, is selected to control the envelope fluctuations of the QAM signal. In SQAM, the maximum envelope fluctuation changes from 0.7 dB (for $A = 0.7$) to 3.0 dB (for $A = 1.0$). The IJF-OQPSK signal described in the previous section is a special case of SQAM signals for $A = 1.0$.

Illustrative measured eye patterns of SQAM baseband signals, measured at the transmit superimposed baseband processor output, prior to modulation, are given in Fig. 7.24. Note that the transmit (premodulation) **vertical eye openings** (VEO), for different values of $A$, have full amplitudes at every sampling instant. Thus in a linearly amplified system, the SQAM signals would be IJF and would have a negligible $P(e)$ performance degradation. In a nonlinearly amplified system, the demodulated VEO for $A < 1.0$ is wider than that for $A \geq 1.0$; hence it is reasonable to expect that the $P(e)$ performance for $A < 1.0$ is better than that for $A \geq 1.0$. Figure 7.25 compares the power spectrum of the SQAM ($A = 0.8$)

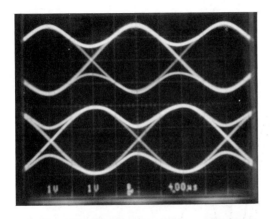

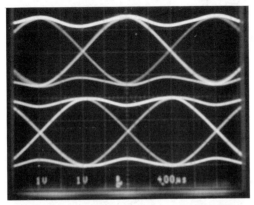

**Figure 7.24** Experimental measured eye patterns (I- and Q-channel) of SQAM baseband signals of a 128-kb/s modem in a linear channel. (a) $A$ = 0.7. (b) $A$ = 0.85. (c) $A$ = 1.0. These eye patterns were measured at the transmitter, prior to modulation (Vertical:1 V/division; horizontal: 4 μs/division.) (Courtesy of J. So Seo, Ph.D Candidate, University of Ottawa [Seo, 7.33; 7.35].)

signal to those of MSK, QBL, IJF-OQPSK (or SQORC), and TFM in a hard-limited channel. Note that the SQAM signal has spectral advantages over QPSK and MSK signals and comparable spectral properties to TFM signals [Austin, 7.19; Amoroso, 7.20; DeJager, 7.18].

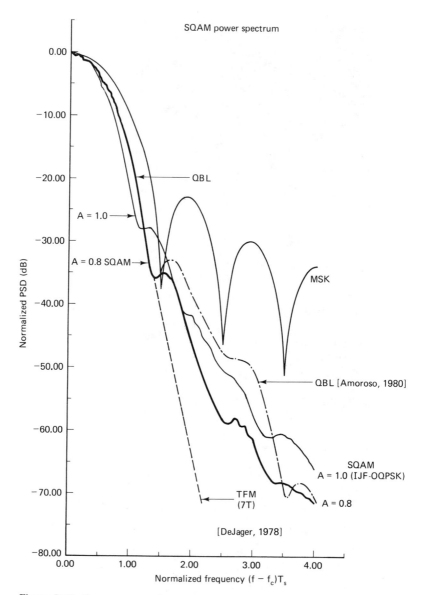

**Figure 7.25** Power spectra (one-sided) of SQAM, MSK, QBL, IJF-OQPSK (SQORC), and TFM modulated signals in a nonlinear (hard-limited) channel. [Seo 7.33; 7.35].

## 7.6.3 Performance of SQAM and Other Types of Modems in an ACI Environment

The ACI environment of three satellite Earth station modulators is illustrated in Fig. 7.21. As stated earlier, in many practical applications, the interference caused by immediate adjacent channels has the most significant impact on the performance degradation, whereas the impact of the more distant channels is not as harmful. Thus it is sufficient to analyze the impact of two adjacent modems on the BER performance of the desired modem.

For the computation of the average error probability, eight values of random phases and eight values of random symbol timings are taken for each interfering signal. Let $P_e(\theta_u, \theta_l, \tau_u, \tau_l)$ be the marginal error probability assuming phases $\theta_u$, $\theta_l$ and symbol timing $\tau_u$, $\tau_l$ of the interfering adjacent channels, where subscripts $u$ and $l$ represent the upper adjacent channel and lower adjacent channel, respectively. The average error probability is defined as

$$P(e) = \tfrac{1}{8}[P_e(\theta_{u1}, \theta_{l1}, \tau_{u1}, \tau_{l1}) + P_e(\theta_{u2}, \theta_{l2}, \tau_{u2}, \tau_{l2}) + \cdots + P_e(\theta_{u8}, \theta_{l8}, \tau_{u8}, \tau_{l8})]$$

The optimum sampling instant is chosen by calculating error probabilities for a given $E_b/N_O$ for each of the sampling points in a symbol interval, until the minimum $P(e)$ is obtained. Fixing this point as the optimum sampling point for this symbol, further sampling points are offset by the symbol duration for the entire PRBS sequence [Seo, 7.33].

Computer simulations have been extensively used in the performance evaluation of various nonlinearly amplified modems which are operated in an ACI environment. The following **assumptions are used in our simulations:**

1. For SCPC satellite systems and other radio applications, an illustrative input data bit rate $f_b = 64$ kb/s (or $f_s = 32$ kbaud data symbol rate) is used. Evidently other rates could also be used.

2. The interfering signals have the same modulation format as the desired main signal.

3. The carrier phases and symbol timings of the two interfering signals are randomized separately over the interval $(0, 2\pi)$ and $(-T_s/2, T_s/2)$, respectively, to avoid the coherence and synchronization between the interfering and main channels.

4. The radio transmit amplifiers are operated in a fully saturated mode, and they are approximated by ideal hard-limiters. As illustrated in Fig. 7.15, an ideal hard-limiter presents a good first-order approximation to a saturated amplifier.

5. Only the desired signal is attenuated. This simulates a flat fade or a lower uplink power of the main channel. The signals in the adjacent channels are not attenuated.

6. Two interfering adjacent channels are equally spaced. The main channel has a carrier frequency $f_c$, and the adjacent channels have carrier frequencies $f_{\pm 1} = f_c \pm \Delta F$, where $\Delta F$ represents the channel frequency separation (or channel spacing).

7. In the SQAM modem, only receive filters (fifth-order Butterworth LPFs) are used.

In [Seo, 7.35] a detailed performance study of SQAM, MSK, IJF-OQPSK, and conventional OQPSK modems, operated in an ACI environment, is presented.

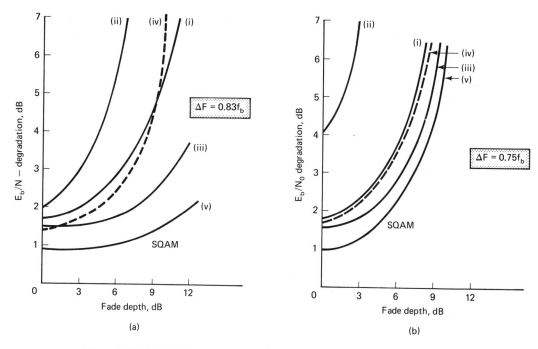

**Figure 7.26** $E_b/N_O$ degradation against fade depth of the desired channel in hard-limited multichannel system at $P_e = 10^{-4}$ [Seo 7.33; 7.35]: (i) MSK with *Tx* and *Rx* filters; (ii) MSK with *Rx* filter only; (iii) IJF-OQPSK (SQORC) [Feher, 7.1; 7.3]; (iv) OQPSK; (v) SQAM: $A = 0.85$.

Here, we present two illustrative results, which demonstrate the potential advantages of SQAM modems. In Fig. 7.26, we assume that the desired main channel signal suffers a flat fade, varying from 0 to 14 dB. The $P(e)$ performance degradation as a function of the **fade depth** (FD) is shown in Fig. 7.26(a) and (b) for $\Delta F = 0.83f_b$ and $\Delta F = 0.75f_b$, respectively. Note that the SQAM modem outperforms the other modems. This can be explained from the fact that the out-of-band energy of SQAM signals for linearly and nonlinearly amplified systems is much lower than those of other signals. Thus the SQAM signals cause less ACI compared to other signals. Also, in a narrow channel spacing or at low levels of FD, the OQPSK modem outperforms the MSK modem. This is because in MSK the side-lobe roll-off is faster, but its main-lobe bandwidth is 50% wider than that of OQPSK.

### 7.6.4 XPSK: Crosscorrelated QPSK Modulation Technique

Following our literature review of a class of power-efficient modems, we described a number of novel, simple modulation techniques such as IJF, TSK-OQPSK, and SQAM. These techniques have one common feature; namely, the pulse shaping is always performed independently in the I and Q baseband channels and there is no correlation between these baseband input signals. In an interesting modem,

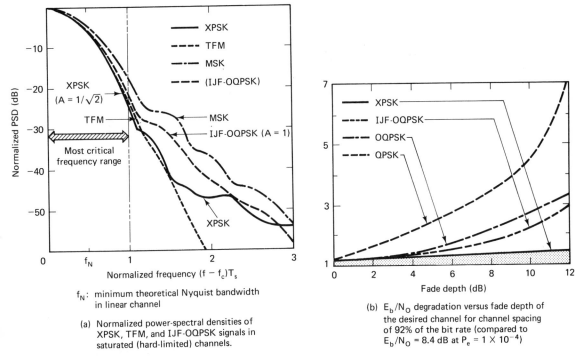

$f_N$: minimum theoretical Nyquist bandwidth
in linear channel

(a) Normalized power-spectral densities of
XPSK, TFM, and IJF-OQPSK signals in
saturated (hard-limited) channels.

(b) $E_b/N_O$ degradation versus fade depth of
the desired channel for channel spacing
of 92% of the bit rate (compared to
$E_b/N_O = 8.4$ dB at $P_e = 1 \times 10^{-4}$)

**Figure 7.27** XPSK, TFM, and IJF-OQPSK power spectral density and performance in an
ACI environment (With permission of the IEEE [Kato, 7.38].)

introduced by Kato and Feher [Kato, 7.29; 7.38], an intentionally controlled cross-correlation between the I and Q transmit baseband input signals (rails) is introduced. This crosscorrelation X reduces the envelope fluctuation of IJF-OQPSK signals and further improves the performance of these systems. The PSD and the $P_e = f(E_b/N_O)$ performance (degradation) curve as a function of fade depth of X-PSK, TFM, and IJF-OQPSK modems is illustrated in Fig. 7.27.

The X-PSK and IJF-OQPSK demodulators are conventional OQPSK demodulators, whereas the TFM demodulator requires a more-complex baseband signal processor. Due to space limitation, we do not describe the details of the X-PSK modulator operation. For information on X-PSK see [7.29].

### 7.6.5 Commercial Satellite Earth Station Applications of New Modem Techniques

Years of intensive R&D on power-efficient modulation techniques led to a number of industrial applications of new generations of modems. We highlight here two Canadian satellite and Earth-station communications systems, which employ new generations of (non-QPSK) modems. These systems are the first commercial systems described in the open literature [Sewerinson, 7.21; Feher, 7.37; 7.40].

Microtel Pacific Research Limited of Burnaby, B.C., developed a preassigned and demand assigned subscriber satellite communications system. The MICRO-TEL system is configured to meet the following objectives:

1. Provide the necessary functions for both the end user and the network operator.
2. Allow for mixture of voice, data, and broadcast services within the same framework.
3. Allow for cost-effective hardware and efficient transponder utilization.
4. Be tamper-proof.
5. Minimize the possibility of interference to other users or systems through misuse or malfunction.
6. Be transportable and accept harsh environmental conditions.
7. Be easily expandable.
8. Provide terminals for single- or multichannel operation to meet communications requirements.
9. Be upgradable to take advantage of new technological developments.

The MICROTEL satellite earth station system uses an in-house developed **constant-envelope QPSK** (CE-QPSK) modulation technique. The performance of the MICROTEL modem, based on available data (spectrum and BER), is practically identical with the previously described constant-envelope hard-limited IJF-OQPSK [Sewerinson, 7.21]. The first operational satellite system using MICROTEL's CE-QPSK modems was installed in Western Canada by the British Columbia Telephone Company during 1983. This satellite system provides loop-extension services throughout the province.

Spar Aerospace Limited developed the SPARCOM system, which employs our patented IJF-OQPSK modulator. The primary application of SPARCOM voice-data channel unit is in $\frac{14}{12}$-GHz and $\frac{6}{4}$-GHz SPARCOM satellite Earth stations (See Fig. 7.28).

For point-to-point and point-to-multipoint data and/or voice satellite transmission, these Earth stations, models 103C and 103D, provide high-quality service (with a typical availability of 0.99995 for a fully redundant system).

SPARCOM digital satellite Earth-station terminals improve upon the data transmission rate restrictions and BER performance of the analog terrestrial telephone system. Small antenna size permits on-premises installation of antennas, often eliminating the need for any telephone company–provided services.

The SPARCOM digital terminal is a complete system from subscriber data interfaces to antenna. The system develops from one to six digital trunks, each capable of operating in the range of 2.4 kb/s to 256 kb/s.

A typical block diagram of the SPARCOM digital thin-route Earth station is illustrated in Fig. 7.29. A number of low-speed (2.4 kb/s to 9.6 kb/s) modem data rates and a digitized voice channel may be multiplexed in the asynchronous and synchronous digital MUX equipment. A **forward-error-correcting encoder/decoder** (FEC codec) may be inserted to improve the BER performance of the

**Figure 7.28** SPAR Aerospace Limited's small-antenna Earth-station subsystem for SPARCOM digital terminals.

data source.   Optional encryption subsystems achieve data communications security.   A large-scale integration (LSI) chip adaptive delta modulator provides good-quality voice in the 20-kb/s to 40-kb/s range.

The IJF-OQPSK modem combined with the frequency synthesizer accepts the multiplexed (or single) data stream and converts it to the appropriate IF frequency in the 50- to 90-MHz range.   The modulator contains a low-power IF hard-limiter, which produces a 70-MHz output that is fed to the upconverter and RF power amplifier with a constant envelope and a reduced spectrum.   The AM/AM and AM/PM conversion, inherent in all saturated HPAs has no effect on the constant-envelope, bandlimited signal.   (Fully saturated HPA operation enables a power-efficient utilization of the RF amplifier).

The significantly lower sidebands of SPARCOM's IJF-OQPSK modulator lead to a performance advantage such as illustrated in Fig. 7.30, where ACI in a faded channel is compared for the QPSK and IJF-OQPSK systems.

The SPARCOM IJF-OQPSK Earth station is equipped with a half-rate convolutionally encoded–soft-decision decoded subsystem.   The performance of this commercial Earth station is illustrated in Fig. 7.31.

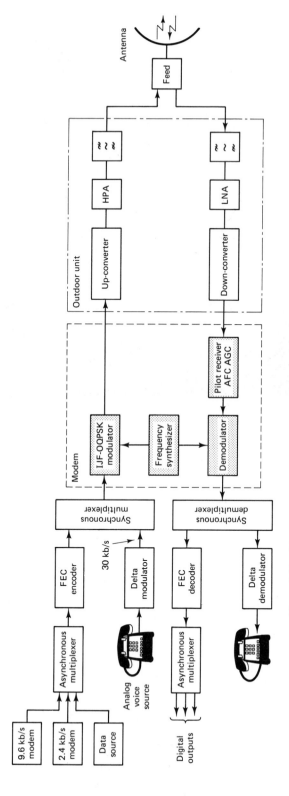

**Figure 7.29** SPARCOM digital thin-route SCPC satellite Earth-station terminal—a possible system configuration. This system is using Feher's patented IJF-OQPSK modem, manufactured by SPAR Aerospace Limited, a world-leading manufacturer of Satellites and Earth-station equipment.

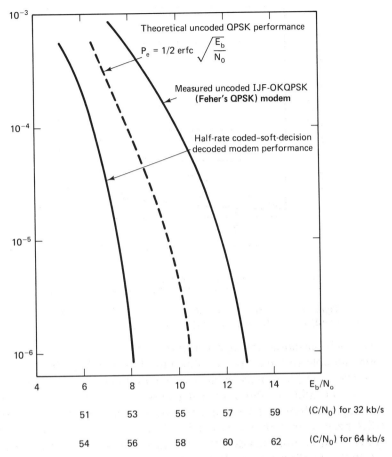

Note: $E_b/N_O$ is the uncoded source bit-energy-to-noise-density ratio; $C/N_O$ is the corresponding $C/N_O$ ratio required for the transmission of 32/kb s and 64/kb s source-rate signals. Single channel back to back performance was measured through a saturated amplifier.

**Figure 7.30**  $P_e$ performance of uncoded and coded IJF-OQPSK systems in SPAR-COM satellite Earth stations [Feher, 7.31; 7.40; Le-Ngoc, 7.42].

# 7.7 DIFFERENTIALLY COHERENT MODEMS: DMSK AND DCTPSK

## 7.7.1 Why Use Differentially Coherent Demodulators?

Most power- and bandwidth-efficient digital communication systems employ coherent detection. Coherent systems perform well in the presence of Gaussian noise but are not very tolerant of other link disturbances, such as multipath fading, shadowing, Doppler shifts, or phase noise. These effects have become more important in recent years, particularly with the proliferation of digital mobile communications systems. Differential detection avoids carrier recovery and achieves

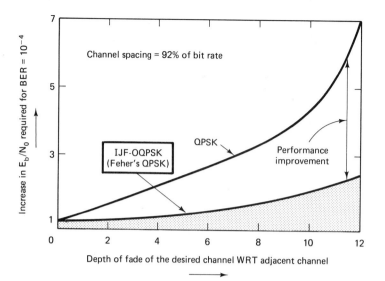

Figure 7.31 Comparison of ACI performance degradation due to channel fading, QPSK versus IJF-OQPSK SPARCOM system.

fast synchronization and resynchronization. Therefore, differential detection is more suitable to bursty traffic transmitted over fading channels, such as TDMA systems, or in phase-noise-controlled environments.

Application of differential detection ot known power-efficient modulation techniques results in DPSK and DMSK systems. Differential DBPSK has good BER performance, but it is not bandwidth efficient. A differentially detected system that is more bandwidth efficient but less power efficient than DBPSK is DQPSK. DQPSK is suitable for single-carrier linear-channel applications. However, when operated in a nonlinear channel, the filtered side lobes of DQPSK signal are restored. Consequently, in a multicarrier ACI environment, DQPSK cannot efficiently utilize the available bandwidth. The structure of DBSK and of DQPSK modulators is practically the same as that of coherent BPSK and QPSK modulators, respectively. Differentially detected binary and quadriphase demodulators are illustrated in Fig. 7.32. Note that instead of a carrier recovery circuit used for coherent demodulation, a one-symbol delay element is used for comparison (i.e., differential detection). The principles and applications of these modems are described in numerous books including [Feher, 7.2].

A recently introduced technique is differentially detected MSK. As a constant-envelope modulation technique filtered MSK suffers very little from side-lobe restoration. But MSK is basically a wideband system that requires large channel spacings and consequently cannot achieve high packing density.

In the following sections we present the principles of operation of DMSK–differentially detected MSK and of nonredundant error correction of DMSK sig-

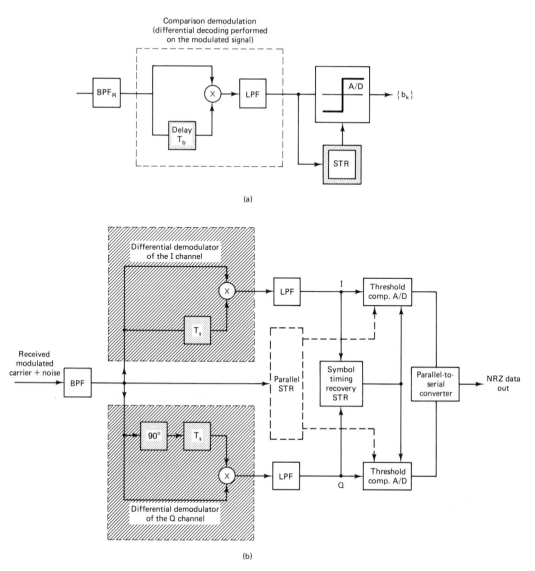

**Figure 7.32** Differential binary (DBSK) and quadriphase demodulator (DQPSK) block diagrams. (a) DBPSK demodulator. (b) DQPSK demodulator.

nals. Then, we describe a modem that considers the modulator jointly with differential detection and uses spectral shaping techniques to achieve bandwidth and power efficiency. This new technique is called *differentially continuous-transition PSK* (DCTPSK) [Yongacoglu, 7.25; 7.26]. Chapter 10 presents premodulation filtered MSK systems that are also suitable for noncoherent demodulation and have a narrower spectrum than conventional MSK systems.

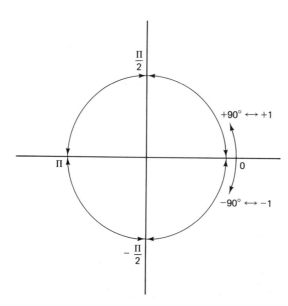

**Figure 7.33** Signal-state diagram of MSK. At the end of each bit interval only ±90° phase shifts corresponding to data polarity (±1) are permitted. The phase changes uniformly during the bit interval.

## 7.7.2 DMSK Demodulation with Nonredundant Error Correction

### Principles of Operation

MSK, also known as FFSK, is a modulation technique that can be represented as an offset QPSK system or as a digital FM system. In the offset QPSK model a sinusoidally shaped baseband filter is used (sinusoidal in the time domain), whereas in the FFSK model, a binary FM modulator having a modulation index of exactly 0.5 is employed. These models are theoretically identical. At the modulator output a continuous-phase constant-envelope signal is obtained [Feher, 7.2].

Unlike some other offset QPSK-modulated signals, a MSK signal can be differentially detected on a bit-by-bit basis. In Fig. 7.33 the signal-state space diagram of an unfiltered MSK signal is illustrated. During each bit interval, phase changes of only ±90° degrees are permitted, as opposed to 0°, ±90°, and 180° in QPSK. Therefore, the phase detector during the phase comparison of two time slots checks to see if the phase has advanced by +90° or −90°. This means that only one binary-phase detector output in which +1 represents a +90° phase change and −1 represents −90° phase change is sufficient. In QPSK two binary-phase detector outputs representing the four possible states are required. This hardware simplification, in the case of DMSK demodulation, also extends to other circuits in the decoder, such as the syndrome-calculating unit of the nonredundant error-correction subsystem. Instead of performing modulo-4 arithmetic, as in QPSK, modulo-2 arithmetic, which is easily realized by exclusive-OR, gates is sufficient. In this respect an MSK decoder is the same as a BPSK decoder. An illustrative DMSK demodulator block diagram and corresponding error-correction circuit diagram are shown in Fig. 7.34.

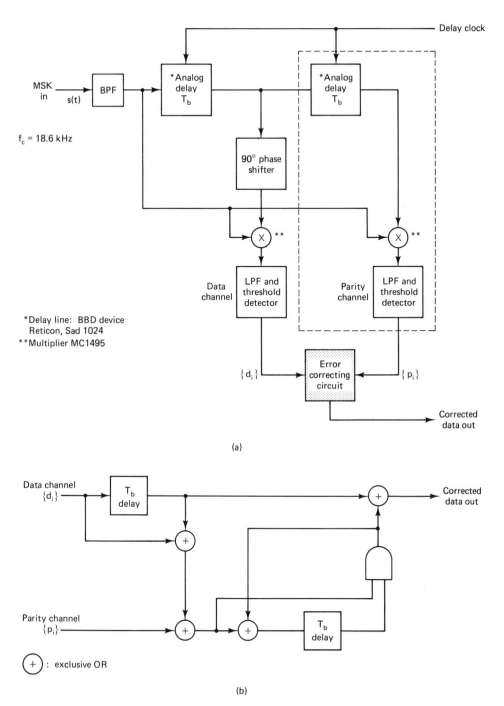

**Figure 7.34** An illustrative experimental DMSK demodulator block diagram and corresponding nonredundant error correcting circuit [Tezcan, 7.65]. (a) Block diagram of an experimental 2.4-kb/s demodulator. (b) Error-correcting circuit.

The FSK representation of an MSK signal is given by

$$s(t) = \cos\left(w_c t + d_k \frac{\pi t}{2T_b} + \psi_k\right) \qquad kT_b \le t \le (k+1)T_b \qquad (7.40)$$

where

$w_c$ = carrier frequency

$d_k$ = data, $\pm 1$

$\psi_k$ = constant phase for the $k$th bit duration

The second term in the parentheses represents the frequency deviation due to data polarity. To satisfy the phase-continuity requirement, we have to satisfy

$$\psi_k + d_k \frac{\pi k t}{2T_b} = \psi_{k-1} + d_{k-1} \frac{\pi k t}{2T_b} \qquad (7.41)$$

which can be rewritten as:

$$\psi_k = \psi_{k-1} + (d_{k-1} - d_k)\frac{\pi k}{2} \qquad (7.42)$$

The multiplier followed by a LPF, illustrated in Fig. 7.34(a), performs the phase-comparison operation between two consecutive symbols (bits). At the data CM output we have

$s(t)s(t - T_b)$

$$= \cos\left[w_c t + \frac{d_k \pi t}{2T_b} + \psi_k\right]\cos\left[w_c(t - T_b) + \frac{d_{k-1}\pi(t - T_b)}{2T_b} + \psi_{k-1}\right]$$

$$= \frac{1}{2}\left\{\cos\left[w_c(2t - T_b) + \frac{\pi}{2T_b}(d_k t + d_{k-1}t - d_{k-1}T_b) + \psi_k + \psi_{k-1}\right] \qquad (7.43)\right.$$

$$\left. + \cos\left[w_c T_b + \frac{\pi}{2T_b}(d_k t - d_{k-1}t + d_{k-1}T_b) + \psi_k + \psi_{k-1}\right]\right\}$$

The LPF removes the higher products, and assuming $\psi_{k-1} = 0$, we have

$$\frac{1}{2}\cos\left\{w_c T_b + \frac{\pi}{2T_b}[t(d_k - d_{k-1}) + d_{k-1}T_b] + \psi_k\right\} \qquad (7.44)$$

as the comparator output, that is, at the input of the threshold detector.

For the four possible 2-consecutive-bit combinations, $\psi_k$ is given by

| Bit Combination | | Phase | |
|---|---|---|---|
| $d_k$ | $d_{k-1}$ | $\psi_k$ | |
| $-1$ | $-1$ | $0$ | |
| $-1$ | $1$ | $-\pi$ | $(7.45)$ |
| $1$ | $-1$ | $\pi$ | |
| $1$ | $1$ | $0$ | |

These four combinations correspond to the following four demodulated waveforms:

$$(d_{k-1}, d_k) = (-1, -1) \quad \frac{1}{2}\cos\left(w_c T_b - \frac{\pi}{2}\right)$$

$$= (-1, \ \ 1) \quad \frac{1}{2}\cos\left(w_c T_b + \frac{\pi t}{T_b} + \frac{\pi}{2}\right)$$

$$kT_b \leq t \leq (k+1)T_b \quad (7.46)$$

$$= (1, \ -1) \quad \frac{1}{2}\cos\left(w_c T_b - \frac{\pi t}{T_b} - \frac{\pi}{2}\right)$$

$$= (1, \ \ \ 1) \quad \frac{1}{2}\cos\left(w_c T_b + \frac{\pi}{2}\right)$$

The term $w_c T_b$ is dependent upon the carrier frequency $f_c$ and bit duration $T_b$; for this reason the circuit is designed such that $(w_c T_b)$ modulo $2\pi = -\pi/2$. With this design value the first and last waveforms in equation (7.48) become constants, having values of $-\frac{1}{2}$ and $+\frac{1}{2}$, respectively. The second waveform corresponds to a half-sinewave going from $+\frac{1}{2}$ to $-\frac{1}{2}$, and the third one is a half-sinewave going from $+\frac{1}{2}$ to $-\frac{1}{2}$. These waveforms combine to form the eye diagram illustrated in Fig. 7.35. Sampling is done at the end of each bit duration. An illustrative DMSK waveform is given in Fig. 7.36.

Similar processing is performed in the second-order detector (see Fig. 7.25). Since the first-order detector is used to detect the data and the second-order detector is used for error correction, from now on we will call them the **data** and **parity** channels, respectively. After detection the outputs are connected to the decoder. The decoder used here corrects errors by generating a syndrome.

A detailed study of the principles, performance, and applications of nonre-

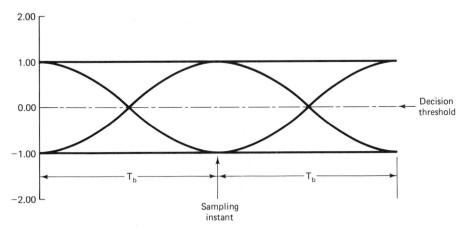

**Figure 7.35** Computer-generated eye diagram of DMSK signal at the output of the phase comparator. During each bit interval there are four possible traces: $[-\frac{1}{2}, \cos(\pi t/T)2, \cos(\pi t/T - \pi)/2, +\frac{1}{2})$. The traces given in the eye diagram are slightly distorted because of the ISI introduced by the harmonic removal filter.

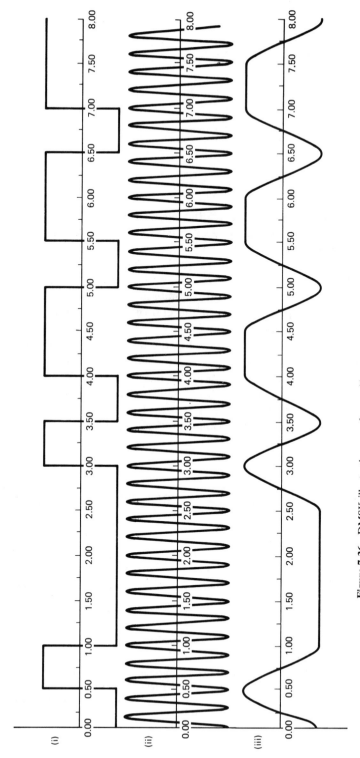

**Figure 7.36** DMSK-illustrative waveforms: (i) segment of an NRZ waveform input to modulator; (ii) modulated, transmitted waveform; (iii) differentially demodulated waveform.

dundant error-correction subsystems in DMSK and in DQPSK systems is given in [Samejima, 7.66; Masamura, 7.67]. Here we present a brief description of the principles of nonredundant error correction and, by means of an illustrative timing diagram, highlight the error-correction capability of this system.

In coherent demodulation an error is caused when the noise shifts the phase of the signal into the error region. In differential demodulation, such as DMSK, an error is caused when the *sum* of noise phases in two consecutive bit intervals shifts the differential signal into the error region. The nonredundant error correction circuitry exploits the fact that if the sum of noise samples in two consecutive intervals causes an error, the sum of noise samples in two alternate intervals does not necessarily cause an error. Hence, if we compare not only two consecutive symbols but also two alternate symbols, then the results can be used for error detection and correction. The nonredundant error correction process in a DMSK system is illustrated in Fig. 7.37. The signal patterns correspond to the DMSK block and circuit diagrams of Fig. 7.35. From the corrected signal pattern we note that

1. Single errors in the data channel are corrected.
2. Single errors in the parity channel are ignored.
3. Double errors in the parity channel cause single errors at the decoder output when there is no error at the data channel detector output.

### Performance in an ACI and AWGN Environment

Most practical data-transmission systems operate in a combined ACI and AWGN environment. The importance of the ACI and its impact on a number of coherently demodulated systems is described in Section 7.6. Here we describe the performance of DMSK systems in an AWGN and ACI environment [Tezcan, 7.65].

Measured and simulated $P_e = f(E_b/N_O)$ curves in an AWGN environment, with no ACI, are given in Fig. 7.38. The performance of a theoretical coherent demodulator is included as a reference. The measured improvement with non-redundant error correction is about 1 dB. A purpose of the ACI investigation is to determine the specifications of the BPF that lead to the "best-possible" performance. (Here best-possible implies best performance within practical hardware and cost constraints.) See Fig. 7.35 and Table 7.2.

In Fig. 7.39 fade depth is defined as the ratio (difference in decibel) of each one of the adjacent channels to the power of the main channel. The channel spacing is defined as the frequency spacing between the carrier center frequencies ($\Delta f$) normalized to the bit rate ($f_b$).

In the single-channel AWGN case, a $BT_b$ of 0.7 was found to give good results. Here we compare this $BT_b$ with a narrower $BT_b$ ($BT_b = 0.6$). The results indicate that a narrower filter has a somewhat worse performance at weak fades, but leads to better performance than the wider filter as the fade is increased. A comparison of Figures 7.39(a) and 7.39(b) shows that for channel spacing 2.25, the performance of $BT_b = 0.6$ filter exceeds the performance of the $BT_b = 0.7$ filter

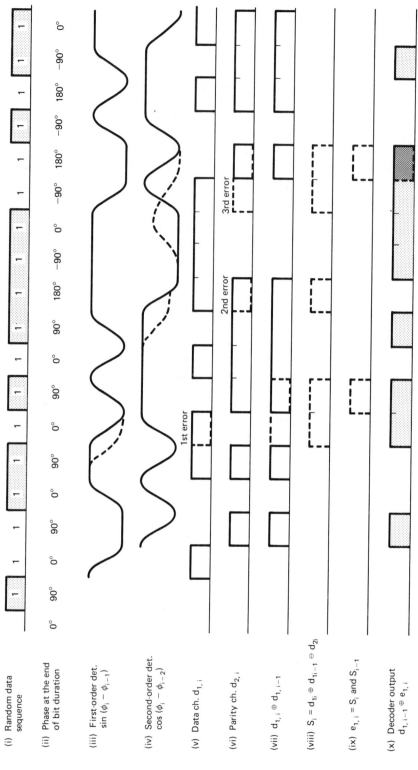

(i) Random data sequence

(ii) Phase at the end of bit duration

(iii) First-order det. $\sin(\phi_i - \phi_{i-1})$

(iv) Second-order det. $\cos(\phi_i - \phi_{i-2})$

(v) Data ch. $d_{1,i}$

(vi) Parity ch. $d_{2,i}$

(vii) $d_{1,i} \oplus d_{1,i-1}$

(viii) $S_i = d_{1i} \oplus d_{1i-1} \oplus d_{2i}$

(ix) $e_{1,i} = S_i$ and $S_{i-1}$

(x) Decoder output $d_{1,i-1} \oplus e_{1,i}$

**Figure 7.37** Illustration of the error-correction process in DMSK. Line (i) presents a random data sequence, (ii) illustrates the phase of the signal at the end of the bit durations. Lines (iii) and (iv) are the outputs of the first- and second-order detectors, respectively. The dotted lines indicate possible changes in the signal level due to noise lines. (v) and (vi) show the regenerated first and second-order detector outputs, respectively; here the dotted lines indicate changes in the bits due to noise. Lines (vii) to (ix) show various points in the decoder. Line (x) is the decoder output. In this example we have three errors. First is a single error in the data channel ($d_{1,i}$); therefore, it is corrected. The second is in the parity channel ($d_{2,i}$); hence, it is ignored. The third is a double error in the parity channel; therefore, a single error is created at the decoder output when there was no error at the first-order detector output.

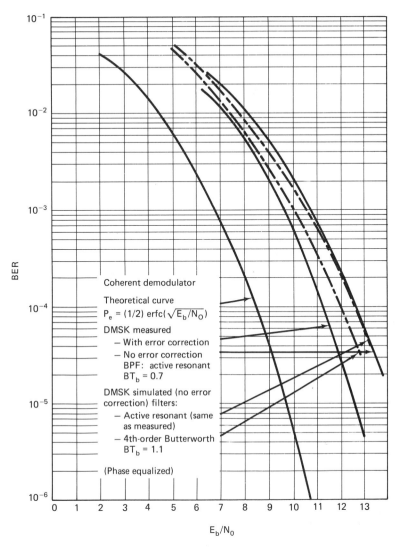

**Figure 7.38** Experimental and computer simulation $P_e = f(E_b/N_O)$ results of a DMSK system operated in an AWGN environment [Tezcan, 7.65].

at a fade of only 4 dBs. Considering that in terrestrial mobile radio and $\frac{14}{12}$-GHz satellite systems large fades are common, the importance of narrower filtering becomes apparent.

Another important point to be noted is that in some cases smaller channel spacings have better performance than larger channel spacings. An example of this is the normalized channel spacings 2.00 and 1.75 at deep fades. When the BER is plotted as a function of normalized channel spacing, the phenomenon may be noticed from the curves of Fig. 7.40. The BER fluctuates with channel spacing; the cause of the fluctuation is the power of the adjacent channel side lobes falling into the desired receiver band. When a side lobe falls into the main-channel filter

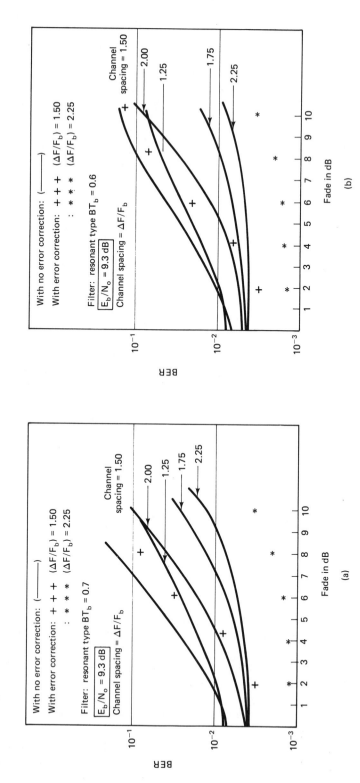

**Figure 7.39** Measured performance of DMSK as a function of fade depth. Fade depth: attenuation of main channel with respect to the adjacent channels. [Tezcan, 7.65].

**TABLE 7.2** Summary of bandwidth bit duration product ($BT_b$) and degradation over ideal coherent detection of measured and simulated DMSK in an AWGN channel (no ACI). Degradation at a BER of $10^{-3}$. Bandwidth (B) is the double-sided 3-dB bandwidth of the BPF [Tezcan, 7.65]

|  | Measured | | Simulated | |
|---|---|---|---|---|
|  | $BT_b$ | Degradation | $BT_b$ | Degradation |
| Resonant type filter non-phase equal-ized | 0.7 | 3.9 | — | — |
| Resonant type filter phase equal-ized | — | — | 0.7 | 3.8 |
| Resonant type filter with non-redun-dant error correc-tion | 0.7 | 2.8 | — | — |
| Fourth-order Butterworth filter phase equal-ized | — | — | 1.1 | 3.2 |

pass band, the interference increases. When a null of the ACI falls within the desired band the performance of the desired channel improves.

### 7.7.3 DCTPSK: Differentially Detected Controlled Transition PSK Systems

**Design Objectives**

For nonlinearly amplified power-efficient radio system applications which require differential detection and a more-efficient spectrum utilization than achievable with DMSK and DQPSK systems, we describe a new technique called DCTPSK. [Yongacoglu, 7.25; 7.26].

For an efficient digital modulation system, the **design objectives** may be summarized as follows:

1. The modulated hard-limited spectrum should be compact to enable a spectral-efficient operation.
2. A good $P_e = f(E_b/N_O)$ is required to enable power-efficient operation.

Bearing these objectives in mind, combined with the constraint that differ-

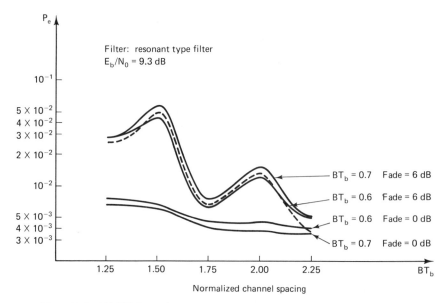

**Figure 7.40** DMSK $P_e$ performance function of channel spacing for different fade depths and filter bandwidths: Solid line indicates measured results; dotted line indicates computer simulation results for $BT_b = 0.7$, fade $= 6.0$ dB [Tezcan, 7.65]

ential detection may be required, we developed the DCTPSK modem. To our knowledge, this is the only Q-modulation technique that constructs the signals particularly for differential detection and also constrains the spectral regrowth after a nonlinearity without using a delay element; that is, it is not an offset keyed or staggered modem.

### DCTPSK-Conceptual Design Procedure

Our starting point is the **signal-space diagram** (SSD), also known as a constellation diagram. The transitions should be controlled such that the amplitude fluctuations are limited (phasor magnitude is never below a certain minimum). Let us illustrate this by an example. The SSD of conventional unfiltered QPSK and of DQPSK modems is drawn in Fig. 7.41(a). Assuming that the signal is at state $A$, then $AB$, $AC$, $AD$, and $AA$ are all legitimate transitions. During the $AA$, $AB$, and $AC$ transitions, the magnitude of the phasor fluctuates at most 3 dB. However, during the $AD$ transition, the phasor passes through zero magnitude, which results in 100% amplitude fluctuation.

In Fig. 7.41(b) we have drawn another SSD, where again $AA$, $AB$, $AC$, and $AD$ transitions are possible. However, in this case, when the $AD$ transition takes place, the trajectory is AEFD rather than AOD. By doing this, the amplitude fluctuation is limited to 3 dB. Consequently, the spectral regrowth after a nonlinearity is kept at a low level.

It should be noted that the SSD indicates only the transition paths. Another important characteristic, the speed of transition from one state to another state, is not conveyed in this diagram. Hence different signals, such as conventional QPSK

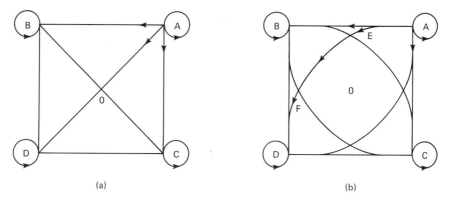

**Figure 7.41** QPSK, DQPSK and DCTPSK signal-space (constellation) diagram [Yongacoglu, 7.25; 7.26] (a) SSD of conventional unfiltered QPSK and DQPSK modems. (b) SSD of DCTPSK.

and IJF-QPSK signals (described in Sections 7.5 and 7.6), may have the same constellation diagrams. In QPSK the transitions are instantaneous, whereas in IJF-QPSK the transitions are smooth and take one symbol duration. The smoother phase transitions of IJF-type signals result in a more-compact spectrum. This same property is exploited for DCTPSK.

Choosing transitions with smooth functions and also limiting the amplitude fluctuation is only one dimension of the signal-design objective. The other dimension is the suitability to differential detection, which requires smooth phase transitions and large phase differentials at the sampling instants.

The phase characteristics of a large number of candidate signal constellation diagrams similar to the one sketched in Fig. 7.41(b) were studied. Through an iterative process signals that satisfy the previously stated design objectives were selected. Next, we describe the encoding rules of these signal elements [Yongacoglu, 7.25; 7.26].

### Generation of Signal Elements

For the random input sequences of $\{a_k\}$ and $\{b_k\}$, to obtain the Q output waveforms $i(t)$ and $q(t)$, the encoding rules can be formulated as follows:

1. If $a_k = a_{k-1}$ and $b_k = b_{k-1}$, then transmit dc, that is,

$$i(t) = \begin{cases} 1 & \text{for } a_k = 1 & 0 \le t \le T_s \\ -1 & \text{for } a_k = -1 & 0 \le t \le T_s \end{cases} \qquad (7.47)$$

and similarly

$$q(t) = \begin{cases} 1 & \text{for } b_k = 1 & 0 \le t \le T_s \\ -1 & \text{for } b_k = -1 & 0 \le t \le T_s \end{cases} \qquad (7.48)$$

where    $T_s$ = the symbol duration (i.e., $T_s = 2T_b$).

2. If $a_k = a_{k-1}$ and $b_k = -b_{k-1}$, then, in channel I transmit dc and in channel Q transmit a segment of a sinusoid. In other words,

$$i(t) = \begin{matrix} 1 & \text{for } a_k = \phantom{-}1 & 0 \le t \le T_s \\ -1 & \text{for } a_k = -1 & 0 \le t \le T_s \end{matrix} \qquad (7.49)$$

and

$$q(t) = -b_k \cos\left(\frac{\pi t}{T_s}\right) \qquad 0 \le t \le T_s \qquad (7.50)$$

Note that $\{a_k\}$ and $\{b_k\}$ take on the values $+1$ and $-1$.

3. If $a_k = -a_{k-1}$ and $b_k = b_{k-1}$, then

$$i(t) = -a_k \cos\left(\frac{\pi t}{T_s}\right) \qquad 0 \le t \le T_s \qquad (7.51)$$

and

$$q(t) = \begin{cases} 1 & \text{for } b_k = \phantom{-}1 & 0 \le t \le T_s \\ -1 & \text{for } b_k = -1 & 0 \le t \le T_s \end{cases} \qquad (7.52)$$

4. If $a_k = -a_{k-1}$ and $b_k = -b_{k-1}$, then the $i(t)$ and $q(t)$ signals must change signs at different times. Therefore, for this case we transmit

$$i(t) = \begin{cases} -a_k & 0 \le t \le \dfrac{T_s}{4} \\[2mm] -a_k \cos\left[\dfrac{4\pi}{3}\left(\dfrac{t}{T_s} - \dfrac{1}{4}\right)\right] & \dfrac{T_s}{4} \le t \le T_s \end{cases} \qquad (7.53)$$

and

$$q(t) = \begin{cases} -b_k \cos\left(\dfrac{4\pi t}{3T_s}\right) & 0 \le t \le \dfrac{3T_s}{4} \\[2mm] -b_k & \dfrac{3T_s}{4} \le t \le T_s \end{cases} \qquad (7.54)$$

Based on these encoding rules, the input and output waveforms of the encoder for an illustrative sequence are shown in Fig. 7.42.

### DCTPSK Modem Block Diagram and Performances

A DCTPSK modulation system is obtained if the signal processors described in the previous subsection are incorporated in a quadrature modulator that has a conventional differential detection receiver structure (see Fig. 7.43). In our implementation [Yongacoglu, 7.25; 7.26], for reasons of simplicity, we use a fourth-order Butterworth predetection BPF with a $BT_s = 1.5$. (*Note:* $T_s = 12T_b$.)

We use Monte Carlo techniques in our computer simulations. The $P_e = f(E_b/N_O)$ performance (or simply BER-performance) of filtered and hard-limited DQPSK, DMSK, and DCTPSK systems is illustrated (in Fig. 7.44). For DQPSK

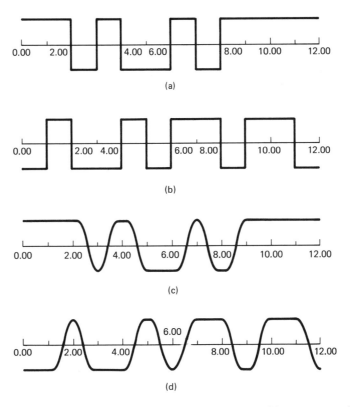

**Figure 7.42** Input and output waveforms of a controlled transition processor. (a) Input I-channel, (b) input Q-channel, (c) output I-channel, and (d) output Q-channel.

and DCTPSK, the 3-dB bandwidth times the bit duration (i.e., $BT_b$) of the transmit and predetection filters are both equal to 0.65. This value is used because Butterworth filters with $BT_b$ in the range of 0.55 to 0.75 yield good BER performance for DQPSK systems.

The BER performance of DMSK is greatly degraded with narrow transmit and predetection filters. For example, when the transmit and predetection filter $BT_b$s are equal to 0.9, in Fig. 7.44 the upper curves (DMSK$_1$) are obtained. Hence DMSK requires wider transmit and predetection filters and consequently results in lower packing density (i.e., wider channel spacing) than DPSK or DCTPSK. In Fig. 7.44 the BER performance of DMSK is also plotted when the transmit and predetection filter both have $BT_b = 1.3$ (curve DMSK$_2$). This indicates that *DMSK requires 1.5 to 2 times wider channel spacing than DQPSK and DCTPSK to achieve a comparable BER performance.*

From Fig. 7.44 we note that in a single-carrier environment, DQSPK has better BER performance than both DCTPSK and DMSK. However, in a hard-limited channel, DQPSK has high spectral components outside of its main lobe (see Figs. 7.45 and 7.46). Consequently, in a multicarrier environment, the ACI becomes a major source of impairment and *DQPSK suffers more degradation than*

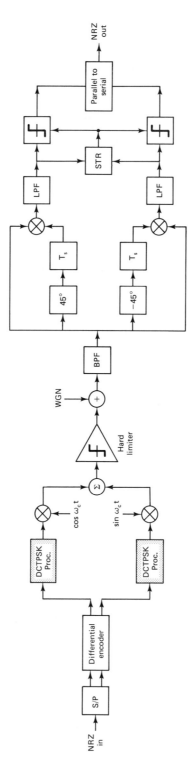

**Figure 7.43** DCTPSK MODEM [Yongacoglu, 7.25; 7.26]

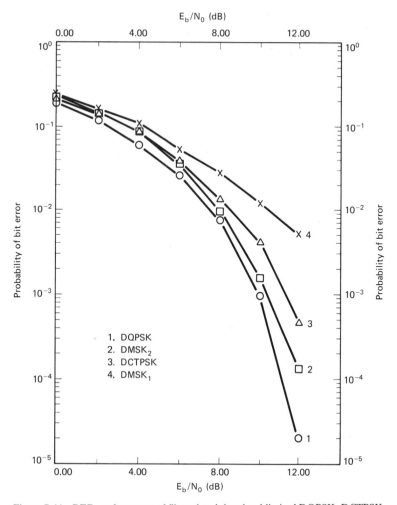

**Figure 7.44** BER performance of filtered and then hard-limited DQPSK, DCTPSK, and DMSK signals (no ACI). For DQPSK and DCTPSK, the transmit and pre-detection filters are fourth-order Butterworth with $BT_b = 0.65$; for DMSK$_1$, the $BT_b$s are 0.9, and for DMSK$_2$, $BT_b$s are 1.3.

*DCTPSK and DMSK.* We observe from Fig. 7.46 that the hard-limited DCTPSK signal has a compact main lobe and lower spectral components than the DMSK and DQPSK signals. For many applications this signal can be employed without any filtering at the transmitter.

## 7.8 SPECTRALLY EFFICIENT MODULATION TECHNIQUES

### 7.8.1 Spectral Efficiency and C/N Requirements

Spectrally efficient modems are used in bandlimited channels (as opposed to the power-limited channels described in the previous sections) in which the available

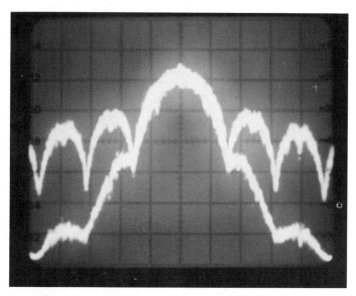

**Figure 7.45** Measured spectra of DQPSK (upper trace) and DCTPSK (lower trace) (vertical: 10 dB/division; horizontal: 20 kHz/division; bit rate = 64 kb/s, carrier frequency = 20 MHz).

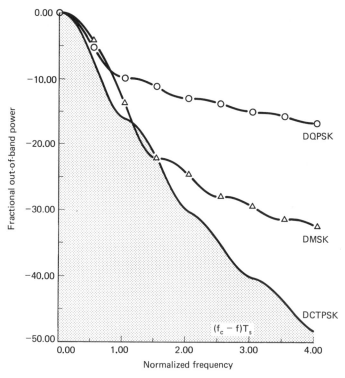

**Figure 7.46** Fractional out-of-band power of unfiltered DQPSK, DCTPSK and DMSK signals in a hard-limited channel [Yongacoglu, 7.25; 7.26].

**TABLE 7.3** Spectral efficiency and C/N requirement of wideband modems

| Modulation technique | Nyquist rate theoretical efficiency bits/s/Hz | Practical efficiency bits/s/Hz | C/N Required for $P_e = 10^{-8}$ (theoretical) |
|---|---|---|---|
| QPSK (4-QAM) | 2 | 1.2–2 | 15 |
| 9-QPRS | 2 | 2  –2.8 | 17.5 |
| 8-PSK | 3 | 2.5–3 | 20.5 |
| 16-QAM | 4 | 2.5–3.5 | 22.5 |
| 49-QPRS | 4 | 2.5–4.3 | 24.5 |
| 64-QAM | 6 | 4.5–5 | 28.5 |
| 128-QAM | 7 | 4.5–5.5 | 31.5 |
| 225-QPRS | 6 | 5.7–6.3 | 31 |
| 256-QAM | 8 | 5  –7 | 34.5 |
| 512-QAM | 9 | 5.5–7.5 | 37.5 |
| 961-QPRS | 8 | 6  –8 | 37 |
| 1024-QAM | 10 | 7  –9 | 40.5 |

Detailed $P = f(C/N)$ charts are given in Fig. 7.5. The spectral efficiency is illustrated in Fig. 7.3. *Note:* Entries are accurate within 1 dB: Noise is specified in the double-sided Nyquist bandwidth [Feher, 7.2; 7.3; Kucar, 7.23].

carrier-to-noise ratio (CNR) is high enough, that is, the transmission system can support a rate of more than 2 bits/s/Hz. The telephone channel has been used extensively for some of the most spectrally efficient modem applications. The reasons have to do with the commercial importance of such channels and with the fact that they can be modeled to first order as linear time-invariant channels, the available CNR is typically more than 30 dB and the relatively low bandwidth, typically 300 Hz to 3000 Hz, permits the utilization of reasonable cost-powerful digital processing elements [Forney, 7.68].

A number of advanced modulation techniques were developed for telephony modem applications and applied several years later to broader-band channels such as LOS microwave, satellite systems, and coaxial systems. During the mid-1980s a number of highly spectrally efficient techniques were developed directly for broader-band radio system applications. In this section we describe such modems, that is, modems with a spectral efficiency in the 2-bit/s/Hz to 10-bit/s/Hz range (see Table 7.3). Particularly, 64-QAM and 256-QAM systems found wide applications in terrestrial microwave systems.

QPRSs are also of high interest. A number of companies have developed 9-QPRS and 49-QPRS radio systems and have been considering the design of 225-QPRS radio systems. QPRSs, due to the inherent correlation properties between adjacent states, might be operated somewhat above the theoretical Nyquist rate. A new QPRS system, which has a better speed tolerance than conventional QPRS systems, is described in Section 7.12 after the discussion of QAM systems.

The principles of operation of spectrally efficient modems are reviewed in

Section 7.2. Additional material, which could be useful for a better comprehension of the following sections, is given in [Feher, 7.3; 7.2]. For example, the theory and systems applications of 8 PSK, 9 QPRS, 16 QAM, and other radio systems that preceded the implementation of 256 QAM are described in [Feher, 7.3].

## 7.9 256-QAM AND 64-QAM RADIO SYSTEMS

QAM modulator and demodulator conceptual block diagrams are described in Section 7.2 (see Figs. 7.1 and 7.4, respectively). A somewhat more-detailed modern QAM system, equipped with digital **baseband adaptive transversal equalizers** (BATE) and FEC subsystems, is illustrated in Fig. 7.47. The following description is for a standard 1.544-Mb/s rate (known as DS-1 rate) radio application. Evidently this illustrative block diagram is also suitable for much-higher transmission rates.

### 7.9.1 Description of a 256-QAM Modem

In the 256-QAM modulator digital interface unit, which extracts a 1.544-MHz clock from the incoming DS-1 signal, the 1.544-Mb/s bipolar signal is converted into a unipolar signal. The unipolar signal is then scrambled by a pseudorandom generator. In case of failure of the incoming DS-1 signal, an all-one **alarm-indicator signal** (AIS) is generated and scrambled for transmission to the demodulator.

To achieve random- and burst-error correction, redundant bits are added to the 1.544-Mb/s scrambled signal. The interleaved FEC-encoded signal is transmitted at a bit rate of 1.6 Mb/s, so a very low redundancy codec is used. Together with a 4-kb/s control and/or system monitor data stream, the data rate for the modem is 1.6 Mb/s. This aggregate data stream is series-to-parallel converted into two 4-bit streams, which, after differential/Gray encoding, are D/A converted into two 16-level analog signals. These two 16-level signals each pass through spectral shaping optimized LPFs. The two bandlimited signals are modulated by an I- and a Q-IF carrier, respectively. For carrier recovery, two low-level pilots may be inserted at the band edges of the modulated signals [Feher, 7.22].

Because of the sharp filtering of the baseband 16-level signals (in case of the 256-QAM modulator, we have 16-levels in the I- and Q-channels, respectively) at the band edges, there is practically no interference between the data and the low-level pilots.

The received 256-QAM signal from the radio system is passed through a protection filter, or *roofing filter*, which removes most spurious signals from adjacent channels. An IF adaptive equalizer compensates for frequency selective fades and time-variable channel distortions. An **automatic gain-control** (AGC) circuit provides a constant output carrier to the demodulator. The demodulated baseband signals are low-pass filtered. The overall modulator-demodulator channel-filtering strategy satisfies Nyquist raised-cosine filters and leads to an optimum performance.

The in-phase baseband signal is also used to extract the symbol-timing clock. This clock samples the two baseband signals and enables A/D conversion. The

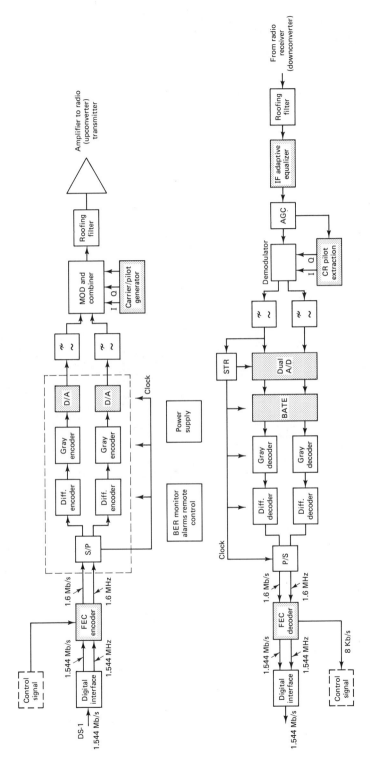

**Figure 7.47** 256-QAM Modem Equipped with forward-error-correction (FEC), digital base-band adaptive transversal equalization (BATE), IF adaptive equalization, automatic gain-control (AGC) and control/monitor subsystems Courtesy of Karkar Electronics, Inc., San Francisco, [7.22].

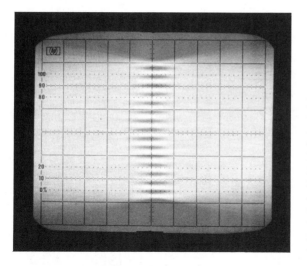

**Figure 7.48(a)** Measured 256-QAM demodulated I-channel eye diagrams. I-channel symbol rate $f_s$ = 200 kBaud corresponds to modem bit rate of 1.6 Mb/s (64 QAM). Raised cosine channel Nyquist filters ($f_n$ = 100 kHz) having a roll-off parameter $\alpha$ = 0.2 and a 55-dB attenuation beyond 120 kHz, designed by Karkar Electronics, Inc., are used in this experiment. Unequalized filters have a sinusoidal group delay distortion. [Wu, 7.34]

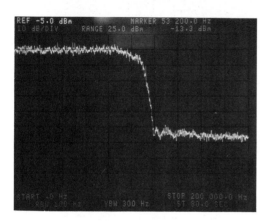

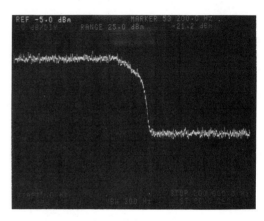

After transmit filter          After cascade of TX and RX filters

**Figure 7.48(b)** Power spectral density of a 1.6-Mb/s, 256-QAM modem (baseband). Measured after an $\alpha$ = 0.2 raised-cosine filter, designed by Karkar Electronics. Corresponding eye diagram is shown in Fig. 7.48(a). 1.6 Mb/s is used here (instead of 1.544 Mb/s) in order to enable the insertion of low-redundancy FEC (error-correction) bits and of a digitized service channel. Measurement performed in the Digital Communications Research Laboratory, University of Ottawa.

BATE automatically equalizes in the time domain the impairments caused by the transmission system and channel filters in general.

After Gray/differential decoding, the equalized I- and Q-signals are parallel-to-serial converted into a 1.6-Mb/s signal. With the 1.6-MHz clock, this 1.6-Mb/s signal passes through the de-interleaved FEC decoder for random- and burst-error corrections and becomes a 1.544-Mb/s data stream. The 8-kb/s control-monitor signal is extracted for further processing. The 1.544-Mb/s unipolar data stream is descrambled and then converted back to the standard DS-1 bipolar format.

**Figure 7.48(c)** The first successful $\alpha = 0.2$ filtered 256-QAM microwave transmission results are reported in [Feher, 7.22]. These systems, having a practical spectral efficiency of 6.66 bits/s/Hz, operate over FDM-FM and high-capacity SSB radio systems manufactured by Karkar Electronics, Inc. Several thousand SSB radio systems carrying data are operational in the United States.

In case a scrambled all-one signal is received, an AIS alarm indicates that a failure has occurred upstream at the modulator.

Illustrative measured eye and spectral diagrams of 256-QAM and 64-QAM systems are shown in Fig. 7.48(a) and (b). A hardware photograph of racks of SSB (analog) radio systems, which carry 256-QAM-1.544 Mb/s data, is shown in Fig. 7.48(c).

### 7.9.2 Performance Degradations

The sensitivity of higher-order modulation techniques to different impairments increases significantly with the number of states. Here we assume that the radio amplifiers, upconverters, downconverters, AGC, and all other hardware components are linear and that there is no AM/AM- and AM/PM-caused degradation in the system. To achieve this "linear" radio system operation, a very high output-backoff in the order of 10 dB is used in most radio systems. For example, a

$+40$-dBm (10-W) output amplifier is operated at $+30$ dBm (1 W), that is, at ten times lower power than the saturated output power. Evidently this leads to a significant loss of power efficiency.

To investigate the effects of selective fading and/or radio system hardware imperfections (linear imperfections), we assume that there is a residual group-delay and/or amplitude response distortion after the adaptive equalizers. This residual imperfection can be interpreted as a deviation from the ideal raised-cosine Nyquist channel characteristics described in Section 7.2.

The residual **group-delay distortions** are defined as follows:

$$D(f) = \begin{cases} L_D \cdot f & \text{for linear group delay} \\ P_D \cdot f^2 & \text{for parabolic group delay} \\ S_D \cdot \sin\left(\dfrac{2\pi Kf}{2f_{BW}}\right) & \text{for sinusoidal group delay} \end{cases} \qquad (7.55)$$

where

$$f_{BW} \overset{\triangle}{=} (1 + \alpha)f_N$$

$$f_N = \text{Nyquist bandwidth in baseband}$$

This $(1 + \alpha)f_N$ bandwidth definition (for group-delay distortion) is more appropriate than $f_N$, as it includes the critical effect of the group delay at the edge of the filter attenuation band. In the case of sinusoidal group delay, we present results only for $K = 4$. The symbol error rate $P_s$ versus average CNR is computed. Typical results are shown in Fig. 7.49. We define maximum group delay $\tau_m$ in the filter bandwidth $(2f_{BW})$ as follows:

$$\tau_m \overset{\triangle}{=} \begin{cases} L_D(2f_{BW})\text{ns} & \text{for linear group delay} \\ P_D(f_{BW})^2\text{ns} & \text{for parabolic group delay} \\ S_D \text{ ns} & \text{for sinusoidal group delay} \end{cases} \qquad (7.56)$$

The degradation of $C/N$ as a function of $\tau_m$ relative to the case with no distortion, for $P_s$ of $10^{-4}$, is shown in Fig. 7.50. Note that here we simulate a 120 Mb/s system with $\alpha = 0.4$. We also include the performance of a 90-Mb/s 64-QAM system with $\alpha = 0.4$ in Fig. 7.50. For a given value of maximum group delay $\tau_m$ in the filter bandwidth, linear group delay causes the most severe degradation to the system's performance as compared to parabolic or sinusoidal group-delay distortion [Wu, 7.34].

In the study of residual **amplitude distortions** and their impact on system performance, we assume that the group-delay distortion is equalized, that is, $D(f)$ is equal to a constant. Three different characteristics for the amplitude distortion are defined as follows:

$$A(f) = \begin{cases} L_A \cdot f & \text{for linear amplitude distortion} \\ P_A \cdot f^2 & \text{for parabolic amplitude distortion} \\ S_a \cdot \sin\left(\dfrac{2\pi Kf}{2f_{BW}}\right) & \text{for sinusoidal amplitude distortion.} \end{cases} \qquad (7.57)$$

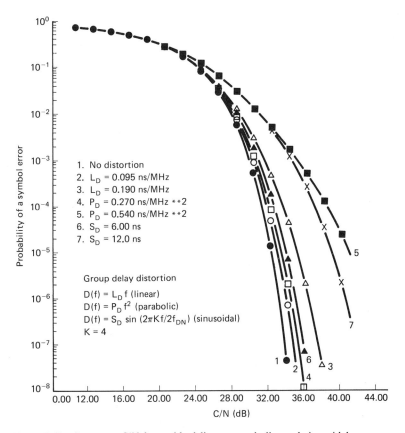

**Figure 7.49** $P_s$ versus $C/N$ for residual linear, parabolic, and sinusoidal *group-delay* distortions for 256 QAM with a bit rate of 120 Mb/s (15 Mbaud and $\alpha = 0.4$). Noise is defined in the double-sided Nyquist bandwidth; that is, the equivalent noise bandwidth is 15 MHz. [Wu, 7.34]

The symbolic error rate $P_s$ versus $C/N$ is computed [Wu, 7.34]. A typical result is shown in Fig. 7.51. Degradations of $C/N$ for 256 QAM and 64 QAM with 15 Mbaud and $\alpha = 0.4$ versus maximum amplitude distortion $A_m$ in the filter bandwidth are plotted in Fig. 7.52, where

$$A_m \triangleq \begin{cases} L_A(2f_{BW}) & \text{for linear amplitude distortion} \\ P_A(f_{BW})^2 & \text{for parabolic distortion} \\ S_A & \text{for sinusoidal distortion} \end{cases} \quad (7.58)$$

We note that for a given value of maximum amplitude distortion $(A_m)$, linear amplitude distortion causes the least degradation, followed in order of increasing degradation by parabolic and sinusoidal amplitude distortions.

CCI is one of the major sources of performance impairments in a frequency-reuse radio communication system. Recently 64-QAM and 256-QAM systems for frequency-reuse system applications have been considered [Saito, 7.71]. For this

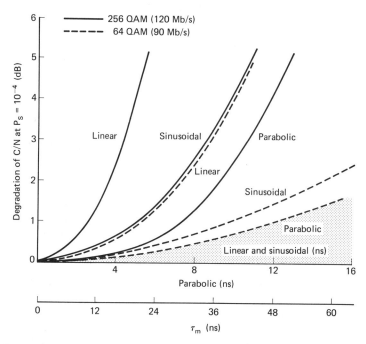

**Figure 7.50** Degradation of *C/N* versus $\tau_m$ for linear, parabolic, and sinusoidal *group-delay* distortions for 256 QAM and 64 QAM with symbol rate of 15 Mbaud and $\alpha = 0.4$ raised-cosine filters.

reason here we present illustrative results of our study in regards to the sensitivity of these systems to CCI [Wu, 7.70]. An in-depth interference analysis of QAM systems is given in Chapter 9. If the radio system is equipped with CCI cancellation circuits, then the CCI power, designated by the desired **carrier-to-interference power** ratio *C/I*, should be interpreted as the *residual* CCI after interference cancellation.

The CCI system model is shown in Fig. 7.53 [Wu, 7.70]. In this study the transmit and receive filters are assumed to be ideal square roots of raised-cosine filters with *x*/sin *x* equalization in the transmitter so that the whole system satisfies Nyquist first criterion. An $\alpha = 0.2$ system is used, where $\alpha$ denotes the ratio of the excess bandwidth to Nyquist bandwidth. In other words, here we assume that the residual group-delay and amplitude distortion are negligible. We assume that the interferer and the desired signal are of the same format but statistically independent of each other. We consider 64 QAM with a 64-QAM CCI, 225 QPRS with a 225-QPRS CCI, and the like. The probability of error performance of a 256-QAM system in the presence of CCI and AWGN is illustrated in Fig. 7.54. The degradation in *C/N* for 64-, 256-QPRS systems due to CCI is illustrated in Fig. 7.54(b). Note that the 256 QAM is the most sensitive. For a *C/N* degradation, that is, fade margin degradation of 1 dB, a residual *C/I* of about 40 dB is required.

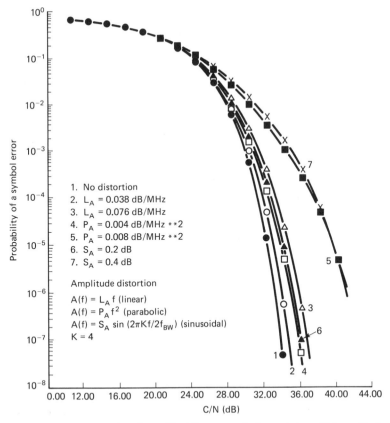

**Figure 7.51** $P_s$ versus $C/N$ for residual linear, parabolic, and sinusoidal *amplitude distortions* for 256 QAM with a bit rate of 120 Mb/s (15 Mbaud and $\alpha = 0.4$). Noise is defined in the double-sided Nyquist bandwidth; that is, the equivalent noise bandwidth is 15 MHz. [Wu, 7.34].

### 7.9.3 FEC-Coded 256-QAM Performance

Illustrative $P_e = F(C/N)$ performance curves for uncoded and FEC-coded 256-QAM modems are given in Fig. 7.55. In this system, the uncoded source rate is 1.544 Mb/s, whereas the transmission rate is 1.6 Mb/s (see Fig. 7.47) and the respective coded modem description is given in Section 7.9. At $P_e = 10^{-8}$, the uncoded 256-QAM modem has a $C/N = 37.5$-dB requirement. Thus the modem, due to channel and hardware imperfections (including residual group delay after adaptive equalization) is 37.5 dB $-$ 34.2 dB $=$ 3.3 dB worse than the theoretical uncoded 256-QAM modem. The FEC-encoded 256-QAM modem requires $C/N$ of only 32.4 dB for $P_e = 10^{-8}$.

From this example we note that the approximately 3% redundancy, which

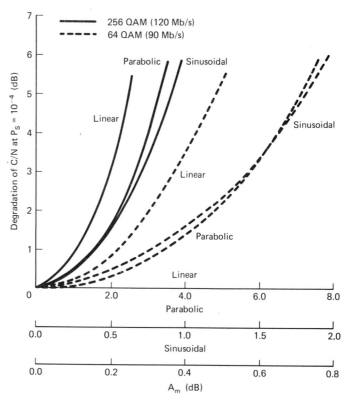

**Figure 7.52** Degradation of *C/N* versus $A_m$ for linear, parabolic, and sinusoidal *amplitude distortions* for 256 QAM and 64 QAM with symbol rate for 15 Mbaud and $\alpha = 0.4$ raised-cosine filters.

was added for FEC encoding bits, reduces the measured *C/N* requirement by 5.1 dB [Feher, 7.22].

## 7.10 NLA-QAM: NONLINEARLY AMPLIFIED HIGH-POWER QAM SYSTEMS

In radio systems, in order to overcome poor detection due to signal loss as a result of long path length or fading, the generation of high transmitter output power is often desirable. In general, output power is most easily maximized by using as a power amplifying element an amplitude-limiting device. Examples of such devices are **traveling wave tubes** (TWTs) and GaAsFET amplifiers in saturation, class-C operated bipolar transistor amplifiers, and Gunn and Impatt diode **injection locked amplifiers** (ILAs).

Here we present a generalized technique for generating QAM signals that permits nonlinear transmitter output power amplification and, hence, higher output power than in methods considered to date. This technique has been designated

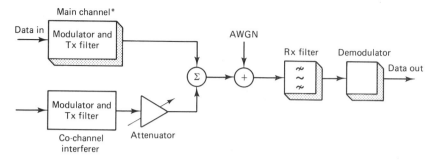

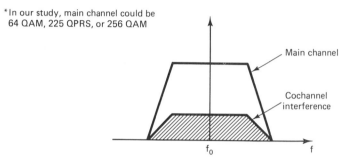

Illustrative frequency spectra of the desired and interfering signals

**Figure 7.53**  CCI system model of 64-QAM, 225-QPRS, and 256-QAM systems [Wu, 7.70].

**nonlinearly amplified quadrature amplitude modulation** (NAL-QAM).  With it, a $2^{2n}$-state QAM signal is generated by combining $n$ unfiltered, nonlinearly amplified, QPSK signals, $n$ being a positive integer.   As 64 QAM is of considerable interest, a 64-state NLA-QAM modulator is described.

One attractive feature of this modulation method is that despite significant differences with conventional QAM, the same straightforward demodulation techniques apply to both [Morais, 7.69; Hill, 7.39].

A block diagram of the NLA 64-state QAM modem is shown in Fig. 7.56. NLA 64-state QAM, as its name implies, permits nonlinear transmit power amplification.   The QPSK-modulated signals are unfiltered prior to the nonlinear amplifiers, so they each contain only one power level and, as such, are unaffected by the AM/AM and AM/PM conversion characteristics of the nonlinear amplifiers. Hence, the $P_e$ performance of NLA 64-state QAM is identical to that of conventional 64-state QAM, assuming equivalent filters are used.   The modulator employs a parallel modulation technique wherein QPSK modulators 1, 2, and 3 operate in parallel.   The modulated QPSK1 signal is added to the QPSK2 and QPSK3 signals, which are 6 dB and 12 dB below the signal power level of QPSK1.   The signal vectors for each of the three QPSK modulators are depicted in Fig. 7.57(b)–(d) with the resulting 64-state signal constellation diagram (Fig. 7.57(a)).

Hill and Feher [Hill, 7.39] have studied the effect of selective fading and/or system filter imperfections on the performance of NLA-64-QAM systems.   They have shown that the effects of modulator imperfections in the power level combining

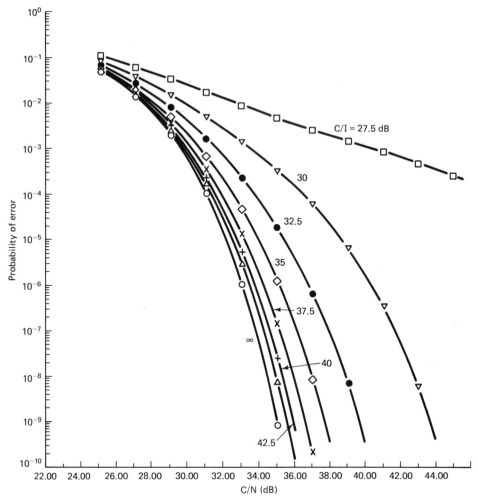

**Figure 7.54(a)** Error probability for 256-QAM ($\alpha = 0.2$) with a 256-QAM ($\alpha = 0.2$) CCI [Wu, 7.70].

on the $P_e$ performance degradation may be a major weakness of this high-power QAM system.

Despite significant differences between NLA-QAM and conventional QAM modulators, the same straightforward demodulation techniques apply to both, and the $P_e$ versus $C/N$ performances are essentially identical. In the straightforward 16-state version, NLA-QAM has been shown to result in a transmitter output power advantage that is on the order of 5 dB compared to the conventional method, when TWTs or GaAsFET amplifiers are used. An even larger advantage results when 16-state NLA-QAM using class-C bipolar amplifiers is compared to conventional 16-QAM using a linear bipolar or GaAsFET amplifier of similar complexity to the class-C types. We note, however, that any power advantage is achieved at the expense of one additional output power amplifier in the case of 16-state NLA-QAM and two such amplifiers in the case of 64-state NLA-QAM [Morais, 7.69].

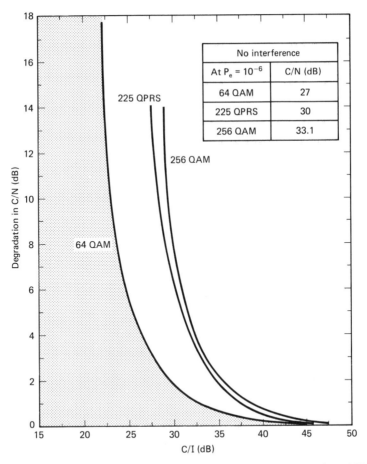

| No interference | |
|---|---|
| At $P_e = 10^{-6}$ | C/N (dB) |
| 64 QAM | 27 |
| 225 QPRS | 30 |
| 256 QAM | 33.1 |

**Figure 7.54(b)** Degradation in *C/N* for 64-QAM, 256 QAM ($\alpha$ = 0.2), and 225 QPRS with an amplitude-modulated CCI, which has the same format as the desired signal, such as 64 QAM with a 64-QAM CCI, and so on. The ordinate represents the increase in *C/N* to maintain an error rate of $10^{-6}$ [Wu, 7.70].

## 7.11 16-SQAM: A SUPERPOSED SPECTRALLY EFFICIENT QAM MODEM FOR NONLINEARLY AMPLIFIED RADIO SYSTEMS

In the previous section we described NLA-QAM systems, which have a significant transmit power advantage (power efficiency advantage) compared to conventional QAM systems. However, as Fig. 7.57 shows, NLA-QAM modulators require post-modulation spectral shaping BPFs in order to bandlimit the bandwidth of the parallel QPSK-modulated signals. As the individual QPSK modulators are amplified by saturated amplifiers, the spectrum prior to the BPF is wide. As pointed out in Section 7.4, for many applications it is not practical to design the spectral shaping BPFs following the nonlinear HPA because

1. It is very difficult to design a Nyquist BPF for very high carrier frequency-to-bit-rate ratios.

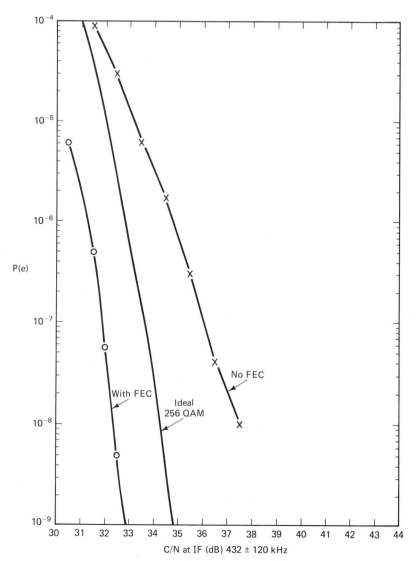

**Figure 7.55** Performance measurement results of 256-QAM, 1.544-Mb/s modem (uncoded) and FEC coded (1.6 Mb/s) modem. In this terrestrial microwave measurement, performed over 900-km nonregenerative sections over AT&T and MCI networks, 256-QAM modems, equipped with adaptive equalizers and manufactured by Karkar Electronics, Inc., of San Francisco, are used.

**2.** For frequency agile system applications—for example, TDMA systems—a bank of filters instead of one filter would be required.

**3.** The post-HPA power handling capability and insertion loss of the radio-frequency BPF might be critical.

In SQAM systems there is no need for post-HPA filtering. SQAM signal processors assure that the nonlinearly amplified spectrum is narrow; thus a spec-

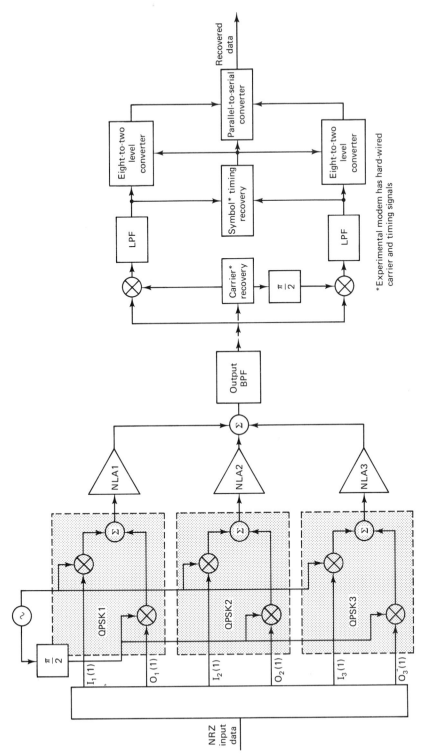

**Figure 7.56** NLA 64-QAM model block diagram.

*Experimental modem has hard-wired carrier and timing signals

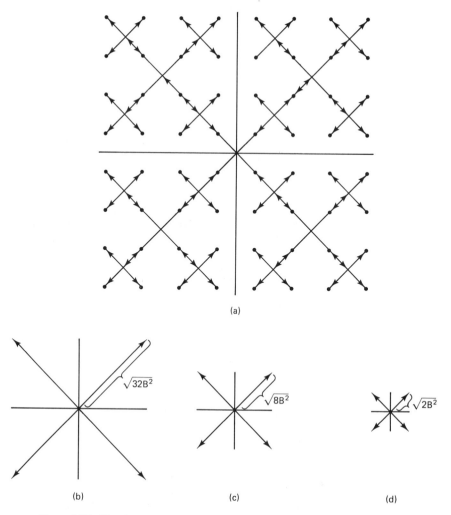

**Figure 7.57** Signal state of "parallel modulation." (a) 64-state QAM. (b) First path signal. (c) Second path signal. (d) Third path signal.

trally efficient nonlinearly amplified signal *without* the need of BPFs is obtained. Four-state (2-level baseband) SQAM systems are described in Section 7.6.2 and in references [Seo, 7.28; 7.33; 7.35]. Here, 16-SQAM signals are generated by a straightforward extension of the 4-SQAM principles to 16 states [Seo, 7.72]. In this case in the I- and Q-channel basebands there are four independent states instead of two. From equations (7.37)–(7.39) we obtain the modulated 16-SQAM signal $p(t)$ given by:

$$p(t) = \sum_k [a_k s(t - kT_s)(\cos 2)\pi f_c t + b_k s\left(t - kT_s - \frac{T_s}{2}\right)(\sin 2 \, \pi)f_c t] \quad (7.59)$$

where

$\{a_k\}, \{b_k\} = \{\pm 1, \pm 3\}$, independent and equiprobable

$\quad T_s$ = symbol duration ($T_s = 4T_b$)

$\quad f_c$ = carrier frequency

$\quad s(t)$ = impulse response of SQAM baseband signals defined as:

$$s(t) = \frac{1}{2}\left(1 + \cos\frac{\pi t}{T_s}\right) - \frac{1 - A}{2}\left(1 - \cos\frac{2\pi t}{T_s}\right) \qquad (7.60)$$

$$0.5 \leq A \leq 1.5, \quad -T_s \leq t \leq T_s$$

Here, $A$ is an amplitude parameter of the SQAM signals to be selected to control the power spectrum and envelope fluctuation of the QAM signals [Seo, 7.72]. Note that the impulse response, equation (7.60), contains two terms. This is the reason that we used the term *superposed* for this modulation scheme.

A block diagram of the 16-SQAM modulator is illustrated in Fig. 7.58. The corresponding demodulator is practically the same as a conventional offset 16-QAM demodulator [Feher, 7.2; 7.3]. The input NRZ data enters 1-to-4 serial-to-parallel converter which is split into four parallel NRZ signals, $I_1$, $Q_1$, $I_2$ and $Q_2$. The Q signals $Q_1$ and $Q_2$ are offset by half a symbol interval ($T_s/2$) from the in-phase signals $I_1$ and $I_2$. These four 2-level signals are then wave-shaped (processed) and Q-modulated through two parallel SQAM modulators. For detailed generation of 2-level SQAM signals refer to Section 7.6.2 and [Seo, 7.35].

The SSDs for 16-SQAM and 16-state **multiamplitude MSK** (MAMSK) [Weber, 7.73] signals are illustrated in Fig. 7.59. We note that for the illustrated amplitude parameters $A = 0.7$ and $A = 0.8$, different SSDs and also different $P_e$ performance results and spectra are obtained.

From the nonlinearly amplified out-of-band-to-total-power ratios, illustrated in Fig. 7.60, we note that conventional 16 QAM has the least desirable spectrum (very wide band). MAMSK has a somewhat better spectrum, and SQAM has the narrowest spectrum (smallest out-of-band energy) among the techniques consid-

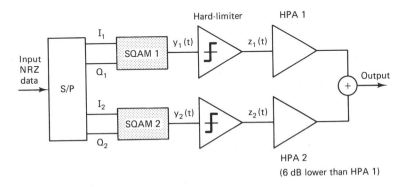

**Figure 7.58** Block diagram of 16-SQAM modulator.

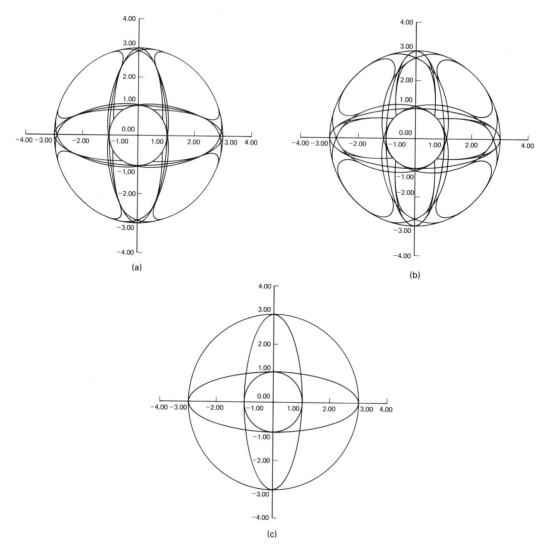

**Figure 7.59**  Signal space diagrams of 16 SQAM and MAMSK. (a) 16 SQAM (A = 0.7) (b) 16 SQAM (A = 0.8) (c) MAMSK. [Seo, 7.72]

ered. The MAMSK modulation technique has been extensively described by Weber and others [Weber, 7.73].

The $P_e = f(E_b/N_o)$ curves given in Fig. 7.61 indicate that the $A = 0.7$ to $A = 0.9$ 16-SQAM has a considerably better $P_e$ performance than an illustrative conventional 16 QAM, even though the 16-QAM signal requires a 9-dB HPA input backoff, whereas the 16-SQAM signal is amplified by fully saturated amplifiers.

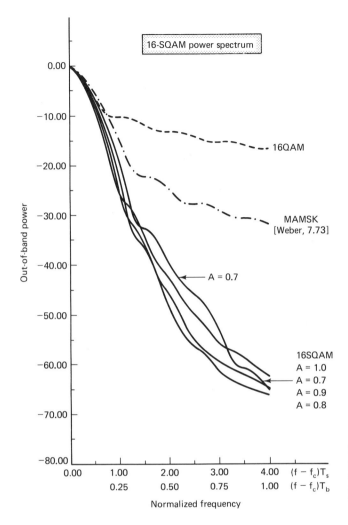

**Figure 7.60** Out-of-band-to-total-power ratios of 16-SQAM, MAMSK and 16-QAM signals in a nonlinear (hard-limited) channel. [Seo, 7.72; Weber, 7.73]

## 7.12 QPRS RADIO SYSTEMS OPERATED ABOVE THE NYQUIST RATE

An improved efficiency partial-response signaling (PRS) and quadrature-partial-response signaling (QPRS) data-filtering strategy, which enables a higher transmission rate (more symbols per second per hertz) than feasible with conventional PRS and QPRS systems, is described in this section. With the use of our technique [Wu, 7.32; Feher, 7.27] a smaller number of signal levels (states) can be chosen for specified spectrally efficient applications, thus an increase in signal-to-noise ratio ($S/N$) immunity is obtained with a lower cost and simpler hardware implementation.

PRS systems are speed tolerant [Lender, 7.6], that is, it is possible to transmit at a rate higher than the Nyquist rate. Thus one may think of using PRS systems

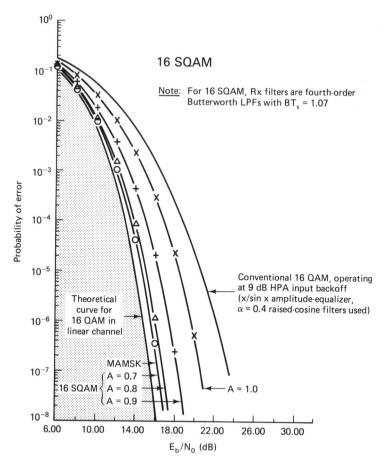

**Figure 7.61** $P_e$ performance of 16-SQAM modems through saturated amplifiers. [Seo, 7.72]

to achieve higher spectral efficiency. For example, for highly spectral-efficient applications, such as data-in-voice supergroup modem applications, we require the transmission of a T-1 carrier (1.544 Mb/s) plus overhead bits, that is, a total rate of 1.6 Mb/s or higher in a bandwidth of 256 kHz (this includes the guard bands of the CCITT Standard frequency division multiplexed (FDM) scheme). Hence, an efficiency of 6.25 bits/s/Hz or more is required. Ideal 225-QPRS modems achieve the Nyquist rate of 6 bits/s/Hz. To achieve 6.25 bits/s/Hz by 225-QPRS we require 6.25 : 6 = 1.042, that is, 4.2% higher than the Nyquist rate. Thus it is of interest to study the performance of PRS and QPRS above the Nyquist rate. These systems could offer an attractive alternative to 256-QAM modems described in Section 7.9.

### 7.12.1 System Models

PRS and QPRS system block diagrams are shown in Fig. 7.62(a) and (b). An illustrative baseband example would be the transmission of an $f_b$ = 800-kb/s rate

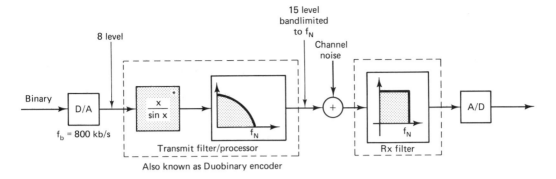

(a)

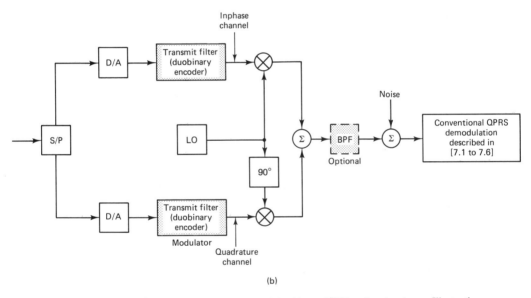

(b)

**Figure 7.62** (a) Block diagram of a conceptual duobinary PRS baseband system. Illustrative example: Source rate $f_b = 800$ kb/s; symbol rate of D/A output, assuming an 8-level output is used, is $f_s = 266,000$ symbols/s. In this case $f_N = 133$ kHz. Thus the spectral efficiency of the 8-level PAM signal, which gets converted to 15-level PRS, is 800 kb/s : 133 kHz = 6 b/s/Hz. [Wu, 7.32; Lender 7.6]. (b) QPRS modem block diagram. The duobinary encoders in the I-and Q-channels are identical to the ones in Fig. 7.62(a)

with a 15-level system filtered at the Nyquist frequency of 133 kHz, that is, 6 bits/ s/Hz. In the corresponding 225-QPRS modem, which consists of modulating in quadrature two baseband 15-level PRS signals, the source rate at 1.6 Mb/s is serial- to-parallel converted and fed into the I- and Q-channels.

Transmission above the Nyquist rate by PRS can be achieved by increasing the signaling rate with the transmission characteristic being kept fixed, as is the

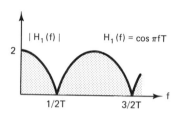

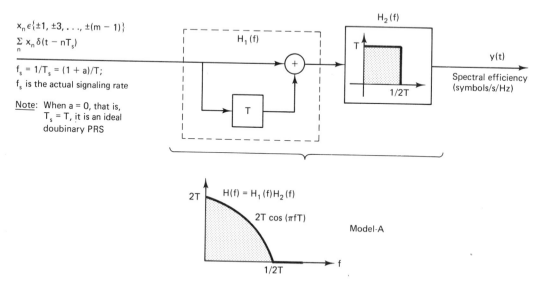

**Figure 7.63** *Model-A*: Conventional system model for duobinary PRS above the Nyquist rate and cascaded transfer function. The spectral efficiency of this system is $2(1 + a)$ symbols/s/Hz, that is, $a\%$ higher than the Nyquist rate of 2 symbols/s/Hz.

case in which speed tolerance is described in numerous references, including [Lender, 7.6; 7.64]. This conventional method is called **Model-A**, as depicted in Fig. 7.63. Model-A transmits $f_s = (1 + a)/T$ symbols per second over a bandwidth of $1/2T$ hertz, resulting in an efficiency of $2(1 + a)$ symbols per second per hertz, where $a$ represents the percentage above the Nyquist rate. Note that spectral efficiency is defined at the infinite attentuation point (or 50 dB in practice), not the 3-dB point. In fact, Model-A is designed to force a symbol rate through a duobinary filter that is higher than the rate for which the filter is designed. As such, it introduces serious undesired ISI and results in very significant performance degradations.

An improved scheme called **Model-B** was introduced by Feher and others [Feher, 7.27; Wu, 7.32]. This model is illustrated in Fig. 7.64. Our Model-B consists of defining a new (infinite) partial-response system in the following manner:

1. Define the ideal duobinary response for the new higher symbol rate.
2. Truncate this response to achieve the desired spectral efficiency using an ideal brickwall filter ("chopping"), which has a cutoff frequency at $f_N/(1 + a)$ below the Nyquist frequency $f_N = 1/2T_s$.

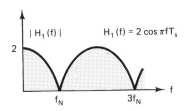

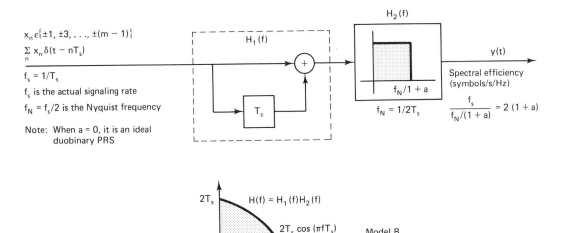

**Figure 7.64** *Model-B*: New system model for duobinary PRS above the Nyquist rate a corresponding cascaded transfer function. The spectral efficiency of this system is $2(1 + a)$ symbols/s/Hz, that is, $a\%$ higher than the Nyquist rate of 2 symbols/s/Hz. [Feher, 7.27; Wu, 7.32].

Intuitively we expect that Model-B performs better than Model-A because in Model-B most of the original frequency components are retained and the power loss is significantly smaller than in the Model-A case.

### 7.12.2 Error-Floor (Dribbling-Error) Problems

Here we demonstrate that multilevel partial-response systems, operated above the Nyquist rate, have an **error floor**, also known as **dribbling errors**. We highlight multilevel partial-response systems and QPRS. However, note that the methods used in the error-floor calculations could be also used for any other system, including QAM systems [Wu, 7.32]. Error floors have been noticed in many operational microwave systems. A practical solution to reduce the error floor is the application of FEC coding to systems that have dribbling errors. As we demonstrate in this section, an error floor is introduced even if the system has an infinite (practically very high) CNR.

Referring to Figs. 7.63 and 7.64, the *M*-ary input symbol sequence is rep-

**TABLE 7.4** Characteristics of duobinary PRS at and above the Nyquist rate

| | Frequency response $H(f)$ | Impulse response $h(t)$ |
|---|---|---|
| Model-A (conventional) | $2T \cos{(\pi f T)}$<br>$\|f\| \leq \dfrac{1}{2T}$ | $\text{sinc}\left(\dfrac{t}{T} + \dfrac{1}{2}\right) + \text{sinc}\left(\dfrac{t}{T} - \dfrac{1}{2}\right)$ |
| Model-B (new scheme) | $2T_s \cos(\pi f T_s)$<br>$\|f\| \leq \dfrac{f_N}{1+a}$ | $\dfrac{1}{1+a}\left\{\text{sinc}\left[\dfrac{1}{1+a}\left(\dfrac{t}{T_s} + \dfrac{1}{2}\right)\right] + \text{sinc}\left[\dfrac{1}{1+a}\left(\dfrac{t}{T_s} - \dfrac{1}{2}\right)\right]\right\}$ |

$f_s = 1/T_s = (1 + a)/T$ is the signaling rate.
$f_N = f_s/2$ is the Nyquist frequency of the signal.

resented by $\sum_n x_n \delta(t - nT_s)$, where $1/T_s$ is the signaling rate and the $x_n$'s are equally possible values of $\{-(m - 1), - (m - 3), \ldots, (m - 1)\}$. Expressions for $H(f)$, the overall transfer function, and $h(t)$, the corresponding impulse response, for the two models are given in Table 7.4, in which the center of the impulse response has been chosen as the time origin, that is, with $a = 0$ for the ideal duobinary PRS. At the Nyquist rate, the only two nonzero samples are at $t = -T_s/2$ and $T_s/2$, respectively. For $M$-ary inputs, the output $y(t)$ has $2m - 1$ levels if $a = 0$. With $a > 0$, that is, transmission above the Nyquist rate, the nominally zero pulse samples at $t = \pm(2n + 1)T_s/2$, $n = 1, 2, 3, \ldots$, are no longer zeros. These nonzero samples present undesired ISI to the system. The peak eye-closure criterion [Kabal, 7.74; Lucky, 7.75] is commonly used to determine eye opening for a system in the presence of ISI. To apply this criterion, we must first evaluate the peak distortion $D_P$ due to the worst data sequence, which is defined as follows:

$$D_P = 2(m - 1) \sum_{n=1}^{\infty} \left| h\left(\frac{2n + 1}{2} T_s\right) \right| \qquad (7.61)$$

By using (7.61) and the peak eye-closure criterion, it can be shown [Wu, 7.32] that neither Model-A nor Model-B gives open eyes for $M$-ary inputs, $m > 4$ and $a > 0.04$. In other words, both of these models introduce an error floor even for very high $C/N$ ratio. This is due to inherently higher ISI in multilevel systems. However, the probability of occurrence of the worst sequence (which closes the eye diagram and is the cause of error floor) may be relatively low in some cases. Instead of the peak distortion, *we introduce the concept of a truncated peak distortion $D_t$ defined by*

$$D_t = 2(m - 1) \sum_{n=1}^{N} \left| h\left(\frac{2n + 1}{2} T_s\right) \right| \qquad (7.62)$$

The difference between (7.62) and (7.61) is that in (7.61) only a finite number of undesired ISI samples are considered. Neglecting all the undesired ISI, for the moment, the output levels at the optimum sampling instant would be 0, $\pm 2h(T_s/2)$, $\pm 4h(T_s/2), \ldots, \pm 2(m - 1)h(T_s/2)$, resulting in a decision distance

**TABLE 7.5** Comparison of the shortest length of the worst-data sequence and the corresponding error floor for model-A and model-B with 8-level inputs (15-level outputs)

| Percent above Nyquist rate | Model-A (conventional model) | | Model-B (new model) | |
|---|---|---|---|---|
| | $L$ | $E_f = 1/L^8$ | $L$ | $E_f = 1L^8$ |
| 3 | 22 | $1.3 \times 10^{-20}$ | 686 | 0 |
| 4 | 12 | $1.4 \times 10^{-11}$ | 206 | 0 |
| 5 | 8 | $6 \times 10^{-8}$ | 92 | $8.2 \times 10^{-84}$ |
| 6 | 6 | $3.8 \times 10^{-6}$ | 60 | $6.5 \times 10^{-55}$ |
| 7 | 6 | $3.8 \times 10^{-6}$ | 34 | $1.9 \times 10^{-31}$ |
| 8 | 6 | $3.8 \times 10^{-6}$ | 28 | $5.1 \times 10^{-26}$ |
| 9 | 6 | $3.8 \times 10^{-6}$ | 22 | $1.3 \times 10^{-20}$ |

of $h(T_s/2)$. Now summing up the undesired ISI for $N$ previous and $N$ forthcoming symbols under the worst case, there may exist a minimum $N$, denoted by $N_{\min}$, such that

$$D_t \geq h\left(\frac{T_s}{2}\right) \qquad \text{for } N \geq N_{\min} \tag{7.63}$$

When (7.63) is met, the eye is closed. We define $L = 2(N_{\min} + 1)$ to be the *shortest length of the worst sequence* which closes the eye. The probability of occurrence of this sequence is $E_f = 1/L^m$. This probability may be regarded as an *error floor*, for even without noise the system can not have a probability of error lower than $E_f$. Though $E_f$ is not a tight bound for the average BER performance, it can serve as a means to predict how good a system would be. As a rule of thumb, a system with lower error floor will be preferred. The problem of finding $L$ and $E_f$ for the two models with 8-level inputs was programmed on a computer. Results are summarized in Table 7.5. From Table 7.5, we see that the error floor of Model-B is much lower than that of Model-A. Thus we reason that our new scheme (Model-B) is better than the conventional method for a practical realization.

### 7.12.3 Computer-Generated and Measured Multilevel Partial-Response Eye Diagrams

Since eye diagrams show a good indication of system performance, we compare a few computer-generated and measured eye diagrams. In the case of 49 QPRS with 10% increase above the Nyquist rate, Fig. 7.65(b) (Model-B) shows a robust eye opening, whereas in Fig. 7.65(a) (Model-A) the eye is about to close. This indicates that Model-B is better than Model-A as the number of input levels increases. This trend becomes clearer as we look into 225-QPRS with 4.2% increase above Nyquist rate. Whereas Fig. 7.66 (Model-B) shows a reasonably good eye opening, in Fig. 7.66(a) (Model-A) the eye is completely closed. To confirm the simulated results we designed an above-the-Nyquist-rate system using our Model-B for a 15-level PRS [Feher, 7.27]. The measured eye diagram and corresponding

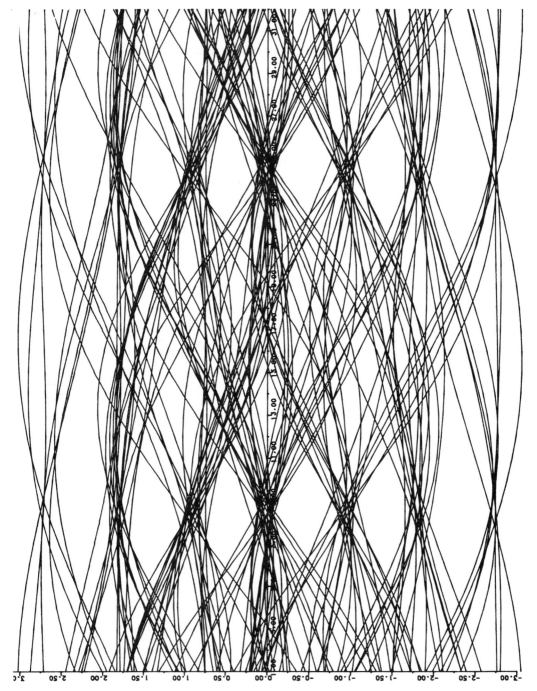

**Figure 7.65(a)** Eye diagram of 7-level/49-QPRS with 10% increase above the Nyquist rate using Model-A. Spectral efficiency: 4.4 bits/s/Hz.

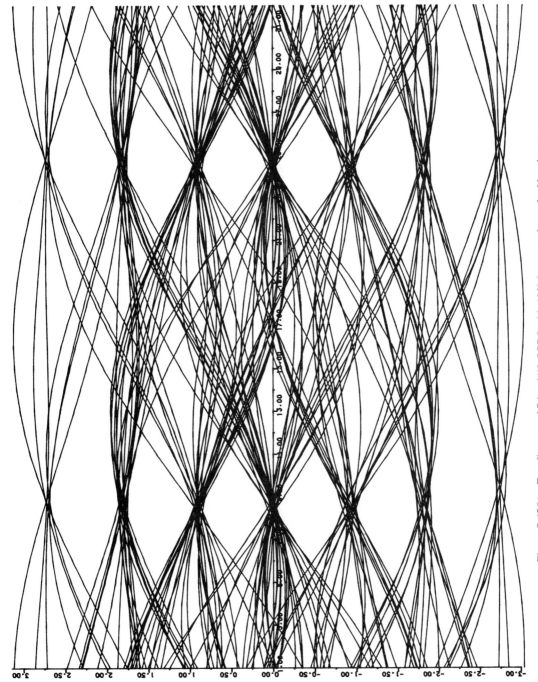

**Figure 7.65(b)** Eye diagram of 7-level/49-QPRS with 10% increase above the Nyquist rate using Model-B [Wu, 7.32]. Spectral efficiency: 4.4 bits/s/Hz.

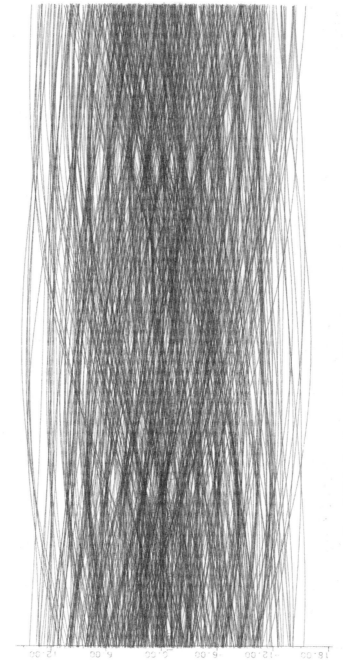

**Figure 7.66(a)** Eye diagram of 15-level PRS/225-QPRS with 4.2% increase above the Nyquist rate using Model-A. Spectral efficiency: 6.25 bits/s/Hz.

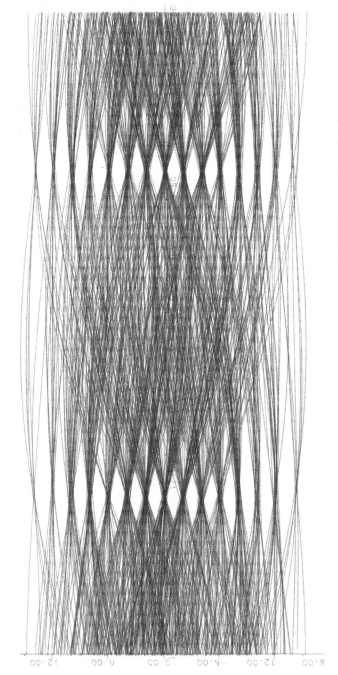

**Figure 7.66(b)** Eye diagram of 15-level PRS/225-QPRS with 4.2% increase above the Nyquist rate using Model-B. Spectral efficiency: 6.25 bits/s/Hz.

417

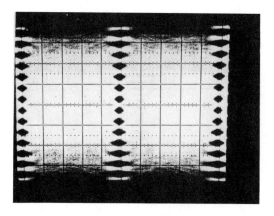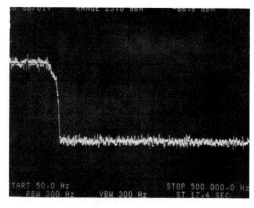

Horizontal: 500 ns/division
Vertical: 100 mV/division

Horizontal: 50 kHz/division
Vertical: 10 dB/division

**Figure 7.66(c)** Baseband eye diagram and corresponding power spectral density of a 15-level PRS. This eye diagram corresponds to the I- (or Q-) channel of a 225-QPRS modem operated at $f_b$ = 1.6 Mb/s. The spectral efficiency of the baseband system and of the quadrature modem is 800 kb/s: 128 kHz = 6.25 bits/s/Hz, that is, a 4.2% higher rate than the theoretical 6-bit/s/Hz Nyquist rate. This efficiency is defined at the 50-dB attenuation point. In these measurements Karkar Electronics, Inc., patented α = 0.01 filter is used [U.S. patent 3,271,705] See [Feher 7.27], patent disclosure by Feher et al.

power spectral density are shown in Fig. 7.66(a). We note that the measured eye diagram is open as predicted in the computer-simulation result of Fig. 7.66(b).

### 7.12.4 225-QPRS BER Performance Above the Nyquist Rate

To achieve an improved spectral efficiency and a lower-cost hardware configuration, we assume that the duobinary PRS filter shaping is fully at the transmitter, whereas the receive filter merely bandlimits the channel noise. This would result in about 1 dB degradation in $C/N$ with respect to the case in which the filter shaping is optimally divided between the transmitter and receiver [Feher, 7.2]. A perfect

**TABLE 7.6** Required C/N* (decibels) at $P_e$ = $10^{-9}$ for 225-QPRS above the Nyquist rate [Wu 7.32; Feher 7.27]

| Percent above Nyquist rate | Spectral efficiency (bits/s/Hz) | Model-A (conventional) | Model-B** (new scheme) |
|---|---|---|---|
| 0 | 6 | 32 dB | 32 dB |
| 3 | 6.18 | 43.3 | 32.4 |
| 4 | 6.24 | Not possible | 34. |
| 5 | 6.3 | Not possible | 34.9 |
| 6 | 6.36 | Not possible | 35.3 |

*The average $CN$ is specified in the minimum Nyquist bandwidth.

**Based on patent disclosure of Feher and others [Feher, 7.27; Wu, 7.32].

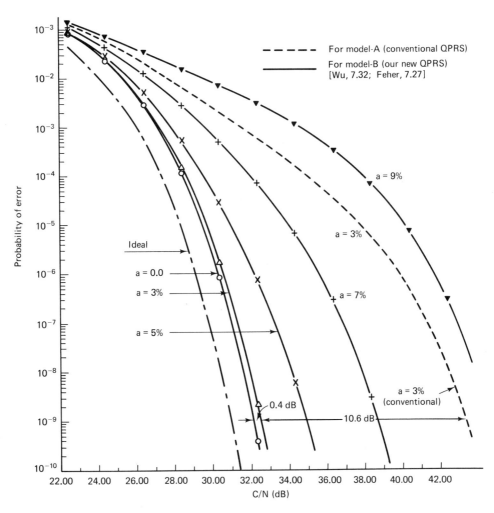

**Figure 7.67** Probability of error of 225-QPRS above the Nyquist rate using Model-A or Model-B versus *C/N*. In the figure, *a* represents the number (in percentage) of the actual transmission rate above the Nyquist rate [Wu, 7.32].

synchronization is assumed and symbol-by-symbol detection is used in the simulation. Results are plotted in Fig. 7.67. [Wu, 7.32].

Table 7.6 summarizes the required *C/N* at $P_e = 10^{-9}$ for 225-QPRS above the Nyquist rate using Model-A and Model-B for comparison.

The significant advantage of Model-B is evident from Fig. 7.68 and Table 7.6. For example, for 225-QPRS transmission, 3% above the Nyquist rate with conventional Model-A yields an 11-dB degradation, whereas with our new Model-B only a 0.4 dB degradation is observed.

The performance of multilevel **modified duobinary** systems [Linder, 7.6] operated above the Nyquist rate can be improved by an extension of our Model-B to modified duobinary systems. A detailed description of these improved modified duobinary systems is given in [Wu, 7.76].

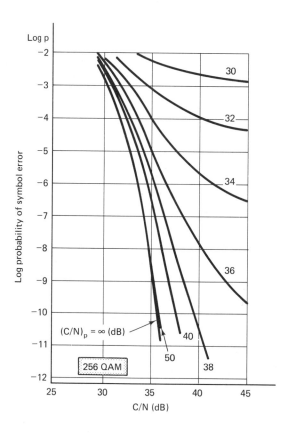

**Figure 7.68** The probability of error-performance curves of 256-QAM system versus the carrier-to-thermal-noise ratio and a carrier-to-phase-noise ratio as a parameter. Double-sided Nyquist bandwidth [Kucar 7.23].

## 7.13 PHASE NOISE AND CARRIER AND SYMBOL TIMING RECOVERY CAUSED DEGRADATIONS

Phase noise introduced by the transmission system and/or by the demodulator synchronization subsystems may cause a significant performance degradation. Phase noises of local oscillators, frequency synthesizers, upconverters, and downconverters are among typical contributors to the total system-caused phase noise. Imperfect carrier recovery circuits may also contribute to phase noise accumulation.

Many authors considered phase-noise-caused degradations of QPSK systems. A detailed study of QPSK-related phase-noise problems and an extensive list of references is presented in Chapter 7 of [Wu, 7.2]. However, an exact mathematical solution or practical guidelines for higher-state modulation schemes were not published until 1985 [Feher, 7.1; Kucar, 7.23; Sasase, 7.77].

Here, as first-order approximation we assume that the probability density function (pdf) of the noise is Gaussian. Computer simulation and experiments confirm that our first-order approximation leads to reasonably accurate results illustrated in Fig. 7.68 through Fig. 7.71. *M*-ary QAM (MQAM), *M*-ary PSK (MPSK) and *M*-ary-QPR (MQPR) system performance curves are given.

In the computation of phase-noise-caused degradations we assume that in adddition to the Gaussian pdf of the phase noise, the additive channel noise has

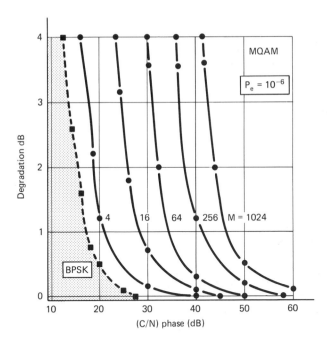

**Figure 7.69** The degradation of MQAM systems versus the carrier-to-phase-noise ratio for $P_e = 10^{-6}$. Double-sided Nyquist bandwidth [Kucar 7.23].

also a Gaussian pdf. The total carrier-to-noise power is

$$\left(\frac{C}{N}\right) = \frac{1}{(N/C_{CH}) + (N/C_P)} \tag{7.64}$$

where

$\left(\dfrac{C}{N_T}\right)$ = total $C/N$ (caused by channel noise plus phase noise)

$\left(\dfrac{C}{N_{CH}}\right)$ = $C/N$ of the channel, that is, $C/N$ caused by the additive Gaussian noise of the channel (known also as *thermal noise*)

$\left(\dfrac{C}{N_P}\right)$ = carrier-to-*phase* noise

We interpret $(C/N)_P$ as the **integrated phase noise** in the total bandwidth of the receiver. This bandwidth frequently equals the double-sided Nyquist bandwidth, that is, the symbol rate bandwidth of the receiver [Feher, 7.2]. Total or integrated phase noise could be also considered as the residual phase noise left in the system after a CRC which may enhance or cancel the phase noise [Feher, 7.1; 7.2; Kucar, 7.23].

In Fig. 7.72, the $C/N$ degradation as a function of Gaussian phase noise (expressed in degrees rms) and of static carrier phase error for 64-QAM and 256-QAM systems at $P_e = 10^{-8}$ is illustrated. Here an $\alpha = 0.2$ filtered ideal raised-cosine Nyquist channel is assumed [Sasase, 7.77]. A straightforward conversion of rms degrees of phase noise into $(C/N)_P$ expressed in dB confirms that the results reported in Fig. 7.68 are practically the same as that of Fig. 7.72. For example

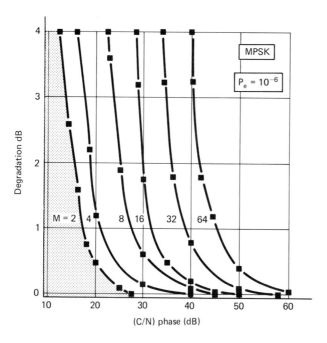

**Figure 7.70** The degradation of MPSK systems versus the carrier-to-phase-noise ratio for $P_e = 10^{-6}$. Double-sided Nyquist bandwidth [Kucar, 7.23].

$0.573°$ of rms Gaussian noise degrades a 256-QAM system by 1.3 dB at a $P_e = 10^{-8}$ (see Fig. 7.72). We recall that $0.573°$ corresponds to $0.573/57.3 = 0.01$ radians, or to $20 \log 0.01 = -40$ dB on a logarithmic scale. From Fig. 7.68 note that a $(C/N)_P = 40$ dB (noise is $-40$ dB in reference to carrier) is causing a 1.5 dB degradation at $10^{-8}$; that is, a good agreement between the two representations

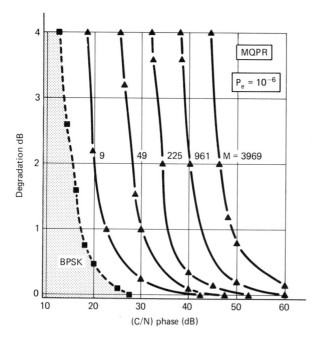

**Figure 7.71** The degradation of MQPR systems versus the carrier-to-phase-noise ratio for $P_e = 10^{-6}$. Double-sided Nyquist bandwith [Kucar, 7.23].

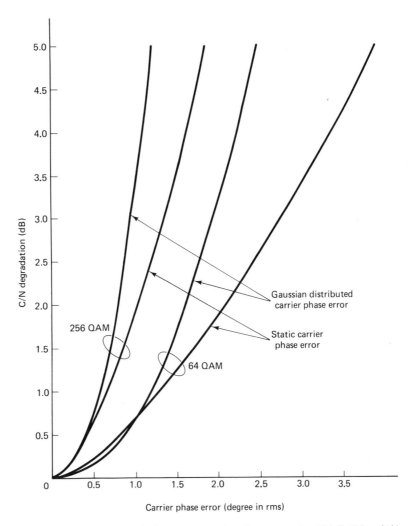

**Figure 7.72** *C/N* degradation versus carrier phase error for 256 QAM and 64 QAM. Parameter, $P_e = 10^{-8}$, $\alpha = 0.2$ [Sasase, 7.77].

has been demonstrated. In general, the accuracy of these curves is good within about $\pm 0.5$ dB for degradations in the range of 0.5 dB to 2.5 dB. For larger degradations our first-order approximations are not valid.

**Symbol timing recovery** (STR) circuit imperfections are also among frequent causes of performance degradations. In Fig. 7.73, the effect of Gaussian, uniform, sinusoidal, and static deviation of the STR circuitry on the *C/N* degradation for a $P_e = 10^{-8}$ in 64-QAM systems is illustrated [Sasase, 7.77]. We note that the $\alpha = 0.2$ raised-cosine filtered 256-QAM system is very sensitive to sampling deviations—that is, clock jitter or clock imperfections—of the STR *timing recovery* circuits. Thus to have a tolerable degradation (in the order of 0.5 dB), the sampling deviation or jitter should be less than $0.75°$ rms. In general, higher-order systems are more sensitive to STR imperfections than lower-order systems. Also, systems

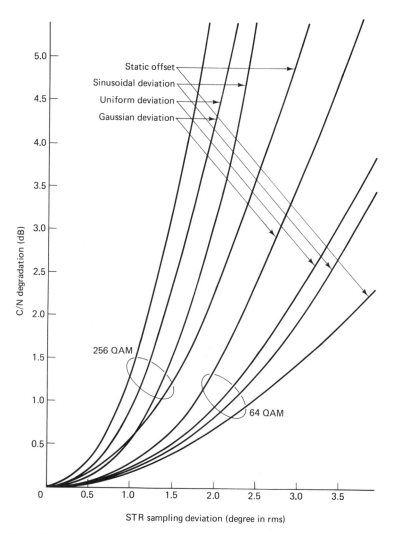

**Figure 7.73** *C/N* degradation versus sampling deviation of the STR circuitry for 256-QAM and 64-QAM systems $P_e = 10^{-8}$, raised-cosine system having $\alpha = 0.2$. [Sasase, 7.77].

that require steeper filters (smaller $\alpha$ in terms of raised-cosine filters) are more sensitive to clock jitter than filtered channels corresponding to larger $\alpha$.

## REFERENCES

[7.1] Feher, K. and Engineers of Hewlett-Packard. *Telecommunications Measurements, Analysis, and Instrumentation*, Prentice-Hall, Englewood Cliffs, N.J., 1986.

[7.2] Feher, K. *Digital Communications: Satellite/Earth Station Engineering*, Prentice-Hall, Englewood Cliffs, N.J., 1983.

[7.3] Feher, K. *Digital Communications: Microwave Applications*, Prentice-Hall, Englewood Cliffs, N.J., 1981.

[7.4] Feher, K. "Digital Modulation Techniques in an Interference Environment," Encyclopedia on EMC, Vol. 9, Don White Consultants, Inc., Gainesville, Va., 1977.

[7.5] Proakis, J. G. *Digital Communications*, McGraw-Hill, N.Y., 1983.

[7.6] Lender, A. "Correlative (partial response) techniques and applications to digital radio systems," Chapter 7 in K. Feher, *Digital Communications: Microwave Applications*, Prentice-Hall, Englewood Cliffs, N.J., 1981.

[7.7] Wilson, S. G., H. A. Sleeper, P. J. Schottler, and M. T. Lyons. "Rate 3/4 convolutional coding of 16-PSK: code design and performance study," *IEEE Trans. Commun.*, December, 1984.

[7.8] Feher, K. and R. deCristofaro, "Transversal filter design and application in satellite communications," *IEEE Trans. Commun.*, November 1976.

[7.9] Bhargava, V. K., D. Haccoun, R. Matyas, and P. P. Nuspl. *Digital Communications by Satellite*, John Wiley, New York, 1981.

[7.10] Le-Ngoc, T. "A class of power and bandwidth efficient transmission techniques for digital radio communications systems" Ph.D. Thesis, Electrical Eng. Dept., University of Ottawa, Ottawa, Canada, 1983.

[7.11] Ungerboeck, G. "Channel coding with multi-level/phase signals," *IEEE Trans. Information Theory*, Vol. IT-28, January, 1982, pp. 55–66.

[7.12] Ungerboeck, G. and I. Csajka. "On improving data link performance by increasing the channel alphabet and introducing sequence copy," presented at Int. Symp. Inform. Theory, Ronneby, Sweden, 1976.

[7.13] Lebowitz, S. A. and S. A. Rhodes. "Performance of coded 8-PSK signalling for satellite communications, *Int. Conf. Commun. Rec.*, Denver, Colo., June, 1981, pp. 47.4–47.4.8.

[7.14] Taylor, D. P., and H. C. Chan. "A simulation study of two bandwidth efficient modulation techniques," *IEEE Trans. Commun.*, Vol. COM-29, March 1981, pp. 267–275.

[7.15] Heller, J. A., and I. M. Jacobs. "Viterbi decoding for satellite and space communications," *IEEE Trans. Commun. Technol.*, Vol. COM-19, October 1971, pp. 835–848.

[7.16] Lindsey, W. C., and M. K. Simon. *Telecommunication Systems Engineering*, Prentice-Hall, Englewood Cliffs, N.J., 1973.

[7.17] Prabhu, V. K., "PSK-type modulation with overlapping baseband pulses," *IEEE Trans. Commun.*, Vol. COM-25, September, 1977.

[7.18] DeJager, D., and C. B. Dekker. "Tamed frequency modulation, a novel method to achieve spectrum economy in digital transmission," *IEEE Trans. Commun.*, Vol. COM-29, March, 1981.

[7.19] Austin, M. C., and M. V. Chang. "Quadrature overlapped raised-cosine modulation," *IEEE Trans. Commun.*, Vol. COM-29, March, 1981.

[7.20] Amoroso, F. "The bandwidth of digital data signals," *IEEE Commun. Mag.*, Vol. 18, No. 6, November, 1980.

[7.21] Sewerinson, A. "A remote subscriber satellite system," *Proceedings of the First Canadian Domestic and International Satellite Communications Conference*, Vol. SCC-1983, Ottawa, Canada, June, 1983.

[7.22] Feher, K., E. M. Karkar, and J. C. Y. Huang. "On 6.66 b/s/Hz—256-QAM and

225-QPRS modems for T1/SG data-in-voice (DIV) applications," *Proc. of the IEEE International Conference on Commun.*, ICC-1985, Chicago, June, 1985.

[7.23] Kucar, A., and K. Feher. "Performance of multi-level modulation systems in the presence of phase noise," *Proc. of the IEEE International Conf. on Commun.*, ICC-1985, Chicago, June, 1985.

[7.24] Sasase, I., K. Feher, and K. T. Wu. "Comparison of improved efficiency 225-QPRS and 256-QAM in distorted channels." *Proceedings of the IEEE International Conf. Commun.*, ICC-1985, Chicago, June, 1985.

[7.25] Yongacoglu, A., and K. Feher. "DCTPSK: An efficient modulation technique for differential detection," *Proc. of the IEEE International Conf. on Commun.*, ICC-1985, Chicago, June, 1985.

[7.26] Yongacoglu, A., and K. Feher. "DCTPSK: An efficient modulation technique for differential detection," Patent Disclosure, Canadian Patent and Development Limited, January, 1985.

[7.27] Feher, K., K-T Wu, J. C. Y. Huang, and D. E. MacNally. "Improved efficiency data transmission technique," U.S. and Canadian patent disclosure CPDL Case No. 265-8148-1, October 1, 1984.

[7.28] Seo, J. S., and K. Feher. "SQAM (superposed QAM) baseband signal processor," Canadian Patent and Development Limited (CPDL–March 1984), File No. 265-8148-1. U.S. Patent Filed June 12, 1984.

[7.29] Kato, S., K. Feher. "Correlated signal processor," U.S. Patent No. 4,567,602, issued Jan. 28, 1986.

[7.30] Feher, K. "In service jitter measurement technique," United States Patent No. 4,350,879, issued September 21, 1982.

[7.31] Feher, K. "Filter: nonlinear digital," U.S. Patent No. 4,339,724, issued July 13, 1982, Canada No. 1130871, August 31, 1982 (*already used by SPAR in satellite earth stations; license requested also by Korea Electrotechnology and Telecommunications*).

[7.32] Wu, K. T. and K. Feher. "Multilevel PRS/QPRS Above the Nyquist Rate," *IEEE Trans. on Commun.*, July, 1985.

[7.33] Seo, J. S., and K. Feher. "Performance of SQAM Systems in a Nonlinearly Amplified Multichannel Interference Environment," *IEE Proceedings-F, Commun. Radar and Signal Processing*, June, 1985.

[7.34] Wu, K. T., and K. Feher. "256-QAM modem performance in distorted channels," accepted for the *IEEE Trans. on Commun.*, May, 1985.

[7.35] Seo, J. S., and K. Feher. "SQAM: A new superposed QAM modem technique," *IEEE Trans. Commun.*, March, 1985, pp. 296–300.

[7.36] Pham Van, H., and K. Feher. "TSI-OQPSK for multiple carrier satellite systems," *IEEE Trans. Commun.*, July, 1984, pp. 818–825.

[7.37] Feher, K. "SCPC satcom systems for voice and data services," *Telecommun.* (Horizon House), Vol. 17, No. 6, June, 1983, pp. 80–86.

[7.38] Kato, S., and K. Feher. "XPSK: A new cross-correlated phase-shift-keying modulation technique," *IEEE Trans. Commun.*, May, 1983, pp. 701–706.

[7.39] Hill, T. and K. Feher. "A performance study of NLA-64 state QAM," *IEEE Trans. Commun.*, June, 1983, pp. 821–826.

[7.40] Feher, K. "Digital modulation techniques for new earth stations," *Satellite Commun.*, February, 1983, pp. 36–42.

[7.41] Le-Ngoc, T., and K. Feher. "Performance of an IJF-OPQPSK modem in cascaded nonlinear and regenerative satellite systems," *IEEE Trans. Commun.*, February, 1983, pp. 296–301.

[7.42] Le-Ngoc, T., and K. Feher. "Performance of IJF-OQPSK modulation schemes in a complex interference environment," *IEEE Trans. Commun.*, January, 1983, pp. 137–144.

[7.46] Doelz, M. L., and E. A. Heald. "Minimum-shift data communication system," U.S. Patent No. 2,977,417 (assigned to Collins Radio Company), March 28, 1961.

[7.47] Pelchat, M. G., R. C. Davis, and M. B. Luntz, "Coherent demodulation of continuous phase binary FSK signals" *Proc. Int. Telemetering Conf.*, Washington, D.C., 1971, pp. 187–190.

[7.48] Osborne, W. P., and M. B. Luntz. "Coherent and noncoherent detection of CPFSK," *IEEE Trans. Comm.*, Vol. COM-22, August, 1974, pp. 1023–1036.

[7.49] Fang, R. J. F. "Quaternary transmission over satellite channels with cascaded nonlinear elements and adjacent channel interference," *IEEE Trans. Commun.*, Vol. COM-29, May, 1981, pp. 567–581.

[7.50] Robinson, G., O. Shimbo, and R. Fang. "PSK signal power spectrum spread produced by memoryless nonlinear TWTs," *COMSAT Tech. Rev.*, Vol. 3, Fall, 1973, pp. 227–256.

[7.51] Jones, M. and M. Wachs. "Optimum filtering for QPSK in bandwidth-limited nonlinear satellite channels," *COMSAT Tech. Rev.*, Vol. 9, Fall, 1979, pp. 465–507.

[7.52] Amoroso, F. "Pulse and spectrum manipulation in the minimum (frequency) shift keying (MSK) format," *IEEE Trans. Commun.*, Vol. COM-24, March, 1976, pp. 381–384.

[7.53] Simon, M. K. "A generalization of minimum-shift keying (MSK)-type signalling based upon input data symbol pulse shaping," *IEEE Trans. Commun.*, Vol. COM-24, August, 1976, pp. 845–856.

[7.54] Rabzel, M., S. Pasupathy. "Spectral shaping in minimum shift keying (MSK)-type signals," *IEEE Trans. Comm.*, Vol. COM-26, January, 1978, pp. 189–195.

[7.55] Bazin, B. "A class of MSK baseband pulse formats with sharp spectral roll-off," *IEEE Trans. Comm.*, Vol. COM-27, May, 1979, pp. 826–829.

[7.56] Boutin, N., S. Morisette, and L. Dussault. "Constant amplitude PSK signal with minimum out-of-band energy," *NTC 78*, December, 1978.

[7.57] Deshpande, G. S., and P. H. Wittke. "Optimum pulse shaping in MSK-type signals," *NTC 79*, December, 1979, pp. 33.4/1–7.

[7.58] Aulin, T., and C-E Sundberg. "Continuous phase modulation–Part I: full-response signalling," *IEEE Trans. Commun.*, Vol. COM-29, March, 1981, pp. 196–209.

[7.59] Aulin, T., N. Rydbeck, and C-E Sundberg. "Continuous phase modulation—Part II: Partial-response signalling," *IEEE Trans. Commun.*, Vol. COM-29, March, 1981, pp. 210–225.

[7.60] Anderson, J. G., and D. P. Taylor. "A bandwidth efficient class of signal space codes," *IEEE Trans. Inf. Theory,* V-1. IT-24, November, 1978, pp. 703–712.

[7.61] Muilwijk, D. "Correlative phase shift keying—A class of constant envelope modulation techniques," *IEEE Trans. Commun.*, Vol. COM-29, March, 1981, pp. 226–236.

[7.62] Rhodes, S. "FSOQ, a new modulation technique that yields a constant envelope," *NTC 80*, December, 1980, pp. 51.1/1–7.

[7.63] Greenstein, L. J. "Spectra of PSK signals with overlapping baseband pulses," *IEEE Trans. Commun.*, Vol. COM-25, May, 1977, pp. 523–530.

[7.64] Lender, A. "Correlative level coding for binary data transmission," *IEEE Spectrum*, Vol. 3, February, 1966, pp. 104–115.

[7.65] Tezcan, T. "Performance evaluation of differential MSK with nonredundant error correction in AWGN and ACI environment," M. Eng. Report, University of Ottawa, Elect Eng. Dept., August, 1984 and *IEEE Trans. Commun.*—jointly with K. Feher, July 1986.

[7.66] Samejima, S., et al. "Differential PSK systems with nonredundant error correction," *IEEE J. on Selected Areas in Commun. SAC-1*, January, 1983, pp. 74–82.

[7.67] Masamura, T., et al. "Differential detection of MSK with nonredundant error correction," *IEEE Trans. Commun.*, Vol. COM-27, June, 1979, pp. 912–917.

[7.68] Forney, G. D., R. G. Gallager, G. R. Lang, F. M. Longstaff, and S. U. Qureshi. "Efficient modulation for band-limited channels," *IEEE J. on Selected Areas in Commun. SAC-2*, September, 1984.

[7.69] Morais, D. H., and K. Feher. "NLA-QAM: A method for generating high-power QAM signals through nonlinear amplification," *IEEE Trans. Commun.*, March, 1982.

[7.70] Wu, K. T., and K. Feher. "Error rate considerations for 64-256-QAM and 225-QPRS with co-channel interference," IEEE Transaction Communic.—1986 (in press).

[7.71] Saito, Y. "Feasibility considerations of high level multicarrier systems," *IEEE Internat. Conf. on Commun.*, ICC-1984, Amsterdam, May, 1984.

[7.72] Seo, J. S., and K. Feher. "Bandwidth compressive 16-state SQAM modems through saturated amplifiers," IEEE Trans. Commun. 1986 (in press.).

[7.73] Weber, W. J., P. H. Stanton, and J. T. Sumida. "A bandwidth compressive modulation system using multiamplutide minimum shift keying (MAMSK)," *IEEE Trans. Commun.*, Vol. COM-26, May, 1978.

[7.74] Kabal, P., and S. Pasupathy. "Partial-response signaling," *IEEE Trans. Commun.*, Vol. COM-23, September, 1975, pp. 921–934.

[7.75] Lucky, R. W., J. Salz, and E. J. Weldon, Jr. *Principles of Data Communication*, McGraw-Hill, New York, 1968, Chapter 4.

[7.76] Wu, K. T., I. Sasase, and K. Feher. "Improved efficiency 15-level modified duobinary PRS above the Nyquist rate," Proc. of the *IEEE International Commun. Conference*, ICC-1985, Chicago, June, 1985.

[7.77] Sasase, I., and K. Feher. "Effect of carrier recovery and symbol timing recovery error on the performance of 256-QAM systems," *IEEE Trans. Commun.* 1986 (in press).

[7.78] Le-Ngoc, T., K. Feher, and H. Pham Van. "New modulation techniques for low-cost earth stations," *IEEE Trans. Commun.*, January, 1982.

# 8

---

# CORRELATIVE CODING: BASEBAND AND MODULATION APPLICATIONS

**DR. SUBBARAYAN PASUPATHY**

*Professor, Electrical Engineering*
*University of Toronto*
*Toronto, Ontario*
*Canada M5S 1A4*

## 8.1 INTRODUCTION

Correlative coding, or partial-response signaling (PRS) was introduced by Lender [Lender, 8.1] in the 1960s as a technique for data communication. The signaling method differs from the conventional pulse amplitude modulation (PAM) system in that a controlled amount of intersymbol interference (ISI) is introduced to attain certain beneficial effects, such as convenient spectral shapes. Such PRS systems also possess, in general, the property of being relatively insensitive to channel imperfections and to variations in transmission rate [Lucky, 8.2].

Kretzmer [Kretzmer, 8.3] categorized the characteristics of many PRS schemes and compared them on the basis of several performance measures. A unified view of PRS systems is presented in [Kabal, 8.4]. (See [Kabal, 8.4] for a list of many important theoretical and applications-oriented papers in the area, spanning the period 1960–1975.) At present, the importance of this signaling scheme is so well recognized that many textbooks on communication systems, such as [Schwartz, 8.5; Stremler, 8.6; Haykin, 8.7; Lathi, 8.8], cover PRS in detail. It is assumed that the reader is familiar with some of the material in these references or surveys such as [Pasupathy, 8.9; Lender, 8.10]. The purpose of this chapter is twofold; the first is to present a brief tutorial introduction to PRS and the second is to discuss in detail the structures and performances of various PRS decoders and applications reported in the literature. The emphasis throughout the chapter will be on those concepts not treated in detail in previous tutorial surveys or textbooks. The references should not be considered as exhaustive but rather as representative sources for further information on the topics.

## 8.2 NYQUIST CRITERION

The difference between PRS and conventional PAM becomes clear after a study of Nyquist's classic work on pulse shape design [Nyquist, 8.11] and one of his criteria for eliminating ISI. In fact, such a study will lead us to the conclusion that PRS is a natural extension and generalization of Nyquist's work.

The communication system studied by Nyquist [Nyquist, 8.11] can be modeled by the baseband synchronous PAM scheme shown in Fig. 8.1(a). The results are easily extended to many other linear modulation schemes. The $\{a_k\}$ are independent $M$-ary symbols taking on the equally likely values $\{0, 1, \ldots, m - 1\}$. $H_T(f)$ and $H_R(f)$ are the transmitter and receiver filters; the channel is modeled as a linear filter $H_C(f)$ with an additive white Gaussian noise (AWGN) $\eta(t)$ having a one-sided spectral density $N_o$. In the absence of noise, the system is characterized by the overall frequency transfer function $G(f) = H_T(f)H_C(f)H_R(f)$; note that $H_R(f)$ can also include an equalizer, if there was one. The noiseless receiver output is

$$y(t) = \sum_k a_k g(t - kT) \qquad (8.1)$$

The principle underlying PAM is that the noiseless received samples should correspond to the data values—that is, $y(nT)$ should equal $a_n$. In other words, the PAM system, represented by $G(f)$ combined with the sampler, should essentially behave as a memoryless digital channel. From (8.1) we have

$$y(nT) = a_n g(O) + \sum_{k \neq n} a_k g[(n - k)T] \qquad n \text{ an integer} \qquad (8.2)$$

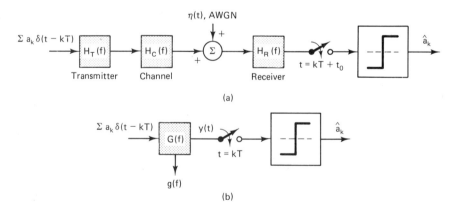

(a)

(b)

**Figure 8.1** (a) Baseband PAM model. (b) Noiseless equivalent model.

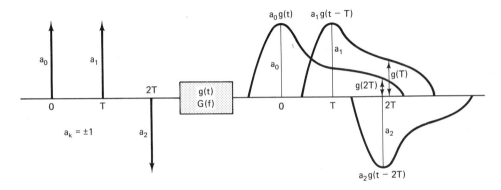

**Figure 8.2** ISI due to overlapping of pulses.

The first term in (8.2) represents the desired datum $a_n$, whereas the second term is the ISI due to the pulse $g(t)$ having nonzero values at the sampling instants $t = mT$, $m \neq 0$. This is illustrated in Fig. 8.2; for example, the desired value of $a_2 = -1$ at $t = 2T$ and the ISI from the previous pulses is $g(T) + g(2T)$.

Nyquist [Nyquist, 8.11] considered the problem of designing the pulse shape $g(t)$ such that no ISI results—that is, for an arbitrary message sequence, $y(nT)$ should equal $a_n$. From (8.2) and Fig. 8.2, it is obvious that this can happen only when

$$g(kT) = \begin{cases} 1 & k = 0 \\ 0 & k \text{ integer, } k \neq 0 \end{cases} \qquad (8.3)$$

Equation (8.3) is known as the **Nyquist's I criterion** for zero ISI. The frequency-domain statement of the Nyquist's I criterion follows by applying (8.3) to the Poisson-sum formula [Franks, 8.12]:

$$\sum_k g(kT)e^{-j2\pi kfT} = \frac{1}{T} \sum_n G\left(f - \frac{n}{T}\right) \qquad (8.4)$$

and we get

$$\sum_n G\left(f - \frac{n}{T}\right) = T \qquad (8.5)$$

In order to maximize the symbol rate in a given bandwidth, we are interested in the $g(t)$ which satisfies (8.3) or (8.5) and has the **minimum bandwidth** (MB). It is clear from (8.5) that such a MB solution is unique and equals the ideal low-pass spectrum

$$G(f) = \begin{cases} T & |f| \leq \dfrac{1}{2T} \\ 0 & \text{elsewhere} \end{cases} \qquad (8.6a)$$

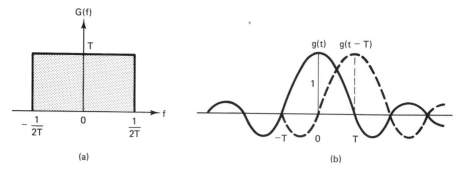

**Figure 8.3** Minimum bandwidth Nyquist I pulse in (a) frequency domain, and (b) time domain.

with the corresponding impulse response

$$g(t) = \text{sinc}\left(\frac{t}{T}\right) \triangleq \frac{\sin(\pi t/T)}{(\pi t/T)} \tag{8.6b}$$

Figure 8.3 illustrates the MB solution. The pulse (8.6b) has zero crossings at all multiples of $T$, except at $t = 0$, as required by (8.3), so that pulses transmitted at the rate $1/T$ do not interfere with each other (see Fig. 8.3(b)).

The implication of the MB solution (8.6) can also be stated as: The *MB required for signaling at the symbol rate* $(1/T)$ *without ISI is* $(1/2T)$ *Hz*; this is the famous relation between symbol rate and bandwidth and $(1/2T)$ is known as the **Nyquist bandwidth**. Thus the maximum theoretical symbol rate packing (the symbol rate per hertz of available bandwidth) is 2 symbols/s/Hz, using the pulse of (8.6). Unfortunately, (8.6) represents an impractical system due to the discontinuity at the band edge; the basic problem is that $\Sigma|g(t - kT)|$ does not converge if $t \ne kT$. It follows that $y(t)$ may take unbounded values between sampling points. Thus the slowly decreasing tail of $g(t)$ causes excessive ISI if any timing perturbations in the system (e.g., a small timing offset in the sampler) occur. Thus the theoretically maximum rate packing of 2 symbols/s/Hz cannot be achieved in practice by Nyquist-type pulse designs.

Nyquist also considered the above problem and showed that **nonminimum bandwidth** (NMB) solutions to (8.5) provide practical filters with smoother transitions at band edge and hence are more tolerant of timing perturbations. Practical filters are nearly always limited to no more than double the Nyquist bandwidth, i.e.,

$$G(f) = 0 \qquad |f| > \frac{1 + \alpha}{2T}$$

where $0 \le \alpha \le 1$ is the roll-off parameter indicating excess bandwidth. The most popular class of NMB solutions satisfying (8.5) has the cosine roll-off characteristic given by

$$G(f) = \begin{cases} T & |f| \le \dfrac{(1-\alpha)}{2T} \\ \dfrac{T}{2}\left\{ 1 - \dfrac{\sin[(|\omega| - \pi/T)T]}{2\alpha} \right\} & \dfrac{(1-\alpha)}{2T} \le |f| \le \dfrac{(1+\alpha)}{2T} \\ 0 & |f| > \dfrac{(1+\alpha)}{2T} \end{cases} \quad (8.7)$$

and the corresponding pulse shape

$$g(t) = \mathrm{sinc}\left(\frac{t}{T}\right) \cdot \frac{\cos(\alpha\pi t/T)}{1 - 4\alpha^2 t^2/T^2}$$

Thus practical symbol-rate packing in Nyquist type schemes is $2/(1 + \alpha)$, with $\alpha$ typically between 0.15 and 1.00.

It was Lender [Lender, 8.1; 8.10] who, at this time, showed that by relaxing the zero-ISI criterion of Nyquist, one can achieve the maximum rate with practical filters. By allowing a controlled amount of ISI or, equivalently, by introducing correlation between the transmitted pulse amplitudes and by changing the detection procedure, Lender could achieve the symbol rate of 2W baud in a bandwidth of W hertz. Thus Lender's correlative codes, generalized later as PRS schemes by Kretzmer [Kretzmer, 8.3], could be regarded as a practical means of achieving the theoretically maximum symbol-rate packing of Nyquist, using realizable and per-turbation-tolerant filters.

## 8.3 A PRS SYSTEM MODEL

The basic concept behind correlative schemes can be illustrated by the example of the duobinary scheme (or class 1 partial response) for binary transmission [Pasu-pathy, 8.9]. The generalization to other schemes and $M$-ary transmission is straightforward.

Assume that a sequence $\{a_k\}$ of binary symbols ($\pm 1$) is to be transmitted at the rate of $(1/T)$ symbols per second over the ideal MB Nyquist channel of band-width $(1/2T)$ hertz (see Fig. 8.4(a)). Let the digits first pass through a digital filter, as shown in Fig. 8.4(a). The discrete linear filter simply adds to the present digit the value of the previous digit: Thus every digit interferes with the next digit to be transmitted. The symbol sequence at the channel input is

$$y_k = a_k + a_{k-1}$$

Hence, unlike the zero-memory system, the input amplitudes $\{y_k\}$ are no longer independent, and this dependency or correlation between transmitted levels can also be thought of as artificially introducing ISI. Since the ideal Nyquist filter transmits all input amplitudes (at $T$ second intervals) without distortion in the

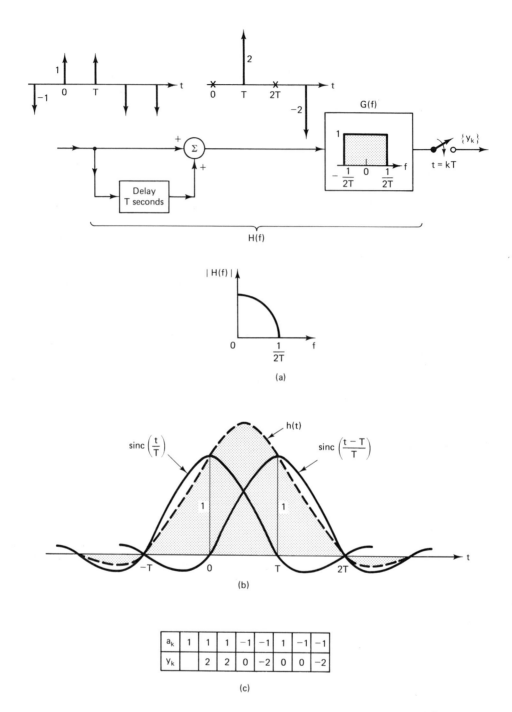

**Figure 8.4** Duobinary system: (a) a system model; (b) the impulse response $h(t)$; (c) sample input and noiseless output sequence.

absence of noise, we once again get the sequence $\{y_k\}$ at the sampler output. Note, however, that the ISI introduced by the digital filter is of a controlled amount, that is, the interference in determining $a_k$ comes only from the preceding symbol $a_{k-1}$. The controlled amount of interference represents a finite-memory system and is the key behind correlative, or partial-response, signaling. Whereas signaling in zero-memory systems is based on eliminating ISI, signaling in correlative or finite-memory systems is based on a controlled amount of ISI. However, the data can still be detected correctly. If $a_{k-1}$ has been correctly decided, its effect on $y_k$ can be eliminated by subtraction and $a_k$ can be decided. This demodulation technique is essentially an inverse of the operation of the digital filter at the transmitter.

The beneficial aspect of this correlation can now be explored. The cascade of the digital filter and the ideal channel $G(f)$ in Fig. 8.4(a) can be shown to be equivalent to

$$|H(f)| = 2 \cos(\pi f T) \qquad |f| \leq \frac{1}{2T}$$

This is because a delay element has a transfer function $e^{-j2\pi fT}$, the digital filter has the transfer function $1 + e^{-j2\pi fT}$, and hence

$$H(f) = G(f)(1 + e^{-j2\pi fT})$$
$$= 2G(f) \cos(\pi fT) \cdot e^{-j\pi fT}$$

Thus $H(f)$ has a gradual roll-off to the band edge and can be implemented by practical and realizable analog filtering. The corresponding impulse response $h(t)$ becomes

$$h(t) = \operatorname{sinc}\left(\frac{t}{T}\right) + \operatorname{sinc}\left(\frac{t - T}{T}\right)$$

as shown in Fig. 8.4(b); we can see that there are only two nonzero samples (at $T$-second intervals), giving rise to controlled ISI from the adjacent bit.

A sample input sequence $\{a_k\}$ and the resulting three-level ($\pm 2$, 0) output sequence $\{y_k\}$ for the duobinary system is shown in Fig. 8.4(c). Apart from the higher symbol-rate packing achieved by the practical and perturbation tolerant filter shape occupying the MB, another advantage of the correlative scheme is its error-detecting capability. From Fig. 8.4(c), we can see that because of the correlation between digits, only certain types of transitions are allowed. The following constraints apply to the three-level duobinary scheme:

1. A positive (negative) level at one sampling time cannot be followed by a negative (positive) level at the next sampling instant.
2. If a positive (negative) peak is followed by a negative (positive) peak, they must be separated by an odd number of zero-value samples.
3. If a positive (negative) peak is followed by another positive (negative) peak, they must be separated by an even number of zero-value samples.

Thus the duobinary waveform can constantly be checked by a level violation monitor, so that any violation of these rules results in an error indication. Although certain combinations of bit errors will be compensatory in nature and will not show up as three-level violations, the three-level error rate will be a good indication of the actual error rate. Similar predetermined rules can be laid out for other partial response formats, and violations of these patterns can be used for error detection.

We shall now examine a model for a general PRS system. For a PRS system, the impulse response $h(t)$ is required to pass through 0 at regular $T$-second intervals with the exception of a period of $NT$ second, $N$ being the smallest number of samples that span all the nonzero samples.

The noiseless PRS system is characterized by the samples of the desired impulse response $h(t)$. If $\{x_n\}$, $n = 0, 1, 2, \ldots, N - 1$ are $N$ sample values of $h(t)$ and $N$ is the smallest number of contiguous samples that span all the nonzero samples, the PRS system is characterized by the system polynomial

$$x(D) = \sum_{k=0}^{N-1} x_k D^k, \qquad x_0, x_{N-1} \neq 0$$

where $D$ is the delay operator corresponding to a $T$-second delay [Kobayashi, 8.13]. The PRS system can be modeled by a cascade of a tapped delay line with coefficients $\{x_k\}$ (a discrete nonrecursive finite impulse response (FIR) filter) and a filter $G(f)$, as shown in Fig. 8.5(a).

The periodic frequency response of the transversal filter is

$$X(f) = x(D)\big|_{D = e^{-j2\pi fT}} \tag{8.8a}$$

$$= \sum_{k=0}^{N-1} x_k e^{-j2\pi fkT}$$

and the impulse response

$$x(t) = \sum_{k=0}^{N-1} x_k \delta(t - kT) \tag{8.8b}$$

The PRS system impulse response thus is

$$h(t) = \sum_{k=0}^{N-1} x_k g(t - kT) \tag{8.9}$$

corresponding to $H(f) = X(f)G(f)$. The following theorem [Kabal, 8.4] throws light on the relation between PRS and Nyquist's criterion.

THEOREM. $h(t)$ has the sample values $\{x_k\}$ if and only if $g(t)$ satisfies Nyquist's I criterion.

*Proof.* $H(f)$ has samples $\{x_k\}$ if and only if by Poisson sum formula [Franks, 8.12]

$$\sum_k x_k e^{-j2\pi fkT} = \frac{1}{T} \sum_n H\left(f - \frac{n}{T}\right)$$

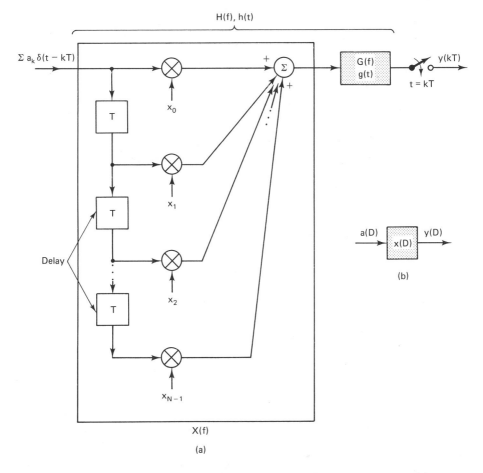

**Figure 8.5** (a) Generalized PRS system model. (b) Discrete system model.

$$= \frac{1}{T} \sum_n \left[ G\left(f - \frac{n}{T}\right) \sum_m x_m e^{-j2\pi(f - n/T)mT} \right]$$

$$= \frac{1}{T} \sum_n G\left(f - \frac{n}{T}\right) \sum_m x_m e^{-j2\pi fmT}$$

which can be true if and only if $G(f)$ satisfies the Nyquist criterion (8.5).

Thus the PRS system can be visualized as a cascade of two parts: $X(f)$ forces the desired sample values but is periodic, whereas the Nyquist filter $G(f)$ preserves the sample values but can be used to bandlimit the resulting system. For a given system polynomial $x(D)$, different choices of $G(f)$ (both MB and NMB) result in different $H(f)$, all of which have identical sampled responses. This separation makes it easier to understand the frequency domain behavior of the PRS system and its spectral properties such as bandwidth and nulls. Also, from Fig. 8.5(a), we see that if $x_k = 0$, $k \neq 0$, the system simplifies to that of a Nyquist-type system. Thus the PRS system can be seen as satisfying a **generalized Nyquist I criterion**: Whereas the Nyquist criterion demands *only one* nonzero sample for $g(t)$, the PRS scheme demands that $h(t)$ have a *finite* number (greater than 1) of nonzero samples. The nonzero samples $h(kT) = x_k$, $k \neq 0$ create a controlled amount of ISI or finite memory. Thus PRS represents a *finite-memory* system and is a natural extension of the *zero-memory* system of Nyquist.

The PRS system can also be understood in the context of a discrete system representation such as Fig. 8.5(b). The information sequence is represented by

$$a(D) = \sum_{k=1}^{\infty} a_k D^k \qquad (8.10)$$

and the encoded output sequence, by

$$y(D) = \sum_{k=1}^{\infty} y_k D^k \qquad (8.11)$$

determined by

$$y(D) = x(D)a(D)$$

or

$$y_k = \sum_{i=0}^{N-1} x_i a_{k-i}, \qquad \text{for all } k \qquad (8.12)$$

Among the $N$ coefficients of $x(D)$ the leading coefficient $x_0$ represents the signal value, whereas the other $N - 1$ coefficients represent controlled ISI terms. Equation (8.12) also illustrates the similarity between correlative coding and convolutional coding as has been pointed out by Kobayashi [Kobayashi, 8.14] and Forney [Forney, 8.15]. A correlative coder can be viewed as a linear finite-state machine defined over the real-number field as opposed to a Galois field over which a convolutional encoder is defined. From this similarity, we shall see later that the maximum-likelihood decoding algorithm devised by Viterbi [Kobayashi, 8.14; Forney, 8.15] for decoding convolutional codes is also applicable to PRS decoding.

## 8.4 CANDIDATE PRS SYSTEMS

We shall now briefly study the effect of a few PRS system parameters and how they influence the choice of a PRS system.

### 8.4.1 System Bandwidth

From Fig. 8.5(a) we see that the bandwidth is controlled by the choice of $G(f)$. In order to maximize the data rate in the available bandwidth, many PRS systems use the MB $G(f)$ given by (8.5). NMB $G(f)$ are also possible and used, as for example, in data-under-voice systems [Seastrand, 8.16; Baker, 8.17].  As we shall see, NMB $G(f)$ also allow the use of system polynomials (such as $(1 - D)$), which are unsuitable for MB systems.

### 8.4.2 Spectral Null at f = 1/2T

It is well known that if $H(f)$ (such as in Fig. 8.3(a)) is discontinuous, $h(t)$ decays asymptotically as $1/|t|$ [Bennett, 8.18].  Recall that it was the slow decay of $g(t)$ in (8.6) that made the MB Nyquist-type system impractical to use and intolerant to timing perturbations.  Hence, similar considerations demand that the MB PRS transfer function $H(f)$ be at least continuous for faster decay.  This implies that $X(f)$ must have a zero at $f = 1/2T$ (where $G(f)$ is discontinuous) for $H(f)$ to be a continuous function.  From (8.8) we see that $X(f)$ has a spectral null at $f = 1/2T$ if and only if $x(D)$ has $(1 + D)$ as a factor.  The resulting MB $H(f)$ with $x(D) = (1 + D)$, shown in Fig. 8.6(a), is the duobinary or class I PRS system (Fig. 8.4) and is a popular choice in many systems, such as T1D [Maurer, 8.19] and digital radio [Lender, 8.10].

For NMB systems, the continuous nature of NMB $G(f)$ (see (8.7)) automatically makes NMB $H(f)$ continuous, and hence a spectral null at $f = 1/2T$ is not necessary to ensure the continuity of $H(f)$; but even in such systems, a null at $1/2T$ is still useful—a pilot tone inserted at this point can be used for clock recovery [Baker, 8.17].

### 8.4.3 Spectral Null at f = 0

Reduced low-frequency components in the spectrum are desirable in systems such as transformer-coupled circuits, dc powered cables, single-sideband (SSB) modems [Becker, 8.20], and carrier systems with carrier pilot tones.  For a null at $f = 0$, it can be seen from (8.8) that $D = e^{-j2\pi fT} = 1$ for $f = 0$ or $1 - D$ must be a factor of $x(D)$.  The MB $(1 - D)$ system frequency response is shown in Fig. 8.6(b). As mentioned before, the system is not practical due to the discontinuity in $H(f)$ and can be used only with NMB $G(f)$.  We will show later that a $(1 - D)$ system

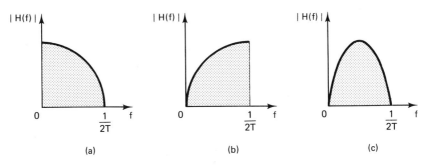

**Figure  8.6**  Minimum-bandwidth PRS system spectra for (a) duobinary $(1 + D)$; (b) bipolar $(1 - D)$; and (c) modified duobinary $(1 - D^2)$.

with precoding is the popular bipolar or **alternate mark inversion** (AMI) code used in the T1 carrier system, a pioneer in digital communication systems [Aaron, 8.21; Kobayashi, 8.13].

### 8.4.4 The Number of Output Levels

An $M$-ary PRS system with $N$ nonzero pulse samples will have $L$ output levels, where

$$(m - 1)N + 1 \leq L \leq m^N \tag{8.13}$$

with the minimum value being obtained when the pulse samples have the same magnitude. Complexity of implementation dictates that the number of output levels be kept to a minimum. It has also been shown [Kabal, 8.22] that if $x(D)$ has $(1 \pm D)$ as a factor, some of the output levels coalesce.

Thus if we restrict $L$ to the minimum value of 3 for binary input, require $1 + D$ and/or $1 - D$ to be a factor of $x(D)$, and do not want any nulls or severe ripples in the middle of the passband, the three systems that emerge as good candidates are $1 + D$, $1 - D$, and $1 - D^2$. The frequency response of MB $1 - D^2$ (modified duobinary or class-4 PRS) as shown in Fig. 8.6(c); the two spectral nulls make this an attractive choice for many applications, such as SSB modulation [Becker, 8.20] and T1-type PCM carrier systems [Lender, 8.23; Chou, 8.24]. Note that $(1 - D^2)$ can be derived by interleaving two $(1 - D)$ streams [Croisier, 8.25]; this is how T1C carrier data is derived from the two T1 (bipolar) data streams [Chou, 8.24]. The NMB $(1 - D^2)$ scheme is also used in data-under-voice schemes [Seastrand, 8.16; Baker, 8.17].

### 8.5 BIT-BY-BIT PRS DECODERS

The sampled output of an ideal noiseless PRS (Fig. 8.5) is given by

$$y_m = x_0 a_m + \sum_{i=1}^{N-1} x_i a_{m-i} \tag{8.14}$$

The second term represents the controlled ISI introduced by the PRS system. The receiver can recover the data $a_m$ by subtracting the effect of previous input symbols.

In practice, only estimates of the data are available, and these are used to cancel the controlled ISI in the decision feedback decoder structure (Fig. 8.7). When additive noise is present, the input to the threshold detector is

$$r_m = \frac{1}{x_0} \left( y_m + n_m - \sum_{i=1}^{N-1} x_i \hat{a}_{m-i} \right) \tag{8.15}$$

where

$$n_m = [\eta(t) * h_R(t)]_{t = mT} \tag{8.16}$$

and where $*$ denotes convolution. The input $r_m$ is then quantized to the nearest

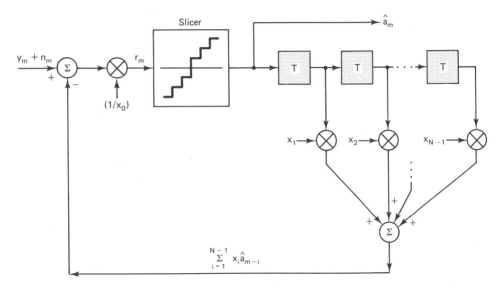

**Figure 8.7** Decision-feedback decoder for PRS.

allowable data level to give the next data estimate, $\hat{a}_m$. If we define a decision error $e_m = a_m - \hat{a}_m$, then from (8.14) and (8.15)

$$r_m = a_m + \frac{1}{x_0}\left(n_m + \sum_{i=1}^{N-1} x_i e_{m-i}\right) \tag{8.17}$$

We see from (8.17) that past decision errors can adversely effect subsequent decisions. We shall first evaluate the probability of error, $P_e$, taking the effect of error propagation into account, and then discuss a technique introduced by Lender [Lender, 8.1] called *precoding*, which prevents error propagation.

### 8.5.1 Probability of Error (No Precoding)

Let us evaluate the probability of error for the $(1 + D)$ system with binary inputs; the method can be easily extended to $M$-ary input and other system polynomials.

For analytical ease, we shall assume that all the spectral shaping is done at the transmitter ($H_T(f) = H(f)$ and $H_R(f)$ is an ideal low-pass filter (LPF)). Since $\eta(t)$ is a zero-mean, white Gaussian noise, $n_i$ is a zero-mean Gaussian random variable with variance

$$\sigma^2 = N_0 \int_0^\infty |H_R(f)|^2 df = \frac{N_0}{2T} \tag{8.18}$$

and $n_i$'s are uncorrelated and hence independent.

If the spectral shaping is divided in other ways, the noise samples $\{n_i\}$ will, in general, be correlated, and the probability of error has to be calculated using a numerical approach [Newcombe, 8.26].

For the $(1 + D)$ system, $x_0 = x_1 = 1$ and the input to the quantizer in Fig. 8.7

is

$$r_i = a_i + n_i + e_{i-1} \tag{8.19}$$

The estimate is made according to the rule

$$\hat{a}_i = \pm 1 \quad \text{if} \quad r_i \gtrless 0 \atop -1 \tag{8.20}$$

or, equivalently, if

$$n_i \gtrless -(a_i + e_{i-1}) \atop -1 \tag{8.21}$$

where $e_i$ can assume the values of 0 (corresponding to no error) and $(-2a_i)$ (denoting an error).

From symmetry,

$$P(e_i = 2) = P(e_i = -2) \tag{8.22}$$

and

$$P_e = P(e_i \neq 0) = P(e_i = 2) + P(e_i = -2) = 2P(e_i = 2) \tag{8.23}$$

$$P(e_i = 2) = \sum_{k=-1}^{1} A_{2k} P(e_{i-1} = 2k) \tag{8.24}$$

where

$$\begin{aligned}
A_{2k} &= P(e_i = 2/e_{i-1} = 2k) \\
&= P[n_i > -(-1 + 2k)] P(a_i = -1) \tag{8.25} \\
&= \frac{1}{2} \int_{-(-1+2k)}^{\infty} \frac{\exp(-x^2/2\sigma^2) \, dx}{\sqrt{2\pi} \, \sigma} \\
&= \frac{1}{2} Q\left(\frac{1 - 2k}{\sigma}\right)
\end{aligned}$$

$$Q(x) = \left(\frac{1}{\sqrt{2\pi}}\right) \int_{x}^{\infty} \exp\left(\frac{-u^2}{2}\right) du \tag{8.26}$$

Then using the constraint $\sum_{k=-1}^{1} P(e_i = 2k) = 1$ and assuming stationary error states $P(e_i) = P(e_{i-1})$, (8.22) through (8.24) result in

$$P_e = \frac{2A_0}{1 + 2A_0 - (A_{-2} + A_2)} \tag{8.27}$$

A Markov-chain analysis can also be used to arrive at analytical and numerical solutions for $P_e$ [Kabal, 8.4; 8.22].

Note that

$$2A_0 = Q\left(\frac{1}{\sigma}\right) \tag{8.28}$$

represents a lower bound for $P_e$, assuming zero error propagation ($e_i = 0$). Hence, we see that error propagation has increased the error probability, as given by the division of (8.27) by (8.28). In fact, we can show [Kabal, 8.4] that for $M$-ary input and $N$ nonzero PRS pulse samples, the increase is at most a factor of $m^{N-1}$; thus for large $m$ and $N$, this effect can be drastic.

The previous analysis was for the $1 + D$ system. Kobayashi [Kobayashi, 8.14] has shown that $1 + D$ has the same performance as $1 - D$. The argument is as follows: Consider the system model in Fig. 8.8(a) with the dc-free binary input ($\pm1$) sequency $\{a_k\}$. Clearly, the performance of the system in Fig. 8.8(a) does not change when an alternating sequence $(-1)^k$ is multiplied at both the input and output. Thus the systems of Fig. 8.8(a) and (b) are equivalent. Figure 8.8(b) can now be redrawn as Fig. 8.8(c).

We shall now show why $(1 \pm D)$ have the same performance. From Fig. 8.8,

$$a(D) = \Sigma a_k D^k \tag{8.29}$$

and

$$a'(D) = \Sigma(-1)^k a_k D^k = \Sigma a'_k D^k \tag{8.30}$$

$$y(D) = (1 + D)a(D) = \Sigma(a_k + a_{k-1})D^k = \Sigma y_k D^k \tag{8.31}$$

$$y'(D) = (1 + D)a'(D) = \Sigma y'_k D^k = \Sigma[(-1)^k a_k + (-1)^{k-1}a_{k-1}]D^k \tag{8.32}$$

$$\bar{y}(D) = \Sigma \bar{y}_k D^k = \Sigma(-1)^k y'_k D^k$$

$$= \Sigma[(-1)^{2k}a_k + (-1)^{2k-1}a_{k-1}]D^k$$

$$= \Sigma(a_k - a_{k-1})D^k \tag{8.33}$$

$$= (1 - D)a(D)$$

Thus Fig. 8.8(c) simplifies to Fig. 8.8(d). If the noise sequence $\{n_k\}$ is uncorrelated (as it is in our model), the systems $1 \pm D$ are equivalent. Also, the systems $x(D)$ and $x(D^k)$ have equivalent performances, since the latter is a $k$-fold interleaved version of the former. Thus the three important PRS systems considered here, $1 + D$, $1 - D$, and $1 - D^2$, all have identical performances.

### 8.5.2 Precoding

Precoding is a technique used [Lender, 8.1; 8.10] to eliminate the error-propagation problem of the decision-feedback decoder of Fig. 8.7. Error propagation was caused by the fact that previous symbols were not available at the receiver and only their estimates were known. The precoder eliminates the effect of previous symbols at the source where they are known precisely. Decisions at the output are then made independently of one another.

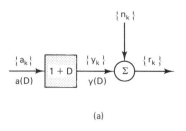

(a)

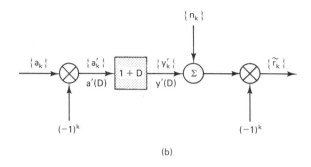

(b)

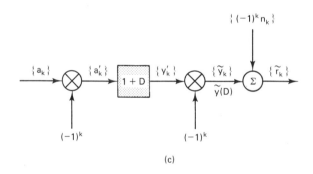

(c)

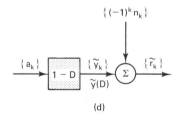

(d)

**Figure 8.8** Equivalence of $(1 + D)$ and $(1 - D)$ schemes [Kobayashi 8.14].

The concepts of error propagation and precoding can be understood for a general PRS system by considering the discrete system representation, such as Fig. 8.9(a). The encoder output ($L$-ary alphabet) $y(D) = x(D)a(D)$ is sent over a channel with an additive noise sequence $n(D)$. The receiver output is denoted by

$$r(D) = y(D) + n(D) \tag{8.34}$$

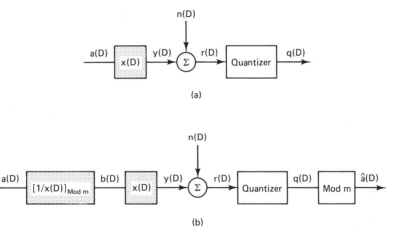

**Figure 8.9** (a) Discrete system PRS model. (b) Precoded PRS system.

At the receiving end, $r(D)$ is first quantized. Assuming hard decisions, the quantizer assigns to each $r_k$ one of $L$ (equal to 3 for our examples) possible values. If no errors are introduced in the channel and quantizer, then $q(D) = y(D)$, and thus the information sequence $a(D)$ can be recovered simply by passing $q(D)$ through the inverse filter $1/x(D)$. An obvious drawback is that if $q(D)$ contains an error, the effect of the error tends to propagate in the decoded sequence. This is clear from the fact that $1/x(D)$ can be expanded, in general, into an infinite power series in $D$. Precoding devised by Lender avoids this error propagation.

Figure 8.9(b) shows a precoded PRS system. The precoder can be described in terms of a discrete filter with transfer function $[1/x(D)]_{\mathrm{mod}\,m}$. With an $M$-ary input $a(D)$, the $M$-ary precoded output $b(D)$ is given by

$$b(D) = \frac{a(D)}{x(D)} \quad \mathrm{mod}\ m \tag{8.35}$$

or, equivalently,

$$x(D)b(D) = a(D) \quad \mathrm{mod}\ m \tag{8.36}$$

The correlative coder transforms $b(D)$ into $y(D)$ according to the relation

$$x(D)b(D) = y(D) \tag{8.37}$$

From (8.36) and (8.37) it follows that

$$y(D) = a(D) \quad \mathrm{mod}\ m$$

or

$$\tag{8.38}$$

$$y_m = a_k \quad \mathrm{mod}\ m \text{ for all } k$$

Thus given the hard decision output $q(D)$, we merely perform modulo $m$ (mod $m$) operation on $q(D)$ to get $\hat{a}(D)$, the estimated input sequence. Since the mod $m$ operation depends only on the present receiver sample $r_k$, propagation of errors in the output $\hat{a}(D)$ is eliminated.

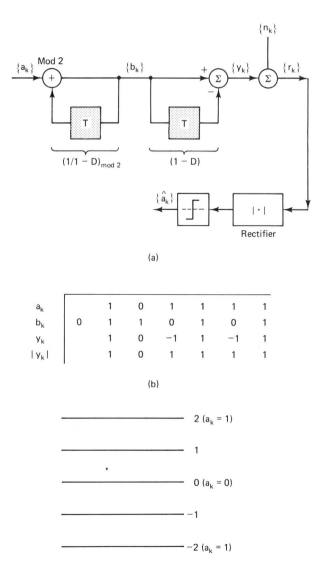

(a)

| $a_k$ | | 1 | 0 | 1 | 1 | 1 | 1 |
|---|---|---|---|---|---|---|---|
| $b_k$ | 0 | 1 | 1 | 0 | 1 | 0 | 1 |
| $y_k$ | | 1 | 0 | −1 | 1 | −1 | 1 |
| $|y_k|$ | | 1 | 0 | 1 | 1 | 1 | 1 |

(b)

─────────────── 2 ($a_k = 1$)

─────────────── 1

─────────────── 0 ($a_k = 0$)

─────────────── −1

─────────────── −2 ($a_k = 1$)

(c)

**Figure 8.10** (a) Precoded $(1 - D)$ system (bipolar). (b) A sample sequence for the bipolar scheme. (c) Detection scheme for bipolar code.

Figure 8.10(a) shows a precoded $(1 - D)$ system for binary input, and Fig. 8.10(b) shows the transformation of a sample sequence through the system. Note that the system, where the binary symbol 0 is represented by no signal and 1 is represented alternately by positive and negative pulses, is the well-known AMI or bipolar code [Aaron, 8.21; Kobayashi, 8.13]. We also note the inherent error-detecting capability in a PRS system from this example; if 1s in the quantizer output $q(D)$ do not alternate, we immediately know that an error has been made. It has the effect of reducing dc wander, since a pulse of one polarity is certain to be followed eventually by a pulse of the opposite polarity. Note also that the mod 2 detector at the receiver is a simple rectifier (absolute value operation). The system $x(D) = 1 - D^2$ with precoding is mathematically equivalent to an inter-leaved form of the bipolar coding system.

The performance of the precoded $(1 - D)$ system can be easily evaluated. From Fig. 8.10(a)

$$r_k = y_k + n_k \tag{8.39}$$

where $y_k = \pm 1, 0$ and the noise sample (sampler output of an ideal LPF at the receiver) is zero-mean Gaussian with variance $\sigma^2$.

The probability of error, then, is (Fig. 8.10(c))

$$P_e = P(y_k = -1)P(n_k > 1) + P(y_k = 0)P(|n_k| > 1)$$
$$+ P(y_k = 1)P(n_k < -1) - P(y_k = -1)P(n_k > 3) \tag{8.40}$$
$$- P(y_k = 1)P(n_k < -3)$$

The two negative terms in (8.40) represent the fact that a large noise sample at extreme levels can still lead to a correct decision, since $y_k = \pm 1$ represents $a_k = 1$.

Using $P(y_k = \pm 1) = \frac{1}{4}$ and $P(y_k = 0) = \frac{1}{2}$,

$$P_e = \frac{3}{2} P(n_k > 1) - \frac{1}{2} P(n_k > 3)$$
$$\approx \frac{3}{2} P(n_k > 1) \tag{8.41}$$
$$= \frac{3}{2} Q\left(\frac{1}{\sigma}\right)$$

A comparison of (8.41) with (8.27) shows the advantage of precoding in reducing probability of error.

## 8.6 ERROR DETECTION

The system shown in Fig. 8.9(b), though extremely simple in structure, does not take advantage of the inherent redundancy in the $L$-level sequence $y(D)$ and will not detect any errors. Lender [Lender, 8.27] proposed an error-detection scheme for the $(1 - D)^2$ scheme, where $q(D)$ is monitored by logic circuits, to check existence of any unallowable patterns (e.g., consecutive 1s of same polarity in the bipolar example).

A general method of determining whether the estimated output sequence $q(D)$ is allowable is to pass it through an inverse linear filter with response $1/x(D)$ and check to see whether an allowable input sequence $\hat{a}(D)$ (i.e., only $\pm 1$ in a binary system) comes out. Figure 8.11 shows such a receiver structure, which has been reported independently by many authors [Forney, 8.15; Gunn, 8.28; Kobayashi, 8.29]. For an $M$-ary input with levels $[0, 1, \ldots, m - 1]$, any detectable error in $q(D)$ will result in an inverse filter output $b(D)$ with coefficients other than the allowable levels $[0, 1, \ldots, m - 1]$. Error-correction schemes, based on such information, have also been reported [Gibson, 8.30; Forney, 8.15].

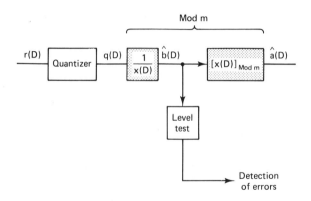

Mod m

**Figure 8.11** A general error-detection scheme.

Detection of errors

## 8.7 VITERBI RECEIVER FOR PRS

Because of the desirable spectral shape and other nice features and simple implementation, PRS schemes have been widely used in data transmission. However, as has been shown [Lucky, 8.2], the increase in the number of output levels seems to require a higher SNR (e.g., 2.1 to 3.00 dB for the duobinary scheme, depending on the filter apportioning between the transmitter and the receiver) to perform as well as the ideal antipodal binary PAM system. This was seen as the inevitable penalty to be paid in exchange for the increase in data rate and the insensitivity to channel perturbations and variations in transmission rate. However, Kobayashi [Kobayashi, 8.14] and Forney [Forney, 8.15] showed that this apparent poorer performance is not an inherent drawback of PRS but is due to the nonoptimality of the conventional bit-by-bit detection method. They showed that **maximum likelihood sequence estimation** (MLSE), using the Viterbi algorithm, can be used to recover almost all the SNR loss. We now discuss the structure of the Viterbi receiver, following the development in [Kobayashi, 8.13].

Let us consider $x(D) = 1 - D$ the precoded bipolar example (Fig. 8.10(a)) without binary input $a_k = (0, 1)$ in AWGN. The input $\{a_k\}$, precoded sequence $\{b_k\}$, and the correlative coded input $\{y_k\}$ are related by

$$b_k = a_k \oplus b_{k-1} \qquad b_0 = 0 \tag{8.42}$$

and

$$y_k = b_k - b_{k-1} \tag{8.43}$$

where $\oplus$ refers to modulo-2 addition. Assume that $\{a_k\}$ is of finite length $N$; thus there are $2^N$ different vectors representing $\{y_k\}$.

If $r_k = y_k + n_k$ represents the noisy sample, then it is well known that the MLSE decision criterion is equivalent to the minimum distance decision rule based on the measure

$$D(\hat{\mathbf{y}}) = \sum_k (r_k - \hat{y}_k)^2$$
$$= \|\mathbf{r} - \hat{\mathbf{y}}\|^2 \tag{8.44}$$

where $\mathbf{r}$ and $\hat{\mathbf{y}}$ are the vectors representing $\{r_k\}$ and $\{\hat{y}_k\}$ and $\|\mathbf{x}\|$ is the norm of vector $\mathbf{x}$. The optimum receiver picks the $\hat{\mathbf{a}}$ that minimizes $D(\hat{\mathbf{y}})$ where $\hat{\mathbf{y}}$ and $\hat{\mathbf{a}}$ are related through (8.42) and (8.43).

$D(\hat{\mathbf{y}})$ can be written as

$$D(\hat{\mathbf{y}}) = -2\langle \mathbf{r}, \hat{\mathbf{y}} \rangle + \|\hat{\mathbf{y}}\|^2 + \|\mathbf{r}\|^2 \tag{8.45}$$

where $\langle \, , \, \rangle$ refers to vector inner product.

Since the last term does not depend on $\hat{\mathbf{y}}$, the decision criterion is to maximize

$$J(\hat{\mathbf{y}}) = \langle \mathbf{r}, \hat{\mathbf{y}} \rangle - \frac{1}{2}\|\hat{\mathbf{y}}\|^2$$

$$= \Sigma r_k(\hat{b}_k - \hat{b}_{k-1}) - \frac{1}{2}\Sigma(\hat{b}_k - \hat{b}_{k-1})^2 \tag{8.46}$$

A brute-force procedure would be to compute $J(\hat{\mathbf{y}})$ for $2^N$ different patterns of $y$ (using $2^N$-different matched filters) and select the largest. Since $N \rightarrow \infty$, this clearly is an impractical approach. The Viterbi algorithm overcomes this dimensionality problem by applying the discipline of dynamic programming [Omura, 8.31].

Let $\mathbf{y}_k$ denote the first $k$ elements of $\mathbf{y}$:

$$\mathbf{y}_k = [y_1, \ldots, y_k] \tag{8.47}$$

and let $J_k$ denote $J(\hat{\mathbf{y}}_k)$. Then

$$J_k = \sum_{n=1}^{k} [r_n(\hat{b}_n - \hat{b}_{n-1}) - \tfrac{1}{2}(\hat{b}_n - \hat{b}_{n-1})^2] \tag{8.48}$$

and can be written in an iterative way as

$$J_k = J_{k-1} + (\hat{b}_k - \hat{b}_{k-1})r_k - \tfrac{1}{2}(\hat{b}_k - \hat{b}_{k-1})^2 \tag{8.49}$$

With

$$\mu_k(i) \stackrel{\triangle}{=} \max_{\{\hat{b}_{k-1}, \hat{b}_k = i\}} \{J_k\} \qquad i = 0, 1 \tag{8.50}$$

substituted into (8.49), we obtain $\mu_k(i)$ recursively as

$$\mu_k(i) = \max_{\{\hat{b}_{k-1}\}} \{J_{k-1} + (i - \hat{b}_{k-1})r_k - \tfrac{1}{2}(i - \hat{b}_{k-1})^2\}$$

$$= \max_{j=0,1} \{\mu_{k-1}(j) + (i - j)r_k - \tfrac{1}{2}(i - j)^2\} \qquad i = 0, 1 \tag{8.51}$$

Equivalently,

$$\mu_k(0) = \max \begin{cases} \mu_{k-1}(0) \\ \mu_{k-1}(1) - r_k - \tfrac{1}{2} \end{cases} \tag{8.52}$$

and

$$\mu_k(1) = \max \begin{cases} \mu_{k-1}(0) + r_k - \tfrac{1}{2} \\ \mu_{k-1}(1) \end{cases} \tag{8.53}$$

Note that (8.52) and (8.53) are not independent. If $\mu_k(0) = \mu_{k-1}(1) - r_k - \tfrac{1}{2}$, then we can verify $\mu_k(1) = \mu_{k-1}(1)$. Similarly, if $\mu_k(1) = \mu_{k-1}(0) + r_k - \frac{1}{2}$, then $\mu_k(0) = \mu_{k-1}(0)$.

Thus $\mu_k(i)$ represents a distance measure (or metric) of the most-likely sequence among all possible candidates with the constraint $\hat{b}_i = i$. Using the initial condition $b_0 = 0$, we start from

$$\mu_0(i) = \begin{cases} 0 & i = 0 \\ -\infty & i = 1 \end{cases} \tag{8.54}$$

The repeated use of (8.52) and (8.53) for $k = 1, 2, \ldots, N$ uniquely determines the MLSE solution.

The performance of such Viterbi receivers can be evaluated using transfer-function bounding techniques and other methods [Kobayashi, 8.14; Forney, 8.15]. The results show that a PRS system with a Viterbi receiver has almost the same performance as that of a binary system.

## 8.8 CORRELATIVE CODING AND CONTINUOUS PHASE MODULATION

Recently, correlative coding has been applied to **continuous phase modulation** (CPM), resulting in improved spectral properties at the expense of only a small penalty in performance [deJager, 8.32; Deshponde, 8.33; Aulin, 8.34; Muilwijk, 8.35]. In particular, the combination of the coding polynomial $(1 + D)^2$ and minimum shift keying (MSK) [Pasupathy, 8.36], known as tamed frequency modulation (TFM) [deJager, 8.32] seems to stand out as the most promising scheme among these. We shall briefly summarize this particular application of correlative coding in this section.

It is well known [Pasupathy, 8.36] that MSK, an instance of binary continuous-phase frequency-shift keying (CPFSK), can be represented as an instance of binary offset quadrature phase-shift keying (OQPSK). Thus a seemingly nonlinear modulation can be optimally demodulated using a simple linear in-phase–quadrature (I-Q) receiver. The spectral properties of MSK can be improved by changing the data pulse shapes, while still ensuring that the signal has constant envelope. Such generalized MSK signals, in general, are no longer exactly equivalent to OQPSK signals. Hence the optimal demodulators for these signals are nonlinear (i.e., Viterbi receiver) and more complicated than the MSK receiver. Fortunately, in many instances, a linear receiver can still demodulate such generalized MSK signals without much loss in performance. The TFM scheme belongs to such a class.

The CPFSK version of MSK (Fig. 8.12) generates the signal

$$s(t) = A \cos(\omega_c t + \phi(t) + \phi_0) \tag{8.55}$$

where $\omega_c$ is the carrier frequency, $\phi_0$ is the starting angle, the excess phase $\phi(t)$ is

$$\phi(t) = \int_0^t f(\tau) \, d\tau \tag{8.56}$$

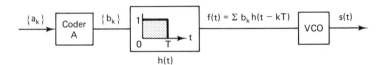

**Figure 8.12** A CPFSK scheme generating an MSK signal. Voltage controlled oscillator (VCO) generates frequencies $f_c \pm 1/4T$.

with

$$f(t) = \frac{\pi}{2T} \Sigma b_k h(t - kT) \tag{8.57}$$

$$h(t) = \begin{cases} 1 & 0 \le t \le T \\ 0 & \text{elsewhere} \end{cases} \tag{8.58}$$

where $T$ is the symbol duration, $\{b_k\}$ is an encoded version of the data stream $\{a_k\}$, and $[a_k, b_k \in (\pm 1)]$. As an OQPSK signal (Fig. 8.13) MSK is given by

$$\begin{aligned}
s(t) = &A \sum_{k\,\text{odd}} c_k p(t - kT) \cos (\omega_c t + \phi_0) \\
&+ A \sum_{k\,\text{even}} c_k p(t - kT) \sin (\omega_c t + \phi_0)
\end{aligned} \tag{8.59}$$

where

$$p(t) = \sin \left\{ \frac{\pi}{2T} \int_0^T [h(\tau) + h(\tau - T)] \, d\tau \right\} \tag{8.60a}$$

$$= \begin{cases} \sin \dfrac{\pi t}{2T} & 0 \le t \le 2T \\ 0 & \text{elsewhere} \end{cases} \tag{8.60b}$$

and $\{c_k\}$ (equal to $\pm 1$) is derived from $\{a_k\}$ by the coding scheme B. Actually, the scheme of Fig. 8.12 with $a_k = b_k$ is known as fast-frequency-shift keying (FFSK)

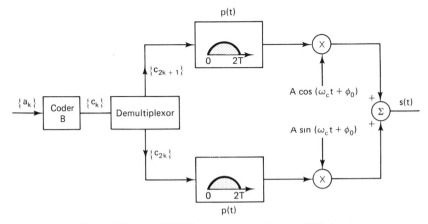

**Figure 8.13** An OQPSK system generating an MSK signal.

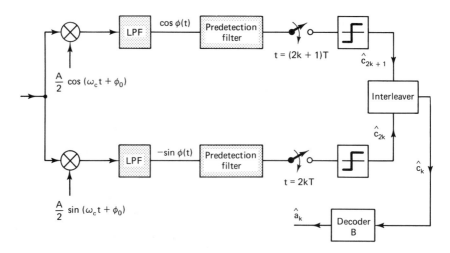

**Figure 8.14** An I-Q receiver for MSK (LPF is low-pass filter).

[deBuda, 8.37] and the original MSK description [Doelz, 8.38] is actually the scheme of Fig. 8.13 with $a_k = c_k$.

In order that the signals of (8.55) through (8.58) and (8.59) through (8.60) be identical for the same $\{a_k\}$, it can be shown [Galko, 8.39] that $\{b_k\}$ and $\{c_k\}$ must be related by the coding rule

$$b_k = (-1)^{k+1} c_k c_{k-1} \tag{8.61}$$

Thus MSK ($a_k = c_k$ in Fig. 8.13) can be described by the system of Fig. 8.12 using the coding rule (8.61). The equivalence of MSK to OQPSK indicates that an MSK signal corrupted by AWGN is optimally demodulated (with respect to $\{c_k\}$) by the I-Q receiver of Fig. 8.14, with predetection filter matched to $p(t)$ (8.60). (In practice, the LPF and the predetection filter may be combined into one filter.)

The MSK-OQPSK equivalence can be extended to any $h(t)$ restricted to [0, $T$] (but satisfying $\int_0^T h(t)\, dt = T$) with pulse shape (8.60a) replacing the sinusoidal shape. Since the excess phase $\phi(nT) = \pi/2 \int_0^{nT} f(\tau)\, d\tau$ is independent of the particular $h(t)$, then we can also demodulate these MSK-like signals by the I-Q receiver of Fig. 8.14.

In the MSK signal of Fig. 8.12,

$$\phi(nT) = \frac{\pi}{2} \sum_{k=0}^{n-1} b_k \equiv \begin{cases} 0 & \text{mod } \pi,\ n \text{ even} \\ \dfrac{\pi}{2} & \text{mod } \pi,\ n \text{ odd} \end{cases} \tag{8.62}$$

Thus

$$\cos \phi(nT) = \begin{cases} \pm 1 & n \text{ even} \\ 0 & n \text{ odd} \end{cases} \tag{8.63}$$

$$\sin \phi(nT) = \begin{cases} \pm 1 & n \text{ odd} \\ 0 & n \text{ even} \end{cases} \tag{8.64}$$

Since the excess phase $\phi(t)$ typically varies smoothly between the phase values at $t = nT$, it follows that in such MSK-like schemes $\cos \phi(t)$ and $\sin \phi(t)$ will be very similar to PAM signals with similar eye diagrams, that is, PAM signals with no ISI at the sampling points. Thus the I-Q receiver responds to the polarities of the I-Q signals at the sampling points and the data can be recovered from these polarity changes.

It follows, then, that one way of guaranteeing detectability by an I-Q receiver is to require $\phi(nT)$ to coincide with the excess phase of MSK. This may be interpreted as a zero-ISI condition and corresponds to the requirement.

$$\int_{nT}^{(n+1)T} h(t) \, dt = \begin{cases} T & n = 0 \\ 0 & n \neq 0 \end{cases} \tag{8.65}$$

Equation (8.65) is known as **Nyquist's third criterion** [Pasupathy, 8.40] and translates in the frequency domain to the condition

$$\sum_n H\left(f - \frac{n}{T}\right) e^{j\pi(fT-n)} \frac{\sin \pi(fT - n)}{\pi(fT - n)} = T \tag{8.66}$$

where $H(f)$ is the Fourier transform of $h(t)$. (The TFM scheme [deJager, 8.32] uses a pulse that approximates the minimum bandwidth solution of (8.66).)

If we require only detectability by a linear receiver, then the zero-ISI condition (8.65) is clearly overly stringent, and relaxing it is likely to result in signals with significantly decreased bandwidths, as was done in correlative coding schemes. Thus such generalizations, resulting in *generalized* MSK [Galko, 8.39], can be thought of as MSK schemes with ISI.

A simple way to weaken the zero-ISI conditions is to allow the use of correlative coding, or controlled ISI. Consider a pulse satisfying

$$\int_{mT}^{(m+1)T} h(t) \, dt = \begin{cases} x_m & m = \pm 1, 0 \\ 0 & m \geq 2 \end{cases} \tag{8.67}$$

and $x_{-1} + x_0 + x_1 = 1$. For such a pulse shape, the excess phase at $t = mT$ becomes

$$\phi(mT) = \frac{\pi}{2} \sum_{k=0}^{m-1} b_k + \frac{\pi}{2T}(x_{-1}b_m - x_1 b_{m-1}) + \frac{\pi}{2T}(x_1 b_{-1} - x_{-1} b_0) \tag{8.68}$$

The last term in (8.68) may be regarded as a fixed phase offset and can be incorporated into the starting angle $\phi_0$ (replacing $\phi_0$ by $\hat{\phi}_0 - \pi/2T(k_1 b_{-1} - x_{-1}b_0)$). Thus in an I-Q receiver using the phase offset $\hat{\phi}_0$, the new excess phase $\hat{\phi}(mT)$ at the sampling points becomes [Galko, 8.41]

$$\hat{\phi}(mT) = \phi_{\text{MSK}}(mT) + \frac{\pi}{2T}(x_{-1}b_m - x_1 b_{m-1}) \tag{8.69}$$

Thus provided the latter term in (8.69), representing controlled ISI, is always less than $\pi/2$, the polarities of the samples of $\cos \hat{\phi}(t)$ and $\sin \hat{\phi}(t)$ will be the same as in MSK. This will permit an I-Q receiver to demodulate the signal (though with decreased performance due to reduction in the difference in the sample levels).

The required condition, then, is when

$$|x_{-1}| + |x_1| < T \qquad \sum_i x_i = T \qquad (8.70)$$

with the eye opening at the sampling point given by $2 \cos (\pi/2T)(|x_{-1}| + |x_1|)$.

CPM schemes employing correlative coding schemes using the polynomials $\sum_i x_i D^i$ satisfying (8.70) have been considered in the literature. Among these, two schemes stand out as being the most useful. These are duobinary MSK (using the coding $1 + D$) or frequency-shift offset quadrature (FSOQ) modulation [Rhodes, 8.42] and TFM scheme, which uses $(1 + D)^2$ [deJager, 8.32]. We shall now derive the expression for the predetection filter used in TFM; the analysis can easily be extended to other schemes [Galko, 8.43].

The basis of the approach in [deJager, 8.32] for filter design is to *approximate* the I- and Q-channel signals by the PAM signals $\sum c_n g(t - nT)$, where $g(t)$ is some suitably chosen pulse shape. If the approximation is adequate, the filter should produce near-optimal performance. The approximation used in [deJager, 8.32] utilized a raised-cosine pulse

$$g(t) = \begin{cases} \dfrac{1}{2\sqrt{2}} \left( 1 + \cos\left(\dfrac{\pi t}{2T}\right) \right) & |t| \leq 2T \\ 0 & \text{elsewhere} \end{cases} \qquad (8.71)$$

The filter-transfer function $R(f)$ that minimizes the probability of error subject to a zero-ISI (Nyquist I) constraint for the PAM signals is known to be [Lucky, 8.2, pp. 109–110]

$$R(f) = \frac{k_1 G(f)}{\sum |G(f - k/2T)|^2} \qquad (8.72)$$

where $G(f) = F \cdot T \cdot [g(t)]$ and $k_1$ is a constant determined by normalization requirements. Since $g(t)$ is of finite duration, the denominator of (8.72) may be determined explicitly using the Poisson sum formula [Franks, 8.12]. Noting that $F \cdot T \cdot [g(t)*g(-t)] = |G(f)|^2$, (8.72) becomes [Galko, 8.43]

$$R(f) = \frac{k_2 G(f)}{\alpha + \beta \cos 4\pi f T} \qquad (8.73)$$

where

$$\alpha = \int_{-2T}^{2T} g^2(t) \, dt \qquad (8.74)$$

$$\beta = 2 \int_0^{2T} g(t)g(t - 2T) \, dt \qquad (8.75)$$

and $k_2$ is some constant determined by the constraint $\int_{-\infty}^{\infty} |R(f)|^2 \, df = 1$ (a function of $T$).

For the pulse (8.71), the result becomes

$$R(f) = \frac{(\sqrt{T}/2^{1/4})G(f)}{(3T/4) + (T/4)\cos 4\pi f T} \tag{8.76}$$

with the filter impulse response given by

$$r(t) = \frac{2^{1/4}}{\sqrt{T}} \sum_n (2\sqrt{2} - 3)^{|n|} g(t - 2nT) \tag{8.77}$$

Equations (8.76) and (8.77) show the well-known result that such an "optimum" filter is a cascade of a matched filter and a transversal equalizer. Both these results agree with the reported numerical results in [deJager, 8.32]. Such I-Q receivers using a filter designed on a PAM approximation for TFM seem to perform very close to the optimal (Viterbi) receivers. Similar results for other correlatively coded CPM schemes [Galko, 8.44] lead us to believe that linear receivers can give rise to near-optimal performance in many cases.

## 8.9 RELATED WORK

The following are some of the other concepts and research related to correlative coding that have not been covered in this chapter:

1. The concept of monitoring error rates in digital communication systems, in general, and in partial-response systems, in particular. The idea of pseudoerrors is a particularly useful one in this context [Newcombe, 8.45].
2. Pulses that simultaneously satisfy Nyquist and partial-response criteria and their use in enhancing performance and error monitoring [Sousa, 8.46].
3. Suboptimal receivers for PRS schemes with performance and complexity lying between conventional and Viterbi receivers [Eggers, 8.47; Grami, 8.48].

One can conclude by observing that the concept of correlative coding, since it deals with the use of memory or correlation to improve spectral properties and/or performance, will continue to play a vital role in many digital communication systems.

## REFERENCES

[8.1] Lender, A. "The duobinary technique for high speed data transmission," *IEEE Trans. Commun. Electron.*, Vol. 82, May 1963, pp. 213–218.

[8.2] Lucky, R. W., J. Salz, and E. J. Weldon, Jr. *Principles of Data Communication*, McGraw-Hill, New York, 1968, pp. 83–90.

[8.3] Kretzmer, E. R. "Generalization of a technique for binary data communication," *IEEE Trans. Commun Technol.*, Vol. COM-14, February, 1966, pp. 67–68.

[8.4] Kabal, P., and S. Pasupathy. "Partial-response signaling," *IEEE Trans. Commun.*, Vol. COM-23, September, 1975, pp. 921–934.

[8.5] Schwartz, M. *Information Transmission, Modulation, and Noise*, 3rd ed., McGraw-Hill, New York, 1980.

[8.6] Stremler, F. *Introduction to Communication Systems*, 2nd ed., Addison-Wesley, Reading, Mass., 1982.

[8.7] Haykin, S. *Communication Systems*, 2nd ed., John Wiley, New York, 1983.

[8.8] Lathi, B. P. *Modern Digital and Analog Communication Systems*, Holt, Rinehart and Winston, New York, 1983.

[8.9] Pasupathy, S. "Correlative coding: a bandwidth efficient signaling scheme," *IEEE Commun. Society Magazine*, Vol. 15, No. 4, July, 1977, pp. 4-11. Also reprinted in *Tutorials in Modern Communications*, eds. V. B. Lawrence et al., Computer Science Press, 1982.

[8.10] Lender, A. "Correlative (partial response) techniques and applications to digital radio systems," in K. Feher, *Digital Communications: Microwave Applications*, Prentice-Hall, Englewood Cliffs, NJ, 1981.

[8.11] Nyquist, H. "Certain topics in telegraph transmission theory," *Trans. AIEE*, Vol. 47, April, 1928, pp. 617–644.

[8.12] Franks, L. E. *Signal Theory*, Prentice-Hall, Englewood Cliffs, NJ, 1969, p. 73.

[8.13] Kobayashi, H. "A survey of coding schemes for transmission or recording of digital data," *IEEE Trans. Commun. Technol.*, Vol. COM-19, pp. 1087–1100, Dec. 1971.

[8.14] Kobayashi, H. "Correlative level coding and maximum-likelihood decoding," *IEEE Trans. Inform. Theory*, Vol. IT-17, September, 1971, pp. 586–594.

[8.15] Forney, G. D., Jr. "Maximum-likelihood sequence estimation of digital sequences in the presence of intersymbol interference," *IEEE Trans. Inform. Theory*, Vol. IT-18, May, 1972, pp. 363–378.

[8.16] Seastrand, K. L., and L. L. Sheets. "Digital transmission over analog microwave systems," in *ICC Conf Rec.*, 1972, pp. 29.1–29.5.

[8.17] Baker, D. M. "Analog/digital hybrid radio," *Can. Electron. Eng.*, February, 1973, pp. 34–38.

[8.18] Bennett, W. R. *Introduction to Signal Transmission*, McGraw-Hill, New York, 1970, p. 16.

[8.19] Maurer, R. E., and W. J. Maybach. "Engineering the T1D line," *Proc. NTC/80*, Houston, Texas, December, 1980, pp. 39.2.1–39.2.8.

[8.20] Becker, F. K., E. R. Kretzmer, and J. R. Sheehan. "A new signal format for efficient data transmission," *Bell Syst. Tech. J*, Vol. 45, May-June, 1966, pp. 755–758.

[8.21] Aaron, M. R. "PCM transmission in the exchange plant," *Bell Syst. Tech. J.*, Vol. 41, January, 1962, pp. 99–141.

[8.22] Kabal, P., and S. Pasupathy. "Partial-response signalling," *Comm. Tech Report*, Dept. of Electrical Engineering, University of Toronto, January, 1975.

[8.23] Lender, A., and V. Stalik. "Engineering of the duobinary repeated line," *ICC/77 Conf. Rec.*, June, 1977, pp. 32.4–306 — 32.4–309.

[8.24] Chow, P., and M. McDonnell. "The T1C: a 48-channel digital carrier system," *Telesis*, No. 3, 1982, pp. 9–13.

[8.25] Croisier, A. "Introduction to pseudoternary transmission codes," *IBM J. Res. Develop.*, Vol. 14, July, 1970, pp. 354–367.

[8.26] Newcombe, E. A., and S. Pasupathy. "Effects of filtering allocation on the performance of a modified duobinary system," *IEEE Trans. Commun.*, Vol. COM-28, May, 1980, pp. 749–752.

[8.27] Lender, A. "Correlative data transmission with coherent recovery using absolute reference," *IEEE Trans. Commun. Tech.*, Vol. COM-16, No. 1, February, 1968, pp. 108–115.

[8.28] Gunn, J. F., and J. A. Lombardi. "Error detection for partial response systems," *IEEE Trans. Tech.*, Vol. COM-17, No. 6, December, 1969, pp. 734–736.

[8.29] Kobayashi, H., and D. T. Tang. "Application of partial-response channel coding to magnetic recording," *IBM J. Res. Develop.*, Vol. 14, July, 1970, pp. 368–375.

[8.30] Gibson, E. D. "Error correction with little or no redundancy," *International Commun. Conf. Rec.*, 1973, pp. 40.32–40.40.

[8.31] Omura, J. K. "On the Viterbi decoding algorithm," *IEEE Trans. Inf. Theory*, Vol. IT-15, January, 1969, pp. 177–179.

[8.32] de Jager, F., and C. B. Dekker. "Tamed frequency modulation, a novel technique of achieving spectral economy," *IEEE Trans. Commun.*, Vol. COM-26, May, 1978, pp. 534–542.

[8.33] Deshpande, G. S., and P. H. Wittke. "Correlative digital FM," *IEEE Trans. Commun.*, Vol. COM-29, February, 1981, pp. 156–162.

[8.34] Aulin, T., N. Rydbeck, and C. E. W. Sundberg. "Continuous phase modulation—Part II: Partial response signaling," *IEEE Trans. Commun.*, Vol. COM-29, March, 1981, pp. 210–225.

[8.35] Muilwijk, D. "Correlative phase shift keying—a class of constant envelope techniques," *IEEE Trans. Commun.*, Vol. COM-29, March, 1981, pp. 226–236.

[8.36] Pasupathy, S. "Minimum shift keying: a spectrally efficient modulation," *IEEE Commun. Magazine*, Vol. 17, No. 4, July, 1979, pp. 14–22. Also reprinted in *Tutorials in Modern Communications*, eds. V. B. Lawrence et al., Computer Science Press, 1982.

[8.37] de Buda, R. "Coherent demodulation of frequency-shift keying with low deviation ratio," *IEEE Trans. Commun.*, Vol. COM-20, June, 1972, pp. 429–435.

[8.38] Doelz, M. L., and E. T. Heald. Minimum-Shift Data Communication System, U.S. Patent 2,977,417, March 29, 1961.

[8.39] Galko, P. "Generalized MSK," Ph.D. Thesis, Dept. of Electrical Engineering, University of Toronto, Toronto, 1983.

[8.40] Pasupathy, S. "Nyquist's third criterion," *Proc. IEEE*, Vol. 62, June 1974, pp. 860–861.

[8.41] Galko, P., and S. Pasupathy. "On a class of generalized MSK," *ICC'81 Conf. Rec.*, Denver, June 15–18, 1981, pp. 2.4.1–2.4.5.

[8.42] Rhodes, S. A. "FSOQ, a new modulation technique that yields a constant envelope," *NTC'80 Conf. Rec.*, Houston, December, 1980, pp. 51.1.1–51.1.7.

[8.43] Galko, P., and S. Pasupathy. "Linear receivers for correlatively coded MSK," *IEEE Trans. Commun.*, Vol. COM-33, No. 4, April, 1985, pp. 338–347.

[8.44] Galko, P., and S. Pasupathy. "Performance evaluation of generalized MSK," *Globecom'82 Conf. Rec.*, Miami, November 29–December 2, 1982, pp. B3.7.1–B.3.7.5.

[8.45] Newcombe, E. A. and S. Pasupathy. "Error rate monitoring in a partial-response system," *IEEE Trans. Commun.*, Vol. COM-28, No. 7, July, 1980, pp. 1052–1061.

[8.46] Sousa, E. S., and S. Pasupathy. "Enhanced receivers for nonoptimally allocated filtering," *IEEE Trans. Commun.*, Vol. COM-31, No. 7, July, 1983, pp. 879–885.

[8.47] Eggers, M. D., and J. H. Painter. "Optimal symbol-by-symbol detection for duo-binary signaling," *IEEE Trans. Commun.*, Vol. COM-31, No. 9, September, 1983, pp. 1077–1084.

[8.48] Grami, A., and S. Pasupathy. "Optimal two-stage detection For PRS," *ICC/86 Conf. Rec.*, June 22–25, 1986, pp. 671–675.

# 9

# INTERFERENCE ANALYSIS AND PERFORMANCE OF LINEAR DIGITAL COMMUNICATION SYSTEMS

**DR. VASANT K. PRABHU**

*AT&T Bell Laboratories*
*Holmdel, New Jersey 07733*

## 9.1 INTRODUCTION

The field of public telecommunications has experienced an unprecedented growth, expansion, and development during the last several years on both the theoretical and applied fronts. In this area of communications, microwave radio, from its introduction in the late 1940s to the present, has become one of the primary media for transmitting information from point to point and from a point into a given area. The advent of satellite communication technology in 1962 and consequent sharing of bands between satellite and radio relay coupled with the explosive growth of microwave radio routes at these frequencies has led to increased sharing of frequency spectrum and to generation of increased mutual interference. This added interference is playing a dominant role in limiting the capacity, efficiency, reliability, and cost of communication systems. In addition, the advances in low-noise receiver technology have made the ubiquitous thermal noise, which used to limit the performance of earlier communication systems, of secondary importance in system design and engineering.

With the expected increase in congestion of frequency spectrum, use of satellites, various frequency reuse techniques, and migration to use of higher bands, the role played by interference is likely to increase in the future. Interference is a ubiquitous property of the environment in which various communication systems coexist either in the same or adjacent frequency bands and is also caused by nonideal mechanisms utilized in the process of communications.

Until now, the system designer almost always assumed that the limiting corrupting signal has Gaussian characteristics—the characteristics of thermal noise. With the advent of low-noise receivers and congestion in the radio frequency (RF) bands, this assumption can no longer be justified, and interference—which may not have Gaussian characteristics—may very well limit the performance of our present and future communication systems. The method of analysis used to determine the effect of thermal noise on communication systems cannot, therefore, be used to determine the effect of interference on our new and evolving communication systems and to design system components.

Hence, a completely new methodology is required to determine the performance of communication systems subject to non-Gaussian interference and to optimize their performance. Since thermal noise, however small, is almost always present, it is necessary that we consider the joint effect of Gaussian-distributed thermal noise and non-Gaussian distributed interference. This makes our task doubly difficult.

This chapter gives a comprehensive analysis of new methods to determine the joint effect of thermal noise and non-Gaussian interference and presents new insights into their performance. Some of the material covered in the chapter is very new, has been reported in the literature during the last few years, and has been sought after extensively by system designers.

We only consider the additive form of interference in this chapter. This includes cochannel interference (CCI), adjacent-channel interference (ACI), and intersymbol interference (ISI). The interference due to the spurious signals caused by nonlinear mechanisms, imperfect carrier recovery, and so on, will not be covered.

Communication systems can be divided naturally into two categories, analog and digital signal transmission, for the treatment of interference. This book deals only with digital transmission, and we shall give a generalized treatment of interference in digital communication systems corrupted by Gaussian noise and additive interference. The additive interference can be caused by a single source such as a single cross-polarized signal or a set of sources such as cochannel interference, ACI, and ISI.

The treatment of interference is necessarily vast, and we do not claim to treat it exhaustively in this chapter. Our intention is to familiarize the reader with the available techniques and to develop some basic concepts and tools so that any linear digital communication system can be analyzed. Since the performance of a canonical binary system can be used as a building block to determine the performance of more complex systems, we only treat the performance of a basic binary system in this chapter. The interested reader is referred to several references given at the end of this chapter for additional methods and techniques that may be used to consider specific modulation systems and interferences.

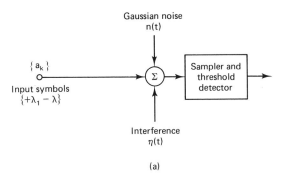

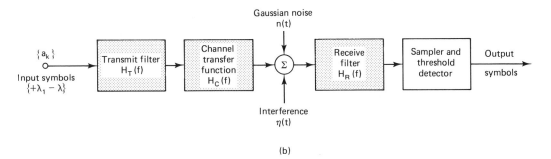

**Figure 9.1** (a) Representation of Gaussian noise and interference in binary canonical system. (b) Portion of a baseband binary canonical system.

## 9.2 ANALYSIS OF INTERFERENCE

In a digital communication system, the symbol error probability often serves as a measure of its performance, and most other performance measures can often be expressed in terms of this probability. For this reason, methods and techniques are often developed to evaluate this probability of error.

Further, the probability of error of M-ary PAM, M-ary QAM, CPSK, DCPSK, and several other modulation schemes can be evaluated or bounded if we can determine the error probability of a binary *canonical* system shown in Fig. 9.1. Input symbols, chosen with a priori probability from a binary alphabet $\{-\lambda, \lambda\}$ are corrupted by additive zero-mean Gaussian noise $n(t)$ of variance $\sigma^2$ and independent additive interference $\eta(t)$, such as ISI, CCI, ACI, and so on. ISI is the residual effect caused by symbols other than the one being detected and is generated by bandlimiting filters used in transmission.

Figure 9.1a is the representation of the method of detection and is not an actual receiver. The actual receiver, including a transmit filter $H_T(f)$, a channel transfer function $H_C(f)$, and a receive filter $H_R(f)$ is shown in Fig. 9.1b. The effect of ISI generated by $H_T(f)$, $H_C(f)$, and $H_R(f)$ as well as CCI or ACI is assumed to be represented by $\eta(t)$. In narrow-band systems, this additive inter-

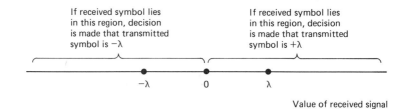

Figure 9.2 Decision regions for binary canonical system. If the received signal vector lies in the right half-plane, we assume that signal vector λ was sent. If it lies in the left half-plane, the signal vector − λ was sent.

ference is more likely to be dominated by ISI caused by bandlimiting; CCI may be the major source of impairment in dual-pol transmission systems, and so on. In any case, $\eta(t)$ represents the total composite interference corrupting the binary signal and can be caused by either a single source (such as a cross-polarized signal) or multiple sources.

We assume maximum-likelihood detection and assume, without loss of generality, that the decision threshold of the receiver is zero. The signal constellation and the decision regions for such a receiver are shown in Fig. 9.2.

In such a case, the average probability $Pe$ of symbol error of the canonical binary system is

$$P_e = \tfrac{1}{2}\{Pr[n + \eta > \lambda] + Pr[n + \eta < -\lambda]\} \tag{9.1}$$

Since λ represents the signal magnitude of the binary canonical system, and it takes values $\pm\lambda$ with equal probability, the mean-squared value $S$ of the binary system is

$$S = \lambda^2 \tag{9.2}$$

Assume that $T$ is the signaling interval of the binary system.

Since the two terms in (9.1) are similar, we consider just the evaluation of $P$, where

$$P = Pr[n + \eta < -\lambda] \tag{9.3}$$

From (9.3)

$$P = \frac{1}{2} E\left[ \operatorname{erfc}\left\{ \frac{\lambda}{\sigma\sqrt{2}} + \frac{\eta}{\sigma\sqrt{2}} \right\} \right] \tag{9.4}$$

where $E[g(z)]$ represents the mathematical expectation of an arbitrary function $g(z)$,

$$E[g(z)] = \int_{-\infty}^{\infty} g(z)dF(z),$$

$$F(z) \text{ is the distribution function of } z \tag{9.5}$$

and

$$\text{erfc}\,(a) = \frac{2}{\sqrt{\pi}} \int_a^\infty e^{-t^2}\, dt \tag{9.6}$$

Since erfc $(x)$ can be expanded into a Taylor series for any $x$, $-\infty < x < \infty$,

$$\text{erfc}\,(x + a) = \text{erfc}\,(x) \tag{9.7}$$

$$+ \frac{2}{\sqrt{\pi}} \exp\,(-x^2) \sum_{k=1}^\infty (-1)^k H_{k-1}(x) \frac{a^k}{k!}$$

where $H_n(x)$ represents the Hermite polynomial of order $n$ [Morse, 9.1],

$$H_j(x) = j! \sum_{m=0}^{[j/2]} \frac{(-1)^m}{m!(j - 2m)!} (2x)^{j-2m} \tag{9.8}$$

where $[b]$ = the largest integer contained in $b$

$$H_{n+1}(x) = 2xH_n(x) - H_{n-1}(x), \qquad n \geq 1,\, H_0(x) = 1 \tag{9.9}$$

In our case,

$$x = \frac{\lambda}{\sigma\sqrt{2}};\, a = \frac{\eta}{\sigma\sqrt{2}} \tag{9.10}$$

Hence,

$$\text{erfc}\left\{ \frac{\lambda}{\sigma\sqrt{2}} + \frac{\eta}{\sigma\sqrt{2}} \right\} = \text{erfc}\left\{ \frac{\lambda}{\sigma\sqrt{2}} \right\}$$

$$+ \frac{2}{\sqrt{\pi}} \exp\left( -\frac{\lambda^2}{2\sigma^2} \right) \sum_{j=1}^\infty (-1)^j H_{j-1}\left( \frac{\lambda}{\sigma\sqrt{2}} \right) \left( \frac{1}{\sigma\sqrt{2}} \right)^j \frac{\eta^j}{J!} \tag{9.11}$$

Also, since a Taylor series is uniformly convergent in any interval that lies together with its end points inside the interval of convergence, and since a uniformly convergent series can be integrated termwise within its interval of convergence [Goodman, 9.40], averaging over the random variable $\eta$,

$$P = \frac{1}{2} \text{erfc}\left( \frac{\lambda}{\sigma\sqrt{2}} \right)$$

$$+ \frac{1}{\sqrt{\pi}} \exp\left( -\frac{\lambda^2}{2\sigma^2} \right) \sum_{k=1}^\infty (-1)^k H_{k-1}\left( \frac{\lambda}{\sigma\sqrt{2}} \right) \left( \frac{1}{\sigma\sqrt{2}} \right)^k \frac{E[\eta^k]}{k!} \tag{9.12}$$

where $E[\eta^n]$ represents the $n$th moment of $\eta$. It therefore follows that the series in equation (9.12) converges, and the error rate $P$ can be computed with any desired accuracy by calculating the moments of $\eta$. We assume that all moments of $\eta$ exist, are finite, and can be calculated.

We shall give some examples. Since, in practice, only a finite number of terms in the series can be used to estimate P, we shall give an expression for the error when we use the first N terms to determine P.

### 9.2.1 Series Truncation Error

If the range $(-\Omega, \Omega)$ of $\eta$ is finite, then

$$E[\eta^{k+s}] = \int_{-\infty}^{\infty} \eta^{k+s} dF(\eta)$$

$$\leq \int_{-\infty}^{\infty} |\eta|^{k+s} dF(\eta)$$

$$= \int_{-\Omega}^{\Omega} |\eta|^{k+s} dF(\eta) \qquad (9.13)$$

$$\leq \Omega^s \int_{-\Omega}^{\Omega} |\eta|^k dF(\eta)$$

$$= m_k \Omega^s, \qquad k \geq 0, \, s \geq 0,$$

where

$$m_k = \int_{-\Omega}^{\Omega} |\eta|^k dF(\eta) \qquad (9.14)$$

and is called the $k$th absolute moment of $\eta$.

If the first $N$ terms in the series are used in estimating P from (9.12), the truncation error $T_N$ is given by

$$T_N = \frac{1}{\sqrt{\pi}} \exp\left(-\frac{\lambda^2}{2\sigma^2}\right) \sum_{k=N+1}^{\infty} (-1)^k H_{k-1}\left(\frac{\lambda}{\sigma\sqrt{2}}\right)\left(\frac{1}{\sigma\sqrt{2}}\right)^k \frac{E[\eta^k]}{k!}. \qquad (9.15)$$

Since it can be shown [Abramowitz, 9.3] that

$$|H_n(t)| \leq p\, 2^{n/2} \sqrt{n!} \, \exp\left(\frac{t^2}{2}\right), \qquad p \approx 1.086435 \qquad (9.16)$$

(9.15) becomes

$$|T_N| < \left(\frac{p}{\sqrt{2\pi}}\right) \exp\left(-\frac{\lambda^2}{4\sigma^2}\right) \frac{m_N}{\sigma^N} \frac{(\Omega/\sigma)}{(N+1)\sqrt{N!}}$$

$$\cdot \left[1 - \frac{\Omega}{\sigma\sqrt{N+1}}\right]^{-1}, \qquad \frac{\Omega}{\sigma\sqrt{N+1}} < 1 \qquad (9.17)$$

Since $T_N$ goes to zero as $N \to \infty$, it follows that $P$ can be estimated with any given accuracy.

### 9.2.2 Recursive Relation to Evaluate Moments of $\eta$

Most often, the interference variable $\eta$ is the sum of a large number of other random variables, which are often statistically independent. Since the probability of error $P$ is the function of moments of $\eta$ and it is easier to evaluate the moments of the components of $\eta$, we develop a recursive method to evaluate the $n$th moment of $\eta$ in terms of the moments of the individual components of $\eta$.

Let $\Theta_k$ denote the partial sum $\Sigma_{i=1}^{k} \eta_i$, where

$$\eta = \sum_{i=1}^{N} \eta_i \qquad (9.18)$$

and $\eta_i$'s are statistically independent random variables. We can write

$$\mu_n = E[\eta^n] = E[\Theta_N^n] = \Omega_n(N) \qquad (9.19)$$

where

$$\Omega_n(i) = E[\Theta_i^n], \qquad n \geq 1, \quad \Omega_0(i) = 1 \qquad (9.20)$$

Now

$$\Omega_n(k) = E[(\Theta_{k-1} + \eta_k)^n], \quad k > 1 \qquad (9.21)$$

or

$$\Omega_n(k) = \sum_{p=0}^{n} \binom{n}{p} \Omega_p(k-1)\alpha_{n-p}(k), \qquad k > 1 \qquad (9.22)$$

where

$$\alpha_{n-p}(k) = E[\eta_k^{n-p}], \qquad \alpha_0(k) = 1, \quad k \geq 1 \qquad (9.23)$$

Since we assume that we can evaluate moments of all order for the component random variables $\eta_k$'s, all $\alpha_{n-p}(k)$'s are known or can be determined, and we have a recurrence relation to compute the moments of all order for the composite variable $\eta$.

In many cases, the $\eta_k$'s are even random variables, and in this case, we can show that

$$\mu_{2l+1} = 0, \qquad l \geq 0, \qquad (9.24)$$

$$\mu_{2n} = \Omega_{2n}(N) \qquad (9.25)$$

$$\Omega_{2n}(k) = \sum_{p=0}^{2n} \binom{2n}{2p} \Omega_{2p}(k-1)\alpha_{2n-2p}(k) \qquad (9.26)$$

The last recurrence relation for the sum of even random variables contains only the sum of positive terms (which reduces possible computation errors) and has successfully been used to compute moments of all order of the composite random variable $\eta$.

If $\eta$ is an even random variable, note that

$$E[\eta^{2k+1}] = 0, \qquad k \geq 0 \qquad (9.27)$$

and

$$P = \frac{1}{2} \mathrm{erfc}\left(\frac{\lambda}{\sigma\sqrt{2}}\right)$$
$$+ \frac{1}{\sqrt{\pi}} \exp\left(-\frac{\lambda^2}{2\sigma^2}\right) \sum_{k=1}^{\infty} H_{2k-1}\left(\frac{\lambda}{\sigma\sqrt{2}}\right)\left(\frac{1}{\sigma\sqrt{2}}\right)^{2k} \frac{E[\eta^{2k}]}{(2k)!} \qquad (9.28)$$

### 9.2.3 Some Examples

**Binary CPSK with a Single Cochannel Interferer**

If the communication system shown in Fig. 9.1 is a binary CPSK system corrupted by cochannel interference and Gaussian noise, the total received signal $s(t)$ may be represented by

$$
\begin{aligned}
s(t) = \; &\text{Re} \, [A \exp \, (j2\pi f_c t + \phi) \\
&+ v + R \exp \, (j2\pi f_c t + \phi_i + \theta)], \quad KT < t \le (K+1)T
\end{aligned}
\tag{9.29}
$$

where $A^2/2$ is the power in the signal, $R^2/2$ is the power in the single cochannel interferer, $f_c$ is the carrier frequency, and $v$ is the zero-mean (complex) thermal noise with variance $\sigma^2$. Further, $\phi$ and $\phi_i$ represent, respectively, statistically independent data symbols transmitted over desired and interfering carriers and take values of $\pm\pi/2$ with equal probability. Since desired and interfering carriers often originate from two different sources, we assume that the interference is statistically independent of the signal and that $\theta$ is uniformly distributed over $[0, 2\pi]$.

Identifying different terms, the probability of error $Pe$ can be shown to be given by (9.28) where $\lambda = A$, and $\eta = R \cos \theta$.

It may be shown that

$$
E[(R \cos \theta)^{2k}] = R^{2k} \frac{(2k)!}{2^{2k} \cdot (k!)^2}
\tag{9.30}
$$

Since we know all the terms in (9.28), $Pe$ can be computed for any given signal-to-noise ratio (SNR; in decibels $= 10 \log (A^2/2\sigma^2)$) and **signal-to-interference ratio** (SIR; in decibels $= 10 \log (A^2/R^2)$). The results are shown in Fig. 9.3.

**Multiple Cochannel Interferers**

In this case,

$$
\begin{aligned}
s(t) = \; &\text{Re} \, [A \exp \, (j2\pi f_c t + \phi) \\
&+ v + \sum_{k=1}^{M} R_k \exp \, (j2\pi f_c t + \phi_k + \theta_k], \quad KT < t \le (K+1)T
\end{aligned}
\tag{9.31}
$$

and

$$
\eta = \sum_{i=1}^{M} R_i \cos \theta_i
\tag{9.32}
$$

Since we assume that $\theta_i$'s are independently distributed random variables uniformly distributed over the range $[0, 2\pi]$, we can show that

$$
E[\eta^{2k+1}] = 0, \quad k \ge 0
\tag{9.33}
$$

and use the recursive relations given in Section 9.2.2 to compute the moments $\mu_{2n} = E[\eta^{2n}]$

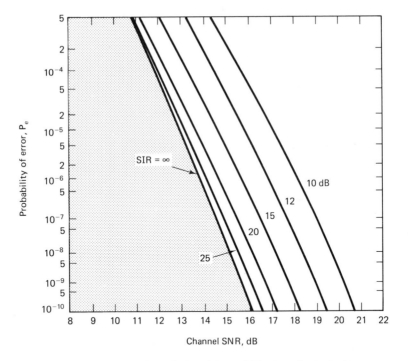

**Figure 9.3** Error rate for binary CPSK with one CCI. SNR (in decibel) is defined as $10 \log_{10} (A^2/2\, \sigma^2)$ and S/R (in decibel) is defined as $10 \log_{10} (A^2/R^2)$.

In this case,

$$\alpha_{2i}(j) = E[(R_j \cos \theta_j)^{2i}] \tag{9.34}$$

$$= R_j^{2i} \frac{(2i)!}{2^{2i}(i!)^2}$$

All even-order moments of $\eta$ can hence be calculated by using the above recurrence relations, and $Pe$ can be computed from

$$Pe = \frac{1}{2} \operatorname{erfc} \left( \frac{A}{\sigma\sqrt{2}} \right)$$

$$+ \frac{1}{\sqrt{\pi}} \exp \left( -\frac{A^2}{2\sigma^2} \right) \sum_{j=1}^{\infty} H_{2j-1} \left( \frac{A}{\sigma\sqrt{2}} \right) \left( \frac{1}{\sigma\sqrt{2}} \right)^{2j} \frac{\mu_{2j}}{(2j)!} \tag{9.35}$$

The challenge is to evaluate the last equation numerically. Even though the given series converges and the error rate can be computed with any desired accuracy, we would like to point out some difficulties that may be encountered. First, $H_j(x)$ may oscillate widely between large positive and negative values, and the alternate terms in the series may be of the same order and of different signs leading to computational problems. For specific cases, the reader may be able to innovate new techniques for summing the series. We point out one such technique.

Let

$$R_{\max} = \max_k \{R_k\} \tag{9.36}$$

Define

$$\eta_r = \frac{\eta}{R_{\max}} \quad \text{and} \quad \mu_r^k = E[\eta_r^k] \tag{9.37}$$

Rearranging the terms, we can show that

$$Pe = \frac{1}{2} \operatorname{erfc}\left(\frac{A}{\sigma\sqrt{2}}\right) + \exp\left(-\frac{A^2}{2\sigma^2}\right) \sum_{k=1}^{\infty} D_{2k}\left(\frac{A}{\sigma\sqrt{2}}\right) \cdot \mu_r^{2k} \tag{9.38}$$

where

$$D_n(x) = \frac{x^n R_{\max}^n H_{n-1}(x)}{n!} \tag{9.39}$$

$$D_n(x) = \frac{2x^2 R_{\max}}{n}\left[D_{n-1}(x) - \frac{(n-2)R_{\max}}{(n-1)} D_{n-2}(x)\right] \tag{9.40}$$

$$D_1(x) = xR_{\max}, \qquad D_2(x) = x^3 R_{\max}^2$$

We found that $D_n(x)$ behaves much more reasonably than $H_n(x)$ and the series converges faster.

Some results for binary phase-shift keying (PSK) when two or more cochannel interferers are corrupting the signal are shown in Fig. 9.4.

Another method of solving this computational problem is to use the Gaussian quadrature rules. In the next section, we outline this method.

### 9.2.4 Computation by Gaussian Quadrature Rules

In (9.4), the probability of error $P$ has been expressed as

$$P = \frac{1}{2} \int_{-\infty}^{\infty} \operatorname{erfc}\left(\frac{\lambda + \eta}{\sigma\sqrt{2}}\right) dF(\eta) \tag{9.41}$$

Using Gaussian quadrature rules, we can approximate the integral in (9.41) by

$$P \approx \sum_{j=1}^{L} w_j \operatorname{erfc}\left\{\frac{\lambda_j + \eta}{\sigma\sqrt{2}}\right\} \tag{9.42}$$

where the $\lambda_j$ are called the abscissas of the formula and the $w_j$ are called the weights. The set of numbers $\{w_i, \eta_i\}_{i=1}^{L}$ is called a **quadrature (Q) rule** and it has been shown that they can be obtained from the $2L + 1$ moments of the random variable $\eta$. This method of computation is described in great detail in [Benedetto, 9.29]. The probability of error can be evaluated with any desired accuracy since the error involved in the approximation can be shown to be bounded above by a quantity that tends to zero as the number $L$ goes to infinity.

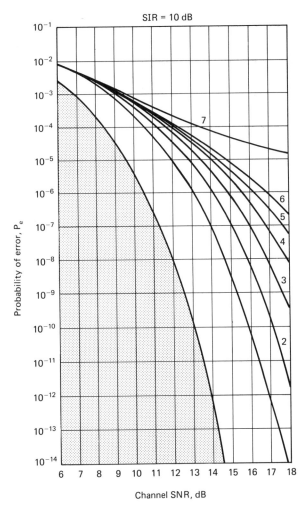

**Figure 9.4** Error rate for binary CPSK with two or more cochannel interferers. SNR (in decibels) is defined as $10 \log_{10} (A^2/2\sigma^2)$ and S/R (in decibels) as $10 \log_{10} (A^2/\Sigma_{k=1}^M R_k^2)$. The number $M$ represents the total number of interferers.

### 9.2.5 Analysis of M-ary Systems

The analysis given in this chapter can be extended to determine the interference performance of M-ary PAM systems, $M > 2$. Such extensions may be found in [Ho, 9.14; Prabhu, 9.22; Prabhu, 9.58].

### 9.2.6 Analysis of M-ary PSK and QAM

In M-ary phase-shift keying (MPSK) and QAM, we can show that the decision probability of a symbol can be expressed as the joint probability of decision of symbols along two axes, not necessarily orthogonal (as in MPSK).

In particular, the probability of error $P_M$ can be expressed as

$$\begin{aligned} P_M &= Pr[E_x \cup E_y] \\ &\le Pr[E_x] + Pr[E_y] \end{aligned}$$

(9.43)

where $Pr[E_x]$ and $Pr[E_y]$ are the probabilities of error of equivalent canonical systems discussed in this chapter. Note that the inequality follows from the union bound of probabilities.

These relations are discussed extensively in [Prabhu, 9.8] for MPSK and in [Prabhu, 9.58] for QAM.

## 9.3 MODIFIED CHERNOFF BOUNDS

Because of the complexity of the exact computation of the $P$, various upper and lower bounds have been derived to compute the error probability of a digital system subject to additive interference and Gaussian noise. Since it is possible to include all the available bounds, we have opted to discuss only two such bounds, which, in our opinion, typify the methods of derivation and also seem to have wide applicability in communication system analyses.

The first such bounds we shall discuss are called **modified Chernoff bounds** and are derived by using principles of analytic continuation and contour integration. The bounds, which are simple functions, require only the evaluation or bounding of the moment-generating function of the additive interference. The bounds are expressed in terms of a parameter that is the unique solution of an equation containing the derivative of the moment generating function of the interference.

The bounds are tight for high SNRs, the region of interest in most communication systems. For low SNR, the series approach is more convenient and can be used.

### 9.3.1 Probability of Error for a Canonical System

As before, the probability of error $P$ can be written as

$$P = Pr(n + \eta > \lambda) \tag{9.44}$$

From (9.43),

$$P = \int_\lambda^\infty P_{n+\eta}(\mu)\, d\mu$$
$$= \frac{1}{2\pi} \int_\lambda^\infty d\mu \int_{-\infty}^\infty \exp\left(-j\rho\mu\right)\Phi_{n+\eta}(\rho)d\rho, \qquad \text{Im } \rho = 0, \tag{9.45}$$

where $P_y(\cdot)$ denotes the probability density function (pdf) of $y$ and

$$\Phi_y(\cdot) = E[\exp\{j(\cdot)y\}] = \langle\exp\{j(\cdot)y\}\rangle \tag{9.46}$$

is its characteristic function.

Assuming that $\Phi_{n-\eta}(\rho)$ for complex $\rho$ can be obtained by the analytic continuation of its values on the real axis,

$$P = \frac{1}{2\pi} \int_\lambda^\infty \int_{-\infty}^\infty \exp\left(-j\mu\rho\right)\Phi_{n+\eta}(\rho)\, d\rho\, d\mu, \qquad \text{for } all\ \rho \tag{9.47}$$

Interchanging the order of integration and using the method of contour integration,

$$P = \frac{1}{2\pi} \int_c \frac{1}{j\rho} \exp(-j\lambda\rho)\Phi_{n+\eta}(\rho)\,d\rho \tag{9.48}$$

for *all* $\rho$ *such that* Im $\rho < 0$

and C is any contour from $-\infty + j0$ to $\infty + j0$ in the lower half of the complex $\rho$-plane. Since the integrand is analytic in the entire $\rho$-plane except for the point $\rho = 0$, and since for any $\rho$ with Im $\rho < 0$ and $|\rho| \to \infty$, we assume that the integrand goes to zero, we choose $\rho = x - j\beta$ and $x$ and $\beta$ *real*; and get

$$P = \frac{1}{2\pi} \int_{-\infty}^{\infty} \frac{1}{\beta + jx} \exp(-\lambda\beta) \exp(-j\lambda x)\Phi_{n+\eta}(x - j\beta)\,dx \qquad \beta > 0 \tag{9.49}$$

Since noise $n$ is a zero-mean Gaussian random variable with variance $\sigma^2$ and is independent of $\eta$

$$\begin{aligned}\Phi_{n+\eta}(\rho) &= \Phi_n(\rho)\Phi_\eta(\rho) \\ &= \exp(-\sigma^2\rho^2/2)\Phi_\eta(\rho)\end{aligned} \tag{9.50}$$

and

$$P = \exp(-\lambda\beta + \beta^2\sigma^2/2)\frac{1}{2\pi} \int_{-\infty}^{\infty} \frac{\exp\{-jx(\lambda - \beta\sigma^2)\}\exp(-x^2\sigma^2/2)}{\beta + jx}\Phi_\eta(x - j\beta)\,dx \tag{9.51}$$

or

$$P = \exp\left(-\lambda\beta + \frac{\beta^2\sigma^2}{2}\right)I \tag{9.52}$$

where

$$I = \frac{1}{2\pi} \int_{-\infty}^{\infty} \frac{\exp\{-jx(\lambda - \beta\sigma^2)\}\exp(-x^2\sigma^2/2)}{\beta + jx}\Phi_\eta(x - j\beta)\,dx \tag{9.53}$$

Equation (9.50) gives an expression for $P$ valid for any real $\beta > 0$.

### 9.3.2 Upper Bound on Error Probability

Since the absolute value of an integral is upper bounded by the integral of the absolute value of the integrand,

$$I = |I| \le \frac{1}{2\pi} \int_{-\infty}^{\infty} \frac{\exp(-x^2\sigma^2/2)}{(\beta^2 + x^2)^{1/2}} |\Phi_\eta(x - j\beta)|\,dx \tag{9.54}$$

From (9.45),

$$\Phi_\eta(x - j\beta) = \langle \exp(j\eta x) \exp(\eta\beta) \rangle \tag{9.55}$$

and

$$|\Phi_\eta(x - j\beta)| \le \langle \exp(\eta\beta) \rangle \stackrel{\triangle}{=} \Lambda_\eta(\beta) \tag{9.56}$$

where $\Lambda_\eta(\cdot)$ is the moment-generating function of $\eta$. Equations (9.51), (9.53) and (9.55) yield

$$
\begin{aligned}
I &\leq \frac{\Lambda_\eta(\beta)}{2\pi} \int_{-\infty}^{\infty} \frac{\exp(-x^2\sigma^2/2)}{(\beta^2 + x^2)^{1/2}} \, dx \\
&\leq \frac{\Lambda_\eta(\beta)}{2\pi\beta} \int_{-\infty}^{\infty} \exp(-x^2\sigma^2/2) \, dx \\
&= \frac{\Lambda_\eta(\beta)}{\sqrt{2\pi}\beta\sigma}
\end{aligned}
\tag{9.57}
$$

and

$$
\begin{aligned}
P &\leq \frac{1}{\sqrt{2\pi}\beta\sigma} \exp(-\Lambda\beta + \sigma^2\beta^2/2)\Lambda_\eta(\beta) \\
&\stackrel{\triangle}{=} Q_u > 0
\end{aligned}
\tag{9.58}
$$

Since the bound is valid for any positive $\beta$, choose $\beta$ to minimize the bound. Since the optimization of $Q_u$ with respect to $\beta$ leads to a complicated equation, we choose to minimize $\beta Q_u$. The quantity $\beta Q_u$ reaches its minimum when $\beta$ is a solution of

$$
\lambda = \beta\sigma^2 + \frac{\Lambda_\eta'(\beta)}{\Lambda_\eta(\beta)} = \beta\sigma^2 + \frac{d}{d\beta} \ln \Lambda_\eta(\beta)
\tag{9.59}
$$

For $\lambda > 0$, Section 9.36 shows that the unique solution of (9.58), which is a strictly monotonic function of $\beta$, lies between 0 and $\lambda/\sigma^2$.

Bound $Q_u$ multiplied by the factor $\sqrt{2\pi}\,\beta\sigma$ is identical to a quantity $Q_c$ that is obtained by Chernoff bounding techniques. However, for $\eta \to 0$, $\beta \to \lambda/\sigma^2$ and the Chernoff bound $Q_c$ goes only to $\exp(-\lambda^2/2\sigma^2)$; our bound $Q_u$ goes to $(1/\sqrt{2\pi}\,(\lambda/\sigma)) \exp(-\lambda^2/2\sigma^2)$, which is asymptotically equal to $\frac{1}{2}$ erfc $[\lambda/\sigma\sqrt{2}]$, the exact probability of error due to noise alone of the canonical system. For this reason, we choose to call our bound the modified Chernoff bound rather than Chernoff bound.

In most cases of interest, $\beta\sigma$ is large for SNR and even for moderately large interference bound $Q_u$ is tight.

### 9.3.3 Lower Bound on Error Probability

From (9.52) and (9.54),

$$
\begin{aligned}
I = \frac{1}{2\pi} E\Bigg[ &\exp(\eta\beta) \\
&\cdot \int_{-\infty}^{\infty} \frac{\beta \cos\{x[\eta - (\lambda - \beta\sigma^2)]\} + x \sin\{x[\eta - (\lambda - \beta\sigma^2)]\}}{(\beta^2 + x^2)} \\
&\cdot \exp(-x^2\sigma^2/2) \, dx \Bigg]
\end{aligned}
\tag{9.60}
$$

Since

$$\cos(\theta) \geq 1 - \frac{\theta^2}{2} \qquad \text{for } all \; \theta \qquad (9.61)$$

and

$$\sin(\theta) \geq -1 \qquad \text{for } all \; \theta \qquad (9.62)$$

$$I \geq \frac{\Lambda_\eta(\beta)}{2\pi} \int_{-\infty}^{\infty} \frac{\beta(1 - A_2 x^2) - |x|}{\beta^2 + x^2} \exp(-x^2\sigma^2/2) \, dx$$

$$\geq \Lambda_\eta(\beta) \left[ (1 + \beta^2 A_2) \frac{1}{2} \exp\left(\frac{\beta^2\sigma^2}{2}\right) \text{erfc}(\beta\sigma/\sqrt{2}) \right. \qquad (9.63)$$

$$\left. - \frac{\beta^2 A_2}{\sqrt{2\pi}\beta\sigma} - \frac{1}{\pi} \frac{1}{(\beta\sigma)^2} \right]$$

where

$$A_2 = \frac{1}{2} \frac{1}{\Lambda_\eta(\beta)} \{\Lambda_\eta''(\beta) - 2(\lambda - \beta\sigma^2)\Lambda_\eta'(\beta) + (\lambda - \beta\sigma^2)^2\Lambda_\eta(\beta)\} \qquad (9.64)$$

Because

$$\text{erfc}(t) \geq \frac{2}{\sqrt{\pi}} \frac{\exp(-t^2)}{t + \sqrt{t^2 + 2}}, \qquad t \geq 0 \qquad (9.65)$$

Equations (9.51), (9.62), and (9.64) yield

$$P \geq Q_1 \qquad (9.66)$$

where

$$Q_l = Q_m = \frac{1}{\sqrt{2\pi}\beta\sigma} \exp\left(-\beta\lambda + \frac{\beta^2\sigma^2}{2}\right) \Lambda_\eta(\beta)$$

$$\left\{ \frac{2}{1 + [1 + 4/(\beta\sigma)^2]^{1/2}} - \beta^2 A_2 \frac{[1 + 4/(\beta\sigma)^2]^{1/2} - 1}{[1 + 4/(\beta\sigma)^2]^{1/2} + 1} - \frac{2}{\sqrt{2\pi}\beta\sigma} \right\}$$

$$= Q_u \left\{ \frac{2}{1 + [1 + 4/(\beta\sigma)^2]^{1/2}} - \beta^2 A_2 \frac{[1 + 4/(\beta\sigma)^2]^{1/2} - 1}{[1 + 4/(\beta\sigma)^2]^{1/2} + 1} - \frac{2}{\sqrt{2\pi}\beta\sigma} \right\}$$

if $Q_m \geq 0$

$$Q_1 = 0, \qquad \text{otherwise} \qquad (9.67)$$

We shall assume that $\beta\sigma$ is such that $Q_m > 0$ so that $Q_l = Q_m$. From (9.66), for $\beta\sigma \rightarrow \infty$ and bounded $A_2$,

$$\frac{Q_u}{Q_l} \rightarrow 1 \qquad (9.68)$$

and the bounds are exponentially tight.

Again the lower bound is valid for any $\beta > 0$. The lower bound or the difference between the upper and lower bounds can be optimized to yield the tightest results; but this leads to a complex equation. We choose to use $\beta$ in (9.58) to simplify the bounds. For this $\beta$

$$
\begin{aligned}
0 \leq A_2 &= \frac{1}{2} \frac{\Lambda_\eta''(\beta)\Lambda_\eta(\beta) - \Lambda_\eta'^2(\beta)}{\Lambda_\eta^2(\beta)} \\
&= \frac{1}{2} \frac{d^2}{d\beta^2} \ln \Lambda_\eta(\beta).
\end{aligned}
\tag{9.69}
$$

### 9.3.4 The Use of Upper and Lower Bounds

The bounds discussed above are simple functions that involve only the moment generating function $\Lambda_\eta(\beta)$ of the interference. If $\Lambda_\eta(\beta)$ and its first two derivatives can be determined, we can simply evaluate upper and lower bounds after computing optimized $\beta$ from (9.58). Since the unique solution of (9.58) lies between 0 and $\lambda/\sigma^2$, conventional methods such as bisection, Newton-Raphson, and so on, can be used to evaluate $\beta$. If we use

$$
P_{\text{approx}} = \frac{Q_u + Q_l}{2} \overset{\triangle}{=} Q_d
\tag{9.70}
$$

to estimate the error probability,

$$
\begin{aligned}
Q_u &= \frac{1}{\sqrt{2\pi}\beta\sigma} \exp\left(-\beta\lambda + \beta^2\sigma^2/2\right)\Lambda_\eta(\beta) \\
&\geq P \\
&\geq Q_u \left\{ \frac{2}{1 + [1 + 4/(\beta\sigma)^2]^{1/2}} \right. \\
&\quad \left. - \beta^2 A_2 \frac{[1 + 4/(\beta\sigma)^2]^{1/2} - 1}{[1 + 4/(\beta\sigma)^2]^{1/2} + 1} - \frac{2}{\sqrt{2\pi}\beta\sigma} \right\} \\
&= Q_l
\end{aligned}
\tag{9.71}
$$

the fractional error, $\Delta_d$, relative to $Q_d$ is

$$
\begin{aligned}
\Delta_d &= \frac{Q_d - Q_l}{Q_d} \\
&= \frac{Q_u - Q_l}{Q_u + Q_l}
\end{aligned}
\tag{9.72}
$$

Note that

$$
P = Q_d(1 \pm \Delta_d)
\tag{9.73}
$$

and the maximum percentage error in approximating $P$ by $Q_d$ can be estimated from the value of $\Delta_d$. For bounded $A_2$, $\Delta_d \to 0$ and the bounds become very tight for $\beta\sigma \to \infty$.

### 9.3.5 Some Examples

**Probability of Error Due to CCI**

The CCI $\eta$ in a canonical binary system with $K$ independent interferers is

$$\eta = \sum_{i=1}^{K} R_i \cos \theta_i \tag{9.74}$$

where $R_i$ is the amplitude of the $i$th cochannel interferer and $\theta_i$ is uniformly distributed over the range $[0, 2\pi]$. From (9.55),

$$\Lambda_\eta(\beta) = \left\langle \exp \left\{ \beta \sum_{i=1}^{K} R_i \cos \theta_i \right\} \right\rangle = \prod I_0(\beta R_i) \tag{9.75}$$

where $I_n(\cdot)$ is the modified Bessel function of the first kind and of order $n$. The parameter $\beta$ is the unique solution of

$$\lambda = \beta \sigma^2 + \sum_{i=1}^{K} \frac{I_1(\beta R_i)}{I_0(\beta R_i)} \tag{9.76}$$

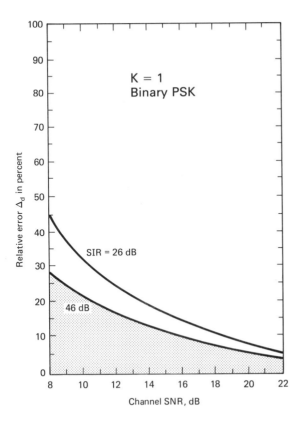

**Figure 9.5** For binary system with CCI, the maximum relative error in approximating probability of error by arithmetic mean of upper and lower bounds.

and

$$A_2 = \frac{1}{2} \sum_{i=1}^{K} R_i^2 \frac{I_0^2(\beta R_i) - I_1^2(\beta R_i) - I_0(\beta R_i)I_1(\beta R_i)/(\beta R_i)}{I_0^2(\beta R_i)} \qquad (9.77)$$

For a single interferer, we plot the maximum fractional error $\Delta_d$ in Fig. 9.5. Note that $\Delta_d$ decreases rather rapidly for increasing values of the SNR and $Q_d$ becomes a very good approximation to $P$.

### Multiple Interferers in 64 *QAM*

As pointed out in Section 9.2.6, the analysis given in this chapter can be extended to analyze the performance of M-ary QAM systems, $M \geq 4$. The details of these extensions are given in [Prabhu, 9.58].

For a 64-QAM signal with $K$ independent but equal sinusoidal interferers, the probability of error $Px$ on one rail (sine or cosine channel) can be shown to be

$$Px = 2\left(1 - \frac{1}{L}\right) P, \qquad L = 2 \qquad (9.78)$$

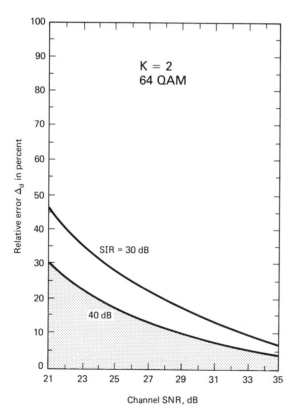

**Figure 9.6** For a 64-QAM system with two CCIs, the maximum relative error in approximating probability of error by arithmetic mean of upper and lower bounds.

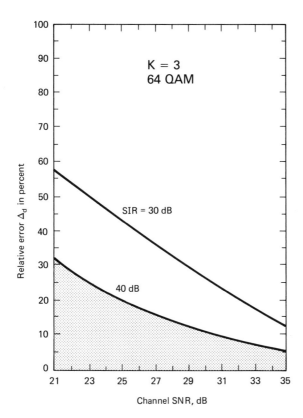

**Figure 9.7** For a 64-QAM system with three CCIs, the maximum relative error in approximating probability of error by arithmetic mean of upper and lower bounds.

where

$$\frac{\lambda}{\sigma\sqrt{2}} = \left(\frac{3 \cdot 10^{\text{SNR}/10}}{2 \cdot (L^2 - 1)}\right)^{1/2}, \qquad \text{SNR given in dB} \qquad (9.79)$$

$$R_m = \left(\frac{2(L^2 - 1)}{3 \cdot 10^{\text{SIR}_m/10}}\right)^{1/2}, \qquad \text{SIR}_m \text{ given in dB} \qquad (9.80)$$

The parameter SNR (in decibels) is the signal-to-noise ratio of the receiver and $\text{SIR}_m$ (in decibels) is the signal-to-interference ratio of each of the individual interferers.

For $K = 2$, 3, and 4, we plot $\Delta_d$ in Fig. 9.6, 9.7, and 9.8. Notice that $\Delta_d$ decreases rapidly for increasing values of the SNR, a very desirable feature. Also, observe the change in curvature of $\Delta_d$ as $K$ increases.

### 9.3.6 Optimum Solution of Parameter β

We need to solve (9.58) to determine the optimum value of β. The solution of (9.58) is also the solution of the equation

$$G(\beta) \stackrel{\triangle}{=} \beta\sigma^2 + \frac{\Lambda'_\eta(\beta)}{\Lambda_\eta(\beta)} - \lambda = 0 \qquad (9.81)$$

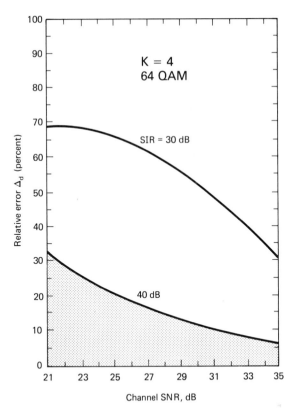

**Figure 9.8** For a 64-QAM system with four CCIs, the maximum relative error in approximating probability of error by arithmetic mean of upper and lower bounds.

Now, $\beta > 0$; and since we assume that the mean of $\eta$ is zero,

$$G(0) = \frac{\Lambda'_\eta(0)}{\Lambda_\eta(0)} - \lambda = -\lambda \tag{9.82}$$

Also,

$$G'(\beta) = \sigma^2 + \frac{\Lambda_\eta(\beta)\Lambda''_\eta(\beta) - \Lambda'^2_\eta(\beta)}{\Lambda^2_\eta(\beta)} \tag{9.83}$$

and

$$\Lambda'_\eta = E[\eta \exp(\beta\eta)] \tag{9.84}$$

$$\Lambda''_\eta = E[\eta^2 \exp(\beta\eta)] \tag{9.85}$$

Since

$$\left[ \eta - \frac{E[\eta \exp(\beta\eta)]}{E[\exp(\beta\eta)]} \right] \geq 0, \quad \text{for all real } \beta \tag{9.86}$$

and

$$\exp(\beta\eta) \geq 0, \text{ for all real } \beta \tag{9.87}$$

$$E[\eta^2 \exp(\beta\eta)] - \frac{E[\eta \exp(\beta\eta)]}{E\exp(\beta\eta)} \geq 0 \tag{9.88}$$

From (9.82)–(9.84) and (9.87)

$$\frac{\Lambda_\eta(\beta)\Lambda_\eta'' - \Lambda_\eta'^2(\beta)}{\Lambda_\eta^2(\beta)} \geq 0 \tag{9.89}$$

and

$$G'(\beta) > \sigma^2 \tag{9.90}$$

The function $G(\beta)$ is, therefore, a strictly monotonic increasing function of $\beta$. Since $G(0)$ is negative and $G(\lambda/\sigma^2)$ is positive, $G(\beta)$ has a unique solution between 0 and $\lambda/\sigma^2$.

## 9.4 UPPER BOUND WITH PEAK-LIMITED INTERFERENCE
[*Glave, 9.25*]

To compute the probability of error in a binary canonical system by the series method, we need to compute all moments of interference, $\eta$. To use modified Chernoff bounds, we need to determine (or bound) the moment-generating function. Since the evaluation of moments or the determination of the moment-generating function can be quite complex and requires detailed knowledge about the interference, we next develop a simple bound, called the peak-limited bound. It is only an upper bound on $P$ and requires two simple parameters for its evaluation.

The concept of the peak-limited bound is quite simple. It is to determine, of all admissible interference probability distribution functions that have given rms and peak values, the one that produces the largest $P$. It has been shown that a maximizing distribution can be found, and it is simple enough so that $P$ is easily evaluated for that distribution to produce the bound.

### 9.4.1 Optimization Theory

As before, the probability of error $P$ can be written as

$$P = Pr(n + \eta > \lambda) \tag{9.91}$$

Let us assume that interference $\eta$ is peak-limited and never exceeds the value $A$, or

$$|\eta| \leq A \tag{9.92}$$

Also, the variance of $\eta$ is

$$E[\eta^2] \leq \sigma_\eta^2 \tag{9.93}$$

Usually, the value of $A$ can be estimated and the value of $\sigma_\eta^2$ can be determined either from the power spectrum or from measurements.

We can derive that

$$P = Pr[n + \eta > \lambda] \le Pr[|n + \eta| > \lambda] \triangleq P_g \qquad (9.94)$$

From (9.93) and performing the integration over the Gaussian random variable, we can show that

$$P_g = \int_{-\infty}^{\infty} \frac{1}{2} \left\{ \text{erfc} \left[ \frac{\lambda - \eta}{\sigma\sqrt{2}} \right] + \text{erfc} \left[ \frac{\lambda + \eta}{\sigma\sqrt{2}} \right] \right\} dF(\eta) \qquad (9.95)$$

where $dF(\cdot)$ is the probability distribution function of $\eta$.

Let

$$q(\eta) = \frac{1}{2} \left\{ \text{erfc} \left[ \frac{\lambda - \eta}{\sigma\sqrt{2}} \right] + \text{erfc} \left[ \frac{\lambda + \eta}{\sigma\sqrt{2}} \right] \right\} \qquad (9.96)$$

so that

$$I(F) = \int_{-\infty}^{\infty} q(\eta) \, dF(\eta) \qquad (9.97)$$

Let $\zeta$ be the class of all distribution functions whose support is contained in $[-A, A]$. Then $I(F)$ is a functional on the space $\zeta$ to the real numbers $R$.

Our maximization problem is to find that $F_o \in \zeta$ which maximizes $I(F)$, if a maximum exists, over all $\zeta$ subject to the constraint

$$\int_{-\infty}^{\infty} \eta^2 \, dF(\eta) \le \sigma_\eta^2 \qquad (9.98)$$

The maximization of $I(F)$ over all allowable distribution functions can be solved by the methods of calculus of variations, but the problem can be solved elegantly by two theorems given by Smith [Smith, 9.16].

We shall need the following.

THEOREM 1. Let

$$P_g = \sup_{F \in \zeta} I(F) \qquad (9.99)$$

If $I$ is a convex cap weakly differentiable continuous mapping from $\zeta$ into the reals, then $P_g$ is achieved by a random variable $\eta_o$ with distribution function $F_o \in \zeta$. A necessary and sufficient condition for $F_o$ to achieve $P_g$ is that for some constant $k > 0$,

$$\int_{-\infty}^{\infty} [q(\eta) - k\eta^2] \, dF(\eta) \le I(F_O) - k\sigma_\eta^2, \qquad \text{for } all \ F \in \zeta \qquad (9.100)$$

COROLLARY 1. Let $F_o$ be an arbitrary distribution function in $\zeta$ such that

$$\int_{-\infty}^{\infty} \eta^2 \, dF_O(\eta) \le \sigma_\eta^2 \qquad (9.101)$$

Let $E_o$ denote the points of increase of $F_o$ on $[-A, A]$. Then $F_o$ is optimal if and only if for some $k > 0$

$$q(\eta) \le I(F_o) + k(\eta^2 - \sigma_\eta^2), \qquad \text{for } all \ \eta \ \epsilon [-A, A] \qquad (9.102)$$

$$q(\eta) = I(F_o) + k(\eta^2 - \sigma_\eta^2), \qquad \text{for } all \ \eta \ \epsilon \ E_o \qquad (9.103)$$

THEOREM 2.   If $q(\eta)$ is analytic on $[-A, A]$ then $E_o$ is a finite set of points.

A discussion and proof of these theorems and their proofs are given by Smith. We will not repeat them here.   The importance of the results lies in the fact that the random variable that maximizes $I(F)$ is discrete and takes on only a finite number of values.   This makes the search for the maximizing distribution at least a tractable computational problem.

### 9.4.2 Peak-Limited Upper Bound

Glave [Glave, 9.25] has shown that we can intuitively pick a candidate for $F_o$ and verify that the necessary and sufficient condition of Corollary 1 are satisfied.   Further, he has proved Theorem 3.

THEOREM 3.   If $n$ is a Gaussian random variable with zero-mean and variance $\sigma^2$, then $Pr\,[|n + \eta| \ge \lambda]$ is a maximum over all random variables $\eta$ satisfying conditions (i) and (ii) when $\eta$ has a distribution given by

$$F_o(\eta) = \frac{\sigma_\eta^2}{2A^2} u_{-1}(\eta - A) + \left(1 - \frac{\sigma_\eta^2}{A^2}\right) u_{-1}(\eta) + \frac{\sigma_\eta^2}{2A^2} u_{-1}(\eta + A) \quad (9.104)$$

provided

$$(\lambda - A) \ge \sqrt{3}\sigma \qquad (9.105)$$

In (9.103), $u_{-1}$ represents the unit step function.   The corresponding bound on $Pr\,[|n + \eta|]$ is then

$$Pr\,[|n + \eta| \ge \lambda] \ge \frac{\sigma_\eta^2}{2A^2}\left[\text{erfc}\left(\frac{\lambda - A}{\sigma}\right) + \text{erfc}\left(\frac{\lambda + A}{\sigma}\right)\right]$$
$$+ \left(1 - \frac{\sigma_\eta^2}{A^2}\right)\text{erfc}\left(\frac{\lambda}{\sigma}\right) \qquad (9.106)$$

Note that $(\lambda - A)$ is essentially the eye opening and hence that condition (9.104) requires that the effective eye opening be almost two standard deviations of the noise.   Normalizing $\lambda$ to unity and assuming, say a 50% opening, this condition requires that the peak SNR be greater than 11 dB.   Most systems operate at SNRs greater than this and thus condition (9.104) does not seem to be unduly restrictive.

*Proof.*   Given the distribution function of (9.103) and using Corollary 1 we must show that for some $k > 0$

$$\frac{1}{2}\left\{\text{erfc}\left(\frac{\lambda - \eta}{\sigma}\right) + \text{erfc}\left(\frac{\lambda + \eta}{\sigma}\right)\right\} \leq \frac{\sigma_\eta^2}{2A^2}\left[\text{erfc}\left(\frac{\lambda - A}{\sigma}\right) + \text{erfc}\left(\frac{\lambda + A}{\sigma}\right)\right]$$

$$+ \left[1 - \frac{\sigma_\eta^2}{A^2}\right]\text{erfc}\left(\frac{\lambda}{\sigma}\right) + k(\eta^2 - \sigma_\eta^2)$$

$$\text{for } all \ \eta \in [-A, A] \qquad (9.107)$$

with equality at the points $\eta = 0, \pm A$. We can pick a value of $k$ by setting $\eta = 0$. Thus

$$k = \frac{1}{2A^2}\left[\text{erfc}\left(\frac{\lambda - A}{\sigma}\right) + \text{erfc}\left(\frac{\lambda + A}{\sigma}\right) - \text{erfc}\left(\frac{\lambda}{\sigma}\right)\right] \qquad (9.108)$$

Using this value for $k$, we must show that

$$\frac{1}{2}\left[\text{erfc}\left(\frac{\lambda - \eta}{\sigma}\right) + \text{erfc}\left(\frac{\lambda + A}{\sigma}\right) - 2\ \text{erfc}\left(\frac{\lambda}{\sigma}\right)\right] \leq \frac{\eta^2}{2A^2}\left[\text{erfc}\left(\frac{\lambda - A}{\sigma}\right)\right.$$

$$\left. + \text{erfc}\left(\frac{\lambda + A}{\sigma}\right) - 2\ \text{erfc}\left(\frac{\lambda}{\sigma}\right)\right], \qquad \text{for all } \eta \in [-A, A]$$

$$(9.109)$$

with equality at the points of increase of $F(\eta)$, namely, at the points $\eta = -A, 0, +A$. This inequality constraint can be verified by inspection. Thus it remains to verify the inequality for all other points in $[-A, A]$. Since each side of (9.106) is even about $\eta = 0$, it suffices to verify the inequality for $\eta \in [0, A]$.

Let

$$g(\eta) = \frac{1}{2}\ \text{erfc}\left(\frac{\lambda - \eta}{\sigma}\right) + \frac{1}{2}\ \text{erfc}\left(\frac{\lambda + \eta}{\sigma}\right) - \text{erfc}\left(\frac{\lambda}{\sigma}\right) \qquad (9.110)$$

Hence, we must show that

$$g(\eta) \leq \frac{\eta^2}{A^2}\ g(A), \qquad \text{for all } \eta \in [0, A] \qquad (9.111)$$

Differentiating $g$ four times we find

$$g^{(3)}(\eta) = \frac{d^3 g(\eta)}{d\eta^3}$$

$$= \frac{1}{\sqrt{2\pi}\sigma^3}\left\{\left[\left(\frac{\lambda - \eta}{\sigma}\right)^2 - 1\right]\exp\left[-\frac{(\lambda - \eta)^2}{2\sigma^2}\right]\right. \qquad (9.112)$$

$$\left. - \left[\left(\frac{\lambda + \eta}{\sigma}\right)^2 - 1\right]\exp\left[-\frac{(\lambda + \eta)^2}{2\sigma^2}\right]\right\}$$

$$g^{(4)}(\eta) = \exp\left[-\frac{(\lambda - \eta)^2}{2\sigma^2}\right] \frac{(\lambda - \eta)}{\sqrt{2\pi}\sigma^5}\left[\left(\frac{\lambda - \eta}{\sigma}\right)^2 - 3\right]$$
$$+ \exp\left[-\frac{(\lambda + \eta)^2}{2\sigma^2}\right] \frac{(\lambda + \eta)}{\sqrt{2\pi}\sigma^5}\left[\left(\frac{\lambda + \eta}{\sigma}\right)^2 - 3\right] \tag{9.113}$$

Clearly, $g^{(4)}$ will certainly be positive for all $\eta \in [0, A]$, provided

$$\lambda - A \geq \sqrt{3}\,\sigma \tag{9.114}$$

Hence, under this condition, $g^{(3)}(\eta)$ is monotonically increasing on $[0, A]$ and since $g^{(3)}(0) = 0$, we have

$$g^{(3)}(\eta) \geq 0, \qquad \eta \in [0, A] \tag{9.115}$$

Using the integral remainder form of the Taylor series expansion for $g(\eta)$,

$$g(\eta) = \frac{\eta^2}{2} g^{(2)}(0) + \int_0^\eta \frac{(\eta - t)^2}{2} g^{(3)}(t)\, dt \tag{9.116}$$

and the inequality which we must verify becomes

$$\int_0^\eta \frac{(\eta - t)^2}{2} g^{(3)}(t)\, dt \leq \frac{\eta^2}{A^2} \int_0^A \frac{(A - t)^2}{2} g^{(3)}(t)\, dt, \qquad \eta \in [0, A] \tag{9.117}$$

Since $g^{(3)}(t) \geq 0$, for all $\eta$ in $[0, A]$,

$$\int_0^\eta \frac{(\eta - t)^2}{2} g^{(3)}(t)\, dt \leq \int_0^\eta \frac{\eta^2}{2}\left(1 - \frac{t}{A}\right)^2 g^{(3)}(t)\, dt$$
$$\leq \int_0^A \frac{\eta^2}{2}\left(1 - \frac{t}{A}\right)^2 g^{(3)}(t)\, dt \tag{9.118}$$
$$= \int_0^A \frac{\eta^2}{A^2} \frac{(A - t)^2}{2} g^{(3)}(t)\, dt$$

Hence, $g(\eta) \leq [\eta^2/A^2]\, g(A)$ for all $\eta$ in $[0, A]$.

This completes the proof of Theorem 3.

Since the use of this bound is rather obvious, we shall not illustrate it by giving examples. If the interference is small, the bound is tight. The reader is referred to [Rosenbaum, 9.41; and Glave, 9.47] for their use in CPSK systems.

## REFERENCES

[9.1] Morse, P. M., and H. Feshback. *Methods of Theoretical Physics*, McGraw-Hill, New York, 1953, pp. 786–787.

[9.2] Saltzberg, B. R. "Error probabilities for a binary signal perturbed by Intersymbol

Interference and Gaussian Noise," *IEEE Trans. Commun. Systems*, Vol. CS-12, March, 1964, pp. 117–120.

[9.3]   Abramowitz, M., and I. A. Stegun. *Handbook of Mathematical Functions*, National Bureau of Standards, Washington, D.C., March, 1965.

[9.4]   Aaron, M. R., and D. W. Tufts. "Intersymbol interference and error probability," *IEEE Trans. Inform. Theory*, Vol IT-12, January, 1966, pp. 26–34.

[9.5]   Saltzberg, B. R. "Intersymbol interference error bounds with applications to ideal bandlimited signaling," *IEEE Trans. Inform. Theory*, Vol. IT-14, July, 1968, pp. 563–568.

[9.6]   Saltzberg, B. R., and M. K. Simon. "Data transmission error probabilities in the presence of low-frequency removal and noise," *Bell Syst. Tech. J.*, Vol. 48, January, 1969, pp. 255–273.

[9.7]   Rosenbaum, A. S. "PSK error performance with Gaussian noise and interference," *Bell Syst. Tech. J.*, Vol. 48, No. 2, February, 1969, pp. 413–442.

[9.8]   Prabhu, V. K. "Error rate considerations for coherent phase-shift keyed systems with co-channel interference," *Bell Syst. Tech. J.*, Vol. 48, No. 3, March, 1969, pp. 743–767.

[9.9]   Prabhu, V. K. "Error probability upper bound for coherently detected PSK signals with co-channel interference," *Electronics Letters*, Vol. 5, No. 16, August 7, 1969, pp. 383–385.

[9.10]  Lugannani, R. "Intersymbol interference and probability of error in digital systems," *IEEE Trans. Inform. Theory*, Vol. IT-15, November, 1969, pp. 682–688.

[9.11]  Rosenbaum, A. S. "Binary PSK error probability with multiple co-channel interferences," *IEEE Trans. Commun.*, Vol. COM-18, No. 3, June, 1970, pp. 241–253.

[9.12]  Ho, E. Y., and Y. S. Yeh. "A new approach for evaluating the error probability in the presence of intersymbol interference and additive Gaussian noise," *Bell Syst. Tech J.*, Vol. 49, November, 1970, pp. 2249–2266.

[9.13]  Rosenbaum, A. S. "Error performance of multiphase DPSK with noise and interference," *IEEE Trans. Commun.*, December, 1970, pp. 821–824.

[9.14]  Ho, E. Y., and Y. S. Yeh. "Error probability of a multilevel digital system with intersymbol interference and Gaussian noise," *Bell Syst. Tech. J.*, Vol. 50, March, 1971, pp. 1017–1023.

[9.15]  Shimbo, O., and M. I. Celebiler. "The probability of error due to intersymbol interference and Gaussian noise in digital communication systems," *IEEE Trans. Commun.*, Vol. COM-19, April, 1971, pp. 113–119.

[9.16]  Smith, J. G. "The information capacity of amplitude- and variance-constrained scalar Gaussian channels," *Information and Control*, Vol. 18, April, 1971, pp. 203–219.

[9.17]  Calandrino, L., G. Corazzo, G. Crippa, and G. Immovilli. "Intersymbol, interchannel and co-channel interferences in binary and quaternary PSK systems," *Alta Frequenza*, Vol. XL, May, 1971, pp. 407–420.

[9.18]  Prabhu, V. K. "Performance of coherent phase-shift keyed systems with intersymbol interference," *IEEE Trans. on Inform. Theory*, Vol. IT-17, No. 4, July, 1971, pp. 418–431.

[9.19]  Hill, F. S. "The computation of error probability for digital transmission," *Bell Syst. Tech. J.*, Vol. 50, July-August, 1971, pp. 2055–2077.

[9.20]  Goldman, J. "Multiple error performance of PSK Systems with co-channel interference and noise," *IEEE Trans. on Commun.*, Vol. COM-19, No. 4, August 1971, pp. 420–430.

[9.21] Yeh, Y. S., and E. Y. Ho. "Improved intersymbol interference error bounds in digital systems," *Bell Syst. Tech J.*, Vol. 50, October, 1971, pp. 2585–2598.

[9.22] Prabhu, V. K. "Some considerations of error bounds in digital systems," *Bell Syst. Tech. J.*, Vol. 50, December, 1971, pp. 3127–3151.

[9.23] Forney, G. D. "Lower bounds on error probability in the presence of large interference," *IEEE Trans. Comm.*, Vol. COM-20, February, 1972, pp. 76–77.

[9.24] Falconer, D. D., and R. D. Gitlin. "Bounds on error pattern probabilities for digital communication systems," *IEEE Trans. Commun.*, Vol. COM-20, April, 1972, pp. 132–139.

[9.25] Glave, F. E. "An upper bound on the probability of error due to intersymbol interference for correlated digital signals," *IEEE Trans. Inform. Theory*, Vol. IT-18, No. 3, May, 1972.

[9.26] Yao, K. "On minimum average probability of error expression for a binary pulse communication system with intersymbol interference," *IEEE Trans. Inform. Theory*, Vol. IT-18, July, 1972, pp. 528–531.

[9.27] Shimbo, O., R. Fang, and M. Celebiler. "Performance of M-ary PSK system in Gaussian noise and intersymbol interference," *IEEE Trans. Inform. Theory*, Vol. IT-19, No. 1, January, 1973, pp. 44–58 and Vol. IT-19, No. 3, May, 1973, p. 365 (corrections).

[9.28] Prabhu, V. K. "Error probability performance of M-ary CPSK systems, with intersymbol interference," *IEEE Trans. Commun.*, Vol. COM-21, No. 2, February, 1973, pp. 97–109.

[9.29] Benedetto, S., G. de Vincentis, and A. Luvison. "Error probability in the presence of intersymbol interference and additive noise for multilevel digital signals," *IEEE Trans. Commun.*, Vol. COM-21, March, 1973, pp. 181–190.

[9.30] Shimbo, O., and R. Fang. "Effects of co-channel interference and Gaussian noise in M-ary PSK system," *COMSAT Tech. Review*, Vol. 3, No. 1, Spring, 1973, pp. 183–207.

[9.31] Hill, F. S., and M. A. Blanco. "Random geometric series and intersymbol interference," *IEEE Trans. Inform. Theory*, Vol. IT-19, May, 1973, pp. 326–335.

[9.32] Aein, J. M., and R. D. Turner. "Effect of co-channel interference on CPSK carrier," *IEEE Trans. Commun.*, Vol. COM-21, July, 1973, pp. 783–790.

[9.33] Mathews, J. W. "Sharp error bounds for intersymbol interference," *IEEE Trans. Inform. Theory*, Vol. IT-19, No. 4, July, 1973, pp. 440–447.

[9.34] McGee, W. F. "A modified intersymbol interference error bound," *IEEE Trans. Commun.*, Vol. COM-21, July, 1973, pp. 862–863.

[9.35] Korn, I. "Probability of error in binary communication systems with causal bandlimited filters—Part 1: nonreturn-to-zero signal," *IEEE Trans. Commun.*, Vol. COM-21, August, 1973, pp. 878–890.

[9.36] Korn, I. "Probability of error in binary communication systems with causal bandlimited filters—Part 2: split-phase signal," *IEEE Trans. Commun.*, Vol. COM-21, August 1973, pp. 891–898.

[9.37] Benedetto, S., E. Biglieri, and V. Castellani. "Combined effects of intersymbol, interchannel and co-channel interferences in M-ary CPSK systems," *IEEE Trans. Commun.*, Vol. COM-21, No. 3, September, 1973, pp. 997–1008.

[9.38] Fang, R., and O. Shimbo. "Unified analysis of a class of digital systems in additive noise and interference," *IEEE Trans. Commun.*, Vol. COM-21, No. 10, October, 1973, pp. 1075–1091.

[9.39] Prabhu, V. K. "Intersymbol interference performance of systems with correlated digital signals," *IEEE Trans. Commun.*, Vol. COM-21, No. 10, October, 1973, pp. 1147–1152.

[9.40] Goodman, A. W. *Analytic Geometry and the Calculus,* MacMillan, New York, 1974, pp. 619–627.

[9.41] Rosenbaum, A. S., and F. E. Glave. "An error-probability upper bound for coherent phase-shift keying with peak-limited interference," *IEEE Trans. Commun.*, Vol. COM-22, No. 1, January 1974, pp. 6–16, and 356–363.

[9.42] Korn, I. "Bounds to probability of error in binary communication systems with intersymbol interference and dependent or independent symbols," *IEEE Trans. Commun.*, Vol. COM-22, No. 2, February, 1974, pp. 251–254.

[9.43] Vanelli, J. C., and N. M. Shehadeh. "Computation of bit error probability using the trapezoidal integration rule," *IEEE Trans. Commun.*, Vol. COM-22, March, 1974, pp. 331–334.

[9.44] McLane, P. J. "Lower bounds for finite intersymbol interference error rates," *IEEE Trans. Commun.*, Vol. COM-22, June, 1974, pp. 853–857.

[9.45] Korn, I. "Improvement to sharp error bounds for intersymbol interference," *Proc. IEE,* London, Vol. 122, March, 1975, pp. 265–267.

[9.46] McLane, P. J. "Error rate lower bounds for digital communications with multiple interference," *IEEE Trans. Commun.*, Vol. COM-23, May, 1975, pp. 539–543.

[9.47] Glave, F. E., and A. S. Rosenbaum. "An upper bound analysis for coherent phase-shift keying with co-channel, adjacent-channel and intersymbol interference," *IEEE Trans. Commun.*, Vol. COM-23, No. 6, June, 1975, pp. 586–597.

[9.48] Yao, K., and R. M. Tobin. "Moment space upper and lower bounds for digital systems with intersymbol interference," *IEEE Trans. Inform. Theory*, Vol. IT-22, January, 1976, pp. 65–74.

[9.49] Spowage, K. A. E. "An upper bound on the error probability for differential phase-shift keying in the presence of additive Gaussian white noise and peak-limited interference," *IEEE Trans. Commun.*, Vol. COM-24, No. 11, November, 1976, pp. 1276–1279.

[9.50] Tobin, R. M., and K. Yao. "Upper and lower error bounds for coherent phase-shift-keyed (CPSK) systems with coherent interference," *IEEE Trans. Commun.*, Vol. COM-25, No. 2, February, 1977, pp. 281–287.

[9.51] Jeruchim, M. C. "A survey of interference problems and applications to geostationary satellite networks," *Proc. IEEE,* Vol. 65, No. 3, March, 1977, pp. 317–331.

[9.52] Jenq, Y. C., B. Liu, and J. B. Thomas. "Probability of error in PAM systems with intersymbol interference and additive noise," *IEEE Trans. Inform. Theory*, Vol. IT-23, September, 1977, pp. 575–582.

[9.53] Prabhu, V. K. "PSK modulation with overlapping baseband pulses," *IEEE Trans. Commun.*, Vol. COM-25, No. 9, September, 1977, pp. 980–990.

[9.54] Celebiler, M. I., and G. M. Coupe. "Effects of thermal noise, filtering and co-channel interference in binary coherent PSK systems," *IEEE Trans. Commun.*, Vol. COM-26, No. 2, February, 1978, pp. 257–267.

[9.55] Krishnamurthy, J. "Bounds on the probability of error in digital communications," *IEEE Trans. on Aerospace and Electronic Systems*, Vol. AES-14, No. 2, March, 1978, pp. 284–291.

[9.56] Yue, O. C. "Saddle point approximation for the error probability in PAM systems with noise and interference," *IEEE Trans. Commun.*, Vol. COM-27, October, 1979, pp. 1604–1609.

[9.57] Stavroulakis, P. *Interference Analysis of Communication Systems*, IEEE Press, New York, 1980.

[9.58] Prabhu, V. K. "Detection efficiency of 16-ary QAM," *Bell Syst. Tech. J.*, Vol. 59, No. 4, April, 1980, pp. 639–656.

[9.59] Prabhu, V. K. "MSK and offset QPSK modulation with bandlimiting filters," *IEEE Trans. on Aerospace and Electronic Systems*, Vol. AES-17, No. 1, January, 1981, pp. 2–8.

[9.60] Prabhu, V. K. "Co-channel interference immunity of high capacity QAM," *Electronics Letters,* Vol. 17, No. 9, September, 1981, pp. 680–681.

[9.61] Feher, K., and T. LeNgoc. "Performance of IJF-OQPSK modulation schemes in the presence of noise, interchannel and co-channel interference," *Conf. Record, National Tel. Conf.*, New Orleans, La., November 29–December 3, 1981.

[9.62] Prabhu, V. K., and J. Salz. "On the performance of phase-shift-keying systems," *Bell Syst. Tech. J.*, Vol. 60, No. 10, December, 1981, pp. 2307–2343.

[9.63] Prabhu, V. K. "Modified Chernoff bounds for PAM systems with noise and interference," *IEEE Trans. Inform. Theory*, Vol. IT-28, No. 1, January, 1982, pp. 95–100.

[9.64] Prabhu, V. K. "Error rate bounds for differential PSK," *IEEE Trans. Commun.*, Vol. COM-30, No. 12, December, 1982, pp. 2547–2550.

[9.65] Coco, R. A. "Bit error rate curves for M-QAM signals with multiple sinusoid interferers," (in press).

[9.66] Coco, R. A. "Bit error rate curves for M-QAM signals with multiple co-channel interferers," (in press).

# 10

# MOBILE-RADIO COMMUNICATIONS

**DR. KENKICHI HIRADE**

*Head, Mobile Communications Applications Section*
*Yokosuka Electrical Communication Laboratory*
*Nippon Telegraph and Telephone Corporation*
*1-2356, Take, Yokosuka-shi, Kanagawa-ken, 238-03 Japan*

## 10.1 INTRODUCTION TO MOBILE RADIO

### 10.1.1 Historical Review

The ultimate objective of communications is to enable anyone to communicate instantly with anyone else from anywhere. This can be achieved only by mobile radio communications. Mobile radio communications started with the early experiments of radio pioneers towards the end of the nineteenth century. A well-known historical event that clearly showed the importance of mobile radio communications was the distress of the Titanic in 1912. In these early mobile radio communications, radio telegraph was dominant, in which Morse-coded on-off keying was used as the modulation scheme. It can, therefore, be seen that mobile radio communications started with digital technology. Radio telegraph was extensively used for ship-to-shore transmission during World War I. However, after World War I radio telephone, or analog voice transmission, began to play an important role along with radio telegraph, or digital data transmission.

This dichotomy of digital for telegraph and analog for telephone continued until the mid-1970s when digital voice transmission became more pervasive. Since the advent of mobile radio telegraph, various technological advances have brought the appearance of many other mobile radio communications systems, such as radio telephone, radio paging, emergency dispatch, navigation control, status reporting, and so on [Bowers, 10.1]. Demand for mobile radio communication services has steadily increased. Current mobile radio communication systems are limited to services for specialized groups because of the limited frequency spectrum allocated for mobile use.

In order to meet such increasing demands with a limited mobile radio-frequency (mobile RF) spectrum, it is necessary to alleviate the spectrum limitation by exploiting new higher frequency bands with inherently wider bandwidth. The 900-MHz band has emerged as a major mobile radio frontier of our technological age. Many advanced technologies and techniques have been developed to facilitate implementation of the new 900-MHz-band mobile radio communication systems with reasonable cost. In consequence, advanced high-capacity **cellular** mobile radio telephone systems using the 900-Mhz band have been developed in various countries [10.2]-[10.8]. [Ito, 10.2; Young, 10.3; Remy, 10.4; Calvert, 10.5; Haug, 10.6; Pfannschmidt, 10.7; Kammerlander, 10.8]. It is expected that the 900-MHz band will be the home for future portable radio-telephone systems [Mikulski, 10.9; Fisher, 10.10; Sakamoto, 10.11].

Cellular mobile radio systems enable high-density geographical cochannel reuse and are effective for achieving efficient spectrum utilization. These systems are adopted in advanced mobile radio system plans. Digital technologies are also effectively applied to achieve not only high-speed, highly reliable data transmission but also high-grade, highly flexible system control. However, in general, digital voice transmission has not been adopted yet for mobile radio systems.

Since the mid-1970s, the application of digital technology in terrestrial telecommunications networks has been expanding rapidly owing to innovative large-scale integration (LSI) circuit technologies. Consequently, the dichotomy of analog for voice transmission and digital for nonvoice transmission is now broken down. As described in Chapter 2, the digital revolution is now surging to construct the integrated services digital network (ISDN) on a global scale. It will also surge over mobile radio communication systems for providing voice and nonvoice services. Therefore, it is expected that future mobile radio communication systems will be integrated into the digital telecommunications network and a variety of effective services will be provided.

### 10.1.2 Digital Applications to Mobile Radio

The common objective of mobile radio systems is to provide high-quality communication services between a fixed land base or satellite station and a moving vehicle with reasonable cost and spectral economy. Digital applications to mobile-radio systems intend to achieve this objective by applying advanced digital technologies [CCIR, 10.12]. Let us now discuss the application of digital technologies to mobile-radio systems.

#### Efficient Spectrum Utilization

Efficient spectrum utilization is a problem on which most developmental studies of radio engineering have focused. Many technologies have been intensively studied and developed for utilizing efficiently the three components of spectrum: **frequency**, **time**, and **space**. Narrow-band transmission, multichannel access, and small cell layout are the three major solutions to conserving spectrum. Narrow-band transmission is achieved by low bit rate (LBR) coding, bandwidth-efficient modulation, and stabilized carrier source techniques. Multichannel access, which enables time-shared use of a channel, requires an electronic central processor using a stored-program-control scheme and microprocessor-controlled frequency-variable mobile-radio transceivers. Small cell layout, which makes possible geographical cochannel reuse with high density, is achieved by a high-grade system control. Diversity and error control, which enable improvement of the cochannel interference performance, are effective for achieving small cell layout.

### Highly-Secure Voice Transmission

In advanced cellular mobile radio telephone systems, the same frequency is reused in different geographic areas. Those areas are located sufficiently far away from each other so as to make the effect of cochannel interference (CCI) negligible. Nevertheless, intelligible crosstalk may occur due to the peculiarity of the mobile radio propagation path. Such crosstalk may cause serious problems for a number of mobile radio services. While the occurrence of crosstalk can surely be reduced by using a well-designed cochannel-assignment scheme, it is more desirable to adopt a *highly secure* voice-transmission technique which prevents intelligible interception. Digital voice transmission is effective for this purpose [Bettinger, 10.13; Sachs, 10.14].

### High-Speed Data Transmission

The ever-increasing use of computers, computer applications, and, in general, the large number of recently discovered digital signal processors have increased the demand for mobile, terrestrial and satellite data transmission services. The digital format is inherently effective for achieving a relatively high-speed data-transmission capability within a limited spectrum. For instance, many routine messages, such as identification code, status information, and location information can be transmitted reliably in less time when a digital format is used [Brenig, 10.15]. Thus channel congestion may be mitigated with the use of digital techniques. Moreover, mobile data-transmission techniques enable many users direct access to computer files to obtain the required information. They also provide means for nonvoice applications, such as alphanumeric and visual services that can easily be displayed or recorded. It is essential to ensure highly reliable, or virtually error-free, data transmission because wrong information is worse than no information.

### Highly Reliable Data Transmission

In the usual land mobile radio environment, signal transmission between a fixed base station and a moving vehicle is performed through random multiple propagation paths due to the effects of reflection, scattering, and diffraction. Therefore, as the vehicle moves, fast and deep fading—which is generally called **multipath fading**—appears on the received carrier. This fading degrades the signal-transmission performance. In order to design highly reliable mobile data-transmission systems, it is necessary to adopt auxiliary techniques such as error detection or correction coding. These techniques can be regarded as a certain kind of time-diversity technique peculiar to digital transmission [Lucky, 10.16; Peterson, 10.17; Mabey, 10.18]. However, we note that error correction-detection techniques require the addition of redundant bits into the original data stream, thus leading to a wider bandwidth requirement.

### High-Quality Voice Transmission

For digital voice transmission, a more-efficient time diversity technique, without any bandwidth expansion, can be applied. The erroneous part of the received voice signal is first removed and then replaced by an appropriate substitute adaptively generated by using the error-free parts of the signal. The erroneous part can easily be detected by monitoring the level of the received carrier subjected to multipath fading. Such an advanced time-diversity approach is achieved with the aid of digital signal processing (DSP) techniques [Duttweiler, 10.19; Murakami, 10.20].

### High-Grade System Control

In advanced cellular mobile radio-telephone systems, a high-grade and highly flexible distributed system control architecture is employed. The **distributed system functions** are partitioned into three major subsystems, the **mobile radio units**, the **cell site**, and the **central systems controller**, to perform the following functions:

1. Interconnection with the existing telephone network
2. Channel set-up for the land- or mobile-originated call
3. Hand-off for the vehicular movement between adjacent cells
4. Location registration from the roaming vehicle
5. Disconnection

Common-channel signaling and control, which are based on high-speed highly reliable data transmission and stored-program control, are indispensable for performing the above functions. Different signals may be simultaneously transmitted on a single carrier by using time division multiplexing (TDM) techniques peculiar to digital transmission. When both voice and data signals are present, then the in-service radio channel can easily be monitored and controlled without interruption, and the functions of system control can be upgraded.

## 10.2 DIGITAL MODULATION FOR MOBILE RADIO

In most commercial VHF and UHF (150–900 MHz) mobile radio systems, single channel per carrier (SCPC) analog FM transmission techniques are used for voice transmission. One of the most-important objectives of emerging digital voice transmission systems is to achieve the same bandwidth efficiency as that of conventional analog FM systems. Bandwidth-efficient digital voice-transmission systems, which use standard 25-kHz channel spacing, require a voice coding method having a bit rate of less than 16 kb/s and a bandwidth-efficient modulation technique with a spectral efficiency of about 1 bit/s/Hz. Low bit rate (LBR) voice-coding techniques are described in Chapter 3. Here we describe bandwidth-efficient modulation techniques for digital mobile radio systems applications.

### 10.2.1 Requirements for Mobile Use

For mobile radio applications, bandwidth-efficient modulation techniques should satisfy the following requirements [Jenks, 10.21]:

1. *Compact output power spectrum*: Out-of-band power radiated into the adjacent channel should be 60 to 80 dB below that in the desired channel. This is necessary to avoid severe adjacent channel interference (ACI), which causes performance degradation. As the vehicle moves, a fading of 40 to 60 dB may occur. Even under such deep fades a carrier-to-interference ratio (CIR) of about 20 dB should be maintained.

2. *Good probability-of-error* ($P_e$) *performance*: This implies high immunity to thermal noise and interference. Thus a low-transmitter power requirement and high-density geographical cochannel reuse may be obtained.

3. *Class*-C *nonlinear power amplifier*: This is necessary to save supplied dc power and improve the efficiency of the output stage and is especially important for miniaturizing mobile or portable radio units.

4. *Frequency-variable carrier source*: This is necessary to enable access to any one of the assigned mobile radio channels. A phase-locked frequency synthesizer with a programmable center frequency is usually used for this purpose [Manassewitsh, 10.22; Gardner, 10.23].

In order to satisfy the first requirement, the power spectrum of the modulated RF signal output must be appropriately filtered. However, this cannot be performed at the final RF stage in multichannel transceivers satisfying the fourth requirement, because no tunable RF filter is available that would satisfy the system requirements. Therefore, intermediate-frequency (IF) or baseband filtering with frequency upconversion is used. However, when a filtered IF signal is upconverted and amplified by a nonlinear class-C power amplifier, then spectral restoration (regrowth) occurs. In order to avoid spectral restoration, constant-envelope bandwidth-efficient modulation techniques, having compact-spectrum and a good $P_e$ performance property, have to be employed.

### 10.2.2 Premodulation-Filtered MSK

A number of power- and spectrally efficient modulation techniques are presented in Chapters 7 and 8. Here we focus attention on minimum shift-keying (MSK) modulation systems and particularly on Gaussian premodulation-filtered MSK (GMSK) systems. MSK, which is a digital FM or a continuous-phase frequency shift keying (FSK) modulation technique having a modulation index of 0.5, pos-

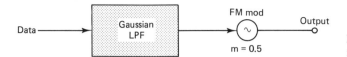

Data → Gaussian LPF → FM mod (m = 0.5) → Output

**Figure 10.1** Premodulation-filtered MSK modulator.

sesses the following attractive properties: constant envelope, fast roll-off of higher-order sidebands, compact spectrum, and coherent detection capability [Buda, 10.24; Pasupathy, 10.25]. However, it does not satisfy the properties of narrow main lobe with low out-of-band radiation. Since MSK signals can be generated by direct FM modulation, a narrow output power spectrum can be obtained by means of a premodulation low-pass filter (LPF), while keeping the constant envelope property; see Fig. 10.1. In order to obtain a compact output power spectrum, the premodulation LPF should meet the following conditions:

**1.** Narrow bandwidth and sharp cutoff
**2.** Low overshoot impulse response
**3.** Preservation of filter output pulse area to assure a $\pi/2$ phase shift

Conditions 1, 2, and 3 are needed to suppress higher-order frequency components, to prevent excessive instantaneous frequency deviation, and to assure the applicability of coherent detection, respectively. However, introduction of a premodulation LPF violates the minimum frequency spacing constraint and the fixed-phase transition constraint of simple MSK modems. Fortunately, these two constraints are not indispensable requirements for the applicability of coherent detection. Premodulation-filtered MSK signals can be detected coherently, if their pattern-averaged phase-transition trajectory does not deviate from that of simple MSK systems. Therefore, premodulation-filtered MSK may be regarded as an *advanced or generalized MSK*.

Sinusoidal FSK (SFSK), which employs a sinusoidal waveform shaping to smooth the phase-transition trajectory, is an example of advanced MSK systems [Amoroso, 10.26]. However, the output power spectrum of SFSK does not satisfy the strict requirement on the out-of-band radiation limits of mobile radio systems. Tamed FM (TFM) requires a specific premodulation LPF having a modified partial-response transfer function. This modulation technique [deJager, 10.27] as well as intersymbol-jitter-free (IJF) offset-quadrature PSK (IJF-OKQPSK) (Chapter 8) are among the likely candidates for applications of mobile-radio systems. For an in-depth description of additional constant-envelope and compact-spectrum modulation techniques, see [Feher, 10.28; Aulin, 10.29; Aulin, 10.30; Muilwijk, 10.31; Honma, 10.32; Akaiwa, 10.33; Asakawa, 10.34].

### 10.2.3 GMSK Modulation

A Gaussian low-pass filtered (LPF) MSK modulation, GMSK, which satisfies conditions 1, 2, and 3, was experimentally shown to be an attractive technique for digital mobile-radio systems applications [Murota, 10.35]. Some of the most important properties of the GMSK modulation technique are highlighted in the following sections.

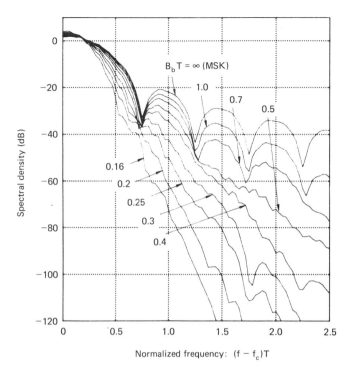

**Figure 10.2** Power spectral density of GMSK. $B_bT$ is the normalized 3-dB bandwidth of a premodulation Gaussian LPF; $T$ is the unit bit duration. (Printed by permission of the IEEE [Murota, 10.35]. © 1981 IEEE.)

### Output Power Spectrum

Figure 10.2 illustrates the output power spectrum of a GMSK modulator as a function of the normalized frequency difference from the carrier center frequency $(f - f_c)T$. The normalized 3-dB bandwidth of the premodulation Gaussian LPF, $B_bT$, is a parameter, where $T$ is the unit bit duration. An optimum condition for maximizing the spectral efficiency of GMSK cellular land mobile radio systems has been shown to be $B_bT \approx 0.25$ [Murota, 10.36].

Figure 10.3 illustrates the fractional power in the desired channel versus the normalized bandwidth of a predetection rectangular band-pass filter (BPF), $B_iT$. Table 10.1 shows the occupied bandwidth for a specified percentage of power, where $B_bT$ is a variable parameter. For comparison, the occupied bandwidth of TFM is also presented. We note that the GMSK system with $B_bT = 0.25$ requires only a slightly wider bandwidth than the more-complex TFM system.

Figure 10.4 illustrates computed results of the ratio of out-of-band radiation power falling into the adjacent channel to the total power in the desired channel. A normalized channel spacing $\Delta f \cdot T$ is used as the abscissa; both channels are assumed to have ideal rectangular band-pass characteristics with $B_iT = 1$ and are assumed to have equal power. The situation of $\Delta f \cdot T \le 1.5$ and $B_iT = 1$ corresponds to the case of $f = 25$ kHz and $B_i = 16$ kHz when $f_b = 1/T = 16$ kb/s. From this figure, we conclude that GMSK with $B_bT = 0.25$ can be adopted for VHF and UHF mobile-radio systems. In such SCPC systems an out-of-band radiation power to total power ratio of less than 60 dB is specified [CCIR, 10.37].

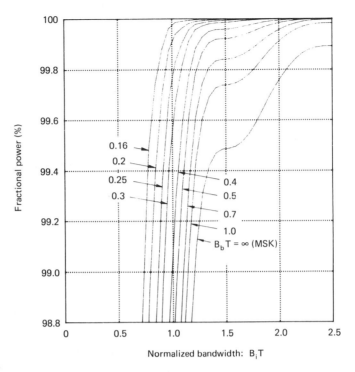

**Figure 10.3** Fractional power ratio of GMSK. $B_i$ is the bandwidth of an ideal rectangular predetection BPF; $B_bT$ is the normalized 3-dB bandwidth of a premodulation Gaussian LPF; $T$ is the unit bit duration. (Printed by permission of the IEEE [Murota, 10.35]. © 1981 IEEE.)

### $P_e$ Performance of Coherent Systems

Let us now consider the theoretical $P_e$ performance of the GMSK modulation technique. Coherent detection in the presence of additive white Gaussian noise (AWGN) is assumed.

**TABLE 10.1**  Occupied bandwidth normalized by bit rate for specified percentage power

| $B_bT$ | Power% 90 | 99 | 99.9 | 99.99 |
|---|---|---|---|---|
| 0.2 | 0.52 | 0.79 | 0.99 | 1.22 |
| 0.25 | 0.57 | 0.86 | 1.09 | 1.37 |
| 0.5 | 0.69 | 1.04 | 1.33 | 2.08 |
| MSK | 0.78 | 1.20 | 2.76 | 6.00 |
| TFM | 0.52 | 0.79 | 1.02 | 1.37 |

Occupied bandwidth

Printed by permission of the IEEE [Murota, 10.35]. At the end of this chapter, details of the reference are given. © 1981 IEEE.

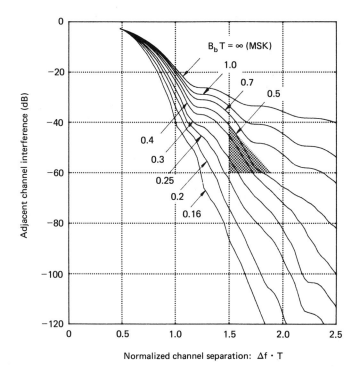

**Figure 10.4** ACI of GMSK systems. Shaded area indicates current specification limit. $B_iT = 1$ is assumed. (Printed by permission of the IEEE [Murota, 10.35]. © 1981, IEEE.)

Since GMSK is a binary digital modulation technique, its $P_e$ performance bound, for a high signal-to-noise ratio (SNR) may be approximately represented by [Wozencraft, 10.38]

$$P_e = \frac{1}{2} \, \text{erfc} \left( \frac{d_{\min}}{2\sqrt{N_0}} \right) \tag{10.1}$$

where $N_0$ is the power spectral density of the AWGN and erfc(·) is the complementary error function defined by

$$\text{erfc}(x) = \frac{2}{\sqrt{\pi}} \int_x^\infty \exp(-u^2) \, du \tag{10.2}$$

Furthermore, $d_{\min}$ is the minimum value of signal distance $d$ between mark and space in Hilbert space observed during the time interval from $t_1$ to $t_2$, where $d$ is defined by

$$d^2 = \frac{1}{2} \int_{t_1}^{t_2} |u_m(t) - u_s(t)|^2 \, dt \tag{10.3}$$

In (10.3), $u_m(t)$ and $u_s(t)$ are the complex signal waveforms corresponding to the mark and space signals, respectively. The $P_e$ performance bound given by (10.1) is attained only when an ideal maximum-likelihood detection scheme is employed.

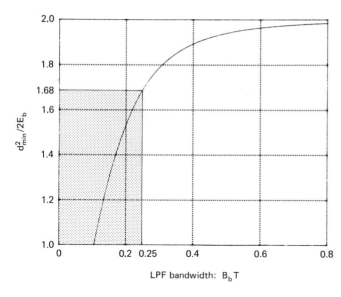

**Figure 10.5** Minimum signal distance of GMSK. (Printed by permission of the IEEE [Murota, 10.35]. © 1981, IEEE.)

However, we may use this equation as an approximate solution for the ideal $P_e$ performance of coherently detected GMSK systems.

Figure 10.5 illustrates computed values of $d_{\min}$ of the GMSK signal as a function of $B_b T$, where $E_b$ is the signal energy per bit defined by

$$E_b = \frac{1}{2} \int_0^T |u_m(t)|^2 \, dt = \frac{1}{2} \int_0^T |u_s(t)|^2 \, dt \qquad (10.4)$$

The case of $B_b T \to \infty$ corresponds to simple MSK that is MSK without a premodulation LPF. An ideal antipodal transmission system (conventional MSK) has $d_{\min} = 2\sqrt{E_b}$; see the top right-hand corner of Fig. 10.5. Due to the ISI effect of the phase transitions, the meaningful observation time interval for the GMSK signal $t_2 - t_1$ may be made longer than $2T_b$ [Murota, 10.39].

Substituting the computed values of $d_{\min}$ into (10.1), the $P_e$ performance of the coherently detected GMSK system is obtained [Murota, 10.39]. Figure 10.6 illustrates the performance degradation of the GMSK system from an ideal antipodal transmission system. This degradation is due to the ISI effect of the premodulation Gaussian LPF. From this figure, we conclude that the performance degradation is small and that the required increase of $E_b/N_0$ of GMSK system with $B_b T = 0.25$ is not more than 0.7 dB compared with that of an ideal antipodal (MSK) transmission system.

## 10.3 PROPAGATION CHARACTERISTICS OF LAND MOBILE RADIO

### 10.3.1 Preface

A typical model of a VHF and UHF land mobile-radio transmission link consists of an elevated base station antenna free from the influence of local scatterers and

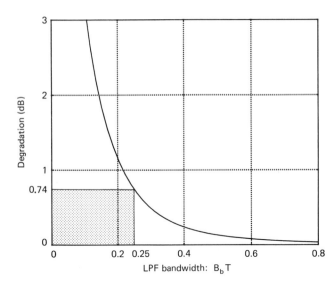

Figure 10.6 BER performance degradation of GMSK in reference to an ideal coherently demodulated antipodal system, such as MSK or BPSK. (Printed by permission of the IEEE [Murota, 10.35]. © 1981, IEEE.)

reflectors, a line-of-sight (LOS) propagation path to the general location of a moving vehicle, and a mobile antenna mounted on the vehicle. No direct LOS propagation path exists between the base and mobile antennas due to natural and constructed obstacles in the neighborhood of the vehicle. In such an environment the transmission link may be modeled as random multiple propagation paths, varying with the vehicle movement. As described in detail in [Jakes, 10.40; Lee, 10.41], VHF and UHF land mobile-radio propagation is characterized by three approximately separable effects, which are known as *multipath fading*, *shadowing*, and *path loss*, respectively. Multipath fading can further be divided into *envelope fading* (nonselective), *Doppler spread* (time-selective), and *time-delay spread* (frequency-selective). Next, we describe the underlying principles of these phenomena.

### Envelope Fading

When the received signal is measured over a distance of a few hundreds of wavelengths, fast and deep fluctuations appear on the signal envelope. It has been theoretically shown that the fluctuated signal envelope exhibits a Rayleigh distribution when the number of waves propagating randomly in different directions is sufficiently large, which is the case in urban areas [Ossana, 10.42; Gilbert, 10.43; Clarke, 10.44; Gans, 10.45]. The distribution is such that fades of 30 dB or more below the root-mean-square (rms) value of the signal envelope occur 0.1% of the time. Moreover, a fading rate is shown to be proportional to the vehicle speed and the carrier center frequency. For instance, the fading rate becomes 30 Hz when the vehicle speed is 36 km/h and the carrier center frequency is 900 MHz. These characteristics agree well with the field test results reported in [Young, 10.46; Okumura, 10.47; Nyland, 10.48].

### Doppler Spread

It has also been shown that fast Rayleigh envelope fading is accompanied by fast phase change, which induces random FM noise on the received carrier. The baseband spectrum of random FM noise extends to about twice the maximum Doppler frequency [Jakes, 10.40]. At 36 km/h and 900 MHz, the random FM noise spectrum extends to about 75 Hz. This effect, which can be regarded as a **temporal decorrelation** effect of the multipath fading and is sometimes called **time-selective fading** effect, may impose a limit on the $P_e$ performance of digital angle modulated transmission systems. The $P_e$ performance cannot be improved only by increasing the transmitter power. *Coherence time*, which is usually defined as the time interval to obtain an envelope correlation of 0.9 or less, is inversely proportional to the maximum Doppler frequency, that is, Doppler spread in such a land mobile radio channel. **Doppler spread** is a measure of the spectral width of a received carrier when a single sinusoidal carrier is transmitted through the multipath fading channel [Bello, 10.49; Schwartz, 10.50; Kennedy, 10.51].

### Time-Delay Spread

Based on the dual relationship between time and frequency we define **coherence bandwidth** of a land mobile radio channel as the required frequency spacing to have an envelope correlation of 0.9 or less. This bandwidth is inversely proportional to the rms value of time-delay spread. **Time-delay spread** is a measure for the temporal width of a received multiple-impulsive carrier if transmitted through a multipath fading channel [Bello, 10.49; Schwartz, 10.50; Kennedy, 10.51]. The characteristics of coherence bandwidth and time-delay spread in the VHF and UHF urban land mobile radio channels have been experimentally clarified by many authors [Turin, 10.52; Cox, 10.53; Cox, 10.54; Bajwa, 10.55]. The effect of time-delay spread, which can be regarded as a **frequency-selective fading** effect, causes waveform distortions in the detected signal and imposes an upper limit on the $P_e$ performance of high-speed digital transmission systems. This effect can, therefore, be neglected in a narrow-band SCPC digital transmission system with a bit rate of 16 kb/s and a channel spacing of 25 kHz. However, it must be taken into account when considering the $P_e$ performance of higher-speed digital transmission systems such as **spread-spectrum** (SS) and time division multiple access (TDMA) mobile radio systems.

### Shadowing

Even after the fast multipath fading is removed by averaging over distances of a few tens of wavelengths, **nonselective shadowing** still remains. Shadowing is caused mainly by terrain features of the land mobile radio propagation environment. It imposes a slowly changing average upon the Rayleigh fading statistics. Although there is no comprehensive mathematical model for shadowing, a log-normal distribution with a standard deviation of 5 to 12 dB has been found best to fit the experimental data in a typical urban area [Okumura, 10.47; Egli, 10.56; Black, 10.57].

### Path Loss

An average value of log-normal shadowing, which is also called the **area average**, is determined by the path loss, which varies with the distance between the fixed base station and the general location of a moving vehicle. Variation of the path loss has been shown experimentally to obey the inverse cubic to fourth-power law [Young, 10.46; Okumura, 10.47; Black, 10.57]. A useful empirical formula has been presented in [Hata, 10.58] for the path loss in the VHF and UHF land mobile radio environment.

### 10.3.2 Multipath Fading

Let us now derive a mathematical model for multipath fading observed in VHF and UHF land mobile radio systems. Since we focus attention in this chapter on narrow-band SCPC digital land mobile radio transmission systems, we have to take the effects of nonselective envelope fading and time-selective random FM noise into account while neglecting those of frequency-selective time-delay spread. The following derivation is obtained by using the Gans' power spectral theory [Gans, 10.45] based on the works of Ossana, Gilbert, and Clarke [Ossana, 10.42; Gilbert, 10.43; Clarke, 10.44]. The Gans' theory was generalized by Lin and Aulin [Lin, 10.59; Aulin, 10.60].

#### Mathematical Model—Unmodulated Carrier

Let us consider single sinusoidal (unmodulated) carrier transmission through random multiple propagation paths between a fixed base station and a constantly moving vehicle. Assuming that the vehicle speed $v$ is sufficiently small compared to the product of the carrier center frequency $f_c$ and carrier wavelength $\lambda$, that is $v << f_c\lambda$, the **received faded carrier** $e(t)$ can be represented as

$$e(t) = \sum_{n=1}^{N} e_n(t) = Re\left[\sum_{n=1}^{N} z_n(t)\, e^{j2\pi f_c t}\right] \qquad (10.5)$$

where $e_n(t)$ is an elementary propagation wave, $Re[\cdot]$ is the real part of $[\cdot]$, and $j = \sqrt{-1}$. Furthermore, $z_n(t)$ is the complex random modulation of $e_n(t)$, which is caused by the random variation of propagation paths due to the vehicle movement.

Provided that $N$ is sufficiently large and that all the $|z_n(t)|$ are equal, $e(t)$ is given by the following narrow-band Gaussian process:

$$e(t) = Re\,[z(t)\, e^{j2\pi f_c t}] \qquad (10.6)$$

where $z(t)$ is a complex zero-mean stationary baseband Gaussian process having the following properties:

$$\langle z(t)\rangle = 0$$

$$\tfrac{1}{2}\,\langle z(t)z*(t - \tau)\rangle = \psi(\tau) \qquad (10.7)$$

$$\tfrac{1}{2}\langle z(t)z(t - \tau)\rangle = 0$$

In the previous equation, $\langle \cdot \rangle$ represents an ensemble average, $(\cdot)*$ is the complex

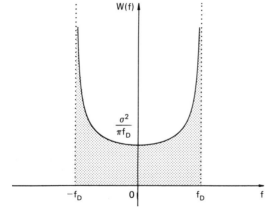

**Figure 10.7** Typical example of power spectrum of a faded single sinusoidal (unmodulated) carrier. $\sigma^2$ is average power of the received faded carrier $e(t)$. $f_D = v/\lambda$ is the maximum Doppler frequency.

conjugate of $(\cdot)$, and $\psi(\tau)$ is the autocorrelation function of $z(t)$ given by

$$\psi(\tau) := \int_{-\infty}^{\infty} W(f)\, e^{j2\pi f\tau} df. \tag{10.8}$$

$W(f)$ is the faded signal spectrum.

In a typical system, the moving vehicle has an antenna with an omni-directional pattern in the horizontal plane and the propagation angle of each elementary wave is uniformly distributed. The faded signal spectrum $W(f)$ is given by

$$W(f) = \begin{cases} \dfrac{\sigma^2}{\pi\sqrt{f_D^2 - f^2}} & |f| \le f_D \\[2mm] 0 & |f| > f_D \end{cases} \tag{10.9}$$

where $\sigma^2$ is the average power of $e(t)$ and $f_D$ is the maximum Doppler frequency, given by $f_D = v/\lambda$. Figure 10.7 shows $W(f)$. Substituting (10.9) into (10.8), $\psi(\tau)$ may be written as

$$\psi(\tau) = \sigma^2 J_0(2\pi f_D \tau) \tag{10.10}$$

where $J_0(\cdot)$ is a zero-order Bessel function of the first kind. Several types of faded-signal spectra other than (10.10) have been observed in land mobile radio environments. Their properties are investigated in more detail in [Aulin, 10.60].

From (10.6), (10.7), and (10.8), it can be shown that multipath fading can be regarded as a multiplicative complex stationary Gaussian process characterized by $W(f)$ and $\psi(\tau)$. As a consequence, a fading simulator for testing land mobile radio systems can be simply constructed as shown in Fig. 10.8 [Arredondo, 10.61; Hirade, 10.62; Ralpho, 10.63]. In this simulator $x(t)$ and $y(t)$ are the real and imaginary parts of $z(t)$, respectively. This construction method can be extended to the case of multiple faded signal generators having variable crosscorrelation coefficients [Hattori, 10.64]. It can also be extended to the case of frequency-selective faded signals observed in land mobile radio environments having severe

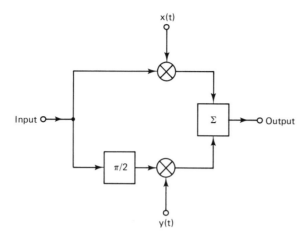

**Figure 10.8** Model of a fading simulator used for testing of land mobile radio systems. $x(t)$ and $y(t)$ are the real and imaginary parts of $z(t)$. $z(t)$ is a complex zero-mean stationary Gaussian process.

time-delay spreads [Caples, 10.65; Arnold, 10.66]. IF measurements of the $e(t)$ waveform illustrate a typical faded carrier. See Fig. 10.9.

**Envelope and Phase**

The received faded carrier $e(t)$, given by (10.6), can also be represented as

$$e(t) = R(t)\cos\{2\pi f_c t + \theta(t)\} \tag{10.11}$$

where $R(t)$ and $\theta(t)$ are the envelope and phase of $e(t)$ given, respectively, by

$$\begin{cases} R(t) = |z(t)| \\ \theta(t) = \angle z(t) \end{cases} \tag{10.12}$$

Here $\angle z(t)$ denotes the argument of $z(t)$. Since at some particular instant $z = z(t)$ becomes a complex Gaussian random variable, the joint probability density function (pdf) of $R = R(t)$ and $\theta = \theta(t)$ can be written as [Rice, 10.67]

$$p(R, \theta) = \frac{R}{2\pi\sigma^2} \exp\left(-\frac{R^2}{2\sigma^2}\right) \qquad R \geq 0, \quad -\pi \leq \theta \leq \pi \tag{10.13}$$

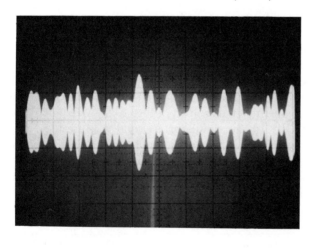

**Figure 10.9** Waveform of a faded unmodulated carrier measured at an intermediate frequency of $f_c = 1$ MHz. The maximum Doppler frequency is $f_D = 40$ Hz.

Consequently, the pdf's of $R$ and $\theta$ are given, respectively, by

$$\begin{cases} p(R) = \dfrac{R}{\sigma^2} \exp\left(-\dfrac{R^2}{2\sigma^2}\right) & R \geqq 0 \\[4mm] p(\theta) = \dfrac{1}{2\pi} & -\pi \leqq \theta \leqq \pi \end{cases} \qquad (10.14)$$

The above equations indicate that the received faded carrier $e(t)$ has a **Rayleigh-distributed** envelope $R(t)$ and a **uniformly distributed** phase $\theta(t)$. The fading model described above is, thus, called the *Rayleigh fading model*, and the modeling has been confirmed experimentally by a number of field tests [Okumura, 10.47; Nyland, 10.48].

The cumulative pdf of $R$—that is, the probability that $R$ does not exceed a specified level $R_s$—can be obtained as

$$\text{Prob}[R \leqq R_s] = \int_0^{R_s} p(R)\,dR = 1 - \exp\left(-\frac{R_s^2}{2\sigma^2}\right) \qquad (10.15)$$

Figure 10.10 shows computed results. From this figure we note that the envelope

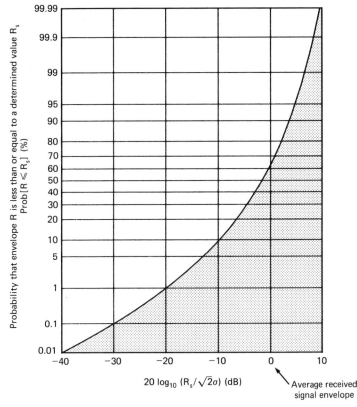

**Figure 10.10** Cumulative probability distribution of the envelope of a Rayleigh faded carrier. $R$ is a random variable representing the Rayleigh-distributed envelope of the faded carrier $e(t)$. $\sigma^2$ is the average power of the faded carrier $e(t)$.

Sec. 10.3   Propagation Characteristics of Land Mobile Radio   **503**

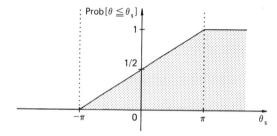

**Figure 10.11** Cumulative probability distribution of the phase of a Rayleigh faded carrier. $\theta$ is a random variable representing the uniformly distributed phase of the faded carrier $e(t)$.

of a Rayleigh faded carrier fluctuates within a dynamic range of about 40 dB with a probability of 99.9%. The pdf of $\theta$, that is, the probability that $\theta$ does not exceed a specified value $\theta_s$ can be described by

$$\text{Prob}[\theta \leq \theta_s] = \int_{-\pi}^{\theta_s} p(\theta)\,d\theta = \frac{\theta_s}{2\pi} \tag{10.16}$$

Figure 10.11 depicts computed results of (10.16). Figure 10.12 shows the envelope and phase variation of a faded carrier generated by a 900-MHz band fading simulator. A two-dimensional vector locus observed on the polar display of a network analyzer is shown in Fig. 10.13.

### Crossing Rate and Fade Duration

In the design of high-speed digital mobile radio transmission systems, it is important to estimate or evaluate the characteristics of multipath fading that induce burst errors. Provided that burst errors always occur when the signal envelope fades below a specific threshold value, the level crossing rate can be used as an appropriate measure for the **burst error occurrence rate**. The fade duration can also be used to estimate or evaluate approximately the **burst error length**.

The level crossing rate, which is generally defined as the expected rate at which the envelope $R$ crosses a specified level $R_s$ in the positive direction, is given

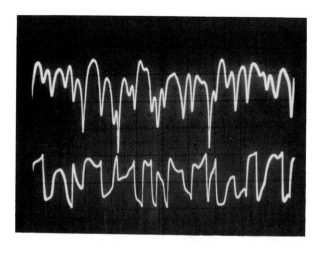

**Figure 10.12** Waveforms of envelope and phase of a Rayleigh faded carrier measured at 900 MHz by a vector voltmeter. Channel 1 is the envelope (Vertical:10 dB/division, horizontal:50 ms/division). Channel 2 is the phase (vertical:$\pi$rad/division, horizontal:50 ms/division).

**Figure 10.13** Two-dimensional network analyzer display of the vector locus of a 900-MHz Rayleigh-faded carrier. The maximum Doppler frequency is $f_D = 40$ Hz.

by

$$N_{R_s} = \int_0^\infty \dot{R}\, p(R_s, \dot{R})\, d\dot{R} \tag{10.17}$$

where the dot denotes a time derivative, and $p(R_s, \dot{R})$ is the joint pdf of $R$ and $\dot{R}$ for $R = R_s$. For the Rayleigh fading model, $p(R_s, \dot{R})$ can be written as [10.67]

$$p(R_s, \dot{R}) = p(R_s)p(\dot{R}) \tag{10.18}$$

where $p(R)$ is the pdf of $R$ given by (10.14). According to Rice's result [Rice, 10.67], the pdf of $\dot{R}$, $p(\dot{R})$, is given by

$$p(\dot{R}) = \frac{1}{2\pi f_D \sigma \sqrt{\pi}} \exp\left[-\left(\frac{\dot{R}}{2\pi f_D \sigma}\right)^2\right] \tag{10.19}$$

Therefore, the level crossing rate $N_{R_s}$ is given by

$$N_{R_s} = \sqrt{2\pi}\, f_D \left(\frac{R_s}{\sqrt{2\pi}}\right) \exp\left(-\frac{R_s^2}{2\sigma^2}\right) \tag{10.20}$$

Computed results of this equation are illustrated in Fig. 10.14.

The statistical distribution function of the fade duration of the Rayleigh faded channel has not yet been derived in a closed form. However, the average value of the fade duration can be expressed by

$$\langle T_f \rangle = \frac{1}{N_{R_s}} \text{Prob}[R \leqq R_s] \tag{10.21}$$

Substituting (10.15) and (10.21) yields

$$\langle T_f \rangle = \frac{1}{\sqrt{2\pi}(R_s/\sqrt{2}\sigma)f_D} \left[\exp\left(\frac{R_s^2}{2\sigma^2}\right) - 1\right] \tag{10.22}$$

Computed results of this equation are shown in Fig. 10.15.

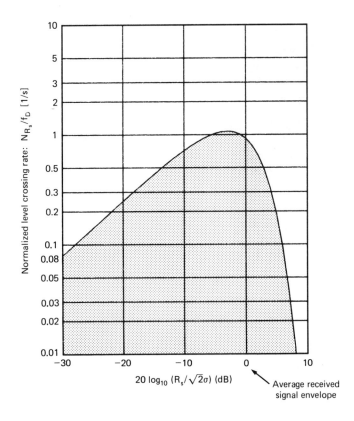

**Figure 10.14** Level crossing rate $N_{R_s}$ of a Rayleigh-faded carrier.

y-axis: Normalized level crossing rate: $N_{R_s}/f_D$ [1/s]

x-axis: $20 \log_{10} (R_s/\sqrt{2}\sigma)$ (dB)

Average received signal envelope

**Example 10.1**

When the fading rate is $f_D = 40$ Hz, that is, $v = 48$ km/h for $f_c = 900$ MHz, and the average signal level is 30 dB higher than the threshold level, that is, $20 \log_{10} (R_s/\sqrt{2}\sigma) = -30$ dB, then the level crossing rate and the average fade duration become $N_{R_s} \approx 3.2$ l/s and $\langle T_f \rangle = 0.316$ ms, respectively.

### 10.3.3 Empirical Formula for Path Loss

Path loss, which varies with the distance between a fixed base station and the general location of a moving vehicle, is determined from the received field strength and the effective radiation power. Of the many technical reports that are concerned with path loss prediction methods for VHF and UHF land mobile-radio systems [Young, 10.46; Okumura, 10.47; Egli, 10.56; Black, 10.57; Barsis, 10.68; Durkin, 10.69; Allsebrook, 10.70], Okumura's report is perhaps the most comprehensive one [Okumura, 10.47]. In his report, many useful curves to predict a *median value* of the received field strength are presented with the following parameters: carrier frequency, $f_c$ (MHz), base station antenna height, $h_b$(m), and mobile-station antenna height, $h_m$(m). Based on Okumura's prediction method,

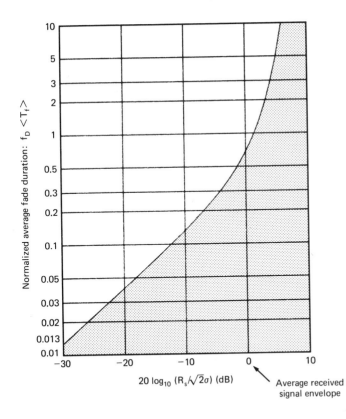

Normalized average fade duration: $f_D \langle T_f \rangle$

$20 \log_{10} (R_s/\sqrt{2}\sigma)$ (dB)

Average received signal envelope

**Figure 10.15** Average fade duration of a Rayleigh-faded carrier.

an empirical formula for the median path loss $L_p$ (decibels) between two isotropic base and mobile antennas has been obtained. It is given by [Hata, 10.58]:

$$L_p = \begin{cases} A + B \log_{10}(r) & \text{for urban area} \\ A + B \log_{10}(r) - C & \text{for suburban area} \\ A + B \log_{10}(r) - D & \text{for open area} \end{cases} \qquad (10.23)$$

where $r$ (kilometers) is the distance between base and mobile stations, and $A$, $B$, $C$, and $D$ are given, respectively, by

$$A = A(f_c, h_b, h_m) = 69.55 + 26.16 \log_{10}(f_c) - 13.82 \log_{10}(h_b) - a(h_m)$$

$$B = B(h_b) = 44.9 - 6.55 \log_{10}(h_b) \qquad (10.24)$$

$$C = C(f_c) = 2 \left[ \log_{10} \left( \frac{f_c}{28} \right) \right]^2 + 5.4$$

$$D = D(f_c) = 4.78 \left[ \log_{10}(f_c) \right]^2 - 19.33 \log_{10}(f_c) + 40.94$$

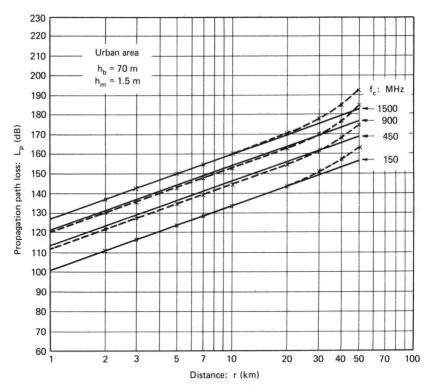

**Figure 10.16** Propagation path loss in urban area—$f_c$ is a parameter.

where

$$a(h_m) = \begin{cases} [1.1 \log_{10}(f_c) - 0.7]h_m - [1.56 \log_{10}(f_c) - 0.8] \\ \qquad\qquad\qquad\qquad\qquad\qquad \text{for medium or small city} \\ \\ \begin{cases} 8.28[\log_{10}(1.54\, h_m)]^2 - 1.1 & \text{for } f_c \geq 200 \text{ MHz} \\ 3.2[\log_{10}(11.75\, h_m)]^2 - 4.97 & \text{for } f_c \leq 400 \text{ MHz} \end{cases} \\ \qquad\qquad\qquad\qquad\qquad\qquad \text{for large city} \end{cases} \quad (10.25)$$

Equation (10.23) can be used if the following conditions are satisfied:

$$\begin{cases} f_c: & 150\text{--}1500 \text{ (MHz)} \\ h_b: & 30\text{--}200 \text{ (m)} \\ h_m: & 1\text{--}10 \text{ (m)} \\ r: & 1\text{--}20 \text{ (km)}. \end{cases} \quad (10.26)$$

Figures 10.16 and 10.17 show the path loss $L_p$ (decibel) in an urban area, where the carrier frequency $f_c$ (megahertz) and the base station antenna height $h_b$ (meters) are variable parameters, respectively. The solid lines are obtained by the empirical

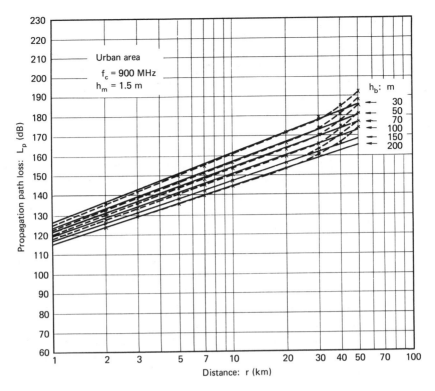

**Figure 10.17** Propagation path loss in urban area—$h_b$ is a parameter.

formula and the dashed lines by the Okumura's prediction method. These figures show that the maximum error is only about 1 dB within the distance range of $r = 1-20$ (km).

## 10.4 DIGITAL FM TRANSMISSION PERFORMANCE IN LAND MOBILE RADIO

Of the many kinds of digital modulation schemes, digital FM, which is also called continuous-phase FSK (CP-FSK) is among the most useful for digital mobile-radio transmission systems because of its constant-envelope property and small RF band-width requirement. (Other modulation techniques having desirable properties for potential digital mobile-radio systems applications are described in Chapters 7 and 8). Three kinds of detection schemes, **discriminator**, **differential**, and **coherent detection** schemes are used to detect digital FM signals. Discriminator detection can be applied irrespective of the value of the modulation index. Differential detection is limited to the case of low modulation index of less than 1.0. Coherent detection can be applied only in the special case of a discrete modulation index of 0.5 or 0.25. Taking into account the effect of fast and deep multipath fading modeled in the previous section, we analyze the $P_e$ performance of digital FM mobile-radio transmission systems. Our analysis, based on a unified method [Hirade, 10.71; 10.72], applies to discriminator and differential detection schemes. Gaussian noise generated in the receiver front end and cochannel interference caused by geographical reuse of the same frequency are taken into consideration.

### 10.4.1 *Mathematical Model of Discriminator and Differential Detectors*

A model of a digital FM receiver is shown in Fig. 10.18. It is assumed that the desired signal and the undesired cochannel interference are received simultaneously and that both of them are subjected to mutually independent fast Rayleigh fading. The receiver front end generates an AWGN. A predetection band-pass filter (BPF) having symmetrical band-pass characteristics precedes the discriminator or differential detector. Practical systems may have a postdetection LPF. The ISI caused by the predetection BPF and the postdetection LPF is assumed to be negligible. The detector output is synchronously sampled by the recovered bit timing clock, and the transmitted baseband pulse signals are sequentially regenerated. Ideal sampling, timing recovery, and decision circuits are assumed.

Based on our system model, the desired signal, $s(t)$, the undesired cochannel interference, $i(t)$, and the additive Gaussian noise, $n(t)$, at the predetection BPF output can be represented, respectively, as the following narrow-band Gaussian processes:

$$\begin{cases} s(t) = \text{Re}\{z_s(t)\exp j[2\pi f_c t + \phi_s(t)]\} \\ i(t) = \text{Re}\{z_i(t)\exp j[2\pi f_c t + \phi_i(t)]\} \\ n(t) = \text{Re}\{z_n(t)\exp j[2\pi f_c t]\} \end{cases} \qquad (10.27)$$

where $\text{Re}\{\cdot\}$ denotes the real part of $\{\cdot\}$, $f_c$ is the nominal center frequency of the carrier, and $\phi_s(t)$ and $\phi_i(t)$ are the modulation phase changes of $s(t)$ and $i(t)$, respectively. Furthermore, $z_s(t)$, $z_i(t)$, and $z_n(t)$ are the mutually independent complex zero-mean stationary baseband Gaussian processes. The detector input

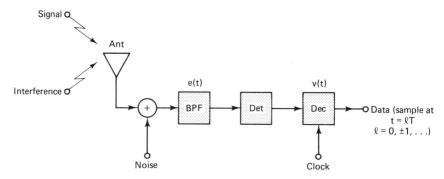

Det: Detector (discriminator or differential detector)
Dec: Decision circuit

**Figure 10.18** Digital FM receiver model.

signal $e(t)$ is

$$e(t) = s(t) + i(t) + n(t) = \text{Re}[z(t) \exp j(2\pi f_c t)] \tag{10.28}$$

where $z(t)$ is

$$z(t) = z_s(t)\exp j[\phi_s(t)] + z_i(t)\exp j[\phi_i(t)] + z_n(t) \tag{10.29}$$

Considering that the frequency discriminator output is given by the time derivative of phase change of $e(t)$ and that the differential detector for digital FM signals operates as a quadrature product detector of $e(t)$ and $e(t - T)$ [Stein, 10.73], the respective detector outputs can be written as

$$v(t) = \begin{cases} \dfrac{\text{Re}[-jz^*(t)\,\dot{z}(t)]}{|z(t)|^2} & \text{for discriminator detection} \tag{10.30a} \\[3ex] \text{Re}[-jz(t)z^*(t - T)] & \text{for discriminator detection} \tag{10.30b} \end{cases}$$

where the symbols of asterisk and dot are the complex conjugate and the time derivative, respectively, and $T$ is the signaling period, that is, unit bit duration.

### 10.4.2 Derivation of Average $P_e$

Provided that the detector output $v(t)$ is sampled at some particular instant $t = lT$, where $l = 0, \pm 1, \pm 2, \dots$, and that the decision about whether a mark or a space was sent is based on the polarity of $v(lT)$, the average $P_e$ is

$$P_e = pP_e(M) + (1 - p)P_e(S) \tag{10.31}$$

where $p$ is the probability of mark transmission in $s(t)$, and $P_e(M)$ and $P_e(S)$ are the conditional probabilities of $v(lT) < 0$ for mark transmission and that of $v(lT) > 0$ for space transmission, respectively.

According to Stein's result [Stein, 10.73], $P_e(M)$ and $P_e(S)$ can be obtained, respectively, as

$$\begin{aligned} P_e(M) &= \text{Prob}(|\zeta_1|^2 < |\zeta_2|^2) = \text{Prob}(|\zeta_1| < |\zeta_2|) \\ P_e(S) &= \text{Prob}(|\zeta_1|^2 > |\zeta_2|^2) = \text{Prob}(|\zeta_1| > |\zeta_2|) \end{aligned} \tag{10.32}$$

where $\zeta_1$ and $\zeta_2$ are given, respectively, by

$$\begin{aligned} \zeta_1 &= \tfrac{1}{4}[(z_1 + z_2)(1 + K) + (z_1 - z_2)(1 - K)e^{j\lambda}] \\ \zeta_2 &= \tfrac{1}{4}[(z_1 + z_2)(1 - K) + (z_1 - z_2)(1 + K)e^{j\lambda}] \end{aligned} \tag{10.33}$$

$K$ and $e^{j\lambda}$ are

$$K = \left[ \frac{(\sigma_1^2 + \sigma_2^2) + \sqrt{(\sigma_1^2 - \sigma_2^2)^2 + 2(\sigma_1\sigma_2\rho_s)^2}}{(\sigma_1^2 + \sigma_2^2) - \sqrt{(\sigma_1^2 - \sigma_2^2)^2 + 2(\sigma_1\sigma_2\rho_s)^2}} \right]^{1/2}$$

$$e^{j\lambda} = \frac{(\sigma_1^2 - \sigma_2^2) - 2j\sigma_1\sigma_2\rho_s}{\sqrt{(\sigma_1^2 - \sigma_2^2)^2 + (2\sigma_1\sigma_2\rho_s)^2}} \tag{10.34}$$

$\sigma_1$, $\sigma_2$, and $\rho = \rho_c + j\rho_s$ are given by

$$\sigma_1^2 = \tfrac{1}{2}\langle |z_1 - \langle z_1 \rangle|^2 \rangle$$

$$\sigma_2^2 = \tfrac{1}{2}\langle |z_2 - \langle z_2 \rangle|^2 \rangle \qquad (10.35)$$

$$\sigma_1\sigma_2\rho = \tfrac{1}{2}\langle (z_1 - \langle z_1 \rangle)^* (z_2 - \langle z_2 \rangle) \rangle$$

where $z_1$ and $z_2$ are given by

$$\begin{aligned} z_1 &= jz(lT) \\ z_2 &= \dot{z}(lT) \end{aligned} \qquad \text{for discriminator detection} \qquad (10.36a)$$

$$\begin{aligned} z_1 &= jz(\overline{l - l}T) \\ z_2 &= z(lT) \end{aligned} \qquad \text{for differential detection} \qquad (10.36b)$$

Since the $z_1$ and $z_2$ are zero-mean complex Gaussian random variables and (10.31) is a linear transformation, the joint pdf of $|\zeta_1|$ and $|\zeta_2|$ is obtained as

$$p(|\zeta_1|, |\zeta_2|) = \frac{4|\zeta_1||\zeta_2|}{\langle|\zeta_1|^2\rangle\langle|\zeta_2|^2\rangle} \exp\left[ -\left( \frac{|\zeta_1|^2}{\langle|\zeta_1|^2\rangle} + \frac{|\zeta_2|^2}{\langle|\zeta_2|^2\rangle} \right) \right] \qquad (10.37)$$

where

$$\begin{Bmatrix} \langle|\zeta_1|^2\rangle \\ \langle|\zeta_2|^2\rangle \end{Bmatrix} = \frac{1}{2}\left( 1 \pm \frac{\rho_c}{\sqrt{1 - \rho_s^2}} \right)\left[ (\sigma_1^2 + \sigma_2^2) + \sqrt{(\sigma_1^2 - \sigma_2^2)^2 + (2\sigma_1\sigma_2\rho_s)^2} \right] \qquad (10.38)$$

Therefore, $P_e(M)$ and $P_e(S)$ are expressed by

$$\begin{cases} P_e(M) = \displaystyle\int_0^\infty d|\zeta_1| \int_{|\zeta_1|}^\infty p(|\zeta_1|, |\zeta_2|)\, d|\zeta_2| = \frac{1}{2}\left( 1 - \frac{\rho_{c+}}{\sqrt{1 - \rho_{s+}^2}} \right) \\[12pt] P_e(S) = \displaystyle\int_0^\infty d|\zeta_1| \int_{|\zeta_2|}^\infty p(|\zeta_1|, |\zeta_2|)\, d|\zeta_1| = \frac{1}{2}\left( 1 + \frac{\rho_{c-}}{\sqrt{1 - \rho_{s-}^2}} \right) \end{cases} \qquad (10.39)$$

where the subscripts of $+$ or $-$ for $\rho$ correspond to mark or space transmission, respectively. Substituting (10.39) into (10.31) yields

$$\boxed{ P_e = \frac{p}{2}\left( 1 - \frac{\rho_{c-}}{\sqrt{1 - \rho_{s+}^2}} \right) + \frac{(1 - p)}{2}\left( 1 + \frac{\rho_{c-}}{\sqrt{1 - \rho_{s-}^2}} \right) } \qquad (10.40)$$

### 10.4.3 Derivation of $\sigma_1^2$, $\sigma_2^2$, and $\rho$

Let us consider the case that the mutually independent complex zero-mean stationary baseband Gaussian processes $z_s(t)$, $z_i(t)$, and $z_n(t)$ have symmetrical power spectra. Provided that the sampling instant $t = lT$ is perfectly synchronized with the timing phase of $\phi_s(t)$ and the timing sources between $\phi_s(t)$ and $\phi_i(t)$ are asynchronous with each other, $\sigma_1^2$, $\sigma_2^2$, and $\rho_\pm = \rho_{c\pm} + j\rho_{s\pm}$ can be obtained to be

$$
\begin{cases}
\sigma_1^2 &= \sigma_s^2 + \sigma_i^2 + \sigma_n^2 \\[2mm]
\sigma_2^2 &= \sigma_s^2 \left(\dfrac{m_s\pi}{T}\right)^2 + \sigma_i^2 \left(\dfrac{m_i\pi}{T}\right)^2 - \sigma_s^2\ddot{\rho}_s(0) - \sigma^2\ddot{\rho}_i(0) - \sigma_n^2\ddot{\rho}_n(0) \quad (10.41a) \\[2mm]
\sigma_1\sigma_2\rho_\pm &= \pm \sigma_{sp}^2 \left(\dfrac{m_s\pi}{T}\right) \qquad \text{for discriminator detection}
\end{cases}
$$

$$
\begin{cases}
\sigma_1^2 &= \sigma_s^2 + \sigma_i^2 + \sigma_n^2 \\[2mm]
\sigma_2^2 &= \sigma_s^2 + \sigma_i^2 + \sigma_n^2 \qquad\qquad\qquad\qquad\qquad\qquad (10.41b) \\[2mm]
\sigma_1\sigma_2\rho_\pm &= -j\left[\sigma_s^2\rho_s(T)\,e^{\pm jm_s\pi} + \sigma_i^2\rho_i(T)\,\xi_i(T) + \sigma_n^2\rho_n(T)\right]
\end{cases}
$$

$$\text{for differential detection}$$

where $m_s$ and $m_i$ are the modulation indices of $s(t)$ and $i(t)$, which are defined, respectively, by

$$
\begin{cases}
m_s = 2\Delta f_{d_s} T \\
m_i = 2\Delta f_{d_i} T
\end{cases}
\qquad (10.42)
$$

where $\Delta f_{d_s}$ and $\Delta f_{d_i}$ are the frequency deviations of $s(t)$ and $i(t)$, respectively. The symbols $+$ or $-$ again correspond to the cases of mark or space transmission, respectively. Here $\sigma_s^2$, $\sigma_i^2$, and $\sigma_n^2$ are the average powers of $s(t)$, $i(t)$, and $n(t)$, respectively, and $\rho_s(\tau)$, $\rho_i(\tau)$, and $\rho_n(\tau)$ are the normalized autocorrelation functions of $z_s(t)$, $z_i(t)$, and $z_n(t)$, which are given by

$$
\begin{cases}
\sigma_s^2\rho_s(\tau) = \displaystyle\int_{-\infty}^{\infty} W_s(f)\,e^{j2\pi f\tau}\,df \\[3mm]
\sigma_i^2\rho_i(\tau) = \displaystyle\int_{-\infty}^{\infty} W_i(f)\,e^{j2\pi f\tau}\,df \\[3mm]
\sigma_n^2\rho_n(\tau) = \displaystyle\int_{-\infty}^{\infty} W_n(f)\,e^{j2\pi f\tau}\,df
\end{cases}
\qquad (10.43)
$$

where $W_s(f)$, $W_i(f)$, and $W_n(f)$ are the power spectra of $z_s(t)$, $z_i(t)$, and $z_n(t)$, respectively. The second derivatives, $\ddot{\rho}_s(\tau)$, $\ddot{\rho}_i(\tau)$, and $\ddot{\rho}_n(\tau)$, are given by

$$
\begin{cases}
-\sigma_s^2\ddot{\rho}_s(\tau) = \displaystyle\int_{-\infty}^{\infty} (2\pi f)^2\,W_s(f)\,e^{j2\pi f\tau}\,df \\[3mm]
-\sigma_i^2\ddot{\rho}_i(\tau) = \displaystyle\int_{-\infty}^{\infty} (2\pi f)^2\,W_i(f)\,e^{j2\pi f\tau}\,df \\[3mm]
-\sigma_n^2\ddot{\rho}_n(\tau) = \displaystyle\int_{-\infty}^{\infty} (2\pi f)^2\,W_n(f)\,e^{j2\pi f\tau}\,df
\end{cases}
\qquad (10.44)
$$

Using Pelchat's result (Pelchat, 10.74], $\xi_i(T)$ is obtained as

$$\xi_i(T) = \frac{1}{2}\left(\cos m_i\pi + \frac{\sin m_i\pi}{m_i\pi}\right) \tag{10.45}$$

### 10.4.4 Average $P_e$ Performance

In particular, when $W_s(f)$, $W_i(f)$, and $W_n(f)$ in (10.43) and (10.44) are assumed to have even symmetry with respect to $f$, then $\rho_s(T)$, $\rho_i(T)$, $\rho_n(T)$, $\ddot{\rho}_s(0)$, $\ddot{\rho}_i(0)$, and $\ddot{\rho}_n(0)$ are real functions. Therefore, the average $P_e$ of our system becomes

$$P_e = \frac{1}{2}\left(1 - \frac{\Gamma\Lambda(m_s\pi)}{(\Gamma\Lambda + \Gamma + \Lambda)^{1/2}\{\Gamma\Lambda[(m_s\pi)^2 - \ddot{\rho}_s(0)T^2]}{} \\ {} + \Gamma[m_i\pi)^2 - \ddot{\rho}_i(0)T^2] - \Lambda\ddot{\rho}_n(0)T\}^{1/2}}\right)$$

for discriminator detection $\tag{10.46a}$

$$P_e = \frac{1}{2}\left(1 - \frac{\Gamma\Lambda\rho_s(T)\sin m_s\pi}{\{(\Gamma\Lambda + \Gamma + \Lambda)^2 - [\Gamma\Lambda\rho_s(T)\cos m_s\pi} \\ {} + \frac{\Gamma}{2}\rho_i(T)\left(\cos m_i\pi + \frac{\sin m_i\pi}{m_i\pi}\right) + \Lambda\rho_n(T)]^2\}^{1/2}}\right)$$

$$\tag{10.46b}$$

for differential detection

where $\Gamma$ and $\Lambda$ are the average carrier-to-noise ratio (CNR) and the average CIR, respectively defined by

$$\Gamma = \frac{\sigma_s^2}{\sigma_n^2} = \text{average CNR}^*$$
$$\Lambda = \frac{\sigma_s^2}{\sigma_i^2} = \text{average CIR} \tag{10.47}$$

When the CCI can be neglected, that is, $\Lambda \to \infty$, then (10.45) reduces to

$$P_{e_{\Lambda\to\infty}} = \frac{1}{2}\left(1 - \frac{\Gamma(m_s\pi)}{(\Gamma + 1)^{1/2}\{\Gamma[(m_s\pi)^2 - \ddot{\rho}_s(0)T^2] - \ddot{\rho}_n(0)T^2\}^{1/2}}\right) \tag{10.48a}$$

for discriminator detection

$$P_{e_{\Lambda\to\infty}} = \frac{1}{2}\left(1 - \frac{\Gamma\rho_s(T)\sin m_s\pi}{\{(\Gamma + 1)^2 - [\Gamma\rho_s(T)\cos m_s\pi + \rho_n(T)]^2\}^{1/2}}\right) \tag{10.48b}$$

for differential detection

*Since the bandwidth of a predetection BPF is usually set as $BT = 1$, the average CNR $\Gamma = E_b/N_0 BT = E_b/N_0$.

Letting $\Gamma \to \infty$ and $\Lambda \to \infty$ in equation (10.46) yields

$$\left\{ \begin{array}{l} P_{e_{\Gamma,\Lambda \to \infty}} = \dfrac{1}{2} \left\{ 1 - \dfrac{m_s \pi}{[(m_s \pi)^2 - \ddot{\rho}_s(0) T^2]^{1/2}} \right\} \qquad (10.49a) \\[1em] \text{for discriminator detection} \\[1em] P_{e_{\Gamma,\Lambda \to \infty}} = \dfrac{1}{2} \left\{ 1 - \dfrac{\rho_s(T) \sin m_s \pi}{[1 - \rho_s^2(T) \cos^2 m_s \pi]^{1/2}} \right\} \qquad (10.49b) \\[1em] \text{for differential detection} \end{array} \right.$$

which corresponds to the average $P_e$ due to *random FM noise*.

For the special case of $m_s = m_i = 0.5$, which corresponds to **simple MSK modulation**, (10.47b) reduces to

$$P_{e_{\Lambda \to \infty}} = \frac{1}{2} \left\{ 1 - \frac{\Gamma \rho_s(T)}{[(\Gamma + 1)^2 - \rho_n^2(T)]^{1/2}} \right\} \qquad (10.50)$$

For $\rho_n(T) = 0$, we obtain

$$P_{e_{\Lambda \to \infty}} = \frac{\Gamma[1 - \rho_s(T)] + 1}{2(\Gamma + 1)} \qquad (10.51)$$

This is identical to Voelcker's result [Voelcker, 10.75], which was obtained for binary phase-shift keying (BPSK) with differential detection, assuming that the temporal correlation effect of thermal noise during the signaling interval is negligible. Differential detection of MSK signals is equivalent to that of BPSK signals when the temporal correlation effect of thermal noise is negligible. For such a detection scheme, (10.51) coincides with Voelcker's BPSK result.

Thus to obtain numerical results for the average $P_e$ performance, we assume that the multipath fades appearing on $s(t)$ and $i(t)$ are mutually independent Rayleigh fades having the following power spectra:

$$W_s(f) = \left\{ \begin{array}{ll} \dfrac{\sigma_s^2}{\pi(f_D^2 - f^2)^{1/2}} & |f| \leq f_D \\[1.5em] 0 & |f| > f_D \end{array} \right. \qquad (10.52)$$

$$W_i(f) = \left\{ \begin{array}{ll} \dfrac{\sigma_i^2}{\pi(f_D^2 - f^2)^{1/2}} & |f| \leq f_D \\[1.5em] 0 & |f| > f_D \end{array} \right.$$

where $f_D$ is the maximum Doppler frequency. We also assume that an ideal rectangular BPF having a center frequency $f_c$ and a bandwidth of $B$ is used as the predetection BPF. The power spectrum $W_n(f)$ is then given by

$$W_n(f) = \begin{cases} \dfrac{\sigma_n^2}{B} & |f| \leq \dfrac{B}{2} \\ 0 & |f| > \dfrac{B}{2} \end{cases} \tag{10.53}$$

From (10.52), (10.53), $\rho_s(T)$, $\rho_i(T)$, $\rho_n(T)$, $\ddot{\rho}_s(0)$, $\ddot{\rho}_i(0)$, and $\ddot{\rho}_n(0)$, we obtain

$$\begin{cases} \rho_s(T) = \pi_i(T) = J_0(2\pi f_D T) \\ \rho_n(T) = \dfrac{\sin(\pi BT)}{(\pi BT)} \end{cases} \tag{10.54}$$

$$\begin{cases} \ddot{\rho}_s(0) = \ddot{\rho}_i(0) = -2(\pi f_D T)^2 \\ \ddot{\rho}_n(0) = -\dfrac{\pi^2}{3} B^2 \end{cases} \tag{10.55}$$

Since the bandwidth of a predetection BPF is usually set as $BT = 1$, $\rho_n(T)$ reduces to zero. Therefore, (10.46), (10.48), and (10.49) are given, respectively, by

$$\begin{cases} P_e = \dfrac{1}{2}\left[ 1 - \dfrac{\Gamma\Lambda m_s}{(\Gamma\Lambda + \Gamma + \Lambda)^{1/2}\left\{\Gamma\Lambda[m_s^2 + 2(f_D T)^2] + \Gamma(m_i^2 + 2(f_D T)^2] + \Lambda/3\right\}^{1/2}} \right] \\ \qquad \text{for discriminator detection} \tag{10.56a} \\[4ex]
P_e = \dfrac{1}{2}\left\{ 1 - \dfrac{\Gamma\Lambda J_0(2\pi f_D T)\sin m_s \pi}{\left[ (\Gamma\Lambda + \Gamma + \Lambda)^2 - \Gamma^2 J_0^2(2\pi f_D T) \left[\Lambda \cos m_s\pi + \frac{1}{2}\left(\cos m_i\pi + \dfrac{\sin m_i\pi}{m_i\pi}\right)\right]^2\right]^{1/2}} \right\} \\ \qquad \text{for differential detection} \tag{10.56b} \end{cases}$$

$$\begin{cases} P_{e_{\Lambda\to\infty}} = \dfrac{1}{2}\left( 1 - \dfrac{\Gamma m_s}{(\Gamma + 1)^{1/2}\left\{\Gamma[m_s^2 + 2(f_D T)^2] + 1/3\right\}^{1/2}} \right) \\ \qquad \text{for discriminator detection} \tag{10.57a} \\[4ex]
P_{e_{\Lambda\to\infty}} = \dfrac{1}{2}\left( 1 - \dfrac{\Gamma J_0(2\pi f_D T)\sin m_s \pi}{\left\{(\Gamma + 1)^2 - \Gamma^2 J_0^2(2\pi f_D T)\cos^2 m_s\pi\right\}^{1/2}} \right) \\ \qquad \text{for differential detection} \tag{10.57b} \end{cases}$$

$$\begin{cases} P_{e_{\Gamma,\Lambda\to\infty}} = \dfrac{1}{2} \left\{ 1 - \dfrac{m_s}{[m_s^2 + 2(f_D T)^2]^{1/2}} \right\} & \text{(10.58a)} \\[2em] \quad \text{for discriminator detection} \\[2em] P_{e_{\Gamma,\Lambda\to\infty}} = \dfrac{1}{2} \left\{ 1 - \dfrac{J_0(2\pi f_D T)\sin m_s \pi}{[1 - J_0^2(2\pi f_D T)\cos^2 m_s \pi]^{1/2}} \right\} & \text{(10.58b)} \\[2em] \quad \text{for differential detection} \end{cases}$$

In the case of MSK modulation, that is, $m_s = m_i = 0.5$, the computed results of average $P_e$ versus CNR performance are shown in Figs. 10.19–10.22 where the

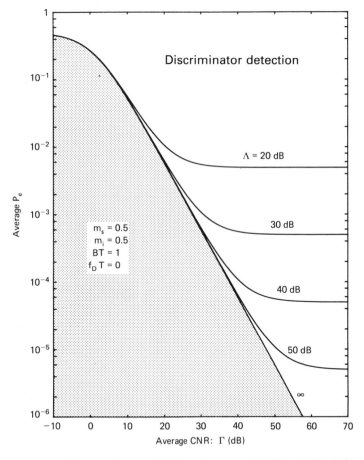

**Figure 10.19** Average $P_e$ versus CNR performance of MSK with discriminator detection (1). Average CIR $\Lambda$ is a parameter. $f_D T$ is normalized maximum Doppler frequency.

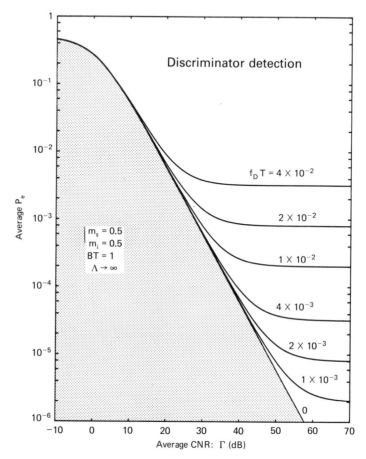

**Figure 10.20** Average $P_e$ versus CNR performance of MSK with discriminator detection (2). Normalized maximum Doppler frequency $f_D T$ is a parameter. $\Lambda$ is average CIR.

average CIR $\Lambda$ and the normalized maximum Doppler frequency $f_D T$ are taken as variable parameters, respectively. Figures 10.23–10.26 illustrate the computed results of average $P_e$ versus CIR performance with parameters of $\Lambda$ and $f_D T$, respectively.

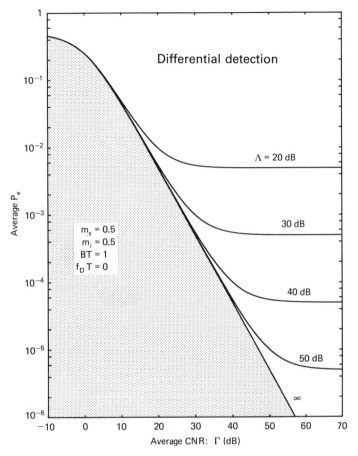

**Figure 10.21** Average $P_e$ versus CNR performance of MSK with differential detection (1). Average CIR $\Lambda$ is a parameter. $f_D T$ is normalized maximum Doppler frequency.

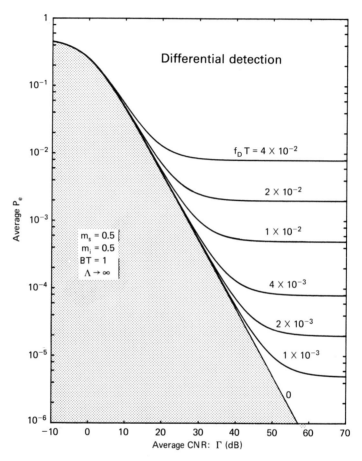

**Figure 10.22** Average $P_e$ versus CNR performance of MSK with differential detection (2). Normalized maximum Doppler frequency $f_D T$ is a parameter. $\Lambda$ is average CIR.

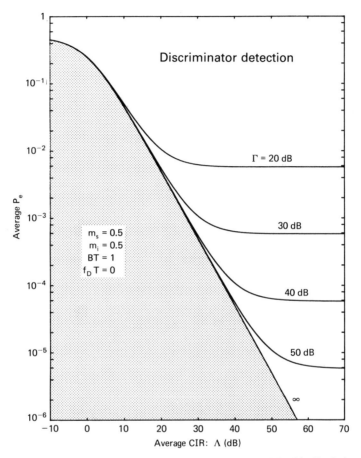

**Figure 10.23** Average $P_e$ versus CIR performance of MSK with discriminator detection (1). Average CNR $\Gamma$ is a parameter. $f_D T$ is normalized maximum Doppler frequency.

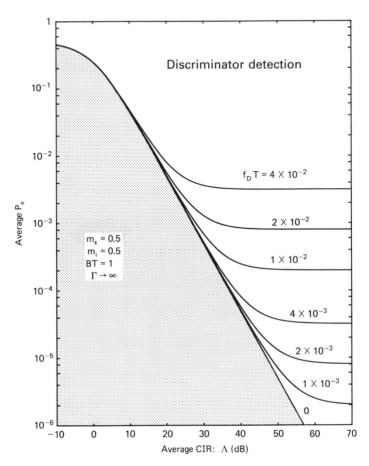

**Figure 10.24** Average $P_e$ versus CIR performance of MSK with discriminator detection (2). Normalized maximum Doppler frequency $f_D T$ is a parameter. $\Gamma$ is average CNR.

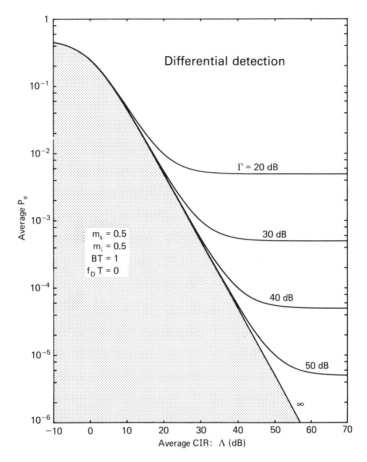

**Figure 10.25** Average $P_e$ versus CIR performance of MSK with differential detection (1). Average CNR $\Gamma$ is a parameter. $f_D T$ is normalized maximum Doppler frequency.

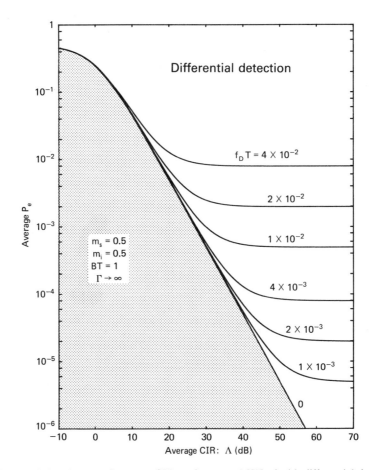

**Figure 10.26** Average $P_e$ versus CIR performance MSK of with differential detection (2). Normalized maximum Doppler frequency $f_D T$ is a parameter. $\Gamma$ is average CNR.

### 10.4.5 Comparison

As mentioned previously, digital FM signals can be detected by three kinds of detection schemes, discriminator, differential, and coherent detection schemes. It would be useful if we could compare the $P_e$ performance of these detection schemes. However, the theoretical $P_e$ performance of coherent detection scheme has not been derived for the fast Rayleigh fading environment of our interest. It is fairly difficult to analyze the tracking performance of a carrier recovery circuit if the temporal variation of a faded carrier phase has to be taken into account. Therefore, in the following discussion we compare the $P_e$ performance of discriminator and differential detection schemes.

Let us first consider the case of $f_D T \to 0$ and $\Lambda \to \infty$, which corresponds to the case of a much-higher signaling rate than the maximum Doppler frequency, and where the CCI is negligible. In this case, the Rayleigh envelope fading is the predominant cause of errors. Letting $f_D T \to 0$ in (10.57) yields

$$
\begin{cases}
P_{e_{\Lambda \to \infty}} = \dfrac{1}{2}\left[1 - \dfrac{\sqrt{3}\Gamma m_s}{(\Gamma + 1)^{1/2}\,(1 + 3\Gamma m_s^2)^{1/2}}\right] \cong \dfrac{1}{4\Gamma}\left(1 + \dfrac{1}{3m_s^2}\right) & \text{(10.59a)} \\[4pt]
\qquad\qquad\qquad \text{for discriminator detection} \\[8pt]
P_{e_{\Lambda \to \infty}} = \dfrac{1}{2}\left\{1 - \dfrac{\Gamma \sin m_s\pi}{[(\Gamma + 1)^2 - \Gamma^2\cos^2 m_s\pi]^{1/2}}\right\} \approx \dfrac{1}{2\Gamma \sin^2 m_s\pi} & \text{(10.59b)}
\end{cases}
$$

If we compute specific numerical values, then for differential detection from these equations we conclude that the average $P_e$ versus CNR performance of differential detection in the quasi-static Rayleigh fading environment is slightly superior to that of discriminator detection unless $m_i = m_s \geq 0.558$.

As for the special case of MSK modulation, that is, $m_i = m_s = 0.5$, the $P_e$ performance of coherent detection in the quasi-static Rayleigh fading environment can be obtained as

$$
P_{e_{\Lambda \to \infty}} = \frac{1}{2}\left[1 - \frac{\Gamma^{1/2}}{(\Gamma + 1)^{1/2}}\right] \cong \frac{1}{4\Gamma} \qquad \text{for coherent detection} \qquad (10.60)
$$

since the detection mechanism is equivalent to that of BPSK with coherent detection. On the other hand, letting $m_s = 0.5$ in (10.59), the $P_e$ *performance of MSK with discriminator and differential detection* schemes are given, respectively, by

$$
\begin{cases}
P_{e_{\Lambda \to \infty}} = \dfrac{1}{2}\left[1 - \dfrac{\Gamma}{(\Gamma + 1)^{1/2}\,(\Gamma + 4/3)^{1/2}}\right] \approx \dfrac{7}{12\Gamma} & \text{(10.61a)} \\[4pt]
\qquad\qquad\qquad \text{for discriminator detection} \\[8pt]
P_{e_{\Lambda \to \infty}} = \dfrac{1}{2(\Gamma + 1)} \approx \dfrac{1}{2\Gamma} & \text{(10.61b)}
\end{cases}
$$

$$
\text{for differential detection}
$$

Equations (10.60) and (10.61) indicate that coherent detection has the best $P_e$ performance for MSK transmission in the quasi-static Rayleigh fading environment. The required values of average CNR for obtaining an average $P_e$ of $1 \times 10^{-3}$ are 24.0 dB, 27.0 dB, and 27.7 dB for coherent, differential, and discriminator detection schemes, respectively.

Let us then consider the case of $\Gamma \to \infty$ and $\Lambda \to \infty$, where the random FM noise caused by the random phase change of a received faded carrier is the predominant cause of error. In this case, the irreducible $P_e$ performances of discriminator and differential detection schemes have already been obtained as (10.57), which can be approximated as

$$
\begin{cases}
P_{e_{\Gamma,\Lambda \to \infty}} \approx \dfrac{1}{2}\left(\dfrac{f_D T}{m_s}\right)^2 & \text{for discriminator detection} \qquad \text{(10.62a)} \\[8pt]
P_{e_{\Gamma,\Lambda \to \infty}} \approx \dfrac{1}{2}\left(\dfrac{\pi f_D T}{\sin m_s\pi}\right)^2 & \text{for discriminator detection} \qquad \text{(10.62b)}
\end{cases}
$$

where

$$J_0(z) \approx 1 - \left(\frac{z}{2}\right)^2 \tag{10.63}$$

(which is an approximation formula) was used for deriving (10.62). Equation (10.62) indicates that the irreducible $P_e$ performance of discriminator detection for high $C/N$ and $C/I$ is superior to that of differential detection irrespective of the values of modulation index. For instance, in the special case of MSK modulation ($m_s = 0.5$), the average $P_e$ of differential detection becomes nearly equal to 2.5 times that of discriminator detection. This result can be understood from the fact that the discriminator, whose detection mechanism is considered to be equivalent to that of a differential detector using a delay line with infinitesimal time delay, is able to track the fast phase change more quickly than the differential detector. In order to analyze the cause of the irreducible $P_e$ performance of coherent detection for such $C/N$ and $C/I$, it is necessary to solve the tracking behavior of a carrier recovery circuit. Though this problem has not been solved yet, a qualitative comparison can be made. Since the reference carrier is usually recovered from the received digital FM signal itself by the use of a narrow-band phase-locked loop with decision-directed channel measurement, its tracking behavior against the fast random phase change, that is, random FM noise, is considered to be inferior to that of the delay line in a differential detector. Thus the irreducible $P_e$ performance of coherent detection in such a condition may be inferior to that of differential detection. Consequently, we conclude that the irreducible $P_e$ performance of the coherent detection scheme is worse than that of the other two detection schemes.

Finally, let us consider the case of $f_D T \to 0$, $\Gamma \to \infty$, and $m_s = m_i = m$, where the CCI predominates in causing errors. In this case, letting $f_D T \to 0$, $\Gamma \to \infty$, and $m_s = m_i = m$ in (10.55) yields

$$P_{e_{\Gamma \to \infty}} = \frac{1}{2(\Lambda + 1)} \qquad \text{for discriminator detection} \tag{10.64a}$$

$$P_{e_{\Gamma \to \infty}} = \frac{1}{2}\left[ 1 - \frac{\Lambda \sin m\pi}{\left\{ (\Lambda + 1)^2 - \left[ \Gamma \cos m\pi + \frac{1}{2}\left( \cos m\pi + \frac{\sin m\pi}{m\pi} \right) \right]^2 \right\}^{1/2}} \right]$$

for differential detection $\tag{10.64b}$

The above equations indicate that the average $P_e$ versus CIR performance of discriminator detection is slightly inferior to that of differential detection unless $m \approx 0.5$. Moreover, as for the special case of MSK modulation, $m_s = m_i = m = 0.5$, the average $P_e$ versus CIR performance of the considered detection schemes in a quasi-static Rayleigh fading environment is given by

$$P_{e_{\Gamma \to \infty}} = \frac{1}{2(\Lambda + 1)} \approx \frac{1}{2\Lambda} \qquad \text{for discriminator detection} \qquad (10.65a)$$

$$P_{e_{\Gamma \to \infty}} = \frac{1}{2}\left[1 - \frac{\Lambda}{\{(\Lambda + 1)^2 - (1/\pi)^2\}^{1/2}}\right] \approx \frac{1}{2\Lambda} \qquad (10.65b)$$

$$\text{for differential detection}$$

$$P_{e_{\Gamma \to \infty}} = \frac{1}{2}\left[1 - \sqrt{\frac{\Lambda}{\Lambda + 1}}\right] \approx \frac{1}{4\Lambda} \qquad (10.65c)$$

$$\text{for coherent detection}$$

### 10.4.6 Effect of Time-Delay Spread (Frequency-Selective Fading)

Possible performance degradation caused by the effect of **time-delay spread**, which is also called **frequency-selective fading,** has not been taken into consideration in our analysis. This degradation is typically negligible in narrow-band single channel per carrier (SCPC) digital land mobile radio transmission systems. However, as has been previously mentioned, potential performance degradation caused by frequency-selective fading cannot be neglected in some cases. Though there are a number of publications that deal with the performance analysis of digital transmission systems in a frequency-selective fading environment [Bello, 10.76; Bailey, 10.77; Hummels, 10.78], a complete analysis of digital FM land mobile radio transmission performance, taking into account all the effects of envelope fading (nonselective), random FM noise (time-selective), and time-delay spread (frequency-

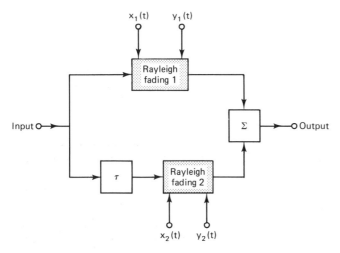

**Figure 10.27** Two-path Rayleigh fading model.

selective) has not been presented yet. Fortunately, for the special case of the MSK modulation technique with differential detection, these three impairments have been analyzed [Hata, 10.79].

A two-path Rayleigh fading model, which is characterized by the sum of two mutually independent Rayleigh fading signals having a time-delay difference $\tau$, as shown in Fig. 10.27, has been assumed.

Provided that each Rayleigh fading signal has identical normalized autocorrelation functions given by $J_0(2\pi f_D T)$ and that the bandwidth of an ideal predetection BPF is set as $BT = 1$, the average $P_e$ performance is presented in the following closed form:

$$P_e = \frac{1}{2}\left\{ 1 - \frac{1}{2}\frac{\Gamma J_0(2\pi f_D T)}{(\Gamma + 1)} \right.$$

$$\left. - \frac{1}{2}\frac{\Gamma J_0(2\pi f_D T) \cos^2\left(\frac{\pi\tau}{2T}\right)}{\left\{ (\Gamma+1)^2 - \left[ \frac{\Gamma}{2} J_0(2\pi f_D T) \sin\left(\frac{\pi\tau}{2T}\right) \right]^2 \right\}^{1/2}} \right\} \qquad (10.66)$$

If $f_D T \ll 1$ and $\tau/T \ll 1$, this equation can be approximately written as

$$P_e \approx P_{e1} + P_{e2} + P_{e3} \qquad (10.67)$$

where

$$\begin{cases} P_{e1} = P_e\left( f_D T = 0, \frac{\tau}{T} = 0 \right) = \frac{1}{2(\Gamma + 1)} \\[2mm] P_{e2} = P_e\left( \Gamma \to \infty, \frac{\tau}{T} = 0 \right) = \frac{1}{2}[1 - J_0(2\pi f_D T)] \\[2mm] P_{e3} = P_e(\Gamma \to \infty, f_D T = 0) = \frac{1}{4}\left[ 1 - \frac{\cos^2\left(\frac{\pi\tau}{2T}\right)}{\left\{ 1 - \frac{1}{4}\sin^2\left(\frac{\pi\tau}{T}\right) \right\}^{1/2}} \right] \end{cases} \qquad (10.68)$$

In these equations, $P_{e1}$, $P_{e2}$, and $P_{e3}$ represent the average $P_e$ due to nonselective Rayleigh envelope fading, time-selective random FM noise, and frequency-selective time-delay spread, respectively. For $\tau/T = 0$, that is, $P_{e3} = 0$, (10.66) coincides with the average $P_e$ formula given by letting $m_s = 0.5$ in (10.57). Computed values of $P_{e1}$, $P_{e1} + P_{e2}$, and $P_{e1} + P_{e3}$ are illustrated by the dashed lines in Fig. 10.28. Values of $P_e$ given by (10.66) are also shown. This figure indicates that there exists a preferable transmission bit rate that is less prone to the effects of both random FM noise and time-delay spread. The preferable transmission bit rate can be determined approximately as the value satisfying $P_{e1} + P_{e2} = P_{e1} + P_{e3}$,

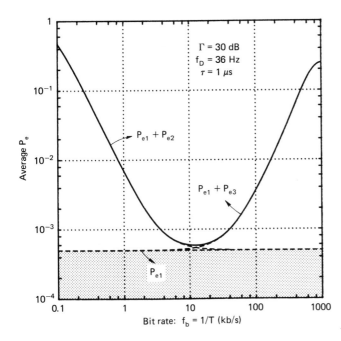

$$\Gamma = 30 \text{ dB}$$
$$f_D = 36 \text{ Hz}$$
$$\tau = 1 \text{ μs}$$

**Figure 10.28** Average $P_e$ versus transmission bit rate of MSK with differential detection. $\Gamma$ is average CNR. $f_D$ is maximum Doppler frequency.

that is, $P_{e2} = P_{e3}$. Since $P_{e2}$ and $P_{e3}$ can be approximated as

$$P_{e2} \approx \frac{\pi^2}{2} (f_D T)^2 \qquad \text{for } f_D T \ll 1 \tag{10.69}$$

$$P_{e3} \approx \frac{\pi^2}{32} (\tau/T)^2 \qquad \text{for } \tau/T \ll 1$$

the preferable transmission bit rate $f_b = 1/T$ is obtained as

$$f_b = \frac{1}{T} = 2 \sqrt{\frac{f_D}{\tau}} \tag{10.70}$$

For example, when $f_D = 36$ Hz and $\tau = 1$, μs, $f_b = 1/T = 12$ kb/s.

## 10.5 DIVERSITY TECHNIQUES FOR DIGITAL LAND MOBILE RADIO

In the previous section we demonstrated that fast multipath fading severely degrades the average $P_e$ performance of digital FM land mobile radio transmission systems. In order to achieve highly reliable digital data transmission without excessively increasing both transmitter power and cochannel reuse distance, it is indispensable to adopt an auxiliary technique that can cope with the fast multipath fading effect. It is well known that diversity reception is one of the most effective techniques for this purpose. Various diversity techniques have been proposed and studied not only for high frequency (HF) radio systems but also for troposcatter and LOS microwave radio relay systems. Operational principles of diversity techniques were discovered in HF radio experiments in the 1930s [Schwartz, 10.80; Sunde, 10.81]. Diversity techniques for VHF and UHF and microwave land mobile radio applications have also been studied for many years [Jakes, 10.82; 10.83; Parsons, 10.84; Lee, 10.85]. Most of them are for analog mobile radio, but they can, in principle, also be applied to digital mobile radio systems. The benefits obtained by diversity techniques are increasing with the increased demand for digital land mobile radio services, since digital transmission is more susceptible to fast multipath fading effects.

### 10.5.1 Branch Construction Methods

A diversity technique requires a number of signal transmission paths, named **diversity branches**, all of which carry the same information but have uncorrelated multipath fadings, and a circuit to combine the received signals. Depending upon the land mobile radio propagation characteristics, there are a number of methods to construct diversity branches. Generally, they are classified into the following five categories: (1) space, (2) angle, (3) polarization, (4) frequency, and (5) time diversity. Each of these methods will be reviewed briefly.

**Space diversity**, which has been most widely used because it can be implemented simply and economically, comprises a single transmitting antenna and a number of receiving antennas. Spacing between adjacent receiving antennas is

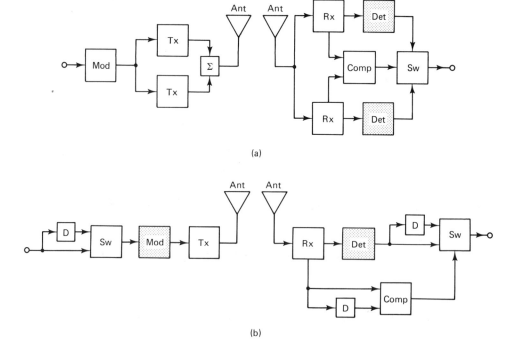

(a)

(b)

**Figure 10.29** Branch construction methods using frequency and time. (a) Frequency diversity. (b) Time diversity.

chosen so that multipath fading appearing in each diversity branch becomes uncorrelated. **Angle diversity**, which is also called direction diversity, requires a number of directive antennas. Each antenna responds independently to a wave propagating with specific angle and direction to produce uncorrelated faded signals. **Polarization diversity**, in which only two diversity branches are available, is effective because the signals transmitted through two orthogonally polarized propagation paths have uncorrelated fading statistics in the usual VHF and UHF land mobile radio environment. Difference in frequency and/or time can also be utilized to construct diversity branches with uncorrelated fading statistics. Figure 10.29 illustrates the block diagrams of two-branch **frequency** and **time diversity** techniques. The required frequency and time spacing can be determined from the characteristics of the time-delay spread and the maximum Doppler frequency. A common advantage of these two techniques compared with space, angle, and polarization diversity techniques is that the number of transmitting and receiving antennas can be reduced to one, and the disadvantage is that a wider bandwidth spectrum is required. **Error-correction coding**, which is peculiar to digital transmission systems, can be regarded as a kind of time diversity technique. Except for polarization diversity, there exists no limit, in principle, to the number of diversity branches.

### 10.5.2 Combining Methods

A number of methods have been studied for many years to combine uncorrelated faded signals obtained from the diversity branches. They are usually classified into the following three categories: (1) maximal-ratio combining, (2) equal-gain combining, and (3) selection [Brennan, 10.86; Feher, 10.87; 10.88]. For the coherent detection case there is no difference whether the combining is carried out in the predetection or in the postdetection stage. However, for noncoherently detected systems, such as FM discriminator or differential detection schemes, a slight difference in performance exists between predetection and postdetection combining methods. Block diagrams of three predetection combining methods are shown in Fig. 10.30.

Assuming ideal operation, predetection stage **maximal-ratio combining** achieves the best performance improvement compared with the other methods. However, it requires cophasing, weighting, and summing circuits, as shown in Fig. 10.30(a) resulting in the most complicated implementation. An **equal-gain combining** system diagram is shown in Fig. 10.30(b). It is similar to maximal-ratio combining, except that the weighting circuits are omitted. The performance improvement obtained by an equal-gain combiner is slightly inferior to that of a maximal-ratio combiner, since interference and noise corrupted signals may be combined with high quality (noise-free) signals. For VHF, UHF and microwave land mobile radio applications, both the maximal-ratio and the equal-gain combining methods are unsuitable. It is difficult to realize a cophasing circuit having a precise and stable tracking performance in a rapidly changing random phase multipath fading environment. Compared with these two combining methods, the **selection** method is more suitable for mobile radio applications due to its simple implementation. In this method, the best diversity branch having the highest signal level is selected

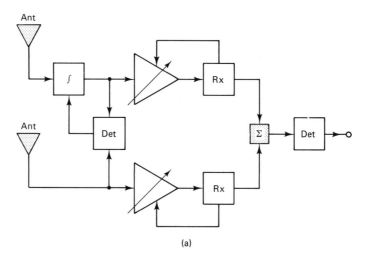

(a)

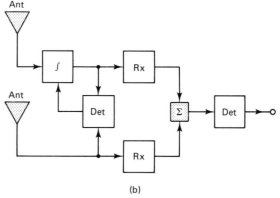

(b)

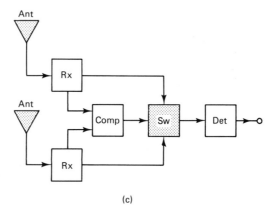

(c)

**Figure 10.30** Typical combining methods. (a) Maximal-ratio. (b) Equal-gain. (c) Selection.

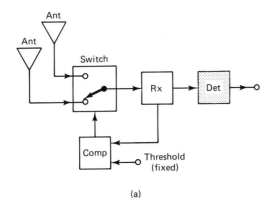

(a)

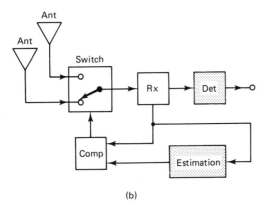

(b)

**Figure 10.31** Switching methods with fixed and variable threshold. (a) Fixed threshold. (b) Variable threshold.

(see Fig. 10.30(c)). In addition, stable operation is easily achieved even in the fast multipath fading environments. It has been shown that the performance improvement obtained by the selection method is only slightly inferior to the ideal one achieved by the maximal-ratio combining method [Feher, 10.87].

The major disadvantage of this method is that continuous monitoring of the signals requires the same number of receivers as the number of diversity branches. This can be alleviated by the use of a **switching** or **scanning** receiver [Rustako, 10.89; Shortall, 10.90]. Figure 10.31 illustrates such a receiver. In Fig. 10.31(a) switching from one branch to the other occurs when the signal level falls below a threshold. The threshold can be set at a fixed value within a small area, but it is not necessarily the best in the whole service area. Therefore, the threshold value should be controlled adaptively with the vehicle movement, as shown in Fig. 10.31(b). The performance improvement obtained by the switching method depends upon the choice of the threshold value and the amount of feedback time delay caused by monitoring, estimation, decision, and switching. However, envelope and phase transients of a carrier, which may be caused by switching in the predetection stage, reduce the performance improvement. In angle modulation systems, such as digital FM systems, phase transients cause errors in the detected data stream, whereas envelope transients can be removed by a predetection band-pass limiter.

Figure 10.32 illustrates a simple implementation of the switching method, in

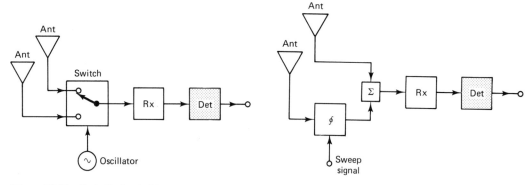

**Figure 10.32** Periodical switching method.

**Figure 10.33** Combining method using phase-sweeping scheme.

which the diversity branches are selected periodically by using a conventional free-running oscillator. This method is used in low-speed large-deviation digital FM systems in which phase transients caused by **periodic switching** can be suppressed. The switching rate, which is the only selectable parameter, is chosen to be equal or larger of the double of the signaling bit rate, so that the signal of the best branch can be received in every signaling period. The same performance improvement as that of a conventional switching method can be achieved by using an FM discriminator followed by a suitable LPF. The effectiveness of this method is experimentally demonstrated in a thermal noise and (CCI) environment [Adachi, 10.91; 10.92]. However, the performance improvement in an adjacent-channel interference (ACI) environment may be reduced since periodic switching in the predetection RF stage could cause the adjacent-channel spectrum to fold over into the desired channel band. This could be solved by an increase of the adjacent-channel selectivity of the receiver.

Another effective variation of the switching method using a single receiver is the **phase-sweeping** method shown in Fig. 10.33. Provided that the sweeping rate is higher than twice the highest frequency of the modulation signal, the same diversity improvement effect as that obtained by the periodic switching method can be achieved [Parsons, 10.93; Rogers, 10.94]. The phase-sweeping method can be regarded as a **mode-averaging** method using spaced antennas with electronically scanned directional patterns [Kazal, 10.95; Villard, 10.96]. Most of the studies related to this method are concerned with AM systems. However, as shown in [Rogers, 10.94], this method can also be applied to FM systems. A **frequency-offset** method such as the one illustrated in Fig. 10.34 is also useful, since the same improvement can be achieved as in the case of the linear phase-sweeping method. This method becomes even more attractive when there are multiple diversity branches.

### 10.5.3 Statistical CNR and CIR Performance Improvement

Performance improvements achieved by various diversity techniques are assessed measurable reduction of the fading dynamic range of the CNR and the CIR of the combined signals. We now discuss the diversity improvement effects on the sta-

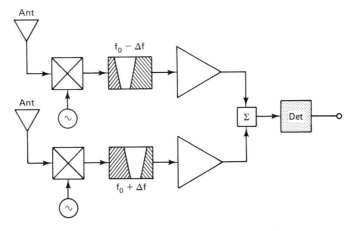

**Figure 10.34** Combining method using frequency-offset scheme.

tistics of CNR and CIR, respectively. In the following model we investigate two-branch predetection diversity systems. These systems are very effective and useful for practical mobile radio applications.

Let us first examine the statistics of CNR and CIR, without diversity, assuming a Rayleigh fading model. Since the instantaneous CNR $\gamma$ is in proportion to the square of a Rayleigh-distributed signal envelope, the pdf of $\gamma$, where $\gamma \geqq 0$, is given by the following exponential distribution:

$$p(\gamma) = \frac{1}{\Gamma} e^{-\gamma/\Gamma} \tag{10.71}$$

where $\Gamma$ denotes the average value of $\gamma$. The instantaneous CIR $\gamma$ is given by the ratio of squared signal envelope to squared interference envelope. These envelopes are mutually independent Rayleigh-distributed random variables. The pdf of $\lambda$, where $\lambda \geqq 0$, is then obtained as the following $F$-distribution:

$$p(\lambda) = \frac{\Lambda}{(\lambda + \Lambda)^2} \tag{10.72}$$

when $\Lambda$ denotes the average value of $\lambda$.

Letting $\gamma_i$ $(i = 1, 2)$ and $\gamma$ be the instantaneous CNR of each diversity branch and of the combined branch, respectively, the three combining methods may be described by

$$\begin{cases} \gamma = \gamma_1 + \gamma_2 & \text{for maximal-ratio combining} \tag{10.73a} \\[2mm] \sqrt{\gamma} = \sqrt{\frac{\gamma_1}{2}} + \sqrt{\frac{\gamma_2}{2}} & \text{for equal-gain combining} \tag{10.73b} \\[2mm] \gamma = \begin{cases} \gamma_1 \cdots \gamma_1 > \gamma_2 \\ \gamma_2 \cdots \gamma_1 < \gamma_2 \end{cases} & \text{for selection} \tag{10.73c} \end{cases}$$

Assuming that the two Rayleigh fades that appear on the diversity branches are mutually independent and that both Rayleigh faded signals have an equal average power, that is, $\Gamma_1 = \Gamma_2 = \Gamma$, where $\Gamma_i$ ($i = 1, 2$) denotes the average value of $\gamma_i$ ($i = 1, 2$), the pdf's of $\gamma$ for the above three combining methods have been expressed [Jakes, 10.83] by

$$p(\gamma) = \frac{\gamma}{\Gamma^2} e^{-\gamma/\Gamma} \approx \frac{\gamma}{\Gamma^2} \qquad \text{for maximal-ratio combining} \qquad (10.74a)$$

$$p(\gamma) \approx \frac{4}{3}\frac{\gamma}{\Gamma^2} \qquad \text{for equal-gain combining} \qquad (10.74b)$$

$$p(\gamma) = \frac{d}{d\gamma}\{(1 - e^{-\gamma/\Gamma})^2\} \approx 2\frac{\gamma}{\Gamma^2} \qquad \text{for selection} \qquad (10.74c)$$

where the approximation holds for $\gamma \ll \Gamma$. The outage rates of $\gamma$—that is, the probabilities that $\gamma$ does not exceed a specified value $\gamma_s$—can be obtained by using the following integral:

$$\text{Prob}[\gamma \leq \gamma] = \int_0^{\gamma_s} p(\gamma)\, d\gamma \qquad (10.75)$$

Computed results of (10.75) for (10.71) and (10.74) are shown in Fig. 10.35. We conclude from this figure that the fading dynamic range of $\gamma$ may be remarkably reduced by the use of the diversity techniques and that there is only a slight difference between the performance improvements of the above combining methods. The pdf's of $\lambda$ can be obtained in a similar manner as derived in [Arredondo, 10.61]. They are

$$p(\lambda) = \frac{d}{d\lambda}\left\{\left(\frac{\lambda}{\lambda + \Lambda}\right)^2\right\} \qquad \text{for maximal-ratio combining} \qquad (10.76a)$$

$$p(\lambda) \approx \frac{d}{d\lambda}\left[\left(\frac{\gamma}{\gamma + \dfrac{\sqrt{3}}{2}\Lambda}\right)^2\right] \qquad \text{for equal-gain combining} \qquad (10.76b)$$

$$p(\lambda) = \left\{\frac{2\Lambda}{(\lambda + \Lambda)^2} - \frac{2\Lambda}{(2\lambda + \Lambda)^2}\right\} \qquad \text{for selection} \qquad (10.76c)$$

where perfect-pilot cophasing scheme is assumed for both the maximal-ratio combining and the equal-gain combining methods and the larger desired signal power algorithm is assumed for the selection method. The mutual correlation among the four Rayleigh fades appearing on the desired signals and the undesired interferences obtained from the respective branches is assumed to be negligible. It is also assumed that the average CIRs of the branches are equal to each other—that is, $\Lambda_1 = \Lambda_2 = \Lambda$. Computed values of the outage rate given by

$$\text{Prob}[\lambda \leq \lambda_s] = \int_0^{\lambda_s} p(\lambda)\, d\lambda \qquad (10.77)$$

are graphically shown in Fig. 10.36. This figure shows that the fading dynamic range of $\lambda$ can also be reduced remarkably by diversity techniques. There is only

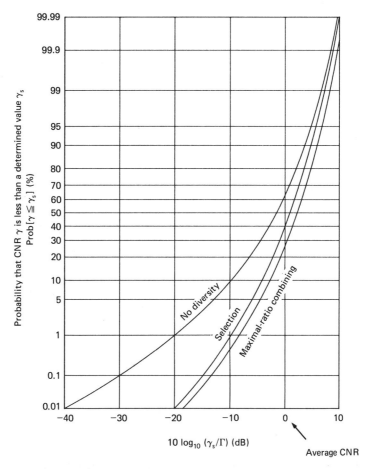

**Figure 10.35** Cumulative probability distribution of CNR $\gamma$. $\Gamma$ is the average CNR of each branch. $\gamma_s$ is a specified $\gamma$.

a slight difference in the performance improvement of the various combining methods.

### 10.5.4 Average $P_e$ Performance Improvement

The performance improvement of a diversity technique can also be assessed by its ability to reduce, for a specific $P_e$, the permissible values of CNR and CIR. Diversity improvement effects on the average $P_e$ performance of digital radio transmission systems in multipath fading environments have been extensively studied [Pierce, 10.97; Lindsey, 10.98; Schwartz, 10.99; Sunde, 10.100; Prapinmongkolkarn, 10.101]. For digital FM land mobile radio transmission systems, diversity improvement has been analyzed by taking into account the effects of nonselective envelope fading and of time-selective random FM noise [Adachi, 10.102; 10.103], whereas the effect of frequency-selective time-delay spread has been neglected. The impact of postdetection selection diversity systems on the average $P_e$ perform-

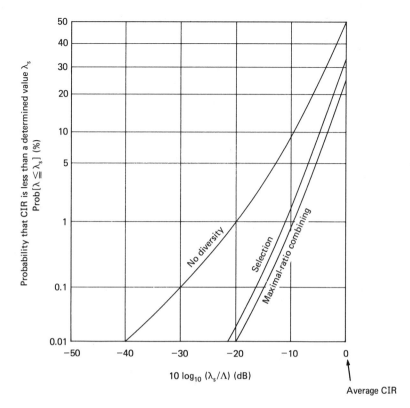

**Figure 10.36** Cumulative probability distribution of CIR $\lambda$. $\Lambda$ is the average CIR of each branch. $\lambda_s$ is a specified $\lambda$.

ance of MSK land mobile radio transmission systems has been analyzed in [Hata, 10.79]. In the following discussions, two-branch predetection diversity improvement effects on the average $P_e$ performance of MSK land mobile radio transmission systems in the quasi-static nonselective Rayleigh fading environment will be considered. The effects of thermal noise and CCI are considered separately.

### Average $P_e$ versus CNR Performance

Let us first consider a case in which the effect of CCI can be disregarded. Provided that an ideal selection method is used, the average $P_e$ versus CNR performance, with and without diversity, is denoted by $P_e^{(2)}(\Gamma)$ and $P_e^{(1)}(\Gamma)$, respectively. It can be obtained from (10.71) and (10.74c) as

$$P_e^{(2)}(\Gamma) = \int_0^\infty P_e(\gamma) \frac{d}{d\gamma} \{(1 - e^{-\gamma/\Gamma})^2\} \, d\gamma \qquad (10.78)$$

$$P_e^{(1)}(\Gamma) = \int_0^\infty P_e(\gamma) \frac{1}{\Gamma} e^{-\gamma/\Gamma} \, d\gamma \qquad (10.79)$$

where $P_e(\gamma)$ denotes the static $P_e$ versus CNR performance in the nonfading condition. Therefore, the following relationship between $P_e^{(2)}(\Gamma)$ and $P_e^{(1)}(\Gamma)$ can be shown to be given by

$$P_e^{(2)}(\Gamma) = 2P_e^{(1)}(\Gamma) - P_e^{(1)}\left(\frac{\Gamma}{2}\right) \tag{10.80}$$

For the maximal-ratio combining method, a similar relationship can be obtained from (10.70) and (10.73a) as

$$P_e^{(2)}(\Gamma) = P_e^{(1)}(\Gamma) - \frac{1}{\Gamma}\frac{\partial}{\partial(1/\Gamma)}\{P_e^{(1)}(\Gamma)\} \tag{10.81}$$

These relationships can be generalized to the case of $M$-branch diversity reception.

By using the previously obtained relationships, (10.60) and (10.61), $P_e^{(1)}(\Gamma)$ for MSK modulation (that is, no diversity case), we have

$$
\begin{cases}
P_e^{(1)}(\Gamma) = \frac{1}{2}\left\{1 - \frac{\Gamma}{[(\Gamma + 1)(\Gamma + \frac{4}{3})]^{1/2}}\right\} & (10.82a) \\[2mm]
\qquad \approx \frac{7}{12\Gamma} \quad \text{for discriminator detection} \\[3mm]
P_e^{(1)}(\Gamma) = \frac{1}{2}\left[1 - \frac{\Lambda}{(\Gamma + 1)}\right] & (10.82b) \\[2mm]
\qquad \approx \frac{1}{2\Gamma} \quad \text{for differential detection} \\[3mm]
P_e^{(1)}(\Gamma) = \frac{1}{2}\left(1 - \sqrt{\frac{\Gamma}{\Gamma + 1}}\right) & (10.82c) \\[2mm]
\qquad \approx \frac{1}{4\Gamma} \quad \text{for coherent detection}
\end{cases}
$$

By substituting (10.82) into (10.80), $P_e^{(2)}(\Gamma)$ for the selection method is obtained as

$$
\begin{cases}
P_e^{(2)}(\Gamma) = \frac{1}{2}\left\{1 - \frac{2\Gamma}{[(\Gamma + 1)(\Gamma + \frac{4}{3})]^{1/2}} + \frac{\Gamma}{[(\Gamma + 2)(\Gamma + \frac{8}{3})]^{1/2}}\right\} & (10.83a) \\[2mm]
\qquad \approx \frac{11}{8\Gamma^2} \quad \text{for discriminator detection} \\[3mm]
P_e^{(2)}(\Gamma) = \frac{1}{(\Gamma + 1)(\Gamma + 2)} & (10.83b) \\[2mm]
\qquad \approx \frac{1}{\Gamma^2} \quad \text{for differential detection} \\[3mm]
P_e^{(2)}(\Gamma) = \frac{1}{2}\left(1 - 2\sqrt{\frac{\Gamma}{\Gamma + 1}} + \sqrt{\frac{\Gamma}{\Gamma + 2}}\right) & (10.83c) \\[2mm]
\qquad \approx \frac{3}{4\Gamma^2} \quad \text{for coherent detection}
\end{cases}
$$

In addition, by substituting (10.82) into (10.81), $P_e^{(2)}(\Gamma)$ for the maximal-ratio combining method is obtained as

$$P_e^{(2)}(\Gamma) = \frac{1}{2}\left\{1 - \frac{\Gamma}{[(\Gamma + 1)(\Gamma + \frac{4}{3})]^{1/2}} - \frac{\Gamma(\frac{7}{6}\Gamma + \frac{4}{3})}{[(\Gamma + 1)(\Gamma + \frac{4}{3})]^{3/2}}\right\}$$

$$\approx \frac{11}{8\Gamma^2} \quad \text{for discriminator detection} \tag{10.84a}$$

$$P_e^{(2)}(\Gamma) = \frac{1}{2(\Gamma + 1)^2}$$

$$\approx \frac{1}{2\Gamma^2} \quad \text{for differential detection} \tag{10.84b}$$

$$P_e^{(2)}(\Gamma) = \frac{1}{2}\left[1 - \sqrt{\frac{\Gamma}{\Gamma + 1}} + \frac{1}{2}\sqrt{\frac{\Gamma}{(\Gamma + 1)^3}}\right]$$

$$\approx \frac{3}{16\Gamma^2} \quad \text{for coherent detection} \tag{10.84c}$$

Curves obtained by computing (10.83) and (10.84) are shown in Figs. 10.37 and 10.38, respectively. Moreover, curves representing (10.82) are also shown in both figures. These figures indicate that in order to improve the average $P_e$ ten times, it is necessary to increase the average CNR by about 5 dB when two-branch diversity techniques are used, whereas it is as much as 10 dB in the case of single-branch reception.

### Average $P_e$ versus CIR Performance

Let us now consider another case in which the effect of CCI predominates in causing errors. In this case, the average $P_e$ versus CIR performances, with and without two-branch selection diversity, are denoted by $P_e^{(2)}(\Lambda)$ and $P_e^{(1)}(\Lambda)$, respectively. It can be obtained from (10.72) and (10.76c) as

$$\begin{cases} P_e^{(2)}(\Lambda) = \int_0^\infty P_e(\lambda)\left[\frac{\Lambda}{(\lambda + \Lambda)^2} - \frac{2\Lambda}{(2\lambda + \Lambda)^2}\right]d\lambda & (10.85) \\[3mm] P_e^{(1)}(\Lambda) = \int_0^\infty P_e(\lambda)\frac{\lambda}{(\lambda + \Lambda)^2}d\lambda & (10.86) \end{cases}$$

where $P_e(\lambda)$ denotes the static $P_e$ versus CIR performance in the nonfading condition. From these equations, the following relationship between $P_e^{(2)}(\Lambda)$ and $P_e^{(1)}(\Lambda)$ can be derived:

$$P_e^{(2)}(\Lambda) = 2P_e^{(1)}(\Lambda) - P_e^{(1)}\left(\frac{\Lambda}{2}\right) \tag{10.87}$$

Furthermore, for the maximal-ratio combining method, a similar relationship can

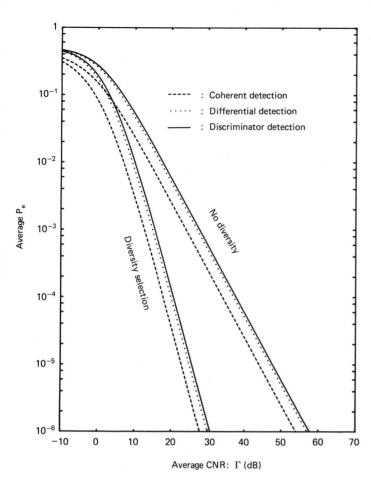

**Figure 10.37** Diversity improvement effect on average $P_e$ versus CNR performance of MSK modulated system in the quasi-static Rayleigh fading environment: Two-branch predetection selection method is used.

be obtained from (10.72) and (10.76a) as

$$P_e^{(2)}(\Lambda) = P_e^{(1)}(\Lambda) + \Lambda \frac{\partial}{\partial \Lambda} \{P_e^{(1)}(\Lambda)\} \tag{10.88}$$

These relationships can be generalized to the case of *M*-branch diversity systems.

As has been given previously by (10.65), $P_e^{(1)}(\Lambda)$ for each of the three detection schemes is

$$
\begin{cases}
P_e^{(1)}(\Lambda) = \dfrac{1}{2(\Lambda + 1)} \approx \dfrac{1}{2\Lambda} \quad \text{for discriminator detection} \tag{10.89a} \\[3mm]
P_e^{(1)}(\Lambda) = \dfrac{1}{2}\left\{ 1 - \dfrac{\Lambda}{[(\Lambda + 1)^2 - (1/\pi)^2]^{1/2}} \right\} \cong \dfrac{1}{2\Lambda} \tag{10.89b} \\[1mm]
\qquad\qquad\qquad \text{for differential detection} \\[3mm]
P_e^{(1)}(\Lambda) = \dfrac{1}{2}\left( 1 - \sqrt{\dfrac{\Lambda}{\Lambda + 1}} \right) \approx \dfrac{1}{4\Lambda} \quad \text{for coherent detection} \tag{10.89c}
\end{cases}
$$

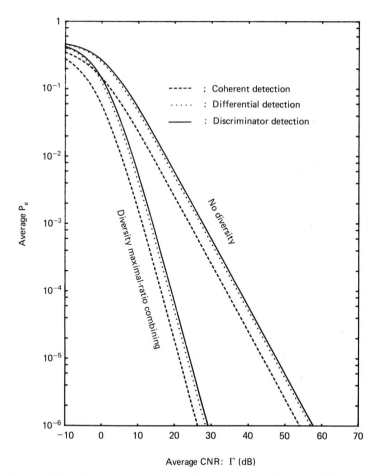

**Figure 10.38** Diversity improvement effect on average $P_e$ versus CNR performance of MSK modulated system in the quasi-static Rayleigh fading environment: Two-branch predetection maximal-ratio-combining method is used.

From (10.87) and (10.89), $P_e^{(2)}(\Lambda)$ for the selection method can be obtained as

$$P_e^{(2)}(\Lambda) = \frac{1}{(\Lambda + 1)(\Lambda + 2)} \approx \frac{1}{\Lambda^2} \qquad \text{for discriminator detection} \qquad (10.90a)$$

$$P_e^{(2)}(\Lambda) = \frac{1}{2}\left\{1 - \frac{2\Lambda}{[(\Lambda + 1)^2 - (1/\pi)^2]^{1/2}} + \frac{\Lambda}{[(\Lambda + 2)^2 - (2/\pi)^2]^{1/2}}\right\} \qquad (10.90b)$$

$$\approx \frac{1}{\Lambda^2}\left(1 + \frac{1}{2\pi^2}\right) \approx \frac{1}{\Lambda^2} \qquad \text{for differential detection}$$

$$P_e^{(2)}(\Lambda) = \frac{1}{2}\left[1 - 2\sqrt{\frac{\Lambda}{\Lambda + 1}} + \sqrt{\frac{\Lambda}{\Lambda + 2}}\right] \qquad (10.90c)$$

$$\approx \frac{3}{4\Lambda^2} \qquad \text{for coherent detection}$$

In addition, by substituting (10.89) into (10.88), $P_e^{(2)}(\Lambda)$ for the maximal-ratio combining method can be obtained as

$$
\begin{cases}
P_e^{(2)}(\Lambda) = \dfrac{1}{2(\Lambda + 1)^2} \cong \dfrac{1}{2\Lambda^2} \qquad \text{for discriminator detection} & (10.91a) \\[4mm]
P_e^{(2)}(\Lambda) = \dfrac{1}{2}\left\{1 - \dfrac{2\Lambda}{[(\Lambda + 1)^2 - (1/\pi)^2]^{1/2}} + \dfrac{\Lambda^2(\Lambda + 1)}{[(\Lambda + 1)^2 - (1/\pi)^2]^{3/2}}\right\} & (10.91b) \\[4mm]
\qquad \approx \dfrac{1}{2\Lambda^2}\left(1 + \dfrac{2}{\pi^2}\right) = \dfrac{1}{2\Lambda^2} \qquad \text{for differential detection} \\[4mm]
P_e^{(2)}(\Lambda) = \dfrac{1}{2}\left[1 - \dfrac{3}{2}\sqrt{\dfrac{\Lambda}{\Lambda + 1}} + \dfrac{1}{2}\sqrt{\dfrac{\Lambda^3}{(\Lambda + 1)^3}}\right] & (10.91c) \\[4mm]
\qquad \approx \dfrac{3}{16\Lambda^2} \qquad \text{for coherent detection}
\end{cases}
$$

Figures 10.39 and 10.40 show curves obtained by computing (10.90) and (10.91), respectively, where calculated results of (10.89) are also shown for comparison. As with the curves in Figs. 10.37 and 10.38, these figures indicate that in order to improve the average $P_e$ ten times, it is necessary to increase the average CIR by about 5 dB when diversity improvement is available, whereas an increase of 10 dB is required in the case of no-diversity reception.

## 10.6 RADIO LINK DESIGN OF DIGITAL LAND MOBILE RADIO

As described in Section 10.3, VHF and UHF land mobile radio propagation characteristics can be approximately modeled by: (1) fast multipath Rayleigh fading; (2) slow log-normal shadowing; and (3) path loss variation with distance. Radio link design, which is one of the most important problems of mobile radio engineers, must be carried out by taking these effects into account. Fundamental parameters of radio link design for cellular land mobile radio systems are transmitter power and cochannel reuse distance. These parameters are determined for a specified *transmission quality* and *allowable outage*. Several design procedures have been proposed and studied [Araki, 10.104; Lundquist 10.105; Yoshikawa, 10.106; Kamata, 10.107; Hansen, 10.108; Gosling, 10.109; MacDonald, 10.110; French, 10.111]. A comprehensive design approach, suitable for VHF and UHF cellular land mobile radio systems, is presented in this section. For the system design, transmission quality requirement is obtained from a knowledge of the transmission performance in a pure multipath Rayleigh fading environment without log-normal shadowing. Thermal noise and CCI are also taken into account. Outage, which is defined as a fraction of the service area over which the required transmission quality cannot be maintained within the service area, is determined from the characteristics of log-normal shadowing and path loss variation with distance.

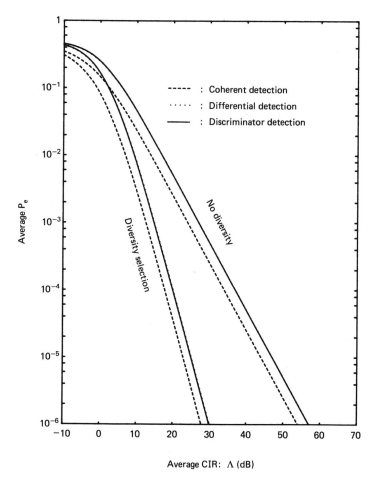

**Figure 10.39** Diversity improvement effect on average $P_e$ versus CIR performance of MSK modulated system in the quasi-static Rayleigh fading environment: Two-branch predetection selection method is used.

### 10.6.1 Outage and Margin

Thermal noise and CCI are two main factors affecting the transmission quality. **Outage**, which takes these two factors into account, can be derived as follows.

Figure 10.41 depicts a simple model of geographical cochannel reuse in cellular land mobile radio systems. Let $X$ and $Y$ be the *local means* of the desired signal and the undesired interference, respectively. Assuming that $X$ and $Y$ are subjected to mutually independent log-normal shadowing, the joint pdf of $X$ and $Y$ is given by

$$p(X, Y) = \frac{1}{2\pi\sigma^2 XY} e^{-(1/2\sigma^2)[\ln^2(X/X_m) + \ln^2(Y/Y_m)]} \tag{10.92}$$

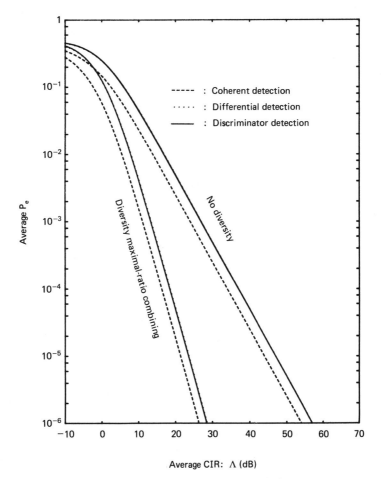

Average CIR: Λ (dB)

**Figure 10.40** Diversity improvement effect on average $P_e$ versus CIR performance of MSK modulated system in the quasi-static Rayleigh-fading enviroment: Two-branch predetection maximal-ratio-combining method is used.

In (10.92), $\sigma$ is the standard deviation, whose value in decibels* is empirically shown as 5 to 12 dB in a typical urban area [Okumura, 10.47; Egli, 10.56; Black, 10.57], and $X_m$ and $Y_m$ are the *area means* of $X$ and $Y$, which are given, respectively, by

$$\begin{cases} X_m = X_m(r_1) = A \cdot r_1^{-\alpha} \\ Y_m = Y_m(r_2) = A \cdot r_2^{-\alpha} \end{cases} \tag{10.93}$$

where $A$ and $\alpha$ are the propagation parameters and the value of $\alpha$ is about 3 to 4 in an urban area [Hata, 10.58]. Moreover, $r_1$ and $r_2$ are the distances from a

---

*Letting the standard deviation in decibels be expressed as $\sigma_0$, the relation between $\sigma_0$ and $\sigma$ is given by $\sigma_0 = 10 \cdot \sigma \cdot \log_{10}e$.

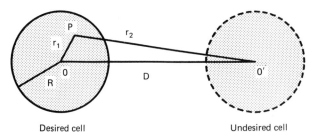

**Figure 10.41** Simple model of geographical cochannel reuse. (Printed by permission of the IEEE [Hansen, 10.108]. © 1982 IEEE).

Desired cell        Undesired cell

moving vehicle to the desired and undesired base stations, respectively. Letting the local means of the CNR and the CIR be $\Gamma$ and $\Lambda$, respectively, and changing the variables in (10.92) such that $X = \Gamma$ and $X/Y = \Lambda$, the joint pdf of $\Gamma$ and $\Lambda$ is given by

$$p(\Gamma, \Lambda) = \frac{1}{2\pi\sigma^2 \Gamma \Lambda} e^{-(1/2\sigma^2)[\ln^2(\Gamma/\Gamma_m) + \ln 2(\Lambda/\Lambda_m \cdot \Gamma_m/\Gamma)]} \tag{10.94}$$

where $\Gamma_m = \Gamma_m(r_1)$ and $\Lambda_m = \Lambda_m(r_1, r_2)$ are the area means of CNR and CIR, respectively.

For the threshold level $\Gamma_{th}$ and $\Lambda_{th}$ specified by the required transmission quality, the probability that $\Gamma \leq \Gamma_{th}$ or $\Lambda \leq \Lambda_{th}$ is expressed as

$$\begin{aligned}\text{Prob}[\Gamma \leq \Gamma_{th} \quad \text{or} \quad \Lambda \leq \Lambda_{th}] \\ = \text{Prob}[\Gamma \leq \Gamma_{th}] + \text{Prob}[\Lambda \leq \Lambda_{th}] - \text{Prob}[\Gamma \leq \Gamma_{th} \quad \text{and} \quad \Lambda \leq \Lambda_{th}]\end{aligned} \tag{10.95}$$

where

$$\begin{cases} \text{Prob}[\Gamma \leq \Gamma_{th}] = \displaystyle\int_0^{\Gamma_{th}} \int_0^{\infty} p(\Gamma, \Lambda)\, d\Gamma\, d\Lambda & (10.96) \\[3mm] \text{Prob}[\Lambda \leq \Lambda_{th}] = \displaystyle\int_0^{\infty} \int_0^{\Lambda_{th}} p(\Gamma, \Lambda)\, d\Gamma\, d\Lambda & (10.97) \\[3mm] \text{Prob}[\Gamma \leq \Gamma_{th} \quad \text{and} \quad \Lambda \leq \Lambda_{th}] = \displaystyle\int_0^{\Gamma_{th}} \int_0^{\Lambda_{th}} p(\Gamma, \Lambda)\, d\Gamma\, d\Lambda & (10.98) \end{cases}$$

Substituting (10.94) into (10.96) through (10.98) and making some modifications, (10.95) becomes

$$\begin{aligned} \text{Prob}&[\Gamma \leq \Gamma_{th} \quad \text{or} \quad \Lambda \leq \Lambda_{th}] \\ &= \frac{1}{2}\text{erfc}\left\{\frac{\ln(\Gamma_m/\Gamma_{th})}{\sqrt{2}\sigma}\right\} + \frac{1}{2}\text{erfc}\left\{\frac{\ln(\Lambda_m/\Lambda_{th})}{\sqrt{2}\sigma}\right\} \\ &\quad - \frac{1}{2\sqrt{\pi}}\int_{-\infty}^{\ln(\Gamma_{th}/\Gamma_m)/\sqrt{2}\sigma} e^{-t^2}\text{erfc}\left\{t + \frac{\ln(\Lambda_m/\Lambda_{th})}{\sqrt{2}\sigma}\right\} dt \end{aligned} \tag{10.99}$$

where erfc is the complementary error function. Equation (10.99) indicates that

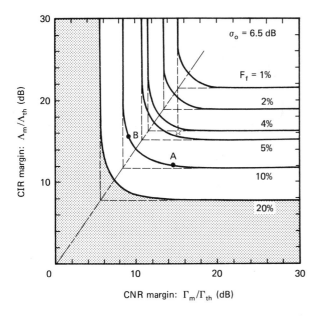

**Figure 10.42** Required margins for CNR and CIR. (Printed by permission of the IEEE [Hansen, 10.108]. © 1982 IEEE).

the outage which takes into account thermal noise and CCI is a function of the required CNR and CIR margins, that is, $\Gamma_m/\Gamma_{th}$ and $\Lambda_m/\Lambda_{th}$. In other words, (10.99) presents the required area means of CNR and CIR $\Gamma_m$ and $\Lambda_m$ for the threshold levels $\Gamma_{th}$ and $\Lambda_{th}$ and the allowable outage Prob$[\Gamma \leq \Gamma_{th}$ or $\Lambda \leq \Lambda_{th}] = F_f$.

Figure 10.42 shows the required margins for CNR and CIR computed from (10.99), where the dashed lines are the asymptotes. The lines parallel to the horizontal axis represent the CIR margin when the CNR margin is infinite, and the lines parallel to the vertical axis represent the CNR margin when the CIR margin is infinite. From this figure, the following conclusions can be drawn.

1. Allotment of the outage for thermal noise and CCI can be made according to the system scale or system grade. For example, when the total allowable outage is 10%, it is possible to make a link design for point *A* which gives priority to thermal noise or for point *B* which gives priority to CCI. The former link design is suitable for realizing large-cell systems, whereas the latter one is for smaller-cell high-capacity systems.

2. When the outage is allotted separately for the respective factors, the required margins for CNR and CIR can be calculated separately. For example, assuming that the total specified outage is 5% and that 1% is allotted for thermal noise and 4% for CCI, the required margins for CNR and CIR are determined from the point marked by the star in this figure. As the point lies a little above the 5% curve, this design requires a slightly larger margin for the specified outage. The same relation generally holds for other allotted values. This procedure leads always to a good performance.

### 10.6.2 Design Procedure

Based on the relationship between outage and margin, the transmitter power and the cochannel reuse distance, both of which are fundamental parameters of a mobile radio link design, can be determined.

Threshold values for the local mean of CNR, $\Gamma_{th}$, and the local mean of CIR, $\Lambda_{th}$, are assumed to be specified separately based on the transmission-quality requirement in a pure Rayleigh fading environment without log-normal shadowing. The allowable outage for thermal noise and CCI can be determined separately. Thus the transmitter power and the cochannel reuse distance can be determined as follows.

**Determination of Transmitter Power**

Let $F_f^1$ be the outage at some point $P$ in the desired cell as shown in Fig. 10.41. The relation between $F_f^1$ and the required CNR margin $\Gamma_m/\Gamma_{th}$ is given by

$$
\begin{aligned}
F_f^1 &= \text{Prob}[\Gamma \le \Gamma_{th}] \\
&= \int_0^{\Gamma_{th}} \frac{1}{\sqrt{2\pi}\sigma\Gamma} e^{-(1/2\sigma^2)\ln^2(\Gamma/\Gamma_m)}\, d\Gamma \\
&= \frac{1}{2}\,\text{erfc}\left\{ \frac{\ln(\Gamma_m/\Gamma_{th})}{\sqrt{2}\sigma} \right\}
\end{aligned}
\tag{10.100}
$$

The obtained numerical values are shown in Fig. 10.43. As $\Gamma_m$ is a function of the distance between the base station and a moving vehicle, $\Gamma_m/\Gamma_{th}$ represents the minimum CNR margin at some point on the cell fringe.

When planning a system, the outage must be specified within the whole cell. Therefore, it is necessary to make clear the relation between the outage at the cell fringe and the outage within the whole cell. Letting $r_1 = r$ and using (10.93), the outage within the whole cell $F_a^1$ can be obtained as

$$
\begin{aligned}
F_a^1 &= \frac{1}{\pi R^2} \int_0^R F_f^1(r) 2\pi r\, dr \\
&= \frac{1}{2}\,\text{erfc}(X_0) = \frac{1}{2} e^{(2X_0 Y_0 + Y_0^2)}\,\text{erfc}(X_0 + Y_0)
\end{aligned}
\tag{10.101}
$$

where

$$
X_0 \equiv \frac{\ln(\Gamma_m(R)/\Gamma_{th})}{\sqrt{2}\sigma} \quad \text{and} \quad Y_0 \equiv \frac{\sqrt{2}\sigma}{\alpha}
\tag{10.102}
$$

The first term of (10.101) is equal to the outage $F_f^1(R)$ at the cell fringe given by (10.100) with $r = R$, and the second term is the correction term. The numerical results obtained by (10.101) are shown in Fig. 10.44. This figure shows that the radio link can be designed based on the outage at the cell fringe. Consequently, the required CNR margin can be computed by (10.100).

The required area mean of CNR $\Gamma_m$ is then given as the sum of $\Gamma_{th}$ (decibel)

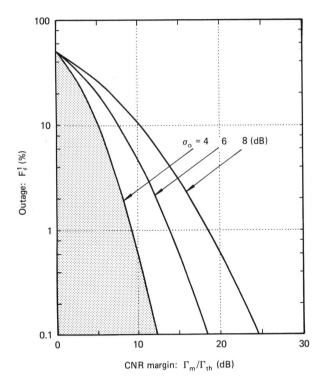

**Figure 10.43** Relation between outage and required CNR margin. (Printed by permission of the IEEE [Hansen, 10.108]. © 1982 IEEE).

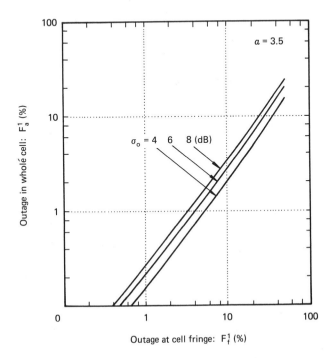

**Figure 10.44** Relation between $F_f^1$ and $F_a^1$. (Printed by permission of the IEEE [Hansen, 10.108]. © 1982 IEEE).

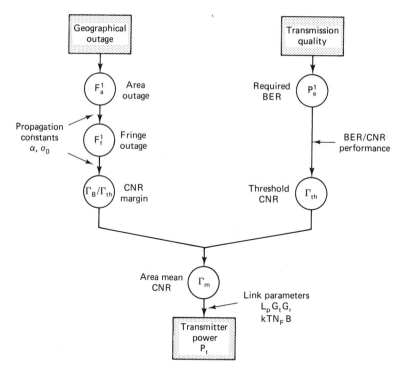

**Figure 10.45** Design procedure to determine the required transmitter power. (Printed by permission of the IEEE [Hansen, 10.108]. © 1982 IEEE).

and margin $\Gamma_m/\Gamma_{th}$ (decibels) corresponding to the outage $F_f^1$. When the path loss with distance $L_p$, which is related to the cell radius, and the receiver noise power, $kTBN_F$, are given, the transmitter power, $P_t$, can be obtained as

$$P_t = \frac{\Gamma_m \cdot kTBN_F \cdot L_p}{G_t \cdot G_r} \qquad (10.103)$$

where $G_t$ and $G_r$ are the antenna gains including line losses at the transmitter and receiver, respectively. A design procedure to determine the transmitter power is illustrated in the form of a flowchart in Fig. 10.45.

### Determination of Cochannel Reuse Distance

Let us assume that there exists a single interfering base station and that the desired signal and the undesired interference are subjected to the mutually independent log-normal shadowing having the same standard deviation. Then the pdf of the local mean CIR $\Lambda$ becomes

$$p(\Lambda) = \frac{1}{2\sqrt{\pi}\sigma\Lambda} e^{-1/4\sigma^2 \ln^2(\Lambda/\Lambda_m)} \qquad (10.104)$$

Letting $F_f^2$ be the outage at the point $P$ in the desired cell, the relationship between outage $F_f^2$ and the required CIR margin $\Lambda_m/\Lambda_{th}$ can be obtained as

$$F_f^2 = \text{Prob}(\Lambda \le \Lambda_{th}]$$

$$= \int_0^{\Lambda_{th}} \frac{1}{2\sqrt{\pi}\sigma\Lambda} e^{-(1/4\sigma^2)\ln^2(\Lambda/\Lambda_m)}\, d\Lambda \qquad (10.105)$$

$$= \frac{1}{2} \text{erfc}\left\{ \frac{\ln(\Lambda_m/\Lambda_{th})}{2\sigma} \right\}$$

The results are shown in Fig. 10.46.

Letting $r_1 = r$ and $r_2 \cong D - r$, the outage within the whole cell $F_a^2$ can be approximately expressed as

$$F_a^2 = \frac{1}{\pi R^2} \int_0^R F_f^2(r)2\pi r\, dr$$

$$= \frac{1}{2}\text{erfc}(X_0') - \frac{1}{2} e^{(2X_0'Y_0' + Y_0'^2)}\text{erfc}(X_0' + Y_0') \qquad (10.106)$$

where

$$X_0' \equiv \frac{\ln[\Lambda_m(R)/\Lambda_{th}]}{2\sigma} \quad \text{and} \quad Y_0' \equiv \frac{2\sigma}{\alpha} \qquad (10.107)$$

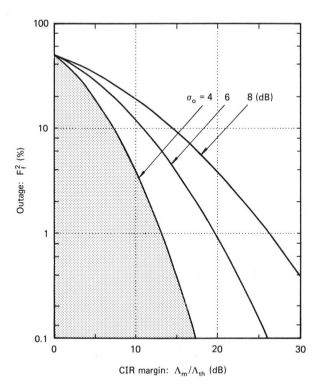

Figure 10.46 Relation between outage and required CIR margin.

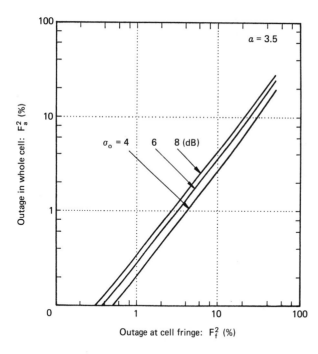

**Figure 10.47** Relation between $F_f^2$ and $F_a^2$. (Printed by permission of the IEEE [Hansen, 10.108]. © 1982 IEEE).

In the above equation, $\Lambda_m$, $\Lambda_{th} \gg 1$ is assumed. The first term of (10.106) is equal to (10.105), and the second term is the correction term. The computed results of (10.106) are shown in Fig. 10.47.

From (10.72), the area mean of CIR $\Lambda_m$ at the worst point on the cell fringe $r = R$ is given by

$$\Lambda_m = \left( \frac{R}{D - R} \right)^{-\alpha} \equiv \Lambda_m(R) \tag{10.108}$$

This can be rewritten as

$$\frac{D}{R} = 1 + \Lambda_m(R)^{1/\alpha} \tag{10.109}$$

and the results obtained are shown in Fig. 10.48. The ratio $D/R$ is the minimum cochannel reuse distance normalized by the cell radius because $\Lambda_m(R)$ is given by the sum of $\Lambda_{th}$ (dB) and the minimum CIR margin $\Lambda_m/\Lambda_{th}$ (decibels), which corresponds to the fixed outage $F_f^2$. A suggested design procedure to determine the cochannel reuse distance is shown in a flowchart form in Fig. 10.49. In the preceding derivation, it is assumed that the desired signal and the undesired interference are subjected to mutually independent log-normal shadowing. In practical land mobile radio propagation, shadowing may, however, be partially correlated, because the shadowing is caused by buildings or the terrain near the vehicle. The pdf of the local mean of CIR, taking the correlation effect into account, can be

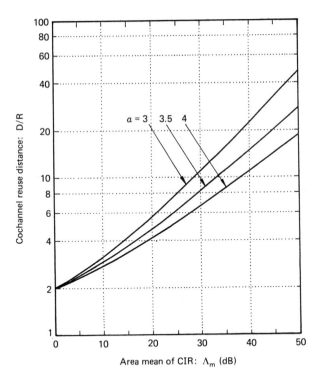

Figure 10.48 Relation between cochannel reuse distance and area mean of CIR. (Printed by permission of the IEEE [Hansen, 10.108]. © 1982 IEEE).

shown to be

$$p(\Lambda) = \frac{1}{2\sqrt{\pi}\sigma \sqrt{1 - \rho\Lambda}} e^{-(1/4\sigma^2(1 - \rho))\ln^2(\Lambda/\Lambda_m)} \qquad (10.110)$$

where $\rho$ is the correlation coefficient. The results are shown in Fig. 10.50. Comparing (10.104) with (10.110), we conclude that the correlation effect is equivalent to decreasing the standard deviation from $\sigma$ to $\sigma\sqrt{1 - \rho}$. Therefore, we may state that Fig. 10.50 presents the worst case interference probability.

### Example of a Mobile Radio Link Design

According to our described procedure, let us determine the transmitter power and the cochannel reuse distance for a digital mobile radio link.

The following conditions are assumed.

1. The frequency band of the system is 900 MHz, and the cell radius is $R = 3$ km. Standard deviation of the log-normal shadowing is $\sigma_0 = 6$ dB, and the propagation constant is $\alpha = 3.5$.

2. An average $P_e$ of $1 \times 10^{-3}$ and an outage $F_a$ of 10% are required, and these values are allotted equally for thermal noise and cochannel interference.

3. MSK with differential detection is adopted as the modem scheme. The

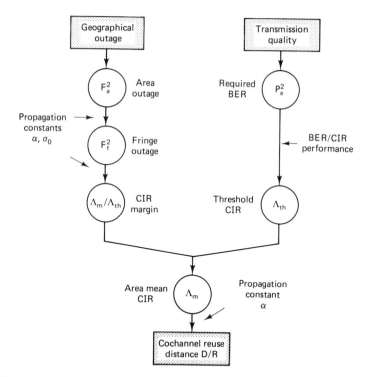

**Figure 10.49** Design procedure to determine the cochannel reuse distance. (Printed by permission of the IEEE [Hansen, 10.108]. © 1982 IEEE).

required transmission bandwidth for a 16-kb/s bit rate is $B = 16$ kHz. A two-branch selection diversity technique is applied.

Figure 10.51 illustrates the derivation procedure and the results obtained according to the flowchart shown in Figs. 10.45 and 10.49. The transmission performance is based on the theoretical performance, taking into account a 2-dB degradation. The path loss with distance is based on the empirical formula given in (10.23). From this procedure, the transmitter power and cochannel reuse distance are determined as $P_t = 1$ W and $D/R = 8.2$, respectively.

### 10.6.3 Discussions

The relation between outage and required margins has been derived by taking into account both thermal noise and CCI. The procedure and the results indicate that the required margins for thermal noise and CCI can be computed separately. A simple and useful procedure to determine the transmitter power and the cochannel reuse distance is presented in a flowchart form. The procedure can be applied to digital and also to analog mobile radio link designs.

To obtain a more accurate specification of required transmitter power, it is necessary to take other degradation factors, such as manufactured noise or terrain factors, into account. These factors should be treated as additional margin or

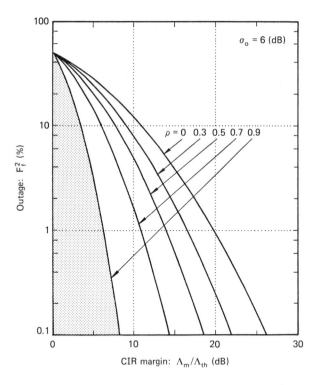

**Figure 10.50** Correlation effect on the relation between outage and required CIR margin.

propagation loss. In a cellular system, it is also necessary to design the cochannel reuse distance more accurately because of the presence of multipath interfering base stations. If we wish to have a conservative (safe) design, then we should add 8 dB, which corresponds to six interfering base stations, to the area mean of the CIR determined by our procedure.

As shown in the mobile radio link design example, the space diversity effect, which mitigates the fast multipath Rayleigh fading, effectively decreases the permissible local means of CNR and CIR. On the other hand, the site diversity effect, which is obtained by the hand-off technique and is effective for mitigating the shadowing, decreases the required CNR and CIR margins.

## 10.7 EFFICIENT SPECTRUM UTILIZATION IN DIGITAL LAND MOBILE RADIO

### 10.7.1 Definition of Spectral Efficiency

Let us first define the spectral efficiency of a cellular land mobile radio system. In this system, the entire service area is covered with many small cells, and the same set of frequencies are geographically reused every cluster of cells, as shown in Fig. 10.52. Let the number of channels assigned to each cell, carried traffic per channel, system bandwidth, and area of a unit cell be $n_{cell}$ (channels/cell), $a_c$ (erl/channel), $W$ (Hz), and $S$ (m$^2$), respectively. The overall spectral efficiency $\eta_T$

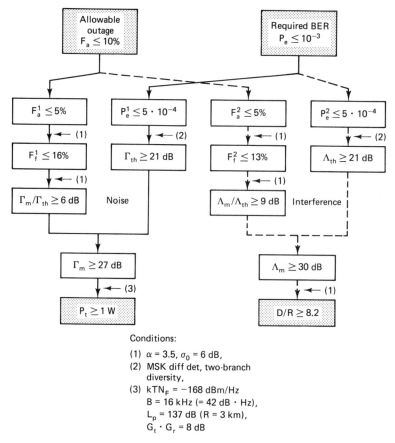

**Figure 10.51** Design example of a digital mobile radio link. (Printed by permission of the IEEE [Hansen, 10.108]. © 1982 IEEE).

(erl/Hz · m²) is defined, in [Murota, 10.36; Suzuki, 10.112; 10.113], by

$$\eta_T = \frac{n_{\text{cell}}\, a_c}{WS} \tag{10.111}$$

This definition indicates that the overall spectral efficiency is given as the spatial traffic density per unit bandwidth. Assuming that every cluster of cells is composed of $N$ cells and that the same number of channels are assigned to each cell, $n_{\text{cell}}$ is given by

$$n_{\text{cell}} = \frac{n_{\text{sys}}}{N} \tag{10.112}$$

where $n_{\text{sys}}$ (channels) is the total number of channels assigned for the system and is given by

$$n_{\text{sys}} = \frac{W}{f_s} \tag{10.113}$$

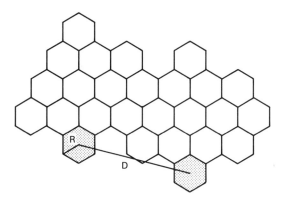

**Figure 10.52** An example of small cell layout (hexagonal cell structure, $N = 13$).

where $f_s$ (Hz/channel) is the channel spacing. By substituting (10.111) and (10.112) into (10.111), $\eta_T$ is obtained as

$$\eta_T = \frac{1}{NS} \cdot \frac{1}{f_s} \cdot a_c \qquad (10.114)$$

This representation means that the overall spectral efficiency $\eta_T$ is given as the product of the following three factors:

$$\eta_s = \frac{1}{NS} \qquad (10.115)$$

$$\eta_f = \frac{1}{f_s} \qquad (10.116)$$

$$\eta_t = a_c \qquad (10.117)$$

where $\eta_s$, $\eta_f$, and $\eta_t$ are the elementary spectral efficiencies with respect to space, frequency, and time, respectively. This decomposition indicates that the definition given by (10.110) is consistent with the convention used by a number of authors [Colavito, 10.114; Berry, 10.115; Hatfield, 10.116; Muilwijk, 10.117].

The number of subscribers accommodated by the system, that is, the system capacity, is another useful measure of efficient spectrum utilization. Assuming that the geographical distribution of traffic is uniform within the service area, the system capacity is proportional to the number of subscribers in a unit area. Letting busy-hour traffic per subscriber and offered traffic per cell be $a_{\text{sub}}$ (erl/sub) and $a_{\text{cell}}$ (erl/cell), respectively, the number of subscribers in a unit area $N_{\text{sub}}$ (sub/m$^2$) is given by

$$N_{\text{sub}} = \frac{a_{\text{cell}}}{a_{\text{sub}}} \cdot \frac{1}{S} \qquad (10.118)$$

The offered traffic per cell, $a_{\text{cell}}$, is related to the carrier traffic per cell, $n_{\text{cell}} \cdot a_c$, by the following equation:

$$a_{\text{cell}} = \frac{n_{\text{cell}} \cdot a_c}{1 - B} \qquad (10.119)$$

where $B$ denotes the blocking rate of calls. Substituting (10.119) into (10.118) and using (10.111), $N_{sub}$ can be obtained as

$$N_{sub} = \frac{W}{a_{sub}(1-B)}\,\eta_T \tag{10.120}$$

Since $W$, $a_{sub}$, and $B$ are already defined system parameters, maximizing system capacity can be reduced to the problem of maximizing $\eta_T$. This problem is considered here in detail.

### 10.7.2 Effective Techniques and Technologies

Equation (10.114) indicates that efficient spectrum utilization can be achieved by increasing $\eta_s$, $\eta_f$, and $\eta_t$. For this purpose, the following three methods are used: (1) high-density geographical cochannel reuse (reducing $N$ and $S$); (2) narrow-band transmission (reducing $f_s$); and (3) demand-assignment multichannel access (increasing $a_c$). They are effective for increasing $\eta_s$, $\eta_f$, and $\eta_t$, respectively. We now consider each method in more detail.

#### Geographical Cochannel Reuse

In the case of an ideal hexagonal layout of cells as shown in Fig. 10.52, the number of cells, $N$, in a cluster of cells is given by [Araki, 10.104]

$$N = \frac{1}{3}\left(\frac{D}{R}\right)^2 \tag{10.121}$$

where $R$ and $D$ are the radius of a unit cell and the cochannel reuse distance, respectively. Under the assumption that both signal and interference are subjected to uncorrelated multipath fading and have a local mean proportional to the inverse $\alpha$th power of the propagation distance, the ratio of $(D/R)$ is determined from the following relationship:

$$\left(\frac{D-R}{R}\right)^\alpha = M_f\,\Lambda_{th} \tag{10.122}$$

where $M_f$ and $\Lambda_{th}$ denote fading margin against allowable geographical outage due to log-normal shadowing and threshold CIR for a specific $P_e$, respectively. Experimental field test results indicate that $\alpha$ equals 3 to 4 in the usual V/UHF land mobile radio environment. Using (10.121) and (10.122), the relationship between $N$ and $M_f\,\Lambda_{th}$, by using the parameter $\alpha$, is obtained as

$$N = \tfrac{1}{3}[1 + (M_f\,\Lambda_{th})^{1/\alpha}]^2 \tag{10.123}$$

where $N$ takes on a discrete value of 3, 4, 7, 9, 12, 13, . . . . Reduction of $N$ is, therefore, achieved by assuring a high CCI protection, that is, by reducing $\Lambda_{th}$.

Space or polarization diversity technique, which can mitigate the multipath fading effect without transmission bandwidth expansion, is the most effective one for this purpose. Forward-error-correction (FEC) coding, which can be regarded as a certain kind of time diversity, is also effective, although it introduces some redundancy into the transmitted data stream and requires an expansion of the transmission bandwidth. Consequently, there exists an optimum coding rate with respect to the maximal spectral efficiency, which is achieved by a trade-off between high-density geographical cochannel reuse and narrow-band transmission.

Moreover, it is self-evident that smaller cell layout with lower transmitter power is effective for reducing the area of a unit cell $S$ in (10.115). However, it is necessary to place more base stations in order to make $S$ smaller. Accordingly, the area of a unit cell $S$ should be optimized from the system cost viewpoint.

### Narrow-Band Transmission

Provided that source coding with bit rate $f_b$, modulation with bandwidth efficiency $m$, and FEC coding with rate $R_c$ ($0 < R_c \leq 1$) are adopted in a digital land mobile radio system, the channel spacing $f_s$ is obtained as

$$f_s = \frac{f_b}{mR_c} + 2\Delta f \qquad (10.124)$$

where $\Delta f$ is the carrier frequency drift. To achieve narrowband transmission—that is, to increase $\eta_f$—low bit rate source coding (reduction of $f_b$), bandwidth-efficient modulation (increase of $m$), high-rate FEC coding ($R_c \to 1$), and stabilized carrier frequency sources ($\Delta f \to 1$) are required.

For efficient spectrum utilization, low bit rate source coding that does not require a very good $P_e$ performance is needed, since an excessively low $P_e$ transmission requirement leads to excessively high CCI protection.

Thus high-density geographical cochannel reuse becomes a difficult requirement. For voice coding, a number of low bit rate (less than 32 kb/s) coding techniques, each of which require only a moderate $P_e$ ($10^{-2}$ to $10^{-4}$) performance to obtain high-quality telephone service, are currently being developed.

It is effective to increase the bandwidth efficiency $m$ of the modulation technique for narrow-band transmission, that is, increasing $\eta_f$. However, this requires higher CCI protection, that is, larger $\Lambda_{th}$ in (10.121), which results in decreasing $\eta_s$. Therefore, bandwidth efficiency $m$ should be optimized to achieve a trade-off between narrow-band transmission (increasing $\eta_f$) and high-density geographical cochannel reuse (increasing $\eta_s$).

To achieve narrow-band transmission, it is desirable not to introduce any FEC coding (not to set $R_c = 1$). However, FEC coding, which can improve the $P_e$ versus CIR performance, or strengthen cochannel interference protection, is effective for achieving high-density geographical cochannel reuse, that is, increasing $\eta_s$ at the sacrifice of transmission bandwidth. Therefore, the coding rate $R_c$ should be optimized in order to maximize the overall spectral efficiency.

Needless to say, carrier frequency drift should be kept as small as possible in order to increase $\eta_f$. It is one of the most important and difficult problems to realize a *miniaturized frequency-agile mobile radio transceiver* having a highly stabilized carrier frequency source. A frequency-agile mobile radio transceiver is necessary for achieving demand-assignment multichannel access, which is effective for increasing $\eta_t$ in (10.117). Phase-locked frequency synthesizers are widely used for this purpose.

**Multichannel Access**

Demand-assignment multichannel access, which enables time-shared use of a channel and increases the traffic $a_c$ carried by a channel, is effective for increasing $\eta_t$ in (10.117). An electronic central processor using a stored-program-control scheme and a microprocessor-controlled frequency-agile mobile radio transceiver may be used for achieving multichannel access. By an application of the well-known Erlang's $B$-formula, it is easy to attain $a_c = 0.7$ to $0.9$ by conventional multichannel access control schemes, while its maximum limit is $a_c = 1$ [Syski, 10.118; Fujiki, 10.119].

### 10.7.3 Optimization

By substituting (10.123) and (10.124) into (10.114), the overall spectral efficiency $\eta_T$ can be obtained as

$$\eta_T = \frac{3a_c}{S[1 + (M_f \Lambda_{th})^{1/\alpha}]^2 [f_b/mR_c + 2\Delta f]} \tag{10.125}$$

If the carrier frequency drift and the log-normal shadowing is negligible—that is, $\Delta f = 0$ and $M_f = 1$—then $\eta_T$ becomes

$$\eta_T = \frac{3a_c \, mR_c}{Sf_b[1 + (\Lambda_{th})^{1/\alpha}]^2} \tag{10.126}$$

In the following discussion, we assume that $\alpha = 3.5$, that $S$, $f_b$, and $a_c$ in equation (10.126) are constants, and that a premodulation Gaussian-filtered MSK (GMSK) modulation technique described in Section 10.2.3 is used. Under these assumptions, bandwidth efficiency $m$ is given as a function of the normalized 3-dB bandwidth $B_bT$ of the premodulation Gaussian LPF. Moreover, the threshold CIR $\Lambda_{th}$ for a specific $P_e$ is given as a function of $B_bT$ and $R_c$. Therefore, the optimization problem can be reduced to the problem of obtaining $B_bT$ and $R_c$ to maximize $\eta_T$. Let us now consider this problem.

**Relationship between $m$ and $B_bT$**

By assuming that no FEC coding is introduced and that the carrier frequency drift can be disregarded, that is, $R_c = 1$ and $\Delta f = 0$, bandwidth efficiency $m$ is derived from (10.123) as the channel spacing normalized by the transmission bit rate ($m = f_s/f_b$). In GMSK-modulated systems $m$ is increased by reducing $B_bT$, where the ACI suppression is a parameter. Figure 10.53 shows computed results of the relationship between $m$ and $B_bT$ for an ACI of $-70$ dB, which is a general requirement for SCPC land mobile radio systems. The ACI is defined as the relative power that falls into an adjacent channel having an ideal rectangular band-pass characteristic whose bandwidth equals the transmission bit rate.

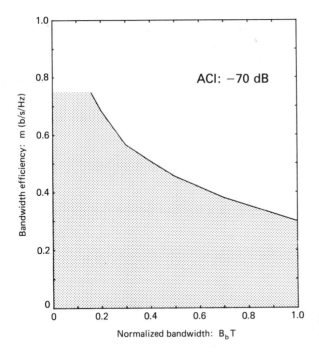

**Figure 10.53** Relationship between $m$ and $B_b T$.

### Relationship between $\Lambda_{th}$ and $B_b T$

In order to clarify the relationship between $\Lambda_{th}$ and $B_b T$, it is necessary to obtain the average $P_e$ versus CIR performance of a GMSK transmission system in a multipath fading environment. To simplify our investigation, let us assume that coherent detection is used. Noise-free and interference-limited conditions are assumed. A GMSK signal is assumed to be subjected not only to ISI due to the premodulation Gaussian LPF but also to CCI caused by geographical reuse of the same frequency, while ISI due to the predetection BPF in the receiver side is not taken into account. Under these assumptions, let us first consider the $P_e$ versus CIR performance in a nonfading environment.

Figure 10.54 shows a vector diagram of the resultant complex envelope for

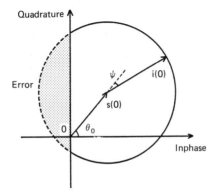

**Figure 10.54** Resultant complex envelope of GMSK signal and interference.

a GMSK signal $s(t)$ and CCI $i(t)$ at a decision instant $t = 0$, where correct decision occurs when the resultant complex envelope $s(0) + i(0)$ falls onto the right half-plane. The carrier phase difference $\psi$ between $s(0)$ and $i(0)$ is a random variable with a uniform distribution. Moreover, $\theta_0 = \theta(0)$ is the modulation phase deflection due to ISI effect of the premodulation Gaussian LPF. For the worst signal pattern, that is . . . , 0, 0, 0, 1, 1, 1, . . . , which is the cause of maximal ISI, the modulation phase change $\theta(t)$ is given by

$$\theta(t) = \frac{\pi}{2T} \left[ \int_0^t \text{erf}(\beta\tau) \, d\tau + \frac{1}{\beta\sqrt{\pi}} \right] \tag{10.127}$$

where $T$ is the signaling period and $\beta$ is given by

$$\beta = \pi B_b \sqrt{\frac{2}{\ln 2}} \tag{10.128}$$

Here, $B_b$ is the 3-dB bandwidth of a premodulation Gaussian LPF. The reference phase is defined as the modulation phase change at $t = 0$ of a simple MSK, that is, $B_b T \rightarrow \infty$, without the ISI effect. By substituting $t = 0$ into (10.126), $\theta_0$ is obtained as

$$\theta_0 = \theta(0) = \frac{\sqrt{\ln 2}}{2\sqrt{2\pi} \, B_b T} \tag{10.129}$$

Letting the instantaneous CIR be $\lambda$, the probability of decision error, $P_e(\lambda)$, is obtained as

$$P_e(\lambda) = \begin{cases} \frac{1}{\pi} \cos^{-1}[\sqrt{\lambda} \cos \theta_0] & \text{for } 0 \leq \sqrt{\lambda} \cos \theta_0 \leq 1 \\ \\ 0 & \text{for } \sqrt{\lambda} \cos \theta_0 > 1 \end{cases} \tag{10.130}$$

Averaging $P_e(\lambda)$ over the fading dynamic range of $\lambda$, the average $P_e$ versus CIR performance in the quasi-static multipath fading environment is obtained as

$$P_e(\Lambda) = \int_0^\infty P_e(\lambda) \, p(\lambda) \, d\lambda \tag{10.131}$$

where $\Lambda$ and $p(\lambda)$ are the average CIR and the pdf of $\lambda$, respectively. Assuming that both the desired GMSK signal and the undesired CCI are subjected to mutually independent quasi-static Rayleigh fadings and that an ideal two-branch diversity technique that has a perfect-pilot maximal-ratio combining method in the predetection stage is adopted, then $p(\lambda)$ may be derived from (10.72) and (10.76a) as

$$p(\lambda) = \begin{cases} \dfrac{\Lambda}{(\lambda + \Lambda)^2} & \text{for no diversity} \\ \\ \dfrac{\lambda\Lambda}{(\lambda + \Lambda)^3} & \text{for diversity} \end{cases} \tag{10.132}$$

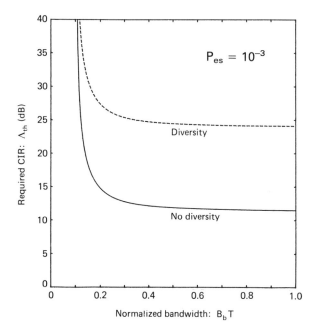

**Figure 10.55** Relationship between $\Lambda_{th}$ and $B_bT$ for $P_{es} = 10^{-3}$.

By substituting (10.129) and (10.131) into (10.130), $P_e(\Lambda)$ is obtained as

$$
P_e(\Lambda) = \begin{cases}
\dfrac{1}{2}\left[1 - \sqrt{\dfrac{\Lambda\cos^2\theta_0}{\Lambda\cos^2\theta_0 + 1}}\right] \approx \dfrac{1}{4\Lambda\cos^2\theta_0} & \text{for no diversity} \\[4mm]
\dfrac{1}{2}\left[1 - \dfrac{3}{2}\sqrt{\dfrac{\Lambda\cos^2\theta_0}{\Lambda\cos^2\theta_0 + 1}} + \dfrac{1}{2}\sqrt{\dfrac{(\Lambda\cos^2\theta_0)^3}{(\Lambda\cos^2\theta_0 + 1)^3}}\right] \\[4mm]
\qquad\qquad\qquad \approx \dfrac{3}{16\Lambda^2\cos^4\theta_0} & \text{for diversity}
\end{cases}
\tag{10.133}
$$

For the particular case of $\theta_0 = 0$, $(B_bT \rightarrow \infty)$, which corresponds to simple MSK, these equations coincide with the previously obtained equations (10.89c) and (10.90c), respectively.

The threshold CIR, $\Lambda_{th}$, for a specific $P_{es}$ can then be obtained in the following approximate form:

$$
\Lambda_{th} \cong \begin{cases}
\dfrac{1}{4\,P_{es}\cos^2\theta_0} & \text{for no diversity} \\[4mm]
\dfrac{\sqrt{3}}{4\,\sqrt{P_{es}}\,\cos^2\theta_0} & \text{for diversity}
\end{cases}
\tag{10.134}
$$

Since $\theta_0$ is given by (10.129) as a function of $B_bT$, the relationship between $\Lambda_{th}$ and $B_bT$ is shown in Fig. 10.55.

**TABLE 10.2** Approximation constant
$c(R_c)$ for self-orthogonal convolutional
code

| $R_c$ | $c(R_c)$ |
|-------|----------|
| $\frac{1}{2}$ | 151 |
| $\frac{2}{3}$ | 807 |
| $\frac{3}{4}$ | 2,227 |
| $\frac{4}{5}$ | 4,569 |
| $\frac{5}{6}$ | 8,192 |
| $\frac{6}{7}$ | 14,000 |
| $\frac{7}{8}$ | 20,000 |

### Relationship between $\Lambda_{th}$ and $R_c$

Now, let us assume that a self-orthogonal convolutional FEC coding technique with a two-bit error correction capability is used along with some auxiliary technique to randomize long burst errors caused by deep fading. Improvement of the average $P_e$ versus CIR performance is presented in [Koga, 10.120], which can be approximated as follows:

$$P_e(\Lambda, R_c) \approx c(R_c)\, P_e^3(\Lambda) \qquad (10.135)$$

where $P_e(\Lambda)$ is given by (10.133), and $c(R_c)$ is an approximation constant dependent upon $R_c$ and is listed in Table 10.2. By using (10.133) and (10.135), the threshold

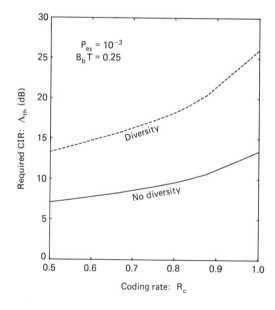

**Figure 10.56** Relationship between $\Lambda_{th}$ and $R_c$ for $P_{es} = 10^{-3}$.

CIR $\Lambda_{th}$ for a specific $P_{es}$ can be obtained as

$$\Lambda_{th} = \begin{cases} \dfrac{\{c(R_c)\}^{1/3}}{4(P_{es})^{1/3}\cos^2\theta_0} & \text{for no diversity} \\[4mm] \dfrac{\sqrt{3}\{c(R_c)\}^{1/6}}{4(P_{es})^{1/6}\cos^2\theta_0} & \text{for diversity} \end{cases} \qquad (10.136)$$

The relationship between $\Lambda_{th}$ and $R_c$ can, therefore, be obtained as shown in Fig. 10.56, where $B_b T = 0.25$, which will be demonstrated later to be near optimum.

**Optimization of $B_b T$ and $R_c$**

Let us first consider a simple case without FEC coding, that is, $R_c = 1$. The relationship of $\eta_T$ versus $B_b T$ for $P_{es} = 10^{-3}$ and $\alpha = 3.5$ is shown in Fig. 10.57, which has been obtained by substituting the calculated results of $m$ shown in Fig. 10.53 and (10.134) into (10.126). The ordinate of this figure is normalized by the value of $\eta_T$ for the simple MSK ($B_b T \to \infty$) without diversity. This figure shows that $B_b T = 0.5$ is a near optimum value in the sense of maximizing the spectral efficiency. An optimum channel spacing for a specific transmission bit rate is usually determined from this optimum $B_b T$ by using the parameter $m$ shown in Fig. 10.53. However, the channel spacing may have to be determined beforehand by some other reasons. In such cases, an optimum transmission bit rate is determined.

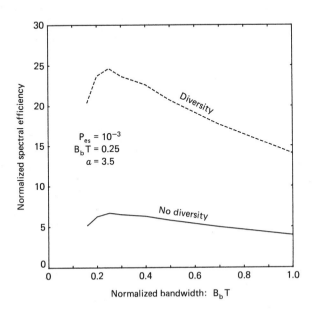

**Figure 10.57** Relationship between $\eta_T$ and $B_b T$.

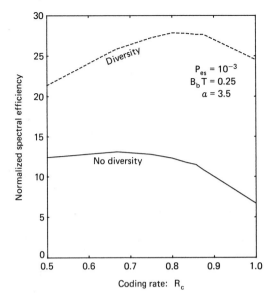

**Figure 10.58** Relationship between $\eta_T$ and $R_c$.

**Example 10.2**

When channels are assigned every 25 kHz, a transmission bit rate of $f_b = 16$ kb/s can be determined as near optimum with $B_bT = 0.25$, where the ACI is suppressed as much as $-70$ dB.

Let us then solve an optimum value of $R_c$ for the FEC coding. The dependency of $\eta_T$ upon $R_c$ for $B_bT = 0.25$, $P_{es} = 10^{-3}$, and $\alpha = 3.5$ is shown in Fig. 10.58, which has been obtained by substituting $m = 0.63$ for $B_bT = 0.25$ and (10.136) into (10.126). The ordinate of this figure is normalized by the value of $\eta_T$ for $R_c = 1$ without diversity. This figure shows that $\eta_T$ has a maximum value at an optimum $R_c$. The optimum values of $R_c$ for diversity and no diversity are given by $R_c = \frac{4}{5}$ and $\frac{1}{2}$, respectively.

## REFERENCES

[10.1] Bowers, R., A. M. Lee, and C. Harshey, Eds. *Communications for a Mobile Society*, SAGE Publications Inc., Beverly Hills, Calif., 1978.

[10.2] Ito, S. and Y. Matsuzaka. "800-MHz band land mobile telephone system—overall view", *Rev. Elec. Commun. Lab.*, Vol. 25, November–December 1977, p. 1147, or *IEEE Trans. Veh. Tech.*, Vol. VT-27, November, 1978, p. 205.

[10.3] Young, W. R. "Advanced mobile phone service; introduction, background, and objectives," *Bell Syst. Tech. J.*, Vol. 58, January, 1979, p. 1, or Blecher, F. H. "Advanced mobile phone service," *IEEE Trans. Veh. Tech.*, Vol. VT-29, May, 1980, p. 238.

[10.4] Remy, J. G., and Y. F. Dehery. "Cellular radiocommunication systems; the French approach," *IEEE ICC 81*, June, 1981, p. 57.3.

[10.5] Calvert, F. J., and R. Uppal. "AURORA—automatic mobile telephone system," *IEEE ICC 81*, June, 1981, p. 57.5.

[10.6] Haug, T. "The nordic mobile telephone system, an extension of the telephone network," *IEEE GCC'83*, November–December, 1983, p. 41.1.

[10.7] Pfannschmidt, H. "MATS E—an advanced cellular radio telephone system," *IEEE GCC 83*, November–December, 1983, p. 41.2.

[10.8] Kammerlander, K. "Presentation of the main system characteristics of the German mobile radio telephone system C," *IEEE GCC 83*, November–December, 1983.

[10.9] Mikulski, J. J. " A system plan for a 900-MHz portable radio telephone," *IEEE Trans. Veh. Tech.*, Vol. VT-26, February, 1977, p. 76.

[10.10] Fisher, R. E. "Portable telephones for 850-MHz cellular systems," *IEEE 30th Veh. Tech. Conf.*, September, 1980, p. E-2.1.

[10.11] Sakamoto, M., S. Kozono and T. Hattori. "Basic study on portable telephone system design," *IEEE 32nd Veh. Tech. Conf.*, May, 1982, p. 279.

[10.12] CCIR Rep. 903. "Digital transmission in the land mobile service," *Report 903 and Question 40-1/8 of the CCIR, 1982*, p. 100 and p. 845.

[10.13] Bettinger, O. "Digital speech transmission for mobile radio service," *Electrical Commun.*, Vol. 47, 1972, p. 224.

[10.14] Sachs, H. M. "Digital voice considerations for the land mobile radio services," *IEEE 27th Veh. Tech. Conf.*, March, 1977, p. 207.

[10.15] Brenig, T. "Data transmission for the mobile radio," *IEEE Trans. Veh. Tech.*, Vol. VT-27, August, 1978, p. 77.

[10.16] Lucky, R. W., J. Salz, and E. J. Weldon, Jr. *Principles of Data Communication*, McGraw-Hill, New York, 1968, Chapters. 10, 11, and 12.

[10.17] Peterson, W. W., and E. J. Weldon, Jr. *Error-Correcting Codes*, 2nd ed., The MIT Press, Cambridge, Mass., 1972.

[10.18] Mabey, P. J. "Mobile radio data transmission—coding for error control," *IEEE Trans. Veh. Tech.*, Vol. VT-27, August, 1978, p. 99.

[10.19] Duttweiler, D. L., and D. G. Messerschmit. "Nearly instantaneous companding and time diversity as applied to mobile radio transmission," *IEEE ICC 75*, June, 1975, p. 40.12.

[10.20] Murakami, H. "Time diversity by pitch-synchronized interpolation," *IEEE Trans. Veh. Tech.*, Vol. VT-29, November, 1980, p. 365.

[10.21] Jenks, F. G., P. D. Morgan, and C. S. Warren. "Use of four-phase modulation for digital mobile radio," *IEEE Trans. Electro-Magnetic Compat.*, Vol. EMC-14, November, 1972, p. 113.

[10.22] Manassewitsh, V. *Frequency Synthesizers Theory and Design*, John Wiley, New York, 1976, Chapter 7.

[10.23] Gardner, F. M. *Phaselock Techniques*, 2nd ed. John Wiley, New York, 1979, Chapter 10.

[10.24] de Buda, R. "Coherent demodulation of frequency shift keying with low deviation ratio", *IEEE Trans. Commun.*, Vol. COM-20, June, 1972, p. 466.

[10.25] Pasupathy, S. "Minimum shift keying: a spectrally efficient modulation," *IEEE Commun. Mag.*, Vol. 17, July, 1979, p. 14.

[10.26] Amoroso, F. "Pulse and spectrum manipulation in the minimum (frequency) shift keying (MSK) format," *IEEE Trans. Commun.*, Vol. COM-24, March, 1976, p. 381.

[10.27] de Jager, F., and C. B. Dekker. "Tamed frequency modulation; a novel method to achieve spectrum economy in digital transmisson," *IEEE Trans. Commun.*, Vol. COM-20, May, 1978, p. 534.

[10.28] Feher, K. *Digital Communications: Satellite/Earth Station Engineering*, Chapter 4, Prentice-Hall, Englewood Cliffs, N.J., 1983.

[10.29] Aulin, T., and C. -E. W. Sundberg. "Continuous phase modulation—Part I: full response signaling," *IEEE Trans. Commun.*, Vol. COM-29, March, 1981, p. 196.

[10.30] Aulin, T., N. Rydbeck, and C. -E. W. Sundburg. "Continuous phase modulation—Part II: partial response signaling," *IEEE Trans. Commun.*, Vol. COM-29, March, 1981, p. 210.

[10.31] Muilwijk, D. "Correlative phase shift keying—a class of constant envelope modulation techniques," *IEEE Trans. Commun.*, Vol. COM-29, March, 1981, p. 226.

[10.32] Honma, K., E. Murata, and Y. Riko. "On a method of constant envelope modulation for digital mobile communication," *IEEE ICC 80*, June, 1980, p. 24.1.1.

[10.33] Akaiwa, Y., I. Takase, S. Kojima, M. Ikoma, and N. Saegusa. "Performance of baseband bandlimited multilevel FM with discriminator detection for digital mobile telephony," *Trans. IECE of Japan*, Vol. 64-E, July, 1981, p. 463.

[10.34] Asakawa, S., and F. Sugiyama. "A compact-spectrum constant-envelope digital phase modulation," *IEEE Trans. Veh. Tech.*, Vol. VT-30, August, 1981, p. 102.

[10.35] Murota, K., and K. Hirade, "GMSK modulation for digital mobile radio telephony," *IEEE Trans. Commun.*, Vol. COM-29, July, 1981, p. 1044.

[10.36] Murota, K., K. Kinoshita, and K. Hirade. "Spectrum efficiency of GMSK land mobile radio," *IEEE ICC 81*, June, 1981, p. 23.8.1.

[10.37] CCIR Rec. 478-3. "Technical characteristics of equipment and principles governing the allocation of frequency channels between 25 and 1000 MHz for the land mobile service," 1982, p. 13.

[10.38] Wozencraft, J. M., and I. M. Jacobs. *Principles of Communication Engineering*, John Wiley, New York, 1965, Chapter 4.

[10.39] Murota, K., and K. Hirade. "Transmission performance of GMSK modulation," *Trans. IECE of Japan*, Vol. 64-B, October, 1981, p. 1123 [in Japanese].

[10.40] Jakes, W. C., Jr., ed., *Microwave Mobile Communications*, John Wiley, New York, 1974, Chapters 1 and 2.

[10.41] Lee, W. C. Y. *Mobile Communication Engineering*, McGraw-Hill, New York, 1982, Chapters 1–7.

[10.42] Ossana, J. F. "A model for mobile radio fading due to building reflections: theoretical and experimental fading waveform power spectra," *Bell Syst. Tech. J.*, Vol. 43, November, 1964, p. 2935.

[10.43] Gilbert, E. N. "Energy reception for mobile radio," *Bell Syst. Tech. J.*, Vol. 44, October, 1965, p. 1779.

[10.44] Clarke, R. H. "A statistical theory of mobile radio reception," *Bell Syst. Tech. J.*, Vol. 47, July–August, 1968, p. 957.

[10.45] Gans, M. J. "A power-spectral theory of propagation in the mobile-radio environment," *IEEE Trans. Veh. Tech.*, Vol. VT-21, February, 1972, p. 27.

[10.46] Young, W. R., Jr. "Comparison of mobile radio transmission of 150, 450, 900, and 3700 MC," *Bell Syst. Tech. J.*, Vol. 31, November, 1952, p. 1068.

[10.47] Okumura, Y., E. Ohmori, T. Kawano, and K. Fukuda. "Field strength and its variability in VHF and UHF land mobile radio service," *Rev. Elec. Commun. Lab.*, Vol. 16, September–October, 1968, p. 825.

[10.48] Nyland, H. W. "Characteristics of small-area signal fading on mobile circuits in the 150-MHz band," *IEEE Trans. Veh. Tech.*, Vol. VT-17, October, 1968, p. 24.

[10.49] Bello, P. A. "Characterization of randomly time-variant linear channels," *IEEE Trans. Commun. Syst.*, Vol. CS-11, December, 1963, p. 360.

[10.50] Schwartz, M., W. R. Bennet, and S. Stein. *Communication Systems and Techniques*, McGraw-Hill, New York, 1966, Chapter 9.

[10.51] Kennedy, R. S. *Fading Dispersive Communication Channels*, John Wiley, New York, 1969, Chapters 2, 3 and 4.

[10.52] Turin, G. L., F. D. Clapp, T. L. Johnston, S. B. Fine, and D. Lavry. "A statistical model of urban multipath propagation," *IEEE Trans. Veh. Tech.*, Vol. VT-21, February, 1972, p. 1.

[10.53] Cox, D. C. "Delay-Doppler characteristics of multipath propagation at 910 MHz in a suburban mobile radio environment," *IEEE Trans. Ant. Prop.*, Vol. AP-20, September, 1972, p. 625.

[10.54] Cox, D. C. "A measured delay-Doppler scattering function for multipath propagation at 910 MHz in an urban mobile radio environment," *Proc. IEEE,* Vol. 61, April, 1973, p. 479.

[10.55] Bajwa, A. S. and J. D. Parsons. "Small-area characterization of UHF urban and suburban mobile radio propagation," *Proc. IEEE,* Vol. 129-F, April, 1982, p. 102.

[10.56] Egli, J. "Radio propagation above 40 MC over irregular terrain", *Proc. IRE,* October, 1957, p. 1383.

[10.57] Black, D. M., and D. O. Reudink. "Some characteristics of radio propagation at 800 MHz in the Philadelphia area," *IEEE Trans. Veh. Tech.*, Vol. VT-21, May, 1972, p. 45.

[10.58] Hata, M. "Empirical formula for propagation loss in land mobile radio services," *IEEE Trans. Veh. Tech.*, Vol. VT-29, August, 1980, p. 317.

[10.59] Lin, S. H. "Statistical behaviour of a fading signal," *Bell Syst. Tech. J.*, Vol. 50, December, 1971, p. 3211.

[10.60] Aulin, T. "A modified model for the fading signal at a mobile radio channel," *IEEE Trans. Veh. Tech.*, Vol. VT-28, August, 1979, p. 182.

[10.61] Arredondo, G. A., W. H. Chriss, and E. H. Walker. "A multipath fading simulator for mobile radio," *IEEE Trans. Veh. Tech.*, Vol. VT-22, November, 1973, p. 241.

[10.62] Hirade, K., K. Abe, T. Hanazawa, and F. Adachi. "Fading simulator for land mobile radio communication," *Trans. IECE of Japan,* Vol. 58-B, September, 1975, p. 449 [in Japanese].

[10.63] Ralphs, J. D., and F. M. E. Sladen. "An HF channel simulator using a new Rayleigh fading method", *The Radio and Electronic Engineer,* Vol. 46, December, 1976, p. 579.

[10.64] Hattori, T., and K. Hirade. "Generation method of mutually correlated multipath fading waves," *Trans. IECE of Japan,* Vol. 59-B, September, 1976, p. 464.

[10.65] Caples, E. L., K. E. Massad, and T. R. Minor. "A UHF channel simulator for digital mobile radio," *IEEE Trans. Veh. Tech.*, Vol. VT-29, May, 1980, p. 281.

[10.66] Arnold, H. W., and W. F. Bodtman. "A hybrid multi-channel hardware simulator for frequency-selective mobile radio paths," *IEEE GTC 82,* November–December, 1982, p. A.3.1.

[10.67] Rice, S. O. "Mathematical analysis of random noise," *Bell Syst. Tech. J.,* Vol. 23, July, 1944, p. 282, and Vol. 24, January, 1945, p. 46.

[10.68] Barsis, A. P. "Determination of service area for VHF/UHF land mobile and broadcast operations over irregular terrain," *IEEE Trans. Veh. Tech.,* Vol. VT-22, May, 1973, p. 21.

[10.69] Durkin, J. "Computer prediction of service area for VHF and UHF land mobile radio services," *IEEE Trans. Veh. Tech.,* Vol. VT-26, November, 1977, p. 323.

[10.70] Allsebrook, K., and J. D. Parsons. "Mobile radio propagation in British cities at frequencies in the VHF and UHF bands,' *IEEE Trans. Veh. Tech.,* Vol. VT-26, November, 1977, p. 313.

[10.71] Hirade, K., M. Ishizuka, and F. Adachi. "Error-rate performance of digital FM with discriminator-detection in the presence of cochannel interference under fast Rayleigh fading environment," *Trans. IECE of Japan,* Vol. 61-E, September, 1978, p. 704.

[10.72] Hirade, K., M. Ishizuka, F. Adachi, and K. Ohtani. "Error-rate performance of digital FM with differential-detection in land mobile radio channel,"*IEEE Trans. Veh. Tech.,* Vol. VT-28, August, 1979, p. 204.

[10.73] Stein, S. "Unified analysis of certain coherent and non-coherent binary communications systems," *IEEE Trans. Inform. Theory,* Vol. IT-10, January, 1964, p. 43.

[10.74] Pelchat, M. G. "The autocorrelation function and power spectrum of PCM/FM with random binary modulation waveforms," *IEEE Trans. Space Electron. Telem.,* Vol. SET-10, March, 1964, p. 39.

[10.75] Voelcker, H. "Phase-shift-keying in fading channels," *Proc. IEEE,* Vol. 107-B, January, 1960, p. 31.

[10.76] Bello, P. A., and B. D. Nelin. "Effect of frequency selective fading on the binary error probability of incoherent and differentially coherent matched filter receivers," *IEEE Trans. Commun. Syst.,* Vol. CS-11, June, 1963, p. 170.

[10.77] Bailey, C. C., and J. C. Lindenlaub. "Further results concerning the effect of frequency selective fading on differentially coherent matched filter receivers," *IEEE Trans. Commun. Tech.,* Vol. COM-16, October, 1968, p. 749.

[10.78] Hummels, D. R., and F. W. Ratchiffe. "Calculation of error probability for MSK and OQPSK systems operating in a multipath fading environment," *IEEE Trans. Veh. Tech.,* Vol. VT-30, August, 1981, p. 112.

[10.79] Hata, M. "Preferable transmission rate of MSK land mobile radio with differential detection," *Trans. IECE of Japan,* Vol. 65-E, August, 1982, p. 451.

[10.80] Schwartz, M., W. R. Bennet, and S. Stein. *Communication Systems and Techniques,* McGraw-Hill, New York, 1966, Ch. 10.

[10.81] Sunde, E. D. *Communication Systems Engineering Theory,* John Wiley, New York, 1969, Chapter. 9.

[10.82] Jakes, W. C., Jr. "A comparison of specific diversity techniques for reduction of fast fading in UHF mobile radio systems," *IEEE Trans. Veh. Tech.,* Vol. VT-20, November, 1971, p. 81.

[10.83] Jakes, W. C., Jr., ed., *Microwave Mobile Communications,* John Wiley, New York, 1974, Chapters 5 and 6.

[10.84] Parsons, J. D., M. Henze, P. A. Ratliff, and M. J. Withers. "Diversity techniques for mobile radio reception," *The Radio and Electronic Engineer,* Vol. 45, July, 1975, p. 357.

[10.85] Lee, W. C. Y.  *Mobile Communication Engineering,* McGraw-Hill, New York. 1982, Chapters 9 and 10.

[10.86] Brennan, D. G.  "Linear diversity combining techniques," *Proc. IRE,* Vol. 47, June, 1959, p. 1075.

[10.87] Feher, K., and D. Chan.  "PSK combiners for fading microwave channels," *IEEE Trans. Commun.,* Vol. COM-23, May, 1975, p. 554.

[10.88] Feher, K.  *Digital Communications: Microwave Applications,* Prentice-Hall, Englewood Cliffs, N.J., 1981, Chapter 10.

[10.89] Rustako, A. J., Y. S. Yeh, and R. R. Murray.  "Performance of feedback and switch space diversity 900 MHz f.m. mobile radio systems with Rayleigh fading," *IEEE Trans. Commun. Tech.,* Vol. COM-21, November, 1973, p. 1257.

[10.90] Shortall, W. E.  "A switched diversity receiving system for mobile radio," *IEEE Trans. Commun. Tech.,* Vol. COM-21, November, 1973, p. 1269.

[10.91] Adachi, F., T. Hattori, K. Hirade, and T. Kamata.  "A periodic switching diversity technique for a digital FM land mobile radio," *IEEE Trans. Veh. Tech.,* Vol. VT-27, November, 1978, p. 211.

[10.92] Adachi, F.  "Periodic switching diversity effect on co-channel interference performance of a digital FM land mobile radio," *IEEE Trans. Veh. Tech.,* Vol. VT-27, November, 1978, p. 220.

[10.93] Parsons, J. D., P. A. Ratliff, M. Henze, and M. J. Withers.  "Single-receiver diversity system", *IEEE Trans. Commun.,* Vol. COM-21, November, 1973, p. 1276.

[10.94] Rogers, A. J.  "A double phase sweeping system for diversity reception in mobile radio," *The Radio and Electronic Engineer,* Vol. 45, April, 1975, p. 183.

[10.95] Kazal, S.  "Antenna pattern smoothing by phase modulation," *Proc. IRE,* Vol. 52, April, 1964, p. 435.

[10.96] Villard, O. G., Jr., J. M. Lomasney, and N. M. Kawachika.  "A mode-averaging diversity combiner," *IEEE Trans. Ant. Prop.,* Vol. AP-20, July, 1972, p. 463.

[10.97] Pierce, J. N.  "Theoretical diversity improvement in frequency shift keying," *Proc. IRE,* Vol. 46, May, 1958, p. 903.

[10.98] Lindsey, W. C.  "Error probabilities for incoherent diversity reception," *IEEE Trans. Inform. Theory,* Vol. IT-11, October, 1965, p. 491.

[10.99] Schwartz, M., W. R. Bennett, and S. Stein.  *Communication Systems and Techniques,* Chapter 9, McGraw-Hill, New York, 1966.

[10.100] Sunde, E. D.  *Communication Systems Engineering Theory,* John Wiley, New York, 1969, Chapter 10

[10.101] Prapinmongkolkarn, P., N. Morinaga, and T. Namekawa.  "Performance of digital FM systems in a fading environment," *IEEE Trans. Aerosp. Electron. Syst.,* Vol. AES-10, September, 1974, p. 698.

[10.102] Adachi, F.  "Selection and scanning diversity effects in a digital FM land mobile radio with discriminator and differential detections," *Trans. IECE of Japan,* Vol. 64-E, June, 1981, p. 398.

[10.103] Adachi, F.  "Postdetection selection diversity effects on digital FM land mobile radio," *IEEE Trans. Veh. Tech.,* Vol. VT-31, November, 1982, p. 166.

[10.104] Araki, K. "Fundamental problems of nationwide mobile radio telephone system," *Rev. Elec. Commun. Lab.,* Vol. 16, May–June, 1968, p. 357.

[10.105] Lundquist, L., and M. M. Peritsky. "Co-channel interference rejection in a mobile radio space diversity system," *IEEE Trans. Veh. Tech.,* Vol. VT-20, August, 1971, p. 68.

[10.106] Yoshikawa, N., and T. Nomura. "On the design of a small zone land mobile radio system in UHF band," *IEEE Trans. Veh. Tech.,* Vol. VT-25, August, 1976, p. 57.

[10.107] Kamata, T., M. Sakamoto, and K. Fukuzumi. "800-MHz band land mobile radio telephone systems," *Rev. Elec. Commun. Lab.,* Vol. 25, November–December, 1977, p. 1157.

[10.108] Hansen, F., and F. I. Meno. "Mobile fading-Rayleigh and lognormal superimposed," *IEEE Trans. Veh. Tech.,* Vol. VT-26, November, 1977, p. 332.

[10.109] Gosling, W. "A simple mathematical model of co-channel and adjacent channel interference in land mobile radio," *The Radio and Electronic Engineer*, Vol. 48, December, 1978, p. 619.

[10.110] MacDonald, V. H. "Advanced mobile phone service: the cellular concept," *Bell Syst. Tech. J.,* Vol. 58, January, 1979, p. 15.

[10.111] French, R. C. 'The effect of fading and shadowing on channel reuse in mobile radio," *IEEE Trans. Veh. Tech.,* Vol. VT-28, August, 1979, p. 171.

[10.112] Suzuki, H. and K. Hirade. "System considerations of *M*-ary PSK land mobile radio for efficient spectrum utilization," *Trans. IECE of Japan,* Vol. 65-E, March, 1982, p. 159.

[10.113] Suzuki, H., and K. Hirade. "Spectrum efficiency of *M*-ary PSK land mobile radio", *IEEE Trans Commun.,* Vol. COM-30, May, 1982, p. 1803.

[10.114] Colavito, C. "On the efficiency of the radio frequency spectrum utilization in fixed and mobile communications systems," *Alta Frequenz,* Vol. XVIII, September, 1974, p. 640.

[10.115] Berry, L. A. "Spectrum metrics and spectrum efficiency: proposed definitions," *IEEE Trans. Electro-Magnetic Compat.,* Vol. EMC-19, August, 1977, p. 254.

[10.116] Hatfield, D. N. "Measures of spectral efficiency in land mobile radio," *IEEE Trans. Electro-Magnetic Compat.,* Vol. EMC-19, August, 1977, p. 266.

[10.117] Muilwijk, D. "Spectrum efficiency in land mobile radio communication," *Philips Telecommun. Rev.,* Vol. 36, April, 1978, p. 53.

[10.118] Syski, R. *Introduction to Congestion Theory in Telephone Systems*, Oliver and Boyd Inc., Lond, 1960.

[10.119] Fujiki, M., and E. Ganbe. *Traffic Theory for Telecommunications*, Chapter 4, ECL Series, Maruzen Inc., Tokyo, 1970 [in Japanese].

[10.120] Koga, K., Y. Yasuda, and T. Muratani. "Bit error rate reduction performance of BCH codes and self-orthogonal convolutional codes," *Trans. IECE of Japan,* Vol. 62-B, February, 1979, p. 117 [in Japanese].

# 11

## ADVANCED CONCEPTS AND TECHNOLOGIES FOR COMMUNICATIONS SATELLITES

**DR. DOUGLAS REUDINK**

*Director, Radio Research Laboratories*
*AT&T Bell Laboratories, Holmdel, New Jersey 07733*

### 11.1 INTRODUCTION

The concept of communication satellites was put forth in the 1940s by science fiction writer Arthur C. Clarke. Later in 1955, J. R. Pierce published a paper entitled "Orbital Radio Relays," estimating technical feasibility. In 1957, Russia launched the first satellite, the low-altitude Sputnik spacecraft. Just a few months later, the United States launched the Explorer satellite. In 1960, a passive reflector, the ECHO balloon was launched by the United States into a circular orbit 1600 km above the earth. This 30-m diameter balloon constructed of lightweight mylar with an aluminized surface provided communications between Bell Laboratories in New Jersey and Jet Propulsion Laboratories in California by bouncing radio signals off the balloon during times of mutual visibility. Large parabolic antennas up to 30 m in diameter, high-power transmitters with several kilowatts output and receivers with low noise maser front ends were necessary to obtain one high-quality voice circuit between the east and west coast of the United States.

The use of active repeaters (devices on board the satellite which would receive weak signals, amplify them, and send them back down towards the earth) was the first of two major advances in technology that accelerated the evolution of communication satellites to the form we know today. The first satellite to use active repeaters with real time transmission and bandwidths sufficient to carry television was the TELSTAR satellite of the AT&T Company. This satellite, launched in 1962, received transmissions and amplified them through a traveling wave tube to an output power of a few watts. It demonstrated for the first time a television link between the United States and Europe.

The SYNCOM satellite, launched in 1963, marked the second major advance in satellite technology. It was the first satellite to be placed in geostationary orbit. To an observer on the earth, the satellite, rotating with a period of 24 h, appears to be stationary some 36,000 km above the equator. From this vantage point, nearly one-half the earth is visible. In contrast to low-altitude orbits, where a dozen or more satellites are required to provide continuous coverage to any point on the earth, geostationary orbits require only three satellites. A major disadvantage of the geostationary orbit is its large distance from Earth. Since the signals travel a much larger distance, they arrive 100 or more times weaker than they would from a low orbit satellite. Since this signal loss can be compensated for with higher gain antennas and more powerful transmitters, it is not of major consequence. More serious, however, is the problem of delays and echoes. When a conversation is held over a geostationary satellite circuit, the talker must wait about half a second longer for a reply compared to a land circuit. Even worse, a partial reflection often occurs near the telephone of the distance user. This causes some of the talker's voice to traverse the return path, causing an echo delayed by about half a second. This echo often confuses the talker. One technique to overcome the echo problem, which also occurs on very long land or undersea routes, is use of echo suppressors, voice-actuated switches which shut off the return signal when either speaker is talking. During active conversations with frequent interruptions and simultaneous talk by both users, the delay before the echo suppressor is activated confuses the users.

The more-recent echo canceler has proved to be especially effective in mitigating the echo problem. An echo canceler adds to the voice at the far end the exact negative value of the reflected signal; thus the echo is canceled, but the far-end speech is transmitted unaltered. With echoes eliminated, the delay factor is tolerable; people throughout the world have found satellite circuits acceptable.

The growth in quantity, quality, and applications of communication satellites has been phenomenal. Since 1965 more than 100 communications satellites have been launched and as seen in Fig. 11.1 the trend continues. Predictions are that the trend will continue to grow, with some studies showing a need for 250 or more 1980 capacity satellites serving the United States by the year 2000. As shown in Fig. 11.2, the growth in all types of telecommunications is very large and satellites are expected to play a significant role in satisfying the demand.

Satellites first found use for international communications. Although North America and Europe were linked in 1956 with voice circuits by submarine cable, the majority of communications between far-separated nations was through the use of low capacity and often erratic High Frequency (HF) radio. It is no wonder that the INTELSAT organization has grown to over 100 member nations. Also, communication satellites have found use for domestic, military, and maritime communications. Additionally, a number of high-technology experimental and special-purpose satellites have been developed. With few exceptions, satellites to date

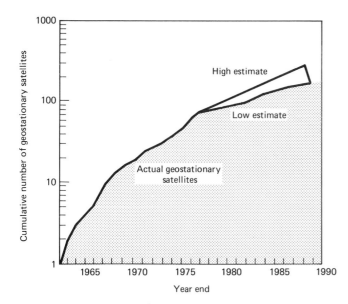

**Figure 11.1** Growth in number of geostationary satellites.

have been essentially tailor made for each application. Typical users order two to four identical satellites to satisfy their special needs. Even though it is far more costly to design and engineer specific application satellites than to employ one of a standard variety, users have found satellites an extremely effective and economical means for communication compared to terrestrial alternatives.

Broadly speaking, communications satellites fall into one of five categories:

1. *International*: International satellite communications is served primarily through the INTELSAT satellite system which serves over 100 nations interconnected through 200 Earth stations. Another system, the Inter-Sputnik satellite system primarily serves Eastern Block Communist countries and Cuba.

2. *Domestic and regional*: More than two dozen countries have domestic satellite systems either operational or in the advanced planning stages. These satellites generally use shaped antenna beams to provide domestic coverage and are employed for telephony, data, and television distribution. Regional satellites are being developed to serve Europe, the Arab States, portions of Africa, South America and Nordic regions with uses and characteristics similar to domestic systems.

3. *Military*: Military communication satellites are typically operated at 7/8 GHz. The U.S.'s **defense satellite communication system** (DSCS) constitutes a series of satellites to provide worldwide military communications. Other networks are employed to provide communications at UHF or VHF to aircraft and naval vessels. The circuit capacities of military satellites are modest compared to international and domestic satellites.

4. *Broadcast and special purpose*: Broadcast satellites are designed for regional or domestic coverage, the intent is to radiate a high-powered television signal

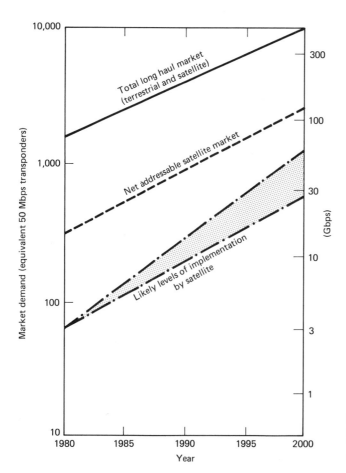

**Figure 11.2** Projected demand for satellite capacity to serve the domestic U.S. market, as estimated by Western Union (1979).

down to a large number of small receiving antennas (one meter diameter or less) using frequencies around 12 GHz. Special purpose satellites include the operational **maritime satellite** (MARISAT). A unique special-purpose satellite is the **tracking and data relay satellite systems** (TDRSS). This satellite enables NASA to track and command all of its low altitude satellites and relay information from the space shuttle to earth. These satellites replace a large number of tracking stations which would be needed for worldwide coverage of low altitude vehicles such as satellites, platforms or the shuttle itself.

5. *Experimental*: Perhaps the best known of the experimental satellites are the NASA developed **applications technology satellites** (ATS). This series began in 1968 and was designed to test new technology concepts for point-to-point, mobile, satellite-to-satellite, and broadcast communications. Each satellite consisted of several experiments operating at frequencies ranging from VHF to 30 GHz. In 1976 **communication technology satellites** (CTS) a joint program with NASA and the Canadian Department of Communications was

launched. Operating in the 11- to 14-GHz frequency band, this satellite was primarily intended to demonstrate a high-powered 200-W movable spot beam for TV applications. Several other satellites have demonstrated advanced capabilities, and many more are being planned. The Japanese, for example, have launched three experimental satellites—one for television broadcast and two for advanced telecommunications experiments. Some satellites like those of the Lincoln Laboratories (LES), the European Space Agency **orbital test satellites** (OTS) and the Japanese (ECS) are designed as preoperational prototypes to gain experience and information prior to embarking on fully operational programs. The propagation of microwaves and millimeter-waves through the atmosphere is of great interest and experiments to measure the effects on propagation have been included on the ATS, CTS, the AT&T COMSTAR, Japanese ECS, and the Italian SIRIO satellites.

## 11.2 BASICS OF SATELLITE COMMUNICATIONS

For the past 20 years, communication satellites have worked by receiving an up-link signal from an Earth station, shifting its frequency to the down-link "transmit" band, amplifying the shifted signal, and retransmitting it back to the ground. The frequency translation is necessary because it is impossible to achieve the necessary isolation between the transmitter output and the receiver input. Without sufficient isolation, the satellite's transmitted signals would block the up-link received signals. By using different up-link and down-link frequency bands, signals from the satellite transmitter are rejected by filters in the receiver and sufficient isolation is achieved.

Commercial satellite communications use a frequency band of 500 MHz bandwidth near 6 GHz for up-link transmissions and another 500 MHz bandwidth near 4 GHz for down-link transmission. Since these frequencies are shared in many countries with terrestrial communications, mutual interference problems must be avoided. Usually the 500-MHz allocation is divided into 12 channels of approximately 40 MHz each. The level of transmit power per 40-MHz channel is typically of the order of 5 to 10 W. This permits each of up to the 12 transponders to carry one TV channel or about 1500 analog FM voice circuits. If digital modulation is used, transponder data rates from 50 to 100 Mb are achievable.

An old modulation technique called single sideband (SSB) is finding popularity in satellite communications because of its greater voice capacity. Theoretically, about 10,000 voice circuits could be carried over a single satellite transponder. In practice, about 7000 voice circuits are achieved.

With traffic demands exceeding available spectrum at $\frac{4}{6}$ GHz, advanced technology satellites are moving to higher-frequency bands for satellite communication. One band with 500 or 1000 MHz bandwidth (depending on the region of the world) has been allocated near 12 GHz for down-links with corresponding up-links near 14 GHz. A third band where extremely high capacities are potentially available is the 20/30-GHz band, where 2.5 GHz bandwidth has been allocated. Much attention will be focused on these bands when we discuss advanced technologies in later sections.

### 11.2.1 Fundamentals of Radio Communications

Energy emanating uniformly in all directions from a point source is called **isotropic radiation**. Antenna gain is the increase in power in the desired direction relative to an isotropic antenna. Gain depends upon the size of the aperture A and the wavelength according to the formula

$$G = \frac{\eta A}{4\pi\lambda^2} \tag{11.1}$$

where $\eta$ represents the antenna efficiency. The quantity $\eta A = A_e$ is often called the **effective area** of the antenna. Depending upon construction, antenna efficiencies usually range from 50 to 80%. A simple and elegant formula which relates the power received to the power transmitted was derived by Friis in 1944:

$$\frac{P_r}{P_t} = \frac{A_t A_r}{d^2\lambda^2} = G_t G_\eta \left(\frac{4\pi}{d\lambda}\right)^2 \tag{11.2}$$

An isotropic radiator has unity gain and an effective aperture $A = \lambda^2/4\pi$. The loss between two isotropic antennas is called the **free space path loss** and is expressed in decibels by

$$\begin{aligned} L_f &= 10 \log_{10}\left[(4\pi d/\lambda)^2\right] \\ &= 92.45 + 20 \log_{10}f + 20 \log_{10}d \end{aligned} \tag{11.3}$$

where $f$ is in gigahertz and $d$ is in kilometers. The free space path loss between Earth and a geostationary satellite at 4 GHz is 195.5 dB. At 12 and 18 GHz, the losses are 205 and 208.5 dB, respectively.

For both practical and political reasons the closest that satellites can be spaced at 4 GHz is about 2° allowing 180 orbit slots worldwide for this frequency band. However, far fewer slots are practical because most applications require satellites to be near or over land masses. Because the same physical size aperture in meters produces a narrower beam at higher frequency, satellites operating at higher frequencies can be placed closer together. However, there is a strong desire to use small antennas in these frequency bands, called **rooftop antennas**, which require satellites to be placed further apart. Thus the challenge of using high-capacity satellites in these high-frequency bands with flexibility—such as that provided by small Earth-station antennas—is where the most significant advances will be made with satellite communications technology in the remainder of this century.

The high-frequency bands are not without problems, however. As mentioned earlier, free space path loss increases with frequency. This problem is not so serious because antenna gain also increases with frequency, and as long as antennas can be built with surface tolerances small compared to a wavelength, increased antenna gain effectively offsets the increased power loss. More serious, however, is the fact that the higher frequencies are attenuated by rain in the atmosphere. This phenomenon has been studied extensively by many researchers over a period of years. At 18 GHz it is not uncommon for radio signals to be attenuated by 20 dB in a heavy thundershower.

### 11.2.2 Noise and Capacity

There are some fundamental relationships which allow us to calculate the ratio of signal and noise and relate this to the capacity of a channel. There are several types of noise—thermal, galactic, and manufactured. The only noise of real concern at microwave frequencies is the ubiquitous thermal noise generated in electronic devices by the thermal motion of electrons. The power of this noise is given by

$$P_n = kTW \tag{11.4}$$

where $P_n$ is the noise power in watts, k is Boltzman's ($k = 1.38 \times 12$ W $\cdot$ s/K), T is the temperature in Kelvin, and W is the bandwidth in hertz. Devices used to amplify or detect low-level signals are characterized by their equivalent noise temperature, $T_s$. A device with a 100 K equivalent noise temperature would double in output power when the signal is the same power as a 100 K noise source. Noise limits the rate at which information can be transmitted.

Shannon's formula gives the capacity of a channel in bits $B$ per second as

$$B = W\log_2\left(\frac{P_r}{P_n + 1}\right) \tag{11.5}$$

The formula does not say how to achieve this capacity; it only says what it is. Typical systems which transmit about 2 bits per hertz and make fewer than one error in a thousand require that the power transmitted be about ten times greater than the noise. The Shannon capacity is also derived for error-free transmission, and it is therefore difficult to judge how close the actual systems come to this limit. When one error in a million is taken to be essentially error free, then some sophisticated systems have been designed which come within about 3 dB of the Shannon limit. The important thing to note about this Shannon formula is that the easy way to increase channel capacity is to increase the bandwidth $W$ and the hard way is to increase power, since capacity only increases with the logarithm of power. However, it is difficult to increase bandwidth because channels are assigned by federal and international bodies, which have deliberated at length to assign frequency allocations among the multitude of potential users. Later we shall examine the technology methods which effectively increase the usable bandwidth for satellite systems to meet the ever-increasing needs for greater capacity.

### 11.2.3 Typical Link Budgets

The first commercial satellite, INTELSAT I employed an Earth-coverage satellite antenna and transmitted only a few watts of power. Using very large 30-m antennas, this satellite was capable of providing 240 intercontinental voice circuits. Generally speaking, satellite performance is limited by the down-link because far more additional power can be provided from an Earth station than from a satellite. This additional power is large enough to allow us to neglect (in a simple analysis) the noise contributions caused in the satellite receiver. Later we shall see that the demodulation of the up-link signals on board the satellite is a powerful technique to reduce Earth station transmitter power and provide advances in signal routing and processing. For the time being, however, we will focus our attention on the satellite down-link, where the capabilities of the satellite transmitter and Earth station receiver are dominant factors. The quality of the greatest interest in satellite design is the down-link signal-to-noise ratio (SNR). This can be expressed in decibels as

$$SNR = P_t + G_t + G_r - L_f - 10 \log(kTW) - L_m \qquad (11.6)$$

where all quantities have been defined before except $L_m$ and are expressed in decibels. $L_m$ consists of the miscellaneous losses such as atmospheric, transmission lines, couplers, and other components between the transmitting amplifier and its antenna and the receiving preamplifier and its antenna. Such losses are usually less than 2 dB in clear air.

### 11.2.4 Impairments Due To Rain

For terrestrial radio applications, the wide bandwidths available motivated propagation studies at millimeter wavelengths. Atmospheric propagation measurements of oxygen and water vapor absorptions showed clear air transmission windows around 30 and 90 GHz. However, large rain attenuation dictated short repeater spacing for terrestrial radio relay.

The effects of oxygen and water vapor in the atmosphere, as well as the cosmic background noise, were determined to be small for microwave satellite communication. For frequencies above 10 GHz, the fundamental obstacle is attenuation by rain. Over the microwave band, the attenuation is considerably greater for rain than for fog. (The reverse is true at optical wavelengths.)

Early experimental measurements on Earth-spce paths began with suntrackers followed by the passive radiometer method. The sum emits large amounts of constant microwave power and thus the attenuation by clouds and rain can be measured directly. The rain medium emits noise in the fashion of a blackbody at some apparent absorber temperature $T_a$. This noise, which increases with the rain attenuation until saturation at about 10 dB, can be utilized to measure attenuation up to this level. Because there is no active source, the measurement method is called **passive radiometry**. The radiometer measurements can be conducted on a fixed earth-space path day and night, whereas the suntracker is limited to an ever-changing path toward the sun in the daytime only. The simplicity of radiometry also facilitates long-term continuous field measurements to collect statistical data on the space diversity schemes.

It is well known that heavy rain showers are often scattered and localized events in contrast with widespread rain. Path diversity, where two or more separated ground stations receive signals from the same satellite, can be used to take advantage of the space inhomogeneity of the heavy rain. Data from radiometers pointing at the same 30° elevation angle and spaced 10, 20, and 30 km was collected. At 19 GHz for a single site, a margin of 30 dB in excess of clear air is needed to guarantee 99.99% availability, as shown in Fig. 11.3. It was found that the percent

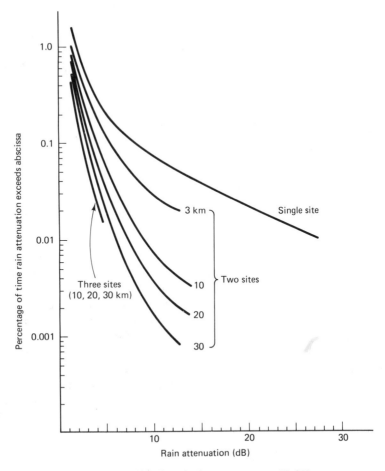

**Figure 11.3** Site diversity improvements at 19 GHz.

of time exceeding 10 dB attenuation decreased more than an order of magnitude by using path diversity. Improvements in link availability are shown in Fig. 11.3 for one, two, and three receiving sites.

Measurements of the 20-GHz beacon on ATS-6 provided initial cross-polarization versus attenuation data on an Earth-space path. These beacons were not continuous but operated in an on-demand mode. They were useful calibration sources but were unable to supply cumulative attenuation and depolarization distribution. Direct measurements of these distributions from satellites were not available until COMSTAR and CTS beacon experiments in 1976.

To measure long-term probability distributions of properties of Earth-space propagation for the planning of future high-capacity satellite communication systems, continuously transmitting 19- and 28-GHz beacons were installed on the COMSTAR communication satellites. This permitted measuring attenuation, depolarization, coherence bandwidth, and scatter of the beacon signals by atmospheric processes. Long-term attenuation and depolarization measurements of an 11.7

Sec. 11.2    Basics of Satellite Communications

**581**

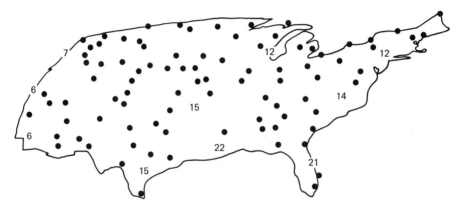

**Figure 11.4** Margins required for 0.01% outage satellite at 70°W, F = 12 GHz.

GHz circularly polarized beacon on the CTS satellites were also made. The attenuation probability distributions were consistent with each other and with the previous suntracker and radiometer data.

Rain rate data, together with COMSTAR measurements, confirmed the previous results that path diversity is necessary for high reliability performance of 18/30 GHz satellite communications. The effect of weather and satellite location strongly impacts the amount of time that an Earth-space link will be out. In Fig. 11.4 we see typical margins (i.e., the number of decibels in excess of clear air) required to insure that the cities will have communications 99.99% of the time. By contrast, Fig. 11.5 shows the required margins for a satellite located at 130°W. Because the cities in the eastern United States must now "look through" much more rain due to an increased slant range, the margins are increased by 10 dB or more. The impact of the rain on the design of the satellite systems will be considered in more detail later.

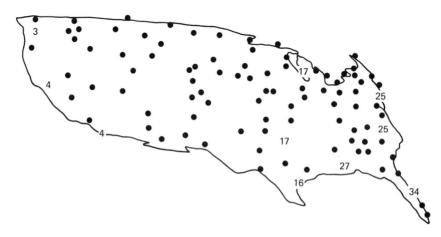

**Figure 11.5** Margins required for 0.01% outage satellite at 130°W, F = 12 GHz.

### 11.2.5 Evolution of Satellite Capabilities

Many of the currently used satellite design parameters began with the INTELSAT I satellite launched in 1965 and are being carried forward through the 1980s. The major improvements in communication satellites have come as a result of increased payload due to ever-increasing launch capabilities. INTELSAT I had an in-orbit mass of only 38 kg. Table 11.1 shows the steady progression in capacity and mass of the INTELSAT satellite series. The space shuttle is capable of lifting nearly 30,000 kg into low earth orbit. Payloads as large as 2,000 kg in geosynchronous orbit are readily achievable and it is projected that payloads as large as 6,000 kg will be available when high-energy upper stages are developed to transport satellites from low earth orbit to geostationary orbit. If this expected trend continues, satellites with circuit capacities of 100,000 or more will be technically and economically feasible.

A key development in the evolution of satellite technology has come about through frequency reuse antennas by use of orthogonal polarizations. Electromagnetic plane waves can be thought of as vectors and devices can be constructed which are sensitive to one vector and not to a vector perpendicular to it. Hence, each vector or polarization can be transmitted and received independently thus reusing (in this example doubling) the allocated frequency.

Other advances that increase the effectiveness of satellites are spot and shaped antenna beams. With increased mass available, satellites can be made physically larger, permitting the use of larger diameter antennas with their narrower antenna beams and the potential for generating several independent beams onboard the satellite. Where a particular region or country is to be served, it is advantageous to shape an antenna beam to follow the contour of the region requiring coverage. This technique was first used in international communications in INTELSAT III and has been especially popular for domestic satellite systems, where beams tailored to cover regions such as Canada, Japan, Indonesia, the Arab States, United States, and Mexico have been designed. The techniques to generate spot and contoured beams are examined in more detail in a later section.

Onboard signal processing is an emerging technology certain to be commonplace in the 1990s. Already antenna beam switching has been demonstrated, and

**TABLE 11.1** Progression of weight and capacity of the INTELSAT series satellites

|  | First launch | Weight in orbit (kg) | Capacity voice circuits |
|---|---|---|---|
| INTELSAT I | 1965 | 38 | 240 |
| II | 1967 | 86 | 240 |
| III | 1968 | 152 | 1200 |
| IV | 1971 | 700 | 4000 |
| IVa | 1975 | 790 | 6000 |
| V | 1979 | 967 | 12,000 |
| VI | 1986 | 1670 | 33,000 |

experimental programs initiated to demonstrate onboard detection, routing, error correction, and multiplexing of digital signals. Additionally, work is progressing on intersatellite links (direct satellite-to-satellite links) to extend coverage around the world, thereby eliminating double-hop satellite circuits (satellite to ground to satellite) and reducing delay.

As mentioned before, satellites today essentially amplify a received signal and redirect it back toward earth. The first modulation method used was FM. In this mode of transmission the signal has no amplitude fluctuation. Thus the transmitter can be operated in its most efficient manner, that is with full power. Often a user does not require the full bandwidth of a satellite transponder. Several signals can be transmitted simultaneously and independently providing they are amplified undisturbed. This requires that the amplifier be linear and necessarily less efficient. Thus weaker signals arrive back to the earth. The advantage, of course, is that multiple users from separated earth stations can simultaneously communicate through the same satellite channel. This technique is called frequency division multiple access (FDMA).

An alternative with greater flexibility is time division multiple access (TDMA). As illustrated in Fig. 11.6 in TDMA users time share a single channel. A user is assigned a time slot in which to transmit. The time slot is measured from a frame marker which repeats at a fixed period. The time slot can be variable in length and can be preassigned or assigned as needed on demand.

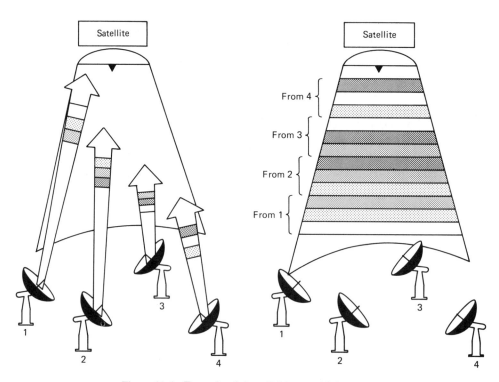

**Figure 11.6** Example of time division, multiple access.

**Figure 11.7** Footprints from a multi-beam satellite.

The effectiveness of **TDMA-demand assignment** (TDMA-DA) is especially dramatic when coupled with satellites using spot antenna beams or scanning beams. Imagine a satellite serving several large cities with spot beams, as illustrated in Fig. 11.7. Each beam is independent of the other, and only Earth stations located within the coverage areas can access the satellite. Adjacent beams use orthogonal polarizations to achieve the required isolation. To utilize this network fully, each spot beam must be connected to every other one. Were this to be done via frequency division, each up-link would have to be connected to ten other down-links. Using frequency division the number of channels needed to connect $N$ spot beams is either $N(N - 1)$ or $N^2$, depending on whether the loop-back path to the same coverage area is required. This implies that a minimum of 110 cross-connections be made onboard the satellite for this example. This arrangement is obviously inefficient and cumbersome and becomes unwieldy as the number of beams become large. On the other hand, a TDMA approach coupled with a switch onboard the satellite greatly simplifies the interconnection problem. A simple matrix switch arrangement as shown in Fig. 11.8 provides access to various beams. Later we shall show under very general circumstances that it is possible to achieve 100% utilization of time slots even though user requirements for communications from beam to beam are different.

In the remainder of this chapter we shall examine in detail those technologies that are enhancing the capability of communication satellites. Of particular importance are the use of the higher frequencies, advances in satellite and Earth-station antennas, the use of digital modulation, and onboard signal routing and processing.

The higher frequency bands offer the potential of greater satellite capacity. The disadvantages are that rain causes signal impairments and that hardware is

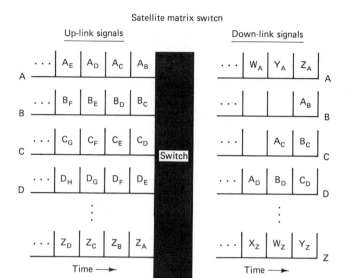

**Figure 11.8** Satellite matrix switch.

less mature and therefore more costly. The strong desire to remain compatible and the huge ground investments in Earth stations work against major technological changes at C-band. As orbit slots fill up at C-band (4 to 6 GHz), users will be forced to exploit the higher frequency bands. It is here that radical new technology will emerge.

## 11.3 MULTIPLE ACCESS AND DEMAND ASSIGNMENT

For broadcast applications, especially television, there is usually a single up-link from the station originating the broadcast. For access to the program, the receiver is merely tuned to the proper transponder channel and the signal is demodulated. For point-to-point applications, matters become more complicated. The first method used to achieve multiple access, that is, several users access the same satellite (usually the same transponder), was FDMA. In this arrangement, each user is assigned a frequency slot in which to transmit. If that frequency slot is made on a permanent basis, the access is called a **fixed assignment**. If these slots are assigned on an as needed basis, that is, on demand, the accesses are called DAs. For small numbers of users, FDMA is adequate, but it has several drawbacks.

First, when the transmissions are on separate carriers, the high-power amplifier in the satellite must be "backed-off" until it is linear. Normally, for constant-envelope, single-carrier transmission, such as TV or multiplexed voice, which is sent on a frequency modulated carrier, the power amplifier can be highly nonlinear, which is the more power-efficient mode of transmission. Intermodulation distortion results in cross talk among the carriers if the amplifier is not linear. Char-

**586**                    Advanced Concepts and Technologies    Chap. 11

acteristics of several amplifiers are shown in Fig. 11.9. With the satellite power reduced, Earth stations must be larger or else a price is paid in terms of reduced SNR. The back-off necessary to obtain a given carrier-to-intermodulation ratio can be calculated. With ordinary traveling wave tubes, large back-offs are required; however with (FET) amplifiers or linearizing circuits, the situation is greatly improved. Figure 11.10 shows the advantage of having linear devices on board the satellite. For digital signals, $C/I + N$ near 20 dB are desirable. Under these conditions, Field Effect Transistor (FET) amplifiers can save a factor of two in power.

Another problem with FDMA is that it is difficult to vary the bandwidth of the assigned frequency slots on demand. In most applications, traffic varies with time but transmission and receiving equipment is usually tailored to a single bandwidth; thus only crude quantization of demand assignment can be made. Finally, for networks where each point is connected to the other in the network, the number of paths and, hence assignments grows as the square of the number of earth stations. While creating large numbers of channels is a formidable task using frequency division, time sharing a single channel among, for instance, 20 users that require 400 time slots is relatively easy. In addition, the length of transmission time, or the time slot, can be adjusted according to user needs. Thus TDMA has the advantages of flexibility in bandwidth and can use the full transponder power with no interference among users.

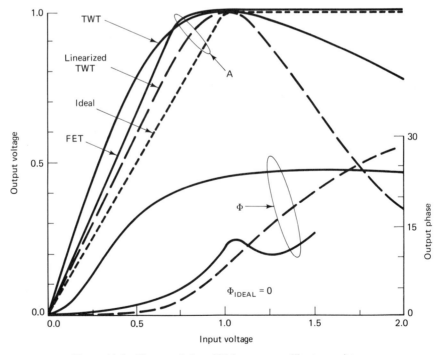

**Figure 11.9** Characteristics of high-power satellite transmitters.

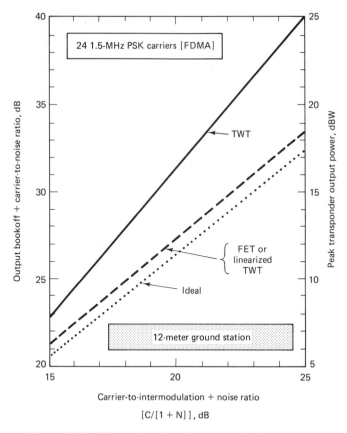

**Figure 11.10** Effect of nonlinearity on transmission.

### 11.3.1 Timing and TDMA Technology

To control and configure a TDMA system, a large time interval called a frame is defined. A frame is subdivided into time **slots**, and a **burst** consists of an integer number of slots. A **guard time** is required between bursts to assure that no overlap occurs. A burst typically consists of a preamble, the message portion, and in some coded systems a postamble, as illustrated in Fig. 11.11.

A **preamble** consists of a signal interval for carrier recovery in coherent demodulating systems, a signal interval for symbol-timing recovery, a **unique word** (UW) for burst synchronization, a **station-identification code** (SIC) and some housekeeping symbols, such as an order wire for voice or data, command and control signaling, or error-monitoring symbols. The **message** or information portion contains the desired data, possibly encoded and modulated onto the carrier. In some applications, a **postamble** is used for decoder quenching for the next burst.

The ratio of useful message information to the total bits transmitted is called the *transmission efficiency*

$$\eta = \frac{rM}{P + M + Q} \qquad (11.7)$$

where $r$ = coding rate of the codec
$\quad\quad M$ = message symbol per burst
$\quad\quad P$ = number of symbols in preamble
$\quad\quad Q$ = number of symbols in postamble

### 11.3.2 TDMA Synchronization Methods

In TDMA systems, several methods are used in synchronizing transmissions from earth stations and the satellite. These several methods fall into four categories; random-access, open-loop, closed-loop, and single sideband-TDMA (SS-TDMA) window methods.

Multiple access in the time domain can be accomplished through **random access** by each station. In the basic form of random access no network timing is present; each station transmits bursts or packets as necessary at random and some bursts overlap. Random access is notably inefficient because retransmission must occur when packets collide.

**Open-loop** methods are characterized by the property that a station transmitting a burst does not monitor its reception; thus the loop is not closed. Without

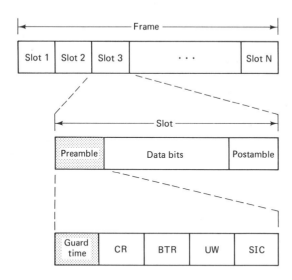

CR  = carrier
BTR = bit-timing recovery
UW  = unique word
SIC = station identification code

**Figure 11.11**  Typical TDMA frame.

any active form of synchronization, it is easy to achieve accuracies of about 5 μs. Based on approximate orbit parameters and free-running clocks, burst positioning has been done to 200-ns accuracy.

Another class of open-loop synchronization involves the use of a reference burst. A **reference burst** is a special preamble used to mark the start of frame. The station transmitting this burst is called the **master station** or **reference station**. The reference bursts are received by each station in the TDMA network and all transmissions are locked to the time base of the reference station. Earth station clocks are synchronized to the received clock, so that stable oscillators are not required.

In **closed-loop** systems the transmitted signals are returned through the satellite transponder to the transmitting station, where the signals are compared and the transmission is adjusted depending on the time of arrival relative to a time base; and early or late burst arrivals are measured and used to control the transmit time. Through careful signal design and processing, precisions of a few nanoseconds are possible.

For satellites with spot beams, connectivity is achieved by a switch in the satellite. Earth stations must synchronize to a switching sequence being followed in the satellite. The switch matrix connections are changed throughout the frame to produce the required interconnection of the earth stations. At the beginning of the frame is the **synchronizing window**, during which signals from each spot-beam zone are looped back to their originating spot-beam zone; this forms the timing reference for all zones and closed-loop synchronization can be established.

### 11.3.3 Time-Slot Assignment

Under very general conditions it has been shown that methods exist that enable frequency reuse via a multibeam satellite system employing identical transponders, so that all transponders are used at maximum efficiency and a uniform grade of service is provided over the service area.

Consider a satellite employing $N$ identical wideband transponders, each with a capacity or throughput of $C$ units, with a number $M$ of distinct footprints needed to provide service. In general, $M$ may be much greater than $N$, but in what follows only $M \geq N$ is required.

The system traffic can be represented by an $M \times M$ matrix $[t_{ij}]$ as shown:

$$[t_{ij}] = \begin{bmatrix} t_{11} & t_{12} & \cdot & \cdot & \cdot & t_{1M} \\ t_{21} & t_{22} & \cdot & \cdot & \cdot & t_{2M} \\ \cdot & \cdot & & \cdot & & \cdot \\ \cdot & \cdot & & & \cdot & \cdot \\ \cdot & \cdot & & & \cdot & \cdot \\ t_{M1} & t_{M2} & \cdot & \cdot & \cdot & t_{MM} \end{bmatrix} \tag{11.8}$$

The element $t_{ij}$ represents the traffic originating in beam $i$ and destined for somewhere in beam $j$. Each footprint might contain several ground stations, so $t_{ij}$

represents the sum of the traffic from all stations within beam $i$ which is directed to stations within beam $j$.

It is not necessary that the traffic matrix be symmetric, and a loop-back feature is possible; that is, we do not require $t_{ij} = t_{ji}$.

Two requirements must be imposed on the traffic matrix $[t_{ij}]$. First, since the total capacity of the satellite is equal to $NC$ ($N$ transponders each capacity $C$), we require that

$$T = \sum_{i=j}^{M} \sum_{j=1}^{M} t_{ij} \leq NC \qquad (11.9)$$

The second requirement is that the traffic originating from or destined for a particular beam should not exceed the capacity of one transponder, that is,

$$\text{Row sum } R_i = \sum_{j=1}^{M} t_{ij} \leq C \qquad i = 1, 2, \ldots, M \qquad (11.10)$$

$$\text{Column sum } S_j = \sum_{i=1}^{M} t_{ij} \leq C \qquad j = 1, 2, \ldots, M \qquad (11.11)$$

The transponders are utilized with 100% efficiency when (11.9) is satisfied as an equality. This equation may be interpreted as establishing the minimum number $N$ of transponders required. Conditions (11.10) and (11.11) are necessary because no two transponders can be connected to a common spot beam (either uplink or downlink) on a noninterfering basis.

If the total offered traffic equals the sum of the transponder capacities, we have the potential for 100% utilization. We will show that it is possible to interconnect the various up-link beams, transponders, and down-link beams such that this is achieved. We do this on a time division basis by enabling each of the $N$ transponders to access any of the $M$ receive (up-link) antenna ports and any of the M transmit (down-link) antenna ports.

It will be shown that all the offered traffic can be allocated among the $N$ transponder on a noninterfering basis; that is, at any instant of time, the $N$ transponder inputs are each connected to a different receive port, and the N transponder outputs are each connected to a different transmit port.

A diagonal of a matrix $[t_{ij}]$ is a $K$-tuple $D = \{d_1, d_2, \ldots, d_K\}$, where each member is a nonzero element of the matrix and no two elements appear in the same row or same column of the matrix. The length of the diagonal is $K$ (the number of elements) and the diagonal is said to cover the $K$ rows and $K$ columns from which the elements are taken.

Assign traffic to the various transponders as follows: Let the TDMA frame consist of $C$ time slots, each representing one unit of traffic. There are $N$ such frames, one belonging to each transponder. In the traffic matrix $[t_{in}]$, find any diagonal of length $N$ which covers all rows and columns summing to $C$ (if any). From these $N$ diagonal elements, extract one unit of traffic from each and assign one unit to each of the $N$ transponders. Since the traffic assigned to the transponders (for this time slot) originates from different up-link beams and directed to different down-link beams, then the traffic has been assigned on a noninterfering basis.

Now since N units of traffic have been removed from the matrix, the reduced matrix has a total traffic of $NC - N = N(C - 1)$ units. Furthermore, each transponder has $C - 1$ units of traffic carrying capacity left, and no row or column of the reduced matrix sums to more than $C - 1$. The latter is true because every row and column that summed to $C$ in the original matrix has had one unit of traffic removed (because of the way the diagonal was constructed).

At this stage, we have the same situation as we started with, except $C - 1$ replaces $C$. By the same technique, we can assign another N units of traffic to the next time slot in each transponder, and end up with a matrix with remaining traffic $N(C - 2)$ in which no row or column sums to more than $C - 2$. Each transponder has then $C - 2$ time slots unallocated. Hence, we can repeat this procedure until all transponder time slots are used and no traffic remains unallocated.

Thus the nonuniform demands of a traffic matrix can be met by N identical transponders each operating at 100% utilization efficiency. Although the method described was for a matrix for which (11.2) was satisfied as an equality (i.e., $T = NC$), it also applies to a matrix for which $T < NC$, because we can always pad such a matrix with dummy traffic until $T = NC$. The assignments corresponding to the dummy traffic can be ignored and simply reflect the fact that the available transponder capacity exceeds the demand.

## 11.4 THE SCANNING SPOT-BEAM CONCEPT

This concept combines the wide-area accessibility of an area coverage system with the high antenna gain of a spot-beam system. It is intimately tied with TDMA. Recalling Fig. 11.6 it is noted that for an area coverage system employing TDMA, the satellite interconnects a single user pair (one transmitting and one receiving) at every point in time. A satellite capable of forming a pair of rapidly movable spot beams (one receiving and one transmitting) can thereby produce the same accessibility as does the area coverage system, provided the beams are scanned synchronously with the TDMA bursts to interconnect the correct user pairs. Operation is illustrated in Fig. 11.12. The up-link (receiving) beam is first pointed at ground station 1, which sequentially transmits bursts intended for a number of receiving ground stations. As these bursts arrive at the satellite, the down-link beam is scanned to distribute the bursts to their appropriate destinations. The up-link beam is then pointed at another ground station, and the process repeats until all ground stations have had an opportunity to transmit.

Implementation of the scanning beam concept requires several technological innovations. An initial question was the method whereby the beam could be made to scan. The original scanning beam concept was based upon a phased array antenna. This is an array of many small radiating elements, each of which is

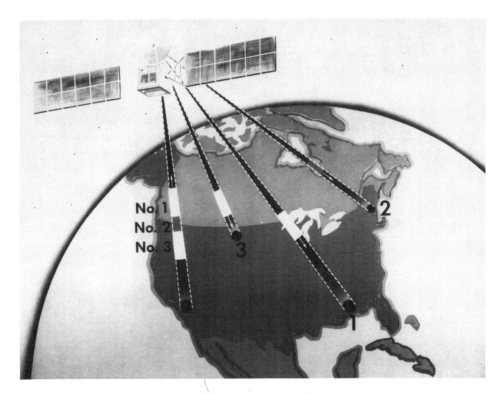

**Figure 11.12** Artist's concept of a scanning beam satellite.

preceded by a device that can shift the phase of the microwave signal passed through it. By imparting the appropriate phase to the wave radiating from each element, the superposition can be made to exhibit a planar wave front, creating a beam focused in a particular direction. This direction can be altered by changing the settings of the individual phase shifters. Such arrays had received earlier attention for communication satellite application. An alternative to the array is an antenna consisting of a large reflector that could be illuminated from different directions by means of multiple feedhorn clusters, as discussed in a later section. By applying the signal to the appropriate cluster, the resulting beam can be made to focus in one of many possible directions. For the NASA 20- to 30-GHz program, advanced communication technology satellite (ACTS), beam switching is achieved by exciting separate clusters of feedhorns.

The feasibility of the scanning spot-beam concept was successfully demonstrated using ground-based hardware simulators. The simulator contained both transmit and receive array elements, phase shifters, and the array controller. The important features of this experiment were to demonstrate

1. Full-band (500-MHz) digital transmission through a satellite channel (up-link, transponder and down-link).
2. Rapid, synchronized scanning of the up-link and down-link beams.

3. Automatic cophasing (or beam forming) of the up-link and down-link beams, concurrent with TDMA transmission.

4. High-speed TDMA burst operation, with demand assignment.

5. Computer control of all of the above.

The experimental system was designed around a transmission scheme using conventional QPSK modulation at 600 Mb/s (300 Mbaud). For TDMA operation a frame length of 25 ms was chosen; this results in 750 symbols/burst (or 1500 bits/ burst with quadrature phase-shift keying (QPSK) modulation); the frame corresponds to approximately 10,000 bursts of 2.5 μs each. With these parameters a channel consisting of one burst per frame has a raw data rate of about 60 Kb/s;

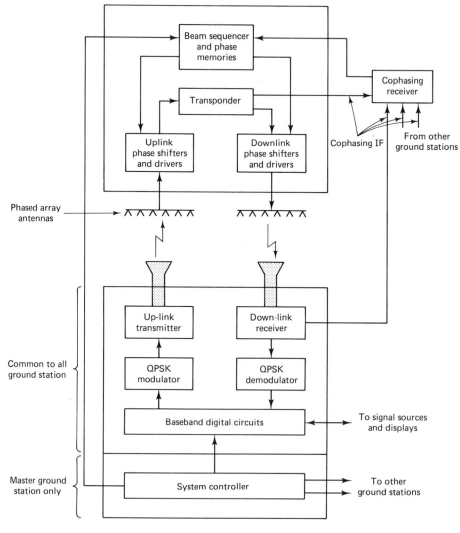

**Figure 11.13** Block diagram of the experimental system.

the usable data is slightly lower, due to guard time between bursts and the overhead required by carrier recovery and other synchronization functions. The choice of a long frame period with a large number of bursts was made to demonstrate the ability of a scanning beam system to efficiently provide service to many small ground terminals, each requiring capacity as low as one single voice circuit.

The five major elements of the experimental facility consisted of (1) the transmission subsystem, which carries 600 Mb/s through a satellite channel; (2) the satellite antenna subsystem, consisting of the up-link and down-link phased arrays and the associated phase shifter/driver circuitry; (3) the cophasing subsystem, which measures and updates the phases needed to form the scanning beams; (4) the baseband digital subsystem, including circuitry for both continuous and burst (TDMA) operation; and (5) the computer-control subsystem, which controls each of the other four and also facilitates the demonstration of the integrated system. Figure 11.13 indicates the interrelationships among the four hardware subsystems.

### 11.4.1 Transmission Subsystem

The transmission subsystem consists of the ground station modulator/transmitter, the satellite transponder, and the ground station receiver/demodulator.

The modulator input is a pair of synchronized 300-Mb/s binary pulse streams delivered by the baseband encoding circuits. These streams are converted to a QPSK signal at 14.25 GHz in one stage, using the quadrature amplitude modulation (QAM) principle. The output is boosted by a GaAsFET amplifier, bandlimited and radiated up-link by a rectangular horn antenna. The satellite transponder (Fig. 11.14) translates the signal to 11.95 GHz and retransmits it on the down-link.

The most-critical part of the receiver/demodulator is the carrier recovery stage, which works on the "modulation wipeoff" principle and performs in near-ideal fashion for both continuous (non-TDMA) and bursty (TDMA) digital signals. The recovered carrier is mixed with the received signal to produce two binary pulse streams, which are further converted to a pair of clocked digital bit streams. These represent the demodulator output, which is sent to the baseband digital subsystem for decoding and signal recovery.

### 11.4.2 Satellite Antenna Subsystem

The satellite antenna subsystem consists of the individual up-link (14.25-GHz) and down-link (11.95-GHz) phased arrays; the phase shifters associated with the array elements and the driver circuits that control the phase shifters.

Each phased array is a linear assembly of 16 pairs of horn elements. The total length of each array is about 40 cm. The array dimensions and scan angle were chosen to achieve a compact array feed structure for a double confocal parabolic reflector antenna system with magnification of about seven (Section 11.6.3).

The azimuth beam position is controlled in each array by setting a phase progression across the elements with the 16 four-bit phase shifters. Each phase shifter consists of four transmission line cells in cascade, the phase change in each cell being determined by a 1-bit control signal that biases a pair of p-i-n diodes.

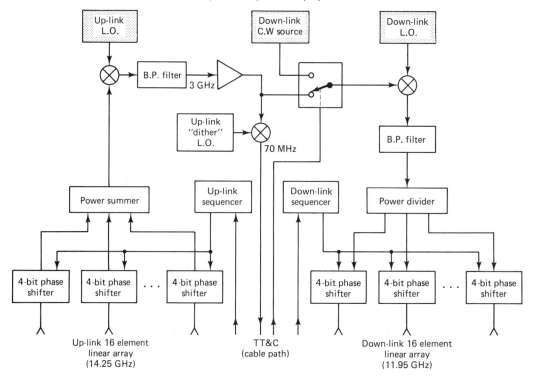

**Figure 11.14** Experimental phased array repeater.

The four cells are designed to obtain 16 uniformly spaced phase-shift values on the interval 0° to 360° (this corresponds to 22.5° steps). This design yields sufficient resolution to achieve the desired control of both beam pointing and beam shape.

Typical data measured for both the 12-GHz and 14-GHz units show the following performance: (1) The switching time between phase states is less than 10 ns. (2) The dc power required per driver circuit at a beam switching frequency of 1 MHz is 36 mW. (3) The radio-frequency (RF) power rating per phase shifter is 800 mW. (4) The RF loss through the phase shifter is 1.6 dB ± 0.2 dB over a 500-MHz bandwidth.

Beam-switching speed is determined by the speed with which the phase shifters can change state. Figure 11.15 shows the beam-switching time as the 12 GHz down-link beam is reoriented to a new position approximately 15° away. Beam-switching times of less than 10 ns (typically 7 to 8 ns) were observed as shown.

### 11.4.3 Cophasing Subsystem

The task of the cophasing subsystem is to determine the values of the phases to be loaded into the phase shifters to form a given up-link or down-link beam. This is done in closed-loop fashion through a simple in-band measurement. The meas-

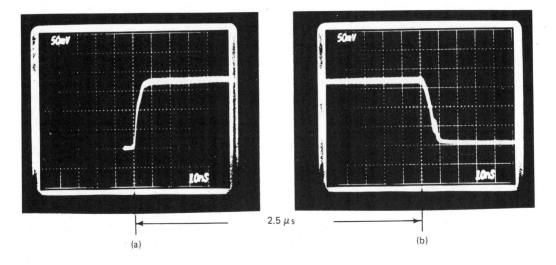

2.5 $\mu$s

(a)                                                                 (b)

**Figure 11.15**  Build up of oriented beam.  (a) Rise time.  (b) Fall time.

urement requires that a carrier-wave CW signal go through the array being cophased (up-link or down-link) while the beam sequencer advances the phases of the array elements one at a time.  The signal that has gone through the array is fed to the "cophasing processor," which uses it to evaluate the correct element phases to be stored in a memory buffer associated with the beam sequencer on the satellite. These values are later retrieved by the beam sequencer and loaded into the phase shifters to form a given beam.

### 11.4.4 Baseband Digital Subsystem

In each ground station the baseband digital subsystem consists of two separate components performing the different functions required by the two possible modes of operation: continuous (non-TDMA) transmission and bursty (TDMA) transmission.  Each component is further subdivided into a transmitter/encoder portion, which provides two digital bit streams for input to the QPSK modulator, and a receiver/decoder portion, which processes the two output bit streams produced by the QPSK demodulator.  The use of emitter coupled logic (ECL) 100 K technology in the high-speed portion of the digital subsystem, together with novel circuit design techniques and careful layout, achieved reliable digital operation at symbol rates well in excess of 300 Mbaud.

The transmitter/encoder for the TDMA-mode converts computer-generated graphics to a bursty (TDMA) digital signal for input to the QPSK modulator. Graphics data, which are already in digital form, are scrambled through addition of a pseudorandom sequence to improve their statistical characteristics, and assembled into short bursts which include the bit sequences required for carrier recovery, phase resolution, and synchronization.  Each ground station included a computer, which performs these functions in software.  The computer transfers the assembled bursts to a high-speed memory buffer, from which they are sent to the QPSK modulator at the full data rate of 600 Mb/s. In the receiver/decoder for the TDMA-mode the bursts arriving from the QPSK demodulator are temporarily stored in another high-speed memory buffer before being processed by the ground station computer under interrupt control. The computer resolves the fourfold phase ambiguity with the aid of special-purpose hardware, and then, through software processing, it unscrambles the data and reconstructs the graphics information into graphics memory for display on a color TV monitor.

### 11.4.5 Computer Control Subsystem

Computer control is involved in each of the four hardware subsystems just discussed. One of the ground stations, designated the *master* station, includes the master control computer. Its task is to control system operation with the aid of some special-purpose hardware. The master computer has output connections to the beam sequencer associated with the phased array, to the cophasing processor, to the transmission subsystems of each ground station and—through a device called *TDMA frame controller*—to the baseband digital subsystem of each ground station. In an actual system most of these connections would not be hard-wired as they are in the simulated system. The connections to the beam sequencer and the on-board cophasing processor would be via the TT&C channel from the master station to the satellite.

Through these connections, the master computer selects the system mode of operation. It controls up-link and down-link beam positions and the beam-switching sequence. It controls cophasing by sending commands to the cophasing controller to select on the links to be cophased. In TDMA mode, it defines the TDMA frame and loads it into the frame controller and the beam sequencer for implementation.

The master station also includes the TDMA frame controller, which implements the TDMA frame defined by the master computer. It has a memory buffer, which is loaded by the master computer with the desired TDMA interconnection pattern to be implemented. The controller then reads the frame data from the memory buffer and generates the appropriate synchronization signals for the various ground stations, enabling transmission or reception of the bursts in the proper time slot within the frame. It also generates synchronization signals for the beam sequencer and the cophasing processor, to insure proper synchronization of the beam-switching pattern and the cophasing sequence with the TDMA frame. The controller advances or delays the various synchronization signals as necessary to compensate for differences in path delay among the ground stations and the satellite repeater. The frame controller also has the task of distributing the system clock to the various subsystems. Through the use of ECL 100 K technology the controller achieves a synchronization resolution of one clock cycle (3.32 ns).

In a scanning spot-beam system, the up-link and down-link beams must be steered in synchronization with the TDMA frame to achieve the desired interconnection pattern. This task is performed by the beam sequencer located on the satellite repeater. It is similar to the frame controller in that it includes a memory buffer, which is loaded by the master computer with the desired beam scanning sequence. The sequencer reads the beam sequence data from the memory buffer, retrieves the corresponding element phases from phase memory (they were put there by the cophasing processor), and loads them into the phase shifters to form the desired beams. As mentioned above, synchronization of the beam sequence with the TDMA frame is established through synchronization signals received from the frame controller.

The primary purpose of the experimental facility sketched in Fig. 11.13 was to demonstrate the feasibility and capabilities of a full-band TDMA system employing scanning spot-beam up-link and down-link antennas. Continuous-mode operation demonstrated high-speed digital transmission at 300 Mbaud in a bandwidth of 500 MHz, up-link and down-link beam switching via the phased array antenna, and the capability for automatic beam forming through a closed-loop in-band measurement, with either a ground-based or satellite-based cophasing processor.

## 11.5 MULTIPLE SCANNING BEAM SYSTEMS

We can imagine using two orthogonal polarizations where beams in one polarization are fixed and the other scans. This offers high capacity, but the satellite resources may be severely mismatched to the nonuniform demand for service across the area of interest. For example, although the capacity of a fixed beam pointed at Denver is identical to that of a fixed beam pointed at New York, it is unreasonable to expect that both cities will offer the same amount of traffic, because their populations are grossly different. This mismatch would imply that valuable satellite resources are being wasted in the Denver beam. By making all the satellite beams scannable, however, it has been shown previously that the total capacity of the multibeam satellite can be used with 100% efficiency.

In principle, a multiple scanning beam satellite can be implemented by passing the signal of each beam through a common array antenna. One phase shifter per beam is needed for each array element. Then each signal can be independently steered by applying the correct phase-shifted commands to the group of phase shifters associated with that signal. Prior to each array element, the signals are combined and since the array is linear, the superposition of all beams is produced. Unfortunately, each element of the array is preceded by a nonlinear device, the final power amplifier, which is operated near saturation to maximize its dc-to-RF conversion efficiency. In Section 11.3.1 it was shown that the resulting intermodulation distortion can produce a degradation of the signal equivalent to that of reducing the available satellite power by more than 5 dB.

To avoid this effective power loss, another concept was proposed which permits each beam to be scanned only over a limited region. This concept is, therefore, less general than that of the fully scannable multiple beams, but implementation is much simpler. The region to be served is divided into a number of parallel strips, each of approximately the same traffic profile as sketched in Fig 11.16. An

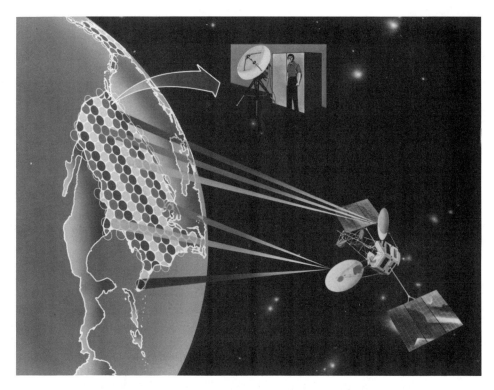

**Figure 11.16** Multiple one-dimensional scanning beams.

SS-TDMA switch is used on the satellite to permit interconnection of the various strips. Within each strip a spot beam is formed which can be scanned only over the strip. Then, to interconnect two users, (1) the up-link spot beam assigned to the strip containing the transmitting ground station is pointed at that ground station, (2) the SS-TDMA switch connects the strip of the transmitting ground station to that of the receiving ground station, and (3) the down-link spot beam assigned to the strip of the receiving ground station is pointed at that ground station. The benefit of this approach is that scanning spot beams can be formed by independent linear phased array antennas, and each array element (along with its corresponding final power amplifier) passes only a single signal. The intermodulation distortion associated with multiple fully scannable beams is therefore avoided. Furthermore, since each array is linear rather than planar, the total number of array elements and phase shifters needed is no greater than that needed to form a single fully scannable beam. Efficient exploitation of this concept is predicated on the ability to segment the coverage region into a number of equal-traffic rectangular strips; any traffic imbalance among the strips will cause some inefficiency in overall satellite utilization. Assuming a population-dependent U.S. traffic model which closely resembles the actual toll traffic demography, it was found that the overall utilization efficiency exceeds 90%.

## 11.6 SPACECRAFT ANTENNAS

Antennas are of major importance in satellite communications and play a significant role in emerging satellite systems. At microwave frequencies relatively small size structures (a meter or so) are capable of directing energy into narrow beams of a few degrees. This is important for earth stations so that their beams do not spread out and interfere with adjacent satellites, and it is important also for satellites so signals can be directed to the intended coverage area.

The first commercial satellites utilized very simple antenna structures designed to illuminate the earth with a circular beam about 18° across. This provided so-called hemispherical coverage in that when placed over the Atlantic Ocean, the satellite illuminated most of the Western Hemisphere. Many advances in antenna technology have occurred in the last 20 years. The technology emphasis has been to shape antenna patterns efficiently to place energy in those places where it is needed and reuse the frequency at the same location by exploiting the orthogonal polarization and at spatially separated locations using spot antenna beams. Other technologies have sought means to get very high gain, especially at lower frequencies where large structures, often unfurlable are required. The scanning beam or time-hop concept, which was described earlier, does not continuously illuminate a single area, but rather the antenna beam is time-shared among several spots. An especially interesting antenna which answers the scanning beam requirement is the phased array, which has great flexibility in that it can be pointed electronically in very short time intervals. The next paragraphs consider many of these developments in more detail.

### 11.6.1 Reflectors

The most-used satellite antennas are reflector antennas, which have the advantage of being simple to analyze, design and construct (see Fig. 11.17). They can be made lightweight and broadband, and the cost per unit area for large sizes is low. A reflector suffers gain loss from aperture blockage or pattern degradation with off-axis feeds. The former is particularly notable for a multibeam paraboloidal reflector antenna using a feed cluster. With a dual-reflector antenna, the feed is close to the primary reflector and thus minimizes the wave guide run to the receiver, and a secondary reflector reduces the overall dimension of the antenna.

Performance efforts for reflector antenna technology are in the area of multifeed and multireflector systems, which have low side-lobe levels. NASA is developing offset feed reflector systems for multiple fixed and scanning beams operating at 30/20 GHz. NASA is also involved in the development of precision deployable large-structure antennas. Narrower beams and greater gain will be achievable with increase in frequency and aperture size of both rigid and deployable designs. Improvements are in perfecting reflector surfaces, thus reducing side-lobe and cross-polarization levels. The largest reflector antenna in orbit to date is a 30-ft (9-m) unfurlable dish on the NASA ATS-6 satellite. It is expected that deployable reflectors, such as used on the ATS-6, could achieve a 25-m diameter

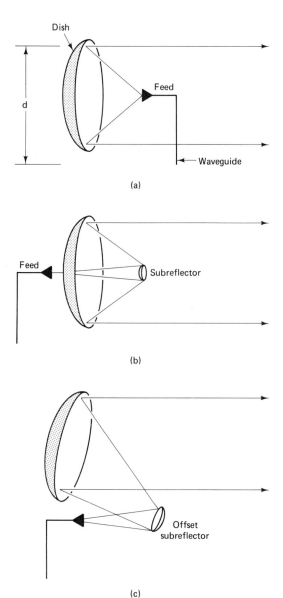

Dish

d

Feed

Waveguide

(a)

Feed

Subreflector

(b)

Offset
subreflector

(c)

**Figure 11.17** Illustration of simple reflector antennas.

or more.  However, because of surface roughness, the very large deployable antennas are expected to be useful only at frequencies below a few gigahertz.

Frequency reuse systems require a multiple-beam antenna design.  Each of
the beams must have low side-lobes in the direction of spatially separated beams
and must be isolated by polarization orthogonality from overlapping beams.  The
beams must generally be contoured to match irregular coverage areas in order to
maximize gain.  When two or more independent beams must be closely spaced
(less than about two beamwidths from center to center), it may be necessary to
increase the edge-of-beam slope to realize adequate isolation.

Since the design of antennas with low sidelobes is a well-known problem in antenna theory, the design problem is to approximate the desired aperture distribution to yield a given beam shape. The symmetrical parabolic or spherical reflector antenna with multiple feed elements is a simple realization; however, the aperture blockage due to the multiple feeds and their structural supports limits the achievable sidelobe level.

Because they are free of aperture blockage, two important multiple-beam configurations continue to be studied: the offset fed reflector antenna and the **multiple beam lens antenna** (MBA). Low side-lobe illuminations are generated by controlled excitation of the multiple-feed elements. First side-lobe levels less than 35 dB have been demonstrated with pencil beams generated by an offset reflector.

The offset-fed reflector is lightweight and relatively easy to deploy or unfurl. Its principal disadvantage is the small asymmetry introduced by the offset. This antenna technology has been applied to several operating satellites, including ANIK, INTELSAT, COMSTAR, and RCA SATCOM as well as the Crawford Hill 7-m radio telescope pictured in Fig. 11.18.

To form a shaped beam the area is covered by a set of constituent beams (each feed corresponds to a beam), which are phase and amplitude weighted and summed to provide the composite coverage. The feed excitations are chosen to satisfy simultaneously a minimum coverage gain and an isolation requirement. A solution in which satisfactory performance is achieved with the minimum reflector size and fewest feed elements is always desirable. To increase the composite beam slope the aperture size is increased beyond that required for normal coverage. The edge-of-beam slope is approximately the slope associated with any one of the constituent beams.

In a well-polarized reflector or homogeneous lens antenna, the surface currents are essentially parallel for a linearly polarized excitation. The cross-polarized currents on the surface of a paraboloidal reflector vanish if the paraboloid is illuminated with a feed having the polarization characteristics of a Huygens source. A simple realization is an open-ended small-aperture circular waveguide, which has the advantage of small electrical size and permits close beam spacing in multiple beam applications while maintaining good polarization characteristics. Peak cross-polarization levels of $-35$ dB or less have been demonstrated with dual mode horns or circular waveguides used as individual feeds.

Most dually polarized antennas now use an overlapping dual reflected system, where the first surface is polarization sensitive. By slightly displacing the back reflector with respect to the front, the feed horn arrays for each polarization are spatially separated, thus simplifying the multiplexing problem.

### 11.6.2 Lenses

The use of a microwave lens eliminates the blockage problem associated with the center-feed reflector configuration. In addition, lenses offer the capability to accomplish scanning over large angles with low distortion, as compared to center-feed or offset-feed reflector systems. Lens technology, however, is the least de-

**Figure 11.18** Seven-meter Crawford Hill offset reflector antenna.

veloped of the antenna types. Lenses involve a substantially higher weight penalty than the other antenna technologies. Only the military is utilizing this technology aimed at focusing on providing designs with wide bandwidths, low sidelobe levels, and polarization isolation for scanning beams up to 50 beamwidths off the antenna boresight without excessive beam shape degradation.

Waveguide lens antennas are used for the DSCS III military communications satellite. The DSCS III receive antenna has a 45-inch aperture and 61 individual beams. An all-electronic **beam-forming network** (BFN) made up of ferrite variable power dividers controls amplitude and phase of each of the individual beams to generate the selective coverage pattern. Amplitude accuracy can be controlled to within 0.4 dB and phase accuracy to within $\pm 3°$. A selective-coverage algorithm has been developed, which programs the BFN to generate patterns varying from the narrow coverage cone to irregular shapes to global coverage, and the G/T varies inversely with the bandwidth. G/T for the receive antenna varies from $-1.0$ dB for a narrow coverage pattern to $-16.0$ dB for full coverage.

The DSCS III transmit antenna subsystem consists of two 19-beam waveguide lenses with apertures of 28 in. Each has its own 19-port BFN optimized for gain (and thus the effective isotropic radiated power (EIRP). No phase setting capability is provided for the transmit multiple beam antenna. The narrow coverage gain including BFN losses is 26.5 dB with greater than 55% efficiency.

### 11.6.3 Phased Arrays

The advantages of a phased array include the fact that the antenna can be designed to form more than one beam. Furthermore, it can theoretically be designed for optimum sidelobe levels. The negative aspects of such a system include complexity, a limited bandwidth, and the fact that its weight and power requirements increase with the number of beams.

Development efforts are focused on beam-forming and control-network components technology at higher frequencies. Examples of current multibeam phased array systems include the NASA TDRSS launched in 1983. A 30-element phased array operating near 2 GHz is capable of forming a steerable beam over a $\pm 13°$ conical field of view. The Japanese have undertaken the research and development of a similar multibeam antenna. INTELSAT V includes a phased array for maritime services.

An especially interesting approach pursued at Bell Laboratories is the use of imaging in conjunction with phased arrays. Phased array antennas are attractive for point-to-point satellite communications because they allow a narrow width beam to be aimed directly at the communicating Earth station or satellite, thereby providing high gain and reducing interference. However, the narrow beamwidth implies a large aperture. A large array is not attractive because of its weight and the loss and complexity of the long interconnections required by the large spacing between array elements.

To obtain the performance of a large-aperture phased array, a small phased array is combined with a large main reflector and an imaging arrangement of smaller reflectors to form a large image of the small array over the main reflector. An electronically scannable antenna with a large aperture is thus obtained, using a small array. An attractive feature of the imaging arrangement is that the main reflector need not be fabricated accurately, since small imperfections can be corrected efficiently by the array.

Many types of phased array feeds may be used with the imaging reflectors. Two-dimensional phased arrays may be used to form a spot beam, or a linear array can be used to scan a fan beam. Multiple linear arrays can be multiplexed to form multiple spot beams, which scan along separate strips.

The magnification, $M$, of main reflector diameter divided by the phased array diameter also represents the range of scan angles of the phased array divided by that of the overall antenna. The magnification allowed for imaging is therefore limited by the maximum scan angle acceptable at the phased array feed, as illustrated in Fig. 11.19.

As an application, the design of an antenna that must transmit at 12 GHz and receive at 14 GHz in a 4.2-m diameter satellite in synchronous orbit at 105° W longitude was considered. A field of view of 3° by 6° (which corresponds

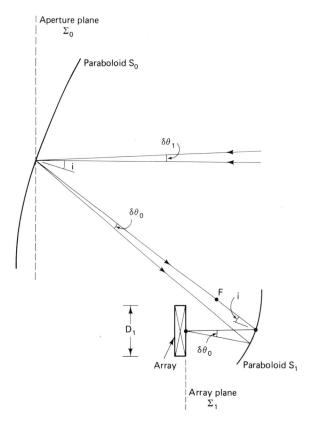

Aperture plane
$\Sigma_0$

Paraboloid $S_0$

$\delta\theta_1$

i

$\delta\theta_0$

F  i

$D_1$

Array

$\delta\theta_0$

Paraboloid $S_1$

Array plane
$\Sigma_1$

**Figure 11.19**  Magnification of an antenna.

approximately to the **contiguous United States** (CONUS) was assumed, with separate arrays employed for transmission and reception. An arrangement suitable for this purpose is shown in Fig. 11.20 using a quasi-optical diplexer between the two arrays and the last reflector.

The design was chosen taking into account the requirement that the arrangement should be free of blockage, it should be efficient and, of course, the array should be reasonably small. It is assumed that in Fig. 11.20 the main reflector can be hinged at the bottom, so that it can be initially stowed in the satellite horizontally. Then once in orbit it would be rotated into its final position, shown in Fig. 11.20.

The scan angle of the phased array is relatively large ($21°$ for an overall antenna azimuth scan angle of $3°$) because of the large magnification, $M = 7$. Thus there is an appreciable variation in illumination over the array aperture for plane waves incident from various directions in the field of view onto the main reflector. This variation in illumination causes a loss in gain. To minimize the loss over the scan region, the feed array is made slightly larger than the image of the main reflector for a boresight beam, and its center is placed slightly lower than the boresight center ray. The antenna gain (relative to a boresight beam with $-10$-dB edge taper and no spillover) versus scan angle was computed. The aperture illumination

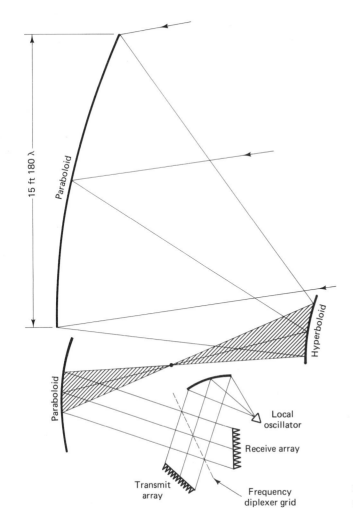

15 ft 180 λ

Paraboloid

Hyperboloid

Paraboloid

Local oscillator

Receive array

Transmit array

Frequency diplexer grid

**Figure 11.20** Antenna with multiple reflectors.

increases due to aberrations for beams scanned near the east and west coast of CONUS so that the antenna directivity is larger than that for a $-10$-dB taper boresight beam. The directivity at center of CONUS is 0.6 dB lower than a $-10$-dB taper boresight beam because the feed array is oversized.

### 11.6.4 Scanning Fan Beam Antenna

As a first step toward the actual realization of scanning beam satellites, a pair of elliptically shaped beams capable of scanning across CONUS in the east-west direction has been tested. The major axis would extend from Canada to Mexico, while the minor axis would be a few hundred miles wide, as sketched in Fig. 11.21. These fan beams (one for the up-link and one for the down-link) would be constructed as a linear array of 16 elements. The expected gain when a beam is directed toward a designated spot on CONUS is 12 dB higher than that of a single

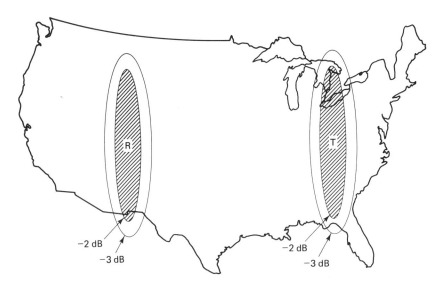

**Figure 11.21** Fan beams transmitting at 12 GHz and receiving at 14 GHz (on-axis gain about 42 dB).

CONUS coverage horn. Of course, the fan beams do not require the full aperture of the antenna that would be available from a shuttle-launched spacecraft and thus could be accommodated as an experimental package on a future operational satellite. However, the beams still provide an impressive on-axis directivity of over 40 dB.

The frequency band is 11.7 to 12.2 GHz for the down-link and 14.0 to 14.5 GHz for the up-link. To cover the CONUS north-south dimension, the 3-dB beamwidth should be about 3° to provide the greatest absolute gain at the north-south limits of the beam. (Although a wider beam would bring the antenna gain at the north-south limits of the beam nearer that at the center, the absolute gain at the center would be reduced.) To observe space on the satellite, it is desirable to use the same antenna aperture for up-link and down-link frequency bands. The 1.2 frequency ratio between these two bands suggested the use of different polarizations in a rectangular horn to keep the north-south beamwidth approximately the same for the two frequency bands. In the east-west direction the beamwidth should be narrow enough to fit about 10 beams on CONUS, i.e., about 0.6° wide.

Even with use of horn lens elements or horn reflector elements the horns are still too bulky. However, the elements can be reduced to an acceptable size with the aid of imaging reflectors. To avoid excessive aberrations when scanning, the focal length of the main reflector should be on the same order as its maximum aperture dimension (about 2 m). Therefore, to make the antenna more compact, it is desirable to divide the width into sections until the width of each section is equal to its height. This results in four subassemblies, each with four horns, as shown in Fig. 11.22

Each subassembly in Fig. 11.22 contains a pair of parabolic reflectors in a Gregorian arrangement with a magnification of five. An inverted five times image

**Figure 11.22** Compact scanning array antenna.

of each four-horn array is produced at the aperture of each main reflector so that side by side they form a 16-element 19-in. by 76-in. array aperture, while the overall antenna is only 23 in. deep. Measurements over the down-link frequency band show negligible variation with frequency. After subtracting out network losses due to electronic phase shifters, line stretchers, coaxial cable, and power dividers, the absolute gain of the RF model referred to the coaxial connector ports of the horn inputs was measured to be 40.7 dB at 11.95 GHz, with an azimuth beamwidth of 0.7° and an elevation beamwidth of 2.8°. The measurements indicate that this design would perform satisfactorily as a scanning fan beam satellite antenna and is relatively insensitive to mechanical alignment errors.

## 11.7 RAIN AND ITS EFFECT ON TRANSMISSION

Some of the fundamental limitations on the performance of satellite communication systems at frequencies greater than 10 GHz result from strong interaction of radio waves with rain and ice in the lower atmosphere. Therefore, the economic design of reliable satellite communication systems in the frequency bands of 12 to 14 GHz and 20 to 30 GHz depends on detailed knowledge of the effects of these interactions.

Direct measurements of the effects of the hydrometeors on radio waves transmitted from satellites are required to provide empirical design data; and as checks on the theoretical models needed for predicting communication system performance in geographical regions where measurements are not available.

Beacons were placed on the COMSTAR communication satellites to satisfy the need for continuously transmitting sources in geostationary orbit. These beacons were specially configured to facilitate comprehensive continuous propagation measurements at 19 and 28 GHz. The single-frequency 19 GHz beacon signal was switched between two orthogonal linear polarizations. The 28-GHz beacon signal included coherent modulation sidebands and was transmitted at a single linear polarization.

Other orbital signal sources for propagation experiments were also available concurrently: at 11 GHz over North America on the CTS satellite; at 11, 14, and 17 GHz over Europe on the OTS and SIRIO satellites; and at 1.7, 11, 20, and 30 GHz over Japan on the ECS II, CS and BS satellites.

The 19- and 28-GHz beacons on the COMSTAR satellites have been used to make unique space-Earth propagation measurements at Crawford Hill, New Jersey. The purpose of these experiments was to provide detailed propagation information not previously available. The major emphasis was on depolarization measurements important to dual-polarized satellite communication systems and on phase- and amplitude-dispersion measurements important to wideband digital systems. The measurements made use of the polarization-switched 19-GHz beacon signals and the coherent modulation sidebands on 28 GHz beacon signals. Many of the most important results are readily extendible to the entire 10- to 30-GHz frequency range and can provide useful parameter estimates for even higher and lower frequencies.

In addition to depolarization and dispersion measurements, space-earth rain attenuation measurements were made over a greater attenuation range than ever before. Measurements over a wide range were needed for checking frequency, elevation angle, and polarization scaling relationships in the 10- to 30-GHz frequency range. Measurements were also made of rain scatter coupling, angle-of-arrival effects, and cloud-produced amplitude scintillation.

Rain attenuation is an important parameter in communication system design. It dominates the system power margin for high-reliability systems operating above 10 GHz. The COMSTAR beacon attenuation measurements confirm the high probabilities of occurrence of severe rain attenuation obtained earlier from radiometer and suntracker measurements. These results reinforce the earlier conclusions that site diversity will be required to meet high availability objectives at 20 to 30 GHz and will also be required to meet these objectives at 11 to 14 GHz for low path-elevation angles at some geographic locations.

Over the 10- to 30-GHz range, attenuation in decibel scales approximately as the square of frequency as can be seen in Fig. 11.23. Although the attenuation variation is a little less than frequency squared, the error in using frequency squared scaling is considerably smaller than the year-to-year variation in measured attenuation distributions.

For path-elevation angles above 10°, rain attenuation in decibel scales linearly with the cosecant of the elevation angle, as illustrated in Fig. 11.24. This is because there is more atmosphere for longer slant paths. This scaling relationship is good for cumulative attenuation distributions spanning at least a year, but it becomes less applicable for shorter time periods and is not generally applicable for scaling

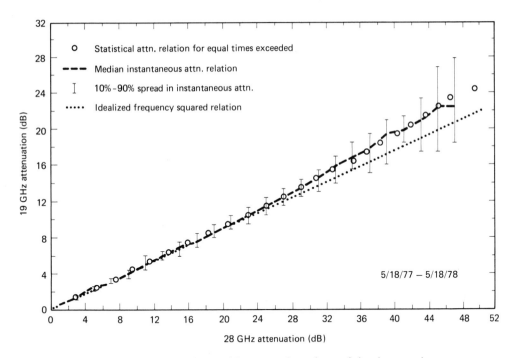

**Figure 11.23** Measured frequency dependence of signal attenuation.

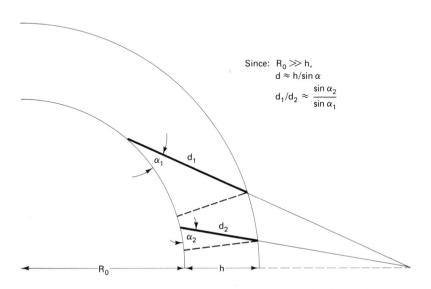

**Figure 11.24** Model for predicting attenuation with elevation angle.

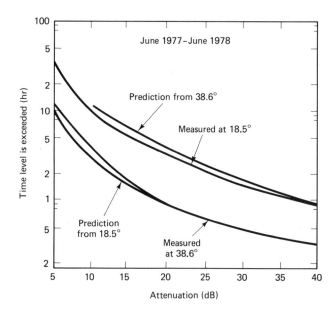

**Figure 11.25** Rain attenuation data at 19 GHz.

attenuation from individual rain storms. As noted in Fig. 11.25, the agreement is quite good for a year's cumulative statistics.

Attenuation probabilities are significantly greater during summer months. The probabilities of severe attenuation (greater than 10 dB) are often more than a factor of ten larger in July and August than for the remainder of the year. The results shown in Fig. 11.26 were obtained in New Jersey, but, since they are influenced by thunderstorm activity, they are probably representative of the eastern half of the United States.

Probabilities of severe attenuations are significantly greater during the business hours and are greater still during the afternoon. As shown in Fig. 11.27, severe attenuation is more than three times as likely between 1 P.M. and 5 P.M. than it is for the remainder of the day. Probabilities of moderate attenuation are more uniformly distributed throughout the day.

From the previous example, we can design a system for an expected outage of 5 h per year, for example, by choosing an appropriate operating margin of −10 dB in a typical case. Another question of importance is: Given a design margin, when an outage occurs how long is it likely to last? Figure 11.28 provides the answer, where smooth curves have been drawn through the histograms of the number of events lasting a particular duration. For the 10-dB fade the average outage is about 7 min. While there are 100 or so total events, only 5 or 6 could be expected to test more than 20 min.

Depolarization is a source of crosstalk (cochannel interference (CCI)) in dual-polarized frequency-reuse communication systems. Depolarization on Earth-space path is caused by nonspherical raindrops and ice particles (snowflakes and small ice crystals). While raindrops strongly attenuate radio signals above 10 GHz, ice particles cause only minimal attenuation. Cumulative distributions of ice depolarization alone and of mixed rain and ice depolarization for Crawford Hill indicate

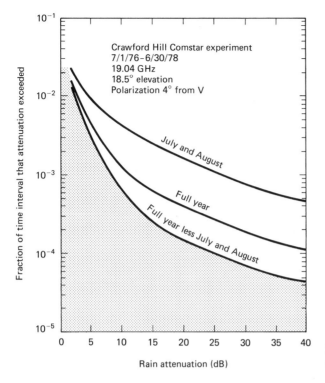

Figure 11.26 Time of year variation in signal attenuation.

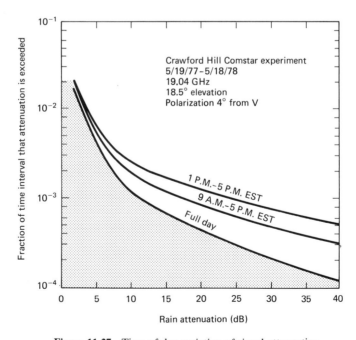

Figure 11.27 Time of day variation of signal attenuation.

Sec. 11.7     Rain and Its Effects on Transmission

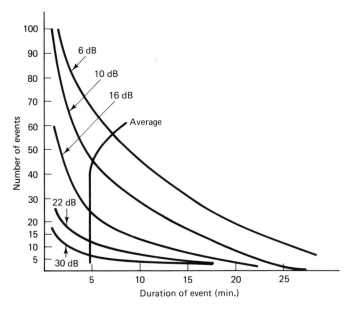

**Figure 11.28** Number of occurrences versus the duration of fade events at 19 GHz.

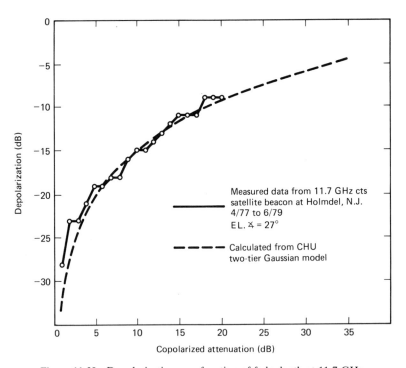

**Figure 11.29** Depolarization as a function of fade depth at 11.7 GHz.

that mixed rain and ice depolarization is about ten times more prevalent than ice depolarization alone. Depolarization is usually strongest for fixed linear polarizations oriented 45° from vertical and horizontal and for circular polarizations. Depolarization is usually minimized for vertical and horizontal polarizations. These minima and maxima occur for linear polarizations because the axes of minimum depolarization for raindrops and ice particles are usually oriented vertically and horizontally. Depolarization data at 11.7 GHz from the CTS satellite are shown in Fig. 11.29.

Phase and amplitude dispersion are potential sources of bit errors in high-rate digital systems. Using the COMSTAR beacons, a comprehensive investigation was made of the dispersive properties of space-earth propagation for frequency separations of 264 MHz, 528 MHz and 9.5 GHz at 28 GHz. No evidence was found of any amplitude or phase dispersion other than the frequency dependence due to the bulk properties of water. No dispersion (frequency selective fading) was found of the type caused by multipath propagation or by resonances in the propagation medium. Therefore, amplitude and phase dispersion should not pose a problem for wideband (on the order of 1 GHz) space-earth communication systems operating above 10 GHz with path elevation angles greater than about 15°.

A potential source of cochannel interference in multisatellite communications systems is rain-scatter coupling into a satellite-earth radio path from a different adjacent communication satellite. Rain-scatter coupling was measured at Crawford Hill for two satellite paths displaced in angle by 0.85°. For rain attenuation up to 15 dB, there was no evidence of rain-scatter coupling and it should not be a significant source of cochannel interference in multisatellite communication systems.

## 11.8 RESOURCE SHARING

As discussed earlier, the major drawback associated with high-frequency satellite systems is the signal attenuation associated with rainfall. At these frequencies, attenuation increases with rain rate, with the result that, over a large portion of the United States, significant power margin must be provided to prevent excessive outage.

An efficient technique to achieve the desired reliability would be to reduce the rain margin to a low value of, for example, 5 to 10 dB, and provide a common pool of resources to be shared among all ground stations and allocated as needed to increase the rain margin only at those stations that might be experiencing a fade depth greater than the built-in margin. One example of pooled resources is demand-assigned TDMA. To increase the rain margin for a particular user for a short interval, we might, for example, consider increasing the radiated power of the transponder serving that user experiencing excess rain attenuation. Such an approach, however, is not very attractive because it requires dual-mode final-power amplifiers and excess battery back-up power and also because the interference produced in adjacent transponders by this high-power mode of operation might tend to become excessive.

A better approach is taken to provide a pool of assignable resources used to increase the rain margin. For this scheme, the pooled resources consist of TDMA burst packet slots. Consider the down-link. In each TDMA frame, let some small number of slots be reserved for use by any ground station experiencing rain attenuation. Then, for each slot occupied by a receiving ground station experiencing a fade, three or four slots would be assigned from the pool and would be encoded at the transmitting ground station using, for example, a rate $r = \frac{1}{3}$ convolutional code. Thus three channel symbols are created for each information symbol. These additional symbols are transmitted during the pooled time slots; there is no increase in the channel data rate. Such an approach might provide 8 to 10 dB of additional fade margin with no increase in either satellite power or satellite or the diameter of the ground station antenna. The overhead associated with the pooled reserve time slots would be low because the probability of several intense simultaneous thunderstorms is low.

A typical TDMA-switching sequence performed at the satellite to interconnect the various footprints of a multibeam satellite system was shown in Fig. 11.8. Within each frame are dedicated time slots used to establish a two-way signaling channel between a master ground station and each remote station in the network. The signaling channels are used to enable TDMA synchronization, as discussed in 11.3.2, and also to distribute system status information, handle new requests for service, assign time slots, and so on. Except for the signaling slots, all other slots can, if needed, be assigned upon demand. At the end of the frame is a pool of unused time slots. For a multiple-spot-beam system with on-board switching, a similar pool of unused time slots might be reserved for each transponder.

When a down-link fade occurs, the carrier-to-noise ratio at the receiving terminal experiencing the fade is no longer sufficient to maintain the desired bit error rate. Thus the capacity into that terminal is reduced. Suppose, for example, the rain attenuation is such that the signal level falls 8 dB below the value required to maintain a voice grade bit error rate (BER) equal to $10^{-3}$. The channel error rate for Gaussian noise is then about 0.1; a lower BER would result if both Gaussian noise and peak-limited interference set the error rate. The BER can still be maintained at $10^{-3}$ or lower if the bit interval is increased by a factor of 7 (i.e., 8 dB) via allocating seven times as much of the TDMA frame and by restructuring the receiver to accept the longer bit interval. Such an approach is unattractive, however, because not only does the longer bit interval involve a great deal of complexity at the receiver, but it also wastes valuable TDMA frame time through inefficient use of the available bandwidth.

Rather, when power measurements indicate that down-link attenuation exceeding the built-in power margin is imminent, use the signaling link to notify the master ground station as well as all transmitting stations communicating with the fade site that a fade is about to occur. Then borrow time slots from the reserve pool and use them as follows. Suppose that, before the fade, the fade site is using the time slot equivalent of $V$ voice circuits. Let us borrow the equivalent of, for instance, $3V$ additional time slots from the pool, thereby providing the equivalent of 4V voice circuits into that ground station. At the originating ground station for each voice circuit, we employ a rate $r = \frac{1}{3}$ convolutional code, which produces three channel bits for each information bit. For both single- and multiple-spot-beam systems, the switching sequence at the satellite is then modified via the signaling link such that each voice circuit packet is transmitted as four contiguous packets, which contain the encoded channel bits (transmitted at the original full bandwidth data rate) plus an extended preamble containing 7 to 10 times the clear air number of bits required to enable carrier and clock recovery at a carrier-to-noise ratio (CNR) as much as 8 to 10 dB below system margin (the bandwidth of the carrier and clock recovery circuits at the receiver are correspondingly reduced by a factor of 7 to 10). At the receiver, the entire extended burst for each voice circuit is serially detected by either a soft-decision or hard-decision detection device and stored in a high-speed buffer. Since the duty cycle of burst arrivals is small, we read out of the buffer during the time interval between bursts arrivals and process the detected channel bits by a relatively slow speed decoder to recover the original information bits.

Figure 30 shows the BER versus channel **symbol-to-noise ratio** ($E_s/N_o$) curves for (1) a constraint length $K = 8, r = \frac{1}{3}$ code used in conjunction with hard-decision Viterbi decoding, and (2) a $K = 4, r = \frac{1}{3}$ code used with 3-bit soft-decision decoding. Both curves assume that Gaussian noise is the only system impairment. We note that, without coding, $E_s/N_o = 7$ dB is required to provide a BER $= 10^{-3}$. Thus the $K = 8$ code can maintain a BER $= 10^{-3}$ with 7.5 dB less power; the $K = 4$ code maintains a BER $= 10^{-3}$ with 9 dB less power. These margins generally increase if both Gaussian noise and peak-limited interference (e.g., CCI and/or intersymbol interference (ISI)) are present. Thus by sharing a small number of TDMA slots among all users, it is possible to provide an additional 8- to 10-dB fade margin at no cost in terms of satellite power, satellite antenna gain, or earth antenna diameter. Figure 11.30 shows that the additional fade margin provided by coding increases if the system BER threshold is reduced to a value less than $10^{-3}$.

When the fade has passed, the extra time slots are returned to the pool to be reassigned as needed to other ground stations in the network.

The primary virtue of this approach is that a relatively small number of equivalent voice circuits can be shared among a large number of users to provide additional rain margin when needed. The additional resources are not wasted by merely retransmitting uncoded data a number of times, but rather the entire trans-ponder bandwidth is exploited to provide additional gain through redundancy cod-ing. Other, lower-rate codes might be used to increase the fade margin still further.

The TDMA time slots reserved for rain fades can be allocated to nonfade sites during periods of high system demand. This possibility provides for an interesting interpretation of rain outage. During clear air conditions, each ground terminal in the network presents an instantaneous demand for some number of equivalent voice circuit packets; the capacity of the satellite is, however, fixed at $C$ two-way voice circuits. Call blockage occurs whenever the total offered load exceeds $C$. A ground station that uses $M$ one-way voice circuit and experiences a fade now demands additional one-way circuits to remain operational; the number

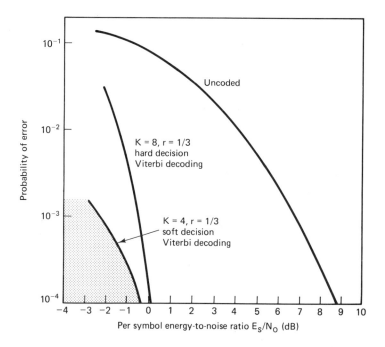

**Figure 11.30** BER performance for coded and uncoded transmission.

of additional circuits required increases with the fade depth, and coding is employed in an attempt to minimize the additional demand. Provided that the additional circuits are available, outage will not occur. Thus rain attenuation can be interpreted as placing additional demands upon the voice circuit resources of the satellite, and outage is interpreted in terms of demand exceeding capacity, that is, blocked calls. Rain outage, then, is more likely to occur during the busy hour, as noted in Fig. 11.27, and would be much less frequent at other times of the day.

For practical reasons, it might be desirable to limit the excess demand for voice circuits due to rain attenuation to a factor of 4 or 5 above the clear air demand. The outage occurs when the attenuation exceeds the additional rain margin provided by these extra circuits. Thus, when designing the network, the offered traffic must be contained to a level such that the desired rain outage and call blockage probability can be achieved by the satellite capacity $C$. Factors affecting this design would include the rain statistics at the various sites, the built-in rain margin, the number of ground stations, the clear air Erlang load of each ground station, and the statistical dependence of rain attenuation in excess of the built-in margin at the various ground stations.

Let us examine the TDMA overhead associated with reserving time slots for rain events. Suppose that $S$ ground stations are in the network, and a total of $N$ one-way voice circuits are available. We reserve $R$ of these for rain events. Thus on the average, each station uses $(N - R)/S$ one-way circuits. The value $R$ is determined by noting that, for each circuit into a given ground station, we need three additional circuits to provide the additional rain margin of 10 dB. We will provide a reserve pool sufficient to accommodate $M$ simultaneous fades. We then obtain the relationship

$$\frac{3M(N - R)}{S = R} = > \frac{R = 3MN}{S + 3M} \tag{11.12}$$

Thus the TDMA inefficiency $\eta$ is given by

$$\eta = \frac{3M}{S + 3M} \tag{11.13}$$

Thus for 100 sites and allowing for two simultaneous fades, the inefficiency or cost is under 6%, assuming that all ground stations carry approximately equal traffic. But, in fact, this 6% "inefficiency" eliminates the need to dedicate several dB of margin to each link. Thus it is not an inefficiency but a portable margin available only as needed, where needed.

If the ground stations of a satellite network exhibit large traffic imbalances, then the rain outage objective for a few high-traffic ground stations might be achieved by more-conventional approaches such as larger antennas or site diversity, with pooled time slots reserved for the exclusive shared usage among a large number of somewhat lower traffic ground stations. In this manner, the shared resource approach can still be applied to provide the outage objective efficiently for most of the ground terminals without requiring a large overhead penalty to protect a small number of high traffic users.

Encoding for the $K = 4$, $r = \frac{1}{3}$ code appears in Fig. 11.31. Data are read into the shift register 1 bit at a time; each time a shift occurs, 3 encoded bits are produced at the outputs of the modulo-2 adders. These encoded bits are augmented by the preamble and start-of-burst unique word, and the entire assembled burst is stored in a buffer awaiting transmission onto the channel. The length of the buffer is about 4000 bits, to be compared against about 1000 bits for clear air bursts.

At the receiver, carrier and clock recovery, demodulation, and bit-by-bit detection (either hard-quantized or soft-quantized) are performed, as shown in

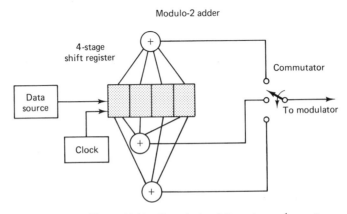

**Figure 11.31** Convolutional $D = 4$, r $= \frac{1}{3}$ encoder.

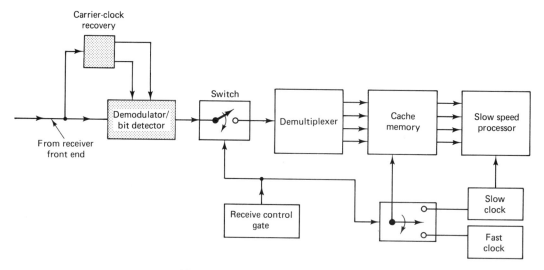

**Figure 11.32** Block diagram of TDMA receiver.

Fig. 11.32. Upon command of the timing circuitry, a window is opened to seize only intended bursts, which are demultiplexed into perhaps eight parallel rails for storage in a cache memory. The size of this memory is about 4000 bits for hard decisions and about 12,000 bits for 3-bit soft decisions. Between burst arrivals, a relatively slow-speed unique detector, shown in Fig. 11.33, locates the beginning of intelligible data; only the sign bit of soft decision 3-bit words is used for this function. The detector consists of digital correlator followed by a comparator, which compares the number of coincidences between the contents of the correlator and the known unique word bit pattern against some preset threshold. The length of the unique word and the threshold (required number of coincidences) are such that reliable detection is possible on a degraded channel. For example, suppose the channel error rate is 0.1, there are 50 bits in the unique word, and we require 30 or more coincidences. Then the probability of missing the start-of-burst unique word is about $4 \times 10^{-9}$.

Having identified the start-of-burst sequence, the remainder of the cache memory is slowly read to the Viterbi decoder. The principles of Viterbi decoding are well known, and the required operations will be only briefly described here. For simplicity, we consider the $K = 3$, $r = \frac{1}{2}$ code shown in Fig. 11.34(a). The decoder is segmented into $2^{K-1} = 4$ states, corresponding to the four possible contents of the initial two states of the shift register. Upon entry of a new data bit into the encoder, permissible state transitions, and the corresponding channel bits generated, are as shown in Fig. 11.34(b). Decoding is in accordance with Fig. 11.34(c). The decoder must correlate the two received words with the channel bits generated for each possible transition, add the appropriate correlation to a metric representing the likelihood of each initial state, and choose which of two merging paths for each state is most likely. The metric of the surviving path for each state is retained and becomes the initial metric for subsequent calculations. Also stored are the surviving paths into each state, to a depth of four or five

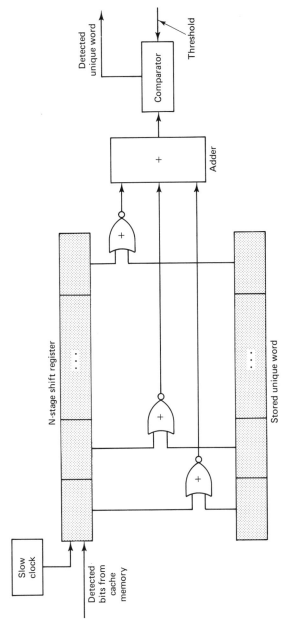

**Figure 11.33** Standard correlator to detect occurrence of a unique word.

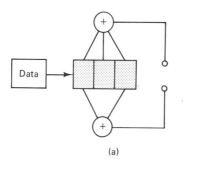

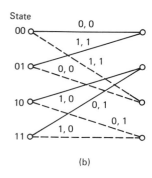

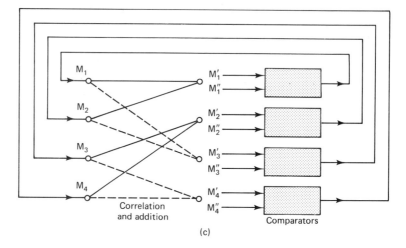

**Figure 11.34** Viterbi decoding for K = 3, 4 = $\frac{1}{2}$ convolutional code.

constraint lengths (about 40 bits for a $K = 8$ code). Thus to implement hard-decision Viterbi decoding for the $K = 8$, $r = \frac{1}{3}$ code, an add-compare-store module to operate on the 1-bit received words and a 40-bit memory must be provided for each state. For soft-decision decoding, the add function consists of adding three 3-bit words (appropriately weighted by $\pm 1$) to the old metric (perhaps a 5-bit word); the comparison is between two 5-bit words, and the storage is a 5-bit word. For the $K = 8$ code, 128 such add-select-store modules, each operating upon 1-bit data, must be provided; for the $K = 4$ code, this number is reduced to 8 modules, but arithmetic operations must be performed upon multibit words.

For the TDMA mode envisioned here, each burst will start with the encoder initially in the all-zeros state, and $K - 1$ zero bits will be stuffed into the encoder after the final data bit; the overhead is small (21 channel bits out of about 4000 for the $K = 8$ code), and ambiguity is prevented at the decoder.

To maintain frame synchronization during rain attenuation conditions, a second, extended frame marker is inserted into each frame marker burst. Recalling that the initial, short-frame marker was provided only to enable rapid acquisition during clear-air conditions, the second, extended frame marker can be used to maintain synchronization after initial acquisition. The function of the second frame marker is analogous to the extended start-of-burst word described earlier, namely, to permit identification via a slow-speed, correlation threshold, as depicted in Fig. 11.32. Since the entire frame-marker burst is stored in cache memory, this memory can be read into the correlator to find the frame marker. Then, by counting the number of elapsed bits until the frame marker is encountered, frame synchronization can be maintained. A shared-resource TDMA approach can increase the down-link rain margin of a digital satellite system by as much as 10 dB. The motivation for such an approach is the observation that dedicated resources (higher transmitter power, large antennas, etc.) are costly means for increasing rain margin, since during the overwhelming clear-air fraction of time, the system would be tremendously overdesigned. By contrast, the TDMA resources of the satellite can be viewed as representing a common pool of resources, which can be dynamically assigned as needed to increase the rain margin only at those sites where they are instantaneously needed. By assigning additional TDMA packets to sites experiencing excessive rainfall and using coding to utilize efficiently the additional bandwidth (time slots) so made available, the fade margin can be increased with no increase in down-link radiated power or satellite or terrestrial antenna size.

## 11.9 TRANSMISSION CAPACITY OF DIGITAL SATELLITES

Orbital crowding is already a problem today. The speculation that future traffic demand might be hundreds of times larger than that currently being carried on satellites (see the introduction) provides the rationale for investigating the theoretical and technical feasibility of satisfying this demand using communication satellites.

The theoretically simple solution to increase capacity is to construct larger, higher-capacity satellites along with larger-aperture Earth stations. Doubling the Earth station antenna aperture halves the antenna beamwidth, permitting twice the number of satellites. Satellites can form four times as many spot beams if the dimensions of the satellite antenna aperture are doubled. However, to make the problem meaningful, it is necessary to constrain certain of these parameters, realizing that capacity can be scaled accordingly. Important factors affecting transmission capacity are signal attenuation due to rain, the peak power limited nature of satellite amplifiers, CCI among neighboring beams, and traffic that is nonuniformly distributed over the terrestrial service region. Additionally, spacecraft-imposed limitations on available RF power and antenna size, constraints on ground terminal equipment, and regulatory constraints upon radiation and bandwidth all affect the achievable capacity. Naturally, as the physical technologies associated with communication satellites continue to advance and as regulatory constraints evolve, the transmission capacity will grow accordingly. Thus mass limitations and regulatory constraints are only parts of overall system considerations.

## 11.9.1 Baseline Assumptions

For purpose of obtaining quantitative examples we choose to fix certain parameters, realizing that recalculating capacity with modified assumptions is fairly straight-forward. The satellite system constraints assumed to permit presentation of quantitative results are summarized as follows:

| | |
|---|---|
| Available radiated power | 500 W |
| Radiation bandwidth | 500 MHz |
| Satellite antenna diameter | 4.8 m |
| Earth station antenna diameter | 5 m |
| Earth station receiver noise temperature | 300 K |
| Usable orbital arc | 40° |

These constraints are consistent with a shuttle-launched 12- to 14-GHz satellite. With multiple shuttle launchers, large space platforms may someday be assembled, equipped with communication apparatus, and carried to geosynchronous orbit. Whether this will happen before the end of this century is not certain; therefore, a more conservative approach will be taken assuming earth-assembled satellites launched by the shuttle. The maximum payload available for the 1980s will be via the **intermediate upper stage** (IUS), which will deliver an on-orbit payload of about 2300 kg. Although in principle, about one-quarter of the 30,000-kg payload of the shuttle could be carried to geosynchronous orbit by an efficient transfer vehicle, an in-orbit satellite mass of 2300 kg is assumed. With this available mass, the satellite radiated power can be calculated using rule-of-thumb estimates. Assume 15% weight available for generating dc power. With advanced state of the art technology, 6 W/kg, including battery backup during eclipse, should be obtainable. Thus 2 kW of dc power is expected. An overall dc to RF efficiency of 25% permits 500 W of radiated power.

A satellite aperture of 4.8 m, the diameter of the space shuttle bay, is assumed. The use of an aperture of this size permits the formation of many simultaneous beams, each capable of simultaneously reusing the available bandwidth. Although unfurled antennas as large as 10 m have been tested in space, the technology has not achieved surface accuracies sufficient for multibeam capability at centimeter wavelengths. The development of this capability will have a tremendous impact on the construction of high-capacity satellites of the future.

While the usable orbital arc for satellite communications is considerably larger than the assumed 40°, this figure represents as much space as most political entities might reasonably expect, and is certainly enough to evaluate orbit spacings for multiple satellite calculations.

Communications through satellites has been demonstrated using Earth station antennas ranging from 0.5 to 30 m. In order to obtain numerical results, some representative earth station antenna size must be chosen. Five-meter antenna diameters are chosen because they are convenient enough to permit the eventual use of hundreds or even thousands of Earth stations for future high-capacity systems and, as will be shown later, their beamwidths are small enough that satellites could be placed less than a degree apart in geosynchronous orbit. A 300-K noise temperature Earth station receiver is assumed because it is within the state of the art and, as will be shown, liquid water in the atmosphere limits system noise temperatures to about 300 K anyway.

A few words need to be said about the satellite frequency bands. Currently, 500-MHz bandwidth is available near $\frac{4}{6}$ and $\frac{12}{14}$ GHz and 2500-MHz bandwidth has been allocated near $\frac{18}{28}$ GHz. Of course, the ultimate capacity system would use all three bands. At $\frac{4}{6}$-GHz growth toward the ultimate capacity is inhibited by the sharing of this band between satellites and terrestrial facilities. At $\frac{18}{30}$ GHz rain severely affects system performance necessitating diversity earth stations for many applications. Since the methodology developed is quite general and can be extended to any frequency range, numerical examples are given only for the $\frac{12}{14}$ GHz frequency band, which has not yet been fully exploited and where rain attenuation for the most part is troublesome but tolerable.

### 11.9.2 Maximum Capacity of a Single Satellite

To overbound the achievable capacity of a single satellite with above parameters, a network is envisioned wherein the ground stations are uniformly spaced and situated such that at the location of any particular ground station, all beams radiated from the satellite exhibit a null* except for the beam serving that ground station. A one-dimensional representation appears in Fig. 11.35. For 0.6° null spacings, about 200-beam footprints can be formed to cover the 3° × 6° span of the continental United States as seen from geosynchronous orbit. Allowing for spectral reuse in orthogonal polarizations, 400 independent beams are formed, serving the 200 optimally located ground stations. A smaller coverage region would permit fewer independent beams and less frequency reuse. Let $\rho_M$ be the resulting clear-air CNR at the location of a single ground station if all available power is allocated to the one beam serving that ground station. The received power is equal to the product of the directive radiated power density (which varies inversely with the square of the distance from the satellite) and the effective area of the receiving antenna aperture. For the assumed parameters, $\rho_M = 41$ dB, which includes 4 dB of implementation loss arising from pointing inaccuracies, antenna inefficiency, waveguide losses, and so on. Then, since the satellite power is divided among $N$ independent beams, the clear-air CNR for a single beam is $\rho_M/N$, and the total available capacity of the $N$ beams is

$$C = NW \log_2\left(1 + \frac{\rho_M}{N}\right) = NW \log_2\left(1 + \frac{P}{NN_0W}\right) \qquad (11.14)$$

From (11.14), it is observed that reusing the frequency band $N$ times has the same effect on the theoretical capacity as does increasing the bandwidth of a single channel by a factor of $N$. For the system parameters chosen, the maximum capacity

---

*Such a pattern could be generated by, for example, a square aperture with uniform illumination.

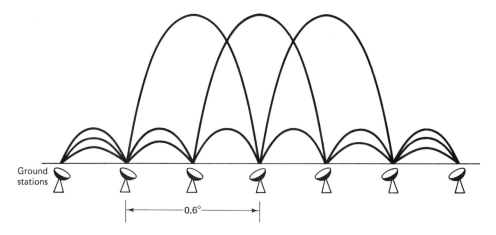

**Figure 11.35** One-dimensional illustration of ground station locations and beam radiation patterns for theoretical capacity calculations.

available from a single satellite is an enormous 1000 G/s. As a basis for comparison, this capacity is about three orders of magnitude larger than that of current satellites.

In practice, several factors limit the achievable transmission capacity of a digital satellite constrained in aperture, power, and bandwidth. In the remaining sections, the effects of two such factors will be addressed. The first is peculiar to satellite systems operating at carrier frequencies above 10 GHz and is associated with signal attenuation from rainfall at these higher frequencies. The second is associated with the nonlinear nature of the final power amplifiers aboard the satellite, which are operated at saturation to maximize the dc to RF conversion efficiency.

### 11.9.3 Capacity Reduction Caused by Rain Attenuation

We saw earlier how rain affects Earth-space transmission. The most significant effect is signal attenuation, which increases with the rain rate, the slant range through the atmosphere, and the operating frequency. From data such as presented in Section 11.2.4, it is seen that to maintain some desired satellite availability, it is generally necessary to provide a significant amount of rain margin in the satellite link budget. Thus only a fraction of the RF power generated at the satellite is available for capacity-bearing purposes; the remainder constitutes the rain margin.

The impact of rain attenuation upon the theoretical satellite capacity depends upon the clear-air CNR at the receiving terminals, the outage objective, the geographical locations of the ground stations, the satellite longitude, and the operating frequency. Recent studies, as illustrated in Fig. 11.29, indicate that cross-polarization isolation is severely degraded under conditions of high rain attenuation. To remove the interference thus produced, cross-polarization cancellation circuits are assumed.

Assuming satellite locations at 100°W and 130°W longitude, 12-GHz atten-

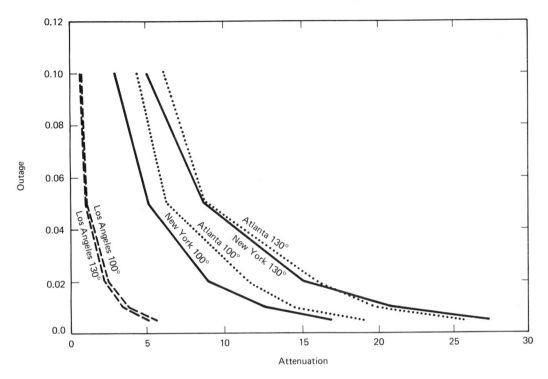

**Figure 11.36** Signal attenuation at 12 GHz for 100° and 130°W.

uation predictions for New York City, Atlanta, and Los Angeles are as shown in Fig. 11.36; these sites may be taken as representative of the northeast, southeast, and west, respectively, as defined in Fig. 11.37.

Let the clear-air CNR for each satellite beam be denoted by

$$\rho = \frac{P}{NN_0W} \tag{11.15}$$

where $N_0 = KT_R$ is the receiver noise spectral density corresponding to a receive temperature $T_R$. In the presence of rain attenuation, the noise spectral density increases to the new value

$$N_0' = K + \left[ T_r \frac{\alpha - 1}{\alpha} T_S \right] = N_0 \left[ 1 + \frac{\alpha - 1}{\alpha} \frac{T_S}{T_R} \right] \tag{11.16}$$

where $\alpha > 1$ is the attenuation caused by rain and $T_S$ is the rain temperature (which may be taken as 300 K). Then, to achieve a given outage objective $\mu$, the capacity per unit bandwidth for the $j$th beam is reduced to

$$C_j = \log_2 \left\{ 1 + \frac{\rho}{\alpha_j(\mu) + [\alpha_j(\mu) - 1]T_R/T_S} \right\} \tag{11.17}$$

where $\alpha_j(\mu)$, the rain attenuation at the geographical location of the $j$th beam which

**Figure 11.37**  Three-zone attenuation map used for capacity calculations.

occurs a fraction μ of the time, is obtained from plots such as presented in Fig. 11.36.

Since an availability of from 99.9 to 99.99% is typically desired, it is concluded from (11.16) that little is to be gained by cooling the receiver below room temperature, since the rain temperature would then predominate. Assuming that $T_R = T_S$, and using the data of Fig. 11.36, results are as shown in Fig. 11.38. In the western region, deep rain fades occur so infrequently that extremely high availability can be maintained with little impact on the capacity per beam, independent of the clear-air CNR and the satellite longitude. In the east, however, the capacity per beam is strongly affected by the rain outage objective. For a favorable satellite longitude (100°W), a capacity reduction of about 50% is incurred to improve the availability from 99.9% to 99.99% if the clear-air CNR is high (28 dB); the capacity is reduced by about 80% if the clear-air CNR is lower (15 dB). For the assumed parameters previously stated, the generation of 400 beams by the satellite will provide a clear-air CNR of 15 dB; if only 20 beams are formed, the clear-air CNR is 28 dB. As the satellite moves westward, the rain attenuation in the east increases because of the greater slant range through the atmosphere between the satellite and ground station, and an order of magnitude capacity reduction is incurred to improve the availability from 99.9% to 99.99% if the clear-air CNR is 28 dB; a reduction of two orders of magnitude is incurred for the lower clear-air CNR of 15 dB.

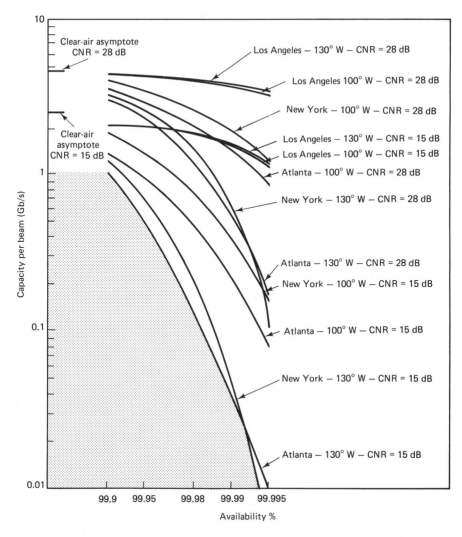

**Figure 11.38** Theoretical capacity per beam versus downlink availability at three sites and two CNRs for two satellite locations.

## 11.9.4 Capacity Reduction Caused by Peak-Limited Power

Another aspect of satellite channels which limits the available capacity is the non-linear nature of the final power amplifiers on board the satellite. The capacity analysis presented thus far has assumed that the satellite is RF average power limited. The true limitation, of course, involves prime dc power, implying that dc-to-RF conversion efficiency affects the satellite capacity. To date, the most-efficient wideband power amplifiers available at frequencies above 10 GHz are

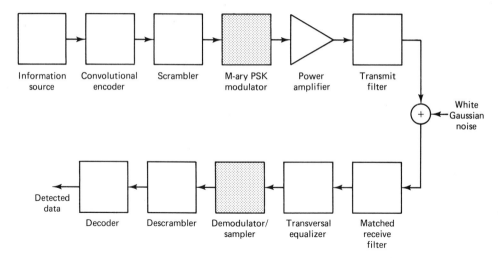

**Figure 11.39**  Communication system model.

operated in the class A mode.   For these class A devices, the dc power drain is constant, and to maximize the efficiency and the RF output power, operation at saturation is indicated.   Thus the satellite channel is more appropriately described by a peak, rather than an average, RF power constraint.   Although the capacity of peak power-limited discrete time channels has been studied, the results have not been extended to the case of a real waveform channel subject to a bandwidth constraint.

The impact of a peak power constraint on the capacity of satellite channels can be investigated in a nonrigorous manner by assuming a constant amplitude form of modulation such as PSK.   The spectrum of the modulated waveform is constrained by imposition of a band-limiting filter (which introduces amplitude modulation) subsequent to the final power amplifier, which is operated at a constant drive to achieve saturation.   The channel model is depicted in Fig. 11.39.   Independent source digits are encoded by a channel encoder, the output of which drives an M-ary PSK modulator, generating one symbol every $T$ seconds.   Subsequent to final power amplification, a four-pole Butterworth filter with a cutoff frequency $f_0$ equal to the allocated bandwidth is inserted such that little power falls outside the allocated band.   Imposition of such a filter causes dispersion of the ideal rectangular input pulses, causing amplitude ripple and overlapping or ISI among the pulses of the modulated stream.   White Gaussian noise is then added in the channel.   The receiver contains a filter matched to the dispersed waveform of an isolated pulse to maximize the resulting SNR.   This is followed by an ideal zero-forcing transversal filter to equalize the ISI, followed by a maximum-likelihood decoder.

For this previously described channel, we can invoke random coding bounds to argue the existence of codes for which the resulting bit error rate can be made arbitrarily small, provided the transmission rate per unit bandwidth does not exceed a CNR-dependent parameter $R_0$.   Here, carrier power is defined as the product of the pulse transmission rate and the energy per rectangular pulse prior to filtering.

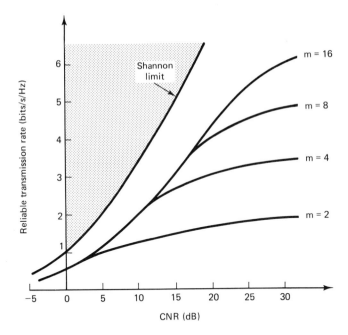

Figure 11.40 Permissible transmission rate versus CNR to achieve reliable communication with PSK.

A plot of $R_0$ versus CNR appears in Fig. 11.40 for $m = 2, 4, 8$, and 16. These plots reflect both loss of signal strength caused by transmit filtering of the ideal rectangular pulses and the noise enhancement resulting from transversal equalization at the receiver. It is seen that for CNRs between $-5$ dB and $+30$ dB, the reduction in capacity caused by the peak-power limitation at the satellite is approximately 50%. To avoid wasting power, higher-order modulations should be used for large CNR. For received CNR less than 15 dB, however, there is little merit in modulation of higher-order than $4\phi$-PSK.

Provided that the received CNR during rain events is within the above range, the combined effects of rain and peak-limiting can be arrived at simply by applying this 50% figure.

### 11.9.5 Frequency Reuse and Interference

The capacity of a single satellite can be greatly increased if the satellite forms multiple directive spot beams and reuses the same frequency band in each beam. Frequency reuse can also be introduced at the ground stations by exploiting the angular discrimination provided by a large aperture ground station antenna to permit the ground station to communicate with a string of satellites in the geosynchronous orbit.

Next, the effects of CCI arising from the antenna side-lobes of adjacent beams are addressed. All beams are assumed to receive the same power, and capacity is found as a function of rain margin. Allocation of different amounts of power to the various beams provides little additional insight to the effects of interference upon capacity and would unnecessarily complicate the analysis.

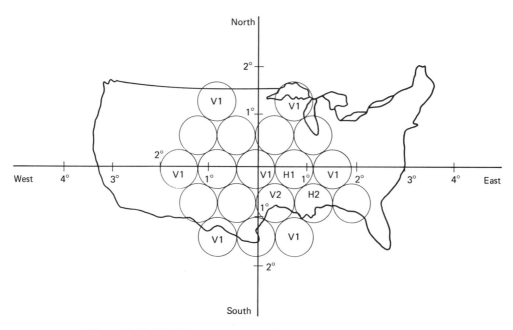

**Figure 11.41** Multibeam coverage of United States by a satellite longitude.

### 11.9.6 Satellite Frequency Reuse

To provide total coverage of CONUS, a satellite which produces contiguous foot-prints, as shown in Fig. 11.41 is assumed. However, unlike the previous section, practical radiation patterns and arbitrary ground station locations are permitted, and CCI may result. Thus the adjacent footprints use frequency plans which do not interfere with each other. In all, four independent signals are obtained by polarization reuse and the division of the available bandwidth into two equal portions. These frequency plans are labeled V1 (vertical, band 1), H1 (horizontal, band 1), V2, and H2. Such footprints can be generated by two parabolic reflector antennas using polarization grid and multiple-feed horns, as discussed in Section 11.6.

For the footprint using frequency plan V1 at the center, the CCIs come from other footprints that are also using V1. The worst location on the ground is a ground station which happens to be at the beam edge and is closest to one of the interfering beams that is also using V1. The received signal power at this ground station is

$$S = \frac{P_s g(A)}{N\alpha} \tag{11.18}$$

where $P_s$ is the power that would be received if $A$ were at beam center and all the satellite power were put into one beam; $g(A)$ is the gain degradation at beam edge (e.g., on the $-3$ dB contour), $g(A) = 0.5$; $N$ is the number of beams produced by the satellite; and $\alpha$ is the down-link power loss ratio caused by rain attenuation.

Advanced Concepts and Technologies    Chap. 11

The received CNR in a 250-MHz bandwidth is

$$\rho_\alpha = \frac{2 \times 10^{\rho M/10} g(A)}{(2\alpha - 1)N} \qquad (11.19)$$

where $\rho_M$ = 41 dB is the clear-air CNR in a 500-MHz bandwidth with all the satellite power put into one beam.

Since down-link signal and interference encounter the same rain attenuation, the down link carrier-to-interference ratio (CIR) is independent of rain fading and is given by

$$\text{CIR} = \frac{g(A)}{\sum_k g_k(A)} \qquad (11.20)$$

where $g_k(A)$ is the side-lobe level at $A$ produced by the $k$th beam that also uses V1. The down-link carrier to interference plus noise level is

$$\rho' = \frac{1}{1/\text{CIR} + 1/\rho_\alpha} \qquad (11.21)$$

The capacity of such an $N$-beam satellite, if the interference is pessimistically treated as Gaussian noise, is

$$C = \frac{WN}{2} \log_2(1 + \rho') \qquad (11.22)$$

To evaluate (11.22), it is assumed that the satellite multibeam antenna surface is illuminated by 15-dB tapered Gaussian illumination. Such an illumination can be produced by a corrugated feed horn located at the focal plane, such that the radiation pattern is circularly symmetric.

Let the beams touch each other at the $-3$ dB points, thus producing 144 beams on CONUS. The aggregated CIR at point $A$ is 27 dB. The capacity is calculated using 11.22 and is presented in Fig. 11.42 as a function of rain margin. Since all beams are assumed to carry the same power, the availability achieved varies with geography. During clear air, the capacity approaches 230 Gb.* With rain fading, the capacity is reduced but still equals 100 Gb, even with a 10-dB rain margin. To see how the packing of beams influences the capacity, beams crossing at the $-1$ dB, $-2$ dB, and $-4$ dB points are considered, and results are also shown in Fig. 11.42. It is interesting to note that, with beam crossings at $-1$ dB, there are more beams, but the total capacity does not increase significantly above a smaller packing density because of interference limitations.

As a comparison, the capacity of a CONUS coverage single-beam system is calculated at its $-3$ dB beam edge with 500 W and a 500 MHz bandwidth. Results are indicated in Fig. 11.42, which shows that the multiple beam satellite yields a 100-fold increase in capacity over a single-beam satellite.

How do practical satellites systems compare with these theoretical capacities?

---

*The reduction of this capacity from the 1000-Gb peak capacity calculated previously illustrates the effect of interference and the arbitrary location of the Earth stations in each antenna footprint.

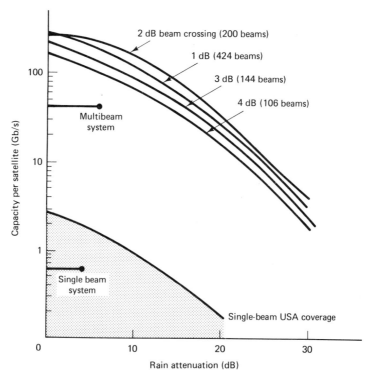

**Figure 11.42** Theoretical capacity limit of a multibeam satellite with 500-MHz bandwidth, 200 W, 4.8-m antenna, 5-m ground station antenna, and 300-K receiver temperature.

As a reference, the following characteristics and definitions for a practical satellite system are selected:

1. Uncoded 4φ PSK modulation with a transmission rate of 1.2 bit/Hz
2. CNR measured in a bandwidth equal to the symbol rate
3. System outage defined at a bit error rate $P_e = 10^{-3}$ and reached when CNR = 13.5 dB†

For CONUS coverage satellites with a contour edge antenna gain of 30 dB, the calculated CNR is 19.2 dB, and this system has 6 dB clear-air margin, which can support about 4 dB rain attenuation. The capacity is 600 Mb per satellite. This characteristic is similar to domestic satellites to be launched in the next few years. Comparing this practical CONUS satellite to the theoretical capacity limit (Fig. 11.42), it is doing quite well. For example, at α = 4 dB, the practical system achieves about 30% of the theoretical limit.

For a practical multibeam satellite, assume that the beams cross at the −3-

---

†The theoretical CNR for $P_e = 10^{-3}$ is 10 dB. We add 3.5 dB to account for nonideal modem performance.

dB point. Since there are 144 beams each with 250 MHz available bandwidth, the capacity is 43.2 Gbit. The calculated beam edge $C/\rho'$, with noise measured in a bandwidth equal to the symbol rate (150 MHz), is 20.7 dB. This system can tolerate about 6 dB rain fading and again is doing quite well as it achieves about 30% of the theoretical capacity limit as shown in Fig. 11.42.

### 11.9.7 Earth Station Frequency Reuse (Multiple Satellites)

To increase the capacity into a particular coverage area,‡ a string of satellites spaced $\theta°$ apart in the geosynchronous orbit may be used, as shown in Fig. 11.43. A ground station $A_0$ located at the edge of the coverage area may communicate with satellite $S_0$ and tolerate the cochannel radiation from the other satellites received by $A_0$ through antenna sidelobes. The situation here is analogous to frequency reuse at the satellite with the exception that, for satellite frequency reuse, the multibeams are used to cover a two-dimensional area whereas, for ground station frequency reuse, multibeams are used to cover a one-dimensional string of satellites. To exploit this dimensional difference and thereby increase ground station frequency reuse, proposals have been made to introduce another dimension in the satellite spacing by, for example, figure-of-eight types of quasi-geostationary orbits. This implies ground tracking of the satellites. In the discussion to follow, however, only one-dimensional frequency reuse is considered.

The down-link CIR is reduced by the presence of multiple satellites. The carrier-to-adjacent-satellite interference is given by

$$\text{CIR}^1 = \frac{1}{\displaystyle\sum_{k \neq 0} G(k\theta)} \tag{11.23}$$

where $G(k\theta)$ is the side-lobe level at $A_0$ toward the $k$th satellite. The total down-link CIR, including the interference from the multiple satellite beams, is

$$\text{CIR}'' = \frac{1}{1/\text{CIR}' + 1/\text{CIR}} \tag{11.24}$$

where CIR is given by (11.20). The down-link carrier to interference plus noise ratio is

$$\frac{C}{p'} = \frac{1}{1/\rho_\alpha + 1/\text{CIR}''} \tag{11.25}$$

where $\rho_\alpha$ is given by (11.19).

The capacity into the coverage area is

$$C_1 = \frac{W}{2} \log_2 1 + \rho' \tag{11.26}$$

With $N$ beams per satellite and $\theta°$ spacing in an orbital arc of 40°, the total capacity

---

‡The coverage area is either the footprint of a multibeam satellite or CONUS in the case of a CONUS coverage satellite.

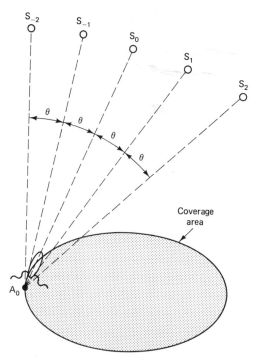

**Figure 11.43** A string of satellites covering the same service area (ground station frequency reuse).

of this multibeam/multisatellite system is

$$C = \frac{40NW}{2\theta} \log_2 1 + \rho' \qquad (11.27)$$

Let the ground station antenna illumination be a 15-dB tapered Gaussian beam with $\theta_{3\,dB} = 0.35°$ and let each satellite have 144 beams whose patterns cross at the $-3$ dB points. The capacities for various satellite spacing are presented in Fig. 11.44. The maximum capacity is seen to be enormous. For example, with $\theta = 0.4°$ the potential capacity is 20,000 Gb. It is also interesting to observe that a 0.3° spacing, although allowing more closely packed satellites, does not yield more capacity than 0.4° and 0.5° spacings because of interference. On the other hand, a 1° spacing has only 50% of the capacity of the 0.5° spacing, an indication that interference is not a factor at this spacing. In comparison, the capacity of CONUS coverage multisatellites is also presented and two orders of magnitude difference in capacity are noted, primarily due to satellite multibeam frequency reuse.

How closely can practical systems approach these capacity limits? Consider a multibeam satellite using $4\phi = $ PSK with beam crossings at $-3$ dB and multisatellites spaced 0.5° apart. The capacity of this system would be

$$C^\rho = \frac{40}{0.5} \times 144 \text{ beams} \times 0.25 \text{ GHz} \times 1.2 \text{ bits/Hz} = 3456 \text{ Gb} \quad (11.28)$$

The calculated clear-air $\rho'$ is 20.2 dB, and it would sustain 5.3 dB rain attenuation,

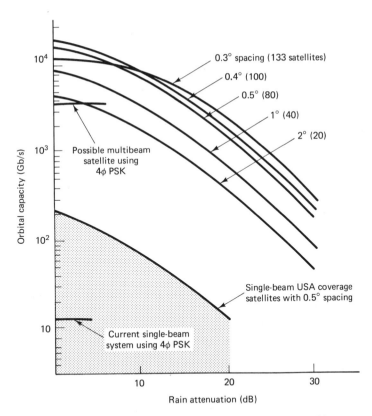

**Figure 11.44** Capacity limit of a multibeam/multisatellite system with power, antenna, and orbital arc constraints.

before dropping below the system threshold of 13.5 dB. The capacity of this system is shown in Fig. 11.44 and is observed to achieve about 30% of the theoretical capacity.

The planned CONUS coverage satellites to be launched in the next few years are typically spaced 2° apart. The capacity of a string of such satellites in the 40° orbital arc is

$$C_\rho = \frac{40}{2} \times 0.5 \text{ GHz} \times 1.2 \text{ bits/Hz} = 12 \text{ Gb} \qquad (11.29)$$

The system, as previously reported, has 19.2 dB clear-air CNR and about 4 dB rain margin. This result is also shown in Fig. 11.44 and it is observed that only 10% of theoretical capacity is achieved.

Although practical considerations limit considerably the ability to achieve theoretical capacities, the calculations here show the enormous potential. Today coding and signal-modulation techniques are such that capacities close (about 3 dB) to the Shannon bound are feasible. The ability to achieve the phenomenal capacities calculated here stems from the expected capability of reusing allocated spectral bands many times over at many orbit slots.

Sec. 11.9    Transmission Capacity of Digital Satellites                **637**

# REFERENCES

[11.1] Pierce, J. R. *Beginnings of Satellite Communications*, San Francisco Press, San Francisco, 1968.

[11.2] Martin, J. *Communications Satellite Systems*. Prentice-Hall, Englewood Cliffs, N. J., 1978.

[11.3] Gould, R. G., and Y. F. Lum, eds. *Communications Satellite Systems; An Overview of the Technology*. IEEE, 1976. Companion volume, *Literature Survey of Communication Satellite Systems and Technology* by J. H. W. Unger.

[11.4] Spilker, J. J. *Digital Communications by Satellite*. Prentice-Hall, Englewood Cliffs, N. J., 1977.

[11.5] Bhargava, V., D. Haccoun, R. Matyas, P. Nespl. *Digital Communications by Satellite: Modulation, Multiple Access and Coding*. John Wiley, New York, 1981.

[11.6] Feher, K. *Digital Communications: Satellite/Earth Station Engineering*. Prentice-Hall, Englewood Cliffs, N. J., 1983.

[11.7] International Conference on Digital Satellite Commun., 4th, 1978, Montreal. Held October 23-25, 1978.

[11.8] 5th International Conference on Digital Satellite Commun., 1981. Genoa, Italy. March 23-26, 1978.

[11.9] Kadar, "Satellite communications systems," 1st ed., *Amer. Inst. Aeronautics and Astronautics*, Vol. 18, 1976.

[11.10] Curtin, D. J., ed. *Trends In Communications Satellites*, Pergamon, 1979.

[11.11] Chu, T. S. "Microwave depolarization of an earth-space path," *Bell Syst. Tech. J.*, 59, July–August, 1980, pp. 987–1007.

[11.12] Gray, D. A. "Earth-space path diversity; Dependence on base line orientation," IEEE G-AP Symposium, Boulder, Colo., August 22–24, 1973, pp. 336–369.

[11.13] Cox, D. C. "An overview of the Bell Laboratories 19 and 28 GHz COMSTAR beacon propagation experiments," *Bell Syst. Tech. J.*, Vol. 57, May–June, 1978, pp. 1231–1255.

[11.14] Cox, D. D., and H. W. Arnold. "Results from the 19 and 28 GHz COMSTAR satellite propagation experiments at Crawford Hill," to be published in IEEE Proceedings, May, 1982.

[11.15] Rustako, A. J., Jr. "An earth-space propagation measurement at Crawford Hill using the 12-GHz CTS Satellite Beacon," *Bell Syst. Tech. J.*, Vol. 57, No. 5, May–June, 1978, pp. 1431–1448.

[11.16] Lin, S. H. "Nationwide long-term rain rate statistics and empirical calculation of 11 GHz microwave rain attenuation," *Bell Syst. Tech. J.*, Vol. 56, No. 9, 1977, pp. 1581–1604.

[11.17] Lin, S. H. "Empirical calculation of microwave rain attenuation distribution on earth satellite paths," *Proc. IEEE Eascon Conference*, Arlington, Va, September, 1978, pp. 372–378.

[11.18] Dragone, C., and Gans M. J. "Imaging reflector arrangements to form a scanning beam using a small array, *Bell Syst. Tech. J.*, Vol. 58, No. 2, 1979, pp. 501–515.

[11.19] Dragone, C., and M. J. Gans. "Satellite phased arrays: Use of imaging reflectors with spatial filtering in the focal plane to reduce grating lobes," *Bell Syst. Tech. J.*, Vol. 59, No. 3, 1980, pp. 449–461.

[11.20] Amitay, N., and M. J. Gans. "Narrow multibeam satellite ground station antenna employing a linear array with a geosynchronous arc coverage of 60°—Part I: Theory," *IEEE Trans. on Antennas and Propagation,* Vol. Ap-30(6), 1982, pp. 1062–1067.

[11.21] Chu, T. S., R. W. Wilson, R. W. England, D. A. Gray, and W. E. Legg. "The Crawford Hill 7-meter millimeter wave antenna," *Bell Syst. Tech. J.,* Vol 57, May-June, 1978, pp. 1257–1288.

[11.22] Dragone, C., and Gans, M. J. "Imaging reflector arrangements to form a scanning beam using a small array," *Bell Syst. Tech. J.,* Vol. 58, No. 2, February, 1979, pp. 501–515.

[11.23] Pierce, J. R. "Orbital radio relays," *Jet Propulsion*, April, 1955.

[11.24] Pierce, J. R., and Kompfner, R. "Transoceanic communications by means of satellite," *Proc. IRE,* Vol. 47, March, 1959.

[11.25] Jakes, W. C., Jr. "Participation of Bell Telephone Laboratories in project echo and experimental results," *Bell Syst. Tech. J.,* July, 1961, p. 975.

[11.26] Reudink, D. O., and Yeh, Y. S. "A scanning spot beam satellite system," *Bell Syst. Tech. J.,* Vol. 56, October, 1977.

[11.27] Viterbi, A. J. "Convolutional codes and their performance in communication systems," *IEEE Trans. Commun. Technology,* Vol. COM-19, October, 1971.

[11.28] Rustako, A. J., G. Vanuci, and C. B. Woodworth. "An experimental scanning spot beam satellite system implementing 600 Mbit/sec. TDMA," *Proc. Sixth International Conference on Digital Satellite Commun.* 11, Phoenix, Az, September 1983.

[11.29] Saleh, A. A. M. "Intermodulation analysis of FDMA satellite systems employing compensated and uncompensated TWTs," *IEEE Trans. on Commun.,* Vol. COM-30, No. 5, May, 1982, p. 1233–1242.

[11.30] Acampora, A. S., C. Dragone, D. O. Reudink. "A satellite system with limited-scan spot beams," *IEEE Trans. Commun.,* Vol. COM-27(10), 1979, pp. 1406–1415.

# 12

# ADAPTIVE EQUALIZATION

## DR. SHAHID U. H. QURESHI

*Senior Director, Research,*
*Transmission Products*
*Codex Corporation*
*20 Cabot Boulevard*
*Mansfield, Mass. 02048*

## 12.1 INTRODUCTION

The rapidly increasing need for computer communications has been met primarily by higher-speed data transmission over the widespread network of voice-bandwidth channels developed for voice communications. A modem (see Chapter 7) is required to carry digital signals over these analog passband (nominally 300 to 3000 Hz) channels by translating binary data to voice-frequency signals and back (Fig. 12.1). The thrust toward common-carrier digital transmission facilities has also resulted in application of modem technology to line-of-sight (LOS) terrestrial radio and satellite transmission and recently to subscriber loops for integrated services digital networks (ISDN, see Chapter 2).

Analog channels deliver corrupted and transformed versions of their input waveforms. Corruption of the waveform—usually statistical—may be additive and/or multiplicative because of possible background thermal noise, impulse noise, and fades. Transformations performed by the channel are frequency translation, nonlinear or harmonic distortion, and time dispersion.

In telephone lines, time dispersion results when the channel frequency response deviates from the ideal of constant amplitude and linear phase (constant delay). Equalization, which dates back to the use of loading coils to improve the characteristics of twisted pair telephone cables for voice transmission, compensates for these nonideal characteristics by filtering.

**640**

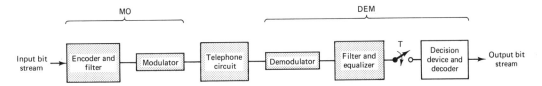

**Figure 12.1** Data transmission system.

A synchronous modem transmitter collects an integral number of bits of data at a time and encodes them into symbols for transmission at the signaling rate. In pulse amplitude modulation (PAM), each signal is a pulse whose amplitude level is determined by the symbol, for example, amplitudes of $-3$, $-1$, 1, and 3 for quaternary transmission. In bandwidth-efficient digital communication systems the effect of each symbol transmitted over a time dispersive channel extends beyond the time interval used to represent that symbol. The distortion caused by the resulting overlap of received symbols is called intersymbol interference (ISI) [Lucky, 12.54]. This distortion is one of the major obstacles to reliable high-speed data transmission over low-background-noise channels of limited bandwidth. In its broad sense, the term *equalizer* applies to any signal-processing device designed to deal with ISI.

It was recognized early in the quest for high-speed (4800 bits/s and higher) data transmission over telephone channels that rather precise compensation, or equalization, is required to reduce the ISI introduced by the channel. In addition, in most practical situations the channel characteristics are not known beforehand. For medium-speed (up to 2400 b/s) modems, which effectively transmit 1 bit/Hz, it is usually adequate to design and use a compromise (or statistical) equalizer, which compensates for the average of the range of expected channel amplitude and delay characteristics. However, the variation in the characteristics within a class of channels, as in the lines found in the switched telephone network, is large enough so that automatic adaptive equalization is used nearly universally for speeds higher than 2400 bits/s. Even 2400-bit/s modems now often incorporate this feature.

Voiceband telephone modems may be classified into one of three categories based on intended application: namely, for two-wire **public switched telephone network** (PSTN), four-wire point-to-point leased lines and four-wire multipoint leased lines. PSTN modems can achieve 2400 bits/s two-wire full-duplex transmission by sending a 4 bits/s symbol and using frequency division to separate the signals in the two directions of transmission. Two-wire full-duplex modems using adaptive echo cancellation are now available for 2400- and 4800-bits/s transmission. Adaptive echo cancellation in conjunction with coded modulation will pave the way to 9600-bit/s full-duplex operation over two-wire PSTN circuits in the near future. At this time commercially available leased line modems operate at rates up to 19.2 kb/s over conditioned point-to-point circuits and up to 9.6 kb/s over unconditioned multipoint circuits. An adaptive equalizer is an essential component of all these modems. (See [Forney, 12.27] for a historical note on voiceband modem development.)

Sec. 12.1    Introduction                                                                      **641**

In radio and undersea channels, ISI is due to multipath propagation [Proakis, 12.87; Siller, 12.98], which may be viewed as transmission through a group of channels with differing relative amplitudes and delays. Adaptive equalizers are capable of correcting for ISI due to multipath in the same way as ISI from linear distortion in telephone channels. In radio channel applications, an array of adaptive equalizers can also be used to perform diversity combining and cancel interference or jamming sources [Brady, 12.26, Monsen, 12.72]. One special requirement of radio channel equalizers is that they be able to track the time varying fading characteristics typically encountered. The convergence rate of the adaptation algorithm employed then becomes important during normal data transmission [Proakis, 12.8]. This is particularly true for 3-kHz-wide ionospheric high-frequency (HF), 3- to 30-MHz radio channels, which suffer from severe time dispersion and relatively rapid time variation and fading. Adaptive equalization has also been applied to slowly fading tropospheric scatter microwave digital radios, in the 4- to 11-GHz bands, at rates up to 200 Mb/s [Murase, 12.82].

In the last decade there has been considerable interest in techniques for full-duplex data transmission at rates up to 144 kb/s over two-wire (nonloaded twisted copper pair) subscriber loops [Agazzi, 12.2; Falconer, 12.22; Messerschmitt, 12.69; Van Gerwen, 12.106; Yasumoto, 12.112] for ISDN (Chapter 2). Two competing schemes for achieving full-duplex transmission are time-compression multiplex or burst mode and adaptive echo cancellation (Chapter 4). Some form of adaptive equalization is desirable, if not indispensible, for these baseband modems due to a number of factors: high transmission rates, especially for the burst mode scheme, attenuation distortion based on the desired range of subscriber loop lengths and gauges, and the presence of bridged taps, which cause additional time dispersion.

The first part of this chapter, intended primarily for those not familiar with the field, is a simplified introduction to ISI and transversal equalizers and an overview of some practical adaptive equalizer structures. In the concluding sections of the first part, we briefly mention other related applications of adaptive filters (such as echo cancellation, noise cancellation, and prediction) and discuss past and present implementation approaches.

Before presenting the introductory material, however, it seems appropriate to summarize the major areas of work in adaptive equalization, with reference to key papers and to sections of this chapter where these topics are discussed. (The interested reader should refer to Lucky [Lucky, 12.59] and Price [Price, 12.83] for a comprehensive survey of the literature and extensive bibliographies of work up to the early 1970s). Unfortunately, use of some as-yet-undefined technical terms in the following paragraphs is unavoidable at this stage.

Nyquist's telegraph transmission theory [Nyquist, 12.115] in 1928 laid the foundation for pulse transmission over band-limited analog channels. In 1960, Widrow and Hoff [Widrow, 12.109] presented a least-mean-squared error (LMSE) adaptive filtering scheme, which has been the workhorse adaptive equalization algorithm for the last decade and a half. However, research on adaptive equalization of PAM systems in the early 1960s centered on the basic theory and structure of zero-forcing transversal or tapped delay line equalizers with symbol interval tap spacing [Lucky, 12.55; 12.56]. In parallel, the theory and structure of linear receive and transmit filters [Smith, 12.116; Tufts, 12.117] were developed to minimize MSE for a time-dispersive additive Gaussian noise channels [George, 12.31]. By the late 1960s LMS adaptive equalizers had been described and understood [Gersho, 12.33; Lucky, 12.54; Proakis, 12.84]. It was recognized that over highly dispersive channels, even the best linear receiver falls considerably short of the matched filter performance bound, obtained by considering the reception of an isolated transmitted pulse [Lucky, 12.54]. Considerable research followed on the theory of optimum nonlinear receiver structures under various optimality criteria related to error probability [Abend, 12.1; Lucky, 12.59; Proakis, 12.87]. This culminated in the development of the maximum likelihood sequence estimator [Forney, 12.24] using the Viterbi algorithm [Forney, 12.24] and adaptive versions of such a receiver [Falconer, 12.17; Lee, 12.51; MacKenchnie, 12.61; Magee, 12.62; Qureshi, 12.88; 12.89; Ungerboeck, 12.103]. Another branch of research concentrated on a particularly simple suboptimum receiver structure known as the decision-feedback equalizer [Austin, 12.3; Belfiore, 12.4; Duttweiler, 12.12; George, 12.32; Monsen, 12.71; Price, 12.83; Salz, 12.93]. Linear feedback, or infinite impulse response (IIR), adaptive filters [Johnson, 12.47] have not been applied as adaptive equalizers due to lack of guaranteed stability, lack of a quadratic performance surface, and a minor performance gain over transversal equalizers [Proakis, 12.87]. As the advantages of double-sideband suppressed-carrier quadrature amplitude modulation (QAM) over single-sideband (SSB) and vestigial sideband (VSB) modulation were recognized, previously known PAM equalizers were extended to complex-valued structures suitable for joint equalization of the in-phase and quadrature signals in a QAM receiver [Falconer, 12.15; 12.16; Kobayashi, 12.49; Proakis, 12.84; CCITT Rec. V. 29; 12.119]. Transversal and decision-feedback equalizers with forward-filter tap spacing that is less than the symbol interval were suggested in the late 1960s and early 1970s [Brady, 12.6; Lucky, 12.58; Monsen, 12.71]. These fractionally spaced equalizers were first used in commercial telephone-line modems [Proakis, 12.86; Logan, 12.53] and military tropospheric scatter radio systems [Monsen, 12.114] in the mid-1970s. Their theory and many performance advantages over conventional "symbol-spaced" equalizers have been the subject of several articles [Gitlin, 12.39; Guidoux, 12.45; Qureshi, 12.91; Ungerboeck, 12.104]. The timing phase sensitivity of the MSE of symbol-spaced [Mago, 12.64], fractionally spaced [Guidoux, 12.45; Salz, 12.91; Ungerboeck, 12.104] and decision-feedback [Salz, 12.94] equalizers has also been a research topic in the 1970s. Recently, interest in a nonlinear decision-aided receiver structure [Proakis, 12.85] now known as an ISI canceler, has been revived by using a fractionally spaced equalizer as a matched filter [Gersho, 12.34; Mueller, 12.80].

In the second part of the chapter we develop the various receiver structures just mentioned and present their important steady-state properties. The first two sections are devoted to the definition of the baseband equivalent channel model and the development of an optimum receive filter, which must precede further linear or nonlinear processing at the symbol rate. The next section on linear receivers shows that while the conventional, matched filter plus symbol-spaced equalizer, and fractionally spaced forms of a linear receiver are equivalent when each is unrestricted (infinite in length), a finite-length fractionally spaced equalizer has significant advantages compared with a practical version of the conventional linear receiver. Nonlinear receivers are presented in the fourth section with a discussion of decision-feedback equalizers, decision-aided ISI cancellation, and adaptive versions of the maximum-likelihood sequence estimator. The final section of this part of the chapter addresses timing phase sensitivity. A few important topics which have been excluded due to space limitations are: adaptive equalization of nonlinearities [Biglieri, 12.5; Falconer, 12.21], diversity combining adaptive equalizer arrays to combat selective fades and interference in radio channels [Brady, 12.6; Monsen, 12.72; 12.114], and a particular passband equalizer structure [Cioffi, 12.11; Mueller, 12.77].

Until the early 1970s most of the equalization literature was devoted to equalizer structures and steady-state analysis [Lucky, 12.59], partly due to the difficulty of analyzing the transient performance of practical adaptive equalization algorithms. Since then some key papers [Gitlin, 12.38; Mazo, 12.65; Ungerboeck, 12.102; Widrow, 12.111] have contributed to the understanding of the convergence of the LMS stochastic update algorithm for transversal equalizers, including the effect of channel characteristics on the rate of convergence. The third part of this chapter is devoted to this subject and a discussion of digital precision considerations [Caraiscos, 12.7; Duttweiler, 12.13; Gitlin, 12.36; 12.38]. The important topic of decision-directed convergence [Macchi, 12.42; Sato, 12.95] and self-recovering adaptive equalization algorithms [Godard, 12.42; Sato, 12.95] has been omitted.

The demand for polled data communication systems using multipoint modems [Forney, 12.26], which require fast setup at the central site receiver, has led to the study of fast-converging equalizers using a short preamble or training sequence. The fourth part of this chapter summarizes three classes of fast-converging equalization algorithms. Some of the early work on this topic was directed toward orthogonalized LMS algorithms for partial response systems [Chang, 12.8; Mueller, 12.75; Qureshi, 12.88]. Periodic or cyclic sequences for equalizer training and methods for fast startup based on such sequences have been widely used in practice [Godard, 12.43; Milewski, 12.70; Mueller, 12.76; Qureshi, 12.90; 12.91]. The third class of fast converging algorithms are self-orthogonalizing [Gitlin, 12.37]. In 1973, Godard [Godard, 12.41] described how the Kalman filtering algorithm can be used to estimate the LMS equalizer coefficient vector at each symbol interval. This was later recognized [Falconer, 12.20] to be a form of **recursive least-squares** (RLS) estimation problem. Development of computationally efficient RLS algorithms has recently been a subject of intense research activity [Honig, 12.46; Morf, 12.73; Mueller, 12.78; 12.79; Proakis, 12.87] leading to transversal [Cioffi, 12.10; 12.11; Falconer, 12.20] and lattice [Ling, 12.52; Makhoul, 12.63; Morf, 12.74; Satorius, 9.96; 9.97; Friedlander, 12.127] forms of the algorithm. Some of these algorithms have been applied to adaptive equalizers for High Frequency (HF) radio modems [Proakis, 12.87; Eleftheriou, 12.126], which need to track a relatively rapidly time-varying channel. However, the extra complexity of these algorithms has so far prevented application to the startup problem of telephone-line modems where periodic equalization [Godard, 12.43] and other cost-effective techniques, for instance, [Eleftheriou, 12.126; Lawrence, 12.125], are applicable.

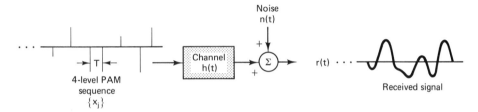

Figure 12.2   Baseband PAM system model.

### 12.1.1 ISI

ISI arises in all pulse-modulation systems, including frequency-shift keying (FSK), phase-shift keying (PSK) and QAM [Lucky, 12.54]. However, its effect can be most easily described for a baseband PAM system. A model of such a PAM communication system is shown in Fig. 12.2. A generalized baseband equivalent model such as this can be derived for any linear modulation scheme. In this model the "channel" includes the effects of the transmitter filter, the modulator, the transmission medium and the demodulator.

A symbol $x_m$, one of $L$ discrete amplitude levels, is transmitted at instant $mT$ through the channel, where $T$ seconds is the signaling interval. The channel impulse response $h(t)$ is shown in Fig. 12.3. The received signal $r(t)$ is the superposition of the impulse responses of the channel to each transmitted symbol and additive white Gaussian noise $n(t)$:

$$r(t) = \sum_j x_j h(t - jT) + n(t)$$

If we sample the received signal at instant $kT + t_o$, where $t_o$ accounts for the channel delay and sampler phase, we obtain

$$r(t_o + kT) = x_k h(t_o) + \sum_{j \neq k} x_j h(t_o + kT - jT) + n(t_o + kT).$$

The first term on the right is the desired signal, since it can be used to identify the transmitted amplitude level. The last term is the additive noise, whereas the middle sum is the interference from neighboring symbols. Each interference term is proportional to a sample of the channel impulse response, $h(t_o + iT)$, spaced a multiple $iT$ of symbol intervals $T$ away from $t_o$, as shown in Fig. 12.3. The ISI is zero if and only if $h(t_o + iT) = 0, i \neq 0$, that is, if the channel impulse response has zero crossings at $T$-spaced intervals.

When the impulse response has such uniformly spaced zero crossings, it is said to satisfy **Nyquist's first criterion**. In frequency domain terms, this condition

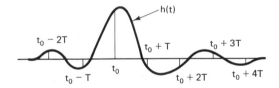

Figure 12.3   Channel impulse response.

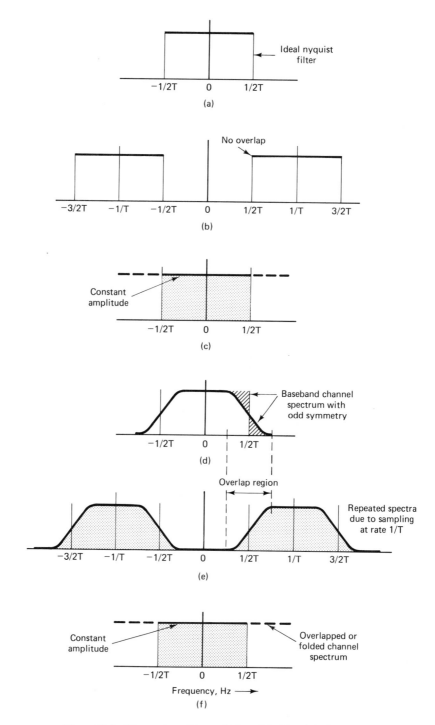

**Figure 12.4**  Linear phase filters which satisfy Nyquist's first criterion.

is equivalent to

$$H'(f) = \sum_n H\left(f - \frac{n}{T}\right) = \text{constant for } |f| \le \frac{1}{2T}$$

$H(f)$ is the channel frequency response and $H'(f)$ is the "folded" (aliased or overlapped) channel spectral response after symbol-rate sampling. The band $|f| \le 1/2T$ is commonly referred to as the **Nyquist**, or **minimum**, **bandwidth**. When $H(f) = 0$ for $|f| > 1/T$ (the channel has no response beyond twice the Nyquist bandwidth), the folded response $H'(f)$ has the simple form

$$H'(f) = H(f) + H\left(f - \frac{1}{T}\right), \qquad 0 \le f \le \frac{1}{T}$$

Figure 12.4(a) and (d) shows the amplitude response of two linear-phase low-pass filters (LPFs): one an ideal filter with Nyquist bandwidth and the other with odd (or vestigial) symmetry around $1/2T$ Hz. As illustrated in Fig. 12.4(b) and (e), the folded frequency response of each filter satisfies Nyquist's first criterion. One class of linear-phase filters, which is commonly referred to in the literature [Lucky, 12.54; Proakis, 12.87; Feher, 12.113] and is widely used in practice [12.23, 12.118] [Fenderson, 12.23; CCITT Rec. V.27, 12.118], is the **raised-cosine** family with cosine roll-off around $1/2T$ hertz.

In practice, the effect of ISI can be seen from a trace of the received signal on an oscilloscope with its time base synchronized to the symbol rate. Figure 12.5 shows the outline of a trace (eye pattern) for a two-level, or binary, PAM system. If the channel satisfies the zero ISI condition, there are only two distinct levels at the sampling time $t_o$. The eye is then fully open and the peak distortion is zero. **Peak distortion** (Fig. 12.5) is the ISI that occurs when the data pattern is such that

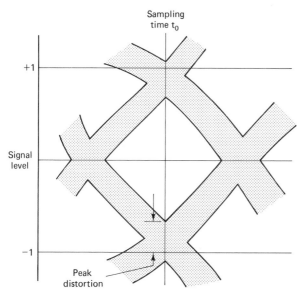

**Figure 12.5** Outline of binary eye pattern.

all ISI terms add to produce the maximum deviation from the desired signal at the sampling time.

The purpose of an equalizer, placed in the path of the received signal, is to reduce the ISI as much as possible to maximize the probability of correct decisions.

### 12.1.2 Linear Transversal Equalizers

Among the many structures used for equalization, the simplest is the transversal (tapped delay-line or nonrecursive) equalizer shown in Fig. 12.6. In such an equalizer the current and past values $r(t - nT)$ of the received signal are linearly weighted by equalizer coefficients (tap gains) $c_n$ and summed to produce the output. If the delays and tap-gain multipliers are analog, the continuous output of the equalizer $z(t)$ is sampled at the symbol rate and the samples go to the decision device. In the commonly used digital implementation, samples of the received signal at the symbol rate are stored in a digital shift register (or memory), and the equalizer output samples (sums of products) $z(t_o + kT)$ or $z_k$ are computed digitally, once per symbol, according to

$$z_k = \sum_{n=0}^{N-1} c_n r(t_o + kT - nt)$$

where $N$ is the number of equalizer coefficients and $t_o$ denotes sample timing.

The equalizer coefficients, $c_n, n = 0, 1, \ldots, N - 1$ may be chosen to force the samples of the combined channel and equalizer impulse response to zero at all but one of the $N$ $T$-spaced instants in the span of the equalizer. This is shown graphically in Fig. 12.7. Such an equalizer is called a **zero-forcing** (ZF) equalizer [Lucky, 12.55].

If we let the number of coefficients of a ZF equalizer increase without bound, we would obtain an infinite-length equalizer with zero ISI at its output. The frequency response $C(f)$ of such an equalizer is periodic, with a period equal to the symbol rate $1/T$ because of the $T$ second tap spacing. After sampling, the

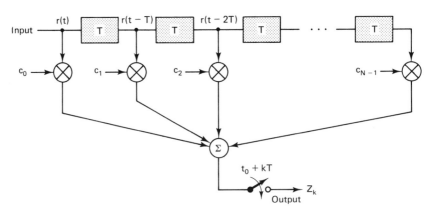

**Figure 12.6** Linear transversal equalizer.

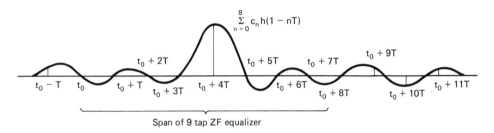

$$\sum_{n=0}^{8} c_n h(1-nT)$$

Span of 9 tap ZF equalizer

**Figure 12.7**   Combined impulse response of a channel and ZF equalizer in tandem.

effect of the channel on the received signal is determined by the folded frequency response $H'(f)$. The combined response of the channel, in tandem with the equalizer, must satisfy the zero ISI condition or Nyquist's first criterion,

$$C(f)H'(f) = 1, \; |f| \leq \frac{1}{2T}$$

From this expression we see that an infinite-length zero-ISI equalizer is simply an inverse filter, which inverts the folded frequency response of the channel. A finite-length *ZF* equalizer approximates this inverse. Such an inverse filter may excessively enhance noise at frequencies where the folded channel spectrum has high attenuation. This is undesirable, particularly for unconditioned telephone connections, which may have considerable attenuation distortion, and for radio channels, which may be subject to frequency selective fades.

Clearly, the ZF criterion neglects the effect of noise altogether. Also, a finite-length ZF equalizer is guaranteed to minimize the peak distortion or worst-case ISI only if the peak distortion before equalization is less than 100% [Lucky, 12.55], that is, if a binary eye is initially open. However, at high speeds on bad channels this condition is often not met.

The **least mean-squared** (LMS) equalizer [Lucky, 12.54] is more robust. Here the equalizer coefficients are chosen to minimize the MSE—the sum of squares of all the ISI terms plus the noise power at the output of the equalizer. Therefore, the LMS equalizer maximizes the signal-to-distortion ratio (S/D) at its output within the constraints of the equalizer time span and the delay through the equalizer.

The delay introduced by the equalizer depends on the position of the main or reference tap of the equalizer. Typically, the tap gain corresponding to the main tap has the largest magnitude.

If the values of the channel impulse response at the sampling instants are known, the $N$ coefficients of the ZF and the LMS equalizers can be obtained by solving a set of $N$ linear simultaneous equations for each case.

Most current high-speed voiceband telephone line modems use LMS equalizers because they are more robust in the presence of noise and large amounts of ISI and superior to the ZF equalizers in their convergence properties. The same is generally true of radio channel modems [Monsen, 12.72; Murase, 12.82; Monsen, 12.114] except in one case [Fenderson, 12.23], where the ZF equalizer was selected due to its implementation simplicity.

### 12.1.3 Automatic Synthesis

Before regular data transmission begins, automatic synthesis of the ZF or LMS equalizers for unknown channels, which involves the iterative solution of one of the previously mentioned sets of simultaneous equations, may be carried out during a training period. (In certain applications, such as microwave digital radio systems, and remote site receivers in a multipoint telephone modem network [Godard, 12.43] the adaptive equalizers are required to bootstrap in a decision-directed mode (see Section 12.1.5) without the help of a training sequence from the transmitter).

During the training period, a known signal is transmitted and a synchronized version of this signal is generated in the receiver to acquire information about the channel characteristics. The training signal may consist of periodic isolated pulses or a continuous sequence with a broad, uniform spectrum such as the widely used maximum-length shift-register or **pseudo noise** (PN) sequence [Chang, 12.9; Lucky, 12.54; Mueller, 12.76; CCITT Rec. V.27, 12.118; CCITT Rec. V.29, 12.119]. The latter has the advantage of much greater average power, and hence a larger received signal-to-noise ratio (SNR) for the same peak transmitted power. The training sequence must be at least as long as the length of the equalizer so that the transmitted signal spectrum is adequately dense in the channel bandwidth to be equalized.

Given a synchronized version of the known training signal, a sequence of error signals $e_k = z_k - x_k$ can be computed at the equalizer output (Fig. 12.8), and used to adjust the equalizer coefficients to reduce the sum of the squared errors. The most popular equalizer adjustment method involves updates to each tap gain during each symbol interval. Iterative solution of the coefficients of the equalizer is possible because the MSE is a quadratic function of the coefficients. The MSE may be envisioned as an $N$-dimensional paraboloid (punch bowl) with a bottom or minimum. The adjustment to each tap gain is in a direction opposite to an estimate of the gradient of the MSE with respect to that tap gain. The idea is to move the set of equalizer coefficients closer to the unique optimum set corresponding to the minimum MSE. This symbol-by-symbol procedure developed by Widrow and Hoff [Widrow, 12.109] is commonly referred to as **stochastic gradient** method because, instead of the true gradient of the MSE,

$$\frac{\delta E[e_k^2]}{\delta c_n(k)}$$

a noisy but unbiased estimate

$$\frac{\delta e_k^2}{\delta c_n(k)} = 2e_k r(t_O + kT - nT)$$

is used. Thus, the tap gains are updated according to

$$c_n(k + 1) = c_n(k) - \Delta e_k r(t_o + kT - nT), \qquad n = 0, 1, \ldots, N - 1,$$

where $c_n(k)$ is the $n$th tap gain at time $k$, $e_k$ is the error signal and $\Delta$ is a positive adaptation constant or step size.

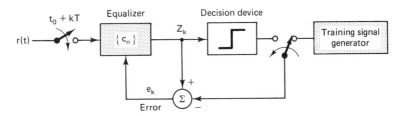

**Figure 12.8**  Automatic adaptive equalizer.

### 12.1.4 Equalizer Convergence

The exact convergence behavior of the stochastic update method is hard to analyze (see Section 12.3.2). However, for a small step size and a large number of iterations, the behavior is similar to the steepest-descent algorithm, which uses the actual gradient rather than a noisy estimate.

Here we list some general convergence properties: (1) Fastest convergence (or shortest settling time) is obtained when the (folded) power spectrum of the symbol-rate sampled equalizer input is flat and when the step size $\Delta$ is chosen to be the inverse of the product of the received signal power and the number of equalizer coefficients; (2) the larger the variation in the folded power spectrum in 1, the smaller the step size must be, and therefore the slower the rate of convergence; (3) for systems where sampling causes aliasing (channel foldover or spectral overlap), the convergence rate is affected by the channel delay characteristics and the sampler phase, because they affect the aliasing. This will be explained more fully later.

### 12.1.5 Adaptive Equalization

After the initial training period (if there is one), the coefficients of an adaptive equalizer may be continually adjusted in a **decision-directed** manner. In this mode the error signal $e_k = z_k - \hat{x}_k$ is derived from the final (not necessarily correct) receiver estimate $\{\hat{x}_k\}$ of the transmitted sequence $\{x_k\}$. In normal operation the receiver decisions are correct with high probability, so that the error estimates are correct often enough to allow the adaptive equalizer to maintain precise equalization. Moreover, a decision-directed adaptive equalizer can track slow variations in the channel characteristics or linear perturbations in the receiver front end, such as slow jitter in the sampler phase.

The larger the step size, the faster the equalizer tracking capability. However, a compromise must be made between fast tracking and the *excess MSE* of the equalizer. The excess MSE is that part of the error power in excess of the minimum attainable MSE (with tap gains frozen at their optimum settings). This excess MSE, caused by tap gains wandering around the optimum settings, is directly proportional to the number of equalizer coefficients, the step size, and the channel noise power. The step size that provides the fastest convergence results in a MSE which is, on the average, 3 dB worse than the minimum achievable MSE. In practice, the value of the step size is selected for fast convergence during the training period and then reduced for fine tuning during the steady-state operation (or data mode).

### 12.1.6 Equalizers for QAM Systems

So far we have discussed only equalizers for a baseband PAM system. Modern high-speed voiceband modems almost universally use PSK for lower speeds, for example, 2400 to 4800 bits/s, and combined phase and amplitude modulation or, equivalently, QAM [Lucky, 12.54], for higher speeds such as 4800 to 9600 or even 19,200 bits/s. At the high rates, where noise and other channel distortions become significant, modems using coded forms of QAM such as **trellis-coded modulation** [Forney, 12.27; van Gerven, 12.105; Wei, 12.107] are being introduced to obtain improved performance. QAM is as efficient in bits per second per hertz as vestigial or SSB amplitude modulation, yet enables a coherent carrier to be derived and phase jitter to be tracked using easily implemented decision-directed carrier recovery techniques [Kobayashi, 12.49]. A timing waveform with negligible timing jitter can also be easily recovered from QAM signals. This property is not shared by vestigial sideband amplitude modulation systems [Lyon, 12.128].

Figure 12.9 shows a generic QAM system, which may also be used to implement PSK or combined amplitude and phase modulation. Two double-sideband-suppressed-carrier (DSB-SC) AM signals are superimposed on each other at the transmitter and separated at the receiver, using quadrature or orthogonal carriers for modulation and demodulation. It is convenient to represent the in-phase (I) and quadrature (Q) channel LPF output signals in Fig. 12.9 by $y_r(t)$ and $y_i(t)$, as the real and imaginary parts of a complex-valued signal $y(t)$. (Note that the signals are real, but it will be convenient to use complex notation.)

The baseband equalizer [Proakis, 12.84], with complex coefficients $c_n$, operates on samples of this complex signal $y(t)$ and produces complex equalized samples $z(k) = z_r(k) + jz_i(k)$, as shown in Fig. 12.10. This figure illustrates more

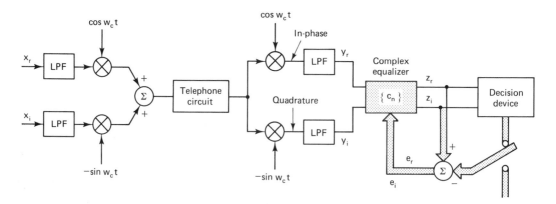

**Figure 12.9** QAM system with baseband complex adaptive equalizer.

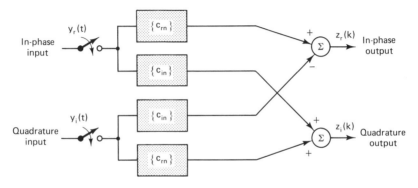

**Figure 12.10** Complex transversal equalizer for QAM modems.

concretely the concept of a complex equalizer as a set of four real transversal filters (with cross-coupling) for two inputs and two outputs. While the real coefficients $c_{rn}$, $n = 0, \ldots, N - 1$, help to combat the ISI in the in-phase and quadrature channels, the imaginary coefficients $c_{in}$, $n = 0, \ldots, N - 1$, counteract the cross interference between the two channels. The latter may be caused by asymmetry in the channel characteristics around the carrier frequency.

The coefficients are adjusted to minimize the mean of the squared magnitude of the complex error signal, $e(k) = e_r(k) + je_i(k)$, where $e_r$ and $e_i$ are the differences between $z_r$ and $z_i$, and their desired values. The update method is similar to the one used for the PAM equalizer except that all variables are complex-valued;

$$c_n(k + 1) = c_n(k) - \Delta e_x y^*(t_o + kT - nT), \qquad n = 0, 1, \ldots, N - 1$$

where $y^*$ is the complex conjugate of $y$. Again, the use of complex notation allows the writing of this single concise equation, rather than two separate equations involving four real multiplications, which is what really has to be implemented.

The complex equalizer can also be used at passband [Brady, 12.6; Falconer, 12.16] to equalize the received signal before demodulation, as shown in Fig. 12.11. Here the received signal is split into its I and Q components by a pair of **phase-splitting filters**, with identical amplitude responses and phase responses that differ by 90°. The complex passband signal at the output of these filters is sampled at the symbol rate and applied to the equalizer delay line in the same way as at baseband. The complex output of the equalizer is demodulated via multiplication by a complex exponential, as shown in Fig. 12.11, before decisions are made and the complex error computed. Further, the error signal is remodulated before it is used in the equalizer adjustment algorithm. The main advantage of implementing the equalizer in the passband is that the delay between the demodulator and the phase-error computation circuit is reduced to the delay through the decision device. Fast phase jitter can be tracked more effectively because the delay through the equalizer is eliminated from the phase correction loop. The same advantage can be attained with a baseband equalizer by putting a jitter-tracking loop after the equalizer.

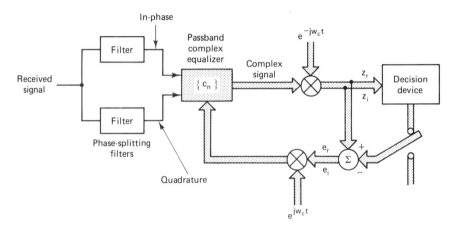

**Figure 12.11**  Passband complex adaptive equalizer for QAM systems.

### 12.1.7 Decision-Feedback Equalizers

We have discussed placements and adjustment methods for the equalizer, but the basic equalizer structure has remained a linear and nonrecursive filter. A simple nonlinear equalizer [Austin, 12.3; Belfiore, 12.4; George, 12.32; Monsen, 12.71; Salz, 12.93], which is particularly useful for channels with severe amplitude distortion, uses decision feedback to cancel the interference from symbols which have already been detected. Figure 12.12 shows such a **decision-feedback equalizer** (DFE). The equalized signal is the sum of the outputs of the forward and feedback parts of the equalizer. The forward part is like the linear transversal equalizer discussed earlier. Decisions made on the equalized signal are fed back via a second transversal filter. The basic idea is that if the value of the symbols already detected are known (past decisions are assumed to be correct), then the ISI contributed by these symbols can be canceled exactly, by subtracting past symbol values with appropriate weighting from the equalizer output. The weights are samples of the tail of the system impulse response including the channel and the forward part of the equalizer.

The forward and feedback coefficients may be adjusted simultaneously to minimize the MSE. The update equation for the forward coefficients is the same as for the linear equalizer. The feedback coefficients are adjusted according to

$$b_m(k + 1) = b_m(k) + \Delta e_k \hat{x}_{k-m}, \qquad m = 1, \dots, M$$

where $\hat{x}_k$ is the $k$th symbol decision, $b_m(k)$ is the $m$th feedback coefficient at time $k$, and there are $M$ feedback coefficients in all. The optimum LMS settings of $b_m$, $m = 1, \dots, M$, are those that reduce the ISI to zero, within the span of the feedback part, in a manner similar to a ZF equalizer. Note that since the output of the feedback section of the DFE is a weighted sum of noise-free past decisions,

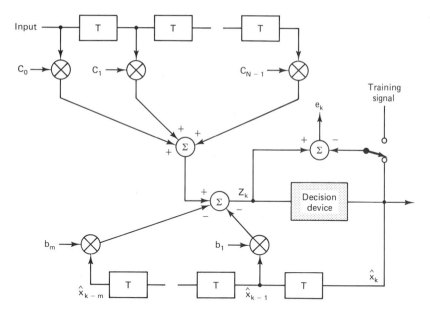

**Figure 12.12** Decision-feedback equalizer.

the feedback coefficients play no part in determining the noise power at the equalizer output.

Given the same number of overall coefficients, does a DFE achieve less MSE than a linear equalizer? There is no definite answer to this question. The performance of each type of equalizer is influenced by the particular channel characteristics and sample phase, as well as the actual number of coefficients and the position of the reference or main tap of the equalizer. However, the DFE can compensate for amplitude distortion without as much noise enhancement as a linear equalizer. The DFE performance is also less sensitive to the sampler phase [Salz, 12.94].

An intuitive explanation for these advantages is as follows: The coefficients of a linear transversal equalizer are selected to force the combined channel and equalizer impulse response to approximate a unit pulse. In a DFE, the ability of the feedback section to cancel the ISI, because of a number of past symbols, allows more freedom in the choice of the coefficients of the forward section. The combined impulse response of the channel and the forward section may have nonzero samples following the main pulse. That is, the forward section of a DFE need not approximate the inverse of the channel characteristics and so avoids excessive noise enhancement and sensitivity to sampler phase.

When a particular incorrect decision is fed back, the DFE output reflects this error during the next few symbols as the incorrect decision traverses the feedback delay line. Thus there is a greater likelihood of more incorrect decisions following the first one, that is, error propagation. Fortunately, the error propagation in a

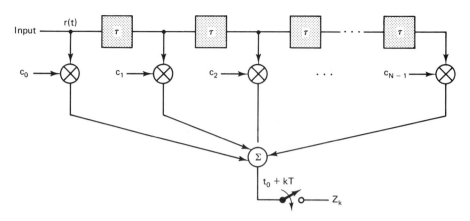

**Figure 12.13** Fractionally spaced equalizer.

DFE is not catastrophic. On typical channels, errors occur in short bursts that degrade performance only slightly.

### 12.1.8 Fractionally Spaced Equalizers

A fractionally spaced transversal equalizer [Brady, 12.6; Gitlin, 12.39; Guidoux, 12.45; Lucky, 12.58; Monsen, 12.71; Qureshi, 12.91; Ungerboeck, 12.104] is shown in Fig. 12.13. The delay-line taps of such an equalizer are spaced at an interval $\tau$, which is less than, or a fraction of, the symbol interval $T$. The tap spacing $\tau$ is typically selected such that the bandwidth occupied by the signal at the equalizer input is $|f| < 1/2\tau$; that is, $\tau$-spaced sampling satisfies the sampling theorem. In an analog implementation, there is no other restriction on $\tau$, and the output of the equalizer can be sampled at the symbol rate. In a digital implementation $\tau$ must be $KT/M$, where $K$ and $M$ are integers and $M > K$. (In practice, it is convenient to choose $\tau = T/M$, where $M$ is a small integer, e.g., 2). The received signal is sampled and shifted into the equalizer delay line at a rate $M/T$, and one output is produced each symbol interval (for every $M$ input sample). In general, the equalizer output is given by

$$z_k = \sum_{n=0}^{N-1} c_n\, r\!\left(t_o + kT - \frac{nKT}{M}\right)$$

The coefficients of a $KT/M$ equalizer may be updated once per symbol based on the error computed for that symbol according to

$$c_n(k+1) = c_n(k) - \Delta e_k r\!\left(t_o + kT - \frac{nkT}{M}\right), \qquad n = 0, 1, \ldots, N-1$$

It is well known (see Section 12.2) that the optimum receive filter in a linear modulation system is the cascade of a filter matched to the actual channel, with a transversal $T$-spaced equalizer [Ericson, 12.14; Forney, 12.24; George, 12.31]. The fractionally spaced equalizer, by virtue of its sampling rate, can synthesize the best combination of the characteristics of an adaptive matched filter and a $T$-spaced equalizer, within the constraints of its length and delay. A $T$-spaced equalizer, with symbol-rate sampling at its input, cannot perform matched filtering. A **fractionally spaced equalizer** (FSE) can effectively compensate for more severe delay distortion and deal with amplitude distortion with less noise enhancement than a $T$-equalizer.

Consider a channel whose amplitude and envelope-delay characteristics around one band edge, $f_c - 1/2T$ hertz, differ markedly from the characteristics around the other band edge, $f_c + 1/2T$ hertz, in a QAM system with a carrier frequency of $f_c$ hertz. Then the symbol-rate sampled or folded-channel frequency response is likely to have a rapid transition in the area of spectral overlap. It is difficult for a typical $T$-equalizer, with its limited degrees of freedom (number of taps), to manipulate such a folded channel into one with a flat frequency response. An FSE, on the other hand, can independently adjust the signal spectrum (in amplitude and phase) at the two band-edge regions before symbol-rate sampling (and spectral overlap) at the equalizer output, resulting in significantly improved performance.

A related property of an FSE is the insensitivity of its performance to the choice of sampler noise. This distinction between the conventional $T$-spaced and FSEs can be heuristically explained as follows: First, symbol-rate sampling at the input to a $T$-equalizer causes spectral overlap or aliasing, as explained in connection with Fig. 12.4. When the phases of the overlapping components match they add constructively, and when the phases are 180° apart they add destructively, which results in the cancellation or reduction of amplitude, as shown in Fig. 12.14. Variation in the sampler phase or timing instant corresponds to a variable delay in the signal path; a linear phase component with variable slope is added to the signal spectrum. Thus changes in the sampler phase strongly influence the effects of aliasing; that is, they influence the amplitude and delay characteristics in the spectral overlap region of the sampled equalizer input. The minimum MSE achieved by the $T$-equalizer is, therefore, a function of the sampler phase. In particular, when the sampler phase causes cancellation of the band-edge ($|f| = 1/2T$ hertz) components, the equalizer cannot manipulate the null into a flat spectrum at all, or at least without significant noise enhancement (if the null is a depression rather than a total null).

In contrast, there is no spectral overlap at the input to an FSE. Thus the sensitivity of the minimum MSE, achieved with an FSE with respect to the sampler phase, is typically far smaller than with a $T$-equalizer.

Comparison of numerical performance results of $T$- and $T/2$-equalizers for QAM systems operating over representative voice-grade telephone circuits [Qureshi, 12.91] has shown the following properties: (1) A $T/2$-equalizer with the same number of coefficients (half the time span) performs almost as well or better than a $T$-equalizer; (2) a pre-equalizer receive shaping filter is not required with a $T/2$-equalizer; (3) for channels with severe band-edge delay distortion, the $T$-equalizer performs noticeably worse than a $T/2$-equalizer regardless of the choice of sampler phase.

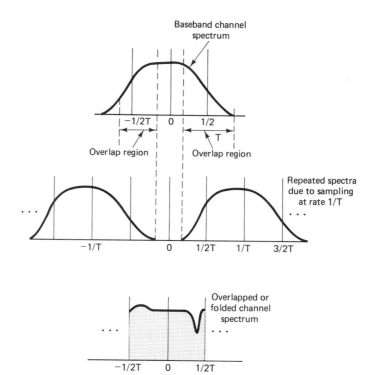

**Figure 12.14**  Spectral overlap at the input to a T-equalizer.

### 12.1.9 Other Applications

While the primary emphasis of this chapter is on adaptive equalization for data transmission, a number of the topics covered are relevant to other applications of adaptive filters.  In this section generic forms of adaptive filtering applications are introduced to help in establishing the connection between the material presented in later sections and the application of interest.

Figure 12.15 shows a general form of an adaptive filter with input signals $x$ and $y$, output $z$, and error $e$.  The parameters of an LMS adaptive filter are updated to minimize the mean-squared value of the error $e$.  In the following paragraphs, we point out how the adaptive filter is used in different applications by listing how $x$, $y$, $z$ and $e$ are interpreted for each case.

#### Equalization

$y$ = received signal (filtered version of transmitted data signal) plus noise uncorrelated with the data signal

$x$ = detected data signal

$z$ = equalized signal used to detect received data

$e$ = residual ISI plus noise

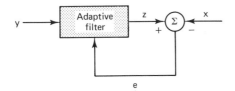

**Figure 12.15** General form of an adaptive filter.

### Echo Cancellation (Chapter 4)

Echo cancellation is a form of a general-system identification problem, where the system to be identified is the echo path linear system. The coefficients of a transversal echo canceler converge in the mean to the echo path impulse response samples.

*Voice*

$y$ = far-end voice signal plus uncorrelated noise

$x$ = echo of far-end voice plus near-end voice plus noise

$z$ = estimated echo of far-end voice

$e$ = near-end voice plus residual echo plus noise

Adaptation is typically carried out in the absence of the near-end voice signal. When double talk is detected (both near- and far-end signals present), update of the echo-canceler coefficients is inhibited [Messerschmitt, 12.69].

*Data*

$y$ = transmitted data signal

$x$ = echo of transmitted data signal plus received signal plus noise

$z$ = estimated echo of transmitted data signal

$e$ = received signal plus residual echo plus noise

Filter adaptation is typically required to be continued in the presence of a large interfering received signal, which is uncorrelated with the transmitted data [Messerschmitt, 12.69; Weinstein, 12.108]. A method proposed in [Falconer, 12.22] involves locally generating a delayed replica of the received signal and subtracting it from $e$ before using the residual for echo-canceler update.

### Noise Cancellation

$y$ = noise source correlated with noise in $x$

$x$ = desired signal plus noise

$z$ = estimate of noise in $x$

$e$ = desired signal plus residual noise

Sec. 12.1    Introduction

One example is that of canceling noise from the pilot's speech signal in the cockpit of an aircraft [Widrow, 12.110]. In this case, $y$ may be picked up from a microphone in the pilot's helmet, and $x$ is the ambient noise picked up by another microphone placed in the cockpit. See [Widrow, 12.110] for a number of other interesting applications of noise and periodic interference cancellation, such as to electrocardiography.

### Prediction (Chapter 3)

$$y = \text{delayed version of original signal}$$

$$x = \text{original signal}$$

$$z = \text{predicted signal}$$

$$e = \text{prediction error or residual}$$

A well-known example is linear predictive coding (LPC) of speech, where the end result is the set of estimated LPC coefficients [Honig, 12.46]. Due to the nonstationary nature of the speech signal, LPC coefficients are typically obtained separately for each new frame (10–25 ms) of the speech signal.

In adaptive differential pulse code modulation (ADPCM) of speech, the purpose of adaptive prediction is to generate a residual signal with less variance so that it can be quantized and represented by fewer bits for transmission [Honig, 12.46]. In this case

$$y = \text{reconstructed speech signal} = \text{quantized residual plus past prediction}$$

$$x = \text{original speech signal}$$

$$z = \text{prediction}$$

$$e = \text{residual to be quantized for transmission}$$

Note that the reconstructed speech signal is used for $y$ instead of a delayed version of the original speech signal, and the predictor coefficients are updated using the quantized residual instead of $e$. Both the reconstructed speech signal and the quantized residual are available at the ADPCM decoder, so the predictor coefficients at the decoder can be adapted in a manner identical to that used at the ADPCM encoder.

### Adaptive Arrays

A further generalization of the adaptive filter of Fig. 12.15 is shown in Fig. 12.16, where a number of input signals are processed through an array of adaptive filters whose outputs are summed together. Such adaptive arrays are useful in diversity combining [Brady, 12.6; Monsen, 114] and in dealing with jamming or spatially distributed interference [Monsen, 12.72; Widrow, 12.110].

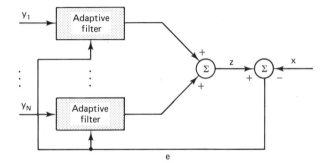

Figure 12.16  Adaptive filter array.

### 12.1.10 Implementation Approaches

We may divide the methods of implementing adaptive equalizers into the following general categories: analog, hard-wired digital, and programmable digital.

Analog adaptive equalizers, with inductor-capacitor (LC) tapped delay lines and switched ladder attenuators as tap gains, were among the first implementations for voiceband modems. The switched attenuators later gave way to field-effect transistors as the variable gain elements. Analog equalizers were soon replaced by digitally implemented equalizers for reduced size and increased accuracy. Recently, however, there is renewed interest in large-scale integrated (LSI) analog implementations of adaptive filters based on switched capacitor technology [Agazzi, 12.2; Yasumoto, 12.112]. Here the equalizer input is sampled but not quantized. The sampled analog values are stored and transferred as charge packets. In one implementation [Agazzi, 12.2], the adaptation circuitry is digital. The variable tap gains are typically stored in digital memory locations and the multiplications between the analog sample values and the digital tap gains take place in analog fashion via multiplying digital-to-analog (D/A) converters. In another case [Yasumoto, 12.112], a five-tap adaptive transversal filter has been fabricated on a single integrated circuit (IC) using an all analog implementation approach combining switched capacitor and **charge-coupled device** (CCD) technologies. The IC can be configured for use as an echo canceler, a linear equalizer, or a decision-feedback equalizer at sampling rates up to 250 kHz. Analog or mixed analog-digital implementations have significant potential in applications, such as digital radio and digital subscriber loop transmission, where symbol rates are high enough to make purely digital implementations difficult.

The most widespread technology of the last decade for voiceband modem adaptive equalizer implementation may be classified as hard-wired digital technology. In such implementations the equalizer input is made available in sampled and quantized form suitable for storage in digital shift registers. The variable tap gains are also stored in shift registers, and the formation and accumulation of products takes place in logic circuits connected to perform digital arithmetic. This class of implementations is characterized by the fact that the circuitry is hard-wired for the sole purpose of performing the adaptive equalization function with a predetermined structure. Examples include the early units based on **metal-oxide**

**semiconductor** (MOS) shift registers and **transistor-transistor logic** (TTL) circuits. Later implementations [Logan, 12.53; Tong, 12.100] were based on MOS LSI circuits with dramatic savings in space, power dissipation and cost.

A hard-wired digital adaptive filtering approach described in [van Gerwen, 12.106] for a 144 kb/s digital subscriber loop modem is based on a **random-access memory** (RAM) table-lookup structure [Halte, 12.124]. Both an echo canceler and a DEF, whose inputs are the transmit and receive binary data sequences, respectively, can be implemented in this way. An output signal value is maintained and updated in the RAM for each of the $2^N$ possible states of an $N$-tap transversal filter with a binary input sequence. Such a structure is not restricted to be linear and, therefore, can adapt to compensate for nonlinearities [Agazzi, 12.123]. A two-chip realization of a digital subscriber loop modem based on a joint echo canceler–DFE RAM structure is described in [Wouda, 12.121].

The most recent trend in implementing voiceband modem adaptive equalizers is toward programmable digital signal processors [Godard, 12.43; Murano, 12.81; Tsuda, 12.101; Qureshi, 12.120]. Here, the equalization function is performed in a series of steps or instructions in a microprocessor or a digital computation structure specially configured to perform efficiently the type of digital arithmetic (e.g., multiply and accumulate) required in digital signal processing. The same hardware can then be time-shared to perform functions such as filtering, modulation, and demodulation in a modem. Perhaps the greatest advantage of programmable digital technology is its flexibility, which permits sophisticated equalizer structures and training procedures to be implemented with ease.

For microwave digital radio systems, adaptive equalizers have been implemented both in the passband at the intermediate frequency (IF) stage and at baseband [Siller, 12.98]. Passband equalizers are analog by necessity, e.g., an amplitude slope equalizer [Fenderson, 12.23] at 70 MHz IF, and a dynamic resonance equalizer using PIN and varactor diodes [Murase, 12.82] at 140 MHz IF. Three-tap $T/2$ transversal equalizers have been implemented in the passband (at 70 MHz IF) using quartz surface acoustic wave filters, analog correlators and tap multipliers for a 4-PSK 12.6 Mb/s digital radio [Monsen, 12.114]. Baseband transversal equalizers have been implemented using a combination of analog and digital or all digital circuitry. A five-tap zero-forcing equalizer using lumped delay elements, hybrid-integrated circuits for variable-gain and buffer amplifiers, and emitter-coupled logic for tap control is described in [Fenderson, 12.23] for a 16-QAM 90-Mb/s digital radio. A five-tap LMS transversal equalizer [Murase, 12.82], and all digital DFEs have also been reported [Siller, 12.98; Monsen, 12.114].

## 12.2 RECEIVER STRUCTURES AND THEIR STEADY-STATE PROPERTIES

### 12.2.1 Baseband Equivalent Model

To set a common framework for discussing various receiver configurations, we develop a baseband equivalent model of a passband data-transmission system. We start from a generic QAM system (Fig. 12.9) since it can be used to implement any linear modulation scheme.

The passband transmitted signal can be compactly written as:

$$s(t) = Re \left[ \sum_n x_n a(t - nT) \exp (j2\Pi f_c t) \right]$$

where $\{x_n\}$ is the complex sequence of data symbols with I (real) and Q (imaginary) components, $x_r$ and $x_i$, respectively, such that $x_n = x_{rn} + jx_{in}$, $a(t)$ is the transmit pulse shape and $f_c$ is the carrier frequency. We shall assume that the baseband transmit spectrum $A(f)$ is bandlimited to $|f| \leq (1 + \alpha)/2T$ Hz where the roll-off factor $\alpha$, between 0 and 1, determines the excess bandwidth over the minimum $|f| \leq 1/2T$. (Note that greater than 100% excess bandwidth is sometimes used in radio and baseband subscriber loop transmission systems).

The received signal is:

$$r(t) = s(t) * h_p(t) + n_p(t)$$

where $h_p(t)$ is the passband channel impulse response, $n_p(t)$ is "passband" Gaussian noise, and the operator $*$ represents convolution. The in-phase and quadrature outputs of the receive LPFs, $y_r(t)$ and $y_i(t)$, may be represented in complex notation as $y(t) = y_r(t) + jy_i(t)$:

$$y(t) = g(t) * [r(t) \exp (-j2\Pi f_c t)]$$

where $g(t)$ is the impulse response of the receive filter. Assume that the receive filter completely rejects the double-frequency signal components produced by demodulation and centered around $2f_c$. Then the baseband received signal may be written as:

$$y(t) = \sum_n x_n h(t - nT) + n(t) \tag{12.1}$$

where

$$h(t) = g(t) * h_b(t) * a(t) \tag{12.2}$$

is the complex-valued impulse response of the baseband equivalent model (Fig. 12.17). The real-valued passband channel impulse response, $h_p(t)$, and the complex-valued baseband channel impulse response, $h_b(t)$, are related according to

$$h_p(t) = Re [h_b(t) \exp (j2\Pi f_c t)]$$

The noise waveform $n(t)$ in (12.1) is also complex-valued, that is,

$$n(t) = g(t) * [n_p(t) \exp (-j2\Pi f_c t)]$$

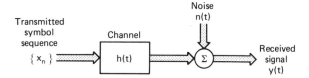

Figure 12.17 General complex-valued baseband equivalent channel model.

The receive LPFs (in Fig. 12.9) typically perform two functions: rejection of the "double-frequency" signal components and noise suppression. The latter function is accomplished by further shaping the baseband signal spectrum. In the baseband equivalent model, only the first of these functions of the receive filters is performed by $g(t)$ and absorbed in $h(t)$ given in (12.2). For simplicity, the noise $n(t)$ in the baseband equivalent model is assumed to be white with jointly Gaussian real and imaginary components.

### 12.2.2 Optimum Receive Filter

Given the received signal $y(t)$, what is the best receive filter? This question has been posed and answered in different ways by numerous authors. Here, we follow Forney's development [Forney, 12.24]. He showed that the sequence of $T$-spaced samples, obtained at the correct timing phase, at the output of a matched filter is a set of sufficient statistics for estimation of the transmitted sequence $\{x_n\}$. Thus such a receive filter is sufficient regardless of the (linear or nonlinear) signal processing that follows the symbol-rate sampler.

For the baseband equivalent model derived in the previous section, the receive filter must have an impulse response $h^*(-t)$, where the superscript $*$ denotes complex conjugate. The frequency response of this matched filter is $H^*(f)$, where $H(f)$ is the frequency response of the channel model $h(t)$.

If the data sequence $\{x_n\}$ is uncorrelated with unit power, that is,

$$E[x_n x_m^*] = \delta_{nm}$$

then the signal spectrum at the matched filter output is $|H(f)|^2$, and the noise power spectrum is $N_o|H(f)|^2$. After $T$-spaced sampling, the aliased or folded signal spectrum is

$$S_{hh}(f) = \sum_n \left| H\left(\frac{f-n}{T}\right) \right|^2, \qquad 0 \le f \le \frac{1}{T}$$

and the noise power spectrum is $N_o S_{hh}(f)$.

If the transmission medium is ideal, then the baseband signal spectrum $H(f)$ at the matched filter input is determined solely by the transmit signal-shaping filters. From the discussion in Section 12.1.1 it is clear that if ISI is to be avoided, the composite of the transmit filter and receive matched filter response must satisfy the Nyquist criterion, that is,

$$S_{hh}(f) = R_o, \qquad 0 \le f \le \frac{1}{T},$$

where, in general, $R_o = T \int_0^{1/T} S_{hh}(f)\, df$. Therefore, the overall Nyquist amplitude response $|H(f)|^2$ must be equally divided between the transmit and receive filters. For instance, each filter may have an amplitude response which is square root of a raised-cosine characteristic [Lucky, 12.54; Feher, 12.113]. For such an ideal additive white Gaussian noise (AWGN) channel, the matched filter, symbol-rate sampler and a memoryless detector comprise the optimum receiver.

Let us now consider the more interesting case of an AWGN channel with linear distortion. For such a channel, the simple linear receiver described above is no longer adequate. The symbol-rate sampled sequence, though still providing a set of sufficient statistics, now contains ISI in addition to noise. The current received symbol is distorted by a linear combination of past and future transmitted symbols. Therefore, a memoryless symbol-by-symbol detector is not optimum for estimating the transmitted sequence. Nonlinear receivers that attempt to minimize some measure of error probability are the subject of Section 12.2.4, where the emphasis is on techniques that combine nonlinear processing with adaptive filters.

It is instructive to first study linear receivers, which attempt to maximize SNR (minimize MSE) prior to memoryless detection.

### 12.2.3 Linear Receivers

We begin by reviewing the conventional linear receiver comprising a matched filter, a symbol-rate sampler, an infinite-length symbol-spaced equalizer, and a memoryless detector. In the following section we show that the matched-filter, sampler, symbol-spaced equalizer combination is a special case of a more-general infinite-length fractionally spaced transversal filter/equalizer. In fact, this general filter may be used as the receive filter for any receiver structure without loss of optimality. Section 12.2.3 presents a contrast between practical forms of the conventional and fractionally spaced receiver structures.

#### Matched Filter and Infinite-Length Symbol-Spaced Equalizer

If further processing of the symbol-rate sampled sequence at the output of a matched receive filter is restricted to be linear, this linear processor takes the general form of a $T$-spaced infinite-length transversal or nonrecursive equalizer followed by a memoryless detector [George, 12.31; Lucky, 12.54]. Let the periodic frequency response of the transversal equalizer be $C(f)$. The equalized signal spectrum is $S_{hh}(f)C(f)$, $0 \le f \le 1/T$. The optimum $C(f)$ is one that minimizes the MSE at its output. The MSE is given by

$$\varepsilon = T \int_0^{1/T} |1 - S_{hh}(f)C(f)|^2 + N_o S_{hh}(f)|C(f)|^2 \, df \tag{12.3}$$

where the first term is the residual ISI power and the second term is the output noise power. Differentiating the integrand with respect to $C(f)$ and equating the result to zero, we obtain the **minimum MSE** (MMSE) equalizer frequency response:

$$C(f) = \frac{1}{N_o + S_{hh}(f)} \tag{12.4}$$

Thus the best (MMSE) matched-filter equalized signal spectrum is given by:

$$S_{hh}\frac{(f)}{N_O + S_{hh}(f)} \tag{12.5}$$

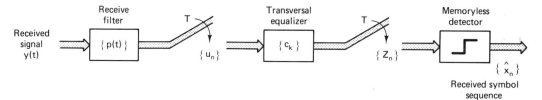

Received
signal
y(t)

Receive
filter

$\{p(t)\}$

$T$

$\{u_n\}$

Transversal
equalizer

$\{c_k\}$

$T$

$\{z_n\}$

Memoryless
detector

$\{\hat{x}_n\}$

Received symbol
sequence

**Figure 12.18**  Conventional linear receiver.

Substituting (12.4) in (12.3), we obtain the following expression for the MMSE achievable by a linear receiver:

$$\varepsilon_{min}(\text{linear}) = T \int_0^{1/T} \frac{N_O}{N_O + S_{hh}(f)} \, df \qquad (12.6)$$

The receiver structure (Fig. 12.18) comprising a matched filter, a symbol-rate sampler, an infinite-length $T$-spaced equalizer, and a memoryless detector is referred to as the **conventional linear receiver**.

If the equalizer response $C(f)$ is designed to satisfy the zero-ISI constraint, then $C(f) = 1/S_{hh}(f)$. The overall frequency response of the optimum ZF linear receiver is given by $H^*(f)/S_{hh}(f)$, which forces the ISI at the receiver output to zero. The MSE achieved by such a receiver is

$$\varepsilon_{zf}(\text{linear}) = T \int_0^{1/T} \frac{N_o}{S_{hh}(f)} \, df \qquad (12.7)$$

$\varepsilon_{zf}(\text{linear})$ is always greater than or equal to $\varepsilon_{min}(\text{linear})$ because no consideration is given to the output noise power in designing the ZF equalizer. At high SNRs the two equalizers are nearly equivalent.

### Infinite-Length Fractionally-Spaced Transversal Filter

In this section, we derive an alternative form of an optimum linear receiver.

Let us start with the conventional receiver structure comprising a matched filter, symbol-rate sampler, $T$-spaced transversal equalizer and a memoryless detector. As a first step, linearity permits us to interchange the order of the $T$-spaced transversal equalizer and the symbol-rate sampler. Next, the composite response $H^*(f)C(f)$ of the matched filter in cascade with a $T$-spaced transversal equalizer may be realized by a single continuous-time filter. Let us assume that the received signal spectrum $H(f) = 0$ outside the range $|f| \leq 1/2T\tau$, $\tau \leq T$, and the flat noise spectrum is also limited to the same band by an ideal anti-aliasing filter with cutoff frequency $1/2\tau$. The composite matched-filter equalizer can then be realized by an infinite-length continuous-time transversal filter with taps spaced at $\tau$ second intervals and frequency response $H^*(f)C(f), |f| \leq 1/2\tau$, which is periodic with period $1/\tau$ hertz. The continuous output of this composite transversal filter may be sampled at the symbol rate without any further restriction on the tap-spacing $\tau$. If the frequency response $C(f)$ is selected according to (12.4), the symbol-rate output of the composite filter is identical to the corresponding output (12.5) in the conventional MMSE linear receiver.

To implement this composite matched-filter equalizer as a fractionally spaced digital nonrecursive filter, the effective tap-spacing must be restricted to $KT/M$, where $K$ and $M$ are relatively prime integers, $K < M$, and the fraction $KT/M \leq \tau$. The desired frequency response $H^*(f)C(f)$ of this fractionally spaced digital transversal filter is periodic with period $M/KT$ hertz, and limited to the band $|f| \leq 1/2\tau < M/2KT$, since $H(f)$ is limited to the same bandwidth. An ideal anti-aliasing filter with a cutoff frequency $M/2T$ Hz is assumed before the rate $M/T$ sampler.

The operation of the digital filter may be visualized as follows. Each symbol interval, $M$ input samples are shifted into digital shift register memory, every $K$th sample in the shift register is multiplied by a successive filter coefficient, and the products are summed to produce the single output required per symbol interval.

Note that the $M/T$ rate input of the fractionally spaced digital filter has the signal spectrum

$$
\begin{cases}
H(f) & |f| \leq \dfrac{M}{2KT} \\[2ex]
0 & \dfrac{M}{2KT} < |f| \leq \dfrac{M}{2T}
\end{cases}
$$

The frequency response of the digital filter is $H^*(f)C(f)$, $|f| \leq M/2KT$. Thus if the output of this filter was produced at the rate $M/T$, the output signal spectrum would be

$$
\begin{cases}
H(f)H^*(f)C(f) & |f| \leq \dfrac{M}{2KT} \\[2ex]
0 & \dfrac{M}{2KT} < |f| \leq \dfrac{M}{2T}
\end{cases}
$$

When the filter output is produced at the symbol rate, it has the desired aliased signal spectrum

$$
\sum_n \left| H\!\left(\frac{f-n}{T}\right) \right|^2 C\!\left(\frac{f-n}{T}\right), \qquad 0 \leq f \leq \frac{1}{T}
$$

Noting that $C(f)$ is periodic with period $1/T$, this output signal spectrum is recognized as $S_{hh}(f)C(f)$, which is the same as for the conventional MMSE linear receiver (12.5), provided $C(f)$ is selected according to (12.4).

This simple development proves the important point that an infinite-length fractionally spaced digital transversal filter is at once capable of performing the functions of the matched filter and the $T$-spaced transversal equalizer of the conventional linear receiver.

Let us further show that the symbol-rate sampled outputs of a fractionally spaced digital filter form a set of sufficient statistics for estimation of the transmitted sequence under the following conditions. The digital filter with tap spacing $KT/M$ has the frequency response $H^*(f)C(f)$, $|f| \leq M/2KT$, where the received signal spectrum $H(f)$ is zero outside the band $|f| \leq M/2KT$, $C(f)$ is periodic with period

$1/T$, and $C(f)$ is information lossless. A sufficient condition for $C(f)$ to be information lossless is that $C(f)$ is invertible, that is, $C(f) \neq 0, 0 \leq f \leq 1/T$. However, it is necessary only that $C(f)/S_{hh}(f)$ is invertible, that is, $C(f)/S_{hh}(f) \neq 0, 0 \leq f \leq 1/T$. In words, this condition implies that $C(f)$ may not introduce any nulls or transmission zeros in the Nyquist band, except at a frequency where the signal (and the noise power spectrum at the matched-filter output) may already have a null.

This result shows that with an appropriately designed $C(f)$, the symbol-rate outputs of a fractionally spaced filter, with frequency response $H^*(f)C(f)$, may be used without loss of optimality, for any linear or nonlinear receiver, regardless of the criterion of optimality.

In a linear receiver, where a memoryless detector operates on the symbol-rate outputs of the fractionally spaced filter, the function $C(f)$ may be designed to minimize the MSE at the detector input. The optimum $C(f)$ is then obtained using the same procedure as outlined in the previous section for the conventional linear receiver. Thus the MMSE $KT/M$-spaced filter frequency response is

$$\frac{H^*(f)}{N_O + S_{hh}(f)}, \qquad |f| \leq \frac{M}{2KT} \tag{12.8}$$

The MMSE achieved by this filter is, of course, the same as $\varepsilon_{\min}(\text{linear})$ derived earlier (12.6) for the conventional receiver structure.

### Fixed Filter and Finite-Length Symbol-Spaced Equalizer

The conventional MMSE linear receiver is impractical for two reasons. First, constraints of finite length and computational complexity must be imposed on the matched filter as well as the $T$-spaced transversal equalizer. Secondly, in most applications, it is impractical to design, beforehand, a filter that is reasonably matched to the variety of received signal spectra resulting from transmissions over different channels or a time-varying channel. Thus the most commonly used receiver structure comprises a fixed filter, symbol-rate sampler, and finite-length $T$-spaced adaptive equalizer (Fig. 12.18). The fixed filter response is either matched to the transmitted signal shape or is designed as a compromise equalizer, which attempts to equalize the average of the class of line characteristics expected for the application. For the present discussion, let us assume that the fixed filter has an impulse response $p(t)$ and a frequency response $P(f)$. Then the $T$-spaced sampled output of this filter may be written as

$$u_k = \sum_n x_n q(kT - nT) + v_k \tag{12.9}$$

where
$$q(t) = p(t) * h(t)$$

$$v_k = \int n(t)p(kT - t)\, dt$$

Denoting the $N$ equalizer coefficients at time $kT$ by the column vector $\mathbf{c}_k$ and the samples stored in the equalizer delay line by the vector $\mathbf{u}_k$, the equalizer output is given by

$$z_k = \mathbf{c}_k^T \mathbf{u}_k \tag{12.10}$$

where the superscript $T$ denotes transpose.

Minimizing the MSE $E[|z_k - x_k|^2]$ leads to the set of optimum equalizer coefficients

$$\mathbf{c}_{\text{opt}} = \mathbf{A}^{-1}\boldsymbol{\alpha} \tag{12.11}$$

where $A$ is an $N \times N$ Hermitian covariance matrix $E[\mathbf{u}_k^*\mathbf{u}_k^T]$, and $\boldsymbol{\alpha}$ is an $N$-element cross-correlation vector $E[\mathbf{u}_k^*x_k]$. Using the assumption that the data sequence $\{x_k\}$ is uncorrelated with unit power, it can be shown that the elements of the matrix $A$ and vector $\boldsymbol{\alpha}$ are given by

$$a_{i,j} = \sum_k q^*(kT)q(kT + iT - jT) + N_o \int p^*(t)p(t + iT - jT)\, dt \tag{12.12}$$

and

$$\alpha_i = q^*(-iT). \tag{12.13}$$

The MMSE achieved by this conventional suboptimum linear receiver is given by

$$\varepsilon_{\min}(\text{con}) = 1 - \boldsymbol{\alpha}^{*T}A^{-1}\boldsymbol{\alpha} \tag{12.14}$$

Alternatively, the $N$ equalizer coefficients may be chosen to force the samples of the combined channel and equalizer impulse response to zero at all but one of the $N$ $T$-spaced instants in the span of the equalizer. The ZF equalizer coefficient vector is given by

$$\mathbf{c}_{zf} = \mathbf{Q}^{-1}\boldsymbol{\delta} \tag{12.15}$$

where $Q$ is an $N \times M$ matrix with elements

$$q_{i,j} = q(iT - jT) \tag{12.16}$$

and $\boldsymbol{\delta}$ is a vector with only one nonzero element, that element being unity.

It is instructive to derive expressions for the equalizer and its performance as the number of coefficients is allowed to grow without bound. Since the $N \times N$ matrices $A$ and $Q$ are Toeplitz, their eigenvalues can be obtained by the **discrete Fourier transform** (DFT) of any row or column as $N \to \infty$ [Grenander, 12.44]. Thus by taking the DFT of the expression (12.11) for $\mathbf{c}_{\text{opt}}$ we obtain the frequency spectrum of the infinite-length $T$-spaced LMS equalizer:

$$C_{\text{opt}}(f) = \frac{Q_{eq}^*(f)}{|Q_{eq}(f)|^2 + N_O S_{pp}(f)} \tag{12.17}$$

where

$$Q_{eq}(f) = \sum_n Q(f - n/T) \tag{12.18}$$

is the aliased spectrum of $q(t)$, and

$$S_{pp}(f) = \sum_n |P(f - n/T)|^2 \tag{12.19}$$

is the aliased power spectrum of $p(t)$. The minimum achievable MSE is given by

$$\varepsilon_{min}(\text{con}) = T \int_0^{1/T} \frac{N_O}{N_O + |Q_{eq}(f)|^2/S_{pp}(f)} \, df \tag{12.20}$$

The corresponding expressions for the ZF equalizer are

$$C_{zf}(f) = \frac{1}{Q_{eq}(f)} \tag{12.21}$$

and

$$\varepsilon_{zf}(\text{con}) = T \int_0^{1/T} \frac{N_o S_{pp}(f)}{|Q_{eq}(f)|^2} \, df \tag{12.22}$$

When $P(f)$ is a matched filter, i.e., $P(f) = H^*(f)$, the above expressions reduce to those given in Section 12.2 because $S_{pp}(f) = Q_{eq}(f) = S_{hh}(f)$.

The smallest possible MSE (zero ISI-matched filter bound) is achieved when in (12.20) we have

$$\frac{|Q_{eq}(f)|^2}{S_{pp}(f)} = S_{hh}(f) = R_o, \qquad 0 \le f \le \frac{1}{T}$$

This occurs when the channel amplitude characteristic is ideal and perfect equalization is achieved by the matched filter. The greater the deviation of $|Q_{eq}(f)|^2 S_{pp}(f)$ from its average $R_o$, the greater $\varepsilon_{min}(\text{con})$. The aliased power spectrum $S_{pp}(f)$, as defined in (12.19), is independent of the phase characteristics of $P(f)$ or the sampler phase. The value of the squared absolute value $|Q_{eq}(f)|^2$ of the aliased spectrum $Q_{eq}(f)$, on the other hand, is critically dependent on the sampler phase in the roll-off region due to aliasing. Thus the minimum MSE achieved by the conventional receiver is dependent on the sampler phase even when the number of $T$-spaced equalizer coefficients is unlimited ($N \to \infty$).

### Finite-Length Fractionally Spaced Transversal Equalizer

This suboptimum linear receiver structure is simply a practical form of the infinite-length structure discussed in Section 12.2.3. We shall restrict our attention to the digitally implemented FSE with tap spacing $KT/M$ (see Fig. 12.19). The input to the FSE is the received signal sampled at rate $M/T$:

$$y\left(\frac{kT}{M}\right) = \sum_n x_n h\left(\frac{KT}{M} - nT\right) + n\left(\frac{kT}{M}\right) \tag{12.23}$$

Each symbol interval, the FSE produces an output according to

$$z(kT) = \sum_{n=0}^{N-1} c_n y\left(kT - \frac{nKT}{M}\right) \tag{12.24}$$

Denoting the $N$ equalizer coefficients at time $kT$ by the vector $\mathbf{c}_k$ and the $N$ most recently received samples (spaced $KT/M$ seconds apart) by the vector $\mathbf{y}_k$, the equal-

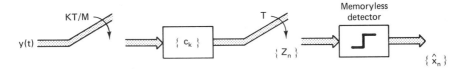

**Figure 12.19** Linear receiver based on a fractionally spaced transversal equalizer.

izer output may be written as

$$z_k = \mathbf{c}_k^T \mathbf{y}_k$$

Minimizing the MSE $E[|z_k - x_k|^2]$ leads to the set of optimum equalizer coefficients

$$\mathbf{c}_{opt} = A^{-1}\boldsymbol{\alpha} \qquad (12.25)$$

where $A$ is an $N \times N$ covariance matrix $E[\mathbf{y}_k^* \mathbf{y}_k^T]$, and $\boldsymbol{\alpha}$ is an $N$-element cross-correlation vector $E[\mathbf{y}_k^* x_k]$.

Using the assumption that the data sequence $\{x_k\}$ is uncorrelated with unit power, it can be shown that the elements of the matrix $A$ and vector $\boldsymbol{\alpha}$ are given by

$$a_{i,j} = \sum_k h^*\left(kT - \frac{iKT}{M}\right) h\left(kT - \frac{jKT}{M}\right) + N_o \delta_{ij} \qquad (12.26)$$

and

$$\alpha_i = h^*\left(-\frac{iKT}{M}\right) \qquad (12.27)$$

The MMSE achieved by the FSE is given by

$$\varepsilon_{min}(\text{FSE}) = 1 - \boldsymbol{\alpha}^{*T} A^{-1} \boldsymbol{\alpha} \qquad (12.28)$$

On the surface, the FSE development is quite similar to that of the conventional $T$-spaced LMS equalizer given in the previous section. There are, however, significant differences. First, unlike the $T$ equalizer, the FSE does not require a fixed receive shaping filter $p(t)$. Secondly, note that whereas the FSE input covariance matrix $A$ is Hermitian, it is not Toeplitz. In fact, each diagonal periodically takes one of $M$ different values. Due to the non-Toeplitz cyclostationary nature of the matrix $A$, it is no longer possible to obtain the eigenvalues of $A$ by simply taking the DFT of one of its rows even as $N \to \infty$. However, it is possible to decompose the set of infinite equations

$$A\mathbf{c} = \boldsymbol{\alpha}$$

into $M$ subsets, each with $M$ Toeplitz submatrices. Using this procedure, it can be shown [Gitlin, 12.39] that as $N \to \infty$, a fraction $(M - K)/M$ of the eigenvalues are equal to $N_O$, and the remaining eigenvalues are of the form $(M/K)S_{hh}(f) + N_o$, where

$$S_{hh}\left(\frac{iM}{NKT}\right) = \sum_n \left| H\left(\frac{iM}{NKT} - \frac{n}{T}\right) \right|^2, \qquad i = 0, 1, 2, \ldots, \left(\frac{NK}{M}\right) - 1 \quad (12.29)$$

The frequency response of the optimum FSE approaches (12.8) as $N \to \infty$, and its MSE approaches $\varepsilon_{min}$(linear) given in (12.6).

As the noise becomes vanishingly small, an infinitely long FSE has a set of zero eigenvalues. This implies that there are an infinite number of solutions which produce the same minimum MSE. The nonunique nature of the infinite FSE is evident from the fact that when both signal and noise vanish in the frequency range $1/2T < |f| \le M/2T$, the infinite FSE spectrum $C(f)$ can take any value in this frequency range without affecting the output signal or MSE.

Gitlin and Weinstein [Gitlin, 12.39] show that for transmission systems with less than 100% excess bandwidth, the matrix $A$ is nonsingular for a finite-length FSE even as the noise becomes vanishingly small. Therefore, there exists a unique set of optimum equalizer coefficients $\mathbf{c}_{opt}$ given by (12.25).

Deviation of the coefficient vector $\mathbf{c}_k$ from the optimum results in the following excess MSE over $\varepsilon_{min}$(FSE) given in (12.28):

$$(\mathbf{c}_k - \mathbf{c}_{opt})^{*T} A (\mathbf{c}_k - \mathbf{c}_{opt})$$

This quadratic form may be diagonalized to obtain $\mathbf{d}_k^{*T} \Lambda \, \mathbf{d}_k$, where the diagonal matrix $\Lambda$ has the eigenvalues of $A$ along its main diagonal, and $\mathbf{d}_k$ is the transformed coefficient deviation vector according to

$$\mathbf{d}_k = V(\mathbf{c}_k - \mathbf{c}_{opt}) \tag{12.30}$$

where the columns of the diagonalizing matrix $V$ are the eigenvectors of $A$. From the analysis of the infinite FSE, we would expect that when the number of FSE coefficients is "large," a significant fraction $(M - K)/M$ of the eigenvalues of $A$ are relatively small. If the $i$th eigenvalue is very small, the $i$th element of the deviation vector $\mathbf{d}_k$ will not contribute significantly to the excess MSE. It is, therefore, possible for coefficient deviations to exist along the eigenvectors corresponding to the small eigenvalues of $A$ without significant impact on the MSE. Thus many coefficient vectors may produce essentially the same MSE.

The most-significant difference in the behavior of the conventional suboptimum receiver and the FSE is a direct consequence of the higher sampling rate at the input to the FSE. Since no aliasing takes place at the FSE input, it can independently manipulate the spectrum in the two roll-off regions to minimize the output MSE after symbol-rate sampling. Thus unlike the $T$-spaced equalizer, it is possible for the FSE to compensate for timing phase as well as asymmetry in the channel amplitude or delay characteristics without noise enhancement. This is discussed further in Section 12.2.5.

### 12.2.4 Nonlinear Receivers

The MMSE linear receiver is optimum with respect to the ultimate criterion of minimum probability of symbol error only when the channel does not introduce any amplitude distortion, that is, $S_{hh}(f) = R_O$, $0 \le f \le 1/T$. The linear receive filter then achieves the matched filter bound for MSE:

$$\varepsilon_{min}(mf) = T \int_0^{1/T} \frac{N_O}{N_O + R_O} \, df$$

and a memoryless threshold detector is sufficient to minimize the probability of error. When amplitude distortion is present in the channel, the output error power spectrum is of a MMSE linear receive filter $N_o/[N_o + S_{hh}(f)]$. Thus noise power is enhanced at those frequencies where $S_{hh}(f) < R_o$. A memoryless detector operating on the output of this receive filter no longer minimizes symbol error probability.

Recognizing this fact, several authors have investigated optimum or approximately optimum nonlinear receiver structures subject to a variety of criteria [Lucky, 12.59]. Most of these receivers use one form or another of the maximum a posteriori probability rule to maximize either the probability of detecting each symbol correctly [Abend, 12.1] or of detecting the entire transmitted sequence correctly. The classical maximum-likelihood receiver [Lucky, 12.54] consists of $m^k$ matched filters, where $k$ is the length of the transmitted sequence whose symbols are drawn from a discrete alphabet of size $m$.

The complexity of the classical receiver, which grows exponentially with the message length, can be avoided by using the Viterbi algorithm. This recursive algorithm, which was originally invented to decode convolutional codes, was recognized to be maximum likelihood sequence estimator (MLSE) of the state sequence of a finite-state Markov process observed in memoryless noise [Forney, 12.25]. Forney [Forney, 12.24] showed that if the receive filter is a whitened matched filter, its symbol rate outputs at the correct sampling times form a set of sufficient statistics for estimation of the information sequence. Thus the transmission system between the data source and the Viterbi algorithm can be considered as a discrete channel, as shown in Fig. 12.20. The state, and hence the input sequence, of the discrete channel can be estimated by the Viterbi algorithm, which

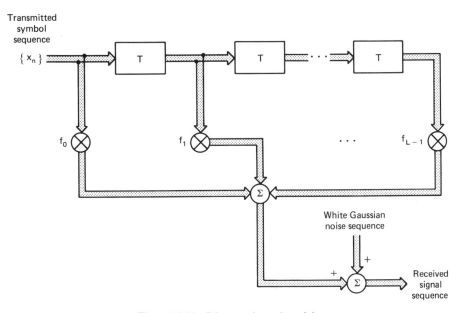

**Figure 12.20** Discrete channel model.

observes the channel output corrupted by AWGN [Forney, 12.24]. The computational complexity of the MLSE is proportional to $m^{L-1}$, the number of discrete channel states, where $L$ is the number of terms in the discrete channel pulse response.

The MLSE maximizes the mean time between error events, a reasonable criterion for practical **automatic repeat request** (ARQ) data communications systems, where efficiency is measured by throughput (i.e., the number of blocks of data correctly received versus the total number of blocks transmitted).

The symbol error probability of the MLSE [Forney, 12.24] is estimated by an expression of the form

$$\Pr(e) \simeq K Q \left[ \frac{d_{\min}}{2\sigma} \right]$$

The minimum Euclidean distance between any two valid neighboring sequences is $d_{\min}^2$, $\sigma^2$ is the mean-squared white Gaussian noise power at the input to the Viterbi algorithm, and $d_{\min}^2/\sigma^2$ is the effective SNR.

$$Q(x) \left( \frac{1}{2\pi} \right) \int_x^\infty exp \left( \frac{-y^2}{2} \right) dy$$

is the Gaussian probability of error function, and the error coefficient $K$ may be interpreted as the average number of ways in which minimum distance symbol errors can occur.

The lower bound on the probability of error for binary transmission over an ideal AWGN channel is given by

$$Q[(2\varepsilon_{\min}(mf))^{-1/2}] \tag{12.31}$$

The MLSE approaches this lower bound at high SNR for all channels except those with extremely severe ISI. For instance, for a class IV partial-response system (Chapter 8) with a discrete channel model of the form $1 - D^2$, the symbol error rate achieved by the MLSE is about four times the lower bound given above. In decibels this difference is small (about 0.5 dB for a symbol error rate of $10^{-4}$) and goes to zero as the effective SNR goes to infinity.

For unknown and/or slowly time-varying channels, the MLSE can be made adaptive by ensuring that both the **whitened matched filter** (WMF) and the channel model used by the Viterbi algorithm adapt to the channel response. Magee and Proakis [Magee, 12.62] proposed an adaptive version of the Viterbi algorithm that uses an adaptive identification algorithm to provide an estimate of the discrete channel pulse response. Structures with adaptive WMFs were proposed in [Mackenchnie, 12.61] and [Qureshi, 12.88].

Forney [Forney, 12.24] derived the WMF as the cascade connection of a matched filter $H^*(f)$, a symbol-rate sampler and a $T$-spaced transversal whitening filter whose pulse response is the anticausal factor of the inverse filter $1/S_{hh}(f)$. The noise at the output of the WMF is white and the ISI is causal. Price [Price, 12.83] showed that the WMF is also the optimum forward filter in a ZF decision-feedback equalizer, where the feedback transversal filter can exactly cancel the causal ISI provided all past decisions are correct.

At this point it is helpful to discuss simpler forms of nonlinear receivers, that is, decision feedback equalization and general decision-aided ISI cancellation, before returning to the topic of adaptive receiver filtering for maximum-likelihood sequence estimation.

### Decision Feedback Equalizers

The MLSE unravels ISI by deferring decisions and weighing as many preliminary decision sequences as the number of states in the discrete channel model. Thus in most cases, the MLSE makes use of all the energy in the discrete channel impulse response to maximize the effective SNR. By contrast, a DFE makes memoryless decisions and cancels all trailing ISI terms. Even when the WMF is used as the receive filter for both the MLSE and the DFE, the latter suffers from a reduced effective SNR, and error propagation, due to its inability to defer decisions.

*Infinite Length.* A DFE takes advantage of the symbols that have already been detected (correctly with high probability) to canel the ISI due to these symbols without noise enhancement. An infinite-length DFE receiver takes the general form (Fig. 12.21) of a forward linear receive filter, symbol-rate sampler, canceler, and memoryless detector. The symbol-rate output of the detector is then used by the feedback filter to generate future outputs for cancellation.

As pointed out by Belfiore and Park [Belfiore, 12.4] a structure equivalent to the DFE receiver of Fig. 12.21 is the structure shown in Fig. 12.22. The latter may be motivated from the point of view that given the MMSE forward filter, for example, an infinite-length FSE, we know that the sequence of symbol-rate samples at the output of this filter form a set of sufficient statistics for estimating the transmitted sequence. Then what simple form of nonlinear processing could further reduce the MSE?

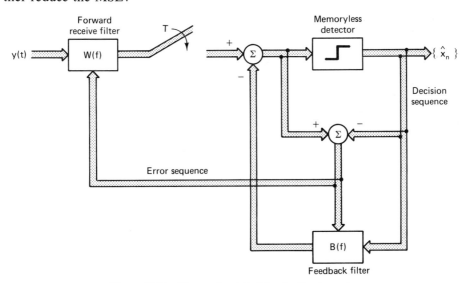

**Figure 12.21** Conventional decision-feedback receiver.

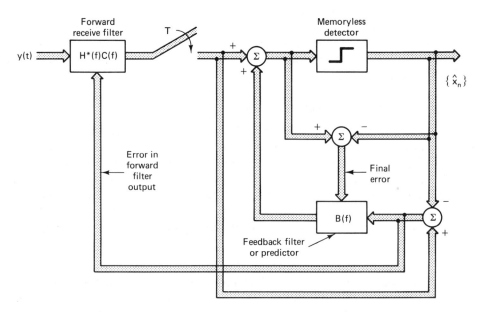

**Figure 12.22** Predictor form of decision-feedback receiver.

The power spectrum of the total distortion or error sequence at the output of the MMSE forward filter is given by the sum of the noise and ISI power spectra, that is,

$$|\hat{E}(f)|^2 = \frac{N_o}{N_o + S_{hh}(f)}, \qquad 0 \le f \le \frac{1}{T} \tag{12.32}$$

The error sequence is white if and only if $S_{hh}(f)$ is a constant; for example, $S_{hh}(f) = R_o, |f| \le 1/2T$, when the channel has no amplitude distortion. In this case further reduction in MSE is not possible. However, for channels with amplitude distortion, the power of the error sequence at the output of the forward filter can be reduced further by linear prediction [Makhoul, 12.122], provided past samples of the error sequence are available. An estimate of these past error samples can be obtained by decision feedback via a memoryless detector, as shown in Fig. 12.22. As the number of predictor coefficients grows without bound, the output MSE can be derived as

$$\varepsilon_{\min}(\text{DFE}) = \exp\left[ T \int^{1/T} \ln |\hat{E}(f)|^2 \, df \right]$$
$$= \exp\left[ -T \int^{1/T_o} \ln \left( 1 + \frac{S_{hh}(f)}{N_o} \right) df \right] \tag{12.33}$$

This expression is identical to the MMSE for an infinite-length conventional DFE [Salz, 12.93], which proves the equivalence of the predictor and conventional DFE structures.

We can write the following equivalence relationship by comparing the two DFE structures shown in Fig. 12.21 and Fig. 12.22:

$$W(f) = H^*(f)C(f)[1 + B(f)] \tag{12.34}$$

After some inspection, it becomes evident that the infinite-length forward filter $W(f)$ in the conventional DFE structure is the cascade of a matched filter and the anticausal factor of the optimum $C(f)$ given in (12.4) [Monsen, 12.71]. So long as the length of the forward filter in each of the two DFE structures is unconstrained, the two structures remain equivalent even when the feedback (or prediction) filter is reduced to a finite length. Note, however, that although the forward filter $W(f)$ in the conventional structure depends on the number of feedback coefficients, the forward filter in the predictor structure is independent of the predictor coefficients.

An alternative to the optimum mean-squared DFE receiver is the ZF formulation [Belfiore, 12.4; Price, 12.83]. The forward filter is again of the form (12.34). However, $C(f)$ is designed to satisfy the zero-ISI constraint, i.e., $C(f) = 1/S_{hh}(f)$. Thus the forward filter $W(f)$ is the cascade of a matched filter and the anticausal factor of the inverse filter $1/S_{hh}(f)$. The noise at the output of $W(f)$ is white and the ISI is causal (as in the case of Forney's WMF). The causal ISI is completely canceled by an infinite-length decision-feedback filter. The final error sequence consists solely of noise which is white and Gaussian, with power

$$\varepsilon_{zf}(\text{DFE}) = \exp\left[-T \int_0^{1/T} \ln\left(\frac{S_{hh}(f)}{N_o}\right) df\right] \tag{12.35}$$

As expected, $\varepsilon_{\min}(mf) \leq \varepsilon_{\min}(\text{DFE}) \leq \varepsilon_{zf}(\text{DFE})$.

***Finite Length.*** Neglecting ZF equalizers, there are four possible structures for finite-length DFE receivers, based on the conventional or FSE forward filters and conventional or predictor forms of feedback filters. Let us denote these as follows:

1. Type 1: conventional forward filter + conventional feedback filter
2. Type 2: conventional forward filter + predictor feedback filter
3. Type 3: FSE forward filter + conventional feedback filter
4. Type 4: FSE forward filter + predictor feedback filter

*Type 1:* To the forward equalizer structure described in Section 12.2.3, we add $N_b$ feedback coefficients. Denoting the latter at time $kT$ by the column vector $\mathbf{b}_k$, the samples stored in the forward equalizer delay line by the vector $\mathbf{u}_k$, and the past $N_b$ decisions by the vector $\mathbf{x}_k$, the equalizer output is given by

$$z_k = \mathbf{c}_k^T \mathbf{u}_k - \mathbf{b}_k^T \mathbf{x}_k$$

Minimizing the MSE with respect to the feedback coefficients leads to

$$\mathbf{b}_k = Q\mathbf{c}_k \tag{12.36}$$

where $Q$ is an $N_b \times N$ matrix with elements given by (12.16). Using (12.36) and proceeding as in Section 12.2.3, it can be shown that the set of optimum forward coefficients is given by

$$\mathbf{c}_{\text{opt}} = \hat{A}^{-1}\boldsymbol{\alpha} \tag{12.37}$$

The $N \times N$ matrix $\hat{A}$ has elements $\hat{a}_{i,j}$ similar to $a_{i,j}$ given by (12.12), except that the summation over $k$ now excludes the set $1 \leq k \leq N_b$ (which is in the span of the feedback filter). Finally,

$$\mathbf{b}_{\text{opt}} = Q\mathbf{c}_{\text{opt}}$$

*Type 2:* In this case, the forward equalizer coefficients can be obtained from (12.11) independently of the predictor coefficients. Let $\hat{\mathbf{e}}_k$ be an $N_b$-element vector at time $kT$ consisting of the $N_b$ most recent error signals before prediction at instants $k, k - 1, \ldots, k - N_b - 1$, with the forward equalizer coefficients at their optimal values. Then the set of optimum predictor coefficients $b_k$ is the solution of the normal equations:

$$E[\hat{\mathbf{e}}_{k-1}^*\hat{\mathbf{e}}_{k-1}^T]\mathbf{b}_k = E[\hat{\mathbf{e}}_k^*\hat{\mathbf{e}}_k] \tag{12.38}$$

*Type 3:* This case is similar to Type 1, except that the forward filter structure described in Section 12.2.3 is used. The set of optimum forward coefficients can be obtained using

$$\mathbf{c}_{\text{opt}} = \hat{A}^{-1}\boldsymbol{\alpha}$$

where the matrix $\hat{A}$ has elements $a_{i,j}$ similar to $\hat{a}_{i,j}$ given by (12.26), except that the summation over $k$ now excludes the set $1 \leq K \leq N_b$. The optimum feedback coefficients are given by

$$\mathbf{b}_{\text{opt}} = H\mathbf{c}_{\text{opt}}$$

where $H$ is an $N_b \times N$ matrix with elements $h_{i,j} = h(iT - jKT/M)$.

*Type 4:* This type of DFE is similar to Type 2. The optimum forward coefficients given by (12.25) still apply and the predictor coefficients can be obtained by solving a set of normal equations similar to that given in (12.38).

For a direct comparison of the performance of the four types of DFE structures, resort must be made to numerical solution for particular channel characteristics, number of equalizer coefficients, and so on. One general comment that can be made is that for an equal number of forward and feedback coefficients, the DFE structures of Types 1 and 3 will always achieve an output MSE at least as low or lower than the MSE achieved by structures of Types 2 and 4, respectively. This is true because unlike Type 1 and 3 structures, independent solution of the forward equalizer and the feedback predictor coefficients in Type 2 and 4 DFE structures does not in general guarantee joint minimization of the final MSE.

One important factor in the practical performance of DFE structures of all types is the effect of error propagation on the final error probability. Assuming correct decisions, the improvement in output signal-to-MSE ratio provided by a DFE reduces the probability of occurrence of the first error. However, once an error occurs it tends to propagate due to incorrect decision feedback. Bounds on the error multiplication factor have been developed by Duttweiler and others [Duttweiler, 12.12] and by Belfiore and Park [Belfiore, 12.4].

**Decision-Aided ISI Cancellation**

The concept of decision feedback of past data symbols to cancel ISI can theoretically be extended to include future data symbols. If all past and future data symbols were assumed to be known at the receiver, then given a perfect model of the ISI process, all ISI could be canceled exactly without any noise enhancement. Such a hypothetical receiver could, therefore, achieve the zero-ISI matched-filter bound on performance. In practice, the concept of decision-aided ISI cancellation can be implemented by using tentative decisions and some finite delay at the receiver. In the absence of tentative decision errors, the ISI due to these finite number of future data symbols (as well as past data symbols) can be canceled exactly before final receiver decisions are made. Such a receiver structure with a two-step decision process was proposed by Proakis [Proakis, 12.85]. Recently Gersho and and Lim [Gersho, 12.34] observed that the MSE performance of this receiver structure could be improved considerably by an adaptive matched filter in the path of the received signal prior to ISI cancellation. In fact the zero-ISI matched-filter bound on MSE could be achieved in the limit, assuming correct tentative decisions. A general theoretical treatment is available in [Mueller, 12.80].

Consider the block diagram of Fig. 12.23, where correct data symbols are assumed to be known to the canceler. Let the forward filter have $N$ coefficients fractionally spaced at $KT/M$-second intervals. Then the final output of the structure is given by

$$z_k = \mathbf{c}_k^T \mathbf{y}_k - \mathbf{b}_k^T \mathbf{x}_k$$

where $\mathbf{c}_k$ and $\mathbf{y}_k$ are the forward filter coefficient and input vectors, respectively, with $N$ elements as defined earlier in Section 12.2.3, $\mathbf{b}_k$ is the vector of the canceler

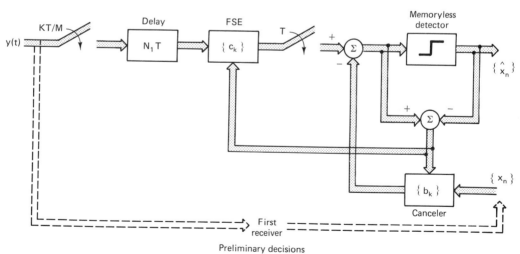

**Figure 12.23**  Decision-aided ISI cancellation.

Sec. 12.2    Receiver Structures and Their Steady-State Properties          **679**

coefficients spaced $T$ seconds apart, and $\mathbf{x}_k$ is the vector of data symbols. Each of these vectors is of length $N_1 + N_2$. The data vector $\mathbf{x}_k$ has elements $x_{k+N_1}, \ldots,$ $x_{k+1}, x_{k-1}, \ldots, x_{k-N_2}$. Note that the current data symbol $x_k$ is omitted. The canceler has $N_1$ noncausal and $N_2$ causal coefficients $b_{-N_1}, \ldots, b_{-1}, b_1, \ldots,$ $b_{N_2}$. To minimize the MSE $E[|z_k - x_k|^2]$, we can proceed as in Section 12.2.4 for the DFE. First, setting the derivative of the MSE with respect to $\mathbf{b}_k$ to zero we obtain

$$\mathbf{b}_k = H\mathbf{c}_k$$

where $\mathbf{H}$ is an $(N_1 + N_2) \times N$ matrix with elements

$$h_{i,j} = h\left(iT - \frac{jKT}{M}\right), \qquad i = -N_1, \ldots, 1, 1, \ldots, N_2$$

$$j = 0, 1, \ldots, N - 1$$

Using this result, it can be shown that the optimum set of forward filter coefficients is given by

$$\mathbf{c}_{\text{opt}} = \hat{A}^{-1}\boldsymbol{\alpha}$$

where the matrix $\hat{A}$ has elements $\hat{a}_{i,j}$ similar to $a_{i,j}$ given by (12.26), except that the summation over $k$ now excludes the set $k = -N_1, \ldots, 1, 1, \ldots, N_2$ (which is in the span of the canceler). The vector $\boldsymbol{\alpha}$ is the same as in (12.27). Finally, the optimum canceler coefficient vector $\mathbf{b}_{\text{opt}} = H\mathbf{c}_{\text{opt}}$ consists of $T$-spaced samples of the impulse response of the channel and the forward filter in cascade, omitting the reference sample.

The role of the forward filter in this structure may be better understood from the following point of view. Given an optimum (LMS) **desired impulse response** (DIR) for the canceler (with the reference sample forced to unity), the forward filter equalizes the channel response to this desired impulse response with least mean squared error. In fact both the DIR, modeled by the canceler coefficients, and the forward filter coefficients are being jointly optimized to minimize the final MSE. The same basic idea was used by Falconer and Magee [Falconer, 12.17] to create a truncated DIR channel for further processing by the Viterbi algorithm. In [Falconer, 12.17] a unit energy constraint was imposed on the DIR, whereas in the ISI canceler structure the reference sample of the DIR is constrained to be unity. Another difference between the two structures is the use of a forward filter with fractional tap spacing rather than a predetermined "matched" filter followed by a $T$-spaced equalizer. The use of a fractionally spaced forward filter for creating a truncated DIR was proposed in [Qureshi, 12.88], noting its ability to perform combined adaptive matched filtering and equalization.

As we allow the lengths of the forward filter and the canceler to grow without bound, the forward filter evolves into a matched filter with frequency response [Gersho, 12.34; Mueller, 12.80]:

$$C(f) = \frac{H^*(f)}{N_O + R_O}$$

The canceler models the $T$-spaced impulse response of the channel and forward filter in cascade, all except the reference sample. The frequency response of the canceler may be written as

$$B(f) = \frac{S_{hh}(f)}{N_O + R_O} - T \int_0^{1/T} \frac{S_{hh}(f)}{N_O + R_O} \, df$$

$$= \frac{S_{hh}(f) - R_O}{N_O + R_O}$$

Note that since all the quantities in the above expression for the canceler frequency response are real, the canceler coefficients must be Hermitian symmetric.

The output noise power spectrum is given by

$$N_o |C(f)|^2 = \frac{N_o S_{hh}(f)}{(N_O + R_O)^2}$$

and the ISI power spectrum may be written as

$$\left| 1 - \frac{S_{hh}(f)}{(N_O + R_O)^2} + B(f) \right|^2 = \frac{N_O^2}{(N_O + R_O)^2}$$

The output MSE, obtained by integrating the sum of the noise and ISI power spectra, is equal to the zero-ISI matched filter bound, that is,

$$\varepsilon_{\min}(mf) = \frac{N_O}{[N_O + R_O)}$$

The critical question regarding decision-aided ISI cancellation is the effect of tentative decision errors on the final error probability of the receiver. Published results answering this question are not yet available. However, reduced MSE has also been reported [Biglieri, 12.5] when a version of the decision-aided receiver structure is used to cancel nonlinear ISI.

### Adaptive Filters for MLSE

Adaptive receive filtering prior to Viterbi detection is of interest from two points of view. First, for unknown and/or slowly time-varying channels, the receive filter must be adaptive in order to obtain the ultimate performance gain from maximum-likelihood sequence estimation. Secondly, the complexity of the MLSE becomes prohibitive for practical channels with a large number of ISI terms. Therefore, in a practical receiver, an adaptive receive filter may be used to limit the time spread of the channel as well as to track slow time variation in the channel characteristics [Falconer, 12.17; Qureshi, 12.89].

By a development similar to that given in Section 12.2.3, it can be shown that a fractionally spaced transversal filter can model the characteristics of the WMF proposed [Forney, 12.24]. However, the constraint on this filter to produce zero anticausal ISI makes it difficult to derive an algorithm for updating the filter coefficients in an adaptive receiver.

The general problem of adaptive receive filtering for MLSE may be approached as follows. We know from Section 12.2.3 that an LMS FSE produces the composite response of a matched filter and an LMS $T$-spaced equalizer, when the MSE is defined with respect to a unit pulse DIR. In general, an LMS FSE frequency response can always be viewed as the composite of a matched filter and an LMS $T$-spaced equalizer response:

$$\frac{H^*(f)G(f)}{N_o + S_{hh}(f)} \tag{12.39}$$

where $G(f)$ is the DIR frequency response with respect to which the MSE is minimized. After symbol-rate sampling, the signal spectrum at the output of the FSE is given by $S_{hh}(f)G(f)/[N_o + S_{hh}(f)]$, $|f| \leq 1/2T$, and the output noise power spectrum may be written as

$$\frac{N_o S_{hh}(f)|G(f)|^2}{[N_o + S_{hh}(f)]^2}, \qquad |f| \leq \frac{1}{2T}$$

The selection of the DIR is therefore the crux of the problem. Fredricsson [Fredricsson, 12.30] has shown that from an effective MSE point of view, best performance is obtained when the DIR is selected such that its power spectrum is

$$|G(f)|^2 = \frac{[N_o + S_{hh}(f)]}{(R_O + N_O)}, \qquad |f| \leq \frac{1}{2T} \tag{12.40}$$

With this optimum DIR and the receive filter selected according to (12.39), the residual ISI power spectrum at the input to the Viterbi algorithm is given by

$$\left| G(f) - \frac{S_{hh}(f)G(f)}{N_o + S_{hh}(f)} \right|^2 = \frac{N_o}{[N_o + S_{hh}(f)](R_o + N_o)}$$

Summing the residual ISI and noise power spectra, we obtain the power spectrum of the combined error sequence at the input to the Viterbi algorithm. The error sequence is found to be white with mean-squared value $N_o/(R_o + N_o)$, which is equal to the matched-filter bound. Note that the Gaussian noise component of the error sequence is approximately white at a moderately high SNR. However, the total error sequence while white is not Gaussian due to residual ISI.

Let us select the DIR such that $G(f)$ is the causal factor of the power spectrum given in (12.40). Then the resulting FSE (12.39) may be recognized as the optimum forward filter of an infinite-length conventional LMS DFE. At moderately high SNR this FSE approaches the WMF. An MLSE receiver using such a receive filter is shown in Fig. 12.24. This receiver can easily be made adaptive by updating both the FSE and the feedback filter coefficients to minimize jointly the mean-squared value of the error sequence. The feedback filter coefficients also provide all but the first DIR coefficient for use by the Viterbi algorithm. The first DIR coefficient is assumed to be unity.

Fredricsson [Fredricsson, 12.30] points out the difficulty of obtaining a general explicit solution for the optimum truncated DIR of a specified finite length. However, when the DIR is limited to two or three terms [Falconer, 12.17; Fredricsson,

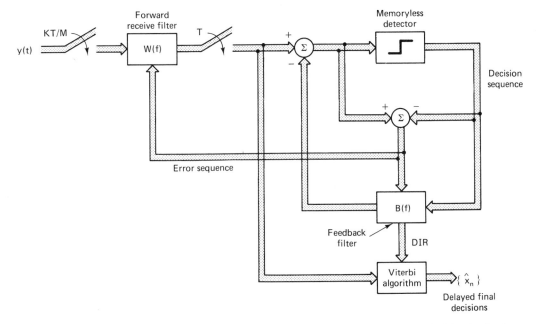

**Figure 12.24** Adaptive MLSE receiver with causal DIR developed via decision-feedback equalization.

12.30], it invariably takes the form of one or the other familiar class of partial-response systems [Kabal, 12.48] with a null at one or both band edges. As mentioned in Section 12.2.4, the Viterbi algorithm is able to recover most of the nearly 3-dB loss which otherwise results from use of bandwidth efficient partial response systems (Chapter 8).

Several methods of jointly optimizing the fractionally spaced receive filter and the truncated DIR are available that minimize the MSE at the input to the Viterbi algorithm. These methods differ in the form of constraint [Falconer, 12.17; Messerschmitt, 12.68] on the DIR which is necessary in this optimization process to exclude the selection of the null DIR corresponding to no transmission through the channel. The general form of such a receiver is shown in Fig. 12.25.

One such constraint is to restrict the DIR to be causal and to restrict the first coefficient of the DIR to be unity. In this case the delay (LT) in Fig. 12.25 is equal to the delay through the Viterbi algorithm and the first coefficient of $\{b_k\}$ is constrained to be unity.

If the causality constraint is removed (as in [Gersho, 12.34] for decision-aided ISI cancellation) but if the reference (or center) coefficient of the DIR is constrained to unity, another form of the receiver, shown in Fig. 12.25, is obtained [Ungerboeck, 12.103]. In this case the delay LT is equal to the delay through the Viterbi algorithm plus $N_1 T$, where the DIR has $2N_1 + 1$ coefficients with its center coefficient constrained to be unity. As before, this structure can easily be made adaptive by updating the FSE and DIR coefficients in a direction opposite to the gradient of the squared error with respect to each coefficient.

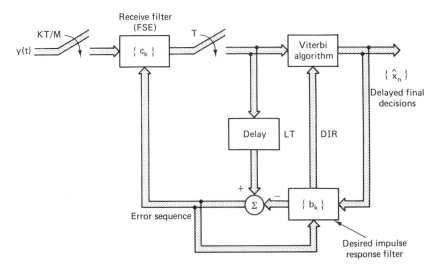

**Figure 12.25** General form of adaptive MLSE receiver with finite-length DIR.

The least restrictive constraint on the DIR is the unit energy constraint proposed by Falconer and Magee [Falconer, 12.17]. This leads to yet another form of the receiver structure shown in Fig. 12.25. However, the adaptation algorithm for updating the desired impulse response coefficients $\{b_k\}$ is considerably more complicated (see [Falconer, 12.17]). Note that the fixed predetermined WMF and $T$-spaced prefilter combination of [Falconer, 12.17] has been replaced in Fig. 12.25 by a general fractionally spaced adaptive filter.

A common characteristic of the just-mentioned truncated DIR suboptimum MLSE receiver structures is that the sample sequence at the input to the Viterbi algorithm contains residual ISI (with respect to the DIR), and that the noise sequence is not white. Bounds on the performance of the Viterbi algorithm in the presence of correlated noise and residual ISI are developed in [Qureshi, 12.89]. In practice, as long as the DIR power spectrum more or less matches the nulls or high attenuation regions of the channel folded-power spectrum, a reasonable length FSE can manipulate the channel response to the truncated DIR without significant noise enhancement or residual ISI. However, the degree to which the noise sequence at the VA input is uncorrelated depends on the constraints imposed on the DIR. As mentioned earlier, as the causal DIR length is allowed to grow the forward filter in the structure of Fig. 12.25 approaches the WMF at moderately high SNR, resulting in uncorrelated noise at the VA input. On the other hand, if the DIR is noncausal, the receive filter approaches a matched filter as the DIR length approaches the length of the original channel impulse response, resulting in the noise to be colored according to the folded-power spectrum of the received signal. For such receiver structures, the modified Viterbi algorithm [Ungerboeck, 12.103] which takes the noise correlation into account, is more appropriate.

### 12.2.5 Timing Phase Sensitivity

As noted in earlier sections, conventional suboptimum receiver structures based on $T$-spaced equalizers suffer from extreme sensitivity to sampler timing phase. The inherent insensitivity of the performance of fractionally spaced equalizers to timing phase was heuristically explained in Section 12.1.8. Here we present an overview of the influence of timing phase on the performance of various receiver structures.

Let us reconsider the conventional linear receiver of Section 12.2.3 in the presence of a timing phase offset $t_O$. Assume that $H(f) = 0$, $|f| > 1/T$, that is, the channel has at most 100% excess bandwidth. The power spectrum, $N_o S_{hh}(f)$, of the noise sequence $\{n_k\}$ is independent of the timing phase $t_O$.

The sampled signal spectrum, however, is a function of $t_O$ according to

$$S_{hh}(f, t_O) = \exp(-j2\pi f t_O)\left[|H(f)|^2 + \left|H\left(f - \frac{1}{T}\right)\right|^2 \exp\left(j2\pi \frac{t_o}{T}\right)\right], \qquad 0 \le f \le \frac{1}{T}$$

Note that when $t_o = 0$, the alias $|H(f - 1/T)|^2$ adds constructively to $|H(f)|^2$, whereas for $t_o = T/2$, destructive aliasing takes place in the foldover region around $1/2T$ hertz. In particular, if the channel power spectrum is the same at the two band edges, a null is created in the sampled signal spectrum at $f = 1/2T$ hertz when $t_O = T/2$.

The MSE corresponding to (12.3) may be written as a function of timing phase:

$$\varepsilon(t_o) = T \int_0^{1/T} |1 - S_{hh}(f, t_o)C(f)|^2 + N_o S_{hh}(f)|C(f)|^2 \, df$$

where $C(f)$ is the periodic frequency response of the $T$-shaped equalizer that follows the sampler. Proceeding to optimize $C(f)$, as in Section 12.2.2, the MMSE may be derived as a function of $t_O$:

$$\varepsilon_{\min}(t_O) = T \int_0^{1/T} \frac{N_O}{N_O + S_{hh}^2(f, t_o)/S_{hh}(f)} \, df \qquad (12.41)$$

Note that if $S_{hh}^2(f, t_o)/S_{hh}(f)$ is small in the foldover region due to poor choice of $t_O$, the integrand becomes relatively large in that range of frequencies. This leads to a larger $\varepsilon_{\min}$. Clearly, timing phase is unimportant when there is no excess bandwidth and therefore no aliasing.

The above development shows that for systems with excess bandwidth, the performance of the conventional MMSE linear receiver with a fixed matched filter is sensitive to choice of timing phase due to the inability of the $T$-spaced equalizer to invert a "null" in the sampled signal spectrum without excessive noise enhancement. This sensitivity can be avoided if the matched filter spectrum $H^*(f)$ is adjusted by a linear phase factor, $\exp(j2\pi f t_O)$, to compensate explicitly for the timing offset $t_O$. An infinite-length fractionally spaced equalizer obtains its insensitivity to timing phase in this way.

For instance, given a timing offset $t_O$, an infinite length $T/2$-equalizer synthesizes the frequency response

$$W(f) = \frac{H^*(f)\exp(j2\pi f t_o)}{N_o + S_{hh}(f)}, \qquad \frac{|f| \le 1}{T}$$

The signal spectrum at the equalizer input is the DFT of the $T/2$-sampled channel impulse response $\{h(kT/2 + t_o)\}$, that is, $H(f, t_o) = H(f)\exp(-j2\pi f t_o)$, $|f| \le 1/T$. Aliasing, due to symbol-rate sampling, takes place at the equalizer output after $W(f)$ has explicitly compensated for the timing offset. This results in the equalized signal spectrum

$$H(f, t_o)W(f) + H(f - 1/T, t_o)W(f - 1/T) = \frac{S_{hh}(f)}{N_o + S_{hh}(f)}, \qquad \frac{0 \le f \le 1}{T}$$

Thus the residual ISI, noise, and MSE at the $T/2$-equalizer output are all independent of the timing phase.

In a practical conventional receiver structure, the matched filter is typically replaced by a filter matched to the transmitted pulse or the received pulse over the average channel. Mazo [Mazo, 12.64] has shown that under some assumptions, the best timing phase for an infinite-length $T$-spaced equalizer is one that maximizes the energy in the band-edge, $f = 1/2T$ hertz, component in the symbol-rate sampled signal at the equalizer input. However, on channels with severe delay distortion and moderately large excess bandwidth, for example, 50%, even the timing phase that maximizes band-edge energy can lead to near nulls elsewhere in the folded spectrum. On such a channel, a $T$-spaced equalizer performs poorly regardless of the choice of timing phase, as explained in Section 12.1.8 (Fig. 12.14). Conditions for amplitude depressions to occur in the folded spectrum are given in [Qureshi, 12.91].

By contrast, a finite-length FSE maintains its ability to compensate for a timing offset in such a way as to equalize with the minimum of noise enhancement.

These results can be extended to decision-feedback receiver structures. Consider a conventional DFE consisting of a matched filter, a symbol-rate sampler, and infinite-length $T$-spaced forward and feedback filters. The MSE of such a receiver can be derived as a function of the timing phase (using the method given in [Salz, 12.94]):

$$\varepsilon_{\min}(t_O) = \exp\left\{ -T\int_0^{1/T} \ln\left[\frac{1 + S_{hh}^2(f, t_o)}{N_o S_{hh}(f)}\right] df\right\} \qquad (12.42)$$

As $t_O$ is varied, the greatest deviation in $\varepsilon_{\min}(t_O)$ occurs for a channel such that

$$|H(f)|^2 = \begin{cases} 1 & |f| \le \dfrac{1 - \alpha}{2T} \\ 0.5 & \dfrac{1 - \alpha}{2T} \le |f| \le \dfrac{1 + \alpha}{2T} \end{cases}$$

where $\alpha$ is the roll-off factor. For this channel

$$S_{hh}(f, t_O)\big|_{t_o = 0} = S_{hh}(f) = 1, \qquad 0 \le f \le \frac{1}{T}$$

and

$$S_{hh}(f, t_O)\big|_{t_o = T/2} = \begin{cases} 1 & |f| \le \dfrac{1 - \alpha}{2T} \\ 0 & \dfrac{1 - \alpha}{2T} \le f \le \dfrac{1 + \alpha}{2T} \end{cases}$$

Using this result in (12.42), we obtain for the DFE

$$\varepsilon_{\min}(t_O)\big|_{t_o = 0} = \frac{1}{1 + 1/N_O}$$

and

$$\varepsilon_{\min}(t_O)\big|_{t_o = T/2} = \frac{1}{1 + 1/N_O} (1 - \alpha)$$

$N_o = 0.01$ (20-dB SNR) and $\alpha = 0.1$, the MSE of the decision-feedback equalizer degrades by 2 dB when $t_O$ is varied from the best choice $t_O = 0$ to the worse choice $t_O = T/2$. The corresponding result for the conventional linear receiver is obtained by using (12.41):

$$\varepsilon_{\min}(t_O)\big|_{t_o = 0} = \frac{N_o}{1 + N_o}$$

and

$$\varepsilon_{\min}(t_O)\big|_{t_o = T/2} = \alpha + \frac{(1 - \alpha)N_o}{1 + N_o}$$

For $N_O = 0.01$ and $\alpha = 0.1$, the MSE of the linear receiver degrades by 10.4 dB as $t_O$ is varied from 0 to $T/2$.

This example illustrates the ability of the conventional DFE to compensate for the spectral null created by poor choice of timing phase with much less noise enhancement than a conventional linear receiver. However, this relative insensitivity may not translate into actual performance insensitivity from an error probability point of view due to error propagation in the DFE. In order to cancel the ISI due to a spectral null created by a bad timing phase, the feedback filter must develop a relatively long impulse response with large-magnitude coefficients. Such a DFE is likely to suffer from severe error propagation.

The performance of a DFE with a fractionally spaced forward filter is, of course, insensitive to timing phase by virtue of explicit compensation of the timing offset by the forward filter before symbol-rate sampling. The same comment applies to a fractionally spaced receive filter used in conjunction with a maximum-likelihood sequence estimator.

## 12.3 LMS ADAPTATION

In this section we expand upon the topics briefly introduced in Sections 12.1.3 through 12.1.5, that is, LMS adaptation algorithms, their convergence properties and excess MSE. The effect of finite precision in digital implementations is also discussed. The results of this section are applicable to other forms of adaptive filters, for example, an echo canceler (see also Chapter 4), with appropriate reinterpretation of terms.

The deterministic gradient algorithm, which is of little practical interest, is presented first to set the stage for a discussion of the LMS or stochastic gradient algorithm.

### 12.3.1 Deterministic Gradient Algorithm

When the equalizer input covariance matrix $A$ and the crosscorrelation vector $\boldsymbol{\alpha}$ (see Section 12.2.3) are known, one can write the MSE as a function of $A$, $\boldsymbol{\alpha}$, and the equalizer coefficient vector $\mathbf{c}_k$ according to

$$\varepsilon_k = \mathbf{c}_k^{*T} A \mathbf{c}_k - 2 \, Re \left[ \mathbf{c}_k^{*T} \boldsymbol{\alpha} \right] + 1 \tag{12.43}$$

Taking the gradient of the MSE with respect to $\mathbf{c}_k$, a deterministic (or exact) gradient algorithm for adjusting $\mathbf{c}_k$ to minimize $\varepsilon_k$ can be written as

$$\mathbf{c}_{k+1} = (I - \Delta A) \, \mathbf{c}_k + \Delta \boldsymbol{\alpha} \tag{12.44}$$

where $\Delta$ is the step-size parameter. (This update procedure is also known as the *steepest descent algorithm*). Using the fact that $\boldsymbol{\alpha} = A \, \mathbf{c}_{\mathrm{opt}}$, from (12.11) or (12.25), and subtracting $\mathbf{c}_{\mathrm{opt}}$ both sides of (12.44), we obtain

$$\mathbf{c}_{k+1} - \mathbf{c}_{opt} = (I - \Delta A) (\mathbf{c}_k - \mathbf{c}_{opt}) \tag{12.45}$$

In order to analyze the stability and convergence of the deterministic gradient algorithm, we use coordinate transformation to diagonalize the set of equations (12.45) so that

$$\mathbf{d}_{k+1} = (I - \Delta \Lambda) \, \mathbf{d}_k \tag{12.46}$$

where the transformed coefficient deviation vector is defined in (12.30).

Since $\Lambda$ is a diagonal matrix, it is clear from (12.46) that the $i$th element of $\mathbf{d}_k$ decays geometrically according to

$$d_{ik} = (1 - \delta \lambda_i)^k \, d_{io}, \qquad i = 0, 1, \ldots, N - 1, \tag{12.47}$$

where $\lambda_i$ is the $i$th eigenvalue of $A$ and $d_{io}$ is the initial value of the $i$th transformed tap gain deviation. Recall from Section 12.2.3 that given $\mathbf{d}_k$, the MSE at step $k$

may be written as the sum of $\varepsilon_{\min}$ and the excess MSE:

$$\varepsilon_k = \varepsilon_{\min} + \mathbf{d}_k^{*T} \Lambda \, \mathbf{d}_k$$

$$= \varepsilon_{\min} + \sum_{i=0}^{N-1} \lambda_i \, (1 - \Delta\lambda_i)^{2k} |d_{io}|^2 \qquad (12.48)$$

Given finite initial deviations, the deterministic algorithm is stable and the MSE converges to $\varepsilon_{\min}$ provided

$$0 < \Delta < \frac{2}{\lambda_{\max}} \qquad (12.49)$$

where $\lambda_{\max}$ is the maximum eigenvalue of $A$. If all the eigenvalues of $A$ are equal to $\lambda$ and $\Delta$ is selected to be $1/\lambda$, the excess MSE will be reduced to zero in one adjustment step of the deterministic gradient algorithm (12.44). For a $T$-spaced equalizer, this condition corresponds to a flat folded-power spectrum at the equalizer input.

When the eigenvalues of $A$ have a large spread, that is, the ratio $p = \lambda_{\max}/\lambda_{\min}$ is large, no single value of the step size $\Delta$ leads to fast convergence of all the tap gain deviation components. When $\Delta = 2/(\lambda_{\max} + \lambda_{\min})$, the two extreme tap gain deviation components converge at the same rate according to $[(p - 1)/(p + 1)]^k$ [Gersho, 12.33]. All other components converge at a faster rate with a time constant at most as large as $(p + 1)/2$ iterations. The impact of the large eigenvalue spread on the convergence of the excess MSE is somewhat less severe because tap gain deviations corresponding to the small eigenvalues contribute less to the excess MSE (see (12.48)).

### 12.3.2 LMS Gradient Algorithm

In practice, the channel characteristics are not known beforehand. Therefore, the gradient of the MSE cannot be determined exactly and must be estimated from the noisy received signal. The LMS gradient [Widrow, 12.109] is obtained from the deterministic gradient algorithm (12.44) by replacing the gradient

$$2(AC_k - \alpha) = 2 \, E \, [\mathbf{y}_k^* \, (\mathbf{y}_k^T \, \mathbf{c}_k - x_k)]$$

by its unbiased but noisy estimate $\mathbf{y}_k^* \, (\mathbf{y}_k^T \, \mathbf{c}_k - x_k)$. The equalizer coefficients are adjusted once in every symbol interval according to

$$\mathbf{c}_{k+1} = \mathbf{c}_k - \Delta \, \mathbf{y}_k^* \, (\mathbf{y}_k^T \, \mathbf{c}_k - x_k) = \mathbf{c}_k - \Delta \, \mathbf{y}_k^* e_k \qquad (12.50)$$

where $\mathbf{y}_k$ is the equalizer input vector, $x_k$ is the received data symbol, and $e_k$ is the error in the equalizer output.

Subtracting $\mathbf{c}_{\mathrm{opt}}$ from both sides of (12.50) allows us to write

$$(\mathbf{c}_{k+1} - \mathbf{c}_{\mathrm{opt}}) = (I - \Delta \, \mathbf{y}_k^* \, \mathbf{y}_k^T) \, (\mathbf{c}_k - \mathbf{c}_{\mathrm{opt}}) - \Delta \, \mathbf{y}_k^* \, e_{k\mathbf{opt}}$$

where $e_{k\mathrm{opt}}$ is the instantaneous error if the optimum coefficients were used. The transformed coefficient deviation vector $\mathbf{d}_k$ is now a random quantity. Neglecting the dependence of $\mathbf{c}_k$ on $\mathbf{y}_k$, we see that the mean of $\mathbf{d}_k$ follows the recursive relationship (12.46), that is,

$$E\left[\mathbf{d}_{k+1}\right] = (\mathbf{I} - \Delta\Lambda)\, E\left[\mathbf{d}_k\right] \qquad (12.51)$$

The ensemble average of the MSE evolves according to

$$\varepsilon_k = \varepsilon_{\min} + E\left(\mathbf{d}_k^{*T} \Lambda\, \mathbf{d}_k\right]$$

where the second term on the right is the average excess MSE,

$$\varepsilon_{\Delta k} = \sum_{i=0}^{N-1} \lambda_i E\left[|d_{ik}|^2\right]$$

This quantity is difficult to evaluate exactly in terms of the channel and equalizer parameters. Using the assumption that the equalizer input vectors $y_k$ are statistically independent [Gitlin, 12.38; Mazo, 12.65; Proakis, 12.84; Ungerboeck, 12.102], the following approximate recursive relationship can be derived:

$$E\left[\mathbf{d}_{k+1}|^2\right] = ME\left[|\mathbf{d}_k|^2\right] + \Delta^2 \varepsilon_{\min}\lambda \qquad (12.52)$$

where $\lambda$ is the vector of eigenvalues of $\Lambda$ and the $N \times N$ matrix $M$ has elements

$$M_{ij} = (1 - 2\,\Delta\lambda_i)\,\delta_{ij} + \Delta^2\lambda_i\lambda_j$$

Let $\rho$ be the ratio of the maximum to the effective average eigenvalue of $A$.

Three important results can be derived [Gitlin, 12.38] using this line of analysis [Ungerboeck, 12.102] and eigenvalue bounds [Gitlin, 12.38]:

1. The LMS algorithm is stable if the step size $\Delta$ is in the range

$$0 < \Delta < \frac{2}{(N\rho\bar{\lambda})} \qquad (12.53)$$

where $\bar{\lambda}$ is the average eigenvalue of $A$ and is equal to the average signal power at the equalizer input, defined according to $E\left[\mathbf{y}_k^T \mathbf{y}_k^*\right]/N$. When $\Delta$ satisfies (12.53), all eigenvalues of the matrix $M$ in (12.52) are less than one in magnitude, permitting mean-squared coefficient deviations in (12.52) to converge.

2. The excess MSE follows the recursive relationship

$$\varepsilon_{\Delta k+1} = \left[1 - 2\Delta\bar{\lambda} + \Delta^2 N\rho(\bar{\lambda})^2\right]\varepsilon_{\Delta k} + \Delta^2\varepsilon_{\min}N\rho(\bar{\lambda})^2 \qquad (12.54)$$

If $\Delta$ is selected to minimize the excess MSE at each iteration, we would obtain

$$\Delta_k = \frac{\varepsilon_{\Delta k}}{(\varepsilon_{\min} + \varepsilon_{\Delta k})\,(N\rho\bar{\lambda})}$$

Initially $\varepsilon_{\Delta k} \gg \varepsilon_{\min}$, so that fastest convergence is obtained with an initial step size

$$\Delta_O = \frac{1}{(N\rho\bar{\lambda})} \qquad (12.55)$$

Note that $\Delta_O$ is half as large as the maximum permissible step size for stable operation.

3. The steady-state excess MSE can be determined from (12.54) to be

$$\varepsilon_\Delta = \frac{\Delta \varepsilon_{min} \, N\rho\bar{\lambda}}{(2 - \Delta N\rho\bar{\lambda})} \tag{12.56}$$

A value of $\Delta = \Delta_O$ results in $\varepsilon_\Delta = \varepsilon_{min}$, that is, the final MSE is 3 dB greater than the minimum achievable MSE. In order to reduce $\varepsilon_\Delta$ to $\gamma\varepsilon_{min}$, where $\gamma$ is a small fraction, $\Delta$ must be reduced to

$$\Delta = \frac{2\gamma}{(1 + \gamma)N\rho\bar{\lambda}} = \frac{2\Delta_0\gamma}{1 + \gamma} \tag{12.57}$$

For instance, a reduction of $\Delta$ to $0.1\Delta_0$ results in a steady-state MSE which is about 0.2 dB greater than the minimum achievable MSE.

Note that the impact of a distorted channel, with an eigenvalue ratio $\rho > 1$, on the excess MSE, its rate of convergence, and the choice of $\Delta$, is the same as if the number of equalizer coefficients N was increased to $N\rho$. For a $T$-spaced equalizer, some of the eigenvalues (and hence $\rho$) depend on the timing phase and the channel envelope delay characteristics in the band-edge regions.

The previous results apply equally to $T$-spaced and fractionally spaced equalizers, except that the recursion for the excess MSE (12.54) for a $kT/M$ spaced equalizer is given by

$$\varepsilon_{\Delta K+1} = \left[1 - 2\Delta\left(\frac{M}{K}\right)\bar{\lambda} + \Delta^2 N\left(\frac{M}{K}\right)P(\bar{\lambda})^2\right] + \Delta^2 \, \varepsilon_{min} \, N\left(\frac{M}{K}\right)P(\bar{\lambda})^2$$

As mentioned earlier, only about $K/M$ of the eigenvalues of the correlation matrix $A$ are significant for a $KT/M$ FSE. For the same average signal power at the equalizer input, equal to the average eigenvalue $\bar{\lambda}$, the significant eigenvalues are generally $M/K$ times larger for an $N$-coefficient $KT/M$ FSE compared with the eigenvalues for an $N$-coefficient $T$-spaced equalizer. Thus the eigenvalue ratio $\rho$ for an FSE should be computed only over the significant eigenvalues of $A$. Note that $A$ and $\rho$ for an FSE are independent of timing phase and channel envelope delay characteristics, and $\rho = 1$ when the unequalized amplitude shape is the square root of Nyquist.

For well-behaved channels ($\rho$ approximately 1), $\varepsilon_{\Delta k}$ for an $N$-coefficient $KT/M$ FSE using the best initial step size given by (12.55) initially converges faster as $(1 - M/KN)^k$ compared with $(1 - 1/N)^k$ for an $N$-coefficient $T$-equalizer using the same best initial step size. Conversely, a $MN/K$-coefficient FSE, using a step size $K/M$ times as large, generally exhibits the same behavior with respect to the convergence and steady-state value of excess MSE as an N-coefficient $T$-spaced equalizer.

As a rule of thumb, the symbol-by-symbol LMS gradient algorithm with the best initial $\Delta$ leads to a reduction of about 20 dB in MSE for well-behaved channels ($\rho$ approximately 1) in about five times the time span $T_{eq}$ of the equalizer. At this time it is desirable to reduce $\Delta$ by a factor of 2 for the next $5T_{eq}$ seconds to permit finer tuning of the equalizer coefficient. Further reduction in excess MSE can be obtained by reducing $\Delta$ to its steady-state value according to (12.57). (For distorted channels, an effective time span of $\rho T_{eq}$ should be substituted in the above discussion.)

### 12.3.3 Digital Precision Considerations

This discussion may suggest that it is desirable to continue to reduce $\Delta$ in order to reduce the excess MSE to zero in the steady state. However, this is not advisable in a practical limited precision digital implementation of the adaptive equalizer. Observe [Gitlin, 12.36] that as $\Delta$ is reduced, the coefficient correction terms in (12.50), on the average, become smaller than half the least-significant bit of the coefficients; adaptation stalls and the MSE levels off. If $\Delta$ is reduced further, the MSE increases if the channel characteristics change at all or if some adjustments made at peak errors are large enough to perturb the equalizer coefficients.

The MSE that can be attained by a digital equalizer of a certain precision can be approximated as follows [Gitlin, 12.38]. Let the equalizer coefficients be represented by a uniformly quantized number of $B$ bits (including sign) in the range $(-1, 1)$. Then the real and imaginary parts of the equalizer coefficients will continue to adapt, on the average, so long as

$$\Delta\left(\frac{\varepsilon\overline{\lambda}}{2}\right)^{1/2} \geq 2^{-B} \tag{12.58}$$

It is desirable to select a compromise value of $\Delta$ such that the total MSE $\varepsilon = \varepsilon_{min} + \varepsilon_\Delta$, with $\varepsilon_\Delta$ predicted by infinite-precision (analog) analysis (12.56), is equal to the lower limit on $\varepsilon$ determined by digital precision (12.58). The required precision can be estimated by substituting $\Delta$ from (12.57) into (12.58):

$$2^B \geq \left[\frac{N\rho(1 + \gamma)}{\sqrt{2}\gamma}\right]\left(\frac{\overline{\lambda}}{\varepsilon}\right)^{1/2} \tag{12.59}$$

where $(\overline{\lambda}/\varepsilon)$ can be recognized as the desired equalizer input-power-to-output-MSE ratio.

As an example, consider a 32-tap $T$-spaced equalizer ($N = 32$) for a well-behaved channel ($\rho = 1$). Select $\gamma = 0.25$ (corresponding to a 1-dB increase in output MSE over $\varepsilon_{min}$) and a desired equalizer output-signal-power-to-MSE ratio of 24 dB (adequate for 9.6-kb/s transmission). Let the input-to-output power scale factor for this equalizer be 2, so that $(\overline{\lambda}/\varepsilon)^{1/2} = 10^{27/20}$. Solving (12.59) for $B$, we find that 12-bit precision is required.

The required coefficient precision increases by 1 bit for each doubling of the number of coefficients and for each 6-dB reduction in desired output MSE. However, for a given $\varepsilon_{min}$, each 6-dB reduction in excess MSE $\varepsilon_\Delta$ requires a 2-bit increase in the required coefficient precision. This becomes the limiting factor in some adaptive filters, such as an echo canceler, which must track slow variations in system parameters in the presence of a large uncorrelated interfering signal [Weinstein, 12.108].

Note that the precision requirement imposed by the LMS gradient adaptation algorithm is significantly more stringent than a precision estimate based on quantization noise due to roundoff in computing the sum of products for the equalizer output. If each product is rounded individually to $B$ bits and then summed, the variance of this roundoff noise is $N2^{-2B}/3$. Assuming an equalizer output signal power of $1/6$, the signal-to-output roundoff noise ratio is $2^{2B}/2N$, which is 54 dB for $B = 12$ and $N = 32$: 30 dB greater than the desired 24-dB signal-to-MSE ratio in our example.

Roundoff in the coefficient update process, which has been analyzed in [Caraiscos, 12.7], is another source of quantization noise in adaptive filters. This roundoff causes deviation of the coefficients from the values they take when infinite precision arithmetic is used. The MSE $\varepsilon_r$ contributed by coefficient roundoff [Caraiscos, 12.7] is approximated by $\varepsilon_r = N2^{-2B}/6\Delta$. Again using an output signal power of $\frac{1}{6}$, the signal-to-coefficient roundoff noise ratio is $\Delta 2^{2B}/N$. For our example, we obtain $\Delta = 0.0375$ from (12.59) using $\gamma = 0.25$, $N = 32$, $\rho = 1$, and $\bar{\lambda} = \frac{1}{3}$. The signal-to-coefficient roundoff noise ratio is 43 dB for $B = 12$, which is 19 dB greater than the desired 24-dB signal-to-MSE ratio, suggesting that the effect of coefficient roundoff is insignificant. Moreover, each time we reduce $\Delta$ by a factor of 2, the coefficient precision $B$ must be increased by 1 bit in order to prevent adaptation from stalling according to (12.58). This reduction in $\Delta$ and the corresponding increase in $B$ further reduces the MSE due to coefficient roundoff by 3 dB. Using the expression for $\varepsilon_r$ given earlier and $\Delta$ from (12.57), we can rewrite (12.58) as

$$\frac{\varepsilon}{\varepsilon_r} \geq \frac{6\rho(1 + \gamma)}{\gamma}$$

Since $\rho \geq 1$ and $\gamma$ is typically a small fraction, the MSE due to coefficient roundoff, $\varepsilon_r$, is always small compared to $\varepsilon$ provided coefficient precision is sufficient to allow adaptation to continue for small desired values of $\gamma$ and $\Delta$. Thus coefficient roundoff noise can be neglected in the process of estimating LMS adaptive filter precision requirements.

## 12.4 FAST-CONVERGING EQUALIZERS

The design of update algorithms to speed up the convergence of adaptive filters has been a topic of intense study for more than a decade. Rapid convergence is important for adaptive equalizers designed for use with channels, such as troposcatter and HF radio, whose characteristics are subject to time variations [Proakis, 12.87]. In voiceband telephone applications, reduction of the initial setup time of the equalizer is important in polling multipoint networks [Forney, 12.26] where the central site receiver must adapt to receive typically short bursts of data from a number of transmitters over different channels.

In this section we present an overview of three classes of techniques devised to speed up equalizer convergence.

### 12.4.1 Orthogonalized LMS Algorithms

Recall from Section 12.3.1 that for the deterministic gradient algorithm, no single value of the step size $\Delta$ leads to fast convergence of all the coefficient deviation components when the eigenvalues of the equalizer input covariance matrix $A$ have a large spread. Using the independence assumption, the same is true regarding the convergence of the mean of the coefficient deviations for the LMS gradient algorithm (see (12.51) in Section 12.3.2). The excess MSE is a sum of the mean-squared value of each coefficient deviation weighted by the corresponding eigenvalues of $A$. Slow decay of some of these mean-squared deviations, therefore, slows down the convergence of the excess MSE. Substituting the best initial $\Delta$ from (12.55) in (12.54), we obtain the recursion

$$\varepsilon_{\Delta k + 1} = \frac{(1 - 1/N_\rho)\varepsilon_{\Delta k} + \varepsilon_{\min}}{N\rho} \tag{12.62}$$

Observe that the initial decay of $\varepsilon_{\Delta k}$ is geometric with a time constant of approximately $N\rho$ symbol intervals. Thus for the same length equalizer, a severely distorted channel ($\rho = 2$) will cause the rate of convergence of the LMS gradient algorithm to be slower by a factor of two compared to that for a good channel. The inadequacy of the LMS gradient algorithm for fast start-up receivers becomes obvious if we consider a 9.6-kb/s, 2400-baud modem with an equalizer spanning 32-symbol intervals. For a severely distorted channel, more than 320 equalizer adjustments over a 133-ms interval would be required before data transmission could begin.

For partial-response systems [Kabal, 12.48; Kretzmer, 12.50], where a controlled amount of ISI is introduced to obtain a desired spectral shape, the equalizer convergence problem is fundamental. It can be shown that $\rho = 2$ for an ideal cosine-shaped spectrum at the equalizer input for a Class IV SSB partial-response [Chang, 12.8] or a Class I QAM partial-response [Qureshi, 12.90] system. Noting this slow convergence, Chang [Chang, 12.8] suggested the use of a prefixed weighting matrix to transform the input signals to the equalizer tap gains to be approximately orthonormal. All eigenvalues of the transformed equalizer input convariance matrix are then approximately equal resulting in faster equalizer convergence.

Another orthogonalized LMS updated algorithm [Mueller, 12.75; Qureshi, 12.88] is obtained by observing that the decay of the mean of all the transformed coefficient deviation components could be speeded up by using a diagonal matrix $\text{diag}(\Delta_i)$ instead of the scaler $\Delta$ in (12.51), such that each element $\Delta_i$ of this matrix is the inverse of the corresponding element $\lambda_i$ of $\Lambda$. Transforming $\text{diag}(\Delta_i)$ back to the original coordinate system, we obtain the orthogonalized LMS update algorithm

$$\mathbf{c}_{k+1} = \mathbf{c}_k - P\,\mathbf{y}_k^*\,e_k \tag{12.63}$$

A practical advantage of this algorithm (12.63) over Chang's structure is that the weighting matrix is in the path of the tap gain corrections rather than the received signal. The computation required, therefore, need not be carried out to as much accuracy.

As we shall see in Section 12.4.3, the fastest-converging algorithms are obtained when $P$ is continually adjusted to do the best job of orthogonalizing the tap gain corrections.

### 12.4.2 Periodic or Cyclic Equalization

As mentioned in Section 12.1.3, one of the most widely used [CCITT Rec. V.27, 12.118; CCITT Rec. V.29, 12.119] methods of training adaptive equalizers in high-speed voiceband modems is based on PN training sequences with periods significantly greater than the time span of the equalizer. Here we discuss the techniques [Milewski, 12.70; Mueller, 12.76; Qureshi, 12.90] that can be used to speed up equalizer convergence in the special case when the period of the training sequence is selected to be equal to the time span of the equalizer.

#### Periodic or Averaged Update

Consider a training sequence $\{x_k\}$ with period $NT$ for a $T$-spaced equalizer with $N$ coefficients. Let the equalizer coefficients be adjusted periodically, every $N$ symbol intervals, according to the following LMS algorithm with averaging

$$c_n(k + 1) = c_n(k) - \Delta \sum_{j=0}^{N-1} e_{kN+j} y^*(kNT + jT - nT), \qquad n = 0, 1, \ldots, N - 1$$

In general, the transformed coefficient deviation vector after the $k$th periodic update is given by

$$\mathbf{d}_k = (I - \Delta N \Lambda_p)^k \mathbf{d}_O$$

where $\Lambda_p$ is a diagonal matrix with eigenvalues $\lambda_{ip}$, $i = 0, 1, \ldots, N - 1$, equal to the coefficients of the discrete Fourier transform (DFT) of the channel periodic autocorrelation function, $R_n$, $n = 0, 1, \ldots, N - 1$. Thus

$$\lambda_{pi} = \int_{n=0}^{N-1} R_n \, exp\left(\frac{-j2\pi ni}{N}\right)$$

$$= | \sum_k H\left(\frac{i/NT - k}{T}\right) |^2, \qquad i = 0, 1, \ldots, N - 1 \qquad (12.64)$$

Using these results, it can be shown that in the absence of noise the perfect periodic equalizer has a frequency response equal to the inverse of the folded channel spectrum at $N$ uniformly spaced discrete frequencies

$$C\left(\frac{n}{NT}\right) = \frac{1}{\sum_k H(n/NT - k/T)}, \qquad n = 0, 1, \ldots, N - 1$$

If the initial coefficients are zero, the excess MSE after the $k$th update is given by

$$\varepsilon_{p\Delta k} = \sum_{i=0}^{N-1} (1 - \Delta N \lambda_{pi})^{2k} \qquad (12.65)$$

Each component of this sum converges provided $0 < \Delta < 2/(N\lambda_{p\max})$.

Fastest convergence is obtained when $\Delta = 2/[N(\lambda_{p\max} + \lambda_{p\min})]$. For an ideal channel, all eigenvalues are unity. Therefore, a single averaged update with $\Delta = 1/N$ results in a perfect periodic equalizer.

### Stochastic Update

So far we have examined the convergence properties of the periodic update or LMS steepest-descent algorithm with averaging. It is more common and, as we shall see, more beneficial to use the continual or stochastic update method, where all coefficients are adjusted in each symbol interval according to (12.50). Proceeding as in Section 12.3.2 and noting that in the absence of noise, $e_{k\text{opt}} = 0$ for a periodic input, we obtain

$$(\mathbf{c}_{k+1} - \mathbf{c}_{p\text{opt}}) = (I - \Delta \mathbf{y}_k^* \mathbf{y}_k^T)(\mathbf{c}_k - \mathbf{c}_{p\text{opt}})$$

Let us define an $N \times N$ matrix $\mathbf{B}_k \triangleq (I - \Delta \mathbf{y}_k^* \mathbf{y}_k^T)$. Note that since $\mathbf{y}_k$ is periodic with period $N$, $B_k$ is a circulant matrix and

$$B_{k+1} = U^{-1} B_k U$$

where the $N \times N$ cyclic shift matrix $U$ is of the form

$$U = \begin{bmatrix} 0 & 1 & 0 & \cdots & 0 \\ 0 & 0 & 1 & \cdots & 0 \\ \cdot & \cdot & \cdot & \cdots & \cdot \\ \cdot & \cdot & \cdot & \cdots & \cdot \\ \cdot & \cdot & \cdot & \cdots & \cdot \\ 0 & 0 & 0 & \cdots & 1 \\ 1 & 0 & 0 & \cdots & 0 \end{bmatrix}$$

Note that $U^T U = U^N = I$. Consider the first $N$ updates from time zero to $N - 1$. Then

$$(\mathbf{c}_N - \mathbf{c}_{p\text{opt}}) = B_{N-1} B_{N-2} \cdots B_0 (\mathbf{c}_O - \mathbf{c}_{p\text{opt}})$$

$$= (U^{-N+1} B_0 U^{N-1})(U^{-N+2} B_0 U^{N-2}) \cdots B_0 (\mathbf{c}_0 - \mathbf{c}_{p\text{opt}})$$

$$= (U B_0)^N (\mathbf{c}_0 - \mathbf{c}_{p\text{opt}})$$

In general,

$$(\mathbf{c}_{kN} - \mathbf{c}_{p\text{opt}}) = (UB_0)^{kN} (\mathbf{c}_0 - \mathbf{c}_{p\text{opt}}) \qquad (12.66)$$

The convergence of the coefficient deviations, therefore, depends on the eigenvalues of $UB_0$. Using the fact that $\mathbf{y}_0^* \mathbf{y}_0^T$ is singular with rank 1, after some manipulation the characteristic equation, $\det(\lambda I - UB_0) = 0$, can be reduced to the form

$$\lambda^N + \Delta N \sum_{n=0}^{N-1} \lambda^n R_n - 1 = 0 \qquad (12.67)$$

Here $R_n = (\mathbf{y}_k^T \mathbf{Y}_{k+n}^*)/N$, $n = 0, 1, \ldots, N-1$, are the coefficients of the periodic autocorrelation function of the equalizer input (defined earlier in terms of the periodic impulse response of the channel). When $R_n = 0$, for $n \geq 1$, corresponding to an ideal Nyquist channel, all roots of (12.67) are equal to the $N$th roots of $(1 - \Delta N R_o)$. Thus perfect equalization is obtained after $N$ updates with $\Delta = 1/(N R_o)$.

Let $\lambda_{\max}$ be the maximum magnitude root of (12.67). Then the coefficient deviations after every $N$ stochastic update are reduced in magnitude according to (12.66) provided $|\lambda_{\max}^N| < 1$. Moreover, from the maximum modulus principle of holomorphic functions [Rudin, 12.92], we have the condition that

$$|\lambda_{\max}^N| \leq \overset{\max}{0 \leq i \leq N-1} \left| 1 - \Delta N \sum_{n=0}^{N-1} R_n \exp\left(\frac{-j 2 \pi n i}{N}\right) \right|$$

or $\qquad\qquad\qquad\qquad\qquad\qquad\qquad\qquad\qquad\qquad\qquad\qquad (12.68)$

$$|\lambda_{\max}^N| \leq \overset{\max}{0 \leq i \leq N-1} | 1 - \Delta N \lambda_{pi} |$$

where $\lambda_{pi}$ are the eigenvalues of $A_p$ in the periodic update method. Since (12.68) holds with equality only when $R_n = 0$ for $n \geq 1$, we reach the important conclusion for periodic training that for the ideal Nyquist channel the stochastic and averaged update algorithms converge equally fast, but for all other channels the stochastic update algorithm results in faster convergence. This behavior has also been observed for equalizer convergence in the presence of random data [Monsen, 12.71].

In the presence of noise, an exact expression for the excess MSE for the stochastic update algorithm is difficult to derive for periodic training sequences. However, assuming zero initial coefficients, the following expression is a good approximation to results obtained in practice for moderately high SNR:

$$\varepsilon_{\Delta kN} = \sum_{i=0}^{N-1} [(1 - \Delta N \lambda_i)^{2k} + \Delta^2 N \lambda_i^2 \, \varepsilon_{pmin}] \qquad (12.69)$$

where $\lambda_i$ are the roots of (12.67) (the effect of noise can be included by defining $R_n = E[\mathbf{y}_k^T \mathbf{y}_{k+N}^*]/N)$. For an ideal channel, $\lambda_i = R_o = 1$ for all $i$, and the excess MSE for periodic training converges to $\varepsilon_{pmin}$ after $N$ adjustments with $\Delta = 1/N$, where $\varepsilon_{pmin}$ is the minimum achievable MSE for periodic training.

The significant difference in the convergence behavior of the stochastic gradient algorithm for random data and periodic training is now apparent by comparing (12.54) and (12.69). The well-controlled correlation properties of periodic training sequences tend to reduce the average settling time of the equalizer by about a factor of two compared to the settling time in the presence of random data.

An important question regarding periodic equalization is: Once the coefficients have been optimized for a periodic training sequence, how close to optimum is that set of coefficients for random data? The answer depends primarily on the selected period of the training sequence (and hence the equalizer span) relative to the length of the channel impulse response. When the period is long enough to contain a sufficiently large percentage (say 95%) of the energy of the channel impulse response, the edge effects in the periodic channel response and equalizer coefficients are small. In frequency domain terms, the discrete tones of the periodic training sequence are adequately dense to obtain representative samples of the channel spectrum. Under these conditions, the excess MSE due to the periodicity of the training sequence is small compared to the excess MSE due to the large value of $\Delta$ that must be selected for fast initial convergence. After rapid initial convergence has been obtained in this manner, it may be desirable to make finer adjustments to the equalizer using a pseudorandom training sequence with a longer period or begin decision-directed adaptation using randomized customer data.

### Application to Fractionally Spaced Equalizers

The averaged and stochastic update methods of periodic training are also applicable to fractionally spaced equalizers [Qureshi, 12.91]. The period of the training sequence is still equal to the time span of the equalizer. Thus a sequence with period $NT$ can be used to train an equalizer with $NM/K$ coefficients spaced $KT/M$ seconds apart.

The equalizer coefficients may be adjusted periodically, every $N$ symbol intervals, according to the averaged update algorithm

$$c_n(k + 1) = c_n(k) - \Delta \sum_{j=0}^{N-1} e_{kN+j} \, y^* \left( \frac{kNT - jT - nKT}{M} \right),$$

$$n = 0, 1, \ldots, \frac{NM}{K - 1}$$

If the channel folded-power spectrum is flat, then a single averaged update with an appropriate value of $\Delta$ results in a matched filter, which also removes ISI. In contrast with $T$-spaced equalizers, the ideal amplitude shape of the unequalized system for fast convergence of a fractionally spaced equalizer is a square root of Nyquist rather than Nyquist. Moreover, the convergence of an FSE is not affected by sampler timing phase or channel delay distortion, since the channel power spectrum is independent of phase-related parameters.

The stochastic update algorithm (12.50) for periodic training of fractionally spaced equalizers can be analyzed along the lines of Section 12.4.2. Like the averaged update algorithm, the convergence of the stochastic update algorithm is also independent of phase-related parameters. When the folded-power spectrum is flat, perfect equalization can be obtained in $N$ adjustments. When the channel power spectrum is not Nyquist, the maximum modulus principle (12.68) applies, ensuring that the stochastic update algorithm will result in faster convergence than that obtained by periodic or averaged update. Comments with regard to the selection of an adequately long period of the training sequence given in Section 12.4.2 still apply.

## Accelerated Processing

One technique for reducing the effective settling time of an equalizer involves performing equalizer coefficient update iterations as often as permissible by the computational speed limitations of the implementation. For instance, for a periodic equalizer, one period of the received sequence of samples may be stored in the equalizer delay line and iterative updates using the averaged or stochastic update algorithm may be made at a rate faster than the usual [Mueller, 12.76]. This update rate may be selected to be independent of the modem symbol rate since the sequence of samples already stored in the equalizer delay line may be circularly shifted as often as required to produce new output samples. Based on each such output new coefficient correction terms can be computed using a circular shift of the locally stored periodic training sequence. Thus after the equalizer delay line has been filled with a set of received samples, the best periodic equalizer for that particular set of received samples can be determined almost instantly by accelerated processing given unlimited computational speed. If this set of received samples is representative of the channel response to the periodic training sequence, then the equalizer obtained by such accelerated processing is a good approximation to the optimum periodic equalizer. However, all sources of aperiodicity, such as initial transients and noise, in these received samples degrade performance, since new received samples are not used in this method to reduce the effect of noise by averaging. Modified versions of the accelerated processing method are possible that reprocess some previously processed samples and then accept a new input sample as it becomes available, thus updating the equalizer coefficients several times per symbol interval.

## Discrete Fourier Transform Techniques

Throughout the discussion on periodic equalization, we have taken advantage of the circulant property of the equalizer input to use the discrete Fourier transform for analysis. As pointed out in Section 12.4.2 in the absence of noise the perfect periodic equalizer has a frequency response equal to the inverse of the folded-channel spectrum at $N$ uniformly spaced frequencies. Therefore, the equalizer coefficients can be directly obtained by transmitting a periodic training sequence and using the following steps at the receiver [Godard, 12.43; Milewski, 12.70; Qureshi, 12.90].

1. Compute the DFT of one period of the equalizer input:

$$Y_i = \sum_{k=0}^{N-1} y_k \exp\left(\frac{-j2\pi ik}{N}\right), \qquad i = 0, 1, \ldots, N - 1$$

2. Compute the desired equalizer spectrum according to

$$C_i = \frac{X_i Y_i^*}{|Y_i|^2}, \qquad i = 0, 1, \ldots, N - 1$$

where $X_i$ is the precomputed DFT of the training sequence.

3. Compute the inverse DFT of the equalizer spectrum to obtain the periodic equalizer coefficients

$$c_n = \left(\frac{1}{N}\right) \sum_{i=0}^{N-1} C_i \exp\left(\frac{j2\pi ni}{N}\right), \quad n - 0, 1, \ldots, N-1$$

A number of modifications may be made to improve performance of this direct computation method in the presence of noise and other distortions, such as frequency translation, which adversely affect periodicity. For instance, when the equalizer input is not strictly periodic with period $NT$ due to channel-induced frequency translation, its DFT at any frequency suffers from interference from adjacent components. The effect of this interference can be minimized by windowing a longer sequence of input samples before taking the DFT. The window function should be selected such that its Fourier transform has reduced side-lobe energy while preserving the property of zero response at $1/NT$-hertz intervals. A $2NT$ second triangular window has both these properties [Godard, 12.43]. A second minor modification can be made to Step 2 by adding a constant estimate of the expected flat noise power spectral components to the denominator.

### 12.4.3 Recursive Least-Squares Algorithms

The orthogonalized LMS algorithms of Section 12.4.1 can provide rapid convergence when the overall received signal spectral shape is known beforehand, as in partial-response systems. In certain voiceband modem applications, special training sequences can be used to design fast equalizer start-up algorithms, such as those in the last section. However, in general, a self-orthogonalizing method, such as one of the **recursive least-squares** (RLS) algorithms described in this section, is required for rapidly tracking adaptive equalizers (or filters) when neither the reference signal nor the input (received) signal (or channel) characteristics can be controlled.

As discussed in earlier sections, the rate of convergence of the output MSE of an LMS gradient adaptive equalizer is adversely affected by the eigenvalue spread of the input covariance matrix. This slow convergence is due to the fundamental limitation of a single adjustable step-size parameter $\Delta$ in the LMS gradient algorithm. If the input covariance matrix is known a priori, then an orthogonalized LMS gradient algorithm can be derived, as in Section 12.4.1, where the scalar $\Delta$ is replaced by a matrix $P$. Most rapid convergence is obtained when $P$ is the inverse of the equalizer input covariance matrix $A$, thus rendering the adjustments to the equalizer coefficients independent of one another.

In [Godard, 12.41], Godard applied the Kalman filter algorithm to the estimation of the LMS equalizer coefficient vector under some assumptions on the equalizer output error and input statistics. The resulting algorithm has since been recognized to be the fastest-known equalizer adaptation algorithm. It is an ideal self-orthogonalizing algorithm [Gitlin, 12.37] in that the received equalizer input signals are used to build up the inverse of the input covariance matrix, which is applied to the coefficient-adjustment process. A disadvantage of the Kalman algorithm is that it requires on the order of $N^2$ operations per iteration for an equalizer with $N$ coefficients.

Falconer and Ljung [Falconer, 12.20] showed that the Kalman equalizer adaptation algorithm can be derived as a solution to the exact least-squares problem without any statistical assumptions. An advantage of this approach is that the "shifting property" previously used for fast RLS identification algorithms [Morf, 12.73] can be applied to the equalizer adaptation algorithm. This resulted in the so-called fast Kalman algorithm [Falconer, 12.20] which requires on the order of $N$ operations per iteration.

A third class of RLS algorithms known as adaptive lattice algorithms [Friedlander, 12.127] were first described for adaptive identification in [Morf, 12.74] and for adaptive equalization in [Makhoul, 12.63; Satorius, 12.96; 12.97]. Like the fast Kalman algorithm, adaptive lattice algorithms are recursive in time, requiring of the order of $N$ operations per iteration. However, unlike the Kalman algorithms, adaptive lattice algorithms are order recursive. That is, the number of equalizer coefficients (and the corresponding lattice filter sections) can be increased to $N + 1$ without affecting the already computed parameters of the $N$th-order equalizer. Low sensitivity of the lattice coefficients to numerical perturbations is a further advantage.

In the remainder of this section, we shall briefly review the least-square criterion, introduce the shifting property used in the fast Kalman algorithm and the structure of the adaptive lattice algorithms, and summarize some important results, complexity estimates, and stability considerations.

### The Least-Squares Criterion

The performance index for RLS algorithms is expressed in terms of a time average instead of a statistical or ensemble average, as in LMS algorithms. The RLS equalizer adaptation algorithm is required to generate the $N$-coefficient vector $\mathbf{c}_n$ at time $n$ that minimizes the sum of all squared errors as if $\mathbf{c}_n$ were used over all the past received signals, that is, $\mathbf{c}_n$ minimizes

$$\sum_{k=0}^{n} |x_k - \mathbf{y}_k^T \mathbf{c}_n|^2 \tag{12.70}$$

In order to permit tracking of slow time variations, a decay factor $w$ with a value slightly less than unity may be introduced. The resulting exponentially windowed RLS algorithm minimizes

$$\sum_{k=0}^{n} w^{n-k} |x_k - \mathbf{y}_k^T \mathbf{c}_n|^2 \tag{12.71}$$

It can be shown [Proakis, 12.87] that given $\mathbf{c}_{n-1}$, the coefficient vector for time $n$ can be generated recursively according to

$$\mathbf{c}_n = \mathbf{c}_{n-1} + \mathbf{k}_n e_n \tag{12.72}$$

where $e_n = x_n - \mathbf{y}_n^T \mathbf{c}_{n-1}$ is the equalizer output error and

$$\mathbf{k}_n = \mathbf{A}_n^{-1} \mathbf{y}_n^* \tag{12.73}$$

is the Kalman gain vector. $A_n$ is the estimated covariance matrix $A_n = WA_{n-1} + \mathbf{y}_n^* \mathbf{y}_n^T$. The presence of the inverse estimated covariance matrix in (12.73) explains the insensitivity of the rate of convergence of the RLS algorithms to the channel characteristics.

In the Kalman algorithm [Godard, 12.41], the inverse matrix $P_n = A_n^{-1}$ and the Kalman gain vector are computed recursively.

The order of $N^2$ complexity of this algorithm is due to the explicit recursive computation of $P_n$. This computation is also susceptible to roundoff noise.

### The Fast Kalman Algorithm

Consider the input vector $\mathbf{y}_{n-1}$ at time $n - 1$ for a $T$-spaced equalizer of length $N$. The vector $\mathbf{y}_n$ at time $n$ is obtained by shifting the elements of $\mathbf{y}_{n-1}$ by one, discarding the oldest sample $y_{n-N}$, and adding a new sample $y_n$. This shifting property is exploited by using least-squared linear prediction. Thus an efficient recursive algorithm can be derived [Falconer, 12.20] for updating the Kalman gain vector $\mathbf{k}_n$ without explicit computation of the inverse matrix $P_n$.

The matrix computations involved in the Kalman algorithm are replaced in the fast Kalman algorithm by recursions, which use forward and backward predictors to update the Kalman gain vector as a new input sample, $y_n$, is received and the oldest sample, $y_{n-N}$, is discarded.

### Adaptive Lattice Algorithms

The Kalman and fast Kalman algorithms obtain their fast convergence by orthogonalizing the adjustments made to the coefficients of an ordinary linear transversal equalizer. **Adaptive lattice** (AL) algorithms, on the other hand, use a lattice filter structure to orthogonalize a set of received signal components [Lyon, 12.127]. The transformed received signal components are then linearly weighted by a set of equalizer coefficients and summed to produce the equalizer output. We shall briefly review the gradient [Satorius, 12.96] and least-squares [Proakis, 12.87; Satorius, 12.97] forms of adaptive lattice algorithms for linear $T$-spaced complex equalizers. See [Mueller, 12.78; Ling, 12.52] for generalization of the least-squares AL algorithm to fractionally-spaced and decision-feedback equalizers.

The structure of an AL gradient equalizer is shown in Fig. 12.26. An $N$-coefficient equalizer uses $N - 1$ lattice filter stages. Each symbol interval a new received sample $y(n)$ enters stage 1. The $m$th stage produces two signals $f_m(n)$ and $b_m(n)$, which are used as inputs by stage $m + 1$. These signals correspond to the forward and backward prediction errors, respectively, of $m$th-order forward and backward linear LMS predictors. The two predictors have identical **reflection coefficients** $k_m$ for the $m$th stage.

The reflection coefficients are updated to minimize the sum of the mean-square value of the forward and backward prediction errors.

One important property of the lattice structure is that as the reflection coefficients converge, the backward prediction errors $b_m(n)$, $m = 0, 1, \ldots, N - 1$, form a vector $\mathbf{b}(n)$ of orthogonal signal components.

A computationally complex (requiring larger number of computations) but

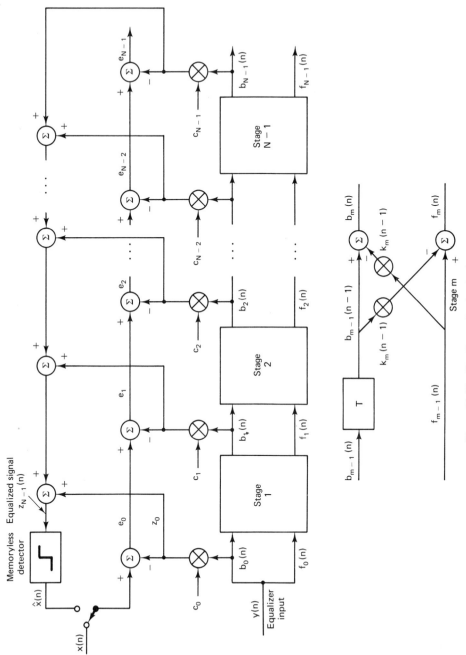

**Figure 12.26** Gradient adaptive lattice equalizer.

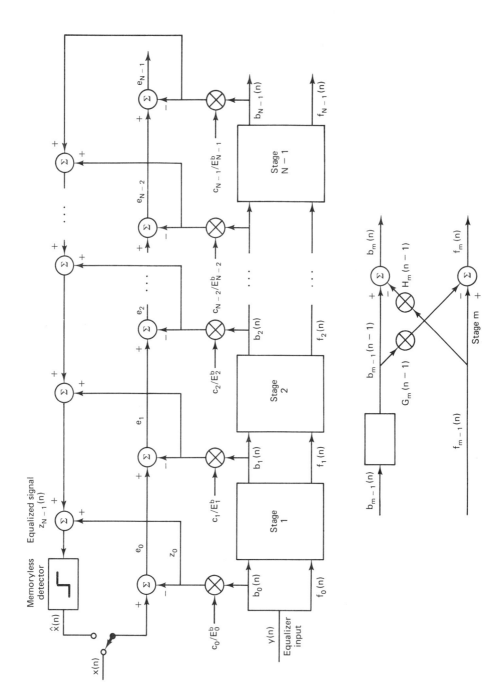

**Figure 12.27** Least-squares adaptive lattice equalizer.

faster-converging least-squares form of the AL equalizer results when the performance index or cost function to be minimized is the exponentially windowed sum of squared errors given in [12.78] instead of the MSE. The structure of the least-squares AL equalizer is shown in Fig. 12.27. Note that the lattice coefficients for forward and backward prediction for any of the lattice stages are no longer equal, each being independently updated to minimize the weighted sum of squared forward and backward prediction errors, respectively.

**Complexity and Numerical Stability**

In the last sections, we have presented an overview of three basic forms of RLS equalization algorithms. Fast RLS algorithms are still being actively studied to reduce computational complexity, specially for "multichannel" (fractionally spaced and decision-feedback) equalizers and to improve stability when limited precision arithmetic is used. Some of the recent results are reported in [Cioffi, 12.10; 12.11; Ling, 12.52].

Accurate counts of the number of multiplications, additions or subtractions, and divisions are hard to determine due to the large number of variations of the RLS algorithms that have been reported in the literature. In the accompanying table the number of operations (multiplications and divisions) required per iteration is listed for $T$ and $T/2$-spaced transversal equalizers of span $NT$ seconds for the LMS gradient, Kalman, fast Kalman/fast transversal, and RLS lattice algorithms. In each case the smallest number of operations is given from the complexity estimates reported in [Mueller, 12.78; Cioffi, 12.10; Ling, 12.52].

| Algorithm | Number of operations per iteration | |
|---|---|---|
| | $T$-equalizer | $\frac{T}{2}$-equalizer |
| LMS gradient | $2N$ | $4N$ |
| Kalman | $2N^2 + 5N$ | $8N^2 + 10N$ |
| Fast Kalman/fast transversal | $7N + 14$ | $24N + 45$ |
| RLS lattice | $15N - 11$ | $46N$ |

The fast Kalman is the most efficient type of RLS algorithm. However, compared to the LMS gradient algorithm, the fast Kalman algorithm is still about four times as complex for $T$-equalizers and six times for $T/2$-equalizers. The RLS lattice algorithm is still in contention due to its better numerical stability and order-recursive structure, despite a twofold increase in computational complexity over the fast Kalman algorithm.

The discussion on RLS algorithms would not be complete without a comment on the numerical problems associated with these algorithms in steady-state operation. Simulation studies have reported the tendency of RLS algorithms implemented with finite precision to become unstable and the adaptive filter coefficients to diverge [Cioffi, 12.10; Honig, 12.46; Ling, 12.52; Mueller, 12.78; Eleftheriou, 12.126]. This is due to long-term accumulation of finite precision errors. Among the different types of RLS algorithms, the fast Kalman or fast transversal type algorithms are the most prone to instability [Honig, 12.46; Mueller, 12.78]. In [Mueller, 12.78] instability was reported to occur when an exponential weighting factor $w < 1$ was used for the fast Kalman algorithm implemented with single precision floating-point arithmetic. The Kalman and RLS lattice algorithms did not show this instability. A sequential processing dual-channel version of the RLS lattice algorithm for a DFE is reported to be stable even for fixed point arithmetic with 10- to 12-bit accuracy [Ling, 12.52]. However, in [Cioffi, 12.10] an "unnormalized" RLS lattice algorithm is shown to become unstable. Normalized versions of fast transversal [Cioffi, 12.10] and RLS lattice [Friedlander, 12.127] algorithms are more stable but both require square roots, the lattice type having greater computational complexity.

# REFERENCES

[12.1] Abend, K., and B. D. Fritchman. "Statistical detection for communication channels with intersymbol interference," *Proc. IEEE*, Vol. 58, May, 1970, pp. 779–785.

[12.2] Agazzi, O., D. A. Hodges, and D. G. Messerschmitt. "Large scale integration of hybrid-method digital subscriber loops," *IEEE Trans. Commun.*, Vol. COM-30, September, 1982, pp. 2095–2108.

[12.3] Austin, M. E. "Decision-feedback equalization for digital communication over dispersive channels," *MIT Lincoln Lab.*, Lexington, Mass., Techn. Rep. 437, August 1967.

[12.4] Belfiore, C. A., and J. H. Park, Jr. "Decision feedback equalization," *Proc. IEEE*, Vol. 67, August, 1979, pp. 1143–1156.

[12.5] Biglieri, E., A. Gersho, R. D. Gitlin, and T. L. Lim. Adaptive cancellation of nonlinear intersymbol interference for voiceband data transmission," *IEEE J. Selected Areas Commun.*, Vol. SAC-2, September, 1984, pp. 765–777.

[12.6] Brady, D. M., "An adaptive coherent diversity receiver for data transmission through dispersive media," *Proc. 1970 IEEE International Conference Commun.*, June, 1970, pp. 21–35 to 21–39.

[12.7] Caraiscos, C., and B. Liu. "A roundoff error analysis of the LMS adaptive algorithm," *IEEE Trans. ASSP*, Vol. ASSP-32, February, 1984, pp. 34–41.

[12.8] Chang, R. W. "A new equalizer structure for fast start-up digital communication," *Bell Syst. Tech. J.*, July–August, 1971, pp. 1969–2014.

[12.9] Chang, R. W., and E. Y. Ho. "On fast start-up data communication systems using pseudo-random training sequences," *Bell Syst. Tech. J.*, Vol. 51, November, 1972, pp. 2013–2027.

[12.10] Cioffi, J. M., and T. Kailath. "Fast, recursive-least-squares, transversal filters for adaptive filtering," *IEEE Trans. ASSP*, Vol. ASSP-32, April, 1984, pp. 304–337.

[12.11] Cioffi, J. M., and T. Kailath. "An efficient exact-least-squares fractionally spaced equalizer using intersymbol interpolation," *IEEE J. Selected Areas Commun.*, Vol. SAC-2, September, 1984, pp. 743–756.

[12.12] Duttweiler, D. L., J. E. Mazo, and D. G. Messerschmitt. "An upper bound on the error probability in decision-feedback equalization," *IEEE Trans. Inform. Theory*, July, 1974, pp. 490–497.

[12.13] Duttweiler, D. L. "Adaptive filter performance with nonlinearities in the correlation multiplier," *IEEE Trans. ASSP*, Vol. ASSP-30, August, 1982, pp. 578–586.

[12.14] Ericson, T. "Structure of optimum receiving filters in data transmission systems," *IEEE Trans. Inform. Theory* (Corresp.), Vol. IT-17, May, 1971, pp. 352–353.

[12.15] Falconer, D. D., and G. J. Foschini. "Theory of minimum mean-square-error QAM system employing decision feedback equalization," *Bell Syst. Tech. J.*, Vol. 53, November, 1973, pp. 1821–1849.

[12.16] Falconer, D. D. "Jointly adaptive equalization and carrier recovery in two-dimensional digital communication systems," *Bell Syst. Tech. J.*, Vol. 55, March, 1976, pp. 317–334.

[12.17] Falconer, D. D. and F. R. Magee, Jr. "Adaptive channel memory truncation for maximum likelihood sequence estimation," *Bell Syst. Tech. J.*, Vol. 52, November, 1973, pp. 1541–1562.

[12.18] Falconer, D. D., and F. R. Magee, Jr. "Evaluation of decision feedback equalization and Viterbi algorithm detection for voiceband data transmission—Parts I and II," *IEEE Trans. Commun.*, vol. COM-24, October, 1976, pp. 1130–1139, and November, 1976, pp. 1238–1245.

[12.19] Falconer, D. D., and J. Salz. "Optimal reception of digital data over the Gaussian channel with unknown delay and phase jitter," *IEEE Trans. Inform. Theory*. January, 1977, pp. 117–126.

[12.20] Falconer, D. D., and L. Ljung. "Application of fast Kalman estimation to adaptive equalization," *IEEE Trans. Commun.*, Vol. COM-26, October, 1978, pp. 1439–1446.

[12.21] Falconer, D. D. "Adaptive equalization of channel nonlinearities in QAM data transmission systems," *Bell Syst. Tech. J.*, Vol. 57, September, 1978, pp. 2589–2611.

[12.22] Falconer, D. D. "Adaptive reference echo cancellation," *IEEE Trans. Commun.*, Vol. COM-30, September, 1982, pp. 2083–2094.

[12.23] Fenderson, G. L., J. W. Parker, P. D. Quigley, S. R. Shepard, and C. A. Siller, Jr. "Adaptive transversal equalization of multipath propagation for 16-QAM, 90-Mb/s digital radio," *AT&T Bell Lab. Tech. J.*, Vol. 63, October, 1984, pp. 1447–1463.

[12.24] Forney, G. D., Jr. "Maximum-likelihood sequence estimation of digital sequences in the presence of intersymbol interference," *IEEE Trans. Inform. Theory*, Vol. IT-18, May, 1972, pp. 363–378.

[12.25] Forney, G. D., Jr., "The Viterbi algorithm," *Proc. IEEE*, Vol. 61, March, 1973, pp. 268–278.

[12.26] Forney, G. D., S. U. H. Qureshi, and C. K. Miller. "Multipoint networks: advances in modem design and control," *National Telecommun. Conference Rec.*, December, 1976, pp. 50-1–1 to 50-1–4.

[12.27] Forney, G. D., Jr., R. G. Gallager, G. R. Lang, F. M. Longstaff, and S. U.

Qureshi. "Efficient modulation for band-limited channels," *IEEE J. Selected Areas Commun.*, Vol. SAC-2, September, 1984, pp. 632–647.

[12.28] Foschini, G. J., and J. Salz. "Digital communications over fading radio channels," *Bell Syst. Tech. J.*, Vol. 62, February, 1983, pp. 429–456.

[12.29] Franks, L. E., ed., *Data Communication*, Dowden, Hutchinson and Ross, 1974.

[12.30] Fredricsson, S. A. "Joint optimization of transmitter and receiver filters in digital PAM systems with a Viterbi detector," *IEEE Trans. Inform. Theory*, March, 1976, pp. 200–2210.

[12.31] George, D. A. "Matched filters for interfering signals." *IEEE Trans. Inform. Theory* (Corresp.), January, 1965, pp. 153–154.

[12.32] George, D. A., R. R. Bowen, and J. R. Storey. "An adaptive decision feedback equalizer," *IEEE Trans. Commun. Technol.*, Vol. COM-19, June, 1971, pp. 281–293.

[12.33] Gersho, A. "Adaptive equalization of highly dispersive channels," *Bell Syst. Tech. J.*, Vol. 48, January, 1969, pp. 55–70.

[12.34] Gersho, A., and T. L. Lim. "Adaptive cancellation of intersymbol interference for data transmission," *Bell Syst. Tech. J.*, November, 1981, pp. 1997–2021.

[12.35] Gitlin, R. D., E. Y. Ho, and J. E. Mazo. "Passband equalization of differentially phase-modulated data signals," *Bell Syst. Tech. J.*, Vol. 52, February, 1973, pp. 219–238.

[12.36] Gitlin, R. D., J. E. Mazo, and M. G. Taylor. "On the design of gradient algorithms for digitally implemented adaptive filters," *IEEE Trans. Circuit Theory*, Vol. CT-20, March, 1973, pp. 125–136.

[12.37] Gitlin, R. D., and F. R. Magee, Jr. "Self-orthogonalizing algorithms for accelerated covergence of adaptive equalizers," *IEEE Trans. Commun.*, Vol. COM-25, July, 1977, pp. 666–672.

[12.38] Gitlin, R. D., and S. B. Weinstein. "On the required tap-weight precision for digitally-implemented adaptive mean-squared equalizers," *Bell Syst. Tech. J.*, Vol. 58, February, 1979, pp. 301–321.

[12.39] Gitlin, R. D., and S. B. Weinstein. "Fractionally-spaced equalization: an improved digital transversal equalizer," *Bell Syst. Tech. J.*, Vol. 60, February, 1981, pp. 275–296.

[12.40] Gitlin, R. D., H. C. Meadors, Jr., and S. B. Weinstein. "The tap-leakage algorithm: an algorithm for the stable operation of a digitally implemented, fractionally spaced adaptive equalizer," *Bell Syst. Tech. J.*, Vol. 61, October, 1982, pp. 1817–1939.

[12.41] Godard, D. N. "Channel equalization using a Kalman filter for fast data transmission," *IBM J. Research and Development*, Vol. 18, May, 1974, pp. 267–273.

[12.42] Godard, D. N. "Self-recovering equalization and carrier tracking in two-dimensional data communication systems," *IEEE Trans. Commun.*, Vol. COM-28, November, 1980, pp. 1867–1875.

[12.43] Godard, D. N. "A 9600-bit/s modem for multipoint communication systems," *National Telecomm. Conference Record*, New Orleans, La., December, 1981, pp. B3.3.1–B3.3.5.

[12.44] Grenander, N., and G. Szego. *Toeplitz Forms and Their Application*, University of California Press, Berkeley, Calif., 1958.

[12.45] Guidoux, L. "Egaliseur autoadaptif a double echantillonnage," *L'Onde Electrique*, Vol. 55, January, 1975, pp. 9–13.

[12.46] Honig, M. L., and D. G. Messerschmitt. *Adaptive Filters; Structures, Algorithms, and Applications*, Kluwer Academic Publishers, Boston, 1984.

[12.47] Johnson, C. R., Jr. "Adaptive IIR filtering: current results and open issues," *IEEE Trans. Inform. Theory*, Vol. IT-30, March, 1984, pp. 237–250.

[12.48] Kabal, P., and S. Pasupathy. "Partial-response signaling," *IEEE Trans. Commun.*, Vol. COM-23, September, 1975, pp. 921–934.

[12.49] Kobayashi, H. "Simultaneous adaptive estimation and decision algorithm for carrier modulated data transmission systems," *IEEE Trans. Commun. Technol.*, Vol. COM-19, June, 1971, pp. 268–280.

[12.50] Kretzmer, E. R. "Binary data communication by partial response transmission," *1965 ICC Conf. Rec.*, pp. 451-455; also, "Generalization of a technique for binary data communication," *IEEE Trans. Commun. Technol.*, February, 1966, pp. 67–68.

[12.51] Lee, W. U., and F. S. Hill. "A maximum likelihood sequence estimator with decision feedback equalization," *IEEE Trans. Commun. Tech.*, Vol. COM-25, September, 1977, pp. 971–979.

[12.52] Ling, F., and J. G. Proakis. "A generalized multichannel least-squares lattice algorithm based on sequential processing stages," *IEEE Trans. ASSP*, Vol. ASSP-32, April, 1984, pp. 381–389.

[12.53] Logan, H. L., and G. D. Forney, Jr. "A MOS/LSI multiple configuration 9600-b/s data modem," *Proc. IEEE International Conference Commun.*, June, 1976, pp. 48-7–48-12.

[12.54] Lucky, R. W., J. Salz, and E. J. Weldon, Jr. *Principals of Data Communication*. McGraw-Hill, New York, 1968.

[12.55] Lucky, R. W. "Automatic equalization for digital communication," *Bell Syst. Tech. J.*, Vol. 44, April, 1965, pp. 547–588.

[12.56] Lucky, R. W. "Techniques for adaptive equalization of digital communication systems," *Bell Syst. Tech. J.*, Vol. 45, February, 1966, pp. 255–286.

[12.57] Lucky, R. W., and H. R. Rudin. "An automatic equalizer for general-purpose communication channels," *Bell Syst. Tech. J.*, Vol. 46, November, 1967, pp. 2179–2208.

[12.58] Lucky, R. W. "Signal filtering with the transversal equalizer," *Proc. Seventh Annual Allerton Conference on Circuits and System Theory*, October, 1969, pp. 792–803.

[12.59] Lucky, R. W. "A survey of the communication theory literature: 1968–1973," *IEEE Trans. Inform. Theory*, Vol. 52, November, 1973, pp. 1483–1519.

[12.60] Macchi, O., and E. Eweda. "Convergence analysis of self-adaptive equalizers," *IEEE Trans. Inform. Theory*, Vol. IT-30, March, 1984, pp. 161–176.

[12.61] MacKenchnie, L. R. "Maximum likelihood receivers for channels having memory," Ph.D. Thesis, Dept. of Electrical Engineering, University of Notre Dame, Notre Dame, Ind., January, 1973.

[12.62] Magee, F. R., Jr., and J. G. Proakis. "Adaptive maximum-likelihood sequence estimation for digital signaling in the presence of intersymbol interference," *IEEE Trans. Inform. Theory* (Corresp.), Vol. IT-19, January, 1973, pp. 120–124.

[12.63] Makhoul, J. "A class of all-zero lattice digital filters: properties and applications," *IEEE Trans. ASSP*, Vol. ASSP-26, August, 1978, pp. 304–314.

[12.64] Mazo, J. E. "Optimum timing phase for an infinite equalizer," *Bell Syst. Tech. J.*, Vol. 54, January, 1975, pp. 189–201.

[12.65] Mazo, J. E. "On the independence theory of equalizer convergence," *Bell Syst. Tech. J.*, Vol. 58, May-June, 1979, pp. 963–993.

[12.66] Mazo, J. E. "Analysis of decision-directed equalizer convergence," *Bell Syst. Tech. J.*, Vol. 59, December, 1980, pp. 1857–1876.

[12.67] Messerschmitt, D. G. "A geometric theory of intersymbol interference: Part I," *Bell Syst. Tech. J.*, Vol. 52, November, 1973, pp. 1483–1519.

[12.68] Messerschmitt, D. G. "Design of a finite impulse response for the Viterbi algorithm and decision feedback equalizer," *Proc. IEEE Int. Conference Communications*, ICC-74, Minneapolis, Minn., June 17–19, 1974.

[12.69] Messerschmitt, D. G. "Echo cancellation in speech and data transmission," *IEEE J. Selected Areas Commun.*, Vol. SAC-2, March, 1984, pp. 283–296.

[12.70] Milewski, A. "Periodic sequences with optimal properties for channel estimation and fast start-up equalization," *IBM J. Resource Development*, Vol. 27, September, 1983, pp. 426–431.

[12.71] Monsen, P. "Feedback equalization for fading dispersive channels," *IEEE Trans. Inform. Theory*, Vol. IT-17, January, 1971, pp. 56–64.

[12.72] Monsen, P. "MMSE equalization of interference on fading diversity channels," *IEEE Trans. Commun.*, Vol. COM-32, January, 1984, pp. 5–12.

[12.73] Morf, M., T. Kailath, and L. Ljung. "Fast algorithms for recursive identification," *Proc. 1976 IEEE Conf. Decision Contr.*, Clearwater Beach, Fla., December, 1976, pp. 916–921.

[12.74] Morf, M., A. Vieira, and D. T. Lee. "Ladder forms for identification and speech processing," *Proc. 1977 IEEE Conference Decision Contr.*, New Orleans, La., December, 1977, pp. 1074–1078.

[12.75] Mueller, K. H. "A new, fast-converging mean-square algorithm for adaptive equalizers with partial-response signaling," *Bell Syst. Tech. J.*, Vol. 54, January, 1975, pp. 143–153.

[12.76] Mueller, K. H., and D. A. Spaulding. "Cyclic equalization—a new rapidly converging equalization technique for synchronous data communication," *Bell Syst. Tech. J.*, February, 1975, pp. 369–406.

[12.77] Mueller, K. H., and J. J. Werner. "A hardware efficient passband equalizer structure for data transmission," *IEEE Trans. Commun.*, Vol. COM-30, March, 1982, pp. 538–541.

[12.78] Mueller, M. S. "Least-squares algorithms for adaptive equalizers," *Bell Syst. Tech. J.*, Vol. 60, October, 1981, pp. 1905–1925.

[12.79] Mueller, M. S. "On the rapid initial convergence of least-squares equalizer adjustment algorithms," *Bell Syst. Tech. J.*, Vol. 60, December, 1981, pp. 2345–2358.

[12.80] Mueller, M. S., and J. Salz. "A unified theory of data-aided equalization," *Bell Syst. Tech. J.*, November, 1981, pp. 2023–2038.

[12.81] Murano, K., Y. Mochida, F. Amano, and T. Kinoshita. "Multiprocessor architecture for voiceband data processing (application to 9600-bps modem)," *Proc. IEEE International Conference on Commun.*, June, 1979, pp. 37.3.1.–37.3.5.

[12.82] Murase, T., K. Morita, and S. Komaki. "200-Mb/s 16-QAM digital radio system with new countermeasure techniques for multipath fading," *Proc. IEEE International Conference on Commun.*, June, 1981, pp. 46.1.1–46.1.5.

[12.83] Price, R. "Nonlinearly feedback-equalized PAM vs. capacity for noisy filter channels," *Proc. 1972 IEEE International Conf. Commun.*, June, 1972, pp. 22-12–22-17.

[12.84] Proakis, J. G., and J. H. Miller. "An adaptive receiver for digital signaling through channels with intersymbol interference," *IEEE Trans. Inform. Theory*, Vol. IT-15, July, 1969, pp. 484–497.

[12.85] Proakis, J. G. "Adaptive nonlinear filtering techniques for data transmission," *IEEE Symposium on Adaptive Processes, Decision and Control*, 1970, pp. XV.2.1–5.

[12.86] Proakis, J. G. "Advances in equalization for intersymbol interference," in A. J. Viterbi, ed., *Advances in Communication Systems*, Vol. 4, Academic Press, New York, 1975, pp. 123–198.

[12.87] Proakis, J. G. *Digital Communications*, McGraw-Hill, New York, 1983.

[12.88] Qureshi, S. U. H. "New approaches in adaptive reception of digital signals in the presence of intersymbol interference," Ph.D Thesis, University of Toronto, Toronto, Ont., Canada, May, 1973.

[12.89] Qureshi, S. U. H., and E. E. Newhall. "An adaptive receiver for data transmission over time-dispersive channel," *IEEE Trans. Inform. Theory*, Vol. IT-19, July, 1973, pp. 448–457.

[12.90] Qureshi, S. U. H. "Fast start-up equalization with periodic training sequences," *IEEE Trans. Inform. Theory*, Vol. IT-23 September, 1977, pp. 553–563.

[12.91] Qureshi, S. U. H., and G. D. Forney, Jr. "Performance and properties of a T/2 equalizer," *National Telecomm. Conf. Record*, December, 1977.

[12.92] Rudin, W. *Real and Complex Analysis*, McGraw-Hill, New York, 1966.

[12.93] Salz, J. "Optimum mean-square decision feedback equalization," *Bell System Tech. J.*, Vol. 52, October, 1973, pp. 1341–1373.

[12.94] Salz, J. "On mean-square decision feedback equalization and timing phase", *IEEE Trans. Commun. Technol.*, Vol. COM-25, December, 1977, pp. 1471–1476.

[12.95] Sato, Y. "A method of self-recovering equalization for multilevel amplitude modulation," *IEEE Trans. Commun.*, Vol. COM-23, June, 1975, pp. 679–682.

[12.96] Satorius, E. H., and S. T. Alexander. "Channel equalization using adaptive lattice algorithms," *IEEE Trans. Commun.*, Vol. COM-27, June, 1979, pp. 899–905.

[12.97] Satorius, E. H., and J. D. Pack. "Application of least squares lattice algorithms to adaptive equalization," *IEEE Trans. Commun.*, Vol. COM-29, February, 1981, pp. 136–142.

[12.98] Siller, C. A., Jr. "Multipath propagation," *IEEE Commun. Mag.*, Vol. 22, February, 1984, pp. 6–15.

[12.99] Schonhoff, T. A., and R. Price. "Some bandwidth efficient modulations for digital magnetic recording," in *Proc. IEEE International Conf. Communications*, ICC-81, Denver, Co., June 15–18, 1981.

[12.100] Tong, S. Y. "Dataphone II service: data set architecture," *National Telecomm. Conference Record*, New Orleans, La., December, 1981, pp. B.3.2.1.–B.3.2.5.

[12.101] Tsuda, T., Y. Mochida, K. Murano, S. Unagami, H. Gambe, T. Ikezawa, H.

Kikuchi, and S. Fujii. "A high performance LSI digital signal processor for communication," *Proc. 1983 IEEE International Conference Commun.*, June, 1983, pp. A5.6.1–A5.6.5.

[12.102] Ungerboeck, G. "Theory on the speed of convergence in adaptive equalizers for digital communication," *IBM J. Research and Development*, Vol. 16, November, 1972, pp. 546–555.

[12.103] Ungerboeck, G. "Adaptive maximum-likelihood receiver for carrier-modulated data-transmission systems," *IEEE Trans. Commun.*, Vol. COM-22, May, 1974, pp. 624–636.

[12.104] Ungerboeck, G. "Fractional tap-spacing equalizer and consequences for clock recovery in data modems," *IEEE Trans. Commun.*, Vol. COM-24, August, 1976, pp. 856–864.

[12.105] Ungerboeck, G. "Channel coding with multilevel/phase signals," *IEEE Trans. Inform. Theory*, Vol. IT-28, January, 1982, pp. 55–67.

[12.106] van Gerwen, P. J., N. A. M. Verhoeckx, and T. A. C. M. Claasen. "Design considerations for a 144-kbit/s digital unit for the local telephone network," *IEEE J. Selected Areas Commun.*, Vol. SAC-2, March, 1984, pp. 314–323.

[12.107] Wei, L.-F. "Rotationally invariant convolutional channel coding with expanded signal space—part II: nonlinear codes," *IEEE J. Selected Areas Commun.*, Vol. SAC-2, September, 1984, pp. 672–686.

[12.108] Weinstein, S. B. "A baseband data-driven echo canceller for full-duplex transmission on two-wire circuits," *IEEE Trans. Commun.*, Vol. COM-25, July, 1977.

[12.109] Widrow, B., and M. E. Hoff, Jr. "Adaptive switching circuits," *IRE Wescon Conv. Rec.*, part 4, August, 1960, pp. 96–104.

[12.110] Widrow, B., J. R. Glover, Jr., J. M. McCool, J. Kaunitz, C. S. Williams, R. H. Hearn, J. R. Zeidler, E. Dong, Jr., and R. C. Goodlin. "Adaptive noise cancelling: principles and applications," *Proc. IEEE*, Vol. 63, December, 1975, pp. 1692–1716.

[12.111] Widrow, B., J. M. McCool, M. G. Larimore, and C. R. Johnson, Jr. "Stationary and nonstationary learning characteristics of the LMS adaptive filter," *Proc. IEEE*, Vol. 64, August, 1976, pp. 1151–1162.

[12.112] Yasumoto, M. T. Enomoto, K. Watanabe and T. Ishihara. "Single-chip adaptive transversal filter IC employing switched capacitor technology," *IEEE J. Selected Areas Commun.*, Vol. SAC-2, March, 1984, pp. 324–333.

[12.113] Feher, K. *Digital Communications: Satellite/Earth Station Engineering*, Prentice-Hall, Englewood Cliffs, N. J., 1983.

[12.114] Monsen, P. "Theoretical and measured performance of a DFE modem on a fading multipath channel," *IEEE Trans. Commun.*, Vol. COM-25, October, 1977, pp. 1144–1153.

[12.115] Nyquist, H. "Certain topics in telegraph transmission theory," *Trans. AIEE*, Vol. 47, April, 1928, pp. 617–644.

[12.116] Smith, J. W. "The joint optimization of transmitted signal and receiving filter for data transmission systems," *Bell Syst. Tech. J.*, Vol. 44, December, 1965, pp. 2363–2392.

[12.117] Tufts, D. W. "Nyquist's problem—the joint optimization of transmitter and receiver in pulse amplitude modulation," *Proc. IEEE*, Vol. 53, March, 1965, pp. 248–260.

[12.118] CCITT Rec. V.27 b. "4800/2400 bits per second modem with automatic equalizer standardized for use on leased telephone-type circuits," *Int. Telegraph and Telephone Consultative Committee*, 1980.

[12.119] CCITT Rec. V.29. "9600 bits per second modem standardized for use on point-to-point 4-wire leased telephone-type circuits," *Int. Telegraph and Telephone Consultative Committee*, 1980.

[12.120] Qureshi, S. U. H., and H. M. Ahmed. "A custom chip set for digital signal processing," in *VLSI Signal Processing*, IEEE Press, 1984, New York, pp. 52–63.

[12.121] Wouda, K. J., S. J. M. Tol, and W. J. M. Reiknkjens. "An ISDN transmission system with adaptive echo cancelling and decision feedback equalization—a two-chip realization," *Proc. IEEE International Conference Commun.*, Amsterdam, The Netherlands, 1984, pp. 685–690.

[12.122] Makhoul, J. "Linear prediction: a tutorial review," *Proc. IEEE*, Vol. 63, April, 1975, pp. 561–580.

[12.123] Agazzi, O., D. G. Messerschmitt, and D. A. Hodges. "Nonlinear echo cancellation of data signals," *IEEE Trans. Commun.*, Vol. COM-30, November, 1982, pp. 2421–2433.

[12.124] Holte, M., and S. Stueflotten. "A new digital echo canceller for two-wire subscriber lines," *IEEE Trans. Commun.*, Vol. COM-29, November, 1981, pp. 1573–1581.

[12.125] Lawrence, V. B., and J. J. Werner. "Low-speed data transmission by using high-speed modems," *Proc. IEEE Globecom*, Atlanta, Ga., November, 1984, pp. 677–682.

[12.126] Eleftheriou, E., and D. D. Falconer. "Restart methods for stabilizing FRLS adaptive equalizers in digital HF transmission," *Proc. IEEE Globecom*, Atlanta, Ga., November, 1984, pp. 1558–1562.

[12.127] Friedlander, B. "Lattice filters for adaptive processing," *Proc. IEEE*, Vol. 70, August, 1982, pp. 829–867.

[12.128] Lyon, D. L. "Envelope-derived timing recovery in QAM and SQAM systems," *IEEE Trans. Commun.*, Vol. COM-23, November, 1975, pp. 1327–1331.

# INDEX

Raised-cosine filter, 321–22
Rate parameter of speech-coding algorithms, 135
Rayleigh fading model, 503
Receiver structures and steady-state properties:
  baseband equivalent model, 662–64
  linear receivers, 665–72
  nonlinear receivers, 672–84
  optimum receive filter, 664–65
Recursive least-squares (RLS), 644, 700–702, 705–6
Redundancies in the speech signal, 135–37
  interpitch processing and, 151–54
  predictive coding and, 140–42
Reference burst, 590
Reference signal for the echo canceler, 190
Reflector antennas, 601–3
Remote switching units (RSUs), 54–55
Residual excited (RELP) vocoders, 158
RFI (radio-frequency interference), 204
RLS (recursive least-squares), 644, 700–702, 705–6
Rooftop antennas, 578
RSUs (remote switching units), 54–55

## S

Sample assignment word (SAW), 265, 268
Sampling, 8
SATCOM (satellite communications), 34, 603
Satellite channels, 17
Satellites, communications:
  antenna gain and, 578
  antennas for, 601–9
  basics of, 577–86
  capabilities of, 583–86
  categories of, 575–77
  early developments, 573–77
  frequency reuse, 632–35
  impairments due to rain, 580–82, 609–10, 612, 615
  ISDN and, 55–56
  link budgets and, 579–80
  modem techniques for, 365–67
  multiple access and demand assignment, 586–92
  multiple scanning beam systems, 599–600
  noise and capacity and, 579
  power-efficient modem techniques for nonlinear, 321
  resource sharing for, 615–20, 622–23
  scanning spot-beam, 592–99
  transmission capacity of digital, 623–37
  use of active repeaters, 573

Saturated power, 26
SAW (sample assignment word), 265, 268
Scalar quantizer, 138–40
Scanning beam systems, satellites and, 599–600
Scanning fan beam antennas, 607–9
Scanning spot-beam, 592–99
Selection method, mobile-radio systems and, 531, 533
Self-routing networks, 82
Series truncation error, 464
Service channels (SCs), 240, 253–54, 256
  DSI connection procedures and, 263–65
  DSI operation and, 262–63
  numbering, 256, 258
Session layer, 58
SFSK (sinusoidal frequency-shift keying), 336, 493
Shadowing, 499
Shannon's formula, 579
Signal-channel transmission techniques, 85
Signaling protocols, 86
Signal-to-distortion ratio, 271–72
Signal to quantization noise, 10
Signal-transfer points (STPs), 42, 94
Single-sideband (SSB) transmission, 207, 577
Sinusoidal frequency-shift keying (SFSK), 336, 493
Slow-scan video, 48
SM (statistical multiplexing), 4
Space diversity, 530–31
SPARCOM system, 366–67
SPEC (speech predictive encoded communications), 18
  analysis of performance, 268–72
  comparison of TASI and implementations, 275, 277
  competitive clipping, 266
  connect clipping, 266
  definition of, 238, 265
  PCM sample prediction, 267–68
  predictor distortion, 265
  sample assignment word, 265, 268
  speech detector clipping, 266
  /TDMA implementation, 272, 274–75
Spectral-efficient modems, 26, 318, 387–90
Spectrum utilization:
  definition of, 555–58
  geographical cochannel reuse and, 558–59
  mobile-radio systems and, 489–90
  multichannel access and, 560
  narrow-band transmission and, 559–60
  optimization and, 560–66
Speech:
  adaptation and nonstationarity in, 142–47
  noise control and, 155–56